Advances in Fig Research and Sustainable Production

FSC
www.fsc.org
MIX
Paper from responsible sources
FSC® C022174

To my late parents, Bella and Katarina, my wife, Hilla, and our children, Naama, Avner, and Amos, who stimulated my life-time passion for fruit trees, especially figs.

Moshe A. Flaishman

To my dearest late husband, Bekir, who used to say that "everything may fade out, yet books live beyond us" and to my beloved children, Uygar and Ayça, who shared my drive for figs, and to my grandchildren, our future.

Uygun Aksoy

Advances in Fig Research and Sustainable Production

Edited by

Moshe A. Flaishman

Institute of Plant Sciences, Agricultural Research Organization, Rishon LeZion, Israel

Uygun Aksoy

Department of Horticulture, Faculty of Agriculture, Ege University, Bornova, İzmir, Turkey

CABI

CABI is a trading name of CAB International

CABI
Nosworthy Way
Wallingford
Oxfordshire OX10 8DE
UK

Tel: +44 (0)1491 832111
E-mail: info@cabi.org
Website: www.cabi.org

CABI
WeWork
One Lincoln St
24th Floor
Boston, MA 02111
USA

Tel: +1 (617)682-9015
E-mail: cabi-nao@cabi.org

A catalogue record for this book is available from the British Library, London, UK.

Library of Congress Cataloging-in-Publication Data

Names: Flaishman, Moshe A., editor. | Aksoy, U. (Uygun), editor.
Title: Advances in fig research and sustainable production / Moshe A. Flaishman, Uygun Aksoy.
Description: Boston : CAB International, [2022] | Includes bibliographical references and index. | Summary: "This book provides a comprehensive review of fig research, both basics and applications. It addresses the recent scientific results in genetics, genomics, metabolomics, pharmaceutical properties, understanding of symbiosis, new management, post-harvest technology and diseases and pest management of figs"--Provided by publisher.
Identifiers: LCCN 2021057091 (print) | LCCN 2021057092 (ebook) | ISBN 9781789242478 (hardback) | ISBN 9781789242485 (ebook) | ISBN 9781789242492 (epub)
Subjects: LCSH: Fig. | Sustainable agriculture.
Classification: LCC SB365 .A38 2022 (print) | LCC SB365 (ebook) | DDC 333.76/16--dc23/eng/20211221
LC record available at https://lccn.loc.gov/2021057091
LC ebook record available at https://lccn.loc.gov/2021057092

References to Internet websites (URLs) were accurate at the time of writing.

ISBN-13: 978 1 78924 247 8 (hardback)
978 1 78924 248 5 (ePDF)
978 1 78924 249 2 (ePub)

DOI: 10.1079/9781789242492.0000

Commissioning Editor: Rebecca Stubbs
Editorial Assistant: Emma McCann
Production Editor: Shankari Wilford

Typeset by SPi, Pondicherry, India
Printed and bound in the UK by Severn, Gloucester

Contents

Contributors

Aksoy, Uygun, Department of Horticulture, Faculty of Agriculture, Ege University, Bornova, İzmir, Turkey. Email: uygun.aksoy@gmail.com

Balcı, Belit, ALARA Tarım, Bursa, Turkey. Email: bbalci@alaraagri.com

Bechar, Avital, Institute of Agricultural Engineering, Agricultural Research Organization, Volcani Center, Rishon LeZion, Israel. Email: avital@volcani.agri.gov.il

Ben-Yakir, David, Department of Entomology, Agricultural Research Organization, Volcani Center, Rishon LeZion 7505101, Israel. Email: benyak@volcani.agri.gov.il

Can, Hakki Zafer, Department of Horticulture, Faculty of Agriculture, Ege University, Bornova, İzmir, Turkey. Email: hakki.zafer.can@ege.edu.tr

Cohen, Yafit, Institute of Agricultural Engineering, Agricultural Research Organization, Volcani Center, Rishon LeZion, Israel. Email: yafitush@volcani.agri.gov.il

Costa, Alain, 8 Gabriel Pérez, Albatera 03340, Spain. Email: alain.costa.ac@gmail.com

Eisikowitch, Dan, Department of Molecular Biology and Ecology of Plants, Tel Aviv University, Israel. Email: dane@tauex.tau.ac.il

Elbeaino, Toufic, Istituto Agronomico Mediterraneo di Bari (CIHEAM-IAMB), Via Ceglie 9, 70010 Valenzano, Bari, Italy. Email: elbeaino@iamb.it

Flaishman, Moshe A., Institute of Plant Science, Agriculture Research Organization, Volcani Center, Rishon LeZion 7528809, Israel. Email: vhmoshea@agri.gov.il

García-Viguera, Cristina, Phytochemistry and Healthy Food Lab (LabFAS), Department of Food Science and Technology, CEBAS-CSIC, Campus de Espinardo, Murcia, 30100, Spain. Email: cgviguera@cebas.csic.es

Ghara, Mahua, Department of Molecular Biology and Ecology of Plants, Tel Aviv University, Israel; Botanical Garden, Tel Aviv University, Ramat Aviv, Tel Aviv 69978, Israel. Email: mahuaghara@post.tau.ac.il

Gündeşli, Muhammet Ali, Department of Plant and Animal Production, Nurdağı Vocational School, Gaziantep University, Gaziantep, Turkey. Email: maligun4646@gmail.com

Ikegami, Hidetoshi, Fukuoka Agriculture and Forestry Research Center Buzen Branch, Yukuhashi, Japan. Email: ikegami@farc.pref.fukuoka.jp

Imrak, Burhanettin, Department of Horticulture, Faculty of Agriculture, University of Çukurova, 01330, Balcali, Adana, Turkey. Email: bimrak@cu.edu.tr

Irget, Mehmet Eşref, Department of Soil Science and Plant Nutrition, Faculty of Agriculture, Ege University, İzmir, Turkey. Email: esref.irget@ege.edu.tr

Kafkas, Nesibe Ebru, Department of Horticulture, Faculty of Agriculture, University of Çukurova, 01330, Balcali, Adana, Turkey. Email: ebruyasakafkas@gmail.com

Kafkas, Salih, Department of Horticulture, Faculty of Agriculture, University of Çukurova, 01330, Balcali, Adana, Turkey. Email: salihkafkas@gmail.com

Lama, Kumar, Department of Life Sciences, School of Science, Kathmandu University, Dhulikhel, Nepal. Email: kumar.lama@ku.edu.np

Lev-Yadun, Simcha, Department of Biology & Environment, Faculty of Natural Sciences, University of Haifa – Oranim, Tivon 36006, Israel. Email: levyadun@research.haifa.ac.il

López-Corrales, Margarita, Fruticultura Mediterránea, Centro de Investigación Finca La Orden-Valdesequera, Junta de Extremadura, Badajoz, Spain. Email: margaritalópez@juntaex.es

Ma, Huiqin, College of Horticulture, China Agricultural University, Beijing 100193, China. Email: hqma@cau.edu.cn

Martín, Alberto, Nutrición y Bromatología, Escuela de Ingenierías Agrarias, Universidad de Extremadura, Badajoz, Spain. Email: amartin@unex.es

Meriç, Mehmet Kamil, Bergama Vocational School, Ege University, İzmir, Turkey. Email: m.kamil.meric@ege.edu.tr

Modi, Arpan, Institute of Plant Science, Agriculture Research Organization, Volcani Center, Rishon LeZion 7528809, Israel. Email: arpanmodi8@live.com

Namdar, Dvora, Institute of Plant Sciences, Agriculture Research Organization, Volcani Center, Rishon LeZion 7528809, Israel. Email: dvoran@volcani.agri.gov.il

Özer, Kamer Betül, Ege University, Faculty of Agriculture, Department of Horticulture, Bornova, İzmir, Turkey. Email: kamer.betul.ozer@ege.edu.tr

Peer, Reut, Institute of Plant Sciences, Agriculture Research Organization, Volcani Center, Rishon LeZion 7528809, Israel. Email: reutp@volcani.agri.gov.il

Piga, Antonio, Dipartimento di Agraria, Università degli Studi di Sassari, Viale Italia 39, 07100 Sassari, Italy. Email: pigaa@uniss.it

Rodov, Victor, Department of Postharvest Science, Agriculture Research Organization, Volcani Institute, Rishon LeZion, Israel. Email: vrodov@volcani.agri.gov.il

Şen, Fatih, Department of Horticulture, Faculty of Agriculture, Ege University, Bornova, İzmir, Turkey. Email: fatih.sen@ege.edu.tr

Serradilla, Manuel Joaquín, Área de Postcosecha, Instituto Tecnológico Agroalimentario de Extremadura (INTAEX), Centro de Investigaciones Científicas y Tecnológicas de Extremadura (CICYTEX), Junta de Extremadura, Badajoz, Spain. Email: manuel.serradilla@juntaex.es

Shirasawa, Kenta, Kazusa DNA Research Institute, Kisarazu, Japan. Email: shirasaw@kazusa.or.jp

Tırkaz, Ece, Aegean Exporters' Associations (EİB), İzmir, Turkey. Email: ece.tirkaz@eib.org.tr

Tsantili, Eleni, Laboratory of Pomology, Agricultural University of Athens, Athens, Greece. Email: ekotsantili@gmail.com

Turanlı, Ferit, Department of Plant Protection, Faculty of Agriculture, Ege University, İzmir, Turkey. Email: ferit.turanli@ege.edu.tr

Valero, Daniel, Department of Food Technology, EPSO, University Miguel Hernández, Alicante, Spain. Email: danielumh@gmail.com

Vera Ponce, Christian, Ficatum-Higos con Historia, Miraflores, Lima, Peru. Email: cveraponce@posteo.de

Zhai, Yanlei, College of Horticulture, China Agricultural University, Beijing 100193, China. Email: zhaiyanlei2012@cau.edu.cn

Ziffer, Irit, Eretz Israel Museum, 2 Haim Levanon Street, Tel Aviv 6997501, Israel. Email: irit.ziffer@gmail.com

Preface

The genus *Ficus* comprises one of the largest and hardiest genera, highly competent to survive and endure unfavorable conditions, consisting of about 40 genera of flowering plants, woody trees, shrubs, vines, and epiphytes, and having as many as 850 species spread throughout the tropics and subtropics worldwide. It is estimated that *Ficus* is a relatively ancient genus, being at least 60–80 million years old. The genus *Ficus* has a remarkable reproductive strategy with a unique fruit, the syconium, hiding the flowers inside its hollow receptacle. It is the site of an obligate mutualism with pollinating wasps, where each *Ficus* species has its own pollinator wasp. Due to this mutualism, the genus *Ficus* produces large crops of fruit at short intervals that maintain continuous development of their wasp mutualists. This combination of large fruit crops and regular fruiting make *Ficus* trees important resources for many frugivores. It has been suggested that the genus *Ficus* is an ecological keystone resource in the world. *Ficus* trees have helped restoration of forests in the tropics. Planting *Ficus* trees accelerated rainforest regeneration in areas where trees have been lost due to logging. Taking a global view, members of the genus *Ficus* have developed important ecosystems. They are used by more than 1200 vertebrate frugivore species, while humans rarely eat "wild fig fruits."

Humans, however, recognized and selected the common fig (*Ficus carica* L.), which produces fruits with delicious sweet flavor. Fig, a pleasant fruit that can be consumed fresh or dry, helped to establish horticulture. It is one of the oldest fruit trees known and appreciated, as it is a vital food source with high energy, which may have helped our ancestors to develop and who, in return, domesticated common figs several thousand years ago. The common fig originated in Asia Minor. The most prominent wild populations of the common fig and its close relatives are widely found all over the Mediterranean Basin, mainly in north-eastern Turkey and northern Iran. The earliest archaeological evidence for exploitation of the wild common fig is from a 780,000-year-old Middle Paleolithic site located in the upper Jordan Valley, Israel. Later, remains of fig trees and seeds were found during excavations in many sites in the Middle East. The common fig is one of the five founder fruit trees that established horticulture. Most likely, it was domesticated in the Chalcolithic period some 6500 years ago. Urukagina, the King of the city states of Lagash and Girsu in Mesopotamia, wrote about them nearly 5000 years ago. The ancient Greeks and Romans said that figs were the fruit of heaven. Figs are used by humans in a variety of other ways. Some fig species are considered sacred and have a symbolic meaning. Fig is known as a "divine" or "sacred" fruit, or a "fruit of paradise" and is accepted as a symbol of wealth, peace, and prosperity in many civilizations and religions. Fig tree bark, leaves, roots, fruits, and latex are used as valuable ethnomedicines. This book will highlight the importance of figs to humans from various perspectives throughout the world.

Fig was well known around the Mediterranean Basin in ancient civilizations. The fig tree appears in archeological records such as on stone tablets and pottery describing the culinary use of figs and many times in holy books. In many religions, including Judaism, Christianity, Islam, Hinduism, and Buddhism, the fig represents fertility, peace, and prosperity. Ancient Olympians earned figs for their athletic achievements. Fig spread mostly after the 12th century. The Crusaders are believed to have introduced fig to the northern European regions. In a source dating back to 1250 CE, a crop called "fegges" or "figges" is mentioned as being regularly imported to Britain from Spain and Portugal. In the 16th and 17th centuries, fig trees were further spread by sailors and church representatives to Far East Asia and the Americas. Today, the leading fig-growing counties are Turkey, Egypt, Iran, Greece, Italy, Morocco, and the USA. In many countries where fig production or consumption dates back centuries, consumers are more acquainted with the fig fruit and its values. However, fig remains a minor crop with an annual global production of about 1.3 million t.

The first comprehensive book on fig was published in 1901 by Eisen. He gathered all the available data on fig production and a worldwide collection of fig varieties. In 1947, I.J. Condit, the world-famous pioneer of fig research, published his book *The Fig*, where he shared his 30 years of experience and research results on figs in California. In 1948, L. Ülkümen and co-workers published a book entitled *Fig and its Diseases*, which included comparison of practices applied in California and Turkey regarding sun drying, packaging, and trade. Another comprehensive book on figs, *Fig Varieties*, was written by N.K. Arendt in 1972. She gathered all her long-term experiences and research results, which were highly valuable in terms of fig production, gained at Yalta, Crimea.

The first International Symposium on Figs was organized by Uygun Aksoy at Ege University in 1997 in Izmir, Turkey, with the aim of establishing the "Fig Club" involving young researchers with world-renowned fig experts. Since then, five additional international symposiums on fig have taken place in Spain, Portugal, Morocco, Italy, and Croatia. The Fig Working Group of the International Society for Horticultural Science (ISHS), starting from its establishment, was chaired by Uygun Aksoy (Turkey), Jose Leitao (Portugal), Moshe Flaishman (Israel), and Carlos H. Crisosto (USA). The Working Group has built up the network, gathered researchers and practitioners, collated presentations that also lead to thorough discussions, and publishes in the journal *Acta Horticulturae*.

Fig is a challenging fruit crop to grow. It is susceptible to many damaging insect pests and diseases, as well as to injuries from abiotic stress during fruit development and ripening. In addition, fig is a very delicate fruit with challenging postharvest procedures. Horticultural fig research has resulted in many achievements; however, various challenges remain for fig growers. One of the main threats for sustainable cultivation of figs is the impact of climate change on natural resources, such as water and nutrient availability, and key physiological processes in fig production. Increased summer temperatures have a direct impact on fruit quality. Climate change is also associated with a higher frequency of extreme events that vary across regions, including drought, hail, and damaging levels of rain. A clear biological and mechanistic understanding of the physiological processes of dormancy, flowering, and fruit cracking is essential for the development of novel growing strategies, including accurate characterization and utilization of fig genetic resources for modifying these traits.

The fig tree has a juvenile period of 2–4 years. There are barriers to interspecific hybridization in fig, but many progeny resulting from artificial fig crosses with *Ficus* species are fertile. It is well documented that there is wide genetic variability in the common fig, *F. carica* L. Most commercially grown varieties have been selected by unknown growers as seed-derived plants that were then maintained and multiplied through vegetative propagation. There are a few cultivars that have been developed from breeding programs via sexual hybridization around the world. As with other fruit-tree breeding programs, classical fig breeding requires a long time period and is expensive. Thus, recent advances in fig genomics, transformation, and genome editing will further improve new cultivar development.

A great deal of modern research on fig has been conducted, but, to date, information has not been comprehensively summarized and presented, and therefore is not available to researchers and horticulturists. The main objective of this book is to summarize the recent advances on tree biology, orchard and postharvest management, pest and diseases, varieties and cultivars, genetics and breeding, health properties, use in ethnomedicine, and fig products and trade. The 29 chapters compiled by leading experts provide reviews on domestication and history, taxonomy, physiology, genetics, production, plant protection, postharvest biology, and global trade of fig. The latest advances in fig science reflect three changes: (i) demand for sustainable and safe products; (ii) the use of new molecular technologies; and (iii) globalization of the markets and of production through adaptation of cultivars to new fig-producing regions.

This book does not aim to provide technical support through a set of guidelines or to give a protocol for fig production. Moreover, the use of specific inputs such as chemicals or pesticides is not included, as countries may differ in product registration for commercial use. The information in this book serves as a reference for current and future practices of fig production, consumption, research, and innovation. We hope that this book will serve as a strong knowledge pool about the common fig (*F. carica* L.), one of the most fascinating species, which everyone can build upon to promote future progress in the global fig industry.

Moshe A. Flaishman, Rishon LeZion, Israel
Uygun Aksoy, Izmir, Turkey

Acknowledgements

The writing of this book started with fruitful discussions of its content almost 3 years ago. More than 40 leading scientists from all over the world have contributed to writing and reviewing of this book. We are grateful to them for their willingness to share their knowledge. We wish to thank them for their efforts, cooperation, and persistence. We thank the growers, processors, research colleagues, technicians, and our students for continued support of our research, which made possible the accumulation of this vast amount of information on fig science and production.

We are grateful to Dr. Reut Peer and Dr. Arpan Modi for their critical reading of the book chapters. Special thanks to Dr. Reut Peer for her creative preparation of the book figures.

This was a challenging project and our home support made its realization possible. Special thanks to our families who were patient and helpful during the years of preparation of this book.

Part I

The History of Fig

1 Fig: from History and Tradition to Scientific Research

Uygun Aksoy[1]* and Moshe A. Flaishman[2]
[1]Department of Horticulture, Faculty of Agriculture, Ege University, Bornova, İzmir, Turkey; [2]Institute of Plant Sciences, Agricultural Research Organization, Rishon LeZion, Israel

1.1 Introduction

The genus *Ficus* constitutes one of the largest of about 40 genera of flowering plants, woody trees, shrubs, vines, and epiphytes, featuring as many as 850 species, and spread throughout the tropics and subtropics worldwide. *Ficus* is a relatively ancient genus estimated to be at least 60–80 million years old, and is able to survive and endure unfavorable conditions.

Mankind recognized and selected the common fig (*Ficus carica* L.), which produces fig fruits with a delicious, sweet flavor that can be consumed fresh or dry. It is one of the oldest fruit trees known and served as a vital, high-energy food source that may have helped our ancestors when it was domesticated several thousand years ago. The common fig originated in Asia Minor. The most prominent wild populations of the common fig are found all over the Mediterranean Basin. Remains of fig trees and seeds have been found during excavations in many sites in the Middle East. The common fig is one of the five founder fruit trees that established horticulture. It was most likely domesticated in the Chalcolithic period some 6500 years ago. Urukagina, the King of the city states of Lagash and Girsu in Mesopotamia, wrote about figs nearly 5000 years ago. In this chapter, we describe the fig journey along with the establishment of human civilization, from history and tradition to current scientific studies.

1.2 Fig from Domestication and Tradition to Current Everyday Life

Figs were well known around the Mediterranean Basin in the ancient civilizations. The fig tree appears many times in holy books. In many religions, including Judaism, Christianity, Islam, Hinduism, and Buddhism, the fig represents fertility, peace, and prosperity. The ancient Greeks and Romans said that figs were the fruit of heaven. The ancient Olympians earned figs for their athletic achievements. Figs were said to be introduced to China during the reign of the T'ang dynasty in the 700s BCE. Another source states that figs were brought to China after emperor Tang Kien's visit to the Turan region in the year 127 BCE. The first Chinese script about the fig belongs to the eighth century. Hia Cheng-Shi in his book entitled *Yu Yang Tsa Tsu* writes about a fruit brought from a country called "Fo-tin" (Palestine). It is called "tin-tin" ("tin" meaning fig in Arabic) and defined as a fruit with no flowers. Today, it is still known as a "fruit without flowers" based on the fruit formation and

* Email: uygun.aksoy@gmail.com

 Advances in Fig Research and Sustainable Production (M.A. Flaishman and U. Aksoy)
DOI: 10.1079/9781789242492.0001

development (Aksoy *et al.*, 2008). In Europe, the fig spread mostly after the 12th century. The crusaders are believed to have introduced fig to the northern European regions. In a source dating back to 1250 CE, a crop called "fegges" or "figges" was mentioned to be regularly imported to Britain from Spain and Portugal. According to another source, however, it was assumed to be first brought by Cardinal Pole in the first half of the 16th century. In England and other northern countries, growing figs in protected backyards was a famous tradition. The fig was also introduced to North America from England and initiated the tradition of planting fig trees in backyards, especially in southern states. The other route was through Latin America in the 16th century by Spanish and Portuguese missionaries.

Throughout history, only dried figs were subject to trade outside its production regions, and fresh figs remained "local" or "exotic" until the end of the 20th century. This was because the fig fruit is highly fragile and could not successfully undergo long-distance transportation. With more resistant fig varieties and improved postharvest handling practices, international fresh-fig trade started increasing. In addition to consumption as fresh and dried fruits, figs were and still are used to make fig wine, "rakı" or "arak", brandy, liqueur, vinegar, jam, marmalade, spreads, sorbet, and non-alcoholic beverages (Sutton, 2014). The strong symbolism attributed to the fig fruit could have appeared as a result of various motives from its origin in the Middle East, which has also given birth to many religions, its high medicinal value, its diversity and abundance in natural flora, its sweet taste and rich flavor, its different fruit form as a receptacle that encloses numerous seeds, or its highly complex life cycle and mutualism with the fig wasp. Possibly, a cumulative effect of all these attributes has made fig fruit special.

Fig is easily propagated vegetatively, so it is possible to find millions of trees of a single variety distributed in different parts of the world, as seen in the example of the Sarılop (= Calimyrna) variety destined for commercial drying in the Meander (western Turkey) and San Joaquin valleys (California, USA). However, if you visit the coast or have a walk in the rural areas of Mediterranean countries in summer, you may find fig fruits ranging from yellow to dark purple in color hanging on wild female and male fig trees dispersed in nature, thanks to the birds who also love to eat the sweet fig fruit and then distribute the seeds as they fly. These fig fruits are consumed locally and bear local names according to the time of ripening, such as "spring," "sickle fig" (because of coinciding with the wheat harvest), or "winter fig," or because of their skin color, such as black, purple, yellow, or white, or from their color together with their origin, such as Roxo de Valinhos, Bursa Black, Brown Turkey, or Fichi di Cosenza. To better explain the names of these well-known varieties, Valinhos is a municipality in the state of São Paulo, Brazil, and Roxo is purple or violet in Portuguese; Bursa is a province in the Marmara Region of Turkey; Brown Turkey cuttings may have been originally taken in the 19th century from Turkey but in today's Turkey, Brown Turkey does not exist; and Cosenza is a province of the Calabria Region in southern Italy famous for its figs. Especially in countries where rapid urbanization took over the agricultural land, districts named after figs are very common. In Turkey, İncirlik, Karaincir, İnciraltı, İncirlibağ, and İncirliova are districts or towns where "incir", the fig, once dominated. In Cyprus, similarly the villages of Aglasyka (meaning excellent figs) and Makrasyka (long figs) are named after figs (Sutton, 2014). Thus, fig gives its name to the location, or the location gives its name to the fig variety, emphasizing the strong association between the fig fruit and the region.

In many countries where fig production or consumption dates back centuries, consumers are more acquainted with the fig fruit and its values but it is still a minor or underutilized fruit. Some authors characterize it as neglected. This is mostly not because of the requirements of the fig tree but due to the very fragile nature of the fig fruit. For centuries, in regions where the climatic conditions present during fruit ripening were suitable, fig fruit was sun dried. It was then preserved as a part of the daily year-round nutrition of the poor as a rich source of energy, sugar, and essential nutrients. Figs were a part of the Mediterranean "garden economy" in the 18th and 19th centuries, which was based on "diversification between wild and cultivated, animal- and plant-resources, intensive and extensive production; on quality of care in production, multiple varieties of animal or plant, processing and packaging for storage and redistribution" (Kolovos, 2019). In the garden economy, the resource-poor families grew vegetables and other annuals in their gardens and planted trees including figs on the borders to supplement their food supply. This diversification has probably formed the strong basis of the healthy Mediterranean diet but also provided resilient systems against ubiquitous risks. Fig was sometimes exchanged with bread, which was more likely to be possessed by the rich (Kolovos, 2019).

Fig is a part of the local culture and traditions, and is thus included in the proverbs of many countries. Some examples are (www.inspirationalstories.com/proverbs/t/about-figs; accessed November 17, 2021):

- There are three tasty morsels: figs, peaches, and melons (Sicily, Italy).
- Better a handful of dried figs and content with that, than to own the gate of peacocks and be kicked in the eye by a broody camel (Morocco).
- When there is shortage of figs, birds eat the fruits of the "mugumo" (Kikuyu, Kenya).
- If you have figs in your knapsack, everyone will want to be your friend (Albania).
- Figs he calls figs, a spade a spade (An honest man who knows what he is speaking of) (Latin).
- Half figs, half raisins (France).
- Drink wine over figs (Spain).
- Sharing the figs can leave you with none at all (Greece).
- Figs after fish and vegetables after meat (Greece).
- Do men gather grapes of thorns, or figs of thistles? (Iran).
- Baklava cannot be made with corn flour, likewise roller pin cannot be made from fig wood (Turkey).

Although well known for thousands of years, fig production is still limited due to its climate dependency. As of 2020, global fig production, which includes fresh, sun-dried, and otherwise-processed figs, is approximately 1.2 million t, whereas apple production is nearly 50 times more and tomato production is 30 times more. Knowledge about the early uses and properties of fig is limited as written documents only increased during the Greek and Roman periods. The common fig, *F. carica* L., originates from the Middle East. It has been present in Asia Minor and Mesopotamia, and was known by Sumerian, Babylonian, Assyrian, Hittite, Phrygian, Trojan, Egyptian, and Persian civilizations prior to the Greeks and Romans but documents are scarce. Fig is mentioned in texts and inscriptions of ancient cities such as Babylon (Babylonian Empire) and Nineveh (an Assyrian city located in today's Mosul, Iraq). The culinary (*c.*2500 BCE) and medicinal (*c.*2750 BCE) uses are reported in Sumerian stone tablets. Excavations in Asia Minor and the Middle East continue to disclose different fig specimens (Sutton, 2014). The species name, *carica*, comes from the Caria region, a historical district located in south-west Turkey along the Aegean Sea. Roman cookery recipes also refer to highly appreciated fig wine made from a fig type called *caricarum* in ancient Rome.

During antiquity, fig was one of the delicacies and an essential element of the banquets during festivities, and was presented as a gift in ancient Greek and Roman periods. During the Thargelia late spring festival, people wore necklaces made of dried figs. The dried-fig trade was an important economic activity in western Turkey and was promoted during the reign of Emperor Tiberius who lived in the years 42 BCE–37 CE. Plinius explained figs and fig varieties including those transferred from Syria to Italy during the reign of Emperor Tiberius. The ease of storage and transportation was instrumental in the spread of figs. Dried fig was among the staple food of sailors during long journeys, although both light- and dark-colored figs were consumed as fresh fruits locally. The experiences gained during production or during consumption are reflected in poems and dramas of the time as a part of their culture. Fig leaf-decorated coins were also an element of trade signifying the importance of fig (Anaç, 2012).

In western Turkey, sun-dried figs have had a high economic impact for centuries. During the Ottoman Period, the first railway privilege was given to the British to build between İzmir and Aydın, which started operation in 1866. During this period, İzmir was one of the most important harbors of the Empire and had great significance as the opening point of a fertile hinterland mainly of dried figs, cotton, tobacco, grapes, and licorice roots. This region was once the main supplier of food for the Ottoman palaces. Dried fruit and nuts including dried figs were transported by boats from the İzmir and Kuşadası ports, and traders sold to the retail shops, which marketed them together with pastries and other confectionary. There were restrictions on the trade of dried fruit and nuts, and priority was given to Istanbul. The freight could go to Alexandria or Thessaloniki only after they had clearance from Istanbul. Dried figs were either shipped or transported by camels until the mid-20th century. Today, dried figs are either shipped to distant markets or transported by trucks to closer markets, whereas fresh figs are distributed by air cargo or by refrigerated trucks.

The Turkish word "incir" is also used as "encir" in Iran and as "incira" in Russian. During the Ottoman Period, they preferred using the term "yemiş," which meant fruit, because of the bad connotation of "incir" in Persia. In western Turkey, fresh figs were traditionally sold in baskets by street vendors shouting "Bardacık" (small glass) and, in fact, the term "bardacık" or "bardakçı" (seller of glass) has been used synonymously for fresh figs as well as "yemiş" (Anaç, 2012).

Figs trees are long lived, but their vegetative growth or vigor and longevity are closely associated with the existing climatic conditions, variety, and management practices. In Topkapı Palace, a monumental fig (*F. carica*) tree is reported to date back to the Ottoman Period (Atasoy, 2002). In southern Peru, ancient fig trees located in Arequipa, Siguas Valley, are estimated to be around 300 years old and continue to be productive, together with the younger fig trees (www.christian-vera.com; accessed November 17, 2021). In some regions of the world, fig trees are grown as multi-trunk bushes providing resistance to cold and sunburn. If the main trunk dies, young suckers replace the old one and contribute to the orchards' longevity. There are many horticultural types, either bearing fruit once, in the spring (breba) or summer (main crop), or twice in both spring and summer, or in some regions there are even everbearing types. The fruit is a syconium bearing many actual drupe-type fruits. The fruit morphology, which is a hollow receptacle enclosing many seeds, is one of the factors that led people to accept fig as a symbol of fertility or prosperity. Male fig trees are easily identified in winter as those with remaining fig fruit on the tree after complete leaf fall during the dormant period. In Mediterranean countries, people pick the male fruits, which have a strong flavor, to prepare jams in early spring before the fig wasp (*Blastophaga psenes*) enters the spring crop of the male trees from overwintering fruit. These people must follow the mutually symbiotic process of fruit development and the life cycle of the fig wasp. For those people living in the Mediterranean, the presence of fig trees and their fruit throughout summer is a part of everyday life, as well as a part of the local history and culture; however, for those from other regions, the fig fruit is an exotic and luxurious holy fruit.

1.3 Fig Fans and Collectors

For many amateur gardeners, the fig is a fruit of interest, and growing figs may become a hobby, starting with having one tree in the home yard. Various information bulletins are available on how to grow a fig tree in your home yard. It is often considered an exotic fruit, and it can be said that "it is so sweet that it sticks to you." Because fig is part of the tradition and culture in various parts of the world, fig fans initiate either private collection gardens, or establish groups to explore, communicate, and exchange material or information. The common target is to prevent the loss of valuable genetic resources. The Fig Database (www.figdatabase.com; accessed November 17, 2021) developed by a group of entrepreneurs is a searchable database for edible fig varieties. It requires input from fig researchers and experts.

The rich diversity of fig genetic resources facilitate the work of fig fans. Among these, Montserrat Pons I Boscana from the Balearic Islands of Spain has gathered a collection of 1000 fig varieties and published a book entitled *Figs of the Balearic Islands*. The book describes 89 fig varieties present at the Son Mut Nou experiment station in Llucmajor, Majorca, in detail and with accompanying photographs (Boscana, 2009, 2015). There is another private collection in Puglia, Italy, and the synonyms, botany, and fruit characteristics of 25 local fig varieties are described in a book with photographs of the leaf and fruit (Minonne *et al.*, 2011). In the USA, Bass Samaan established a fig collection composed of 300 non-commercial varieties in the suburbs of Pennsylvania. He also sells propagation material of these fig varieties to those interested (https://treesofjoy.com; accessed November 17, 2021). In Ontario, another fig fan, Adriano Ferreira, has gathered 200 cold-tolerant fig varieties. These fans and private collectors of figs are expected to increase when fig diversity is taken into consideration.

1.4 Fig in Science

Scientific studies of figs are closely related to the significance of the fruit, whether of economic importance or in terms of genetic resources status. In this respect, the earlier studies appeared toward the end of the 19th and beginning of the 20th century from California, USA, as fig was newly introduced and required scientific information to develop the sector. The importance of the fig crop continued in California. The main research topics focused on fruit physiology, orchard management, pests and diseases, and long-term breeding studies for variety development. Studies in Mediterranean countries such as Italy, Spain, and Portugal started to fade out after the 1970s as fig became an underutilized species. However, research programs were initiated in the major producing countries of Turkey, Egypt, Iran, Tunisia, Morocco, and Israel aiming at solving the problems encountered during production, postharvest handling, and processing. Fields of interest included: (i) pests and diseases, namely viral diseases, aflatoxins, and storage pests; (ii) evaluating genetic resources to develop new varieties resistant to transportation; and (iii) tree architecture, stress physiology, plant nutrition, breeding and postharvest practices such as drying, cold storage, and packaging. More recently, molecular studies were carried out not only in traditional fig-producing countries such as France, Spain, and Israel but also in Japan and China. The medicinal properties of the fig fruit, leaves, and latex are being investigated thoroughly.

1.4.1 Books on figs

The first comprehensive book on figs was published in 1901 by Eisen. He gathered all of the then available data on fig production and the worldwide collection of fig varieties. Some of these varieties have lost their economic importance locally, but it provides insights to explore, especially the eroding fig genetic resources. Condit (1947), the world-famous pioneer of fig research, in the preface of his famous book, *The Fig*, mentions that he has two purposes in writing the book, to collate all the information on fig facts published by other authors in periodicals and to share his 30 years of experience and research results on figs in California. These experiences gained not only in California but also in other regions of the world that he visited or contacted were thus recorded for future generations. Ülkümen *et al.* (1948) published a book entitled *Fig and Diseases* including comparison of practices applied in California and Turkey regarding sun drying, packaging, and trade. Another comprehensive book on figs, *Fig Varieties*, was written by Arendt (1972). She gathered all her long-term experiences and research results, which were highly valuable in terms of fig production, gained at Yalta, Crimea. The book mainly concentrated on fig varieties, floral biology, fruit formation, ripening and composition, resistance to diseases, and suitability of varieties for various methods of processing. As these last two books were in Turkish and Russian, respectively, dissemination of these research results remained more limited. In addition to the aforementioned focus on *F. carica*, many other books, with shorter versions in the local language, were published for practitioners. Pereira (1981) gathered information on fig production in Brazil, whereas Grassi (1991) provided information on fig varieties and fig production from planting to diseases and pests, and provided recipes regarding local utilization of figs. There are two other books with a more limited focus but which have compiled priceless information. Turkey currently supplies 55–60% of the dried figs worldwide, and Aydın province has a share of 65%; the book published by Anaç (2012) gives the historical background of figs and focuses on local traditions and the important effects created by the dried-fig trade throughout the history of Aydın, while Sutton (2014) has given an overview of the history of figs globally including excerpts from traditions involving figs. *Figs: The Genus Ficus*, by Lansky and Paavilainen (2017) focused on traditional and modern medicinal usage of various members of the *Ficus* genus including *F. carica*. The book also provides an in-depth analysis of the biochemistry and bioactive compounds of different *Ficus* species and explores the similarities between species. The book will continue to shed light on the pharmacological exploration of *F. carica* fruit, leaves, latex, and woody material.

1.4.2 Fig networks: how it started

In 1994, both the Food and Agriculture Organization of the United Nations (FAO) and the International Center for Advanced Mediterranean Agronomic Studies (CIHEAM) Mediterranean Institute of Zaragoza (IAMZ) established networks on minor or underutilized fruit species, which included figs. Most experts from the Mediterranean countries who joined the meetings presented the state-of-the-art fig production, the institutions involved and genetic resources. *Descriptors for Fig* published by Bioversity International is an outcome of the IAMZ network. The research institutions working on figs were still very limited, as the role of the Mediterranean countries as prime producers underwent a significant change. Spain, Portugal, and Italy started losing competitiveness in the world dried-fig trade after the 1970s, and parallel with this, research work had faded out. To initiate a working group on figs under the International Society for Horticultural Science (ISHS), the main global horticultural network, required organization of a scientific symposium. Therefore, the First International Symposium on Fig was organized by Uygun Aksoy at Ege University in June 1997 in İzmir, Turkey. The aim was to realize a dream and establish a "Fig Club" where young researchers who wanted to learn more about figs could interact with world-known experts. Some of the participants who attended the training course are now senior researchers and became regular followers of the ISHS fig symposia. The manuscripts of the First International Symposium on Fig were published as volume 480 of *Acta Horticulturae*.

In 1998, after the First International Symposium on Fig, the ISHS Board took the decision to initiate a working group on figs. Within this scientific structure, symposia were organized in Caceres, Spain (2001), Vilamoura, Algarve, Portugal (2005), Meknes, Morocco (2009), Naples, Italy (2015), and Rovinj, Croatia (2019), and the next meeting is planned to be in Weiyuan, Neijiang City, Sichuan Province, China, in 2023. The proceedings of these scientific fig meetings are published as *Acta Horticulturae* volumes 480, 605, 798, 1173, and 1310, respectively.

The Fig Working Group of the ISHS starting from its establishment was chaired by Uygun Aksoy (Turkey), Jose Leitao (Portugal), Moshe A. Flaishman (Israel) and Carlos H. Crisosto (USA). The Fig Working Group continues to build up the network, archive the accumulated knowledge, and pass this knowledge on to coming generations.

1.5 Figs in Business and Research

The fig sector is mostly dispersed, as the history of the fresh-fig trade is comparatively new and relatively small in volume. In contrast, dried figs have been traded for centuries yielding professional organizations of farmers as well as traders.

1.5.1 Fig farmers' cooperatives

Many countries with traditional fig growing have developed fig farmer organizations. In Turkey, the farmers' cooperative has a rather long history in the sun-dried fig sector. TARİŞ İncir (https://tarisincir.com.tr; accessed November 17, 2021) was among the first cooperatives on figs. Due to the monopoly in the Ottoman dried-fig trade, fig farmers started organizing meetings in 1912, and established the first cooperative in 1915 in Aydın, which later became one of the largest cooperative unions in Turkey. TARİŞ is still buying dried figs from its members, and processes, packs, and sells in the domestic and export markets under its own brand name.

SYKIKI is a union of cooperatives that was first established in Kalamata, Greece, in 1935 with fig growers from Messinia and Lakonia and then officially established (www.sykiki.gr/cooperative/sykiki-cooperative; accessed November 17, 2021). The Cooperative aims to promote dried-fig production based on scientific research findings.

The Valley Fig Growers of California (USA) was established in 1959 and is still involved in production and marketing of Mission and

Calimyrna dried figs (https://valleyfig.com; accessed November 17, 2021).

1.5.2 Fig associations

The Dried Fruit Association (DFA) of California

The DFA (https://dfaofcalifornia.com; accessed November 17, 2021) was established in 1908 as a non-profit trade association to support the dried fruit and nut industries in California including figs. Among various services provided by the DFA, dried-fig processors and research institutes use the rapid moisture tester that was developed by the DFA based on electrical conductivity. The DFA still serves the relevant sectors through training, consulting, laboratory testing, and analysis related to food safety issues.

The Aegean Exporters' Association (AEA)

The Aegean Exporters' Association (www.eib.org.tr; accessed November 17, 2021) was founded in 1939 in İzmir, Turkey, as a non-profit organization working closely with the Turkish Ministry of Trade to promote dried fruit production (dried figs, raisins, and apricots) and trade. All dried-fig exporters are obliged to be a member of the AEA. The AEA supports research and training programs and represents the sector at international organizations and events.

The International Nut and Dried Fruit Council Foundation

As the edible nut industry prospered worldwide and organized meetings at an international level, the business people decided to establish an international organization in 1983. The Foundation targeting the nut industry decided to include dried fruit because in business both go together. The aim of the Foundation is to consider the problems at production level and to promote consumption, support research, monitor issues related to international trade, promote meetings, and serve as a knowledge source for its members and the related industries. Within this context, the Foundation brings together the dried-fig industry worldwide.

The Fig Industry Association of China

Commercial-scale fig production is quite new in China. The first organization of fig growers was established in 2012, named the Baoding Fig Industry Association. Baoding is a historical city 150 km to the south-west of Beijing. Annual meetings have been organized in different growing areas in March every year since 2012. Based on the Baoding Fig Industry Association, the China Fig Industry Alliance was organized, and at their annual meeting was upgraded to the Figs Industry Forum of China (FIFC), which includes presentations and discussions on fig varieties, production, processing, and strategic topics in the industry development. At present, the Alliance has more than 500 members including processing companies, cooperatives, growers, and researchers. As well as the annual meeting, the Alliance publishes the *China Fig Year Book*. The "China Fig Group" of WeChat is an active online platform for daily communication between members.

1.5.3 Research institutions on figs

Being an important agricultural country, after the founding of the Republic, the government of Turkey started promoting research by founding thematic research institutes at a regional level. The Fig Research Institute was founded in the heart of the fig production region, Erbeyli, Aydın, in western Turkey. The institute is still performing research on figs.

1.6 Concluding Remarks

The common fig, *F. carica* L., is regaining importance as a healthy and exotic fruit and as a species possessing medicinal properties. Many research studies are adding new information or creating new areas to investigate. The International Symposium on Figs under the auspices of the ISHS brings together valuable results that need to be further analyzed. There are only a few books on figs as a horticultural crop to date, such as Eisen (1901), Condit (1947), Ülkümen *et al.* (1948), and Arendt (1972), published at almost 50-year intervals. Ira Condit, the pioneer of fig research has worked for nearly 60 years and has provided guidance to those who started working with figs through his comprehensive book, *The Fig*. Although the book *Fig Cultivars* (Arendt, 1972) contains very valuable information based on long-term observations, its impact was limited as it is in Russian. Since then, there have many booklets published on figs in different languages, helping the practical aspects at the local level.

Hundreds of research studies have been carried out during the last decades in different parts of the world. This book aims to gather the updated knowledge, reliable research results, and available data on fig production, research, uses, processing, and trade based on the experience of well-known fig experts. Focus is also given to the areas that have newly emerged such as biotechnology, sex inheritance, breeding, the genetic basis of fruit development and ripening, and medicinal properties, as well as on areas where thorough reviews are scarce such as postharvest of fresh and dried figs, sustainable production systems, aroma compounds, and viral diseases.

The knowledge accumulated and collated in the different chapters will bridge the science from Condit's era to the most recent data to lead the way forward. Considering the current and future challenges, as more scientific data is available for fig research, practitioners, and industry, fig production and consumption will be able to grow more sustainably.

References

Aksoy, U., Can, H.Z., Meyvacı, K.B., and Şen, F. (2008) Kuru incir (Dried figs). In: Aksoy, U. (ed.) *Türk Sultanları, Kuru üzüm, Kuru İncir ve Kuru Kayısı*. Can Digital Printing Center, İzmir, Turkey, pp. 51–85.

Anaç, H. (2012) *İncir – Geçmişten Günümüze (Fig – from Past to Present)*. ATOK Publisher, Turkey (in Turkish).

Arendt, N.K. (1972) *Sorta Incira (Fig Varieties)*. The State Nikita Botanical Garden, Yalta, USSR.

Atasoy, N. (2002) *Hasbahçe: Osmanlı Kültüründe Bahçe ve Çiçek (Royal Garden: Gardens and Flower in Ottoman Culture)*. Koç Kültür Sanat ve Tanıtım, Istanbul, Turkey.

Boscana, M.P. (2009) *Les Figueres a Les Illes Balears*. Rústica con solapas, Madrid, Spain.

Boscana, M.P. (2015) Traditional and heritage cultivars of fig trees at the experimental farm of Son Mut Nou. *Acta Horticulturae* 1173, pp. 111–116.

Condit, I.J. (1947) *The Fig*. Chronica Botanica, Waltham, Massachusetts.

Eisen, G. (1901) *The Fig: Its History, Culture and Curing*. Kessinger Publishing, Whitefish, Montana.

Grassi, G. (1991) *The Fig (Il Fico)*. Manuale Pratico, REDA, Rome.

Kolovos, E. (2019) The Mediterranean economies as "garden economies." *Meltem Journal of the Izmir Mediterranean Academy* 5, 98–109.

Lansky, E.P. and Paavilainen, H.M. (2017) *Figs: The Genus Ficus*. CRC Press, Boca Raton, Florida.

Minonne, F., Belloni, P., and de Leonardis, V. (2011) *Fichi di Puglia*. Coop, Ulliside, Castiglione d'Otranto, Puglia, Italy.

Pereira, F.M. (1981) *Fig Culture (Cultura da Figueira)*. Livroceres, Piracicaba, São Paulo, Brazil.

Sutton, D.C. (2014) Figs: a global history. In: Smith, A.F. (ed.) *The Edible Series*. Redaktion Books, London.

Ülkümen, L., Özbek, S., and İleri, M. (1948) *İncir ve Hastalıkları (Fig and its Diseases)*. Yüksek Ziraat Enstitüsü Basımevi, Ankara, Turkey.

2 Remains of the Common Fig (*Ficus carica* L.) in the Archaeological Record and Domestication Processes

Simcha Lev-Yadun*
Department of Biology & Environment, Faculty of Natural Sciences, University of Haifa – Oranim, Tivon, Israel

2.1 Introduction

The common fig (*Ficus carica* L.) is one of the five founder fruit trees or woody vine species that established horticulture in the late prehistoric Near East (Levant) (Zohary and Spiegel-Roy, 1975; Zohary *et al.*, 2012), along with olive (*Olea europaea*), grape vine (*Vitis vinifera*), date palm (*Phoenix dactylifera*), and pomegranate (*Punica granatum*). Defining ancient Near-Eastern horticulture through this group of species is based on reconstructions of the presumed natural distribution of the wild progenitors of these fruit trees and on the appearance of significant quantities of their remains in the archaeological record. The relevant studied remains were found mostly in Chalcolithic sites (mid-seventh to mid-sixth millennia ^{14}C-calibrated years before the present (BP)) in the Near East (e.g. Zohary and Spiegel-Roy, 1975).

All of the five founders of Near-Eastern horticulture can reproduce by seed germination as well as by vegetative reproduction. Some of these trees (common fig, grape wine, and pomegranate) reproduce vegetatively with more ease than others (olive and date palm). For millennia, horticulturists exploited this characteristic and planted orchards with clones expressing the desired fruit characteristics (Zohary and Spiegel-Roy, 1975). In this sense, the most complex figure of this phenomenon can be seen in fig horticulture. While most of the common fig orchards are planted with female clones bearing edible syconia, small numbers of male trees of the common fig are also grown by horticulturists for the sake of pollinating those female varieties where pollination is obligatory for fruit set. However, the majority of modern common fig female varieties are parthenocarpic, which means that in spite of the ability of their main summer crop to be pollinated and produce seeds, in the summer they can produce without pollination commercial amounts of fresh edible figs (Zohary *et al.*, 2012; see also Chapter 23, this volume).

This chapter will refer specifically to the botanical remains of common fig found in archaeological contexts and to fig domestication within its broader cultural significance to prehistoric and historic Near-Eastern agriculture and horticulture.

2.1.1 The wild type of the fig

According to Zohary (1973), the wild type of fig belongs to a swarm of genetically and reproductively compatible forms, the outcome of the

* Email: levyadun@research.haifa.ac.il

 Advances in Fig Research and Sustainable Production (M.A. Flaishman and U. Aksoy)
DOI: 10.1079/9781789242492.0002

spread of feral types of the domesticated common fig among its wild types, and of various combinations of crosses between its domesticated, feral, and wild types. The most prominent wild populations of the common fig are found in the Colchic district of north-eastern Turkey, and in the Hyrcanian district of northern Iran, although populations of what is considered and sometimes proved to be wild common figs are found in many places all over the Mediterranean Basin (Zohary and Spiegel-Roy, 1975; Khadari *et al.*, 2005; Werker, 2006; Melamed *et al.*, 2011, 2016). Moreover, the existence of wild common fig populations in various sites around the Mediterranean Basin provided the pollinating wasp and valuable common fig genetic resources when common fig horticulture spread or arrived from the Near East. Unfortunately, the ancient distribution and routes of spread of the domesticated types are currently not known.

The reproductive biology of the common fig has a bearing on the discussions concerning some critical common fig finds from archaeological excavations, and on the still very partial understanding of the process of common fig domestication, and therefore the relevant aspects will be given briefly below.

The wild type of the common fig is theoretically gynodioecious (female and hermaphroditic plants coexist within a population), but it is functionally dioecious because the hermaphroditic individuals that produce both female and male flowers in their syconia function as males. *F. carica*, being dioecious, thus has two principal tree morphs:

1. The male trees (termed caprificus or goat fig), which produce several generations of non-edible syconia each year, serving throughout the year as a home and a food source for the symbiotic *Blastophaga psenes* wasps, and in the summer as a source of pollen for pollinating the female figs. There are three male synconia generations, mamme, profichi, and mammoni, but only one out of the three (profichi), which ripens in June in Israel and from late May to early June in the Mediterranean region of Turkey (Caliskan *et al.*, 2017), produces significant amounts of pollen and serves as the actual pollen source (Galil and Neeman, 1977).

2. The common fig (the functional female type), to which all the many edible domesticated fig varieties belong.

2.1.2 Types of female common fig varieties

Three types of female fig varieties are known, classified according to their pollination requirement and their reliance upon *Blastophaga* wasps for fruit production, or on their type of parthenocarpy (ripe fruit formation without pollination and without fertilization of ovules). For detailed recent accounts of the classification of fig varieties, see Stover *et al.* (2007), Flaishman *et al.* (2008), Zohary *et al.* (2012), Abbo *et al.* (2015), and Chapter 4 (this volume).

The first group of female common fig varieties consists of those that require pollination for fruit set (classified as non-persistent figs). These are known traditionally as Smyrna figs, due to their abundance in the Smyrna province of Turkey and in adjacent areas, and in modern times, following the establishment of plantations of these varieties in California, as Calimyrna figs (e.g. Stover *et al.*, 2007). To ensure effective pollination of the Smyrna and San Pedro (elaborated below) female that will result in high yield, a population of pollen-loaded female wasps should be maintained. In traditional agrosystems, this is achieved by planting a few wild, feral crosses, or domesticated caprifig (male) trees, or by obtaining branches with ripe profichi-generation syconia at the right season (e.g. Condit, 1947; Kjellberg and Valdeyron, 1984; Flaishman *et al.*, 2008). The dried-fig industry relies mostly on the Smyrna-type varieties, due to their higher sugar content and their capacity to partially dry up while they are still on the tree before they are shed to the ground.

The second group of female common fig varieties is those that require no pollination to produce edible syconia crops. They are often called parthenocarpic common fig varieties, but as the syconium is not a true fruit but rather a pocket made out of a modified branch, a more botanically appropriate term for these varieties is "persistent." The parthenocarpic common fig types have the capacity to develop (rather than drop off) and mature without fertilization of the female flowers, and thus without seed set. However, their main summer crop can be pollinated and produce viable seeds. These parthenocarpic/persistent varieties are the major source for the current fresh-fig industry. In addition to the main summer crop, common fig varieties may produce additional amounts of ripe fruit without pollination, either prior to the main fruiting season, or sometimes even after it. These fruits (known as breba) are seedless, and because they ripen out of the main fig fruiting season, they are highly appreciated (Flaishman *et al.*, 2008).

The third group is known as San Pedro types, and produce both a significant early breba crop and, following pollination, a large main summer crop (Flaishman *et al.*, 2008).

2.2 Archaeological Remains of the Common Fig

2.2.1 General considerations regarding archaeobotanical studies of fig remains

When dealing with the data, hypotheses, and understanding of the significance of archaeological and some genetic finds of the common fig, recent developments made over the last 50 years in the archaeobotanical research in the Levant should be considered. For many decades, archaeobotanical research lagged behind other studies such as those of pottery or architecture. Until 1970, collecting and documenting wood remains found in archaeological excavations in the Levant (mostly as small pieces of charcoal) was anecdotal and their study was non-systematic. Similarly, studying archaeological fruit and seed remains from the Levant gradually became systematic only after the mid-1970s. Therefore, botanical remains were not sampled and thus not studied in critical or even for monumental excavations, such as that of Tel Megiddo carried out by the Oriental Institute of Chicago (1925–1939), or Tel Hazor, which was excavated in the 1950s by the Hebrew University of Jerusalem expeditions.

As fig seeds (pips) are tiny, they can easily be overlooked when they do not appear in compact masses. They are therefore hard to find in archaeological sites that use crude excavation methods. Delicate dry sediment sieving and floating of excavated archaeological sediments in water tanks help in retrieving the minute common fig seeds. It is also important to remember that it is impossible to distinguish between seeds of wild common fig types and those of domesticated or feral types (Zohary *et al.*, 2012). However, with the advance of systematic and careful sampling of archaeological sediments for botanical remains in the last several decades, fruits and seeds of the common fig have been identified in an increasing number of archaeological sites.

Archaeobotanical fig remains appear in several shapes: (i) wood—waterlogged, charred, or dry (Lev-Yadun, 2007); (ii) fruits (Zohary and Spiegel-Roy, 1975); (iii) seeds (Melamed *et al.*, 2011); (iv) pollen (Langgut *et al.*, 2013); and (v) leaf prints (Frumin and Tchekhanovets, 2016). In addition, they are described in religious and art items and texts (see Chapter 3, this volume). Because of its pollination via the symbiotic wasp that transfers small amounts of pollen from closed profichi male syconia to closed female ones, *F. carica* pollen is very rarely found in archaeological and geological sediments; however, it was found in sediments of a Late Bronze Age (mid-14th century BCE) latrine in Megiddo Israel (Langgut *et al.*, 2016), and in the *c.*2400-year-old plaster of the water pool in the Royal Persian garden of Ramat Rahel on the southern outskirts of Jerusalem (Langgut *et al.*, 2013).

2.2.2 Archaeological fruit and seed remains of the common fig

The earliest archaeological evidence for exploitation of the wild common fig is from the 780,000-year-old Middle Paleolithic site at Gesher Benot Ya'aqov, located in the upper Jordan Valley, Israel (Fig. 2.1), where both wood (Werker, 2006) and seeds (Melamed *et al.*, 2011, 2016) were found. Common fig fruits were also found in the 23,000-year-old site of Ohalo II, located on the south-western shore of the Sea of Galilee, Israel (Fig. 2.1; Weiss, 2017). Another site where pre-domestication fig seeds were found is Atlit-Yam (Fig. 2.1), dated to the Pre-Pottery Neolithic C (9200–8500 ^{14}C-calibrated years BP) (Hartmann-Shenkman *et al.*, 2015). These three sites are submerged sites, and their sediments were waterlogged, and therefore the anaerobic environment led to a rare good state of preservation of many delicate botanical remains. Weiss (2015) reviewed the finds of the wild common-fig seed remains in prehistoric sites in the Levant. Fig seeds were found in Pre-Pottery-Neolithic A strata (predating agriculture) at Netiv Hagdud, located in the lower Jordan Valley (Fig. 2.1),

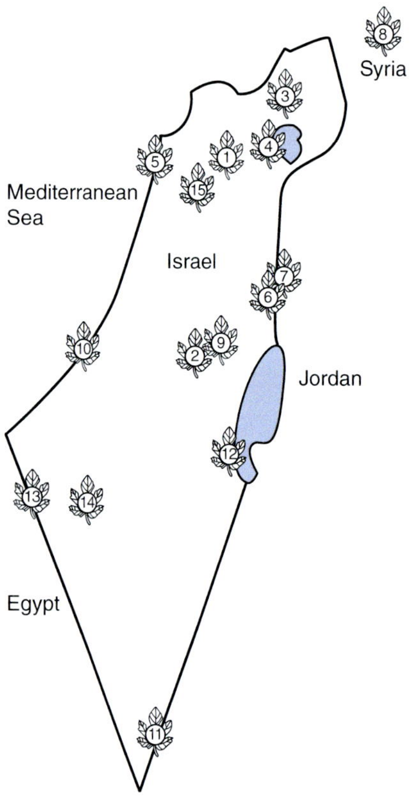

Fig. 2.1. Map of archaeological sites (indicated by fig leaves) where fig botanical remains have been found in the Israel area. The archaeological sites listed are as follows: 1, Zippori; 2, Ramat Rahel; 3, Gesher Benot Ya'aqov; 4, Ohalo II; 5, Atlit-Yam; 6, Netiv Hagdud; 7, Gilgal; 8, Tell Aswad; 9, Jerusalem; 10, Ashkelon; 11, Timna; 12, Masada; 13, Shivta; 14, Nessana; 15, Megiddo.

about 12 km north of Jericho (Kislev, 1997), and in the somewhat later nearby site of Gilgal (Fig. 2.1; Kislev *et al.*, 2006). Fig seeds from the Pre-Pottery Neolithic B, the period of the beginning of field crops agriculture but still predating horticulture by millennia, were found in Tell Aswad, located not far from Damascus, Syria (Fig. 2.1; van Zeist and Bakker-Heeres, 1982). It is very important and even critical to know that the erroneous dating of Tell Aswad to the Pre-Pottery-Neolithic A by its first excavator Henry de Contenson, who excavated it in the early 1970s, as it appears in various publications, was a mistake that caused various misunderstandings concerning the date and place of the beginning of agriculture. Tell Aswad was later redated correctly to the Pre-Pottery Neolithic B (range between 10,450 and 10,150 ^{14}C-calibrated years BP), based on re-examination of its flint tools from the older excavation (Gopher, 1994), followed by a renewal excavation conducted in the early 2000s, where ^{14}C dating was executed (Stordeur *et al.*, 2010). Seeds of common fig were found in various other Near-Eastern Pottery Neolithic sites (reviewed by Weiss, 2015). It is clear that figs were used as food a very long time ago, by Paleolithic pre-*Homo sapiens* but still unidentified hominins (e.g. Melamed *et al.*, 2011, 2016), and that they continued to be used in the Upper Paleolithic and throughout the Neolithic periods, until just before the proposed period of common fig domestication in the Chalcolithic period some 6500 years ago (Zohary and Spiegel-Roy, 1975; Weiss, 2015, 2017).

It seems that after horticulture was established in the southern Levant some 6500 years ago (probably earlier for some species such as olive), figs became a commonly consumed fruit (Zohary and Spiegel-Roy, 1975). However, the location(s) where domestication first occurred is not that easy to decipher. In addition to the complicated archaeological issues, and the luck needed for preservation of plant remains for millennia, as mentioned above, domesticated common fig includes dried-fig-producing varieties, which can be preserved for relatively long periods and can be traded over relatively long distances. Thus, finding common fig fruits or seeds in a given location does not necessarily indicate that it was grown in the environs of the find.

Most of the archaeobotanical remains of domesticated common fig from the Levant, dated to the Chalcolithic period, Early Bronze Age, and later periods, were of their fruit or seeds (see Zohary *et al.*, 2012; Frumin *et al.*, 2015; Weiss, 2015; and citations therein), illuminating the importance of the common fig as a fruit. Charred remains of fig wood were found in Tel Tsaf, northern Israel, dating to the Chalcolithic period (i.e. *c.*7000 years BP; Fig. 2.2). In ancient Israel, seeds or fruit remnants of the common fig were found in several famous sites, such as Jerusalem (Abbasid period, eight to tenth centuries CE; Amichay *et al.*, 2019), Iron Age Ashkelon (Weiss and Kislev, 2004), Iron Age Timna (Ben-Yosef *et al.*, 2017), the early Roman period at Masada (Kislev and Simchoni, 2007), and Early Islamic (mid-seventh to mid-eighth century CE) Shivta, Negev Desert, Israel (Dunseth *et al.*, 2019) (Fig. 2.1).

However, contrary to the situation in the Near East where fig was not a staple, it seems that the common fig served as a staple for the pre-Hispanic Guanches population of the Canary Islands (e.g. Morales *et al.*, 2014).

2.2.3 Archaeological wood remains of the common fig

Compared with the archaeological finds of common fig's fruit and seed remains, its wood remains are quite rare, and when found in archaeological contexts, it usually comprises only a small fraction of the wood assemblage. The wood of the common fig tree is of low quality, so it does not provide long and strong beams, and it is not known to have been traded in the same way as high-quality timbers used for building such as oak, pine, cedar, and cypress. Therefore, when fig tree wood remains are found in an archaeological context, it may be deduced that the fig tree grew in the environs of the site.

Common fig wood remains were found in very small amounts in the 780,000-year-old Middle Paleolithic site of Gesher Benot Ya'aqov, Israel (Werker, 2006). In this case, it certainly represents wild trees, and, importantly, attests to the existence of wild common fig populations outside their main natural distribution in eastern Turkey and northern Iran. Some wood charcoal from wild common fig trees was found in Pre-Pottery Neolithic B strata in the site Nahal Zippori 3, in the Lower Galilee, Israel (Caracuta *et al.*, 2014). Wood charcoal from common fig trees that were probably domesticated was found in several Bronze Age sites in southern Turkey and northern Syria (Deckers and Pessin, 2010). A special find was the wood charcoal of construction timbers of two buildings from the site of Akrotiri on the remains of the island Thera (Santorini), which were preserved under a thick layer of volcanic ash. The houses were repaired after being damaged by severe earthquakes that predated the enormous Plinian eruption of the Thera volcano, *c.*3600 years ago (Late Bronze Age). Most of the repairs were done using high-quality timbers such as oak, pine, and olive, but the inferior timber of the common fig was also used in smaller amounts (Mavromati, 2021). A special find of some fourth-century BC waterlogged wooden items from south-east Spain revealed that common fig wood was sometimes used for the production of delicate furniture parts and handles (Carrión and Rosser, 2010). Common fig wood was also used for construction in the Roman

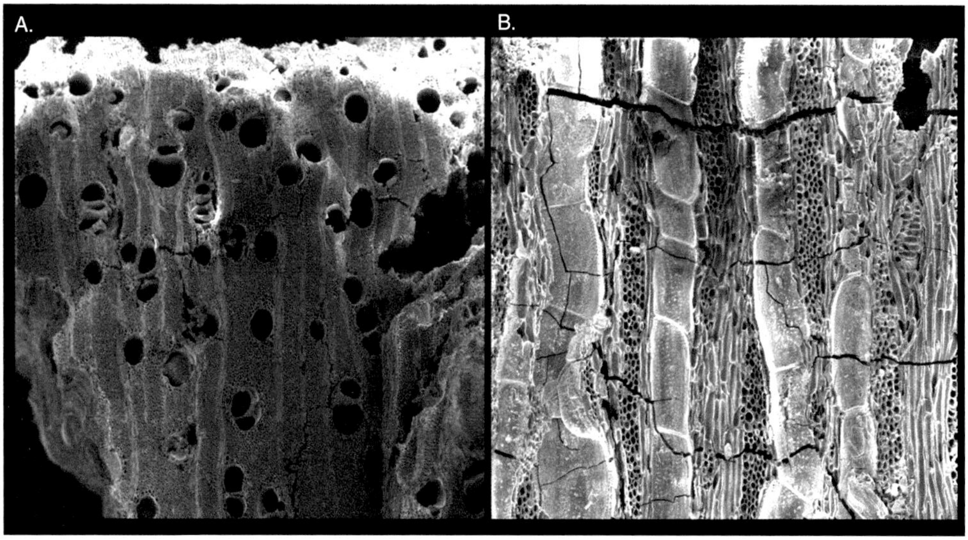

Fig. 2.2. Scanning electron microscopy images of charred wood sections of common fig from Tel Tsaf, Israel, dated to *c.*7000 years ago. A. A tangential longitudinal section, with the typical chains of vessel members. B. A cross-section, with the typical dark vessel members. Images were taken using a Tescan VEGA3 LMH scanning electron microscope. Figure used with permission of Dr. Dafna Langgut.

phase of the Spanish site of the San Chuis hill fort (Badal *et al.*, 2013). Roth *et al.* (2019) suggested that the common fig wood charcoal found in strata of the Early Roman period lower city of Jerusalem represent the use of common fig tree pruning material as domestic firewood. Similarly, common fig wood charcoal found at the Roman period strata of Ses Païsses on Mallorca Island were also classified as firewood remains (Picornell-Gelabert and Carrión Marco, 2017). In the Byzantine (fifth to sixth centuries CE) Negev Desert, Israel, common fig wood charcoal was found in the large sites of Shivta and Nessana, indicating that the common fig was grown there as a fruit tree under irrigation, and that its pruned branches served as firewood (Langgut *et al.*, 2021). The non-typical use of the inferior timber of the common fig as an important construction material along with the high-quality timber of *Pinus canariensis* was discovered in communal granaries of the Guanches, the pre-Hispanic indigenous extinct nation of the Canarian Archipelago (Vidal-Matutano *et al.*, 2021). The common fig was an important component of the diet of the Guanches (Morales *et al.*, 2014), and, in accordance, they used its timber for construction more than in other cultures. However, it seems that overall, common fig wood was usually of minor importance either as firewood or as timber.

2.2.4 Archaeological leaf prints of the common fig on pottery

One of the special archaeological common fig finds is indirect, in the shape of leaf imprints on pottery from Hellenistic Jerusalem (Frumin and Tchekhanovets, 2016), indicating that leaves were pressed into the wet clay by a potter. It is unknown how common this decorative practice was. For further discussion see Chapter 3, this volume.

2.3 Domestication of the Common Fig

2.3.1 Common fig domestication: the basic questions

The scientific attention given to fruit-tree domestication was much less than that given to annual grain crops (Goldschmidt, 2013), and this was also true of domestication of the common fig. Reliable answers to several major questions are required in order to have a good understanding of the domestication process of the common fig: (i) How to define domestication of female common fig? (ii) When was the common fig domesticated? (iii) Where was it domesticated? (iv) Was it domesticated only once and in one place? (v) Was the first domestication phase done via clonal propagating by cuttings, propagation by seeds, or by both methods? (vi) How many genotypes were used the first time it was domesticated? (vii) Were additional genetic resources added to the primary domesticated stock, and if yes, when and where? (viii) When, where, and how many times was the male (caprifig) type domesticated? It is probable that two separated lines of domestication processes occurred, as female trees were probably domesticated first where large populations of figs existed; and, (ix) when and where did the Smyrna, San Pedro, and main summer crop parthenocarpy appear? Many of these questions cannot be addressed with the current state of fig domestication studies, because of the lack of relevant archaeological, botanical, and genetic data.

It is clear that the archaeological and archaeobotanical records per se are silent concerning the genetic aspects of these questions, and that we do not have ancient materials that will provide DNA samples of the first stages of fig domestication. It is hoped that future genetic studies and additional common fig finds from archaeological excavations will at least partly illuminate some of these questions, such as the routes of spread from the Near East.

2.3.2 Common fig domestication: general theoretical considerations

Studying domestication in general, and domestication of the common fig in particular, requires an understanding of the theoretical and practical differences between original or primary domestication versus secondary and repeated domestication. An original/primary domestication has two mandatory components: (i) the idea, or conscious decision of domesticating a new crop or a package of crops; and (ii) the identity of the initial genetic stock that was taken from the wild, which established the crop. In many crops, there are cases of secondary domestications by deliberate addition of wild genotypes, feral types or crosses of domesticated types with wild or feral types, based on the previous original idea of domestication. In the history of domestication, there are various cases in which people from territories outside that of the original site or region of the first domestication, who knew that a certain crop was domesticated, took wild types found where they resided and domesticated them. In such cases, it is not an original domestication from the cultural/knowledge point of view, but rather a secondary one, as the idea emerged earlier in another cultural and genetic setting. Concerning the founders of Near-Eastern horticulture, there are two well-known and genetically proven cases of secondary domestication: (i) in olive, where several wild types were domesticated and sometimes even crossed with the original types that arrived in southern Europe or the Iberian Peninsula from the Near East (Besnard *et al.*, 2013, 2018; Langgut *et al.*, 2019); and (ii) in grape wine, where wild European types that are more resistant to low temperatures and have good winemaking qualities were added to the domesticated types (Myles *et al.*, 2011). There is no cultural, genetic, ecological, or agronomic reason to assume that the situation for wild types, feral types and crosses of common fig types is different.

In addition to the need to distinguish between primary and secondary domestication, there is also a need to distinguish between the vague term "cultivation" and the term "domestication." Collection of female common fig vegetative plant material/seeds from wild trees bearing the desired fruit quality in order to propagate it and establish a clone or a seed-based population in an orchard or a garden would be considered here as an act of domestication. However, the cultural and genetic status of domestication in the early stages of horticulture is not well defined and is actually still unknown concerning all five founder species of Near-Eastern horticulture. Defining "cultivation" and distinguishing it from domestication is not within the scope of this chapter. An in-depth discussion of the differences between these two terms is given by Gopher *et al.* (2021). There is also much confusion in the literature between domestication and crop evolution under domestication (Ladizinsky, 1998; Abbo *et al.*, 2014b). Domesticated crops gather mutations and other genetic changes endogenously, and, in addition, new genetic materials are deliberately added by the farmers (e.g. Ladizinsky, 1998; Abbo *et al.*, 2014b), as well as by chance introgression from the wild (e.g. Cruz *et al.*, 2016), or in the case of the common fig, from other domesticated stocks, by crosses with feral or wild male individuals, or with domesticated caprifigs.

There is a common trend of viewing the domestication of the founder Near-Eastern Neolithic field crops as an unconscious process (e.g. Zohary, 2004; Purugganan and Fuller, 2011). A contradictory view is that their domestication was a conscious and knowledge-based short process (e.g. Lev-Yadun *et al.*, 2000; Abbo *et al.*, 2010a,b, 2013, 2014a; Gopher *et al.*, 2021). While there are contradicting theoretical approaches regarding domestication processes of Near-Eastern Neolithic field crops, as far as fruit trees are concerned, it is agreed that their domestication was a conscious and immediate act (as termed by Gopher *et al.*, 2021).

When trying to understand the origin of domesticated types, there is another principal theoretical difference between annual field grain crops and fruit trees. Cereal grain crops in particular, and legumes possibly in a similar but milder way, were not domesticated as chosen exceptional individuals like fruit trees but rather as large and genetically variable populations in the form of fields. The original cultivated field crop population of a given cereal species was composed of many thousands of seeds, originating in a wild population, and representing a certain part of the species' natural genetic variation (Ladizinsky, 1985). This enables comparison of the genetic composition of the current domesticated population (or ancient ones via fossil DNA) with that of various current wild populations, and finding which current wild population is genetically closest to the domesticated stock. This method allowed identification of the origin of einkorn wheat in the environs of the Karacadağ in south-eastern Turkey (Heun *et al.*, 1997), where, as was found later, several other founder grain crops (emmer wheat, lentil, chickpea, and two out of three barley spike disarticulation types) also originated (Gopher *et al.*, 2021, and citations therein). In the domestication of the Near-Eastern Neolithic field crops, the breakdown of two important wild-type adaptations was involved: seed dispersal (more important in cereals) and seed dormancy (more important in legumes) (Zohary, 1996; Gopher *et al.*, 2021). In the five founder species of fruit trees, there are no such critical characteristics. Therefore, it is much harder to do this with fruit trees, in which exceptional individuals were chosen to establish clones, and when we have no archaeological or genetic information on which clone was taken first, and what the order of domesticating other clones was. It is also harder to study the origin of fruit trees because of the existence of many genetic swarms of wild, domesticated, feral types, and crosses among these types, which coexist simultaneously. Thus, from the genetic point of view, it is certain that the common fig was domesticated more than once.

Because of the potential clonality, there is an inherited problem in defining domestication concerning all five founder fruit trees of the Near East. If, for instance, an exceptional female individual of the common fig, with large, juicy, sweet syconia, was vegetatively cloned by cuttings in order to establish the first orchard of that desired genetic line, that genetic line was simultaneously both wild and domesticated. In order to have a chance of acquiring genetic differences between the wild type and the domesticated form, the trees should either have a very rare spontaneous beneficial somatic mutation, as happened in the 19th century with the mutation that developed into the Jaffa orange, locally known as Shamouti, or should pass some generations of sexual reproduction followed by selection of desired progeny by the early horticulturists. In fruit trees that are propagated by seeds such as the almond, in which a mutation of losing its strong seed toxicity clearly differentiates the wild type from the domesticated (Browicz and Zohary, 1996; Ladizinsky, 1999a), genetic evidence points to a single domestication event (Delplancke *et al.*, 2013). Concerning the common fig, we are still in the dark about all aspects of the early stages of its domestication.

The fact that there are many swarms of domesticated female common figs—feral types, wild types, and various combinations of crosses between them (e.g. Zohary, 1973)—makes it extremely complicated in many cases to decide whether certain common fig populations or individuals are domesticated, feral, wild, or the progeny of a cross. For instance, when I was young, the 1–2 km long and hundreds of meters wide belt of sand dunes of western Herzliyya (my hometown) were what seemed to be the original habitat of many hundreds if not thousands of common fig individuals, including many female trees that produced edible syconia. Some of them were multi-stemmed and shrub-sized, while others were small trees. Many of them were males, but the female ones were both the dark purple and the yellow types, expressing a great variability in syconia size, juiciness, and sweetness, and in tree productivity. I never knew which of them was domesticated or wild. The great genetic heterogenicity of the common fig (e.g. Janick, 2005) makes it difficult to use genetic markers to determine what is domesticated and what is not. Twenty years of studying *Arabidopsis thaliana* after sequencing its genome (Kaul *et al.*, 2000) and about 9500 citations of it left many unsolved issues in *A. thaliana* biology. Therefore, I suspect that even the recent sequencing of the whole genome of the common fig (e.g. Mori *et al.*, 2017) with 31 citations will allow only a partial understanding of fig domestication.

When dried fig fruits were traded over the millennia, some of their seeds germinated and established feral common fig populations in various environs of archaeological sites along trade routes. Sometimes these feral populations were established within habitats that had wild common fig populations (and thus could be colonized by the pollinating wasp), as well as in habitats that did not harbor wild common figs. However, these feral populations could serve as corridors for the spread of the pollinating wasps. They included both female and male trees and should have influenced the genetic composition of both domesticated female and caprifig populations.

At some unknown point in time, female common fig types with fully parthenocarpic main summer crops emerged, and there and then males of the common fig ceased to be essential. This possibly occurred through mutations. Those rare mutated individuals were identified in the wild or among feral trees and crosses, and were probably vegetatively propagated.

The fact that all five founder species of Near-Eastern horticulture can be reproduced vegetatively with more or less ease and produce clones bearing desired characteristics (Zohary and Spiegel-Roy, 1975), does not exclude the possibility of seed propagation during the early stages of domestication. It should be remembered that from the beginning of agriculture 10,500 years ago until the origin of horticulture some 6500 years ago, for almost 4000 years farmers used seeds for growing field crops, such as the eight founder grain crops—emmer wheat, einkorn wheat, barley, pea, lentil, chickpea, bitter vetch, and flax. Even if some of the founder fruit trees of horticulture were domesticated before the Chalcolithic period, (i.e. during the late Pottery Neolithic, which seems to be the case for olive and grape vine), the farmers that did this had a history of some 3000 years of seed-based agricultural practices and crop propagation. Therefore, it is probable that the farmers who sowed their annual grain crops every single year also tried to propagate desired fig types via seeds. Unfortunately, the archaeological record does not indicate whether the common fig was propagated only vegetatively at the beginning of its domestication as is done today, or whether it was also or only by seeds. In nature, common figs may propagate vegetatively in riparian habitats, but elsewhere, propagation is mostly if not solely by seeds. Propagating fruit trees such as the common fig by seeds has its complications, and Janick (2005) described several difficulties emerging when propagating desired female common fig individuals by seeds. First, the common fig was already found long ago to be genetically highly heterogeneous. Therefore, propagation of a desired common fig fruiting type by seeds is usually not a viable practice in modern horticulture because a significant part of the new progeny will not be fertile and half will be males (Rottenberg, 1998). When establishing an orchard from seeds, there will be a juvenile phase of several years. Today, the juvenile phase in common fig is not very long in some of the progeny of the current high-yielding varieties, and significant fruit production can be achieved in some of the progeny grown from seeds of current varieties by the fourth year after sowing (Flaishman *et al*., 2017). However, it should be noted that current common fig varieties are the result of millennia of selection toward high productivity, whereas productivity was not so strong in the early stages of horticulture. Moreover, progenies that produce low-quality syconia that are not in accord with modern customers and market needs are soon discarded. For instance, out of more than 50,000 seedlings, the progeny of 154 cross combinations of 60 common fig female and 26 male parents, 60 promising seedlings were secured, ultimately resulting in 20 new Israeli common fig cultivars (Flaishman *et al*., 2017). It is not known what the original average length or range of juvenility was. There are no data about the proportion of the progeny that produced edible figs in general in such progeny, or whether good productivity was expressed by the juvenile plants some years later. The lack of such data does not allow us to reach a reliable evaluation of the possibility that the common fig trees (and probably other founder species of horticulture) were grown from seeds before vegetative reproduction of desired clones became known to the farmers and dominated horticulture. In any case, it is highly probable that the first domesticated common fig generations were not as productive as the current domesticated fig varieties, and an average juvenile phase of at least 10 years might better represent the ancient domesticated types. However, in the early periods, when agriculture occupied only a small fraction of the land (Gophna *et al*., 1986; Liphschitz *et al*., 1989; Lev-Yadun, 2012), sowing many seeds in order to arrive at some good fruit trees, fig included, could have been a viable practice, even when human life expectancy was much shorter than today (Hassan, 1981).

Similar to the grape vine and pomegranate, fig trees can easily be propagated from branch or stem cuttings. Branch cuttings with several nodes are taken in the late winter, and when planted in wet soil will readily root and resume growth when temperatures rise in the early spring after bud break. Such trees, planted from branch cuttings, are capable of commercial fruiting within 2–3 years, depending on soil fertility and other environmental factors (Stover *et al*., 2007; Flaishman *et al*., 2008). Like many deciduous fruit trees, figs require yearly pruning prior to the termination of their winter dormancy (Flaishman *et al*., 2008). These seasonal pruning operations are imperative to control tree shape, restrain its aggressive growth (and size) and facilitate efficient fruit harvest. This provides common fig growers with ample plant material, enabling both propagation from cuttings, and with firewood. Logically, millennia of clonal common fig propagation via cutting should have selected for types that have a stronger tendency to root than the wild type or earlier domesticated types, but this aspect has not been studied yet.

2.3.3 The timing of common fig domestication

One should note that the beginning of fig (or any crop, or even farm animals) domestication should not be confused with the arrival of a crop to a new territory. An in-depth discussion of this problem is beyond the scope of this chapter.

There are several hypotheses concerning the timing of common fig domestication. The classic and widely accepted hypothesis suggests that the package of five founder fruit trees that established Near-Eastern horticulture was either domesticated or already domesticated in the Chalcolithic period some 6500 years ago (Zohary and Spiegel-Roy, 1975). Weiss (2015), while accepting the major lines of the hypothesis made by Zohary and Spiegel-Roy (1975), suggested that the common fig was already widely domesticated in several countries in the Early Bronze Age but that its domestication was rooted in the Chalcolithic period. A dramatic hypothesis raised by Kislev *et al.* (2006) suggested that the common fig was domesticated in the lower Jordan Valley 11,400–11,200 years ago, already in the Pre-Pottery Neolithic A, thus even before the emergence of grain crop agriculture at *c.*10,500 years ago. However, this hypothesis has been rejected by others (Lev-Yadun *et al.*, 2006; Denham, 2007; Zohary *et al.*, 2012; Abbo *et al.*, 2015; Weiss, 2015).

The above-mentioned three hypotheses were based on the limited information available at the time. When Zohary and Spiegel-Roy (1975) suggested that horticulture emerged in the Chalcolithic period, there were no DNA-based genetic tools to study parenthood and genetic relatedness. Moreover, much of the current botanical data coming from archaeological excavations had not yet been collected. Kislev *et al.* (2006) had much more botanical data from archaeological excavations than Zohary and Spiegel-Roy (1975), but they did not consider the relevant reproductive biology of various domesticated common fig types. Weiss (2015) had a much broader archaeobotanical record at his disposal, and in general corroborated Zohary and Spiegel-Roy's (1975) suggestion. There is evidence that it is possible that Near-Eastern horticulture began as early as the late Pottery Neolithic, but discussing this in depth will be done elsewhere.

Kislev *et al.* (2006) described the find of nine carbonized seedless parthenocarpic figs found at Gilgal I, dating to the Pre-Pottery Neolithic A. In this study, the authors concluded that as modern fig varieties with a parthenocarpic main summer crop produce no seeds, the seedless trees could only have been propagated by cuttings. Thus, Kislev *et al.* (2006) posited that these seedless female common figs represent the first domesticated plant species of the Neolithic agricultural revolution. This means that a fruit tree was the first domesticated plant, rather than several field grain crops, as was assumed prior to their suggestion (Zohary and Hopf, 2000). If Kislev *et al.* (2006) were correct, it would have dramatically changed the understanding of the origin of Near-Eastern agriculture, as it would predate the beginning of the agricultural revolution by 700–900 years, from 10,500 years ago to 11,200–11,400 years ago. Moreover, in this case, it would mean that a conscious rather than an unconscious process led the agricultural revolution, and that domestication was an immediate act rather than a protracted one. The notion that domestication was a conscious and not a protracted act contradicts the hypothesis advocated by the Bar-Ilan archaeobotany research group and by others (e.g. Zohary, 2004; Purugganan and Fuller, 2011). However, Kislev *et al.* (2006) based the grounds for their assumption on a botanical mistake, as all parthenocarpic common fig types produce viable seeds when pollinated (Flaishman *et al.*, 2008), thus making this hypothesis unacceptable.

The hypothesis made by Kislev *et al.* (2006) was immediately criticized. Based on the well-known reproductive biology of the common fig, Lev-Yadun *et al.* (2006) suggested an alternative explanation. Lev-Yadun *et al.* (2006) posited that the three known domesticated female fig tree types were: (i) Smyrna types, which in certain cases produce some seedless breba but must be pollinated in order to produce their main summer crop with its viable seeds; (ii) San Pedro types, which produce significant amounts of breba but also must be pollinated to have a main summer crop with viable seeds; and (iii) common fig varieties, which may produce parthenocarpic breba, a facultative parthenocarpic main summer crop, and occasionally some parthenocarpic autumn figs. The facultative seedless parthenocarpic main summer crop of such common fig varieties produces seeds when male trees producing profichi syconia harboring pollinating wasps are present, or when caprification is practiced. As fig females naturally produce both seed-bearing and seedless syconia, they are thus capable of seed-based spontaneous reproduction. Thus, Lev-Yadun *et al.* (2006) concluded that the unique prehistoric seedless finds described by Kislev *et al.* (2006) neither essentially indicated small-scale common fig cultivation, notwithstanding domestication, nor did they prove that horticulture predated annual grain crops agriculture. Lev-Yadun *et al.* (2006) suggested that it is more parsimonious that the parthenocarpic figs described by Kislev *et al.* (2006) were breba. It is very important to stress that most of the critics of the suggestion given by Kislev *et al.* (2006) that the common fig was the first domesticated crop (i.e. Lev-Yadun *et al.*, 2006; Denham, 2007; Abbo *et al.*, 2015; Weiss, 2015) are not well known and not well cited, and therefore, in too many publications, the misleading hypothesis made by Kislev *et al.* (2006) was considered to be valid, introducing to the scientific literature a mistake that will be very hard to correct. The date of appearance and the region(s) of origin of the San Pedro types and the parthenocarpic summer crop types are still an enigma.

2.3.4 The place of common fig domestication

From genetic studies, it is clear that at least six out of the eight Near-Eastern Neolithic founder grain crops—emmer wheat, einkorn wheat, lentil, chickpea, pea, and flax—were domesticated only once (Zohary, 1996; Fu and Allaby, 2010; Gopher *et al.*, 2021; and citations therein). Moreover, emmer wheat, einkorn wheat, two out of three barley spike disarticulation types, lentil, and chickpea were domesticated in a small region in the northern Levant (e.g. Heun *et al.*, 1997; Ladizinsky, 1999b; Lev-Yadun *et al.*, 2000; Salamini *et al.*, 2002; Gopher *et al.*, 2021). Furthermore, from the archaeological/cultural point of view, all eight founder crops of the Near-Eastern Neolithic agriculture were domesticated together in that small core area and in a short single event (Lev-Yadun *et al.*, 2000; Gopher *et al.*, 2001; Gopher *et al.*, 2021). For fig domestication, the genetic picture is profoundly different. The multiclonality of common fig orchards points toward several or even many genetically (but not culturally) independent domestication events for common fig in the Near East and wherever it was grown in the Mediterranean Basin within the range of the wild type. This long process, occurring over millennia, resulted in thousands of varieties, most of which have never been described (Flaishman *et al.*, 2008). Such non-dated events of adding local genetic material to the domesticated stocks happened in numerous places within the vast area of distribution of the wild common fig. Later, or simultaneously, feral types or crosses between domesticated female clones and local wild or feral male trees added various genotypes to the domesticated gene pool. These additions probably occurred because local wild types and some of the crosses were better adapted to some specific local biotic or abiotic environmental conditions than the domesticated type(s) that had arrived in the area. Another possible explanation would assume that local wild types, feral types, and crosses included individuals with fruit qualities that were preferred and thus preserved by local farmers. Deliberate introductions of domesticated varieties from other countries added to the complex genetic mosaic. For these reasons, it is still unclear whether common fig domestication was an independent innovation in a single place that gradually spread all over the Mediterranean Basin, or whether it was in the shape of multiple culturally unrelated events. As the Near East should be regarded as a monocultural unit from not later than the Pre-Pottery Neolithic A (Gopher, 1994; Gopher *et al.*, 2021), the assumption that the Chalcolithic invention of horticulture could have occurred independently, several times, in this restricted area is unlikely. The clonality of the female common fig orchards, and the lack of anatomical differences between the wild types and the domesticated figs, challenge the attempts to understand the exact mode of fig domestication and the location(s) where it occurred.

2.4 Caprifig Domestication

One of the reasons that caprifig domestication is a very complicated and poorly understood subject is that, although female trees comprise most of the domesticated types and orchards, various types of male trees (caprifigs) were also domesticated in order to provide the essential pollination for many female varieties. It should be remembered that caprifigs were studied much less than the female fig types.

Wild male trees could probably have provided the essential pollination when the common fig was grown in the first stages of female common fig domestication, because the domesticated orchards were still small and located within the region where wild male populations existed. Because, at an unknown stage, the plantations of the common fig reached a size that was too large to be sufficiently pollinated by the wild caprifig populations, domestication of caprifig trees must have started not very long after domestication of the female type. It is not known when the parthenocarpic summer crops emerged; however, if, before that stage, all domesticated female common figs required pollination of their main summer crop, when the crop spread outside its natural habitats, growing of male trees became essential. Where and when this happened is yet not known, nor when male caprifigs ceased to be essential everywhere (when common fig types with fully parthenocarpic main summer crops emerged). All these questions remain understudied.

Caprifigs express significant differences in many characters, including those that matter a lot to farmers, such as seasonality and productivity, both pollen-wise and wasp-wise. The genetic diversity of caprifigs is well manifested in the different genetic stocks that exist in various countries (e.g. Dalkiloç *et al.*, 2011; Caliskan *et al.*, 2017; Bilgin *et al.*, 2020; Mirheidari *et al.*, 2020). As the population genetics of the thousands of fruiting female common fig varieties has not yet been studied within the perspective of the whole wild and domesticated genetic range of the species, it is not a surprise that our partial knowledge about the caprifig is only regional and even more restricted than our limited knowledge about the female type.

2.5 Conclusions

According to botanical remains from archaeological excavations, fruit of the wild common fig were consumed by hominins at least hundreds of thousand years before the emergence of our species (*Homo sapiens*). After the emergence of common fig horticulture *c.*7000 years ago or somewhat earlier, it gradually spread around the Mediterranean Basin and beyond. Genetic material was added to the initial domesticated crop many times from both wild resources and from crosses between types. The details of this millennia-long complicated process of evolution under domestication of both the female and male types are mostly unknown. Common domesticated fig trees provided not only fruit (fresh and dried) and fruit fermentation products but also firewood, timber, and various leaf products.

Acknowledgements

The author thanks Dvory Namdar, Volcani Center, Israel, for her valuable comments on the manuscript, and Dafna Langgut, Head of the Laboratory of Archaeobotany and Ancient Environments at Tel Aviv University, Israel, for the photographs in Fig. 2.2.

References

Abbo, S., Gopher, A., and Lev-Yadun, S. (2015) Fruit domestication in the Near East. *Plant Breeding Reviews* 39, 325–377.

Abbo, S., Lev-Yadun, S., and Gopher, A. (2010a) Agricultural origins: Centers and noncenters; a Near Eastern reappraisal. *Critical Reviews in Plant Sciences* 29, 317–328.

Abbo, S., Lev-Yadun, S., and Gopher, A. (2010b) Yield stability: an agronomic perspective on the origin of Near Eastern agriculture. *Vegetation History and Archaeobotany* 19, 143–150.

Abbo, S., Zezak, I., Zehavi, Y., Schwartz, E., Lev-Yadun, S., and Gopher, A. (2013) Six seasons of wild pea harvest in Israel: bearing on Near Eastern plant domestication. *Journal of Archaeological Science* 40, 2095–2100.

Abbo, S., Lev-Yadun, S., and Gopher, A. (2014a) The 'Human Mind' as a common denominator in plant domestication. *Journal of Experimental Botany* 65, 1917–1920.

Abbo, S., Pinhasi van-Oss, R., Gopher, A., Saranga, Y., Ofner, I., and Peleg, Z. (2014b) Plant domestication versus crop evolution: a conceptual framework for cereals and grain legumes. *Trends in Plant Science* 19, 351–360.

Amichay, O., Ben-Ami, D., Tchekhanovets, Y., Shahack-Gross, R., Fuks, D., and Weiss, E. (2019) A bazaar assemblage: reconstructing consumption, production and trade from mineralised seeds in Abbasid Jerusalem. *Antiquity* 93, 199–217.

Badal, E., Carrión Marco, Y., and Jordá, H.F. (2013) Charcoal analysis at the San Chuis hill fort (Allande, Asturias, Spain). *Sagvntvm Extra* 11, 157–158.
Ben-Yosef, E., Langgut, D., and Sapir-Hen, L. (2017) Beyond smelting: new insights on Iron Age (10th c. BCE) metalworkers community from excavations at a gatehouse and associated livestock pens in Timna, Israel. *Journal of Archaeological Science: Reports* 11, 411–426.
Besnard, G., Khadari, B., Navascués, M., Fernández-Mazuecos, M., El Bakkali, A., *et al.* (2013) The complex history of the olive tree: from Late Quaternary diversification of Mediterranean lineages to primary domestication in the northern Levant. *Proceedings of the Royal Society B* 280, 20122833.
Besnard, G., Terral, J.F., and Cornille, A. (2018) On the origins and domestication of the olive: a review and perspectives. *Annals of Botany* 121, 385–403.
Bilgin, N.A., Misirli, A., Belge, A., and Özen, M. (2020) The pollen and fruit properties of *Ficus carica Caprificus*. *International Journal of Fruit Science* 20 (Suppl. 3), S1696–S1705.
Browicz, K. and Zohary, D. (1996) The genus *Amygdalus* L. (Rosaceae): species relationships, distribution and evolution under domestication. *Genetic Resources and Crop Evolution* 43, 229–247.
Caliskan, O., Bayazit, S., Ilgin, M., and Karatas, N. (2017) Morphological diversity of caprifig (*Ficus carica* var. *caprificus*) accessions in the eastern Mediterranean region of Turkey: potential utility for caprification. *Scientia Horticulturae* 222, 46–56.
Caracuta, V., Weiss, E., van den Brink, E.C.M., Liran, R., Vardi, J., and Barzilai, O. (2014) From natural environment to human landscape: new archaeobotanical data from the Neolithic site of Nahal Zippori 3, Lower Galilee. *Neo-Lithics* 14, 33–41.
Carrión, Y. and Rosser, P. (2010) Revealing Iberian woodcraft: conserved wooden artefacts from south-east Spain. *Antiquity* 84, 747–764.
Condit, I.J. (1947) *The Fig*. Chronica Botanica, Waltham, Massachusetts.
Cruz, C.D., Peterson, G.L., Bockus, W.W., Kankanala, P., Dubcovsky, J., *et al.* (2016) The 2NS translocation from *Aegilops ventricosa* confers resistance to the *Triticum* pathotype of *Magnaporthe oryzae*. *Crop Science* 56, 990–1000.
Dalkiloç, Z., Mestav, H.O., Gunver-Dalkiloç, G. and Kocataş, H. (2011) Genetic diversity of male fig (*Ficus carica caprificus* L.) genotypes with random amplified polymorphic DNA (RAPD) markers. *African Journal of Biotechnology* 10, 519–526.
Deckers, K. and Pessin, H. (2010) Vegetation development in the Middle Euphrates and Upper Jazirah (Syria/Turkey) during the Bronze Age. *Quaternary Research* 74, 216–226.
Delplancke, M., Alvarez, N., Benoit, L., Espíndola, A., Joly, H.I., *et al.* (2013) Evolutionary history of almond tree domestication in the Mediterranean basin. *Molecular Ecology* 22, 1092–1104.
Denham, T. (2007) Early fig domestication, or gathering of wild parthenocarpic figs? *Antiquity* 81, 457–461.
Dunseth, Z.C., Fuks, D., Langgut, D., Weiss, E., Melamed, Y., *et al.* (2019) Archaeobotanical proxies and archaeological interpretation: a comparative study of phytoliths, pollen and seeds in dung pellets and refuse deposits at Early Islamic Shivta, Negev, Israel. *Quaternary Science Reviews* 211, 166–185.
Flaishman, M., Rodov, V., and Stover, E. (2008) The fig: botany, horticulture and breeding. *Horticultural Reviews* 34, 113–196.
Flaishman, M., Peer, R., Freiman, Z.E., Izhaki, Y., and Yablovitz, Z. (2017) Conventional and molecular breeding systems in fig (*Ficus carica* L.). *Acta Horticullturae* 1173, 1–10.
Frumin, S., Maeir, A.M., Kolska Horwitz, L., and Weiss, E. (2015) Studying ancient anthropogenic impacts on current floral biodiversity in the southern Levant as reflected by the Philistine migration. *Scientific Reports* 5, 13308.
Frumin, S.I. and Tchekhanovets, Y. (2016) Plant imprints on pottery reveal fig tree in Hellenistic Jerusalem. *Israel Exploration Journal* 66, 188–201.
Fu, Y.B. and Allaby, R. (2010) Phylogenetic network of *Linum* species as revealed by non-coding chloroplast DNA sequences. *Genetic Resources and Crop Evolution* 57, 667–677.
Galil, J. and Neeman, G. (1977) Pollen transfer and pollination in the common fig (*Ficus carica* L.). *New Phytologist* 79, 163–171.
Goldschmidt, E.E. (2013) The evolution of fruit tree productivity: a review. *Economic Botany* 67, 51–62.
Gopher, A. (1994) *Arrowheads of the Neolithic Levant: A Seriation Analysis*. American Schools of Oriental Research, Winona Lake, Indiana.
Gopher, A., Abbo, S., and Lev-Yadun, S. (2001) The "when", the "where" and the "why" of the Neolithic Revolution in the Levant. *Documenta Prehistorica* 28, 49–62.
Gopher, A., Lev-Yadun, S., and Abbo, S. (2021) *Breaking Ground. Plant Domestication in the Neolithic Levant: The "Core-area – One-event" Model*. Emery and Claire Yass Publications in Archaeology. The Institute of Archaeology, Tel Aviv University, Tel Aviv, Israel.
Gophna, R., Liphschitz, N., and Lev-Yadun, S. (1986) Man's impact on the natural vegetation in the Central Coastal Plain of Israel during the Chalcolithic and the Bronze Ages. *Tel Aviv* 13, 71–84.

Hartmann-Shenkman, A., Kislev, M.E., Galili, E., Melamed, Y., and Weiss, E. (2015) Invading a new niche: obligatory weeds at Neolithic Atlit-Yam, Israel. *Vegetation History and Archaeobotany* 24, 9–18.

Hassan, F.A. (1981) *Demographic Archaeology*. Academic Press, New York.

Heun, M., Sahafer-Pregal, R., Klawan, D., Castagna, R., Accerbi, M., *et al.* (1997) Site of einkorn wheat domestication identified by DNA fingerprinting. *Science* 278, 1312–1314.

Janick, J. (2005) The origins of fruits, fruit growing and fruit breeding. *Plant Breeding Reviews* 25, 255–320.

Kaul, S., Koo, H.L., Jenkins, J., Rizzo, M., Rooney, T., *et al.* (2000) Analysis of the genome sequence of the flowering plant *Arabidopsis thaliana*. *Nature* 408, 796–815.

Khadari, B., Grout, C., Santoni, S., and Kjellberg, F. (2005) Contrasted genetic diversity and differentiation among Mediterranean populations of *Ficus carica* L.: a study using mtDNA RFLP. *Genetic Resources and Crop Evolution* 52, 97–109.

Kislev, M.E. (1997) Early agriculture and paleoecology of Netiv Hagdud. In: Bar-Yosef, O. and Gopher, A. (eds) *An Early Neolithic Village in the Jordan Valley. Part I: The Archaeology of Netiv Hagdud*. Bulletin of the American School of Prehistoric Research, Vol 43, Cambridge, Massachusetts, pp. 209–236.

Kislev, M. and Simchoni, O. (2007) Hygiene and insect damage of crops and food at Masada. In: Aviram, J., Foerster, G., Netzer, E., and Stiebel, G.D. (eds) *Masada VIII. The Yigael Yadin Excavations 1963–1965 Final Reports*. Israel Exploration Society and The Hebrew University of Jerusalem, Jerusalem, Israel, pp. 133–170.

Kislev, M.E., Hartman, A., and Bar-Yosef, O. (2006) Early domesticated fig in the Jordan Velley. *Science* 312, 1372–1374.

Kjellberg, F. and Valdeyron, G. (1984) The pollination of the fig tree (*Ficus carica* L.) and its control in horticulture. *Acta Oecologica Oecologia Generalis* 5, 407–412.

Ladizinsky, G. (1985) Founder effect in crop-plant evolution. *Economic Botany* 39, 191–199.

Ladizinsky, G. (1998) *Plant Evolution Under Domestication*. Kluwer, London.

Ladizinsky, G. (1999a) On the origin of almond. *Genetic Resources and Crop Evolution* 46, 143–147.

Ladizinsky, G. (1999b) Identification of the lentil's wild genetic stock. *Genetic Resources and Crop Evolution* 46, 115–118.

Langgut, D., Gadot, Y., Porat, N., and Lipschits, O. (2013) Fossil pollen reveals the secrets of the Royal Persian Garden at Ramat Rahel, Jerusalem. *Palynology* 37, 115–129.

Langgut, D., Shahack-Gross, R., Arie, E., Namdar, D., Amrani, A., *et al.* (2016) Micro-archaeological indicators for identifying ancient cess deposits: an example from Late Bronze Age Megiddo, Israel. *Journal of Archaeological Science: Reports* 9, 375–385.

Langgut, D., Cheddadi, R., Carrión, J.S., Cavanagh, M., Colombaroli, D., *et al.* (2019) The origin and spread of olive cultivation in the Mediterranean Basin: the fossil pollen evidence. *Holocene* 29, 902–922.

Langgut, D., Tepper, Y., Benzaquen, M., Erickson-Gini, T., and Bar-Oz, G. (2021) Environment and horticulture in the Byzantine Negev Desert, Israel: sustainability, prosperity and enigmatic decline. *Quaternary International* 593–594, 160–177.

Lev-Yadun, S. (2007) Wood remains from archaeological excavations: a review with a Near Eastern perspective. *Israel Journal of Earth Sciences* 56, 139–162.

Lev-Yadun, S. (2012) The environmental impact of Naḥal Zehora II on the Menashe Hills. In: Gopher, A. (ed.) *Village Communities of the Pottery Neolithic Period in the Menashe Hills, Israel. Archaeological Investigations at the Sites of Naḥal Zehora*, Vol. III. Monograph Series 29. The Sonia and Marco Nadler Institute of Archaeology, Tel Aviv University, Tel Aviv, Israel, pp. 1364–1384.

Lev-Yadun, S., Gopher, A., and Abbo, S. (2000) The cradle of agriculture. *Science* 288, 1602–1603.

Lev-Yadun, S., Ne'eman, G., Abbo, S., and Flaishman, M.A. (2006) Comment on "Early Domesticated Fig in the Jordan Valley." *Science* 314, 1683.

Liphschitz, N., Gophna, R., and Lev-Yadun, S. (1989) Man's impact on the vegetational landscape of Israel in the Early Bronze Age II–III. In: Miroschedgji, P. (ed.) *The Urbanization of Palestine in the Early Bronze Age*. British Archaeological Reports International Series, Vol. 527, Oxford, UK, pp. 263–268.

Mavromati, A. (2021) Wood as a structural element in the houses of Akrotiri on Thera, Greece. The anthracological evidence. *Quaternary International* 593–594, 71–84.

Melamed, Y., Kislev, M., Weiss, E., and Simchoni, O. (2011) Extinction of water plants in the Hula Valley: evidence for climate change. *Journal of Human Evolution* 60, 320–327.

Melamed, Y., Kislev, M.E., Geffen, E., Lev-Yadun, S., and Goren-Inbar, N. (2016) The plant component of an Acheulian diet: a case study from Gesher Benot Ya'aqov, Israel. *Proceedings of the National Academy of Sciences USA* 113, 14674–14679.

Mirheidari, F., Khadivi, A., Moradi, Y., and Paryan, S. (2020) Phenotypic variability of naturally grown edible fig (*Ficus carica* L.) and caprifig (*Ficus carica* var. *caprificus* Risso) accessions. *Scientia Horticulturae* 267, 109320.

Morales, J., Rodríguez-Rodríguez, A., del Cristo González-Marrero, M., Martín-Rodríguez, E., Henríquez-Valido, P., and del-Pino-Curbelo, M. (2014) The archaeobotany of long-term crop storage in northwest African communal granaries: a case study from pre-Hispanic Gran Canaria (cal. AD 1000–1500). *Vegetation History and Archaeobotany* 23, 789–804.

Mori, K., Shirasawa, K., Nogata, H., Hirata, C., Tashiro, K., *et al.* (2017) Identification of *RAN1* orthologue associated with sex determination through whole genome sequencing analysis in fig (*Ficus carica* L.). *Scientific Reports* 7, 1–15.

Myles, S., Boyko, A.R., Owens, C.L., Brown, P.J., Grassi, F., *et al.* (2011) Genetic structure and domestication history of the grape. *Proceedings of the National Academy of Sciences USA* 108, 3530–3535.

Picornell-Gelabert, L. and Carrión Marco, Y. (2017) Landscape and firewood procurement at the prehistoric and protohistoric site of Ses Païsses (island of Mallorca, Western Mediterranean). *Quaternary International* 458, 56–74.

Purugganan, M.D. and Fuller, D.Q. (2011) Archaeological data reveal slow rates of evolution during plant domestication. *Evolution* 54, 171–183.

Roth, H., Gadot, Y., and Langgut, D. (2019) Wood economy in Early Roman period Jerusalem. *Bulletin of the American Schools of Oriental Research* 382, 71–87.

Rottenberg, A. (1998) Sex ratio and gender stability in the dioecious plants of Israel. *Botanical Journal of the Linnean Society* 128, 137–148.

Salamini, F., Özkan, H., Brandolini, A., Schäfer-Pregl, R., and Martin, W. (2002) Genetics and geography of wild cereal domestication in the Near East. *Nature Reviews Genetics* 3, 429–441.

Stordeur, D., Helmer, D., Jamous, B., Khawam, R., Molist, M., and Willcox, G. (2010) Le PPNB de Syrie du Sud à Travers les Découvertes Récentes à Tell Aswad. In: al-Maqdissi, M., Braemer, F. and Dentzer, J.M. (eds) *Hauran V. La Syrie du Sud du Néolithique à l'Antiquité Tardive, Recherches Récentes*. Actes du Colloque de Damas 2007, Vol. 1, Beyrouth, Lebanon, pp. 41–68.

Stover, E., Aradhya, M., Ferguson, L., and Crisosto, C.H. (2007) The fig: overview of an ancient fruit. *HortScience* 42, 1083–1087.

van Zeist, W. and Bakker-Heeres, J.A.H. (1982) Archaeobotanical studies in the Levant, 1. Neolithic sites in the Damascus basin: Aswad, Goraife, *Ramad. Paleohistoria* 24, 165–256.

Vidal-Matutano, P., Rodríguez-Rodríguez, A., del Cristo González-Marrero, M., Morales, J., Henríquez-Valido, P., and Moreno-Benítez, M.A. (2021) Woodworking in the cliffs? Xylological and morpho-technological analyses of wood remains in the Prehispanic granaries of Gran Canaria (Canary Islands, Spain). *Quaternary International* 593–594, 407–423.

Weiss, E. (2015) "Beginnings of Fruit Growing in the Old World" – two generations later. *Israel Journal of Plant Sciences* 62, 75–85.

Weiss, E. (2017) Paleolithic vegetal diet in the Southern Levant. In: Enzel, Y. and Bar-Yosef, O. (eds) *Quaternary of the Levant. Environments, Climate Change, and Humans*. Cambridge University Press, Cambridge, UK, pp. 329–335.

Weiss, E. and Kislev, M.E. (2004) Plant remains as indicators for economic activity: a case study from Iron Age Ashkelon. *Journal of Archaeological Science* 31, 1–13.

Werker, E. (2006) 780,000-year-old wood from Gesher Benot Ya'aqov, Israel. *Israel Journal of Plant Sciences* 54, 291–300.

Zohary, D. (1996) The mode of domestication of the founder crops of southwest Asian agriculture. In: Harris, D.R. (ed) *The Origins and Spread of Agriculture and Pastoralism in Eurasia*. Routledge, London, pp. 142–158.

Zohary, D. (2004) Unconscious selection and the evolution of domesticated plants. *Economic Botany* 58, 5–10.

Zohary, D. and Hopf, M. (2000) *Domestication of Plants in the Old World: The Origin and Spread of Domesticated Plants in Southwest Asia, Europe, and the Mediterranean Basin*, 3rd edn. Oxford University Press, Oxford, UK.

Zohary, D. and Spiegel-Roy, P. (1975) Beginning of fruit growing in the Old World. *Science* 187, 319–327.

Zohary, D., Hopf, M., and Weiss, E. (2012) *Domestication of Plants in the Old World: The Origin and Spread of Domesticated Plants in Southwest Asia, Europe, and the Mediterranean Basin,* 4th edn. Oxford University Press, Oxford, UK.

Zohary, M. (1973) *Geobotanical Foundations of the Middle East*. Gustav Fischer, Stuttgart, West Germany.

3 Figs in History and Art

Dvora Namdar[1]* and Irit Ziffer[2]
[1]*Institute of Plant Sciences, Agriculture Research Organization, Volcani Center, Rishon LeZion, Israel;* [2]*Eretz Israel Museum, Tel Aviv, Israel*

3.1 Introduction

Among the earliest fruits used by mankind, figs tell a complex story in history. *Ficus carica* L., generally known as the fig, originates in northern Asia. The fig tree is indigenous to Persia, Asia Minor, and Syria, but is now wild or culture escaped in most Mediterranean countries (Condit, 1947; Ramírez, 1977; Zohary and Hopf, 2000). It expanded with the Greeks and the Romans throughout the Mediterranean region and later to the Americas. Figs are not actually fruit, but an edible part called the "syconium," a false fruit that grows out of a mass of tiny unisexual flowers. All members of the genus share the distinctive syconium—the fig fruit.

The genus *Ficus* (family Moraceae) is among the largest genera of angiosperms (plants that produce seeds within an enclosed space) with approximately 750 species of trees, epiphytes, and shrubs in tropical and subtropical regions worldwide. Figs are a keystone resource for humans and other vertebrates throughout the area in which they occur. Figs are also used by humans in a variety of ways other than nutrition. Some fig species are attributed as sacred, being used as art symbols, employed in ethnomedicine, and utilized into useful objects. In this chapter, we will review the rather rare findings of fig botanical remains and the artistic expression of fig in antiquity.

The fig has been cultivated in the Holy Land, today known as Israel, for more than 5000 years. Its Hebrew name is *te'ena* and it is much the same in other Semitic languages: *tin* in Arabic and *teinta* in Aramaic. The Latin is *ficus*, most probably deriving from a Hebrew source as the unripened fruit is called *paga*. The Latin therefore gives its different names to the fig in European languages: in French *figue*, in German *Feige*, in Italian and Portuguese *fico*, and in Spanish *higo* or *brevo*. Haitians gave it the name *figue France* to distinguish it from the small, dried bananas called *fig* (Condit, 1947).

The fruit qualities and sweetness of the fig were recognized, and selection of seedling trees bearing superior fruits was established, creating fig varieties. In many cultures, the fig was a symbol of peace, and it was believed to distribute wealth and happiness the more it was planted. In antiquity, the fig fruit was inexpensive and provided year-round nourishment, a virtue of considerable economic importance. Figs were sold fresh or dried, and the dried ones were distributed singly, in strings of individual fruits or in cake form, pressed down and squeezed together like a loaf of bread; the loaf type also varied in shape from round or cubic to beehive-like (Fig. 3.1).

* Email: dvoran@volcani.agri.gov.il

 Advances in Fig Research and Sustainable Production (M.A. Flaishman and U. Aksoy)

DOI: 10.1079/9781789242492.0003

Fig. 3.1. Fig cakes and figs on a string. In the Old Testament, figs are known for their nourishing and healing properties. Figs threaded on a string and stored in storage jar were found carbonized at Tel Miqne, Ekron. Israel Antiquities Authority (IAA) 2007 excavation, reference no. 9029. The authors are thankful to Professor Seymour Gitin for granting permission to publish.

Dry figs were the most abundant source of sugar and energy; hence, people travelling long distances consumed dried figs. As travelers ate the fruit, the fig tree was causally transplanted simply by dropping the seeds. Greek and Roman writers describe propagating fig foreign varieties by seeds from the dried fruit. Both Theophrastus (285–370 BCE) in his book entitled *Enquiry into Plants*, Volume II, Chapter I, and Marcus Varro (116–27 BCE) in his book *On Agriculture*, Book I, Chapter XLI, mentioned this custom (Goor, 1965). Mankind also used fig for cooking and medicinally. Under the fig tree shade, they escaped the blazing sun. In addition, they burnt its wood for fuel.

3.2 Archaeobotanical Remains and Archaeological Evidence

Fig was cultivated in the Holy Land more than 7000 years ago. In the excavations of Tell Gezer, situated in the hills of the Shephelah guarding the ancient coastal highway and the pass up to Jerusalem, fig remains were unearthed and recovered, dating to the Neolithic period, fifth millennium BCE. Fig remains, both fruits and seeds, were unearthed in different locations over Israel in controlled archaeological excavations. Table 3.1 summarizes the data regarding these remains that have been published in peer-reviewed journals.

Besides the archaeobotanical physical remains discovered in many sites and over many millennia in Israel (Table 3.1), another type of a direct evidence for the consumption of figs came from analyses of a cesspit in Megiddo, Northern Israel (Langgut *et al.*, 2016). Chemical analysis of archaeological deposits found it to be the most ancient cesspit detected, dated to the mid-14th century BCE, located near the Late Bronze Age palatial area of Megiddo. Pollen analyses of the cesspit remains yielded data on the ancient vegetative diet and the use of medicinal plants by the Megiddo inhabitants, including residues of *F. carica* (common fig) which comprised 16% of the total pollen found in the cesspit. Researchers suggested that the botanical material had been preserved over many millennia thanks to phosphate replacement, i.e. phosphate-rich solutions penetrated the plant material and deposited minerals that replaced the organic components, leading to the preservation of the original microscopic cellular structure of the botanic material (e.g. Green, 1979; McCobb *et al.*, 2001, 2003; Marshall *et al.*, 2008). The unique conditions in the

Table 3.1. Non-exhaustive list of fig remains found in archaeological contexts and excavation sites in Israel. Data collected, arranged, and presented in Table 3.1. by Professor Ehud Weiss, Bar-Ilan University, Israel.

Period	Timespan (in YBP)	Archaeological site	Seeds (*n*)	Fruits (*n*)	Reference(s)
Early Middle Pleistocene	*c.* 1.2–0.5 Ma	Gesher Benot Ya'aqov	79		Melamed *et al.* (2016)
Epipalaeolithic	20,000–10,000	Ohalo II	4		Kislev *et al.* (1992); Weiss (2002)
Middle Epipalaeolithic	18,000–12,000	Jordan River Dureijat	32		Marder *et al.* (2015)
Kebaran Culture	18,000	Nahal Oren	4		Noy *et al.* (1973)
Early Neolithic	11,000	Gilgal I	313	9	Kislev *et al.* (2006a,b,c)
Early Neolithic	11,000	Gilgal III	1	30	
Early Neolithic	11,000	Gesher	36		
PPNA	12,000–9,000	Netiv HaGdud	5013	1	
PPNA	12,000–9,000	Jericho	46		Hopf (1983)
PPNC	8,000	Atlit-Yam	3689		Kislev *et al.* (2004)
PN (Yarmukian Culture)	6400–6000	Nahal Zehora II		2	Kislev *et al.* (2012)
Neolithic	6400–6000	Sha'ar Hagolan	129		Allen (2002)
Chalcolithic	6000–5200	Shiqmim I	2		Kislev (1987)
Chalcolithic	6000–5200	Tel Tsaf	P		Kislev (1979)
Chalcolithic	6000–5200	Jericho	P		Hopf (1969)
EB IB	5100	Tel Beth She'an (Area M)	6		Hopf (1983); Simchoni and Kislev (2012)
EB I	5100	Jericho	5		
EB	5200–4100	Jericho	8		
EB	5200–4100	Tel Gezer	P		Liphschitz and Miroschedji (1989)
EB	5200–4200	Tel Yarmouth	23		Salavert (2008)
EB III	4500	Tel Beth She'an	4		Simchoni and Kislev (2012)
MB	4200–3600	Jericho	2		Hopf (1983)
MB	4200–3600	Tel Megiddo	55		Borojevic (2006)
MB IIB	3700	Tel Beth She'an	1		Simchoni *et al.* (2007)
MB III	3600	Shiloh	31		Kislev (1993)
LB	3600–3200	Tel Miqne		Hundreds (whole dry fruits)	Gitin (1993)
LB IB/IIA	3400	Tel Batash	17		Kislev *et al.* (2006b)
LB IIB	3350	Aphek	107		Kislev and Mahler-Slasky (2009)
LB IIB	3350	Tell es-Safi/Gath	92		Frumin *et al.* (2020)
LB III	3250	Tell es-Safi/Gath	5		Mahler-Slasky and Kislev (2012)
LB–Iron	3200	Tel Burna	P		Orendi *et al.* (2017)
Iron	3200–2586	Ashkelon	5478	6	Weiss *et al.* (2011)

Iron Age	3200–2800	Horbat Rosh Zayit	1		Kislev and Melamed (2000)
Iron I	3200–3000	Shiloh	5		Kislev (1993)
Iron I	3200–3000	Tel Beth-Shemesh		Hundreds	Weiss and Mahler-Slasky (2016)
Iron I	3200–3000	Tell es-Safi/Gath	2		Mahler-Slasky and Kislev (2012)
Late Iron I	3000	Tel Megiddo	1		Borojevic (2006)
Late Iron I–early Iron II	3000–2925	Tell es-Safi/Gath	416		Mahler-Slasky and Kislev (2012)
Iron IB	3000	Tell Qasile	1		Kislev and Hopf (1985)
Iron IB	3000	Aphek	62		Kislev and Mahler-Slasky (2009)
Iron IB	3000	Tell es-Safi/Gath	4		Mahler-Slasky and Kislev (2012)
Iron II	2950	Tel Lachish		2	Liphschitz (2004)
Iron IIA	2950	Aphek	1		Kislev and Mahler-Slasky (2009)
Late Iron IIA	2950	Tell es-Safi/Gath	46		Mahler-Slasky and Kislev (2012)
Iron II	2950	Tel Burna	191		Shai *et al*. (2014)
Iron Age–Persian	2560	Ramat Rahel	~24		Langgut *et al*. (2013)
Persian	2320	Ma'agan Mikhael Ship	2284		Kislev (2003b)
Persian–Hellenistic	2320–2020	Tell Kedesh	540	47	Borojevic (2011)
Roman	2063–1670	Masada		1	Kislev and Simchoni (2007)
Roman	2063–1670	Caves of the Spear	56		Simchoni and Kislev (2004) (in Hebrew)
Late Roman–Early Byzantine	1650	Horvat Raqit	1		Kislev (2003a) (in Hebrew)
Late Roman–Early Byzantine	1650	Huqoq	P		Foley (2015)
Byzantine	1670–1350	Shivta	40		Ramsay and Tepper (2010); Dunseth *et al*. (2019)
Byzantine	1670–1350	Caesarea Maritima	12		Ramsay (2010)
Byzantine	1670–1350	Ein Gedi	1		Melamed and Kislev (2005) (in Hebrew)
Islamic	1100	City of David (Giv'ati parking lot), Jerusalem	7060		Amichay *et al*. (2019)
Islamic	1100	Caesarea Maritima	985		Ramsay and Holum (2015)
Islamic	1100	Caesarea Inner Harbour (Area I14)	~50		Giorgi (1999)
Islamic	1100	Caesarea Maritima	727		Ramsay and Holum (2015)
Islamic	1100	City of David (Giv'ati parking lot), Jerusalem	7060		Amichay *et al*. (2019)
Islamic	1100	Caesarea Maritima	387		Ramsay and Holum (2015)

YBP, years before present; Ma, 1 million years; PPN, Pre-Pottery Neolithic periods A–C; EB, Early Bronze Age; MB, Middle Bronze Age; LB, Late Bronze Age; P, numbers not reported.

cesspit, which was probably characterized by relatively low oxygen levels, led to the good state of the pollen preservation (Langgut *et al.*, 2016).

Pollen is introduced into the intestinal tract through the consumption of flowers, foliage that includes buds and flowers, seeds and fruits of certain types, and drinking teas derived from foliage or flowers (Adams, 1980; Reinhard *et al.*, 1991). Identifying common fig within the pollen assemblage accumulated and preserved in the cesspit indicated that these were indeed ingested pollen grains. The common fig flowers are practically invisible, as they bloom inside the syconium (Galil and Neeman, 1977). Therefore, the common fig is usually extremely under-represented in pollen spectra of ancient samples. Thus, the presence of fig pollen within this assemblage may indicate that the entire fruit was consumed, digested, and then deposited in the cesspit. This provides direct evidence for the consumption of figs, at least by the elite population of Megiddo, as before the Roman period almost all latrine facilities from the region were found in palatial contexts or in buildings that imitate regal luxury (Fink, 2009; McMahon, 2015).

Because of its pollination via the symbiotic wasp that transfers small amounts of pollen from closed male syconia to closed female ones, *F. carica* pollen is very rarely found in archaeological and geological sediments. However, it was found in the analysis of a *c.*2400-year-old plaster wall of the water pool in the Royal Persian Garden of Ramat Rahel, located on the southern outskirts of Jerusalem (Langgut *et al.*, 2013). A rather surprising and very well-preserved pollen assemblage was detected within the layers of a plastered wall. During the construction of a water pool designed to collect rainfall and drainage water to water the Royal Garden, its walls were plastered. While drying, apparently seasonal pollen adhered to the plaster and was sealed there for many decades, until the excavators, to reveal its collection, carefully peeled it off. Surprisingly and despite the specific method of pollination in fig, common fig pollen was detected. Based on the overall assemblage detected and cross-matching of the blooming periods of all the plants represented in this assemblage, the researcher concluded that the maintenance works took place during springtime, by the end of April to early May (Langgut *et al.*, 2013).

Another interesting approach to archaeobotanical remains is looking at plant imprints detected on ceramic vessels (Frumin and Tchekhanovets, 2016). Pottery sherds with plant imprints on them are a common find in archaeological contexts (e.g. Hajnalová, 1986; Wendorf and Schild, 2001). It has also been proposed that the plant imprints may have been used as a substitute for gauze to repair jars that had cracked before firing. Frumin and Tchekhanovets (2016) suggested that the choice of plants used for pottery repair reflects the plant's availability in the potter's living environment, for example plants grown for shade or as hedges (provided the manufacturer's location and the location of the find are near). Therefore, data on plants used for pottery repair may serve as a valuable source of information for this hitherto little-documented aspect of ancient life. Thus, pottery sherds with plant imprints retrieved from Hellenistic strata from the excavation of Ancient Jerusalem were analyzed using three-dimensional scanning and the results showed clear evidence for the intentional use of green leaves for vessel rim repair. Archaeobotanical analysis of these sherds revealed that the leaves used were from a fig tree, *F. carica*, as all of the studied features of the imprints on the pottery sherds analyzed clearly assigned them to various parts of fig tree leaves. These fig leaf impressions are the earliest record of the presence of fig trees in Jerusalem.

In rather rare cases, the content of a vessel is declared on its exterior. In one case, a ceramic jar found at Tel Miqne, Ekron, dated to the 7th century BCE, bore the ancient Hebrew letters *dbl* on the vessel's exterior wall (Fig. 3.2; Gitin, 1993; Ahituv, 2005), *dblt*, dried figs, or fig cakes are listed in a few texts from Ugarit (Watson, 2004), and Hittite sources mention a sort of cake made by mashing up inferior figs served in parts of the Greek Archipelago as a substitute for bread. The term is common in Semitic (Hebrew *dəbēlâh*, Aramaic *dəbêltâ*, Arabic *dublat* for "clump") and also in Egyptian *d3bw* for "figs" and *db* for "sweet fig" (Watson, 2004). At Ugarit, *dblt*, rancid dried figs, were mixed with rancid raisins and groat flour to prepare a medical concoction (Del Olmo Lete and Sanmartín, 2003).

The Hebrew *dblh* (*dəbēlâh*, pl. *dəbēlîm*), or "fig cake," made of a compressed lump of squashed dried figs (Ahituv, 2005), round or rectangular in shape (Mishnah Terumoth 4, 8), recalls the biblical lump of dried figs mentioned in 1 Samuel 25:18: "Then Abigail made haste, and took two hundred loaves, and two bottles of wine, and

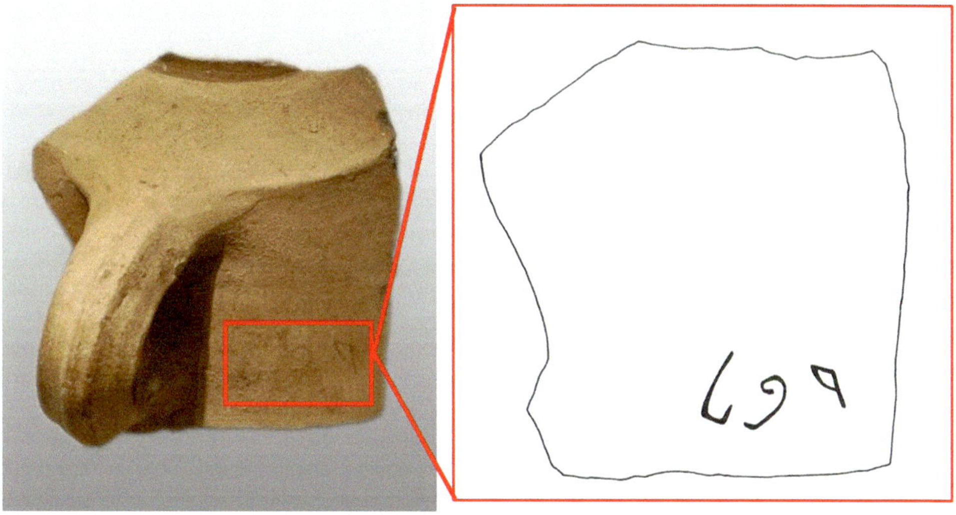

Fig. 3.2. Photograph and line sketch of an inscribed pottery sherd with the inscription *dbl* (Israel Antiquities Authority (IAA) 1996 excavation, reference no. 1315) from the excavation project at Tel Miqne, Ekron. Courtesy of the Israel Museum; curator Eran Arie. The authors are thankful to Professor Seymour Gitin for granting permission to publish. Photograph by Zev Radovan.

five sheep ready dressed, and five measures of parched corn, and a hundred clusters of raisins, and two hundred cakes of figs, and laid them on asses." In 1 Samuel 30:12, it says: "...and they gave him a piece of a cake of figs, and two clusters of raisins; and when he had eaten, his spirit came back to him." In 2 Kings 20:7, we find; "And Isaiah said: 'Take a cake of figs.' And they took and laid it on the boil, and he recovered." An example of a cake of figs is given in Fig. 3.1.

This commodity is also mentioned in later Hebrew texts like the Talmud (see Ta'anit 4, 7). Other pottery sherds bearing the inscription *dblh* or *dbl'* were found in Masada (Yadin *et al.*, 1989). Although in the texts the word is given in plural, on the jar fragment the word is inscribed in singular. The jar from Tel Miqne, Ekron, is the only example where the word is written in the masculine gender (Ahituv, 2005).

3.3 Figs in Different Cultures

3.3.1 Mesopotamia

Ancient Near-Eastern cultures known as Mesopotamian civilization were largely based on Semitic populations that existed between the Tigris and Euphrates River (present-day Iraq), that soon expanded to the area known as the Fertile Crescent including parts of present-day Israel, Jordan, Lebanon, Syria, Iraq, and Iran. During the Bronze Age, the transit from villages to urban centers and the development of a settled agriculture coincided with the beginning of fruit culture.

In the third millennium BCE, northern Mesopotamia consisted of a network of large cities involved in trade with each other and with the neighboring regions (Oates *et al.*, 2007; Kubiak-Martens, 2015). Human agricultural activities in the arid margins of the Fertile Crescent were particularly pronounced during the Early Bronze Age (4900–4000 BP; for comparison with Ancient Israel, see Table 3.1). This was attributed to the climatic amelioration, especially increased rainfall, that occurred over the early Holocene in many areas of the Near East (Wick *et al.*, 2003; Smith, 2005). In Mesopotamia, the emergence of huge urban centers centralizing rich, rainfed agricultural hinterlands corresponded to the emergence of specialized pastoralism and production of pastoral surpluses (Fleming, 2004; Stein, 2004). Four appearances of species *Ficus* were counted in the analysis of charred wood samples found in excavations conducted in Mesopotamia, as summarized in Table 3.2, which shows that during the third millennium BCE, the amount of *Ficus* species present in archaeological contexts increased significantly. Attributed either to the climatic change, the increased amount of rainfall, or to more complex and developed social networking, this appearance is unique and restricted to this period.

The fig tree, *tittu*/*ti'ttu*, has been attested in cuneiform sources since the Akkadian period (the last quarter of the third millennium) among trees of the orchard. At that time, during a "month of figs" (ITI.KAM *tí-i-na-tim*), fig trees and their offshoots were used to plant orchards. The uses of the fig tree were multiple. High-quality figs were offered to the gods at meals and at festivals, and strings of figs were a welcome greeting present. The fruit was used as an ingredient in a remedy applied to the ear to cure hardness of hearing. The leaves, kneaded with milk, were used for a head poultice, and mixed with oil, the preparation was used in poultices for the lungs (Thompson, 1949). According to texts retrieved from the Mari archive (on the Middle Euphrates, dating from the 18th century BCE), its wood was used in the construction of chairs, including the throne of Adad, the storm god. A costly bed with fig-shaped ornaments is mentioned in a Neo-Babylonian text. In literary texts, the fruit was symbolically likened to the breasts of a deity. In addition, it was said that if a man's house smelled like a fig, his wife would be happy. Assyrian kings, ardent botanists, collected specimens of flora. In the stela erected outside the Throne Room of King Assurnasirpal's Northwest Palace at Nimrud (Kalḫu, biblical Calah) upon inauguration of the palace (879 BCE), the sycamore fig tree is mentioned among the trees that the king collected in his travels (Wiseman, 1952).

3.3.2 Egyptian art and the origins of horticulture

The artistic genius engendered by Egyptian civilization, the unique protection provided by the stable environmental conditions that existed in many burial chambers, and the dry surrounding climate have made it possible to reconstruct a detailed history of Egyptian agriculture (Janick, 2002).

In ancient Egypt, the sycamore fig tree (Fig. 3.3) was cultivated for its fig fruits, which were used as sweeteners in baking and as a remedy for various ailments. For the fruit to mature and become aromatic each fig had to be incised by hand shortly before harvesting. A notch in the bottom is recognizable in art. This agricultural practice is described by the prophet Amos, who declares that he was a dresser of sycamore trees (Amos 7:14).

Goor (1965) lists several Egyptian documents in which fig tree and fruits are mentioned. The first document dates to the days of Snefru, the first king of Dynasty 4 (2575–2765 BCE) who fostered the evolution of the highly centralized administration, and who mentions fig as the tree on whose fruit the people in the Land of Canaan grow and flourish. A stela dating to around 2400 BCE tells us about Uni, an army commander of Pepi, who dispatched troops to Canaan to suppress a rebellion: "His troops returned in peace after destroying cities and cutting down the figs and vines." The famous Sinuhe Papyrus (1800 BCE) speaks also of the fertile land of Yaa (Canaan) and of its figs and grapes, barley, emmer, and several fruit trees including olive.

In the Papyrus Anastasi III, Pibesa describes Pelusium, on the borders of Canaan, to his lord Amenemope in the reign of Rameses II (1298–1235 BCE): "And there is a plantation of fig trees" (Erman, 1923); the Papyrus Anastasi

Table 3.2. Charred remains of *Ficus* species wood identified in different sites along the Khabur Basin, Mesopotamia, by period and relative amounts. Data from McCorriston (2007).

Sample no.	Period	Date (BP)	Total amount of identified species (mg)	Amount of *Ficus* sp. identified (mg)	*Ficus* sp. of total identifiable species (%)
94	Neolithic	8600–7400	128	7	5.5
39	Jazira 1–2	4900–4400	83	21	25.3
43	Jazira 3–4	4400–4100	95	67	70.5
45	Jazira 3–4	4400–4100	581	20	3.5

Fig. 3.3. A tree with notched figs, from Thebes, Tomb of Userhat, Dynasty 19, reign of Seti I, 13th century BCE. The tomb owner, his wife, and his mother sit in the shade of the sycamore fig tree. In Egypt, the sycamore fig tree goddess was depicted growing from the tree or, as in this image, carrying the tree on her head. The sycamore goddess gave water and food to the dead to invigorate them in preparation for their journey to the underworld. The goddess Hathor was the "Lady of the Sycamore"; the tree was also associated with Isis and Nut. From the Metropolitan Museum of Art, Norman de Garis Davis, MMA. 30.4.33 Rogers Fund 1930 Public Domain.

IV quotes a letter from Amenemope to Pibesa asking him to prepare "grapes, pomegranates, figs and flowers" (A. Erman, 1923, p. 265) as well as "figs from Kharu (Canaan), pomegranates and apples" for the king (Erman, 1923). In the third and second centuries BCE, the Papyrus of Zenon (59012) lists dried figs, oil, olives, nuts, honey, and pomegranates as commodities brought from "Syria" (Fig. 3.4). Egyptian records suggest that, throughout the region, idol-worshippers offered wine and figs as a sacrifice, which gave both fruits a special sanctity.

3.3.3 Figs in the Hebrew Bible and in biblical times (Iron Age and Persian Period, 1200–445 BCE)

Fruit plays an important role in the biblical narrative. Israel is blessed as the land of Seven Species, "wheat and barley, of vines, figs and pomegranates, a land of olive trees and honey," in Deuteronomy 8:8 (Fig. 3.1). Fruit in the Hebrew Bible was much more than merely a food for the ancient Israelites. Symbolically it appears in the culture's names, laws, proverbs, ritual ceremonies, and traditions. The fig is the first tree mentioned in the Bible, in the story of Adam and Eve, believed to be the tree of knowledge (Midrash, Genesis Rabbah, 15, 8; Babylonian Talmud, Brachot 40, 1; Sanhedrin 70, 1–2). It is also believed that Adam and Eve used fig leaves to cover their nakedness, as written in Genesis 3:7: "And the eyes of them both were opened, and they knew that they were naked, and they sewed fig leaves together and made themselves aprons." Jewish Midrashim tells of Noah taking dried figs and fig cuttings into the Ark (Midrash Raba, 31, 14).

Fig. 3.4. A woman holding a child with her left hand stacks figs in a bowl. Egypt, Luxor West Bank, 670–650 BCE. Brooklyn Museum, Charles Edwin Wilbour Fund 48.74. Creative Commons-BY. Photograph: Brooklyn Museum, CUR.48.74_wwg8.jpg.

The Mediterranean climate suits the growth requirements of the fig tree. Many places around the Holy Land include the word "fig" in their names, which is *tə'ēnâh* in Hebrew. Goor (1965) indicated a selection of such places, such as Taanath-shiloh, "the Fig of Shiloh" in the boundaries of Ephraim (Joshua 16:6); Almon-diblathaim "Almon of Dried Figs" (Numbers 33:46), and Beth-diblathaim, "House of Dried Figs" (Jeremiah 48:22) to be identified with *bt dbltn*, Beth-diblathen, north of Dibon, capital of the kingdom of Moab, in the Mesha stele (Ahituv, 2005: p. 360); Beit Pagi "House of Unripe Figs" (Bethphage in Matthew 21:1, Mark 11:1, Luke 19:29; Tosefta Menahot 8:18; Babylonian Talmud, Pesahim, 63b; Babylonian Talmud, Sota, 45a), Kfar-Pagi, "Village of Unripe Figs" (Tosefta Kilayim 2:12), and Pagi Bethyany, "Unripe Figs of Bethyany" (Tosefta Sheviit 7:14). The Babylonian Talmud in Pesahim p. 53a reports that the figs of Beit Hinni (a place in Samaria) ripen early, but they vanished after the destruction of the Temple.

In biblical times (1200–445 BCE), all over the Holy Land, grape and fig were teamed up. They were planted side by side, with the grape growing as a climber on the fig. The fig's virtue as a foodstuff is emphasized in many passages, such as: "And they gave him a piece of a cake of figs, and two clusters of raisins and when he had eaten, his spirit came again to him" (Samuel 1 30:12). The prophets warned Israel: "If the People walked not in the path of righteousness, I will destroy her vines and her fig" (Hosea 2:12). Restful shade of vine and fig proverbially representing peace, happiness, security, wealth, and plenty: "And Judah and Israel dwelt safely, every man under his vine and under his fig tree, from Dan even to Beersheba, all the days of Solomon" (1 Kings 4:25).

Figure 3.5 shows the Assyrian king Sennacherib sitting on his elaborately decorated throne borne by protective creatures in the military camp facing Lachish, from which he controlled the violent campaign against the city.

Fig. 3.5. The Conquest of Lachish, Gypsum relief, Nineveh, South-west Palace, 700–692 BCE. Drawing by Judith Dekel. The authors are thankful to David Ussishkin for the permission to reproduce the figure from its original publication.

Holding two arrows in his right hand and a bow in his left hand, he receives prisoners of war. David Ussishkin believes that Sennacherib's camp was situated to the south-west of the mound, where the Israeli village of Lachish is now located. The natural landscape of Judah, abounding with grape-laden vines and fruit-bearing fig trees, reflects the bountiful land in times of peace. The violent military scene contrasts with the environment in which it is set, as reflected in the biblical passages, where grapevine and fig tree symbolizing prosperity and well-being contrast with weapons and war, Micah 4:3b–4: "...and they shall beat their swords into plowshares, and their spears into pruning hooks, nation shall not lift up sword against nation, neither shall they learn war anymore;" Zechariah 3:10: "In that day, saith the LORD of hosts, shall ye call every man his neighbor under the vine and under the fig-tree."

Jotham's Parable (Judges 9:8–15) reverses the role of the useful fig and vine for food and shade. The trees, looking for a king, appealed to the olive tree, the fig, and the vine, each refusing to rule the trees with reference to their function in the human world: the oil and the wine were used in cult and in daily life, and the fig for its sweet and delicious fruit. It was the barren bramble (or thornbush) whose fruit is inedible that agreed to wave over the trees inviting them to take refuge in its shade, an ironic description because the small bush provides little shade. Medicinal uses of fig were mentioned several times in the Bible: "And Isaiah said, take a lump of figs. And they took and laid it on the boil, and he recovered" (2 Kings 20:7; see also Isaiah 38:21).

The fig fruit and tree are described several times in the New Testament. A fig tree that Jesus beheld on the outskirts of Jerusalem in the spring was decked in foliage but with no fruit (Matthew 21:18–19), consistent with the fact that in the spring, no fig tree in Israel is in bearing and only a few local varieties have the early crop (Hebrew: bākkûrâh). It may also have been a wild tree or a variety that needed caprification. It is believed that most of fig varieties fruited in the Holy Land without caprification. Later in history, the Greeks introduced into Palestine varieties that needed caprification, and the fig in the New Testament may have been one of them.

3.3.4 Figs in Greek and Roman cultures

The rise of Greek culture (750–450 BCE) had a profound influence on Western civilization. Greek culture (i.e. Hellenism) is associated with the flourishing of the arts and sciences. Greek agriculture has been passed down to us via art in the form of paintings on ceramics and other artifacts. Known as the "forbidden fruit" and the "fruit of the God", figs are nothing new. The Greeks used these sweet drop-like fruits as sign of victory and for ancient Olympic training. Ancient Olympians received figs for their athletic prowess and Pliny the Elder admired the fruit's restorative powers and prescribed the milky juice as well as the still-green leaves for alopecia and the shoots of branches for dog bites (Thompson, 1949). Ancient Greeks were fascinated by the taste and juicy flesh of figs. They created laws against export of high-quality fig fruits to keep them entirely for themselves.

Roman horticulture survives in a rich written tradition and naturalistic images, through paintings and mosaic floors. Roman techniques of pomology are shown in mosaics illustrating grafting, manuring, harvesting, and juice extraction (Janick, 2002). Fig leaves have long been used to conceal the genitals of nude figures in paintings and sculptures. Art curators and exhibitors often added these depictions after the original work was created.

The Roman diplomat Cato the Elder persuaded Rome to go to war over a basket of figs. He explained the threatening distance of the Carthaginian (Carthage, now known as Tunisia) by showing the Senate a fresh fig and claiming it was 3 days old, picked in Carthage. This proved not just how close Carthage was but that it was economically prosperous and blessed with such competitive products. Consequently, this led to the destruction of Carthage by the Romans, in the third Punic war (146 BCE), the last of the wars between the ancient civilizations of the Mediterranean Sea. Romans considered the fig tree as symbolic of the civilization's prosperity. Romulus and Remus, the legendary founders of Rome, were breastfed by a she-wolf under a wild fig tree growing close to the Tiber River.

Pompeii, situated on the shores of the Bay of Naples, Italy, was a Roman town destroyed by the eruption of the volcano Mount Vesuvius in 79 CE. The "Pompeii premise" claims the city, caught underneath the cooling lava, captured a "picture in time" of ancient lives. Charred and preserved remains of bread, figs, and olives were excavated from the ruins of Pompeii, demonstrating this premise (Meyer, 1980).

Archaeobotanical remains of figs dating to the Roman period but excavated outside Italy are not that scarce. There are many discoveries of pips from Roman and later urban and military sites in Britain. These pips were imported figs and not the fruits of locally cultivated trees. There is no proof that the Romans grew the common fig in Britain and the earliest documentary evidence of cultivation is as late as the 15th century AD (Dickson and Dickson, 1996).

Pollen diagrams produced from two sedimentary cores taken from the sea bottom at the harbor of Purtos show that pips of *F. carica* L. (common fig) were abundant, accompanied at times by endocarps of *Rubus* sp. (blackberry), seeds of *Cucumis melo* L. (melon), and grapevine (Pepe *et al.*, 2013). Considering the huge amounts of cereals that were imported and stored in the horrea of Portus (Keay *et al.*, 2005; Keay and Paroli, 2011), it is surprising that the most common crops (cereals and legumes) are completely absent in both cores (Sadori *et al.*, 2015). According to the botanical analysis of shipwrecks from Pisae, ship frames were made of common fig (Giachi *et al.*, 2011).

3.3.5 Figs in the Muslim culture

Figs hold a position of symbolism in many religions, including Islam, Christianity, Hinduism, Buddhism, and Judaism, symbolizing fertility, peace, and victory. In contrast to the frequent appearance of figs in any Jewish written source, the fig appears only once in the Qur'an. The fig is mentioned in the first verse of Surat at-Tin: "[I swear] by the fig and the olive" (Qur'an 95:1). The prophet Muhammad reportedly named the fig as the one fruit he would most wish to see in heaven. The Hadith quotes Muhammad claiming that figs surely "descended from paradise." The Qur'an says: let man consider his food. Nineteen plant names are mentioned in the Qur'an as good nutritional elements and fig is one of them (Azarpour *et al.*, 2014). Muslims believe that continuous consumption of Qur'anic plants, because of their water, sugar, fat, protein, and vitamin components, plays a significant role in prevention and treatment of many diseases (Azarpour *et al.*, 2014).

In the Land of Israel, grapevines and fig trees are harvested (and laid out to dry) simultaneously; hence, the pairing of the two. As in olden times, guests are received and attended to under the trees. An Arab proverb describes the harvest days as times of gain for friends: *yom 'inbak utinak kull en-nas (i)mhibbinak, halas 'inbak utinak kull en-nas 'adauwinak* "during the days of your grapes and figs all people love you; but when your grapes and figs are over then all people become your enemies" (Canaan, 1928).

Figs are found in Muslim graveyards, comprising about 10% of the trees and shrubs found in Muslim burial contexts today. The fig has a symbolic virtue in Muslim passage rites (Dafni *et al.*, 2006).

Leaves of *Ficus septica* Burm. f. var. *septica*, commonly known as septic fig, is used as a medicinal plant by the Muslims of Maranaos, The Philippines, in their healthcare practices (Olowa and Demayo, 2015). The leaves of septic fig are boiled with water and the boiled leaves transferred into a small basin. The sore stomach is placed above the steam while enclosing the whole body (except the head) with a big blanket (the process is called "pauslob" or "tuob"), directing the steam/vapor over the stomach. This treatment is aimed at relieving sickness felt after missing a meal or "pasmo." The roots of septic fig, together with the roots of tobacco are also used to relieve muscle pain, or overfatigue or "bughat" in women. The roots should be cut into small pieces and soaked in lukewarm water or boiled with water, and the soup drunk three times a day (Olowa and Demayo, 2015).

3.3.6 Figs in mosaics of the Roman world

Mosaics form one of the most widespread types of Roman art, found throughout the Roman Empire from Britain and Spain in the west to Jordan and northern Mesopotamia in the east. They were used in public buildings such as Roman baths and marketplaces and, in late antiquity, in places of worship such as synagogues and churches. The majority of mosaics, however, decorated private buildings, both large city houses and villas in the countryside or by the sea. Mosaics were a costly and time-consuming addition to a property, although they would thereafter require little maintenance and could be expected to last a lifetime. Because of the heavy initial investment, considerable care and attention was paid to creating designs that were attractive and appropriate, both to the owner and to the setting.

Despite the diversity in their manner of execution, design, geographical setting, and dating, Roman mosaics display a number of shared or common characteristics. Mosaics are highly characteristic of the Roman world and provide valuable information about the tastes and preoccupations of the time. Figs are represented in mosaic floors found in many ancient synagogues and Roman churches in Israel. In synagogues found in Tiberias, Sepphoris, and Bet Alpha, dating to the sixth century AD, figs are depicted according to the appropriate time of year—autumn (around September–October) and winter (around December–January). In other cases, the figs are depicted embalming the human figure representing "Egypt" (Ben-Sasson, 2013).

The fig is also represented in almost all mosaic floors found in Christian monuments in Israel. In the fifth century AD mosaic found at the church of the birthplace of Jesus in Bethlehem, a pair of figs is depicted. In the Byzantine church built at the top of Masada, probably also dating to the fifth century AD, a cluster of figs is found among other fruits in the floor mosaic. In the mosaic of Bet Loya church, 500 CE, which was deliberately destroyed by later conquests, a basket of fruit full of only figs is detailed. It is possible that in the figures where figs of different sizes are detailed, two types of *Ficus* species are drawn—*Ficus carica* (common fig) and *Ficus sycomorus*, whose fruits were also habitually eaten in this region during this time (Ben-Sasson, 2013). This phenomenon of different *Ficus* species being depicted in one mosaic was also recorded in the John the Baptist Armenian memorial chapel at the olive mount. In this mosaic, two non-identical groups of figs, described at different maturation levels, are seen. Some of the figs in this mosaic are described as very ripped, identified as such by an open and black triangle at their wide end (Ben-Sasson, 2013).

In several chapels of the sixth century AD, figs are found in the mosaics. In the burial chapel of El-Hammam in Bet Shean, 567 CE, figs can be found around the vines of grapes. In the convent chapel of Kursi, 585 CE, on the eastern shores of the Sea of Galilee, figs are placed on a heart-shaped leaf. At the Martyr church of Tel Esteba situated near Bet Shean, inside a geometric pattern, figs and other fruits are depicted, and in the apsis of Hurvat Midras church in the Shephela, a basket of fruits containing one fig.

In the Dominus Flevit Church on the Mount of Olives, figs are also found. This church follows the Byzantine building style, but according to the literature is dated to 675 CE, the Muslim period (Ben-Sasson, 2013).

Note that despite the numerous appearances of trees in different Roman mosaics, a depiction of a fig tree has never been found in a mosaic, neither in Israel nor in Jordan (Ben-Sasson, 2013).

In 1996, a series of mosaic floors measuring approximately 50 × 27 feet were accidentally uncovered during highway construction in Lod, Israel (Fig. 3.6). The mosaics had been laid in *c.*300 CE. In contrast to the above-described mosaics, these probably belonged to a private house and served as the floors for a series of reception or audience halls. Despite its uniqueness, richness, rarity, and beauty depicting all sorts of animals, fish, and boats sailing in and out of shore, fruits are not depicted in this mosaic. However, these mosaics are probably part of a much larger-scale artistic display, of which only a small fraction has currently been found.

3.4 Sacred Figs: Figs in Myth and Art

Out of the 202 taxa of plants identified in ancient art, *F. carica* is one of the seven most represented species due to its strong association with mythological and religious symbolic meaning. Fig fruits and trees make 58 appearances in Roman art objects (Kumbaric, 2013). The most compelling "use" by humans of figs, which is not shared by any other group of animals, is the view of figs as sacred by people worldwide. The tree as a sacred object or a metaphysical construct is not a new concept; however, the prevalence of figs among these is quite striking (Wilson and Wilson, 2013). Many religions talk of "cosmic" trees; while some have no worldly equivalent, others are identified with specific species, e.g. the Indian "sky tree" with the sacred fig (*Ficus religiosa*), the Egyptian "tree of life" with the date or sycamore (*F. sycomorus*), and the mythic world tree of the Hindus with the banyan (*Ficus benghalensis*) (Forlong, 1883; Frese and Gray, 1995; Hamilton, 2002). *F. religiosa* and *F. benghalensis* are considered sacred to Buddhists and Hindus, respectively, worldwide. Buddha reached enlightenment under the bodhi tree, a large and old holy fig tree (*F. religiosa*). Even today, the banyan is India's national tree, with Hindu women flocking to it to receive blessings for a long and happy married life. *F. religiosa*, also known as the pipal tree, has its own religious connotations in its native regions of India, Nepal, Bangladesh, Pakistan, and Sri Lanka.

Another ritual concerning fig sacrifice is specific to Roman harbors. Excavations and

Fig. 3.6. Part of the mosaic of Lod, 300 CE. Modified from the Lod Mosaic Museum website.

sampling conducted at the city of Lattara (Lattes, France) revealed a foundation offering in a small pit situated in a storehouse in the Roman port of the city (25–60 CE). This phenomenon of foundation offering, as opposed to those deposits found in burial contexts, is rare and unique to this region and is unknown during the Roman period. The offering consisted of one date pit, one fig, and one cone of stone pine containing nuts, pips of grape, grains of hulled barley, naked wheat and emmer, grass pea, lentil, and flax. In an attempt of decipher the meaning of this symbolic attribute, the researchers tried to determine whether this assemblage was unique to the foundation offering or whether it was a mundane and common assemblage of fruits and grains (Rovira and Chabal, 2008). In other contemporary ritual offerings, figs are a relatively rare find (Robinson, 2002). The researchers of the Lattara foundation offering reported that most of the fruits and seeds found in this context were also commonly found in contemporary contexts of this site, mainly as consumption or production waste. However, fig was rare, and date was found for the first time. When appreciating the level of attention made in fruit and seed choice for burial, it is worth noting that other fruits and grains frequently used in Roman offerings were absent from this one (Rovira and Chabal, 2008). Olives, walnuts, hazelnuts, almonds, cherries, and peaches, although quite well known at Lattara during the same period (Alonso *et al.*, 2008), were not included in this foundation offering. Thus, the researchers concluded that the choice of the assembled fruits pointed to a voluntary act intended to produce a specific effect. The symbolic and magical properties of the plants burnt in the offering was aimed at producing an effect related principally to abundance, fecundity, and longevity.

Figs are mentioned also for their medicinal virtues. According to an updated evaluation of Dafni and Böck (2019), only five species are mentioned directly in the Bible as medicinal plants: common fig (*F. carica*), nard (*Nardostachys jatamansi*), hyssop (*Origanum syriacum*), balm of Gilead (*Commiphora gileadensis*), and mandrake (*Mandragora officinarum*). These biblical medicinal plants were also known for their beneficial medical properties in Ancient Egypt and Mesopotamia (Zohary *et al.*, 2012; Dafni and Böck, 2019).

In Asian cultures, fig trees were regarded as symbols of power and places of prayer. Figs feature in creation stories, folklore, and fertility rites. The champion is the Indian banyan (*F. benghalensis*), a very large tree that resembles a small forest from afar. *F. religiosa* had stories and poems written about it. The epic poem named *The Mahavamsa* talks about one such tree that had a branch removed on the command of the Indian legendary emperor Ashoka the Great. Ashoka granted holiness to the branch and planted it in a thick-rimmed solid gold vase. He then took the branch over mountains and down the Ganges River to the Bay of Bengal, where his daughter took it and sailed with it to Sri Lanka, to present it to the king. The epic poem tells that Ashoka loved the plant so much he shed tears as he watched it leave.

3.5 Concluding Remarks

In this chapter, we have described some of the ways in which humans used figs in different ancient civilizations. The collection of images of individual figs in the field of art and history highlights both the importance and diversity of this interaction. The rich connection between art and horticultural technology suggests that the systematic collection of fig iconography would be an invaluable resource providing significant information on taxonomy, crop history, and evolution. Our examples span the global distribution of figs and a wide range of different fig practices. The same views and uses of figs occur repeatedly, showing figs as an important food source, and thus they provide vital spiritual, health, and functional roles to humans wherever they are distributed.

References

Adams, K.R. (1980) *Pollen, Parched Seeds, and Prehistory: A Pilot Investigation of Prehistoric Plant Remains from Salmon Ruin, a Chacoan Ruin in Northwest New Mexico*. Eastern New Mexico University Contributions to Anthropology 9, Eastern New Mexico University, Portales, New Mexico.

Ahituv, S. (2005) *HaKetav VeHaMiktav. Handbook of Ancient Inscriptions from the Land of Israel and the Kingdoms Beyond the Jordan from the Periods of the First Commonwealth*. Mosad Bialik, Jerusalem, Israel.
Allen, S.E. (2002). Palaeoethnobotany: preliminary results. In: Sha'ar, H.I., Garfinkel, Y., and Miller, M.A.A. (eds) *Neolithic Art in Context*. Oxbow Books, Oxford, UK, pp. 236–246.
Alonso, N., Buxó, R. and Rovira, N. (2008) Archéobotanique des semences et des fruits de Lattara: bilan des recherches. In: Janin, T. and Py, M. (eds) *La Ville Portuaire de Lattara (Lattes, He´rault) et son Territoire: Nouveaux Acquis, Nouvelles Questions*. Gallia 65 Editions du CNRS, Paris, pp. 193–200.
Amichay, O., Ben-Ami, D., Tchekhanovets, Y., Shahack-Gross, R., Fuks, D., and Weiss, E. (2019) A bazaar assemblage: reconstructing consumption, production and trade from mineralised seeds in Abbasid Jerusalem. *Antiquity* 93, 199–217.
Azarpour, E., Moraditochaee, M., and Bozorgi, H.R. (2014) Nutritional and biochemical compounds of Quranic plants. *Biological Forum* 6, 242–248.
Ben-Sasson, R. (2013) Flora motifs in the art of the Land of Israel from the Roman Period to the end of the Byzantine Era. PhD thesis, Hebrew University of Jerusalem, Jerusalem, Israel.
Borojevic, K. (2006) The archaeobotanical finds. In: Finkelstein, I., Ussishkin, D., and Halpern, B. (eds) *Megiddo IV: the 1998–2002 Seasons*. Monograph Series of the Institute of Archaeology University of Tel Aviv, Vol. 24. Emery and Claire Yass Publications, Tel Aviv, Israel, pp. 519–541.
Borojevic, K. (2011) Interpreting, dating, and reevaluating the botanical assemblage from tell Kedesh: a case study of historical contamination. *Journal of Archaeological Science* 38, 829–842.
Condit, I.J. (1947) *The Fig*. Chronica Botanica, Waltham, Massachusetts.
Dafni, A. and Böck, B. (2019) Medicinal plants of the Bible—revisited. *Journal of Ethnobiology and Ethnomedicine* 15, 1–14.
Dafni, A., Lev, E., Beckmann, S., and Eichberger, C. (2006) Ritual plants of Muslim graveyards in northern Israel. *Journal of Ethnobiology and Ethnomedicine* 2, 38–50.
Del Olmo Lete, J. and Sanmartín, J. (2003) *A Dictionary of the Ugaritic Language in Alphabetic Tradition*. Handbuch der Orientalistik 67. Brill, Leiden, The Netherlands/Boston, Massachusetts.
Dickson, J.H. and Dickson, C. (1996) Ancient and modern occurrences of common fig (*Ficus carica* L.) in the British Isles. *Quaternary Science Reviews* 15, 623–633.
Dunseth, Z.C., Fuks, D., Langgut, D., Weiss, E., Melamed, Y., *et al.* (2019) Archaeobotanical proxies and archaeological interpretation: a comparative study of phytoliths, pollen and seeds in dung pellets and refuse deposits at Early Islamic Shivta, Negev, Israel. *Quaternary Science Reviews* 211, 166–185.
Erman, A. (1923) *Die Literatur der Ägypter*. J.C. Hinrichs, Leipzig, Germany.
Fink, S.A. (2009) Levantine standardized luxury in the Late Bronze Age: waste management at Tell Atchana (Alalakh). In: Fantalkin, A. and Yasur-Landau, A. (eds) *Bene Israel: Studies in the Archaeology of Israel and the Levant during the Bronze and Iron Ages in Honour of Israel Finkelstein, Culture and History of the Ancient Near East*. Brill, Leiden, The Netherlands, pp. 165–195.
Fleming, D.E. (2004) *Democracy's Ancient Ancestors: Mari and Early Collective Governance*. Cambridge University Press, Cambridge, UK.
Foley, A. (2015) Seeds of Roman Galilee: A preliminary archaeobotanical analysis of Huqoq, Israel. *The Spectrum: A Scholars Day Journal* 3, 5.
Forlong, J. (1883) *Rivers of Life*. Bernard Quaritch, London.
Frese, P., and Gray, S. (1995). Trees. In: Eliade, M. (ed.) *The Encyclopaedia of Religion*. Simon & Schuster/ Macmillan, New York, pp. 26–33.
Frumin, S.I. and Tchekhanovets, Y. (2016) Plant imprints on pottery reveal fig tree in Hellenistic Jerusalem. *Israel Exploration Journal* 66, 188–201.
Frumin, S., Melamed, Y., and Maeir, A.M. (2020) Canaan's staples, seasonal vegetables and luxury fruits: finds from Stratum E4. In: Maeir, A.M. and Uziel, J. (eds) *Tell es-Safi/Gath II: Excavations and Studies*. Ägypten und Altes Testament 105. Zaphon, Münster, Germany, pp. 459–488.
Galil, J. and Neeman, G. (1977) Pollen transfer and pollination in the common fig (*Ficus carica* L.). *New Phytology* 79, 163–171.
Giachi, G., Macchioni, N., Pizzo, B., and Capretti, C. (2011) The waterlogged wood of the shipwrecks found in Pisa (Italy). Diagnosis and conservation. In: *Chemistry and Preservation of Waterlogged Wooden Shipwrecks*. CNR IVaLSA, Stockholm, pp. 78–84.
Giorgi, J. (1999) Plant remains from the sediments of the Inner Harbour (area I14). *Journal of Roman Archaeology-Supplementary Series* 35, 353–357.
Gitin, S. (1993) Seventh century BCE cultic elements at Ekron. In: Biran, A. and Aviram, J. (eds) *Biblical Archaeology Today, Proceedings of the Second International Congress on Biblical Archaeology*. Keterpress Enterprises, Jerusalem, pp. 248–258.

Goor, A. (1965) The history of the fig in the Holy Land from ancient times to the present. *Economic Botany* 19, 124–135.
Green, F.J. (1979) Phosphatic mineralization of seeds from archaeological sites. *Journal of Archaeological Science* 6, 279–284.
Hajnalová, E. (1986) Holzkohle Stücken aus dem Gräberfeld in Krásna Ves. In: Budinský-Krička, V. and Veliačik, L. (eds) *Krásna Ves, Gräberfeld der Lausitzer Kultur*. Archäologisches Institut der Slowakischen Akademie der Wissenschaften, Nitra, Slovakia, pp. 137–138.
Hamilton, L. (2002) Forest and tree conservation through metaphysical constraints. *The George Wright Forum* 19, 57–78.
Hopf, M. (1969) Plant remains and early arming in Jericho. In: Ucko, P.J. and Dimbleby, G.W. (eds) *The Domestication and Exploitation of Plants and Animals*. Gerald Duckworth & Co., London, pp. 355–359.
Hopf, M. (1983) Jericho plant remains. *Excavations at Jericho* 5, 576–621.
Janick, J. (2002) Ancient Egyptian agriculture and the origins of horticulture. In: Sansavini, S. and Janick, J. (eds) *International Symposium on Mediterranean Horticulture: Issues and Prospects*. International Society for Horticultural Science, Leuven, Belgium, pp. 23–39.
Keay, S. and Paroli, L. (2011) *Portus and its Hinterland: Recent Archeological Research*. Archaeological Monographs of the British School at Rome, Vol. 18, London.
Keay, S., Millet, M., Paroli, L., and Strutt, K. (2005) *Portus: An Archaeological Survey of the Port of Imperial Rome*. Archaeological Monographs of the British School at Rome, Vol. 15, Rome.
Kislev, M.E. (1979) Tel Saf. *Revue Biblique* 96, 112–115.
Kislev, M.E. (1987) Chalcolithic plant husbandry and ancient vegetation at Shiqmim. In: Levy, T.E. (ed.) *Shiqmim I Studies concerning Chalcolithic Societies in the Northern Negev Desert, Israel (1982–1984)*. British Archaeological Reports, International Series, Oxford, UK, pp. 251–279.
Kislev, M.E. (1993) Food remains. In: Finkelstein, I., Bonimovich, S., and Lederman, Z. (eds) *Shiloh: the Archaeology of a Biblical Site*. Institute of Archaeology of Tel Aviv University, Tel Aviv, pp. 354–361.
Kislev, M. (2003a) Remains of seeds and fruits from Rakit. In: Dar (ed.) *Rakit – The Estate of Marinos in Carmel*, pp. 305–309 (in Hebrew).
Kislev, M.E. (2003b) The botanical remains: a clue to dating the season of the ship's voyage. In: Black, E. (ed.) *The Ma'agan Mikhael Ship – The Recovery of a 2400-year-old Merchantman*. Israel Exploration Society Centennial, Jerusalem, Israel, p. 244.
Kislev, M.E. and Hopf, M. (1985) Food remains from Tell Qasile, with special reference to *Lathyrus sativus/cicera*. *Excavations at Tell Qasile* 2, 140–147.
Kislev, M.E. and Mahler-Slasky, Y. (2009) Food remains. In: Gadot, Y. and Yadin, E. (eds) *Aphek-Antipatris II, the Remains on the Acropolis, The Moshe Kochavi and Pirhiya Beck Excavations*. Tel Aviv University, Tel Aviv, Israel, pp. 499–525.
Kislev, M.E. and Melamed, Y. (2000) Appendix 3. Ancient infested wheat and horsebean from Horbat Rosh Zayit. In: Gal, Z. and Alexandre, Y. (eds) *Horbat Rosh Zayit – An Iron Age Storage Fort and Village*. Israel Antiquity Authority, Jerusalem, Israel, pp. 206–220.
Kislev, M.E. and Simchoni, O. (2007) Hygiene and insect damage of crops and foods at Masada. In: Aviram, J., Foerster, G., and Netzer, E. (eds) *Masada VIII*. Israel Exploration Society and The Hebrew University, Jerusalem, Israel, pp 133–170.
Kislev, M.E., Nadel, D., and Carmi, I. (1992) Epipalaeolithic (19,000 BP) cereal and fruit diet at Ohalo II, Sea of Galilee, Israel. *Review of Palaeobotany and Palynology* 73, 161–166.
Kislev, M.E., Hartmann, A., and Galili, E. (2004) Archaeobotanical and archaeoentomological evidence from a well at Atlit-Yam indicates colder, more humid climate on the Israeli coast during the PPNC period. *Journal of Archaeological Science* 31, 1301–1310.
Kislev, M.E., Hartmann, A., and Bar-Yosef, O. (2006a) Early domesticated fig in the Jordan Valley. *Science* 312, 1372–1374.
Kislev, M.E., Melamed, Y., and Langsam, Y. (2006b) Plant remains from Tel Batash. In: Panitz-Cohen, N. and Mazar, A. (eds) *The Finds from the Second Millennium BCE*. The Hebrew University, Jerusalem, Israel, pp. 295–311.
Kislev, M.E., Simchoni, O., Melamed, Y., and Maroz, L. (2006c) Plant remains. In: Garfinkel, Y. and Dag, D. (eds) *Gesher a Pre-Pottery Neolithic A site in the Central Jordan Valley, Israel*. Ex Oriente, Berlin, pp. 187–191.
Kislev, M.E., Hartmann, A., and Gopher, A. (2012) Food crops from Nahal Zehora II. In: Gopher, A. (ed.) *Village Communities of the Pottery Neolithic Period in the Menashe Hills, Israel*. Tel Aviv University, Tel Aviv, Israel, pp. 1321–1326.

Kubiak-Martens, L. (2015) Plant remains from Tell Ashara (Terqa) and Tell Masaikh in the Middle Euphrates, south-eastern Syria. In: Margueron, J.-C., Rouault, O., Butterlin, P., and Lombard, P. (eds) *Akh Purattim – Les Rives de L'euphrate*. MOM Editions, Lyon, France, pp. 423–442.

Kumbaric, A. (2013) *Biodiversity and phytoiconology of Roman archaeology and influences on Renaissance: meanings for our natural and cultural heritage*. PhD thesis, Department of Sciences, University Roma Tre, Rome.

Langgut, D., Finkelstein, I., and Litt, T. (2013) Climate and the Late Bronze collapse: new evidence from the southern Levant. *Tel Aviv* 40, 149–175.

Langgut, D., Shahack-Gross, R., Arie, E., Namdar, D., Amrani, A., Le Bailly, M., and Finkelstein, I. (2016) Micro-archaeological indicators for identifying ancient cess deposits: an example from Late Bronze Age Megiddo, Israel. *Journal of Archaeological Science: Reports* 9, 375–385.

Liphschitz, N. (2004) The archaeobotanical remains. In: Ussishkin D. (ed.) *The Renewed Archaeological Excavations at Lachish 1973–1994*. Institute of Archaeology, Tel Aviv, Israel, pp. 2230–2247.

Liphschitz, N. and Miroschedji, P.D. (1989) Plant economy and diet in the Early Bronze Age in Israel: a summary of present research. In: Miroschedji, P. (ed.) *L'urbanisation de la Palestine à l'age du Bronze Ancien: bilan et perspectives des recherches actuelles*. BAR International, Oxford, UK, pp. 269–277.

Mahler-Slasky, Y. and Kislev, M.E. (2012) Preliminary archaeobotanical research at Tell es-Safi/Gath – the 1997–2002 seasons. In: Maeir, A.M. (ed.) *Tell es-Safi/Gath I. The 1996–2005 Seasons*. Harrassowitz Verlag, Wiesbaden, Germany, pp. 579–587.

Marder, O., Biton, R., Boaretto, E., Feibel, C.S., Melamed, Y., *et al.* (2015) Jordan River Dureijat – a new Epipaleolithic site in the Upper Jordan Valley. *Journal of the Israel Prehistoric Society* 45, 5–30.

Marshall, L.-J.R., Almond, M.J., Cook, S.R., Pantos, M., Tobin, M.J., and Thomas, L.A. (2008) Mineralised organic remains from cesspits at the Roman town of Silchester: processes and preservation. *Spectrochimica Acta A* 71, 854–861.

McCobb, L.M.E., Briggs, D.E.G., Evershed, R.P., Hall, A.R., and Hall, R.A. (2001) Preservation of fossil seeds from a 10th century AD cess pit at Coppergate, York. *Journal of Archaeological Science* 28, 929–940.

McCobb, L.M.E., Briggs, D.E.G., Carruthers, W.J., and Evershed, R.P. (2003) Phosphatization of seeds and roots in a Late Bronze Age deposit at Potterne, Wiltshire, UK. *Journal of Archaeological Science* 30, 1269–1281.

McCorriston, J. (2007) Cultural and environmental history in archaeological charred woods from the Khabur drainage, upper Mesopotamia. *Publications de l'Institut Français d'Études Anatoliennes* 19, 503–522.

McMahon, A. (2015) Waste management in early urban southern Mesopotamia. In: Mitchell, P.D. (ed.) *Sanitation, Latrines and Intestinal Parasites in Past Populations*. Ashgate Publishing, Farnham, UK, pp. 19–39.

Melamed, Y. and Kislev, M. (2005) Remains of seeds, fruits and pests from the excavations of the village of Ein Gedi. *Atiqot* 49, 89–102 (in Hebrew).

Melamed, Y., Kislev, M.E., Geffen, E., Lev-Yadun, S., and Goren-Inbar, N. (2016) The plant component of an Acheulian diet at Gesher Benot Ya'aqov, Israel. *Proceedings of the National Academy of Sciences USA* 113, 14674–14679.

Meyer, F.G. (1980) Carbonized food plants of Pompeii, Herculaneum, and the Villa at Torre Annunziata. *Economic Botany* 34, 401–437.

Noy, T., Legge, A.J., Higgs, E.S., and Dennell, R.W. (1973) Recent excavations at Nahal Oren, Israel. *Proceedings of the Prehistoric Society* 39, 75–99.

Oates, J., McMahon, A., Karsgaard, P., Al Quntar, S., and Ur, J. (2007) Early Mesopotamian urbanism: a new view from the North. *Antiquity* 81, 585–600.

Olowa, L. and Demayo, C.G. (2015) Ethnobotanical uses of medicinal plants among the Muslim Maranaos in Iligan City, Mindanao, Philippines. *Advances in Environmental Biology* 9, 204–215.

Orendi, A., Smejda, L., McKinny, C., Cassuto, D., Sharp, C., and Shai, I. (2017) The agricultural landscape of Tel Burna: ecology and economy of a Bronze Age/Iron Age settlement in the southern Levant. *Journal of Landscape Ecology* 10, 165–188.

Pepe, C., Giardini, M., Giraudi, C., Masi, A., Mazzini, I., and Sadori, L. (2013) Plant landscape and environmental changes recorded in marginal marine environments: the ancient Roman harbour of Portus (Rome, Italy). *Quaternary International* 303, 73–81.

Ramírez, W. (1977) A new classification of *Ficus*. *Annals of the Missouri Botanical Garden* 64, 296–310.

Ramsay, J. (2010) Trade or trash: an examination of the archaeobotanical remains from the Byzantine harbour at Caesarea Maritima, Israel. *International Journal of Nautical Archaeology* 39, 376–382.

Ramsay, J. and Holum, K. (2015) An archaeobotanical analysis of the Islamic period occupation at Caesarea Maritima, Israel. *Vegetation History and Archaeobotany* 24, 655–671.
Ramsay, J. and Tepper, Y. (2010) Signs from a green desert: a preliminary examination of the archaeobotanical remains from a Byzantine dovecote near Shivta, Israel. *Vegetation History and Archaeobotany* 19, 235–242.
Reinhard, K., Hamilton, D.L., and Hevly, R.H. (1991) Use of pollen concentration in paleopharmacology: coprolite evidence of medicinal plants. *Journal of Ethnobiology* 11, 117–132.
Robinson, M. (2002) Domestic burnt offerings and sacrifices at Roman and pre-Roman Pompeii, Italy. *Vegetation History and Archaeobotany* 11, 93–99.
Rovira, N. and Chabal, L. (2008) A foundation offering at the Roman port of Lattara (Lattes, France): the plant remains. *Vegetation History and Archaeobotany* 17, 191–200.
Sadori, L., Allevato, E., Bellini, C., Bertacchi, A., Boetto, G., *et al.* (2015) Archaeobotany in Italian ancient Roman harbours. *Review of Palaeobotany and Palynology* 218, 217–230.
Salavert, A. (2008) Olive cultivation and oil production in Palestine during the early Bronze Age (3500–2000 BC): the case of Tel Yarmouth, Israel. *Vegetation History and Archaeobotany* 17, 53–61.
Shai, I., Dagan, A., Riehl, S., Orendi, A., Uziel, J., and Suriano, M. (2014) A private stamped seal handle from Tell Bornāṭ/Tēl Burnā, Israel. *Zeitschrift des Deutschen Palästina-Vereins* 130, 121–137.
Simchoni, A. and Kislev, M. (2004) Remains of plants and insects from the Spear Caves in the Judean Desert. *Judea and Samaria Studies* 13, 135–148 (in Hebrew).
Simchoni, O. and Kislev, M.E. (2012) Food and fodder in Early Bronze Age strata in area M. In: Mazar, A. (ed.) *Excavations at Tel Beth-Shean 1989–1996, Vol. IV, The Fourth and the Third Millenia B.C.E.* Israel Exploration Society and Hebrew University Jerusalem, Jerusalem, Israel, pp. 422–429.
Simchoni, O., Kislev, M.E., and Melamed, Y. (2007) Beth-Shean as a trade center for crops in the Bronze Age: botanical and entomological evidence. In: Mazar, A. and Mullins, R.A. (eds) *Excavations at Tel Beth-Shean 1989–1996*, Vol. II, *The Middle and Late Bronze Age Strata in Area R.* The Israel Exploration Society and The Hebrew University of Jerusalem, Jerusalem, pp. 702–715.
Smith, A. (2005) Climate, culture, and agriculture: examining change in the Near East during the Bronze and Iron Ages. PhD thesis, Boston University, Boston, Massachusetts.
Stein, G. (2004) Structural parameters and sociocultural factors in the economic organization of North Mesopotamian urbanism in the third millennium B.C. In: Feinman, G.M. and Nicholas, L.M. (eds) *Archaeological Perspectives on Political Economies*. The University of Utah Press, Salt Lake City, Utah, pp. 61–78.
Thompson, R.C. (1949) *A Dictionary of Assyrian Botany*. The British Academy, London.
Watson, W.G.E. (2004) A botanical snapshot of Ugarit. Trees, fruit, plants and herbs in cuneiform texts. *Aula Orientalis* 22, 107–155.
Weiss, E. (2002) Reconstruction of the human economy and society of the Epipalaeolithic Site Ohalo II inhabitants by the macrofossil botanical remains. Unpublished PhD thesis, Bar-Ilan University, Bar-Ilan, Israel.
Weiss, E. and Mahler-Slasky, Y. (2016) Excursus B: jar of dried figs (Tel Beth-Shemesh). In: Boaretto, E., Brown, E., Gilboa, A., Golani, A., Goren, Y., *et al.* (eds) *Tel Beth-Shemesh: A Border Community in Judah. Renewed Excavations 1990–2000: The Iron Age*. Eisenbrauns, Penn State University Press, University Park, Pennsylvania, pp. 247–256.
Weiss, E., Kislev, M., and Mahler-Slasky, Y. (2011) Plant remains. In: Stager, L.E., Master, D.M., and Schloen, J.D. (eds) *Ashkelon 3: The Seventh Century B.C.* Eisenbrauns, Penn State University Press, University Park, Pennsylvania, pp. 591–613.
Wendorf, F. and Schild, R. (2001) *Holocene Settlement of the Egyptian Sahara*. Springer, New York.
Wick, L., Lemcke, G., and Sturm, M. (2003) Evidence of Late glacial and Holocene climate change and human impact on Eastern Anatolia: high-resolution pollen, charcoal, isotopic and geochemical records from the laminated sediments of Lake Van, Turkey. *The Holocene* 13, 665–675.
Wilson, D. and Wilson, A. (2013) Figs as a global spiritual and material resource for humans. *Human Ecology* 41, 459–464.
Wiseman, D.J. (1952) A new stela of Aššur-naṣir-pal. *Iraq* 14, 24–44.
Yadin, Y., Naveh, J., and Meshorer, Y. (1989) The Aramaic and Hebrew ostraca and jar inscriptions. In: Yadin, Y. and Naveh, J. (eds) *Masada I*. Israel Exploration Society, Jerusalem, Israel, pp. 46–47.
Zohary, D. and Hopf, M. (2000) *Domestication of Plants in the Old World: The Origin and Spread of Domesticated Plants in Southwest Asia, Europe, and the Mediterranean Basin*. 3rd edn. Oxford University Press, Oxford, UK.
Zohary, D., Hopf, M., and Weiss, E. (2012) *Domestication of plants in the Old World: The origin and spread of domesticated plants in Southwest Asia, Europe, and the Mediterranean Basin*. 4th edn. Oxford University Press, Oxford, UK.

Part II

Fig Biology and Orchard Management

4 Fig Morphology and Development

Moshe A. Flaishman*
Institute of Plant Sciences, Agriculture Research Organization, Volcani Center, Rishon LeZion, Israel

4.1 Introduction

Ficus carica is a deciduous tree that can range from open to compact and from pendulous to upright to spreading, depending on the variety. Tree size and density also depend somewhat on variety, although cultural factors and soil quality can have a substantial impact on eventual tree size. Exceptional specimens exist that have attained 9–12 m in height with a 10 m spread, although commercially grown trees are much smaller. Some varieties produce trees that are round-topped and dense with many twiggy lateral spurs. Other varieties have more apically dominant branches with fewer spurs, producing a more open or leggy appearance. The tree grows in dry, sunny places throughout the Mediterranean region, from hillsides to rocky areas in canyons, near springs and close to the sea (Fig. 4.1). Environmental factors such as temperature, photoperiod, and humidity affect the development and yield of the fig tree. This general botany chapter will mainly describe the different fig tree parts, the vegetative and reproductive morphology, with an emphasis on the tree's adaptability to different growing conditions.

4.2 Vegetative Morphology and Development

4.2.1 Root system

The fig tree has a system of fibrous roots that spread out to up to three times the canopy diameter. This system is typically very shallow, with no taproot (Condit, 1947). In some soils, it can reach great depths (Fig. 4.2). Fig plants tolerate average to poor soil and moderate to high salinity (Golombek and Lüdders, 1990). The roots penetrate natural or artificial cracks and can damage urban infrastructure and archaeological monuments. Once the plants are established, they are relatively drought tolerant, probably due to their very extensive and wide-ranging root system. The efficient rooting process of figs is known from ancient times and has been used to establish new orchards from cuttings. A recent study evaluated the establishment performance of *F. carica* trees 3 years after planting. The study employed two types of soil bed—alluvium and terra rossa—and each tree was planted in 6 m^3 of soil (2 m × 2 m × 1.5 m). The results showed fast and efficient establishment of the rooting system. In addition, the roots were not affected

* Email: vhmoshea@agri.gov.il

DOI: 10.1079/9781789242492.0004

Fig. 4.1. *Ficus carica* L. A. A wild fig tree in a dry, rocky area in Israel. B. A fig tree in a commercial orchard.

Fig. 4.2. The root system of a mature fig tree. Figure used with the permission of R. Peer.

by changing soil type in the rhizosphere, suggesting a highly effective root system in figs (Atzmon and Henkin, 1998). Like many species of *Ficus*, the fig tree can produce aerial roots, generally in humid climates. The aerial roots are produced from nodes, internodes, or cuts in the fig tree trunk. Fig root extract contains numerous compounds that have been used for centuries as popular remedies for health disorders (Modi *et al.*, 2012). Fig roots are sensitive to the root-knot nematode, *Meloidogyne incognita*, especially where the trees are planted in light, sandy soils (Condit, 1947). Many fig trees, however, are immune or tolerant to nematodes (Storey, 1975), and nematode-resistant rootstocks are used to overcome the problem. Hosomi *et al.* (2002) tested 20 commercial fig varieties for nematode resistance. Trees grafted on the variety Zidi, which is tolerant to nematodes, were vigorous under various field conditions.

4.2.2 Wood and bark

Fig wood is light in color, soft and pithy, and of little value. Its specific gravity (oven-dried weight/green volume) is 0.43. The bark is smooth, but a very few varieties may have fissures. The trunks and larger branches may have bark tubers: these are swellings or tubers formed from dormant buds whose apex has died but that retain vascular connections with the wood. The young twigs are commonly glabrous and brownish to green, but quickly change to gray as the branch ages. The lenticels become corky, rough, and darker with age. The length of the internode increases toward the median part of the shoot. The wood anatomy of the genus *Ficus* is remarkably homogeneous. It is characterized by abundant axial parenchyma in regular concentric apotracheal bands and narrow vasicentric rings, and by relatively wide vessels. *Ficus* species can be recognized based on these characteristics. Significant anatomical differences have been found between the stem and branch wood of *F. carica*. Fiber length and vessel element length in the branch wood was about 16% and 3% shorter, respectively, than in the stem wood. Vessel frequency in the branch wood was about 52% higher than in the stem wood. Whereas the number of rays per millimeter did not differ in branch and stem wood, ray width was about 18% narrower in the former (Yaman, 2014). To maximize the use of all of the tree material, the fig branch wood can be used in the wood and coal industries, and fig bark extract, which contains a high level of furanocoumarins, has medicinal properties (Rouaiguia-Bouakkaz *et al.*, 2013).

4.2.3 The leaf

Fig has typical large, bright green, single, alternate leaves. They start to develop in the early spring, and growth of new leaves will continue until the temperature drops in autumn. Toward the end of the growing season, environmental conditions such as low temperature, wind, and rain cause leaf fall. Figs exhibit highly pronounced heterophylly, with the leaf lamina differing in size and shape. The alternate leaves are borne on long petioles. The lamina is obovate to nearly orbicular, either undivided or with highly pronounced lobation. Wide variation is found among leaf sizes, lobation, and petioles of the different fig varieties. Kadota leaves have no lobes, whereas others show shallow lobing and yet others, deeply dissected leaves (Fig. 4.3). In some varieties, the basal and distal leaves differ from the middle leaves on a single branch with respect to the number of lobes and sinus depth. The number of lobes varies from none to seven. Additional leaf characteristics that are stable enough to be used in varietal classification and identification are: lobe shape, presence and shape of the teeth, shape of the base (petiole sinus) and leaf venation, dimensions and color, and presence and texture of hairs or spicules.

The dissected fig leaf shows veins and veinless tissue. A close-up of the lower leaf surface shows the minor veins and lithocysts. On the lower leaf surface, the lithocysts and the capitate trichomes are confined to the leaf veins (Fig. 4.3). The leaf petiole consists of the epidermis, a differentiated cortex, and the conductive tissue, the latter consisting of two rings: one of phloem and the other of xylem elements. The petiole and midrib have leptocentric medullary vascular bundles, laticifers containing druse crystals. The leaves of *F. carica* have a compact double-layered

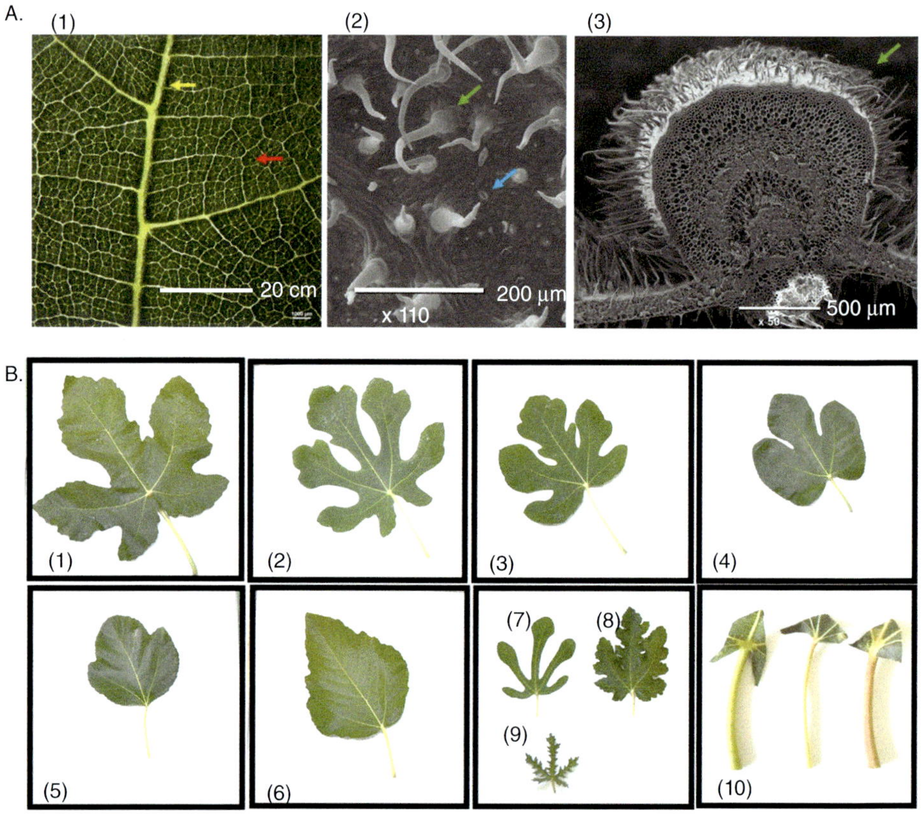

Fig. 4.3. Fig leaves. A. (1) Abaxial side of a fig leaf showing the vein system. The yellow arrow indicates the main vein and the red arrow indicates a lithocyst. (2) Scanning electron micrograph of the abaxial side of the leaf. The green arrow indicates a trichome and the blue arrow shows a stomata. (3) Scanning electron micrograph of a leaf in cross-section, showing the main vein and blade covered with trichomes (green arrow). B. Wide variations in leaf size, lobe presence and shape, and petioles of different fig varieties. (1) Nazareth; (2) Deep Purple; (3) Blanca; (4) Hamadi; (5) Kadota; (6) Hamam; (7) Galope; (8) Ice; (9) Kor; (10) different colored fig petioles.

palisade parenchyma lying below the single-layer adaxial epidermis. The upper epidermis is followed by a one-layered hypodermis with lithocysts and containing hydrated amorphous calcium carbonate bodies termed cystoliths. Numerous idioblasts with crystals, laticifers, and small-sized lithocysts appear within the compact fig leaf mesophyll (Mamoucha *et al.*, 2016).

The fig leaf hairs, trichome secretions, and solid inclusions play significant roles in protecting the plant from external factors, including high-intensity radiation, herbivores, and pathogens. The morphology and ultrastructure of the trichomes, calcium carbonate cystoliths, and silicified structures of the leaves of *F. carica* variety Dottato were investigated with light, confocal, and scanning electron microscopy, and with energy-dispersive X-ray spectroscopy (Giordano *et al.*, 2019). The leaf was found to be covered by non-glandular and capitate glandular trichomes. The non-glandular trichomes were simple, unicellular, and spine-like, and presented different morphologies and sizes. The capitate glandular trichomes were present on the leaf adaxial and abaxial surfaces and consisted of a one-celled stalk and a three- to four-celled spherical head. Histochemical characterization of leaf hairs revealed the presence of flavonoids, whereas glandular trichome head cells showed a complex mixture of alkaloids, essential oils, and flavonoids. Copper and aluminum were found in the constitutive structures, the spike and dome, of the cystoliths (Giordano *et al.*, 2019).

New leaves develop in the spring, starting from the terminal bud. The fig terminal bud generally contains four or five primordial leaves flanked on either side by a scale. Toward the base of the bud, one vegetative and two inflorescence primordia are present. A primordium destined to become vegetative has three or four scales covering the bud axis. During tree growth, the vegetative primordia continue to initiate scales and leaves. As the bud elongates, the cover scales abscize and the apical meristem develops into a shoot, producing leaves and new inflorescences (Fig. 4.4).

Vemmos *et al.* (2013) studied seasonal changes in fig leaf photosynthetic activity and the effects of chlorophyll, nitrogen, specific leaf weight, and leaf anatomy on net photosynthetic rate and carbohydrate concentration. The concentration of chlorophyll (chlorophyll *a* + chlorophyll *b*) increased rapidly until early June, when the fig leaves reached their maximum area, and then started to slowly decrease. The main soluble sugars were predominantly sucrose but also glucose and fructose. Soluble sugar and starch concentrations increased until the leaves reached their maximum area, and then decreased. This period coincided with inflorescence emergence.

The palisade cells below the adaxial surface of the leaf are specialized for photon collection; in many leaves, these cells are responsible for most of the light harvesting. Anatomical strategies to increase light penetration into the lower mesophyll have evolved, using mineral scatterers to introduce light deep in the leaf tissue. *Ficus* species can inhabit very different environments, some of which are characterized by very strong irradiance (Janzen, 1979; Harrison, 2005; Zhang *et al.*, 2016), and some *Ficus* species are known to have among the highest rates of photosynthesis in plants (Zotz *et al.*, 1995; Hao *et al.*, 2011). Leaves from many *Ficus* trees deposit cystoliths just below the leaf surface (Omori and Watabe, 1980; Setoguchi *et al.*, 1989). Pierantoni *et al.* (2020) compared variations in photosynthesis and photosynthetic efficiency in six different *Ficus* species (*F. religiosa*, *F. benghalensis*, *F. microcarpa*, *F. benjamina*, *F. lyrata*, and *F. carica*) with cystoliths deposited adaxially (below the upper surface of the leaf), abaxially (below the lower surface of the leaf), on both sides, or with no deposition at all. Micro-computed tomography three-dimensional cross-sectional perspective views of minerals and soft tissues in *Ficus* leaves showed that in *F. carica*, spherical cystoliths are deposited on the abaxial side, while adaxial cystoliths are absent. *F. carica* was the only species among those studied to deposit abundant silica in the epidermis and in the hairs. It was concluded that cystoliths contribute to optimization of photosynthesis under saturating light (Pierantoni *et al.*, 2020).

As already noted, leaf extracts of *F. carica* have medicinal properties (Badgujar *et al.*, 2014). In addition, a recent study prepared silver nanoparticles using this leaf extract. The silver nanoparticles enhanced cellular membrane

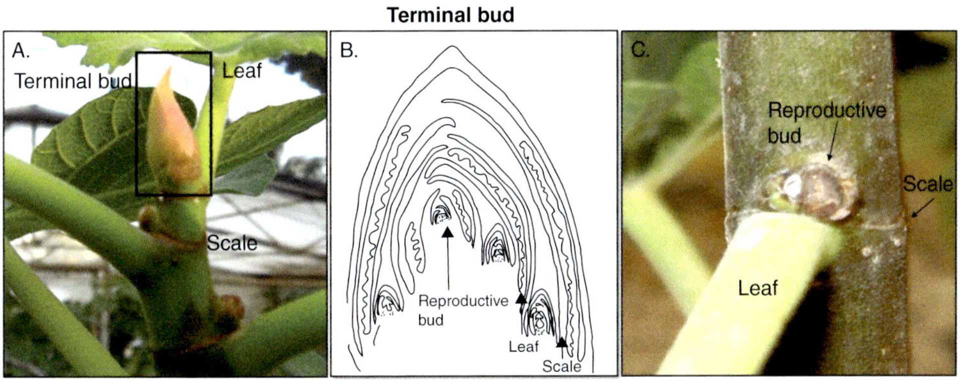

Fig. 4.4. The terminal bud generally contains four or five primordial leaves flanked by a scale on either side, and toward the base of the bud, one vegetative and two inflorescences.

permeability and decreased the effluent flux of the encapsulated drug, resulting in good antimicrobial activity (Begum *et al.*, 2020).

4.3 Reproductive Morphology and Development

As with all *Ficus* species, the fig bears a unique closed inflorescence structure, the syconium, defined as a collective, fleshy, hollow peduncle. It is a false fruit composed entirely of vegetative peduncle, while the true fruits are the pedicellate drupelets found inside. In effect, the syconium is an enlarged, fleshy, involute, hollow, and almost closed receptacle with multiple ovaries on its inner surface. It commonly has a bulbous shape with a small opening, called the ostiole, at the distal end that allows access to pollinators (Fig. 4.5A). Primordia destined to develop into reproductive or vegetative organs are identical at their initial stage of development. In general, four scales cover the bud axis. Following this

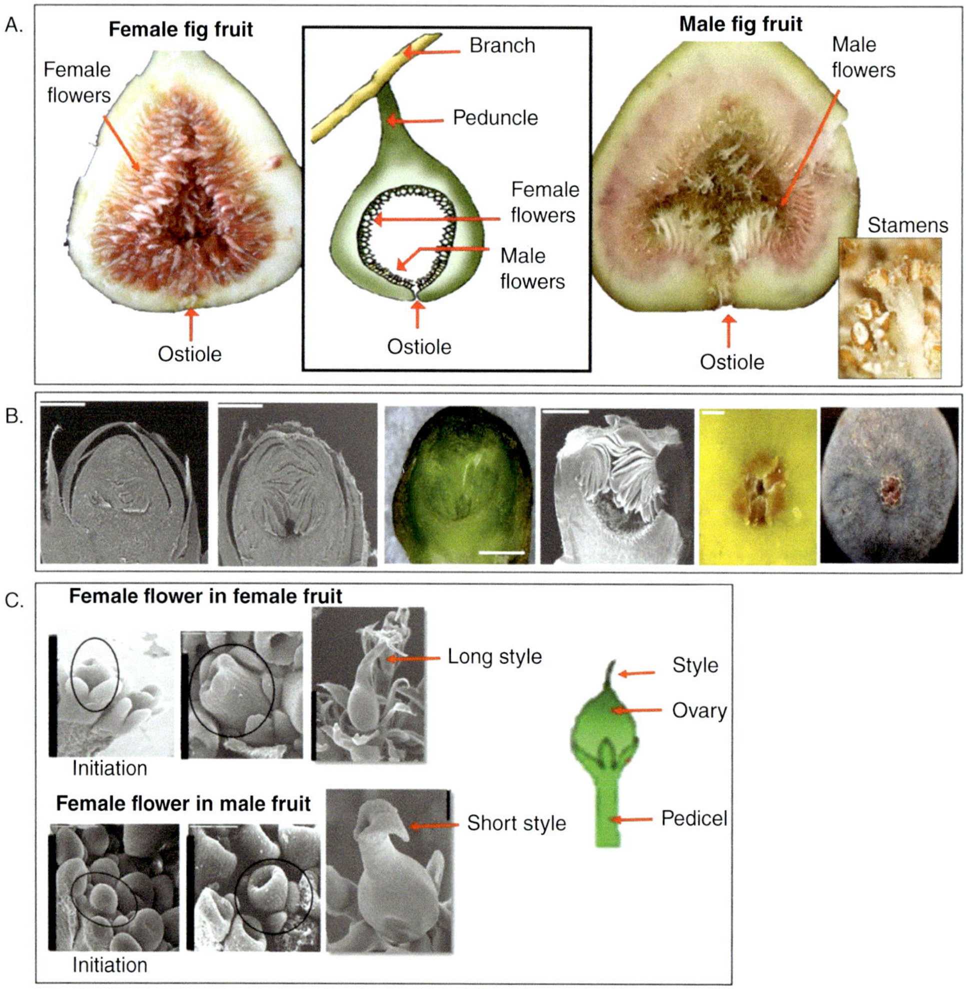

Fig. 4.5. The fig syconium. A. Female (left) and male (right) cross-sectional diagrams of the fig fruit syconia. B. Fig ostiole development. Scale bar, 900 µm. C. Close-up of female long style (top) and male short style (bottom) of the flower. Scale bar, 100 µm. Reproduced and modified from Kamara (2013), with permission.

stage, the vegetative meristem continues to initiate scales and leaves, unlike the reproductive meristem, which broadens and elongates and then begins the initiation of ostiolar scales. Hence, the first visual microscopic evidence of inflorescence differentiation is elongation of the axis and the initiation of scales that eventually surround the ostiole of the syconium (Fig. 4.5B). Further development of the inflorescence primordia consists of continued broadening of the apex and cell division around the periphery, giving rise to many ostiolar scales forming a cup-shaped structure lined with floral primordia (Fig. 4.5B). The formation of the syconium is complete when the apical portion of the cup-shaped structure grows toward the center and forms an ostiole, which is partly closed with numerous scales (Crane and Brown, 1950; Crane and Baker, 1953; Crane, 1986). The multiple fig fruit is composed of small individual drupelets, which develop from the ovaries enclosed in the succulent receptacle to form a single accessory fruit (Storey, 1977). The inner wall of the syconium is lined with flowers and, at maturity, with numerous fruitlets. Botanically, the drupaceous fruitlets are the real fruit. The exocarp and mesocarp stay fleshy until maturation. The endocarp hardens at the end of July/beginning of August. In contrast, the edible fruit is a vegetative peduncle. A passageway leading toward the apical opening (ostiole = eye) links the internal cavity of the fruit to the external atmosphere. The ostiolar channel is surrounded by scales, termed bract leaves (Condit, 1947; Ferguson *et al.*, 1990).

F. carica is functionally dioecious, bearing "male" and "female" figs on separate plants. The edible female fig flowers all have long styles, whereas the caprifig—the male fig tree—carries staminate, short-styled pistillate flowers (Fig. 4.5C).

4.3.1 Fig fruit types

Fig is a gynodioecious species but is functionally dioecious, and thus many varieties require pollination to produce fruit, a process known as caprification. *Blastophaga psenes* wasps carrying pollen emerge from the caprifig to pollinate the long-styled female flowers of the edible fig. The caprifig, commonly termed "male fig", is not edible, and is the commercial planting source for pollen. The caprifig has a fascinating life cycle, which is linked to that of the pollinating wasp, *B. psenes*. The caprifig bears three fig crops annually. In the spring, the profichi crop is initiated from latent buds on the old wood of the previous year's growth. These figs contain mostly female flowers, with a cluster of male flowers near the ostiole. The pollen ripens just as the fig wasps are leaving the fig. The profichi crop is used to pollinate the female figs, and supplies growers with pollen-carrying wasps that will pollinate the Smyrna- and San Pedro-type varieties. The second caprifig crop, known as mammoni, initiates on the concurrent growth in the summer, and matures in the autumn. The third stage is termed mamme crop. These figs remain on the tree during winter and are important as an overwintering crop for the fig wasp, which will be transferred to the profichi in the spring (Fig. 4.6).

The edible fig varieties are divided into three horticultural types according to their pollination, caprification requirement, and cropping season: Smyrna, San Pedro, and common (Storey, 1975). Smyrna and San Pedro types require caprification to bear their main crops; the syconia shrivel and drop off at the end of anthesis if the flowers are not pollinated. Varieties of the Smyrna type usually do not produce the early parthenocarpic crop known as "breba," whereas the San Pedro types yield breba and the pollinated main crop. Fig trees of the San Pedro type have moderate to large breba crops. These develop to maturity without pollination and ripen into edible fruit. A satisfactory second crop of syconia is initiated, but generally only pollinated flowers develop to maturity (Gaaliche *et al.*, 2011; Trad *et al.*, 2012). The third type is the common fig, also called "persistent" fig, which can yield fruit without pollination. Parthenocarpy was probably selected for in fig early in its domestication history, dating back to the Early Neolithic period, and may have been derived as a point mutation favored by humans (Aradhya *et al.*, 2015). Common fig varieties yield one or two crops a year. Common-type fig trees may or may not have a breba crop, but if they do, it develops to maturity without pollination. Most common fig varieties produce a considerable main (second) crop that develops and matures without pollination. Pollination of the different fig types is a consequence of a complex symbiotic relationship with the fig wasp (Galil and

Fig. 4.6. The caprifig bears three crops of figs during the year. Mamme, the winter crop, mammoni, the summer crop, and profichi, the spring crop, which supplies growers with pollen-carrying wasps.

Eisikowitch, 1968; Galil and Neeman, 1977; Flaishman *et al.*, 2008).

4.3.2 Fruit growth and development

In contrast to many other fruit trees, the fig tree exhibits unique flower bud differentiation and fruit-bearing characteristics. Flower bud (inflorescence) differentiation occurs throughout the shoot-elongation period, simultaneously with fruit growth. The entire process of flower bud differentiation to fruit ripening can be completed in one growing season. In the spring, the terminal bud unfolds and growth occurs. The fig fruit are borne in the axils of the leaves, where two inflorescences and one vegetative bud are present at the same lateral position. In varieties such as Mission and Brown Turkey, usually only one inflorescence develops into a multiple fruit, whereas in varieties Kadota and Calimyrna, the two inflorescences develop. Similar to other fruit, fig syconium development has three defined growth periods represented by a double-sigmoid curve (Crane and Brown, 1950; Crane and Baker, 1953). The first, phase I, is characterized by a very rapid increase in diameter and a slower increase in fresh and dry weight, with almost no change in sugar accumulation. Phase II is a quiescent stage marked by almost no change in fruit diameter, dry or fresh weight, or sugar content. Phase III is characterized by an acceleration in the rate of diameter increase, and increases in fresh and dry weights, moisture, and sugar content. Over 70% of the total dry weight and 90% of the total sugar content accumulate in the fruit during this growth phase. Matsuura *et al.* (2001) studied the distribution of ^{13}C at the lower nodes administered to the leaf of a bearing shoot of a fig tree during fruit enlargement and maturation. The ^{13}C-accumulation data revealed that the fruit at phase I has a greater sink strength than those at phase II. When the fruit was treated with a drop of oil placed on the ostiole to induce early maturity, it presented a highly active sink, importing ^{13}C-labeled photosynthates primarily from leaves positioned near the fruit. In varieties that produce fig fruits that develop on the same branch in the same season, phase I growth lasts 5–6 weeks and phase III lasts 3–5 weeks. However, there are great differences between fig varieties in the duration of phase II, the quiescent period of development. In varieties such as Mission and Brown Turkey, the second period lasts 3–4 weeks, whereas in the autumn-producing cultivars Sierra and Autumn Honey, the second period lasts 6–8 weeks.

The pattern of growth and development of breba figs differs from that of the fig syconia that develop on the same branch in the same season. Shortly after initiation, the breba syconium enters winter dormancy. It resumes growth in the spring, for 9–11 weeks, until maturation in June–July. About 2–3 weeks before maturation of the breba fruit, growth rate and sugar accumulation increase significantly. Since all breba fig syconia on the same branch are at similar developmental stages, the fruit harvest is shorter, lasting only 2–3 weeks.

The fruit shape differs with variety, but is usually pyriform–obovoid to globose, and develops from green–yellow to red–violet hues when fully ripe. The fruit flesh is whitish, yellow, amber, pink, or red to purple, and is juicy and sweet when ripe and gummy with latex when unripe. Studying the unique flowering characteristics of fig provides an important basis for its cultivation, flowering regulation, fruit production, and molecular breeding. The molecular mechanism of flower bud differentiation in fig has not been elucidated. Ma *et al.* (2020) isolated a *FLORICAULA/LEAFY* homolog in fig (*FcLFY*), and identified its function and expression pattern in *Arabidopsis*. *FcLFY* was most highly expressed in the apical bud meristem of fig. Transgenic *Arabidopsis* lines overexpressing *FcLFY* flowered 6 days earlier than the wild type. More studies are needed to better characterize the occurrence and function of flower initiation in fig.

4.3.3 Epicuticular fruit wax

Scanning electron microscope studies of fruit surfaces have revealed the presence of epicuticular wax, an important structure related to gas exchange, transpiration, and peel susceptibility to physiological disorders and pathogen attack. In fig, regulation of water loss by surface wax is of great importance when the fruit are to be canned or dried. The quantity of wax extracted per unit surface area varies significantly among the different stages of fig fruit development, with no significant difference between varieties. In phase III, the main accumulated epicuticular wax is in the form of regularly layered platelets. As ripening and senescence proceed, the quantity of wax decreases and the platelet structure changes (Chessa *et al.*, 1992).

4.4 Latex Cells in the Vegetative and Reproductive System

The laticifer is a cell that secretes latex, an exudate composed of several substances, including polyisoprene hydrocarbons, triterpenes, sterols, fatty and aromatic acids, carotenes, phospholipids, proteins, alkaloids, and inorganic compounds, dispersed in a liquid with a distinct refractive index (Fahn, 1979; Ascensão, 2007). Structurally, laticifers can be non-articulated (a single multinucleated cell that grows as the plant develops) or articulated (a row of several cells whose end walls can remain intact, become porous or disappear completely). Both types can branch, or not, producing a complex tube-like system that permeates different tissues of the plant body (Evert, 2006). These structures act in plant defense against herbivores by releasing their toxic or repellent content, which is under strong turgor pressure, in response to injury (Fahn, 1979; Agrawal and Konno, 2009; Konno, 2011). In addition, latex is a coagulant that seals wounds and prevents pathogen entry (Fahn, 1979; Farrell *et al.*, 1991).

In the Moraceae, a family with 40 genera, the non-articulated branched laticifer is considered a synapomorphy. There are numerous studies on the distribution and morphology of laticifers in vegetative organs of species in the family Moraceae (Metcalfe and Chalk, 1950; Fahn, 1979; Evert, 2006). Kajii *et al.* (2014) investigated the distribution and structure of laticifers in an *F. carica* concurrent-year branch and an 8-year-old trunk. Laticifers were found in the cortex, secondary phloem, pith, and secondary xylem. In addition, horizontal laticifers extending from the phloem through the cambium to the xylem were found in some rays. Latex production occurred after wounding of the cortex and phloem, and the latex exuded from the cut site of the laticifers into the surrounding wounded tissue and broken cells.

Fig trees and fruit contain these typical latex-secreting cells, which are found in most parts of the fig. The properties of fig latex have been attributed to the presence of several hydrolytic enzymes, such as protease, lipase, and amylase. The milky latex exudate causes skin irritation because it contains the proteolytic enzyme ficin. Latex is the cytoplasmic fluid of laticiferous tissues, which contain the usual plant cell organelles: nucleus, mitochondria, vacuoles, ribosomes, and Golgi apparatus. Rubber (*cis*-1,4-polyisoprene) is produced in latex as a high-energy cost, and latex secretion is suggested to be a defense against mechanical wounding and/or herbivores such as insects, vertebrates, microorganisms, and fungi (John, 1993). Several lines of evidence suggest that latex production in plants is responsive to environmental conditions, such as soil moisture, soil fertility, and leaf herbivores. The defensive role of latex is based on the fact that it contains a variety of defense-related proteins. Kim *et al.* (2003) identified three rubber particles and latex genes from *F. carica*: a trypsin inhibitor, a chitinase, and a peroxidase. Peroxidase is widely known to participate in a variety of plant defense mechanisms in which hydrogen peroxide is often supplied by an oxidative burst (Lamb and Dixon, 1997; Shigeoka *et al.*, 2002). The transcript level of peroxidase increased in *F. carica* exposed to various types of abiotic stresses or hormones, including wounding, drought, jasmonic acid, and abscisic acid. In contrast, the transcript level of the trypsin inhibitor was markedly increased by wounding treatment and only slightly increased by jasmonic acid treatment. The expression of chitinase was markedly induced by wounding and jasmonic acid treatment. The presence and expression of stress-related proteins on the surface of rubber particles and latex in *F. carica* suggest a role in defense for these particles and latex. A recent study investigated latex enzyme activity in 20 Tunisian fig varieties (Lazreg-Aref *et al.*, 2018). A great disparity was revealed between enzyme activity and latex yield for the different fig varieties during fruit maturation. Protease exhibited the highest activity compared with amylase and lipase activities for all of the varieties. The highest latex yield was produced in the common-type Kahli and San Pedro-type Bither Abiadh and Bither Akhal, coinciding with the increase in protease activity and inactivation of lipase during maturation.

Studies of laticifers in the vegetative organs of the plant body have shown their role in efficiently sealing wounds by coagulation, thereby preventing microorganism invasion and herbivory. However, very little is known about the functions of laticifers in flowers. A recently published study on *Ficus* species showed that laticifers help with the pollination mechanism in fig trees because their arrangement along the syconium seems to favor the mutualism between pollinating wasps and fig trees (Marinho *et al.*, 2018). The laticifers are present in the pedicel and stamens of shorter-styled flowers, and their absence in the pistil suggests that the distribution of this secretory structure in the fig flower was selected for by pressures imposed by the mutualism between the fig and fig wasp. The laticifers in the pedicel of shorter-styled flowers may protect the developing wasp larvae from natural enemies. However, the absence of laticifers in the pistil of most *Ficus* species studied was probably selected for by the mutualistic relationship with the agaonid pollinating wasps, because the latex might interfere with oviposition through the style, larval development of the pollinating fig wasps, and the emergence of pollinator offspring from their galls. The findings of Marinho *et al.* (2018) raise new questions for investigation regarding the mutualism between the fig and fig wasp.

4.5 Concluding Remarks

The fig tree is the only deciduous *Ficus* tree, showing high adaptability to different growing conditions. Today, the tree is distributed all over the world. The general biology of the root, stem, leaf, and fruit illustrates the complex adaptation of *F. carica* trees to development in different climates. The leaf's unique structure encompasses minor nerves, lithocysts, and anatomical strategies to increase light penetration into the lower mesophyll by using mineral scatterers. This is an excellent example of how basic science enables an understanding of a tree's adaptive capabilities. The unique syconium structure of the fig tree and its development in male and female trees are well adapted to the specific mutualism between the fig fruit and the fig wasp (see Chapter 5, this volume). An understanding of the fig's general biology, together with its genetic variation, will provide new answers to current and future challenges.

References

Agrawal, A.A. and Konno, K. (2009) Latex: a model for understanding mechanisms, ecology, and evolution of plant defense against herbivory. *Annual Review of Ecology Evolution and Systematics* 40, 311–331.

Aradhya, M., Preece, J.E., and Velasco, D. (2015) Multivariate analysis of molecular and morphological diversity in fig (*Ficus carica* L.). *Acta Horticulturae* 1173, 23–28.

Ascensão, L. (2007) Estruturas secretoras em plantas. Uma abordagem morfo-anatomica. In: Figueiredo, A.C., Barroso, J.G., and Pedro, L.G. (eds.) *Potencialidades e aplicações das plantas aromaticas e medicinais. Curso teorico-pratico*, 3rd edn. Faculdade da Universidade de Lisboa e Centro de Biotecnologia Vegetal, Lisboa, pp. 19–28.

Atzmon, N. and Henkin, Z. (1998) Establishing forest tree species on peatland in a reflooded area of the Huleh Valley, Israel. *Forestry* 71, 141–146.

Badgujar, S.B., Patel, V.V., Bandivdekar, A.H., and Mahajan, R.T. (2014) Traditional uses, phytochemistry and pharmacology of *Ficus carica*: a review. *Pharmaceutical Biology* 52, 1487–1503.

Begum, M.Y., Alhamhoom, Y., and Roy, A. (2020) Study of antimicrobial and DNA cleavage property of biocompatible silver nanoparticles prepared by using *Ficus carica* L. *Materials Research Innovations* 25, 147–154.

Chessa, I., d'Aquino, S., Cossu, A., and Mannu, L. (1992) Epicuticular wax structure and accumulation on main-crop fig fruit (*Ficus carica* L.) during development. *Advances in Horticultural Science* 6, 47–54.

Condit, I.J. (1947) *The Fig.* Chronica Botanica, Waltham, Massachusetts.

Crane, J.C. (1986) Fig. In: Monselise, S.P. (ed.) *Handbook of Fruit Set and Development.* CRC Press, Boca Raton, Florida.

Crane, J.C. and Baker, R.E. (1953) Growth comparisons of the fruits and fruitlets of figs and strawberries. *Proceedings of the American Society for Horticultural Science* 62, 142–153.

Crane, J.C. and Brown, J.B. (1950) Growth of the fig fruit, *Ficus carica* var. Mission. *Proceedings of the American Society for Horticultural Science* 56, 93–97.

Evert, R.F. (2006) *Esau's Plant Anatomy: Meristems, Cells, and Tissues of the Plant Body: Their Structure, Function, and Development*, 3rd edn. Wiley, Hoboken, New Jersey.

Fahn, A. (1979) *Secretory Tissues in Plants*. Academic Press, London/New York/San Francisco.

Farrell, B.D., Dussourd, D.E., and Mitter, C. (1991) Escalation of plant defense: do latex/resin canals spur plant diversification? *The American Naturalist* 138, 881–900.

Ferguson, L., Michailides, T.J., and Shorey, H. (1990) The California fig industry. *Horticultural Reviews* 12, 409–490.

Flaishman, M.A., Rodov, V., and Stover, E. (2008) The fig: botany, horticulture, and breeding. *Horticultural Reviews* 34, 132–196.

Gaaliche, B., Hfaiedh, L., Trad, M., and Mars, M. (2011) Caprification efficiency of three Tunisian fig (*Ficus carica* L.) cultivars. *Journal of Natural Product and Plant Resources* 1, 20–25.

Galil, J. and Eisikowitch, D. (1968) Flowering cycles and fruit types of *Ficus sycomorus* in Israel. *New Phytologist* 67, 745–758.

Galil, J. and Neeman, G. (1977) Pollen transfer and pollination in the common fig (*Ficus carica* L.). *New Phytologist* 79, 163–171.

Giordano, C., Maleci, L., Agati, G., and Petruccelli, R. (2019) *Ficus carica* L. leaf anatomy: trichomes and solid inclusions. *Annals of Applied Biology* 176, 47–54.

Golombek, S.D. and Lüdders, P. (1990) Effects of short-term salinity on leaf gas exchange of the fig (*Ficus carica* L.). *Plant and Soil* 148, 21–27.

Hao, G.Y., Wang, A.Y., Liu, Z.H., Franco, A.C., Goldstein, G., and Cao, K.F. (2011) Differentiation in light energy dissipation between hemiepiphytic and non-hemiepiphytic *Ficus* species with contrasting xylem hydraulic conductivity. *Tree Physiology* 31, 626–636.

Harrison, R.D. (2005) Figs and the diversity of tropical rainforests. *AIBS Bulletin* 55, 1053–1064.

Hosomi, A., Dan, M., and Kato, A. (2002) Screening of fig varieties for rootstocks resistant to soil sickness. *Journal of the Japanese Society for Horticultural Science* 71, 171–176.

Janzen, D.H. (1979) How to be a fig. *Annual Review of Ecology and Systematics* 10, 13–51.

John, P. (1993) *Biosynthesis of the Major Crop Products*. Wiley, Chichester, New York.

Kajii, C., Morita, T., and Kuroda, K. (2014) Laticifers of *Ficus carica* and their potential role in plant defense. *IAWA Journal* 35, 109–115.

Kamara, I. (2013) Morphological aspects of *Ficus carica*: flower development and characterization of inheritance of morphological properties in intra- and inter-species crossings. MSc thesis, Hebrew University of Jerusalem, Jerusalem, Israel.

Kim, J.S., Kim, Y.O., Ryu, H.J., Kwak, Y.S., Lee, J.Y., and Kang, H. (2003) Isolation of stress-related genes of rubber particles and latex in fig tree (*Ficus carica*) and their expressions by abiotic stress or plant hormone treatments. *Plant and Cell Physiology* 44, 412–414.

Konno, K. (2011) Plant latex and other exudates as plant defense systems: roles of various defense chemicals and proteins contained therein. *Phytochemistry* 72, 1510–1530.

Lamb, C. and Dixon, R.A. (1997) The oxidative burst in plant disease resistance. *Annual Review of Plant Physiology and Plant Molecular Biology* 48, 251–275.

Lazreg-Aref, H., Gaaliche, B., Ladhari, A., Hammami, M., and Hammami, S.O. (2018) Co-evolution of enzyme activities and latex in fig (*Ficus carica* L.) during fruit maturity process. *South African Journal of Botany* 115, 143–152.

Ma, N., An, Y., Li, J., and Wang, L. (2020) Cloning and characterization of a homologue of the *FLORICAULA/LEAFY* gene in *Ficus carica* L., *FcLFY*, and its role in flower bud differentiation. *Scientia Horticulturae* 261, 109014.

Mamoucha, S., Fokialakis, N., and Christodoulakis, N.S. (2016) Leaf structure and histochemistry of *Ficus carica* (Moraceae), the fig tree. *Flora: Morphology, Distribution, Functional Ecology of Plants* 218, 24–34.

Marinho, C.R., Pereira, R.A.S., Peng, Y.Q., and Teixeira, S.P. (2018) Laticifer distribution in fig inflorescence and its potential role in the fig–fig wasp mutualism. *Acta Oecologica* 90, 160–167.

Matsuura, K., Tanabe, K., Tamura, F., and Itai, A. (2001) Distribution of ^{13}C administered to a leaf of a bearing shoot of fig trees during the fruit enlargement and maturation stages at the lower nodes. *Journal of the Japanese Society for Horticultural Science* 70, 466–472.

Metcalfe, C.R. and Chalk, L. (1950) *Anatomy of the Dicotyledons: Leaves, Stem, and Wood, in Relation to Taxonomy, with Notes on Economic Uses*. Clarendon Press, Oxford, UK.

Modi, R.K., Kawadkar, M., Sheikh, S., Kastwar, R., and Tiwari, G. (2012) A review on: comparative studies on ethanolic extract of root and stem bark of *Ficus carica* for analgesic and anti-inflammatory activities. *International Journal of Pharmacy & Life Sciences* 3, 1930–1934.

Omori, M. and Watabe, N. (1980) *The Mechanisms of Biomineralization in Animals and Plants: Proceedings of the Third International Biomineralization Symposium*. Tokai University Press, Tokyo.

Pierantoni, M., Paudel, I., Raphael, B., Tenne, R., Brumfeld, V., *et al.* (2020) Cystoliths in *Ficus* leaves: increasing carbon fixation in saturating light by light scattering off a mineral substrate. *bioRxiv* https://doi.org/10.1101/2020.04.08.030999

Rouaiguia-Bouakkaz, S., Amira-Guebailia, H., Rivière, C., Delaunay, J.C., Waffo-Téguo, P., and Mérillon, J.M. (2013) Identification and quantification of furanocoumarins in stem bark and wood of eight Algerian varieties of *Ficus carica* by RP-HPLC-DAD and RP-HPLC-DAD-MS. *Natural Product Communications* 8, 485–486.

Setoguchi, H., Okazaki, M., and Suga, S. (1989) Calcification in higher plants with special reference to cystoliths. In: Crick, R.E. (ed.) *Origin, Evolution, and Modern Aspects of Biomineralization in Plants and Animals*. Springer, New York, pp. 409–418.

Shigeoka, S., Ishikawa, T., Tamoi, M., Miyagawa, Y., Takeda, T., *et al.* (2002) Regulation and function of ascorbate peroxidase isoenzymes. *Journal of Experimental Botany* 53, 1305–1319.

Storey, W.B. (1975) Figs. In: Janick, J. and Moore, J.N. (eds) *Advances in Fruit Breeding*. Purdue University Press, West Lafayette, Indiana, pp. 568–589.

Storey, W.B. (1977) *The Fig (Ficus carica Linnaeus): Its Biology, History, Culture, and Utilization*. Jurupa Mountains Cultural Center, Riverside, California.

Trad, M., Ginies, C., Gaaliche, B., Renard, C.M., and Mars, M. (2012) Does pollination affect aroma development in ripened fig (*Ficus carica* L.) fruit? *Scientia Horticulturae* 134, 93–99.

Vemmos, S.N., Petri, E., and Stournaras, V. (2013) Seasonal changes in photosynthetic activity and carbohydrate content in leaves and fruit of three fig cultivars (*Ficus carica* L.). *Scientia Horticulturae* 160, 198–207.

Yaman, B. (2014) Anatomical differences between stem and branch wood of *Ficus carica* L. subsp. *carica*. *Modern Phytomorphology* 6, 79–83.

Zhang, S., Huang, W., Zhang, J., and Cao, K. (2016) Differential responses of photosystems I and II to seasonal drought in two *Ficus* species. *Acta Oecologica* 73, 53–60.

Zotz, G., Harris, G., Königer, M., and Winter, K. (1995) High rates of photosynthesis in the tropical pioneer tree, *Ficus insipida* Willd. *Flora* 190, 265–272.

5 *Ficus*–Wasp Mutualism with a Special Emphasis on *Ficus carica*

Dan Eisikowitch,[1*] Mahua Ghara,[1,2] Reut Peer[3] and Moshe A. Flaishman[3]
[1]*Department of Molecular Biology and Ecology of Plants, Tel Aviv University, Israel;* [2]*Botanical Garden, Tel Aviv University, Ramat Aviv, Tel Aviv, Israel;* [3]*Institute of Plant Sciences, Agriculture Research Organization, Volcani Center, Rishon LeZion, Israel*

5.1 Introduction

The genus *Ficus*, with over 750 species, contributes the highest number of species to the family Moraceae (Rohwer, 1993; Datwyler and Weiblen, 2004). Species of *Ficus* grow mostly in the tropics, with a few exceptions in the subtropics (Berg, 1989; www.figweb.org). They vary widely in their growth form, which can range from tall trees of over 10 m in height to shrubby plants of less than a few meters, found in a variety of habitats, from tropical forests to deserts (Hill, 1967; Berg and Wiebes, 1992; www.figweb.org). *Ficus* species are often popularly referred to as figs. Fig-pollinating wasps have an obligate pollinating mutualism with fig trees. The fig–fig wasp system is one of the most dramatic and ancient examples of a compelling herbivore–plant mutualism. Unlike most other insect herbivores, fig wasps are highly mutualistic with the plants on which they feed. Both figs and fig wasps show morphologies that are highly specialized for this mutualism.

To understand this mutualism in *Ficus* species, it is important to know about figs and their different interactive strategies with their pollinator. The aim of this chapter is to shed light on *Ficus*–wasp mutualism, with a focus on *Ficus carica* and its pollinating wasps.

5.2 The Fig

All species of fig have two distinguishing features: (i) a unique inflorescence resembling an urn-shaped structure, commonly called a syconium, containing hundreds to thousands of unisex flowers (Kugler, 1955; Barth, 1985); and (ii) specialized pollination that is carried out only by wasps of the family Agaonidae (superfamily Chalcidoidea). The association between fig and fig wasps dates back approximately 60–100 million years (Rønsted *et al.*, 2005; Cruaud *et al.*, 2012). The interaction between fig and its pollinator is a classic example of mutualism and coevolution.

As the syconium is a closed structure and the flowers are concealed within it, pollinators can only access the flowers through an opening called the ostiole. The ostiole is found at the top of the syconium, leading to a bract-lined tunnel and, eventually, to the syconium cavity.

In nearly half of the known species of *Ficus*, the male flowers, as well as both kinds of female flowers, are found in the same syconium; these are referred to as monoecious (hermaphrodite) figs (Berg, 1989); the remaining species are known as gynodioecious. Gynodioecious species have separate trees that function as pollen donor (male tree) and pollen acceptor (female tree).

* Email: dane@tauex.tau.ac.il

 Advances in Fig Research and Sustainable Production (M.A. Flaishman and U. Aksoy)
DOI: 10.1079/9781789242492.0005

F. carica is an example of a gynodioecious species (Condit, 1947). The female trees bear syconia with only long-styled female flowers, whereas syconia on the male tree have functional male flowers and female flowers with a short style length (Kugler, 1955; Galil, 1973). The long flower styles in the "female" trees tend to prevent wasps from laying their eggs within the ovules, whereas the short-styled flowers of the "male" trees are accessible for egg laying.

5.3 The Fig Wasps

The family Agaonidae is a group of pollinating and non-pollinating fig wasps. They spend their larval stage inside the fig fruit. The pollinating wasps (Agaoninae, Kradibiinae, and Tetrapusiinae) are the mutualistic partners of fig trees. Fig wasps are specialized for pollination of only *Ficus* species (Weiblen, 2002). They are mostly species-specific (i.e. each pollinator species pollinates only one species of *Ficus*), but there are some exceptions (Molbo *et al.*, 2003; Su *et al.*, 2008). Fig wasps are sexually dimorphic; the females have wings, whereas the males are wingless (Berg and Wiebes, 1992). Only the female fig wasps are capable of pollinating the flowers. These female wasps are small, with total body length of up to only a few millimeters. The pollinator of the common fig, *Blastophaga psenes* L., is approximately 1.8 mm in length (Berg and Wiebes, 1992) with an average lifespan of 24–48 h, as inferred from studies under laboratory as well as field conditions (Kjellberg *et al.*, 1988; Ghara and Borges, 2010). Despite their limited lifespan, they can disperse more than 100 km for pollination, as recorded for an African fig wasp (Ahmed *et al.*, 2009). The pollinators are guided to pollen-receptive syconia by volatile signals emitted by these syconia (van Noort *et al.*, 1989; Hossaert-McKey *et al.*, 1994; Ware and Compton, 1994). Upon arriving on a syconium, the pollinator wasps inspect it and then enter it.

Once inside the syconium, the wasps pollinate actively or, in a few systems, passively. Upon pollination, most of the wasps die inside the syconium, although in certain cases they can exit and pollinate other syconia (Gibernau *et al.*, 1996). In return for their pollination services, the wasps oviposit (lay eggs) in some of the flowers by inserting their ovipositor through the style, and the wasp offspring complete their development inside the consequently modified fig flowers, also known as galls. They develop by feeding on the ovary or hypertrophied endosperm (Condit, 1932; Jansen-González *et al.*, 2012). However, because of the constraints of ovipositor length, the wasp cannot oviposit in all flowers (and more specifically, in long-styled flowers) and therefore, the long-styled flowers mostly produce seeds whereas—due to successful oviposition—the short-styled flowers nurture the wasp offspring.

Thus, the fig provides a nursery for the wasp offspring, giving this mutualism the name of nursery pollination mutualism (Sakai, 2002; Dufaÿ and Anstett, 2003). The pollinator is therefore in a mutualistic interaction with its host fig species, where the development and survival of each partner depends on the other (Condit, 1920, 1947). In the wild, figs can produce seeds only if the pollinator exists, and the pollinator offspring can develop only if the host species of fig exists.

Fig-pollinating wasps form obligate symbioses with their fig hosts. This mutualism arose approximately 75 million years ago. Symbionts that retain intimate contact with their hosts often show genome reduction. The female fig wasps fly great distances to transfer horizontally between hosts, whereas the wingless male wasps cannot disperse. It is not clear if the widely dispersing female fig wasps will show genome reduction. To address this question, Xiao *et al.* (2013) compared genomes of both sexes of the fig wasp *Ceratosolen solmsi*. The fig wasp genome generally resembled that of other insects in that its gene content was similar to that of other hymenopterans. However, importantly, the fig wasp genome showed marked reductions in genes associated with environmental sensing and protection. Extreme host specificity and endosymbiosis in a mutualistic host appeared to drive this pattern. They concluded that the complex wasp dispersal is viable with a streamlined chemosensory toolbox, perhaps because overwhelming selection pressure has involved homing in on the correct host during the female's short adult life; the sensing of other aspects of a complex external environment is of little importance, and might even be distracting.

5.4 Pollination in *F. carica*

Pollination in *F. carica* was described for the first time in detail by Condit (1947) and later by Galil and Neeman (1977) in Israel, and by Kjellberg *et al.* (1987) in France. The species-specific pollinator *B. psenes* pollinates *F. carica*. Using a mark–recapture technique with fluorescent dyes, Kjellberg *et al.* (1988) found that the adult pollinators can live up to 48 h in the wild.

To understand the pollination process in the genus *Ficus*, Galil and Eisikowitch (1968) divided its life cycle into five stages (A–E) for monoecious figs. In stage A (pre-female), the syconium is young, prior to opening of the ostiole. In stage B (female), flowers are mature, and the scales on the ostiole loosen; pollinator wasps enter the syconium, pollinate, and oviposit. In stage C (inter-floral), the wasp larvae and fig seeds develop. In stage D (male), flowers are mature; the adult male wasps emerge from the galls, mate with the female wasps, and help the female wasps leave the syconium. In stage E (post-floral), the seeds in the syconia are ready for dispersal.

These stages were also later adapted for the gynodioecious species *F. carica* (Galil and Neeman, 1977; Galil, 1989). Here, we describe the stages of pollination in *F. carica* as observed in Israel. In *F. carica*, the female trees, called the fichi, produce several crops annually; however, it is mainly the spring crop that will be pollinated. The syconia on fichi have only long-styled flowers. The male tree is called caprificus; syconia on caprificus have short-styled flowers. Caprificus produces more than one crop per year, namely the spring crop (profichi), the summer crop (mammoni), and the winter crop (mamme) (Fig. 5.1). These crops have overlapping stages. The number of crops and their nomenclature might vary between different climatic zones.

The stages of the pollination cycle in *F. carica* with respect to the fichi and caprificus are as follows:

- **Stage A (pre-pollination).** Syconia are present as a small protuberance on the axil of the stem. The flowers within the syconia are immature. The ostiole is closed, with overlapping scales.
- **Stage B (pollination).** Syconia on the female trees (fichi) are ready to receive pollen. The adult wasp developing inside the syconium of the profichi crop of caprificus (Fig. 5.2), already swollen from hydration due the wet

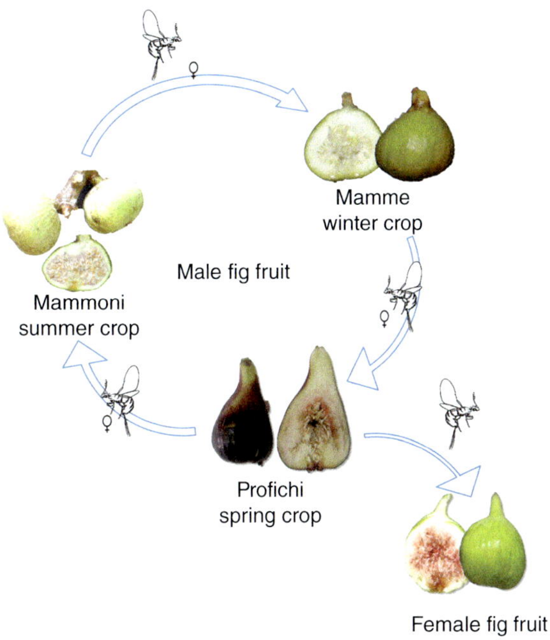

Fig. 5.1. Caprificus produces three crops per year, namely the spring crop, profichi, the summer crop, mammoni, and the winter crop, mamme.

Fig. 5.2. The pollinator of common fig, *Blastophaga psenes*, developing inside the syconium of the profichi crop of caprificus during the spring. Photographs courtesy of R. Peer.

conditions there, encounters the pollen before emerging from the syconium (Galil, 1984). Once outside the syconium environment, where conditions are dry, the wasp dehydrates and shrinks. The pollen on the wasp's body thus becomes concealed in intersegmental and pleural invaginations (Galil and Neeman, 1977). The pollinator enters into the syconium cavity on the female tree through the ostiole (Fig. 5.3). As the ostiole provides only a narrow space for entry, the pollinator wasp attempting to enter loses its wings and part of its antennae (Galil and Neeman, 1977). Once inside, the wasp becomes rehydrated, its body swells, and the concealed pollen is released, thus carrying out passive pollination. Although the wasp attempts oviposition, eggs cannot be laid because the style length is longer than the length of the ovipositor.

Pollination in *Ficus* is mediated by chemical signals—volatiles emitted by the fig host (see section 5.7.1). Some wasps from the profichi crop also enter syconia in stage B of male trees, i.e. the mammoni crop (Fig. 5.3). The wasps oviposit in the short-styled flowers of the mammoni, and the flowers also get pollinated.

- **Stage C (post-pollination).** Development of seeds from pollinated flowers takes place in fichi. In mammoni syconia, the wasp offspring and seeds develop together.
- **Stage D (wasp dispersal).** Due to the absence of developing wasps in fichi, the wasp-dispersal stage is not valid for these trees. Seeds continue to develop in the fichi. On the male tree, the pollinator offspring are ready to emerge from the mammoni. Male wasps developing inside mammoni syconia hatch from their galls and copulate with the females, which remain arrested within the galls. Later on, these fertilized females crawl out from their galls and

Fig. 5.3. Fig–wasp symbiosis during the spring season.

disperse from the natal syconium through the open ostiole (Galil and Neeman, 1977).

- **Stage E (seed dispersal).** The fichi syconia have seeds that are ready for dispersal. The fruit is sweet and fleshy, and it changes color and becomes attractive to seed-dispersal agents (Neeman and Galil, 1978).

To continue the cycle, wasps from mammoni enter the co-occurring receptive stage B syconia of an additional mammoni crop (Fig. 5.3). Mammoni do not produce pollen and are only used for oviposition; seeds are therefore not produced, and mammoni function exclusively as a nursery for the wasps. However, Neeman and Galil (1978) found that artificially pollinated mammoni can produce seeds, indicating that lack of seeds in the wild from mammoni crops is due to the absence of pollen in the previous caprificus crop.

Once the wasps are ready to emerge from mammoni, the syconia of the next mamme crop are in stage B. Thus, wasps from stage D mammoni enter into the stage B mamme. They also do not have any pollen on their body and hence only oviposit, and then develop in the mamme flowers. Stage D of the mamme crop overlaps with the stage B profichi in which, like mamme, there is only wasp development. However, stage D profichi produce viable pollen, which is carried by the wasps to the fichi. Thus, wasp cycling takes place on multiple crops on the caprifig so that the wasps will be ready for fichi pollination and seed production (Fig 5.1).

5.5 Non-pollinating Fig Wasps

Most of the 750 recognized *Ficus* species host a community of non-pollinating fig wasps, in relationships ranging from mutualism to parasitism, which almost never transfer pollen (Cruaud *et al.*, 2011). Most species are of similar size to the pollinators, oviposit from outside the fig, and can be costly to either or both mutualists via a reduction in seeds and/or pollinator offspring. These non-pollinating fig wasps can be gall inducers, kleptoparasites, or parasitoids. Some wasps may, at least facultatively, develop within a seed (Santinelo Pereira *et al.*, 2007). Oviposition in a few groups is achieved after entry through the ostiole, but most non-pollinating fig wasps utilize an extremely long ovipositor to reach the ovules through the thick fig walls. Gall inducers are assumed to lay eggs in pre-receptive to receptive figs, either very early in fig development, before the development of flowers, or later in fig development, in non-galled, healthy flowers. Kleptoparasites lay eggs in the same flowers as their hosts, probably shortly after gall induction, when plant tissue is still available in the flower (Kuttamathiathu, 1958). Parasitoids also oviposit in the same flower as their hosts, but feed on host larval tissues (Tzeng *et al.*, 2008). On this basis, five types of non-pollinating fig wasps can be recognized: (i) early gallers; (ii) gallers ovipositing close to fig receptivity; (iii) fig-entering gallers (not represented in the neotropics); (iv) kleptoparasite; and (v) parasitoids (Cruaud *et al.*, 2011). Most of these species develop in galls that are similar in size to pollinator galls. Some species are large and develop in galls that are obviously larger than the pollinator's galls. Most of these species will have a negative impact on the reproductive success of fig trees because they kill pollinators, consume pollinator larvae, and reduce seed numbers (Kerdelhué and Rasplus, 1996), and fig ovules may be particularly easy to eat because the plant cannot defend them chemically without harming its pollinators (Cook and Rasplus, 2003). In southern Africa, the total fig wasp fauna on different *Ficus* species ranges from three to 30 species, with up to 18 species being recorded on a single crop (Compton and Hawkins, 1992). Limited studies have demonstrated the occurrence of non-pollinating fig wasps in *F. carica* trees. Kuttamathiathu (1958) demonstrated the existence of the non-pollinating fig wasp *Philotrypesis caricae*, a kleptoparasite in *F. carica* (Kuttamathiathu, 1958). Similarly, a study of fig wasps in Turkey found two fig wasp species in the fig syconia: the pollinator wasp *B. psenes* and the additional non-pollinating wasp *P. caricae* (Doğanlar, 2012).

5.6 Can Seed Protection Lead to Dioecy in *Ficus*?

About half of fig trees (*Ficus* species) are monoecious and the remainder are functionally dioecious. Monoecy is ancestral, and dioecy probably evolved once or twice and reverted to monoecy at least twice (Weiblen, 2000; Machado *et al.*, 2001). In monoecious species, the pollinating wasps, pollen, and seeds are all produced in the same fig. An individual flower ovary can become either a wasp gall or a seed; therefore, each wasp offspring costs the host tree one potential seed (Janzen, 1979). In monoecious fig species, a striking property of the mutualism between figs and their pollinating wasps was reported: the wasps consistently oviposit in the inner flowers of the fig syconium, which develop into galls that house developing larvae. Across monoecious fig tree species, approximately 55% of the flowers within individual figs become seeds (Herre, 1989). This is so, even in tree species in which figs contain on average enough wasp eggs to potentially convert all flowers into wasp galls, suggesting that mechanisms are present to restrict wasp galling. Wasps typically do not use the outer ring of flowers, which develop into seeds. To better understand the differences between gall and seed flowers, gene expression was analyzed in fig flowers at the time of oviposition choice and early gall development (Martinson *et al.*, 2015). Significant differences were found in gene expression between gall and seed flowers in receptive syconia prior to oviposition. Transcripts assigned to flavonoids and carbohydrate metabolism were significantly upregulated in gall flowers relative to seed flowers. In response to oviposition, gall flowers significantly upregulated the expression of chalcone synthase, which has been connected to gall formation in other plants. This study provided insight into a possible stability mechanism in the ancient fig–fig wasp association.

In dioecious species, the pollinating wasps and pollen are produced only in the figs of "male" trees, with seeds produced only in the figs of "female" trees. Gynodioecy has thus effectively resolved the conflict occurring in monoecious species between wasp and tree over the flowers' fate (Cook and Rasplus, 2003). The wasps still pollinate and oviposit in "female" figs but produce no offspring, because the wasps cannot gall the flower ovaries. Several ideas have been suggested for the evolutionary advantage of dioecy (Kjellberg *et al.*, 1987; Kjellberg and Maurice, 1989; Kerdelhué and Rasplus, 1996; Kameyama *et al.*, 1999) over the maintenance of monoecy (Herre, 1989). One of these hypotheses suggested that non-pollinating wasps might play a role (Kerdelhué and Rasplus, 1996). Greeff and Compton (2002) showed that non-pollinating wasps that reduce the number of seeds, but not those that reduce the number of pollinators, favor seed protection, a potential initial step toward dioecy. The non-pollinating wasps drain resources from the tree, or directly attack the pollen vectors or seeds. This creates a fitness trade-off for the plant, as facilitating the pollinators results in vulnerability to other wasp species. One evolutionary response to this might simply have been to exclude all wasps from developing in the figs. These fruit would be functionally female because no pollinators are produced but may have a selective advantage if the parasitic load excluded in this way was large enough (Greeff and Compton, 2002).

5.7 *Ficus* Pollination

5.7.1 Pollination mediated by volatiles emitted by the fig host

Pollination in *Ficus* is mediated by chemical signals—volatiles emitted by the fig host. Because the fig crop on any particular tree is generally highly synchronized, the short-lived female wasps must leave their natal tree to find figs that are suitable for oviposition. It is thought that wasps detect a suitable host by identifying the chemical volatiles produced by figs when they are ready for pollination. Gas chromatograms of the fig volatiles of several species of *Ficus* showed them to be species-specific. Age-related changes in the volatile profiles were noted, with extra volatiles being produced when the figs were ready for pollination (Ware *et al.*, 1993).

Female figs emit the same volatiles as male figs (essentially mimicking them), thereby deceiving the pollinators into entering the female fig. Changes in fig volatiles during fruit development have been reported in different fig species (Borges, 2015). The fig–fig wasp interactions are regulated through volatile organic compounds emitted by host figs. Volatile organic compound signals with a specific combination of volatiles act as the attractant during the receptive stage, whereas after pollination, the host species change their volatile organic compound blend and become repellent to pollinating and non-pollinating wasps (Proffit *et al.*, 2018). The qualitative and quantitative differences in various volatile organic compounds during fruit ripening are characterized by the presence of several major compounds; e.g. β-ocimene is dominant in receptive *Ficus racemosa*, whereas the monoterpenes limonene and α-pinene contribute more to the post-pollination stage (Soler *et al.*, 2012).

In *F. carica*, Nawade *et al.* (2020) found significant changes in the accumulation of the volatile organic compound apocarotenoids β-ionone, 6-methyl-5-hepten-2-ol (MHOL), and β-cyclogeraniol during fruit development. The level of MHOL and β-cyclogeraniol during the second week of hand pollination was twofold higher than that found in fruit harvested after the first week of pollination. The accumulation of these compounds decreased during subsequent fruit-development periods. This post-pollination decline may be because excess pollinators are unnecessary after a sufficient number of foundresses have become available for pollination. This suggests the possible role of apocarotenoids in the attraction of pollinating wasps by *F. carica*. Apocarotenoids, such as MHOL and 6-methyl-5-hepten-2-one, were predominant volatiles during the receptive stage of *Ficus curtipes* (Gu *et al.*, 2012), and a behavioral study then confirmed their role in the attraction of fig wasps (Gu *et al.*, 2016). The interaction between figs and their pollinating wasps is regarded as a model system for studying specialized coevolved mutualism. Chemoreception of fig wasps plays an important role in their pollination and odorant-binding protein functions in the first step of odorant detection. A high degree of host specificity is ensured by the fact that female wasps are attracted by their specific fig tree species through the volatile organic compounds released by the figs when they are ready to be pollinated. Zeng *et al.* (2018) studied transcriptome sequencing data from volatile organic compound-treated and control fig wasps. Overall, 16 upregulated genes and 13 downregulated genes were identified. One gene involved in the synoptic vesicle cycle and two genes coding for odorant-binding proteins are likely to have potential impacts on the response of fig wasps to the volatile organic compounds emitted by their host figs. This suggests the presence of a dynamic molecular system of detection of, and hence response to, host plant volatile organic compounds.

Wang *et al.* (2014) studied how odorant-binding proteins are related to the mating process occurring within the fig cavity and the extreme host specificity of the species in the fig wasp *C. solmsi*. They quantified the expression of odorant-binding protein genes in different life stages of female and male fig wasps by real-time quantitative polymerase chain reaction (PCR). The results indicated that *CsolOBP1* and *CsolOBP2* in males might bind to pheromones and play important roles in mate choice, whereas *CsolOBP4* and *CsolOBP5* may function primarily in host localization by females through binding of volatile organic compounds emitted by receptive figs. They concluded that odorant-binding protein genes play important roles in mating and host specificity in the mutualistic fig–pollinator system (Wang *et al.*, 2014).

5.7.2 Environmental effects on pollination

Ficus–wasp mutualism provides an important ecosystem service that is threatened by global warming and climate change. *Ficus* species have a reciprocally obligate mutualism with the tiny, short-lived (1–2 days) fig wasps. The small size and short life of these pollinators is expected to make them more vulnerable to climate change than their larger and longer-lived hosts. Specialized pollination relationships, in which a plant species has few alternative pollinators, are expected to be more vulnerable than generalized ones (Koh *et al.*, 2004; Gilman *et al.*, 2011). In the extreme case of obligate one-to-one relationships (one pollinator to one plant), neither species will survive climate conditions that are unsuitable for the more sensitive partner (Gilman *et al.*, 2011). Four species of adult female fig wasps, *Valisia malayana* (pollinating *Ficus grossularioides*), *Ceratosolen constrictus hewitti* (*Ficus fistulosa*), *Ceratosolen appendiculatus* (*Ficus variegata*) and *Eupristina verticillata* (*Ficus microcarpa*), decrease steadily with increasing temperature (Jevanandam *et al.*, 2013). The latter results suggested that an increase of 3°C or more above the current temperatures experienced across much of the equatorial tropics would markedly decrease the active adult lifespan of all four species. Unless fig wasps can acclimatize or adapt to warmer temperatures with time, these responses may disrupt the mutualism. Warmer-than-usual temperatures can decrease or increase synchrony in the fig–wasp interactions. The current and future responses of *Ficus* trees to climate change are largely unknown, because the fig and wasp are facing an unprecedented human-induced change in climate that has no past representation. Continued in-depth study of the fig–wasp mutualism with climate change is warranted.

5.7.3 Importance of pollination in *F. carica*

The edible fig, *F. carica*, is one of the few species of *Ficus* that has been cultivated. Cultivation of common fig has been known since prehistoric times; the finding of a fig dating back to 11,400–11,200 years ago in the Jordan Valley might suggest early domestication (Kislev *et al.*, 2006; Lev-Yadun *et al.*, 2006). During their long history of cultivation, these figs probably passed through the process of pollination, as seen in native varieties in Turkey, Greece, and other countries in the region. As the young syconia of *F. carica* (stage B) is not hermetically sealed, artificial pollination with the aid of a puffer or syringe is possible. The relative ease of pollinating *Ficus* species without the aid of wasps enabled plant breeders to create new varieties and even new interspecies hybrids of different *Ficus* species (Condit, 1950; Flaishman *et al.*, 2008). However, during the thousands of years of cultivation of the common fig, many parthenocarpic varieties were selected for. The ovaries of these varieties turn into drupelets without fertilization. As no seeds are formed in these parthenocarpic varieties, propagation is totally dependent on stem cuttings leading to clonal varieties, and on traditional agriculture in which the diversity of such clonal varieties is maintained (Achtak *et al.*, 2010).

5.8 Concluding Remarks

Ficus trees occupy large tropical and subtropical areas that are exposed to fluctuating climate. Their pollinating wasps thus go through developmental arrest or multiple generations in different fig crops. Once the female trees are ready for pollination, they trick the pollinators into entering their syconia, thus obtaining pollination services. This interaction is unlike the fig–fig wasp interaction in the monoecious system where the same syconium produces both seeds and wasp offspring. These factors illustrate the complexity of the fig–fig wasp process, regarded as one of the oldest known examples of mutualism. Native to the temperate Mediterranean region, the edible fig, *F. carica*, has been cultivated since prehistoric times. In the wild, the edible fig goes through a complex and interesting cycle of wasp development and wasp cycling processes. The cultivated edible figs produce fruit with the aid of the pollinator wasp *B. psenes*, and propagation is by seed. However, many parthenocarpic varieties have been developed that can propagate without wasp pollination. Finally, artificial pollination with pollen from different donors has enabled the creation of new cultivars and even new hybrids of edible figs. An understanding of fig–fig wasp biology will provide new answers to current and future challenges associated with climate change.

Acknowledgement

We sincerely thank Professor Finn Kjellberg for sharing his knowledge on this topic.

References

Achtak, H., Ater, M., Oukabli, A., Santoni, S., Kjellberg, F., and Khadari B. (2010) Traditional agroecosystems as conservatories and incubators of cultivated plant varietal diversity: the case of fig (*Ficus carica* L.) in Morocco. *BMC Plant Biology* 10, 28.

Ahmed, S., Compton, S.G., Butlin, R.K., and Gilmartin, P.M. (2009) Wind-borne insects mediate directional pollen transfer between desert fig trees 160 kilometers apart. *Proceedings of the National Academy of Sciences USA* 106, 20342–20347.

Barth, F.G. (1985) *Insects and Flowers: the Biology of a Partnership*. Princeton University Press, Princeton, New Jersey.

Berg, C.C. (1989) Classification and distribution of *Ficus*. *Experientia* 45, 605–611.

Berg, C.C. and Wiebes, J.T. (1992) *African Fig Trees and Fig Wasps*. Koninklijke Nederlandse Academic von Wetenschappen Verhadnelingen Afdeling Naturrkunde, Tweede Reeks (Verhandelingen Der Koninkliijke Nederlandse Akademie Van Wetenschappen, Afd. Natuurkunde, Vol. 89). Royal Netherlands Academy, Amsterdam.

Borges, R.M. (2015) How to be a fig wasp parasite on the fig–fig wasp mutualism. *Current Opinion in Insect Science* 8, 34–40.

Compton, S.G. and Hawkins, B.A. (1992) Determinants of species richness in southern African fig wasp assemblages. *Oecologia* 91, 68–74.

Condit, I.J. (1920) Caprigs and caprification. *California Agricultural Experiment Station Bulletin* 319, 341–375.

Condit, I.J. (1932) The structure and development of flowers in *Ficus carica* L. *Hilgardia* 6, 443–481.

Condit, I.J. (1947) *The Fig*. Chronica Botanica Company, Waltham, Massachusetts.

Condit, I.J. (1950) An interspecific hybrid in *Ficus*. *Journal of Heredity* 41, 165–168.

Cook, J.M. and Rasplus, J.-Y. (2003) Mutualists with attitude: coevolving fig wasps and figs. *Trends in Ecology & Evolution* 18, 241–248.

Cruaud, A., Jabbour-Zahab, R., Genson, G., Couloux, A., Yan-Qiong, P., *et al.* (2011) Out of Australia and back again: the world-wide historical biogeography of non-pollinating fig wasps (Hymenoptera: Sycophaginae). *Journal of Biogeography* 38, 209–225.

Cruaud, A., Rønsted, N., Chantarasuwan, B., Chou, L.S., Clement, W.L., *et al.* (2012) An extreme case of plant-insect codiversification: figs and fig-pollinating wasps. *Systematic Biology* 61, 1029–1047.

Datwyler, S.L. and Weiblen, G.D. (2004) On the origin of the fig: phylogenetic relationships of Moraceae from *ndh*F sequences. *American Journal of Botany* 91, 767–777.

Doğanlar, M. (2012) Occurrence of fig wasps (Hymenoptera: Chalcidoidea) in *Ficus caria* and *F. microcarpa* in Hatay, Turkey. *Turkish Journal of Zoology* 36, 721–724.

Dufaÿ, M. and Anstett, M.-C. (2003) Conflicts between plants and pollinators that reproduce within inflorescences: evolutionary variations on a theme. *Oikos* 100, 3–14.

Flaishman, M.A., Rodov, V., and Stover, E. (2008) The fig: botany, horticulture, and breeding. *Horticultural Reviews* 34, 132–196.

Galil, J. (1973) Pollination in dioecious figs. Pollination of *Ficus fistulosa* by *Ceratosolen hewitti*. *The Gardens' Bulletin, Singapore* 26, 303–311.

Galil, J. (1984) Anther dehiscense in fig syconia. In: *Minisymposium Figs and Fig Insects*. Centre de la Recherche Scientifique, Paris, pp. 35–40.

Galil, J. (1989) *Ficus*. In: Halevy, A. (ed.) *Handbook of Flowering*, Vol. 6. CRC Press, Boca Raton, Florida, pp. 331–349.

Galil, J. and Eisikowitch, D. (1968) On the pollination ecology of *Ficus sycomorus* in East Africa. *Ecology* 49, 259–269.

Galil, J. and Neeman, G. (1977) Pollen transfer and pollination in the common fig (*Ficus carica* L.). *New Phytologist* 79, 163–171.

Ghara, M. and Borges, R.M. (2010) Comparative life-history traits in a fig wasp community: implications for community structure. *Ecological Entomology* 35, 139–148.

Gibernau, M., Hossaert-McKey, M., Anstett, M.-C., and Kjellberg, F. (1996) Consequences of protecting flowers in a fig: a one-way trip for pollinators? *Journal of Biogeography* 23, 425–432.

Gilman, R.T., Fabina, N.S., Abbott, K.C., and Rafferty, N.E. (2011) Evolution of plant–pollinator mutualism in response to climate change. *Evolutionary Applications* 5, 2–16.

Greeff, J.M. and Compton, S.G. (2002) Can seed protection lead to dioecy in *Ficus*? *Oikos* 96, 386–388.

Gu, D., Compton, S.G., Peng, Y., and Yang, D. (2012) 'Push' and 'pull' responses by fig wasps to volatiles released by their host figs. *Chemoecology* 22, 217–227.

Gu, D., Yang, D.-R., Yang, P., Peng, Y.-Q., and Wang, Z.-J. (2016) Work division of floral scent compounds in mediating pollinator behaviours. *Chemistry and Ecology* 32, 733–741.

Herre, E.A. (1989) Coevolution of reproductive characteristics in 12 species of New World figs and their pollinator wasps. *Experientia* 45, 637–647.

Hill, D.S. (1967) *Figs (Ficus spp.) of Hong Kong*. Hong Kong University Press, Hong Kong.

Hossaert-McKey, M., Gibernau, M., and Frey, J.E. (1994) Chemosensory attraction of fig wasps to substances produced by receptive figs. *Entomologia Experimentalis et Applicata* 70, 185–191.

Jansen-González, S., Teixeira, S.P., and Pereira, R.A.S. (2012) Mutualism from the inside: coordinated development of plant and insect in an active pollinating fig wasp. *Arthropod–Plant Interactions* 6, 601–609.

Janzen, D.H. (1979) How many parents do the wasps from a fig have? *Biotropica* 11, 127–129.

Jevanandam, N., Goh, A.G., and Corlett, R.T. (2013) Climate warming and the potential extinction of fig wasps, the obligate pollinators of figs. *Biology Letters* 9: 20130041.

Kameyama, T., Harrison, R., and Yamamura, N. (1999) Persistence of a fig wasp population and evolution of dioecy in figs: a simulation study. *Researches on Population Ecology* 41, 243–252.

Kerdelhué, C. and Rasplus, J.Y. (1996) The evolution of dioecy among *Ficus* (Moraceae): an alternative hypothesis involving non-pollinating fig wasp pressure on the fig–pollinator mutualism. *Oikos* 77, 163–166.

Kislev, M.E., Hartmann, A., and Bar-Yosef, O. (2006) Early domesticated fig in the Jordan Valley. *Science* 312, 1372–1374.

Kjellberg, F.E. and Maurice, S. (1989) Seasonality in the reproductive phenology of *Ficus*: its evolution and consequences. *Experientia* 45, 653–660.

Kjellberg, F., Gouyon, P.-H., Ibrahim, M., Raymond, M., and Valdeyron, G. (1987) The stability of the symbiosis between dioecious figs and their pollinators: a study of *Ficus carica* L. and *Blastophaga psenes* L. *Evolution* 41, 693–704.

Kjellberg, F., Doumesche, B., and Bronstein, J.L. (1988) Longevity of a fig wasp (*Blastophaga psenes*). *Proceedings of the Koninklijke Nederlandse Akademie van Wetenschappen* 91, 117–122.

Koh, L.P., Dunn, R.R., Sodhi, N.S., Colwell, R.K., Proctor, H.C., and Smith, V.S. (2004) Species coextinctions and the biodiversity crisis. *Science* 305, 1632–1634.

Kugler, H. (1955) *Einführung in die Blütenökologie*. Gustav Fisher Verlag, Stuttgart, Germany.

Kuttamathiathu, J. (1958) Recherches sur les chalcidiens, *Blastophaga psenes* (L.) et *Philotrypesis caricae* (L.) du figuier (*Ficus carica* L.). *Annales des Sciences Naturelles Zoologie et Biologie Animale* 10, 197–260.

Lev-Yadun, S., Neeman, G., Abbo, S., and Flaishman, M.A. (2006) Comment on early domesticated fig in the Jordan Valley. *Science* 314, 1683.

Machado, C.A., Jousselin, E., Kjellberg, F., Compton, S.G., and Herre, E.A. (2001) Phylogenetic relationships, historical biogeography and character evolution of fig-pollinating wasps. *Proceedings of the Royal Society of London B: Biological Sciences* 268, 685–694.

Martinson, E.O., Hackett, J.D., Machado, C.A., and Arnold, A.E. (2015) Metatranscriptome analysis of fig flowers provides insights into potential mechanisms for mutualism stability and gall induction. *PLoS One* 10: e0130745.

Molbo, D., Machado, C.A., Sevenster, J.G., Keller, L., and Herre, E.A. (2003) Cryptic species of fig-pollinating wasps: implications for the evolution of the fig–wasp mutualism, sex allocation, and precision of adaptation. *Proceedings of the National Academy of Sciences USA* 100, 5867–5872.

Nawade, B., Shaltiel-Harpaz, L., Yahyaa, M., Bosamia, T.C., Kabaha, A., *et al.* (2020) Analysis of apocarotenoid volatiles during the development of *Ficus carica* fruits and characterization of carotenoid cleavage dioxygenase genes. *Plant Science* 290: 110292.

Neeman, G. and Galil, J. (1978) Seed set in the "male syconia" of the common fig *Ficus carica* L. (caprificus). *New Phytologist* 81, 375–380.

Proffit, M., Bessière, J.-M., Schatz, B., and Hossaert-McKey, M. (2018) Can fine-scale post-pollination variation of fig volatile compounds explain some steps of the temporal succession of fig wasps associated with *Ficus racemosa*? *Acta Oecologica* 90, 81–90.

Rohwer, J.G. (1993) Moraceae. In: Kubitzki, K. (ed.) *The Families and Genera of Vascular Plants*. Springer, Heidelberg, Germany, pp. 438–453.

Rønsted, N., Weiblen, G.D., Cook, J.M., Salamin, N., Machado, C.A., and Savolainen, V. (2005) 60 million years of co-divergence in the fig-wasp symbiosis. *Proceedings of the Royal Society B: Biological Sciences* 272, 2593–2599.

Sakai, S. (2002) A review of brood-site pollination mutualism: plants providing breeding sites for their pollinators. *Journal of Plant Research* 115, 161–168.

Santinelo Pereira, R.A., de Pádua Teixeira, S., and Kjellberg, F. (2007) An inquiline fig wasp using seeds as a resource for small male production: a potential first step for the evolution of new feeding habits? *Biological Journal of the Linnean Society* 92, 9–17.

Soler, C.C.L., Proffit, M., Bessière, J.-M., Hossaert-McKey, M., and Schatz, B. (2012) Evidence for intersexual chemical mimicry in a dioecious plant. *Ecology Letters* 15, 978–985.

Su, Z.H., Iino, H., Nakamura, K., Serrato, A., and Oyama, K. (2008) Breakdown of the one-to-one rule in Mexican fig–wasp associations inferred by molecular phylogenetic analysis. *Symbiosis* 45, 73–81.

Tzeng, H.-Y., Tseng, L.-J., Ou, C.-H., Lu, K.-C., Lu, F.-Y., and Chou, L.-S. (2008) Confirmation of the parasitoid feeding habit in *Sycoscapter*, and their impact on pollinator abundance in *Ficus formosana*. *Symbiosis* 45, 129–134.

van Noort, S., Ware, A.B., and Compton, S.G. (1989) Pollinator-specific volatile attractants released from the figs of *Ficus* Burtt-Davyi. *South African Journal of Science* 85, 323–324.

Wang, N., Wang, N.X., Niu, L.M., Bian, S.N., Xiao, J.H., and Huang, D.W. (2014) Odorant-binding protein (OBP) genes affect host specificity in a fig–pollinator mutualistic system. *Insect Molecular Biology* 23, 621–631.

Ware, A.B. and Compton, S.G. (1994) Responses of fig wasps to host plant volatile cues. *Journal of Chemical Ecology* 20, 785–802.

Ware, A.B., Kaye, P.T., Compton, S.G., and van Noort, S. (1993) Fig volatiles: their role in attracting pollinators and maintaining pollinator specificity. *Plant Systematics and Evolution* 186, 147–156.

Weiblen, G.D. (2000) Phylogenetic relationships of functionally dioecious *Ficus* (Moraceae) based on ribosomal DNA sequences and morphology. *American Journal of Botany* 87, 1342–1357.

Weiblen, G.D. (2002) How to be a fig wasp. *Annual Review of Entomology* 47, 299–330.

Xiao, J.H., Yue, Z., Jia, L.Y., Yang, X.H., Niu, L.H., *et al.* (2013) Obligate mutualism within a host drives the extreme specialization of a fig wasp genome. *Genome Biology* 14: R141.

Zeng, J., Yu, H., and Kjellberg, F. (2018) Transcriptome analysis of genes involved in the response of a pollinator fig wasp to volatile organic compounds from its host figs. *Acta Oecologica* 90, 91–98.

6 Fig Propagation

Kamer Betül Özer*
Department of Horticulture, Faculty of Agriculture, Ege University, Bornova, İzmir, Turkey

6.1 Introduction

Fig is a fruit species that has been grown since ancient times. Because it is easily propagated vegetatively by root cuttings, its global commercial production was based on a few varieties, although a high diversity exists in nature. Figs can be propagated by seeds or vegetatively by cuttings. Commercially, the propagation method by cutting is still the most used in fig. Propagation occurs through mechanisms of cell division and differentiation without altering the plant genotype due to mitotic multiplication (Hartmann *et al.*, 2011). Most fig orchards are grown on their own roots (Hartmann *et al.*, 2002). However, other methods can also be used, such as the use of burrs or plunging and grafting techniques and tissue culture (Fig. 6.1). This chapter gives an overview and summarizes the results of the research work carried on propagation of various fig varieties to produce new orchards. Propagation is also of interest for persons multiplying a fig variety appreciated for its form or fruits in their home yards.

6.2 Seeds

Fertilized female fig fruit contains hundreds of seeds. In nature, frugivorous bats and birds feeding on the fig fruit act as the vectors for dispersal of fig seeds. The hard endocarp covering the seed cannot be digested, so remains intact and preserves the viability of the seed. In the natural growing habitats of fig, the Mediterranean countries, seed-borne fig trees are spread all over. This is the main source for the different kinds of figs that have been collected and selected by ancient farmers and that have given rise to the main varieties in use by us today. However, in many countries, seedlings belonging to types requiring pollination for fruit set and produced by seeds are not harvested due to the absence of the fig wasp *Blastophaga psenes* L., the natural pollinator of the fig tree.

Currently, propagation of figs by seeds is used mainly for breeding purposes to develop new varieties. Fig is a dioecious species, and the female fruit needs to receive pollen from male parents to have viable seeds. Moreover, both female and male fig varieties vary according to their fruit set requirements. The drupelets in fig fruit may have fertile seeds only if fertilized, as in the second crop of San Pedro and Smyrna types. The hardened endocarp may not have seeds enclosed, as is the case for brebas and the unfertilized second crop of common-type varieties. Among caprifig (male) varieties, there are types that have persistent-type fruit that do not bear

* E-mail: kamer.betul.ozer@ege.edu.tr

 Advances in Fig Research and Sustainable Production (M.A. Flaishman and U. Aksoy)
DOI: 10.1079/9781789242492.0006

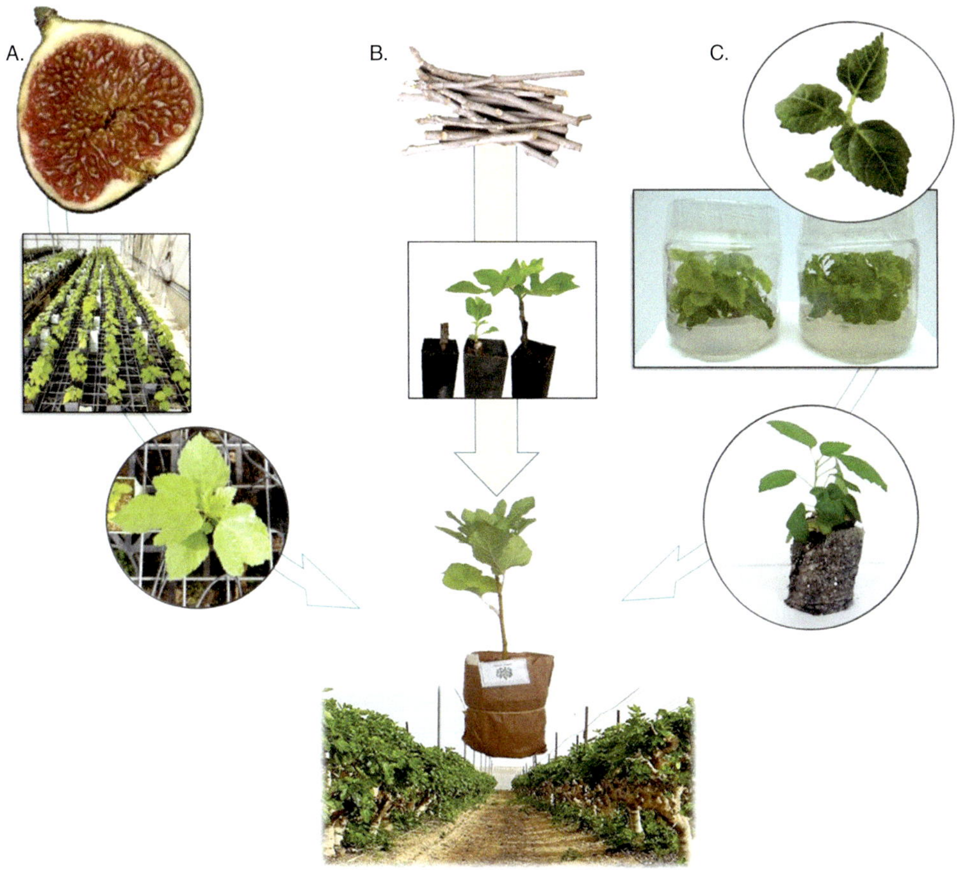

Fig. 6.1. Figs can be propagated A. by seeds or vegetatively by B. cuttings, and C. burrs, plunging, air layering, grafting or micropropagation. Figure used with the permission of R. Peer.

fertile seeds, but upon fertilization, caprifig fruit may also develop viable seeds. When pollinated with preserved pollen from the previous year, short-styled gall flowers in the male fruit may develop viable seeds like the long-styled female flowers pollinated by pollens of the profichi (spring) crop. The germination rate of female fig seeds is higher than that of male fig seeds. If the gall flowers are occupied by wasps, seed formation is not completed in most cases due to mechanical damage created by the ovipositor of the pollinating wasp at the time of pollination or because further growth is hindered later (Neeman and Galil, 1978). Seeds that are not fully developed can be separated from fertile seeds by floating them in water. The heavier fertile seeds sink to the bottom, whereas the empty ones are lighter and float on top.

Germination of fig seeds may be influenced by internal factors such as the composition and thickness of the hard endocarp, or the composition of the endosperm due to the presence of nutrients or compounds that inhibit germination or promote seed dormancy. Although fig seeds can germinate easily compared with many fruit species, physiological dormancy is present. Therefore, various pretreatments can be applied to enhance germination in fig seeds such as stratification, scarification, soaking in water, or pretreatments with chemicals such as potassium nitrate or with growth regulators such as gibberellin. The time required for germination also varies according to the varietal characteristics and the pretreatments. Using gibberellic acid (GA_3), at concentrations of 500 and 1000 mg L^{-1}, breaks seed dormancy in figs and promotes germination and the emergence of new growth (Çalışkan *et al.*, 2012). Most of the globally known dried-fig varieties, such as Sarılop (= Calimyrna), Sabz, Kalamata, and Zidi, are of the Smyrna type, and the pollination and fertilization processes increase the sugar content and acidity flavor, and the seed-bearing achenes add a crunchy texture.

Regarding the formation of apomictic seeds, there are controversial results. Neeman and Galil (1978) reported that female fig varieties were proved not to undergo apomixis. However, Armstrong (2006) suggested that apomictic seeds may be formed in female fig varieties and that there are two main types of apomixes: (i) parthenogenesis (agamogenesis) where a haploid or diploid egg within the embryo sac (or a diploid cell from two fused haploid cells of embryo sac) develops into an embryo; and (ii) agamospermy, when an embryo arises from tissue surrounding the embryo sac. If this involves cells of the nucellus or inner integument, it is called a nucellar embryo, which is chromosomally identical or a clone of the sporophyte parent.

Fig seeds can germinate under warm conditions, and the seedlings can be grown for the first season in a temporary pot and then transferred to the soil medium (Condit, 1947). In germination and emergence experiments, seeds are kept under room conditions (21–27°C), and germinated after 5–25 days (Çalışkan *et al.*, 2012).

6.3 Vegetative Propagation

6.3.1 Cuttings

In prehistoric times, rooting of cuttings may have started when a gardener realized that a fig branch accidentally falling to the ground could develop roots and grow into a tree. At a later stage, this method became widespread in commercial nurseries (Condit, 1947). When the performance of green and hardwood cuttings was compared, the best vegetative propagation method for fig genotypes was reported as hardwood cuttings (Aljane and Nahdi, 2014; Ahmad *et al.*, 2018). Fig cuttings should be taken from healthy annual shoots of middle-aged and strong mother trees. For this purpose, 1-year-old shoots can be taken, as well as 2–3-year-old healthy branches. The thickness of the shoot should be 1–1.5 cm and the length between 15 and 30 cm (Özen *et al.*, 2007; Flaishman *et al.*, 2008). The cuts should be made straight just below the eye and at 45° above the eye at the top end using sharp pruning shears (Condit, 1947). If there is a terminal bud, it is not cut but remains on the branch. Most fig varieties have apical dominance, strengthening the role of the terminal bud in further shoot growth. If there is no terminal bud, the cut is made just above and below the node. However, cuttings with terminal buds are preferred for straight tree growth. As the cuttings that do not have a terminal bud will produce shoots from the lateral buds, they develop slowly and mostly form slender primary branches that will form the main scaffold. Saplings formed from lateral buds do not have strong and straight trunks (Özen *et al.*, 2007). *Ficus carica* is generally propagated by tip cuttings (U. Aksoy, 1990, unpublished data; Lionakis, 1995; F.M. Pereira and J.C. Nachtigal, 1997, unpublished data; Kai *et al.*, 1997). To avoid problems related to transplantation of fig plants that have thin roots from the nursery to orchards, propagation of fig plants in plastic bags or pots may be necessary (Zerhoune, 2003).

The time for preparing cuttings varies according to the growing regions. Some researchers recommend taking cuttings in late autumn or early winter, while others state that they can be taken directly before planting time in January–February after pruning (Hartmann *et al.*, 2002; Özen *et al.*, 2007; Flaishman *et al.*, 2008). If the cuttings are taken in autumn and winter, they can be rooted in sawdust or in another humid medium in the greenhouse or in a closed space during the winter months, or they can be kept under cool conditions (Flaishman *et al.*, 2008). Cuttings are tied in bundles of approximately 50, each wrapped with a damp cloth or folded in a slightly moist sand or perlite medium to preserve their moisture content. The bundles are then placed in plastic cases and stored in cold storage (4°C) at 80–85% relative humidity until planting. Under these conditions, cuttings can be stored for 1–2 months. If the cuttings are taken in early spring, they should be immediately planted in a well-drained and aerated soil mixture to develop roots. In cooler climates, plastic chambers used in propagation with elevated temperatures proved successful in promoting early callus formation in fig cuttings (Pipattanawong *et al.*, 2008). Cuttings are planted in March or April when the soil temperature reaches around 10°C. The soil is tilled with a plow, and rows are established to insert cuttings at 60 cm between rows and 30 cm within the row. An irrigation system is installed accordingly. To protect new growth from early frost risk in autumn, irrigation is rescheduled and decreased starting from early September to ensure adequate growth and resistance (Özen *et al.*, 2007). In addition to young branches, 0.9–1.2 m long hardwood cuttings can be planted directly in the orchard at their permanent places during the dormant period.

To prevent problems arising from soil-borne diseases and pests, namely nematodes in nurseries, sapling production may be carried out by rooting cuttings in substrate cultures designed as beds. In this method, a mixture of 20% perlite, 20% garden soil, 20% coarse sand, 20% forest soil, and 20% animal manure is used after disinfection. In bed culture, fig cuttings are planted with 25 × 25 cm between and within the row spacings. Successful results have been obtained in the substrates perlite or perlite:peat (1:1) mixture as a soil-less medium (Ertan, 2005; Özen *et al.*, 2007). When different growing mediums such as sand, perlite, sawdust, peat, marc, or wood chips were examined for rooting, a mixture of 30:70 perlite:sand was the best medium for the growth performance of hardwood cuttings (Aljane and Nahdi, 2014; Ahmad *et al.*, 2018). When rooting woody cuttings of the fig variety Bursa Black, the highest rooting rate (71%) was obtained in sand used as the medium, followed by peat (31%), perlite (27%), and soil (25%) (Tekintaş and Seferoğlu, 1998). Küden *et al.* (1993) found a wide range of rooting rates for cuttings, between 0% and 90%, based on the effects of cutting collection time, rooting environment, and indole 3-butyric acid (IBA) applications. Substrates composed of different mixtures such as sand, perlite, peat, and sawdust have been tested for rooting fig cuttings (Shamsuddin *et al.*, 2021). Two types of cuttings, semi-hardwood and hardwood, were used in the study, and the results indicated that a 1:1 combination of peat and perlite medium and hardwood cuttings were the most effective rooting media and cutting type for fig propagation, as evidenced by a significantly higher quality of root and shoot growth. Nava *et al.* (2014) evaluated the substrate, cutting collection time, position and cutting depth, and the medium for rooting cuttings of the fig variety Roxo de Valinhos. According to results of the experiment, collecting hardwood cuttings in the first half of July and planting them in sand under plastic cover is recommended in practice. However, the inclination position and cutting depth, and covering the substrate with plastic mulching did not have a marked effect on rooting of cuttings.

Growth regulators are used to promote rooting or survival of cuttings. The effect of different concentrations of IBA on rooting was investigated in a study testing the rooting performances of herbaceous cuttings, taken from various sources such as shoots with leaves or leafless, and the mother plants bearing fruit or not (Nogueira *et al.*, 2007). Herbaceous cuttings taken from the middle part of the branches of the fig variety Roxo de Valinhos were treated with IBA, at concentrations of 0, 150 and 300 mg L^{-1}, planted in a mixture of sand and soil, and placed in plastic bags. After 60 days rooting in the greenhouse, the dry weight of new growth and roots was measured, and no positive effect of IBA on rooting performance of the cuttings was identified. However, the IBA concentrations tested to increase the rooting of fig cuttings taken from different locations on the shoot revealed that cuttings from the basal parts of the shoot, when treated with IBA at 3000 ppm, had the highest survival percentages. The same success was not achieved with the cuttings taken from the upper parts of the branches, possibly due to the difference in the amount of nutrient stocks (Sivaji *et al.*, 2014). Khapare *et al.* (2012) investigated the effects of different IBA and α-naphthalene acetic acid (NAA) concentrations on cutting quality in hardwood and semi-hardwood cuttings. According to the number of sprouts per cutting, leaf area, number of leaves, and root and shoot dry matter, the best result was observed with 2500 ppm IBA + 2500 ppm NAA or 1000 ppm IBA + 1000 ppm NAA combinations. Polat *et al.* (2000) found that cuttings of the Sarılop fig variety formed more roots than the other two varieties.

In addition, cuttings immersed in 1000 ppm IBA rooted better than the control group. With the lengthening of the stratification time applied in hardwood fig cuttings, the rooting rate decreased from 100% to 46%. In another study, the best result in terms of shooting and rooting was found with 100 mg L^{-1} IBA applied to cuttings without stratification and rooted in soil:sand (1:1) mixture (Antunes *et al.*, 2003). The highest rooting rate of fig cuttings (95%) taken every 2 weeks from April to August was found in cuttings taken in April, regardless of the IBA application. IBA application increased the weight of dry matter in shoots and roots, but this effect decreased in late cuttings (Chalfun *et al.*, 2003).

Bud type, shoot age, shoot length, and shoot diameter were analyzed in hardwood cuttings taken from a number of *F. carica* varieties. The percentage of root emergence was also measured. The results showed that there was a high variability among the five varieties analyzed according to the various parameters (Aljane and Nahdi, 2014). In another study, terminal cuttings at three different lengths—long (20–25 cm), medium (10–12 cm), and short (6–8 cm)—were taken from 1-year-old shoots. In addition, a total of five types of cutting were prepared by leaving five nodes from 2-year-old wood at the bottom of medium and short cuttings. The increase in the length of the old wood left at the bottom of the cuttings caused a significant increase in the number of roots (Özeker and İsfendiyaroğlu, 2001). In other studies, it has been determined that when fig cuttings are prepared from 3-year-old shoots, they root better than from 1- and 2-year-old shoots (Aminov, 1973; A.S. Eroğlu, 1977, unpublished data). It is claimed that the bottom cuttings of well-developed 1-year-old shoots give good results, in addition to the older shoots (Hartmann *et al.*, 2002). Some studies reveal that the increase in the length of the cutting has a positive effect on rooting (Seferoğlu *et al.*, 1994; Özeker and İsfendiyaroğlu, 2001). In the reproduction of fig trees in Greece, wood cuttings with a length of 20–30 cm and a thickness of 1–2 cm are generally used, while in Brazil, 1-year-old, 30–40 cm long and 1.5–3 cm diameter cuttings are used (Lionakis, 1995; F.M. Pereira and J.C. Nachtigol, 1997, unpublished data). Ertan *et al.* (2006) determined the effects of using different sizes and diameters of cutting material on sapling growth performance and quality in the production of the Bursa Black fig variety. The cuttings were divided into three size groups of 12–13 cm, 17–18 cm, and 21–23 cm in length, and it was observed that saplings with the longest and thickest shoot were obtained from the cuttings with the lowest diameter and height. In addition, seedlings having the highest value in terms of root length and root number were also obtained from the lowest diameter and lowest height seedlings. Peche *et al.* (2017) applied a lopping system to increase the number of apical shoots and rooting of the cuttings. Rooting percentage and sprouting were higher in cuttings obtained from plants that had been lopped. Application of IBA (2000 ppm) also increased the proliferation of roots and shoots.

The effect of the variety and cutting style on the rooting and vigor of cuttings was investigated using five fig varieties. The results indicated that neither the variety nor the cutting layout affected the percentage of rooting. However, the plant vigor was affected by the variety (Zerhoune, 2003). Karadeniz (2003) found that rooting was affected by the variety and the applied IBA concentration. The highest rooting performance was 58% with 1000 ppm IBA.

6.3.2 Grafting

Grafting is joining a portion of one plant (bud or scion) into or on a stem, root, or branch of another (stock) in such a way that a union will be formed and the partners will continue to grow. When more than two parts are involved, the middle piece is called the "interstock." When the scion consists of a single bud, the process is called budding. Grafting is used when a scion variety is grafted on to seedlings or to change the variety in older trees, repair injured trees, produce dwarf trees, strengthen plants' resistance to certain diseases, enhance adaptation to adverse soil or climatic conditions, ensure pollination and/or produce multi-fruit or multi-flowered plants. Figs are generally propagated by cuttings. Budding or grafting is not practiced widely. Budding may be preferred in nurseries (Hartmann *et al.*, 2002). Grafting is used when a scion variety of choice is grafted on to fig seedlings or when changing the variety in older trees. Where resistant rootstocks develop, for example with resistance to nematodes, grafting on healthy, high-quality, and productive seedlings could be implemented widely (Boliani *et al.*, 2019). The interaction of rootstocks may affect the performance of the scion through dwarfing or invigoration and, in most cases, may alter the precocity and quality.

Budding or grafting in fig is a difficult process due to the milky latex produced by the plant. The basic technique of grafting consists of placing cambial tissues of stock and scion in intimate association, so that the resulting callus tissue produced from the stock and scion interlocks to form a living continuous connection. For grafting, buds on a short branch, taken from 1-year-old strong shoots in figs, are budded on to old branches in the form of T or patch grafts. The bark is lifted away from around an upright or inverted T-shaped incision, and the bud is then inserted under the bark and tied securely in place. The best time for budding as a T or reverse T graft is from April to the end of May. If budding is administered from September to the end of October, then it remains dormant until spring (Condit, 1947). In split bark grafting, two short scions, each containing two or three buds, can be grafted on to a cut branch with a diameter of 10–15 cm. Bark grafting is like budding with the insertion of a scion having buds, instead of placing a single bud as occurs in budding. With an inclined cut at the lower end, the scion is then inserted into the cut made in the rootstock and pushed down slowly until it makes firm contact with the bark over the entire cut (Condit, 1947). The best time for split bark grafting is February and early March under Mediterranean conditions (Özen *et al.*, 2007).

The success or failure of any grafting operation is based on the compatibility of each plant part, closeness of fit, and cambial contact. The union is initially held in place by pressure exerted by the stock, by grafting tape, or by rubber strips applied over the point of union. The performance of fig varieties and the ratio of take may vary. For example, the grafting performance of eight fig varieties grafted on to the Brown Turkey variety varied between 100% and 72%, and the King and Brown Turkey varieties had the highest take ratio (Rattanpal *et al.*, 2014). In another experiment evaluating 13 fig varieties for their T budding performance on Brown Turkey rootstock, maximum budding success was recorded in Brown Turkey (93.4%), while the lowest success was in Black Fig (70.3%) (Rattanpal and Sidhu, 2017).

Two plants that are closely related botanically and that have a continuous cambium can be grafted. Grafts between species are often successful, and between genera, more occasionally. Hosomi (2017) grafted Masui Dauphine figs on to 21 different species to provide resistance against soil disorders. Grafting on *Ficus palmata*, *Ficus septica*, *Ficus pumila*, and *Ficus stipulata* rootstocks, classified in the same subgenus as *Ficus*, was partially successful.

In grafting by cleft graft in the fig variety Roxo de Valinhos, Kotz *et al.* (2011) obtained an adequate percentage of survival and seedling development among grafted individuals, and good compatibility. In grafted fig trees, productivity is essentially related to the rootstock and to the grafting method. Casaroti *et al.* (2015) observed high fruit production mainly during the off-season when the Bonato variety was used as a rootstock and Roxo de Valinhos as a scion variety. In studies related to the development and production of Roxo de Valinhos fig fruits on different rootstocks, it was observed that there was an effect of Turco and Palestino rootstocks on the production of green figs, and Bonato IAC and Caprifigo IAC varieties affected mainly the production of ripe fruits, providing high productivity and quality (E.A.D. Silva, 2010, unpublished data).

6.3.3 Layering

Figs can be grown easily using a few simple methods such as plunging and air layering. This method is an alternative way to multiply fig varieties that have a low rooting ability when propagated with cuttings, such as the Nazareth, Montes, or Brunswick varieties (Bisi *et al.*, 2016). For layering, 1- or 2-year-old bottom shoots growing from the fig trees are buried in the soil in March after the side shoots are cleaned, leaving the tip out. After rooting, they are separated from the main tree, and rooted branches are planted in the spring of the following year. Rooted suckers are also used in propagation, mainly to fill out the gaps in fig orchards. Suckers are separated from the main stem with their roots and planted directly in the orchard as a sapling (Özen *et al.*, 2007). Figs can easily be reproduced by air layering. For air layering, a canister or a pot filled with soil is placed around the base of the new shoots, and after a while, a good root system forms and the entire rooted young tree is then transplanted to the desired location (Condit, 1947).

If 1-year-old branches are layered in early spring, they will usually root in mid-summer. A 2.5 cm belt cut is made just below the point of rooting. The removed part is covered with sphagnum moss or any rooting medium and then wrapped with a polyethylene sheet. The ends are wrapped with tape to prevent rain from entering under the plastic cover. After rooting, the branch is cut just below the rooting medium, the polyethylene is removed, and the new fig tree is ready for planting (Hartmann *et al.*, 2002). For air layering, Daneluz *et al.* (2009) created different types of scar tissues on the branch and tried different IBA concentrations. They also covered the wound with different materials. The best result in the study resulted from the application of 1000 mg L^{-1} IBA in the middle of the branches without injury and covering with *Pinus* substrate. Although fig plants can be reproduced by bottom shoots or layering (Dolgun and Tekintaş, 2008), these methods are not preferred in commercial nurseries because of the limited number of plants obtained, depending on the material (Kabasakal, 1990; Dolgun and Tekintaş, 2008).

6.4 Micropropagation

Sexual propagation using seeds is not preferred because some seeds of fig are non-viable and the fertile ones are not true to type (Qrunfleh *et al.*, 2013). Traditional clonal propagation methods such as rooting cuttings and/or grafting of fruit trees are time-consuming as the resulting plant material is limited, and damage may be created on the mother plant, which is a problem if it is unique. Therefore, reproduction with tissue culture techniques allows high-quality, disease-free, and true-to-type plants in a smaller area independent of environmental conditions (Debergh and Zimmerman, 1991). Tissue culture techniques with high multiplication rates are being developed using various plant parts as the explants to fulfill the increasing demand for fig saplings (Azhar and Zainuddin, 2020). *In vitro* regeneration studies on figs have tested various explants such as shoot tips (Gella *et al.*, 1998; Yancheva *et al.*, 2005), nodal explants (Fráguas *et al.*, 2004), leaves (Kim *et al.*, 2007; Soliman *et al.*, 2010; Dhage *et al.*, 2012), and apical buds (Gella *et al.*, 1998; Kumar *et al.*, 1998; Yakushiji *et al.*, 2003). These studies have been carried out on figs for more than three decades, but the frequency of released articles has increased in recent years.

In all producing countries, fig trees are threatened by viruses that cause chlorosis and damage leaves and fruits, thus reducing productivity and quality, and in some cases causing complete die-out. Shoot-end culture is a suitable method for virus sanitation. Various studies have focused on meristem or shoot-tip culture for large-scale production of high-quality, healthy fig plants (Gella *et al.*, 1998; Günver and Ertan, 1998). Researchers have reported a number of reproducible regeneration methods, but most successful results have been achieved using shoot tips and apical buds (Kumar *et al.*, 1998; Hepaksoy and Aksoy, 2006). *In vitro* studies on figs are limited to the production of virus-free plants. However, *in vitro* mosaic virus-free fig plant production is a faster technique, and when shoot tips and apical buds are used, the chance of success increases. The study of Kumar *et al.* (1998) determined the first protocol for multiple shoot induction and individual plant regeneration. In this study, 68% of the plants had well-developed root systems and were successfully acclimatized to the external environment. In addition to the direct advantages of micropropagation and plant regeneration without antivirus targets, optimization of *in vitro* culture conditions can speed up and facilitate research on transgenic plants (Sahraroo *et al.*, 2019). Günver Dalkılıç *et al.* (2015) used shoot *in vitro* thermotherapy against the fig mosaic virus. Shoot tips (0.7–1.0 cm) of Sarılop and Bursa Black as female, and Ak ilek and Kaba ilek as male fig varieties were cultivated in Murashige and Skoog (MS) medium including 2.0 mg L^{-1} 6-benzylaminopurine (BAP), 0.2 mg L^{-1} gibberellic acid (GA_3), 0.1 mg L^{-1} IBA, and 89 mg L^{-1} phloroglucinol. At the start of the thermotherapy application, the relative humidity and temperature were 70% and 27°C, respectively, and the temperature was increased to 37°C by 2°C daily increments. After this point, the temperature was kept constant. After 3 weeks of thermotherapy application, the shoot tips were transferred to fresh medium and developed new shoots successfully.

According to the results of an *in vitro* study using different fig varieties, the establishment of shoot bud and multiple shoot induction in figs is thought to be quite genotype specific (Dhage *et al.*, 2015). Studies revealed that fig genotypes play a role in the performance of shoot and root formation (Shahcheraghi and Shekafandeh, 2016; Sahraroo *et al.*, 2019). Meristem size may also play an important role in fig meristem culture and micropropagation (Sahraroo *et al.*, 2019). Furthermore, the responses of the varieties used in *in vitro* reproduction studies may differ in terms of parameters analyzed and their genetic background and adaptability to environmental conditions. Kumar *et al.* (1998) used apical buds and applied a protocol for *in vitro* multiple shoot induction and obtained a maximum of 4.8 shoot tips per explant. However, a large-scale *F. carica* production system is still lacking (Sharma *et al.*, 2015). Sharma *et al.* (2015) aimed to develop a clonal production protocol in fig by encapsulation of node segments, which may significantly promote germplasm exchange and conservation programs. Node segments were found to be more effective than shoot tips for shoot multiplication. In *in vitro* studies, fig regeneration was limited to the use of a single shoot tip alongside the apical buds. However, methods have been developed for organogenesis induction from *F. carica* leaf explants by applying different auxin and cytokinin combinations (Yakushiji *et al.*, 2003; Soliman *et al.*, 2010). Li *et al.* (2015), in their *in vitro* studies with the Masui Dauphine fig variety, found that the most suitable medium was modified MS + 1.0 mg L^{-1} BAP + 0.05 mg L^{-1} NAA + 1.0 mg L^{-1} GA_3 + 20 mg L^{-1} sucrose + 7 mg L^{-1} agar at pH 5.8, with axillary buds as the most suitable explant. Bayoudh *et al.* (2015) stated that the most suitable shoot-tip cross-section length for sanitation of local fig varieties is 0.5 mm. Medium containing MS + 0.5 mg L^{-1} BAP + 0.1 mg L^{-1} NAA + 0.1 mg L^{-1} GA_3 gave the most successful results in the proliferation of fig shoots. With this optimized protocol, 90% of the *in vitro* plants rooted in a moist peat environment and the micropropagation time was shortened. Optimum regeneration was observed using MS medium supplemented with 4.0 mg L^{-1} 2,4-dichlorophenoxyacetic acid for callusing, 7 mg L^{-1} thidiazuron and 0.25 mg L^{-1} NAA for shoot production, and 1.0 mg L^{-1} IBA for rooting (Dhage *et al.*, 2015). In *in vitro* culture experiments, the BAP concentration had diverse effects on regeneration of plants. This situation is thought to be related to the genotype specificity of fig varieties (Mustafa and Taha, 2012). While some researchers argue that IBA is necessary for root induction for *in vitro* propagation of figs (Nobre and Romano, 1998; Hepaksoy and Aksoy, 2006; Shahcheraghi and Shekafandeh, 2016; Sahraroo *et al.*, 2019), others have claimed that IBA is effective for *in vitro* root induction and microshoots in figs (Brum, 2001; Fráguas *et al.*, 2004). Brum (2001) determined the type of medium, sucrose concentration, and their interactive effects on optimal multiple shoot induction and plantlet regeneration of the *F. carica* Roxo de Valinhos variety. They observed that a sucrose concentration of 15–20 g L^{-1} tends to stimulate the increase of shoot growth on different culture media. Lethal browning is a major problem arising when working with *in vitro* cultures of *F. carica*, as it kills up to 50% of the leaf segment explants. Lethal browning results from the oxidation of phenolic compounds, released from the cut ends of the explants. The oxidized products, quinones, are known to be highly reactive and inhibit enzyme activity, leading to death of the explants. The best treatment was found to be incubation of the culture in darkness with addition of citric and ascorbic acids, which significantly increased the rates of survival for leaf segment explants (Soliman *et al.*, 2010). Kim *et al.* (2007) in their studies on optimizing conditions for induction of random buds and regeneration of multiple shoots from leaf tissues of *F. carica* found that the addition of phloroglucinol to the medium delayed the exudation of phenolic substances. In addition, thidiazuron added to MS medium with IBA produced more shoots from the cultured wounded leaf explants than the other auxin and cytokinin combinations. In a study investigating *in vitro* surface sterilization and shoot proliferation in figs, the best survival rate was 60%. MS medium supplemented with 0.5 mg L^{-1} BAP and 0.2 mg L^{-1} NAA gave the best results related to average shoot number, average shoot length, and average leaf number

(Moniruzzaman *et al.*, 2017). Volo *et al.* (2017) studied the effects of the cutting shape of explants and various amounts of cytokinin on the micropropagation of figs, and determined that shoots grow faster and have more leaves in explants grown in medium with 0.3 μM kinetin added rather than IBA. Also, more shoot growth was observed on explants where the stem was cut in half, lengthwise.

The genetic stability of *in vitro*-propagated plants should be evaluated as early as possible, especially in perennial woody trees (Marum *et al.*, 2009). Various techniques have been used for this purpose (Breiman *et al.*, 1989; Chowdhury and Vasil, 1993). Many studies have been conducted to investigate the effectiveness of DNA-based molecular markers such as restriction fragment length polymorphisms, amplified fragment length polymorphisms, random amplified polymorphic DNA (RAPD) markers, and inter-simple sequence repeat (ISSR) markers. El-Dessoky *et al.* (2016) aimed to establish an *in vitro* shoot propagation protocol to evaluate the applicability and reliability of RAPD and ISSR markers to determine the *in vitro* genetic stability of regenerated plants. The markers used in their study showed good genetic compatibility with plants grown *in vitro*. Thus, they recommended using the tested methods effectively for rapid microreplication without the risk of instability. Direct shoot development from the apical meristem minimizes the somaclonal variation by bypassing the callus formation phase, thus preserving the genetic structure of the plants obtained (Kaushal *et al.*, 2014).

Tissue culture still encounters some obstacles. Barbosa *et al.* (2008) reported that it took up to 4 years to produce the first fruits in micropropagated figs, probably indicating the occurrence of long juvenility of micropropagated plants via meristem culture. Despite the disadvantages, propagation of the fig tree by tissue culture may become an important technique to produce healthy seedlings (Boliani *et al.*, 2019).

6.5 Fig Nurseries

To establish a vigorous and healthy fig orchard, enough saplings are needed. Fig plant material can be cheaply propagated by growing hardwood stem cuttings directly under field conditions, or by developing saplings in the nursery. Zerhoune (2003), studying propagation of five fig varieties, namely Kadota, Ghouddane, Ournaksi, Chaari, and Ounaq El Hmam, using hardwood stem cuttings under field conditions indicated that neither the variety nor the cutting layout significantly affected the percentage of rooted cuttings. At the end of the first year, most of the cuttings were rooted. However, both the variety and the type of cutting layout affected plant vigor significantly. During the initial development of the cutting under field conditions, sudden changes of temperature and low moisture adversely affect the rooting and growing of cuttings. It is important to prepare the soil and ensure suitable climate conditions to use this method for reproduction. Even the slightest stress conditions faced can adversely affect production costs and quantity. For production of large-scale commercial orchards, fig nurseries are essential. A fig nursery, depending on the size, can produce saplings by using different multiplication methods such as layering, bottom shoots or tip cuttings. The classical layering and bottom shoots methods are not preferred. In commercial fig nurseries, plants are generally propagated by tip cuttings and in some cases in misted rooting units. Recently, the stool bed layering method was used intensively during propagation of many different fruit crops. Dolgun and Tekintaş (2008, 2009) examined fig sapling production with the stool bed layering method, using 1- and 2-year-old branches of the fig variety Sarılop. They concluded that by using stool bed layering, most of the nursery plants had sufficient qualitative and quantitative features and displayed a successful rooting rate.

Pest, disease, and replanting problems have always been a great problem in fig nursery tree production. Şirin *et al.* (2010) studied the best substrate to cultivate nursery fig trees, and analyzed the effects of substrates on the morphological characteristics of fig saplings of the variety Sarılop. They concluded that the use of perlite or peat plus perlite led to increased plant growth and positively impacted the quality of fig nursery trees grown in high tunnels.

To achieve optimum crop yield and quality, or to supply the market demand, orchards may be established with different fig varieties. Therefore, fig nurseries must propagate different varieties, which differ in rooting potential. The propagation method has an influence on the rooting of stem segments of fig tree varieties. To expand the genetic base of fig nurseries, Bisi *et al.* (2016), tested several propagation methods to achieve successful rooting of stem segments of 16 fig varieties. The propagation methods used were layering, hardwood cuttings, nodal segments, herbaceous cuttings originating from the removal of sprouts, and herbaceous cuttings obtained during growth. They concluded that the propagation method influences the rooting of stem segments and that the method utilized must suit the varieties to increase the rooting potential.

6.6 Concluding Remarks

Figs can be propagated by seeds or vegetatively by cuttings, burrs, plunging, air layering, grafting, or micropropagation. It is important to develop propagation techniques that provide producers with high-quality saplings, free of pests or diseases, and with an optimum vigor to allow an orchard that guarantees good productivity for a long period of time. There are many studies that have been carried out to optimize the process of fig tree propagation. Currently, most commercial fig propagation relies on rooting of hardwood or herbaceous cuttings, leading to fig orchards grown on their own roots. With the demand for establishing fig orchards in new regions of the world, propagation of fig varieties has regained interest. The use of resistant varieties and pathogen-free substrates should be prioritized due to the susceptibility of fig trees to nematodes. The combination of micropropagation and grafted fig seedlings could provide virus-free nursery trees with better adaptability to resist soil-borne diseases or abiotic stress conditions.

References

Ahmad, H., Stanica, F., and Stanciu, V. (2018) Cuttings propagation of some fig genotypes (*Ficus carica*). *Scientific Papers Series B: Horticulture* 62, 241–244.

Aljane, F. and Nahdi, S. (2014) Propagation of some local fig (*Ficus carica* L.) cultivars by hardwood cuttings under the field conditions in Tunisia. *Hindawi Publishing Corporation International Scholarly Research Notices* 2014, 2–5.

Aminov, K.H.L. (1973) Some biological and technical aspects of propagating figs from cuttings. *Horticultural Abstracts* 43, 820.

Antunes, L.E.C., Chalfun, N.N.J, Pasqual, M., Dutra, L.F., and Cavalcante-Alves, J.M. (2003) Factors affecting on rooting of fig (*Ficus carica* L.) cuttings. *Acta Horticulturae* 605, 141–146.

Armstrong, W. (2006) Sex determination and life cycle of *Ficus carica*. Available at www2.palomar.edu/users/warmstrong/pljun99b.htm (accessed November 24, 2021).

Azhar, N.A. and Zainuddin, Z. (2020) Tissue culture of *Ficus carica* variety BTM-6. *Malaysian Journal of Sustainable Agriculture* 4, 26–28.

Barbosa, W., Pio, R., de Arruda Veiga, R.F., Chagas, E.A., and Feldberg, N.P. (2008) Efeito de concentrações do AIB no enraizamento *in vitro* de cultivares de figueira. *Bioscience Journal* 24, 1–6 (in Portuguese).

Bayoudh, Ch., Labidi, R., Majdoub, A., and Mars, M. (2015) *In vitro* propagation of caprifig and female fig varieties (*Ficus carica* L.) from shoot-tips. *Journal of Agricultural Science and Technology* 17, 1597–1608.

Bisi, R.B., Locatelli, G., Barbosa, C.M.A., Pio, R., and Balbi, R.V. (2016) Rooting of stem segments from fig tree cultivars. *Acta Scientiarum Agronomy* 38, 379–385.

Boliani, A.C., Arruda Ferreira, A.F., Monteiro, L.N.H., da Silva, M.S.C., and Rombola, A.D. (2019) Advances in propagation of *Ficus carica* L. *Revista Brasileira Fruticultura* 41, 1–13.

Breiman, A., Flisenburg, T., and Galun, E. (1989) Is Nor region variability in wheat invariably caused by tissue culture? *Theoretical and Applied Genetics* 77, 809–814.

Brum, G.R. (2001) Micropropagação da Figueira (*Ficus carica* L.) 'Roxo de Valinhos'. MSc thesis, University of Lavras, Lavras, Brazil (in Portuguese).

Çalışkan, O., Mavi, K., and Polat, A. (2012) Influences of presowing treatments on the germination and emergence of fig seeds (*Ficus carica* L.). *Acta Scientiarum: Agronomy* 34, 293–297.

Casaroti, B.G., Pereira, G.A., and de Souza Corrêa, L. (2015) Efeito de diferentes porta-enxertos sobre o desenvolvimento e produção de figo verde, pela variedade Roxo de Valinhos (Effect of different

rootstocks on the development and production of green fig, by the Roxo de Valinhos variety). *Revista Cultura Agronômica* 24, 225–233 (in Portugese).
Chalfun, N.N.J., Pasqual, M., Norberto, P.M., Dutra, L.F., and Cavalcante-Alves, J.M. (2003) Rooting of fig (*Ficus carica* L.) cuttings: cutting time and IBA. *Acta Horticulturae* 605, 137–140.
Chowdhury, M.K.U. and Vasil, J.K. (1993) Molecular analysis of plants regenerated from embryonic cultures of hybrid sugarcane cultivars (*Saccharum spp.*). *Theoretical and Applied Genetics* 86, 181–188.
Condit, I.J. (1947) *The Fig.* Chronica Botanica, Waltham, Massachusetts.
Daneluz, S., Pio R., Chagas, E.A., Barbosa, W., Ohland, T., and Kotz, T.E. (2009) Propagation of the fig tree 'Roxo De Valinhos' by air layering. *Revista Brasileira de Fruticultura* 31, 285–290.
Debergh, P.C. and Zimmerman, R.H. (1991) Micropropagation. In: Debergh, P.C. and Zimmerman, R.H. (eds) *Micropropagation Technology and Application*. Kluwer Academic Publishers, Dordrecht, The Netherlands, pp. 1–14.
Dhage, S.S., Pawar, B.D., Chimote, V.P., Jadhav, A.S., and Kale, A.A. (2012) *In vitro* callus induction and plantlet regeneration in fig (*Ficus carica* L.). *Journal of Cell and Tissue Research* 12, 3395–3400.
Dhage, S.S., Chimote, V.P., Pawar, B.D., Kale, A.A., Pawar, S.V., and Jadhav, A.S. (2015) Development of an efficient *in vitro* regeneration protocol in fig (*Ficus carica* L.). *Journal of Applied Horticulture* 17, 160–164.
Dolgun, O. and Tekintaş, F.E. (2008) Production of fig (*Ficus carica* L.) nursery plants by stem layering method. *Agriculturae Conspectus Scientificus* 73, 157–160.
Dolgun, O. and Tekintaş, F.E. (2009) Effective use of vegetative material in fig (*Ficus carica* L.) nursery plant production. *African Journal of Agricultural Research* 4, 701–706.
El-Dessoky, D., Attia, O.A., and El-Awady, A.M.M. (2016) An efficient protocol for *in vitro* propagation of fig (*Ficus carica* L.) and evaluation of genetic fidelity using RAPD and ISSR markers. *Journal of Applied Biology & Biotechnology* 4, 57–63.
Ertan, B. (2005) İncir Fidanı Üretiminde Topraksız Tarım Yöntemlerinden Yatak Kültürü Sisteminin Kullanım Olanakları, T.C. Tarım ve Köyişleri Bakanlığı. *Ege Tarımsal Araştırma Enstitüsü Müdürlüğü* 119, S169–S178 (in Turkish).
Ertan, E., Ertan, B., Şirin, U., and Dolgun, O. (2006) Farklı boy ve çapta odun çeliklerinin "Bursa Siyahi" incir çeşidinde fidan gelişim performansı üzerine etkileri. *ADÜ Ziraat Fakültesi Dergisi* 3, 37–44 (in Turkish).
Flaishman, M.A., Rodov, V., and Stover, E. (2008) The fig: botany, horticulture, and breeding. In: Janick, J. (ed.) *Horticultural Reviews*, Vol. 34. Wiley, Hoboken, New Jersey, pp. 113–197.
Fráguas, C.B., Pasqual, M., Ferreira Dutra, L., and Cazetta, J.O. (2004) Micropropagation of fig (*Ficus carica* L.) 'Roxo de Valinhos' plants. *In Vitro Cellular & Developmental Biology – Plant* 40, 471–474.
Gella, R., Marin, J.A., Corrales, M.L., and Toribio, F. (1998) Elimination of fig mosaic from shoot tip cultures by thermotherapy. *Acta Horticulturae* 480, 173–177.
Günver Dalkılıç, G., Ertan, B., Dağ, S., and Görücüoğlu, P. (2015). Effect of *in vitro* thermotherapy application on shoot development in some fig cultivars. *Acta Horticulturae* 1173, 195–198.
Günver, G. and Ertan, E. (1998) A study on the propagation of figs by the tissue culture techniques. *Acta Horticulturae* 480, 169–172.
Hartmann, H.T., Hudson, T., Kester, D.E., Dale, E K., Davies, F.T. Jr., and Geneve, R.L. (2011) *Plant Propagation: Principles and Practices*. Prentice Hall, Boston, Massachusetts.
Hartmann, H.T., Kester, D.E., Davies, F.T., and Geneve, R.L. (2002) Propagation methods and rootstocks for fruit and nut species. In: Davis, F.T., Geneve, R.L. and Wilson, S.B. (eds) *Plant Propagation Principles and Practices*. Pearson Education, Cranbury, New Jersey, pp. 717–757.
Hepaksoy, S. and Aksoy, U. (2006) Propagation of *Ficus carica* L. clones by *in vitro* culture. *Biologia Plantarum* 50, 433–436.
Hosomi, A. (2017) Variation in graft compatibility of wild *Ficus* species as rootstock for common fig trees (*Ficus carica*). *Acta Horticulturae* 1173, 199–205.
Kabasakal, A. (1990) İncir Yetiştiricliği (fig cultivation). In: *Foundation of Agricultural Research and Promotion.* Yalova, Istanbul, Turkey (in Turkish).
Kai, M., Fen, L.Z., Fen, T.Y., and Weibing, J. (1997) The green cutting propagation techniques for fig trees. *China Fruits* 3, 32–38.
Karadeniz, T. (2003) A study on some fruit characteristics and propagations of these by hardwood cuttings of local fig cultivars grown in Ordu (Tukey). *Acta Horticulturae* 605, 107–112.
Kaushal, S., Sidana, A., and Dev, K. (2014) *In vitro* plant production through apical meristem culture of Gentiana kurroo Royle. *Journal of Medicinal Plants Studies* 3, 4–9.
Khapare, L.S., Dahale, M.H., and Bhusari, R.B. (2012) Propagational studies in fig as affected by plant growth regulators. *Asian Journal of Horticulture* 7, 118–120.

Kim, K.M., Kim, M.Y., Yun, P.Y., Chandrasekhar, T., Lee H.Y., and Song, P.S. (2007) Production of multiple shoots and plant regeneration from leaf segments of fig tree (*Ficus carica* L.). *Journal of Plant Biology* 50, 440–446.

Kotz, T.E., Pio, R., Campagnolo, M.A., Chagas, E.A., and Dalastra, I.M. (2011) Enxertia em figueira 'Roxo de Valinhos' por borbulhia e garfagem. *Bragantia* 70, 344–348 (in Portugese).

Küden, A.B., Kaska, N., and Yılmaz, M.A. (1993) Bursa Siyahi ve 01.IN.10 incir çeşitlerinde farklı çelik alma zamanları ile köklendirme ortamları ve IBA uygulamalarının karşılaştırılması. *Çukurova Üniversitesi Ziraat Fakültesi Dergisi* 8, 181–188 (in Turkish).

Kumar, V., Radha, A., and Kumar, C.S. (1998) *In vitro* plant regeneration of fig (*Ficus carica* L. cv. Gular) using apical buds from mature trees. *Plant Cell Report* 17, 717–720.

Li, J., Mi, L., Chen, X., Wan, C., Huo, H., and Chen, B. (2015) *In vitro* rapid propagation of *Ficus carica* 'Masui Dauphine'. *Agricultural Science and Technology* 16, 1360–1362.

Lionakis, S.M. (1995) Present status and future prospects of the cultivation in Greece of the plants: fig, loquat, Japanese persimmon, pomegarante and barbary fig. *Chairs Options Méditerranéennes* 13, 21–30.

Marum, L., Rocheta, M., Maroco, J., Oliveira, M.M., and Miguel, C. (2009) Analysis of genetic stability at SSR loci during somatic embryogenesis in maritime pine (*Pinus pinaster*). *Plant Cell Report* 28, 673–682.

Moniruzzaman, M., Yaakob, Z., and Taha, R.A. (2017) *In vitro* production of fig (*Ficus carica* L.) plantlets. *Acta Horticulturae* 1173, 231–235.

Mustafa, N.S. and Taha, R.A. (2012) Influence of plant growth regulators and subculturing on *in vitro* multiplication of some fig (*Ficus carica*) cultivars. *Journal of Applied Sciences Research* 8, 4038–4044.

Nava, G.A., Júnior, A.W., Mezalira, E.J., Cassol, D.A., and Alegretti, A.L. (2014) Rooting of hardwood cuttings of Roxo de Valinhos fig (*Ficus carica* L.) with different propagation strategies. *Plant Propagation Revista Ceres* 61, 989–996.

Neeman, G. and Galil, J. (1978) Seed set in the 'male syconia' of the common fig *Ficus carica* L. (Caprificus). *New Phytologist* 81, 375–380.

Nobre, J. and Romano, A. (1998) *In vitro* cloning of *Ficus carica* L. adult trees. *Acta Horticulturaae* 480, 161–164.

Nogueira, A.M., Chalfun, N.N.J., Dutra, L.F., and Villa, F. (2007) Propagation of fig tree (*Ficus carica* L.) by cuttings obtained during vegetative period. *Ciência e Agrotecnologia Lavras* 31, 914–920.

Özeker, E. and İsfendiyaroğlu, M. (2001) Çeşme yarımadasında yellopu oluşturan bazı incir tiplerinin çelikle çoğaltılması. *Ege Üniversitesi Ziraat Fakültesi Dergisi* 38, 17–24 (in Turkish).

Özen, M., Çobanoğlu, F., Kocataş, H., Tan, N., Ertan, B., *et al.* (2007) İncir Yetiştiriciliği (Fig Production). In: Özen, M. (ed.) *T.C. Tarım ve Köyişleri Bakanlığı. Tarımsal Araştırmalar Genel Müdürlüğü Erbeyli İncir Araştırma Enstitüsü Müdürlüğü*. Aydın Tuna Matbaacılık, Aydın, Turkey, p. 145 (in Turkish).

Peche, P.M., Moura, P.H.A., Curi, P.N., Melo, E.T., Figueriredo, A.L., *et al.* (2017) Rooting of apical cuttings of fig trees with a lopping system. *Acta Horticulturae* 1173, 217–220.

Pipattanawong, N., Tiwong, S., Thongyean, B., Darak, R., Thamin, P., and Techa, W. (2008) Improvement of propagation by hardwood cuttings with and without using plastic pavilions in fig (*Ficus carica* L.). *Kasetsart Journal – Natural Science* 42, 207–214.

Polat, A.A., Durgaç, C., and Kamiloglu, Ö. (2000) Indol butirik asidin (IBA) incir çeliklerinin köklenmesi üzerine etkisi, *MKU Ziraat Fakültesi Dergisi* 5, 1–6 (in Turkish).

Qrunfleh, I.M., Shatnawi, M.M., and Al-Ajlouni, Z.I. (2013) Effect of different concentrations of carbon source, salinity and gelling agent on *in vitro* growth of fig (*Ficus carica* L.). *African Journal of Biotechnology* 12, 936–940.

Rattanpal, H.S. and Sidhu, G.S. (2017) Studies on propagation of fig (*Ficus carica* L.) cultivars by budding. *Acta Horticulturae* 1173, 173–176.

Rattanpal, H.S., Singh, G., and Singh, G. (2014) Standardization of rapid propagation technique for fig (*Ficus carica* L.) cultivars by budding. *HortFlora Research Spectrum* 3, 94–96.

Sahraroo, A., Zareib, A., and Babalar, M. (2019) *In vitro* regeneration of the isolated shoot apical meristem of two commercial fig cultivars 'Sabz' and 'Jaami-e-Kan'. *Biocatalysis and Agricultural Biotechnology* 17, 743–749.

Seferoğlu, G., Mısırlı, A., Can, Z., and Şahin, N. (1994) Sarılop incir uç çeliklerinin köklenmesi üzerinde araştırmalar, II: alt ısıtmanın etkisi. *Ege Üniversitesi, Ziraat Fakültesi Dergisi* 31, 85–89 (in Turkish).

Shahcheraghi, S.T. and Shekafandeh, A. (2016) Micropropagation of three endemic and endangered fig (*Ficus carica* L.) genotypes. *Advances in Horticultural Science* 30, 129–134.

Shamsuddin, M.S., Shahari, R., Amri, C.N.A.C., Tajudin, N.S., Mispan, M.R., and Salleh, M.S. (2021) Early development of fig (*Ficus carica* L.) root and shoot using different propagation medium and cutting types. *Tropical Life Sciences Research* 32, 83–90.

Sharma, S., Shahzad, A., Mahmood, S., and Saeed, T. (2015) High-frequency clonal propagation, encapsulation of nodal segments for short-term storage and germplasm exchange of *Ficus carica* L. *Trees* 29, 345–353.
Şirin, U., Ertan, E., and Ertan, B. (2010) Growth substrates and fig nursery tree production. *Scientia Agricola* 67, 633–638.
Sivaji, T., Madhavi, K., and Sudha Vani, V. (2014) Effect of type of cuttings and IBA concentrations on the propagation of fig (*Ficus carica*) cv. Poona fig under polyhouse conditions. *International Journal of Tropical Agriculture* 32, 85–88.
Soliman, H.I., Gabr, M., and Abdallah, N. (2010) Efficient transformation and regeneration of fig (*Ficus carica* L.) via somatic embryogenesis. *GM Crops* 1, 47–58.
Tekintaş, F.E. and Seferoğlu, G. (1998) Propagation of fig by hardwood cuttings in the field conditions (*Ficus carica* L.). *Acta Horticulturae* 480, 119–120.
Volo, P., Preece, J.E., and Caruso, T. (2017) The effect of explant cut and cytokinin type on micropropagation of fig (*Ficus carica* L.) 'Brown Turkey'. *Acta Horticulturae* 1173, 227–230.
Yakushiji, H., Mase, N., and Sato, Y. (2003) Adventitious bud formation and plantlet regeneration from leaves of fig (*Ficus carica* L.). *Journal of Horticulture Science and Biotechnology* 78, 717–20.
Yancheva, S.D., Golubowicz, S., Yablowicz, Z., Perl, A., and Flaishman, M.A. (2005) Efficient *Agrobacterium*-mediated transformation and recovery of transgenic fig (*Ficus carica* L.) plants. *Plant Science* 168, 1433–1441.
Zerhoune, M. (2003) Propagation of five fig (*Ficus carica* L.) varieties under field conditions. *Acta Horticulturae* 605, 103–106.

7 Fig Orchard Establishment

H. Zafer Can*
Department of Horticulture, Faculty of Agriculture, Ege University, Bornova, İzmir, Turkey

7.1 Introduction

The fig tree is well adapted to relatively high temperatures and low precipitation regimes. Thus, it has traditionally been cultivated in marginal soil conditions and areas for millennia (Golombek and Lüdders, 1990; Aksoy and Anaç, 1993). Its conventional cultivation has been performed as low input without implementing essential cultural practices, but recently different agrotechniques have been performed. During the last decades, fig cultivation has increasingly been performed under irrigated conditions that increase both fruit yield and quality (Erez *et al.*, 2003; Melgarejo *et al.*, 2007; Galván *et al.*, 2019; Shamin-Shazwan *et al.*, 2019).

Considerable opportunity for phenotypic variability from natural mutations, and its uncomplicated and rapid propagation ability have made it quite easy to obtain new types of fig plants (Flaishman *et al.*, 2008). Beside the caprifig (male) trees, there are three horticultural types of female figs, described as common, San Pedro and Smyrna types (Condit, 1947), and quite different training systems and pruning practices have been performed for each of these groups (Yablowitz *et al.*, 1998; Erez *et al.*, 2003). In Smyrna-type figs, an open-vase shape training method is the usual practice. Canopy architecture becomes very important for breba fruit-bearing fig varieties because breba fruits are born on the previous year's wood, and to improve the yield and regulate the vegetative growth, breba trees should be pruned at different intensities. Training and pruning methods in figs are chiefly variety dependent and may differ according to the harvest method or utilization as fresh or dry.

As well as the natural phenotypic characteristics and variations of fig varieties and their different training and cultivation systems, the climatic and topographic peculiarities of the growing environment also have great importance in terms of site selection and plantation systems for fig orchards. Figs could be an attractive option for poor marginal conditions in arid and semi-arid regions in Mediterranean climates. However, global climate change is threatening overall productivity in these regions. Knowing plant ecophysiology at canopy and orchard levels becomes more and more important, particularly in such marginal areas, in order to obtain the desired fruit yield and quality (León-Sánchez *et al.*, 2016). For these reasons, various factors starting from site selection and environment management to plantation distances have great importance in orchard establishment and management, as fig cultivation is perennial. The fundamental aim of fig orchard planning and a

* Email: hakki.zafer.can@ege.edu.tr

 Advances in Fig Research and Sustainable Production
(M.A. Flaishman and U. Aksoy)
DOI: 10.1079/9781789242492.0007

proper design is to achieve optimum crop yield and quality by using cost-effective cultural practices and optimizing factors affecting physiological processes in plant growth and fruit development. Prior to fig orchard planning, analysis of the training system and orchard management techniques should be conducted to respond to the physiological and economic bases of fig cultivation.

7.2 Production Planning and Analysis

Fig orchard planning includes all the intertwined environmental factors such as climate, geographical location, and soil, in addition to socio-economic factors such as financial capital, labor availability, and machinery use. The environmental factors are closely related to each other and impact management of the farm unit, which is expressed as the growing site, or the living conditions for fig trees. Beside these interactions, there are relationships between the trees and the environmental biotic factors. Selection of the site and preparation of the field for planting are the most important stages in fig production and can make a significant difference to tree growth and farm management. The planting design, spacing, location, and environment can each have a huge effect on fruit yield.

A fig orchard is characterized by a composite of all atmospheric factors such as precipitation, light intensity, temperature, relative humidity, and wind, and all rhizosphere conditions. However, plant growth is also affected by biotic and abiotic stress factors such as drought, light, and salinity, and is impacted by the interaction and cumulative effects of all of these stress factors (Mittler, 2006). According to Fischer and Orduz-Rodríguez (2012), site selection that is not ecologically sophisticated and fit for the prevailing conditions may result in an economically low income. Taking into account all the environmental conditions, site selection and the management system are the key factors for success from planting to harvest. Additionally, global climate change should be taken into account during site selection, because changing and adverse climatic conditions may affect the yield and quality in varying proportions depending on the geographical location (Ramírez and Kallarackal, 2015). In addition, climate change also impacts important pests and diseases and the emergence of new problems (Seidel, 2016).

Selection of the site and variety should be done with a holistic approach because the site should be selected and established according to the targeted variety, type, and market demand. Additionally, fig fruits have multiple uses such as table/dried/processed, and may bear multiple crops as the first (spring) and second (summer) crop. Each variety needs different environmental conditions according to the training system used. Production planning and analysis should be based on a detailed assessment for the specific climate and soil requirements. Environmental factors as prerequisites are more limited during the establishment of a fig orchard for fresh-fig consumption, whereas when sun drying is considered, the conditions affecting yield and especially quality are much more important or more focused on fruit maturation and the drying period.

In addition to variety and site selection, management analysis and production planning are also among the issues that need to be examined in detail. The key performance indicators of successful fig growing are regular bearing and maturation of fruit, return of investment in a short period of time, reduced production costs and labor, resilience or higher adaptability to changing conditions and market demands, and high marketable yield and high fruit quality. Proper planning leads the way to regular and optimum yields and the reduction of production costs per unit area.

The most important factors to be taken into consideration when undertaking accurate planning and analysis of the site are physical and chemical structure, depth, pH, and salinity of the soil; slope, aspect, and risk of erosion; soil organic matter and microorganism activity; previous crops; diseases, pests, and weed potential; precipitation form, amount, and annual distribution; biological diversity; and irrigation water availability and quality. In addition, some market-related factors such as distance and quality of the road connections to the market, market demand, handling and storage conditions, and contracted farming possibilities need to be reviewed. Furthermore, the farm management system and any quality assurance system to be practiced should also be decided because sustainable farming systems such as organic agriculture or good agricultural practices may require additional orchard structures such as a composting area, natural habitats for beneficials, green belts, or a separate storage for inputs and produce. As fig trees are long lived or regenerated easily, proper planning and designing of an orchard saves many years of disappointment and economic loss. A well pre-planned production process and designed plantation ensures the earliest economic return per unit area, optimum yields with high marketable ratios, maximum quality, and low management cost.

During planning and analysis, the most important issues to consider at the initial stage are the climate and soil properties; however, the impact of climate change should never be ignored, and planning must include adaptation strategies for possible future threats. In all climate-change scenarios, nearly all of the current important fig production regions of the world seem to be affected by climate change (Smith *et al.*, 2007; Anon., 2019). Mainly heat and drought stress will be the major problems in the future, and the global climate crisis will particularly affect cultivated perennials. Projected rainfall changes are spatially complex, and the change in precipitation will vary significantly from season to season and across regions (Aksoy and Can, 2012). Additionally, decreases in fertile agricultural land area (Ausubel *et al.*, 2013), soil organic matter (Charman and Roper, 2000), and social issues such as labor populations in agriculture (Anon., 2016) must all be considered. Determination of the status and existing potential of soil and natural resources is essential for farm management planning. Soil and land degradation are among the main factors at greater risk from the effects of climate change, and this impact will increase further. As fig trees are generally planted in marginal and less fertile areas, proper measures should be considered to prevent fig production from being affected by climate change, especially during the initial establishment phase. The proper decisions on land management options will provide not only increased crop yields and quality but also a sustainable agricultural ecosystem.

7.3 Choosing Site and Variety

Site and variety selection must be handled together because the fig crop is utilized in different ways, and some are more dependent on climatic and geographical conditions. The current commercial fig varieties and cultivars have revealed different responses to environmental conditions based on their wide adaptation to different climates and topography.

7.3.1 Variety selection

According to Condit (1955), of the described nearly 600 fig varieties, 78% are common types, 4% are San Pedro types, and 18% are Smyrna types, although the prevalence of a specific type may vary according to the location. All the current commercial fig plantations demonstrate different cultural practices applied to these fig types, which have different fruit-bearing habits. Additionally, these varieties have many phenotypic variations and also vary in terms of plant morphology, maturation period and duration, growth vigor, and fruit quality parameters such as skin and internal color, fruit flavor, sugar content, Brix to titratable acidity, fruit shape, skin thickness and firmness, and ostiole opening (U. Aksoy, 1994, unpublished data).

Different fig types can be found growing in different climatic conditions and at different altitudes in the same areas. Because of this rich diversity, some studies were performed to select the best-performing fig varieties, types, or clones for different climatic and edaphic conditions for the fresh and dried-fig markets. In Turkey, which is the major mainland where fig is grown, 272 promising female fig varieties and 86 Sarılop (= Calimyna) clones were collected by screening different regions (A.S. Eroğlu, 1982, unpublished data) and the collected germplasm was evaluated (U. Aksoy, 1994, unpublished data). Some other adaptation and selection studies have been performed to identify varieties best suited to different regions (Can *et al.*, 1998; Küden and Tanriver, 1998; Kutlu *et al.*, 1998; Kutlu-Kusaksız, 1999, unpublished data).

Botti *et al.* (2003) revealed the effects of climatic conditions on crop yield, fruit quality and production type, and season. Site and variety selection for fresh-fig production as breba or the main crop is more complex due to the diversity available in comparison with dried-fig production. The climatic conditions are more restrictive in dried-fig production due to the necessary sun-drying conditions such as temperature, humidity, and prevailing wind during maturation and drying (Duyar, 1998). Fresh figs can perform well under a wider range of environmental conditions.

As fruit crops are formed on the previous year's or on current-season branches consecutively in both female and male fig trees, the initial growth and maturation periods of the fruits differ significantly. In the northern hemisphere, the ripening and harvesting period of the female fig fruits is generally in June–July for breba fruits and August–November for the summer crop. Fruits ripen on the shoot at 3–7-days intervals starting from the bottom toward the tip (Aksoy, 1981). For this reason, the harvest time also varies according to the varieties. Harvest time is completed in a relatively shorter period: 10–15 days for brebas, and 30–90 days for summer fruits depending on the variety and climatic conditions. Fruit weight and diameter are important quality parameters for both dried and fresh table figs. Additionally, the ratio of total soluble dry matter to titratable acidity is also a major factor used in determining the harvest maturity (Çelikel and Karacali, 1992). For these reasons, accurate variety selection is one of the factors with high impact in terms of sustainable marketing and is also of great importance in breeding because of intermittent ripening and variations in fruit development periods and the intervals between sequential fruits.

There is a dominance of Smyrna-type fig varieties in the global dried market. However, Smyrna-type varieties are also well known in the fresh market, such as Bursa Black. Similarly, Black Mission, which does not require pollination for fruit set, can be dried commercially. The major reason why breba fruits of San Pedro and common fig types are sold as fresh is their comparatively lower sugar or soluble solids contents. Brebas that set on shoots from the previous year generally mature in early summer in the northern hemisphere. These breba crop varieties are quite important for the markets early in the season, although they are more sensitive to transportation due to their higher moisture content (Alayunt *et al.*, 1997a,b).

Fresh fig fruit is highly perishable, and physical hazards and bruising can occur easily during transportation and packing (Çakmak *et al.*, 2010), as well as at the retail stage. Due to its high sugar content and suitability for long-distance transportation, figs are traditionally sun dried and subject to worldwide trade more in the dried form. Fresh figs at full maturity, particularly when they start to soften, are highly sensitive to mechanical damage. Therefore, fresh figs are largely consumed where they are produced (Chessa, 1997). Thus, postharvest and marketing conditions are highly important in terms of choosing the most appropriate variety for a specific location.

The production material should be true to type, healthy, and certified as these factors affect the yield and quality to be harvested from the orchard for many years. It is important to use certified material during the orchard establishment and to keep all documents, as some are only confirmed as the fruits are obtained after some years. Additionally, sapling production should never be done in orchards where disease problems are present and never used as vegetative propagation material because of the risk of spreading diseases. The planting stock should be, at a minimum, healthy, free of all kinds of pests and diseases, including viruses, not damaged, vigorous, uniform, true to type for the variety, and preferably a clone of the target variety and homogeneous in purity.

7.3.2 Site selection

Site selection is one of the most important decisions because every aspect of production including productivity, quality, cultivation practices, and marketing is affected by site conditions. Incorrect site selection is an issue that can cause problems in both the short term and in the longer term during the long productive years.

The orchard site should be selected according to the climatic and edaphic requirements of the fig tree. Besides the climate and soil properties, the most important issues to be considered in site selection are altitude, topography, slope, aspect irrigation availability, water quantity and quality, erosion risk, fluctuating groundwater, and drainage needs. In addition, the history of the land and previous crops, possible pollution and contamination foci, disease and pest risks, and integrated pest management possibilities should be considered. Finally, employment opportunities, marketing, and contracted farming possibilities are important during the initial evaluation.

Although fig is grown as plantations between latitudes 23° and 43° in the northern and southern hemispheres, the fig tree was originally adapted to semi-arid subtropical regions. However, it can be found in the natural flora of mild temperate climates and tropics. While in the tropics, evergreen fig trees can be found at higher altitudes up to 2500 m, commercial orchards are located at lower altitudes in semi-arid subtropical regions. Practically, altitude should not be a limiting factor in fig tree growth. Elevation up to 800 m is not a limiting factor in fig tree growth in milder climates, and altitude enhances the ripening process, as seen in Bouhouli figs in Tunisia (Trad *et al.*, 2013). In Turkey, dried figs are produced at different altitudes ranging from sea level to 800 m above sea level. According to Şen *et al.* (2019), the average fruit weight decreases at higher altitudes, but the share of marketable dried-fruit quality increases, depending on the elevation. Thus, in practical terms, altitude is not a limiting factor in fig tree growing in most fig-producing countries. In Turkey, which is the leading dried-fig producer, most of the fig orchards were located on the fertile alluvial soils of lowland areas until the 1960s, but after establishment of irrigation networks, fig orchards were replaced by irrigated crops. Today, almost 90% of dried figs of the Sarılop (Calimyrna) variety come from the orchards located at altitudes of 500–800 m (Şen *et al.*, 2019).

Fig trees can tolerate a range of extreme site conditions such as rocky lands or hot dry soils (Lansky and Paavilainen, 2011); therefore, fig can be grown in marginal areas where many other commercial crops cannot. According to Flaishman *et al.* (2008), fig plants can be grown on soil with good drainage and a wide range of soil pH (pH 5.5–8.0). Additionally, the fig tree is accepted as moderately tolerant to salinity with negligible salt stress at 6 mS cm^{-1} electrical conductivity (Golombek and Lüdders, 1990; Maas, 1993). Although the fig tree is considered to be tolerant to harsh conditions, adverse climatic and edaphic factors can cause a decline in vegetative and reproductive growth, resulting in yield and quality losses. Subtropical and warm temperate climates provide the optimum climatic conditions for optimum growth, yield, and quality of figs. Sudden temperature decreases in the autumn and early winter period may affect tree growth and yield due to frosts. In warm temperate climates, the most important limiting factor is the low temperature. Temperatures of −3°C in early dormancy (October and November), −9°C in the winter dormancy period, and −1°C and lower after the initiation of bud break in early spring damage fig trees (Aksoy, 1981). However, extreme climatic conditions can be turned to advantage. According to Flaishman and Al Hadi (2002), where the winter temperature is above 6–10°C, such as the Dead Sea area in Israel, Brown Turkey, a common-type variety, starts to act like an everbearing tree and continues to produce fruit from November to May.

There is a significant difference between fig varieties regarding the initiation and ripening of fruits on the shoot. In all cases, fruit ripening occurs at intervals, and this period is also reflected in the length of the harvest period. The harvest period is rather compact in breba fruits, whereas it may last 2 months or longer for the second crop. If the selected region or site is risky in terms of autumn temperatures while entering the cooler period, and if the selected variety forms fruit over a long interval, then there may be problems in terms of fruit ripening, and some fruit may stay unripe. Additionally, high relative humidity or precipitation, or improper irrigation scheduling during the harvest period, depending on the selected fig variety, may lead to increased cracking/splitting in fruits of sensitive varieties. Consequently, fungal diseases may rapidly develop and cause fruit rot (Aksoy, 1981).

Fig trees can be grown without irrigation in places with an annual rainfall of 550–650 mm but the annual total water requirement in sandy soils is reported as 725 mm. Particularly in dry years or when the spring rainfall is insufficient, irrigation is recommended during initial shoot growth as supplementary to winter rains. In the case of drought, the leaves are shed early, before the fruit start ripening, and thus an increase is observed in the amount of extremely dry and sunburned fruits. Additional soil moisture conservation measures must be taken, especially in rainfed orchards located on slopes, to alleviate the impact of drought.

As well as climatic factors such as drought, precipitation, temperature, and soil water reserves, bioclimatic factors also affect fig tree growth, crop yield, and quality. Fig trees are adapted to a wide range of climates from dry to humid. The initial and main distribution areas are in infra- and meso-Mediterranean bioclimates, which are characterized by dry summers with high temperatures. Some climatic indexes linked with vegetation are used in understanding the spreading of different species. For optimum fig yields and conditions, the simple thermicity index, which is based on mean and extreme temperatures, was 18–22, and the annual ombrothermic index based on precipitation and temperature was over 2.5 (Ighbareyeh *et al.*, 2018; López Fernández and López Fernández, 2018).

In general, fig trees can have good vegetative growth under suboptimal soil conditions that are accepted as limiting for most fruit species; however, the yield, but mainly the fruit quality, is highly correlated with the nutritional status of the tree. According to Proebsting and Warner (1954), nitrogen, phosphorus, potassium, calcium, and magnesium are important in terms of plant growth, yield, and fruit quality. In comparison with other deciduous tree crops growing in the same regions, seasonal variations in fig leaf nutrient concentrations are similar to those of other tree crops (Brown, 1994).

Although fig trees may grow on all soil types, even in rocky areas, soil depth accelerates the vegetative growth of fig trees and thus enhances the yield and quality of the fruits. On slopes threatened by erosion, a low soil depth and nitrogen leaching lead to a reduction in tree vigor. Fig trees are not selective in terms of soil properties, but for successful commercial production, deep soils with good drainage, optimum organic matter, and calcium are recommended to obtain high-quality products. Fig trees are sensitive to high sodium and boron levels in the soil, but tolerant of high sodium levels in the soil solution (G. Aydin and S. Seferoğlu, 1999, unpublished data). Fig is resistant to moderate salinity, but there are differences among the varieties in terms of salt resistance. However, a high and/or fluctuating groundwater level is important while establishing a fig orchard, as it negatively affects fruit quality.

Although the use of geothermal sources and energy is considered one of the least environmentally polluting, it may have some negative effects on the environment. Geothermal waters contain more dissolved chemicals and heavy metals than drinking water because these waters are hot and have high salt concentrations, which facilitate the dissolution of all the elements in the soil. Water discharged from geothermal plants is a potential thermal pollutant, as it has a higher temperature than natural water sources. It contains large amounts of carbonate, silica, sulfate, and chloride, along with chemical pollutants such as lithium, boron, arsenic, fluoride, hydrogen sulfide, mercury, lead, zinc, and ammonia. Because of the marked effects of geothermal power plants and/or thermal springs on tree growth, fruit yield, and quality of the figs (S. Dag, 2015, unpublished data; Yilmaz and Kaptan, 2017), analysis of groundwater resources is vitally important during site selection.

Soil fertility and nutritional balance are important for tree vigor, yield, and fresh- and dried-fruit quality; therefore, prior to establishment, the nutritional condition of the site must be analyzed in detail. Depending on the variety, nitrogen is the key element as it affects shoot growth and overall tree vigor (U. Aksoy, 1987, unpublished data). For common-type varieties grown under frost-free conditions, longer shoots favored by nitrogen mean more fruit bearing and, if the heat summation is adequate, higher yields. In varieties destined for sun drying, excess nitrogen may have an adverse effect on dried-fruit quality as it promotes higher moisture content in the fresh fruit leading to longer drying times and a reduction in quality. The research work on Sarılop has shown that under excess nitrogen conditions, dried-fruit size, color, and the ratio of cracked fruits increases. Potassium and calcium and their ratio are also important regarding ostiole-end cracks or splits and sunscald, both accepted as defects in severe cases. As the potassium level was elevated, splitting was promoted, whereas the reverse happened with calcium, and as the calcium increased, the ratio of sun-scalded fruit escalated.

7.4 Establishing the Orchard

7.4.1 Site planning and preparation

Site planning and preparation should be accepted as the fundamental first step of orchard management to maximize production of high-quality fruit and increase grower profitability. It constitutes a base for all future farming practices.

Site preparation requires significant planning and effort in order to grow healthy trees with remarkable fruit quality. It is of great importance to map the site and to make a risk assessment by identifying possible hazards. A risk-management strategy to cope with future risks and proper solutions should be decided and written down, even in very small family farms. Additionally, it is highly important to get preliminary information about the ecophysiological and phenological features, and the ecological requirements of fig trees and the variety to be selected.

In orchard planning, some basic points should be considered. In designing the orchard, roads or alleys are needed to help move harvested fruit or for moving of trucks. Buildings, shelters, and storage areas must first be identified according to the available infrastructure. Suitable social areas equipped with clean running water, toilets, and a hall, must be designed for workers where they can have their meals or have breaks. Storage areas must allow separate storage of both the inputs and unused containers and packaging material from irrigation and fertigation tanks and pumps. The harvested fruit, even if not packed at the farm, goes through a cleaning and sorting process, which also requires a clean and hygienic handling area. For composting of waste on the farm, a composting area should be established away from the main buildings to prevent any unpleasant odors. Where figs are going to be sun dried, a suitable and adequate space must be allocated as a sun-drying area or to establish drying tunnels in dried-fig production. After they are fully dried, figs need to be stored in rooms with protection against stored-product pests. As well as the production plot, it is also of great importance to implement effective environmental management around the orchard and to make the landscaping fit any regional peculiarities.

Optimum distances should be determined according to the slope of the land, soil conditions, and training system to accommodate the maximum number of trees per unit area. However, no space is required for pollinator male fig plants within the edible female fig orchard because male (caprifig) trees are recommended to be planted in separate orchards or locations. If the area to be planted is sloped, planting along the contour lines should be ensured because slopes cause asymmetric root architecture and modification of root morphology (Chiatante *et al.*, 2003), and the soil composition and nutrition levels may vary due to erosion and leaching. After site and variety selection, some preliminary operations such as cleaning the site, deep plowing, leveling and manuring must be done according to the determined planting and management systems. Besides these usual and common horticultural practices, some important operations and processes such as investigating the history of the site, knowing about antagonist effects of previous crops, taking preventive measures for erosion particularly for sloping areas, and environmental management practices are crucial for successful fig growing.

It is important to know the previous use of the site where the fig orchard will be established in case of diseases. The most critical ones are *Rosellinia* root rot (caused by *Rosellinia necatrix*) and *Armillaria* root rot (caused by *Armillaria mellea*) (Papachatzis *et al.*, 2008), which have spread all over the world as root pathogens of forest (Anselmi and Giorcelli, 1990) and orchard (Arora, 1986) trees. Thus, investigating the surrounding environment and neighboring farms is important. When fig orchards are established near forests, these root rots could be of significant importance. As it is not possible to completely eradicate these diseases, it is inevitable that the saplings will be damaged after a few years when the roots reach infected zones. Besides root rots, the site may be investigated for weed density, possible host plants and vectors of prominent pests or diseases, possible pollution, and contamination foci.

During the preparation of the land for planting, some issues need to be carefully considered. As many other commercial crops cannot be successfully grown in degraded marginal areas, fig plantations shifted to these areas due to comparatively greater economic returns. In many fig-producing countries, fig plantations are on high altitudes or slopes that are prone to erosion. For this reason, erosion control measures are quite important. It is known that overland flow due to steep slopes causes erosion, and both flow and erosion are firmly tied to land properties, surface conditions, and soil-management practices (Biddoccu *et al.*, 2017; Ouyang *et al.*, 2018). Conventional short-term soil-management practices have negative effects on land degradation and reduce fig yields by triggering erosion.

Applying sustainable soil-management practices as minimum or no-tillage, cover crops, or strips between fig tree rows may help to avoid runoff, nutrient leaching, and floods (Telak *et al.*, 2020). When planting on slopes, various soil-conservation techniques such as ditches constructed along the contour lines or different types of terracing according to the expected function must be applied following expert recommendation. Additionally, some sustainable management practices such as contour or triangular planting, conservation agriculture (no-till) or minimum tillage, use of cover crops and/or intercrops, or mulching contribute highly to alleviate erosion. However, the possible side effects of the cover crop and intercropping on fig trees regarding water and nutrient competition and hosting pests and diseases must be investigated in detail. Intercropping or agroforestry with an appropriate species is a recommended practice in organic fruit growing. Besides increasing biological diversity, it also provides financial support to producers, especially during the juvenile period.

Before planting, irrigation water quality must be analyzed, and irrigation water should be free of any contaminants and be of acceptable quality. If soil analysis finds any deficiencies in terms of organic matter and nutrients, a strategy should be implemented to overcome this. Potential weeds should be determined in advance, and an appropriate weed-management system should be decided. Green manuring before planting is important in terms of enriching the soil with nitrogen and improving the soil structure. When orchards are on marginal, poor, and sloping areas, soil organic matter content could be very limited, so increasing the organic matter before planting and during the juvenile period is important to maintain long-term soil fertility.

According to the properties of the terrain, the training system and the crop to be marketed (brebas, dried or fresh), additional soil tillage practices may be required, such as deep soil tillage in case of a hardpan, or solving the drainage problem if the groundwater table increases up to 1 m during the rainy season. Furthermore, necessary precautions could be taken if previous factors have antagonistic effects, such as selection and installation of the correct irrigation system (subsurface or above ground), elimination of any buildings blocking the sunlight, selection of appropriate species for windbreaks if there is a wind problem, and timely preparation of the necessary infrastructure and facilities.

7.4.2 Planting density and orchard design

Planting distance, which is the vertical distance between any two trees varies, depends on many factors such as the production system, topography, growth tendency and ecological requisition of the variety, rainfall and irrigation, soil type and fertility, the training system, and open fields or the presence of protection.

Fig can be grown in very different climatic conditions under different training and growing systems. According to Flaishman *et al.* (2008), common-type figs can be grown in hot desert and tropical areas in Israel and Brazil as everbearing because of their low chilling requirement. However, fig can be grown successfully in Mediterranean and temperate climate conditions. Because of the different shoot growth tendencies, different training and pruning systems can be applied according to the tree vigor and growth habits.

During planting, the growth habit, soil characteristics, and the maintenance measures to be applied play an important role in determining the distance between and within the rows. Fig varieties differ in vigor, which is partly related to the strength of the apical dominance. Although both are Smyrna-type fig varieties, Bursa Black develops a strong framework and crown, whereas Sarılop shoots have a drooping growth habit. In contrast, upright and strong growth is observed particularly in common-type fig varieties such as Noire de Coromb and Brown Turkey (U. Aksoy, 1994, unpublished data; Erez *et al.*, 2003; Can, 2008) and also in San Pedro-type varieties as Nazareth (Puebla *et al.*, 2003) because of the shoot-tip dominance. As well as the tree vigor of the fig variety, the training system, topographic conditions, soil properties, and climatic factors are the most important factors to be taken into consideration. The planting method is of great importance, especially in sloping terrain, and the recommended planting method is contour planting, which is defined as using ridges and furrows formed by tillage along the contour lines (Anon., 2007). Contour planting, which is one of the most important soil-conservation management practices, is needed to reduce soil erosion (Long *et al.*, 2010). Triangular planting methods are also quite advantageous in sloping lands in terms of erosion control and biological sustainability. Square and rectangular planting patterns should be avoided on slopes; these can be used in high-density planting orchards on lowlands.

The main determinant of planting density and orchard design is the ultimate canopy size of the trees. In favorable areas, larger distances are required depending on the tree vigor. Spacing between the trees should allow interception of the highest amount of radiant energy as photosynthetically active radiation, although the light-use efficiency of the plants, mainly related to the training system, tree spacing, and planting design, have to be considered. Several orchard designs such as square, offset square, rectangle, and quincunx can be used while establishing a fig orchard, depending on site direction and environmental conditions. The square and rectangle designs allow the same amount of space for the trees, and cultural practices are easily carried out. In the offset square and quincunx designs, the light interception is better, but cultural practices are relatively demanding. Moreover, to reduce the erosion risk in sloping areas, the offset square and quincunx designs should be used with additional cover cropping and mulching to alleviate surface-water flow and soil loss. Determining the plant density per unit area and planting design are mainly important to optimize yield and maximize the fruit quality. Thus, photosynthesis is the key factor for optimum yield and maximum quality. When global climate change is considered, water-use efficiency (WUE), which is the ratio of photosynthesis to transpiration per unit area (Condon and Hall, 1997), stands out as particularly critical for fig trees, which can be grown to add value to marginal areas. To determine the crop performance from leaf to orchard levels, WUE should be considered instead of photosynthesis (Gimenez *et al.*, 1997). However, photoinhibition has adverse effects on photosynthetic carbon metabolism. Besides excessive radiation, other environmental conditions such as high temperatures and water deficit may strongly increase photoinhibition in such marginal conditions (Powles, 1984; Long *et al.*, 1994). According to Can and Aksoy (2007), photosynthesis is probably decreased by both stomatal and non-stomatal resistance mechanisms such as photoinhibition, and the fig tree appears to be a shade-tolerant plant. In contrast, in Tunisia, due to the prevailing climatic conditions, fruit yield and quality was higher in sun-exposed sectors of the canopy than in the shaded canopy parts, and the microclimate inside the canopy was different from that in the periphery (Trad *et al.*, 2013). Therefore, environmental and light conditions must be carefully evaluated when determining planting distances and design.

All gas-exchange parameters and photoinhibition are important in terms of the fig plant's ecophysiology, particularly at the leaf level. At the canopy level or for wider areas such as orchards, the importance of the leaf area index (LAI), calculated as the vertically integrated surface of leaves per unit of ground area (Cournac *et al.*, 2002), cannot be ignored. LAI and WUE are the main stress indicators not only for individual trees but also for wider areas (Waring, 1985; Condon and Hall, 1997). Thus, LAI and especially canopy structure and light-use efficiency should be considered to determine the plant density and design (Norman and Campbell, 1989; Norman and Arkebauer, 1991). Pruning applications such as cutting two-thirds of the fig shoots caused increases in LAI values during the season. The LAI is closely related to the architectural structure of the crown, as well as the total photosynthetically active leaf area (H. Can, 2003, unpublished data).

In recent decades, new growing techniques such as high-density training systems, which provide high fruit yield per area and high quality, have been put into practice in fig growing. As there is no dwarf type or dwarfing rootstocks used commercially in figs, training and pruning are the only tools that can be used to restrict growth and obtain stunted plants, as occurs in olives (Rallo *et al.*, 2013). In Turkey, old Sarılop orchards located on alluvial soils of lowlands were established as 10 × 10 m between and within rows. A 5 × 5 m density is used on slopes where soils are shallow and comparatively yield poor vegetative growth (Aksoy and Anaç, 1993). Planting can be practiced as 5 × 5 m spacing for weak-growing breba fig orchards, and as 6–7 m for vigorous varieties, as the optimum densities.

In intensive orchards, 2–3 m spacing is possible only when vegetative growth and fruit bearing are controlled with pruning practices. The spacing may vary between 1.8 and 7.5 m depending on the variety, soil fertility, and management of the tree architecture (Fig 7.1). Vase training systems are commonly used in fig production; however, a modified open-vase system is used in California (Ferguson *et al.*, 1990), and a V-shaped training system under nets has been used for relatively higher-density plantings (2 × 5 m) in Israel (Maimon, 1998; Erez *et al.*, 2003). Some stub training and pruning systems have also been used in fig production in Israel (Flaishman *et al.*, 2008), Latin America (Botti and Muchnik, 1998; Mendoza-Castillo *et al.*, 2017), and China (Wang *et al.*, 2003) for both the first and second crops. In contrast, a different straight-line training system is used in Japan in glasshouse and open-field conditions (Yamakura *et al.*, 2008a,b; Hosomi *et al.*, 2015). In Malaysia, intensive fig production was conducted under both glasshouse and open-field conditions successfully with a different type of pruning (Shamin-Shazwan *et al.*, 2019). As well as different training systems, some high-density growing systems are used under soil-less greenhouse conditions with a density ranging from 26,000 to 34,000 plants ha^{-1} in Spain (Melgarejo *et al.*, 2007). According to Norberto *et al.* (2018), Roxo de Valinhos fig trees planted at a spacing of 1 × 2.5 m with a density of 4000 plants ha^{-1} provide higher productivity per area. In Spain, at 217 m above sea level, a spacing of 5 × 2 m providing a density of 1000 plants ha^{-1} was tried with some fig varieties under a drip-irrigation system. Among the tested varieties, Calabacita was the most interesting variety with high fruit yield and high-quality fruits. Additionally, easy harvesting and high sanitary quality dried figs could be obtained due to a lower risk of contamination (Galván *et al.*, 2019). In India, trees of the cultivar Deanna and variety Poona were planted under densities of 1000, 800, 666, 571, and 500 plants ha^{-1}, and high-density planting with 1000 or 800 plants ha^{-1} was superior to other planting densities for maintaining tree vigor and fruit yield under continuous cropping. With a decrease in tree density, growth and gas-exchange parameters declined gradually, but photosynthetically active radiation interception by the canopy did not differ significantly among the treatments (Chithiraichelvan *et al.*, 2017).

7.4.3 Planting out

Planting is basically thought as a digging and refilling a pit in the ground; however, some important points should be considered. Pits must be scooped out when the ground is dry, especially if the soil is clay rich, approximately 1 month

Fig. 7.1. A high-density fig orchard. A. Orchard establishment. B. A 5-year-old intensive orchard. Figure used with the permission of R. Peer.

before plantation and should remain empty at least for 10 days for exposure to the sunlight to kill any soil-borne pathogens. Polyethylene covers can be used to increase the solarization effect. Fungicide and insecticide sprays also can be applied to protect the root system. The pits must be filled with well-structured soil and well-composted farmyard manure or fertilizers. It is of great importance to protect the trunk of the saplings from sun or any other mechanical damage by wrapping them in different protector materials after planting. Wrapping provides protection from sunscald and sunburn damage (Harris *et al.*, 2004; Ophardt and Hummel, 2016). Especially in recent years, parallel with climate change, sunburn has been found to occur on the trunks, even if there is no visible damage, and diseases can enter through small cracks caused by excessive dehydration. Wrapping of the trunk can also be done for purposes such as providing earliness with ethylene release or protection from various pests and some wild animals. Another method that can be applied to provide earliness is a stone pavement of the root zone after planting.

7.5 Concluding Remarks

Orchard establishment and previous stages of collecting information for planning are of great importance in fig cultivation because of the variation in geographical and climatic conditions. Additionally, fig is being introduced to quite different environmental conditions and may be exposed to different cultivation systems and consumption purposes. As well as the natural phenotypic variations among fig varieties, different training and cultivation systems and growing environments have great importance in terms of appropriate site selection and plantation systems in fig orchards.

Demand for fig fruit continues to increase in markets and creates good economic revenues, and therefore fig production is increasing and constantly shifting to new production regions. Recently, many producers are choosing to grow figs in high-density systems. For this reason, fig production is undergoing a significant change from non-irrigated traditional farming to super-intensive soil-less greenhouse conditions, creating windows of opportunity in different regions of the world.

Many biotic and abiotic environmental stress factors may seriously threaten fig cultivation. Furthermore, climate crisis on a global scale may multiply the size of the environmental risks. The dimensions of the damage caused by constantly increasing environmental threats can be reduced by appropriate site and variety selection. In addition, implementing good practice management designed for the site-specific conditions is needed.

References

Aksoy, U. (1981) Studies on fruit growth, ripening and storage of Akca, Goklop and Sarilop fig varieties. *Journal of Ege University Faculty of Agriculture* 20, 241–246.

Aksoy, U. and Anaç, D. (1993) Determination of optimum soil and plant nutrients for higher yield and quality in fig production. In: Fragoso, M.A.C. and van Beusichem, M.L. (eds) *Optimization of Plant Nutrition*. Kluwer Academic Publishers, Dordrecht, The Netherlands, pp. 309–313.

Aksoy, U. and Can, H.Z. (2012) Expected effects of climate change on organic agriculture in Turkey. *New Medit* 11, 3–11.

Alayunt, F., Çakmak, B., and Can, H.Z. (1997a) Friction and rolling resistance coefficients of fig. *Acta Horticulturae* 480, 301–304.

Alayunt, F., Çakmak, B., Can, H.Z., and Aksoy, U. (1997b) Vibration damage trial on some fig cultivars. *Acta Horticulturae* 480, 305–310.

Anon. (2007) Conservation practice standard, contour farming. United States Department of Agriculture, Natural Resources Conservation Service. Code 330.

Anon. (2016) *Key Indicators of the Labour Market*. 9th edn. International Labour Office, Geneva.

Anon. (2019) Intergovernmental Panel on Climate Change Report. IPCC, Geneva.

Anselmi, N. and Giorcelli, A. (1990) Factors influencing the incidence of *Rosellinia necatrix* Prill. in poplars. *European Journal of Forest Pathology* 20, 175–183.

Arora, D. (1986) *Mushrooms Demystified*, 2nd edn. Ten Speed Press, Berkeley, California.

Ausubel, J.H., Wernick, I.K., and Waggoner, P.E. (2013) Peak farmland and the prospect for land sparing. *Population and Development Review* 38, 221–242.

Biddoccu, M., Ferraris, S., Pitacco, Cavallo, A., and Temporal, E. (2017) Variability of soil management effects on soil hydrological properties, runoff and erosion at the field scale in a hillslope vineyard, North-West Italy. *Soil and Tillage Research* 165, 46–48.

Botti, C. and Muchnik, E. (1998) Fig production in Latin America. *Acta Horticulturae* 480, 37–42.

Botti, C., Franck, N., Prat, L., Ioannidis, D. and Morales, B. (2003) The effect of climatic conditions on fresh fig fruit yield, quality and type of crop. *Acta Horticulturae* 605, 37–44.

Brown, P.H. (1994) Seasonal variations in fig (*Ficus carica* L.) leaf nutrient concentrations. *Hortscience* 29, 871–873.

Çakmak, B., Alayunt, F.N., Akdeniz, R.C., Aksoy, U. and Can, H.Z. (2010) Assessment of the quality losses of fresh fig fruits during transportation. *Tarım Bilimleri Dergisi: Journal of Agricultural Sciences* 16, 180–193.

Can, H.Z. (2008) Estimation of leaf area index in some breba fig cultivars. *Acta Horticulturae* 798, 109–115.

Can, H.Z. and Aksoy, U. (2007) Seasonal and diurnal photosynthetic behaviour of fig (*Ficus carica* L.) under semi-arid climatic conditions. *Acta Agriculturae Scandinavica Section B: Soil and Plant Science* 57, 297–306.

Can, H.Z., Hepaksoy, S., Aksoy, U., and Kutlu, E. (1998) Leaf characteristics and net gas exchange of fig cultigens adapted to different climatic zones. *Acta Horticulturae* 516, 131–138.

Çelikel, F.G. and Karacali, I. (1992) Effect of harvest maturity and precooling on fruit quality and shelf life of the fig variety "Bursa Siyahi". *Proceedings of the 1st Turkish National Horticultural Congress* 1, 437–441.

Charman, P.E.V. and Roper, M.M. (2000) Soil organic matter (SOM). In: Charman P.E.V. and Murphy, B.W. (eds) *Soils: Their Properties and Management*. Oxford University Press, Oxford, UK, pp. 260–270.

Chessa, I. (1997) Fig. In: Mitra, S. (ed.) *Postharvest Physiology and Storage of Tropical and Subtropical Fruits*. CAB International, Wallingford, UK, pp. 245–268.

Chiatante, D., Scippa, S.G., Di Lorio, A., and Sarnataro, M. (2003) The influence of steep slopes on root system development. *Journal of Plant Growth Regulation* 21, 247–260.

Chithiraichelvan, R., Kurian, R.M., Awachare, C.M., and Laxman, R.H. (2017) Performance of fig (*Ficus carica* L.) under different planting densities. *International Journal of Current Microbiology and Applied Sciences* 6, 2603–2610.

Condit, L.J. (1947) *The Fig*. Chronica Botanica, Waltham, Massachusetts.

Condit, L.J. (1955) Fig varieties: a monograph. *Hilgardia* 23, 323–538.

Condon, A.G. and Hall, A.E. (1997). Adaptation to diverse environments: variation in water use efficiency within crop species. In: Jackson, L.E. (ed.) *Ecology in Agriculture*. Academic Press, USA, pp. 79–116.

Cournac, L., Dubois, M.A., Chave, J., and Riera, B. (2002) Fast determination of light availability and leaf area index in tropical forests. *Journal of Tropical Ecology* 18, 295–302.

Duyar, E. (1998) Impact of ecological changes on fig plantations in Big Menderes (Meander) basin. *Acta Horticulturae* 480, 311–316.

Erez, A., Flaishman, M., and Yablowitz, Z. (2003) Breba production under net: effect of training system. *Acta Horticulturae* 605, 119–123.

Ferguson, L., Michailides, J.T., and Shorey, H.H. (1990) The California fig industry. *Horticultural Reviews* 12, 409–490.

Fischer, G. and Orduz-Rodríguez, J.O. (2012) Ecofisiología en frutales. In: Fischer, G. (ed.) *Manual Para El Cultivo De Frutales En El Trópico*. Produmedios, Bogotá, Colombia, pp. 54–72.

Flaishman, M.A. and Al Hadi, F.A. (2002) *Fig growth in Israel. Alon Hanotea* 56, 56–57 (in Hebrew).

Flaishman, M.A., Rover, V., and Stover, E. (2008) The fig: botany, horticulture, and breeding. *Horticultural Reviews* 34, 113–197.

Galván, A.I., Pérez-Gragera, F., Serradilla, M.J., Córdoba, M.G., Domínguez, G., and López-Corrales, M. (2019) Agronomic behaviour and fruit quality attributes of three dried-fig varieties in super high density system. *Acta Horticulturae* 1310, 217–222.

Gimenez, C., Orgaz, F., and Fereres, E. (1997) Productivity in water-limited environments: dryland agricultural systems. In: Jackson, L.E. (ed.) *Ecology in Agriculture*. Academic Press, USA, pp. 117–143.

Golombek, S.D. and Lüdders, P. (1990) Effects of short-term salinity on leaf gas exchange of the fig (*Ficus carica* L.). *Plant and Soil* 148, 21–27.

Harris, R.W., Clark, J.R., and Matheny, N.P. (2004) *Arboriculture: Integrated Management of Landscape Trees, Shrubs, and Vines*, 4th edn. Prentice Hall, Upper Saddle River, New Jersey.

Hosomi, A., Isobe, T., and Miwa, Y. (2015) Shoot growth and fruit production of the 'Masui Dauphine' variety of fig (*Ficus carica* L.) undergoing renewal long pruning, *Japan Agricultural Research Quarterly* 49, 391–397.

Ighbareyeh, J., Suliemieh, A., Ighbareyeh, M., Daraweesh. A.-Q., Oritz, A.C., and Carmona, E.C. (2018) Impact of bioclimatic and climatic factors on *Ficus carica* L. yield: increasing the economy and maintaining the food security of Jerusalem in Palestine. *Transylvanian Review* 26, 1221–1249.

Küden, A.B. and Tanriver, E. (1998) Plant genetic resources and selection studies on figs in the east Mediterranean and southeast Anatolia regions. *Acta Horticulturae* 480, 49–54.

Kutlu, E., Can, H.Z., Aksoy, U., and Hepaksoy, S. (1998) Evaluation of gas exchange capacity and physiological responses of selected Sarilop (=Calimyrna) fig clones. *Acta Horticulturae* 517, 59–64.

Lansky, E.P. and Paavilainen, H.M. (2011) *Figs: The Genus Ficus*. CRC Press, Boca Raton, Florida.

León-Sánchez, L., Nicolás, E., Nortes, P.A., Maestre, F.T. and Querejeta, J.I. (2016) Photosynthesis and growth reduction with warming are driven by nonstomatal limitations in a Mediterranean semiarid shrub. *Ecology and Evolution* 6, 2725–2738.

Long, D.S., Wuest, S.B., Williams, J.D., Rauwendaal, R., and Bailey, M.J. (2010) Contour planting: a strategy to reduce soil erosion on steep slopes. In: *Proceedings of the 10th International Conference on Precision Agriculture, July, Denver, Colorado*, pp. 18–21.

Long, S.P., Humphries, S., and Falkowski, P.G. (1994) Photoinhibition of photosynthesis in nature. *Annual Review Plant Physiology Plant Molecular Biology* 45, 633–662.

López Fernández, M.L. and López Fernández, M.S. (2018) Worldwide bioclimatology manual and guide. *Naturaleza en Hispania* 5, 1–157.

Maas, E.V. (1993) Testing crops for salinity tolerance. In: Maranville, J.W., Baligar, B.V., Duncan, R.R., and Yohe, J.M. (eds) *Proceedings of a Workshop on Adaptation of Plants to Soil Stress*. University of Nebraska, Lincoln, Nebraska, pp. 234–247.

Maimon, A. (1998) The optimal density and fig tree (*Ficus carica* L.) pruning for a breba crop in Israel. *Acta Horticulturae* 480, 133–136.

Melgarejo, P., Martínez, J.J., Hernández, F., Salazar, D.M., and Martínez, R. (2007) Preliminary results on fig soilless culture. *Scientia Horticulturae* 111, 255–259.

Mendoza-Castillo, V.M., Vargas-Canales, J.M., Calderon-Zavala, G., Mendoza-Castillo, M.D.C., and Santacruz-Varela, A. (2017) Intensive production systems of fig (*Ficus carica* L.) under greenhouse conditions. *Experimental Agriculture* 53, 339–350.

Mittler, R. (2006) Abiotic stress, the field environment and stress combination. *Trends in Plant Science* 11, 15–19.

Norberto, P.M., Alvarenga, A.A., de Andrade, J.C., Rodrigues, F.A., Pereira, L.V., and Gonçalves, E.D. (2018) Cultivation of "Roxo de Valinhos" fig tree in different plant densities for production of green figs for industry in the Region of Campo Das Vertentes-MG. *Agricultural Sciences* 9, 1097–1106.

Norman, J.M. and Arkebauer, T.J. (1991) Predicting canopy photosynthesis and light use efficiency from leaf characteristics. In: Boote, K.J. and Loomis, R.S. (eds) *Modelling Crop Photosynthesis from Biochemistry to Canopy*. CSSA, Madison, Wisconsin, pp 75–94.

Norman, J.M. and Campbell, G.S. (1989) Canopy structure. In: Pearcy, R.W., Ehleringer, J., Mooney, H.A., and Rundel, P.W. (eds) *Plant Physiological Ecology: Field Methods and Instrumentation*. Chapman & Hall, New York, pp. 301–325.

Ophardt, M.C. and Hummel, R.L. (2016) Environmental injury: sunscald and sunburn on trees. *Washington State University Research and Extension Center Bulletin* FS197E, 2–6.

Ouyang, W., Wu, Y., Hao, Z., Zhang, Q., Bu, Q., and Gao, X. (2018) Combined impacts of land use and soil property changes on soil erosion in a mollisol area under long-term agricultural development. *Science of the Total Environment* 613, 298–809.

Papachatzis, A., Eliopoulos, P., Statharas, G., and Vagelas, I. (2008) *Ficus carica* root rot disease caused by *Armillaria mellea* and *Rosellinia necatrix* in Greece. *University of Craiova* 13, 143–148.

Powles, S.B. (1984) Photoinhibition of photosynthesis induced by visible light. *Annual Review of Plant Physiology* 35, 15–44.

Proebsting, E.L. and Warner, R.M. (1954) The effect of fertilizers on yield, quality and leaf composition of figs. *Proceedings of the American Society for Horticultural Science* 63, 10–18.

Puebla, M., Toribio, F., and Montes, P. (2003) Determination of fruit bearing pruning date and cutting intensity in "San Pedro" (*Ficus carica* L.) type fig cultivars. *Acta Horticulturae* 605, 147–157.

Rallo, L., Diego, B., Castro-García, S., Connor, D.J., Gomez-del-Campo, M., and Rallo, P. (2013) High-density olive plantations. *Horticultural Reviews* 41, 303–384.

Ramírez, F. and Kallarackal, J. (2015) *Responses of Fruit Trees to Global Climate Change*. Springer Briefs in Plant Science Book Series. Springer, New York.
Seidel, P. (2016) Impacts of extreme weather events on pests, damage caused by pests and plant protection measures – first evidence. *Journal of Kulturplanzen* 68, 253–269.
Şen, F., Aksoy, U., Özer, K.B., Can, H.Z., and Konak, R. (2019) Effect of altitudes on physical and chemical properties of sun-dried fig (*Ficus carica* cv. Sarılop) fruit. *Acta Horticulturae* 1310, 149–156.
Shamin-Shazwan, K., Shahari, R., Aini Che Amri, C.N., and Tajuddin, N.S.M. (2019) Figs (*Ficus carica* L.): cultivation method and production based in Malaysia. *Engineering Heritage Journal (GWK)* 3, 6–8.
Smith, P., Martino, D., Cai, Z., Gwary, D., Janzen, H., *et al.* (2007) Agriculture. In: Metz, B., Davidson, O.R., Bosch, P.R., Dave, R., and Meyer, L.A. (eds) *Climate Change 2007: Working Group III: Mitigation of Climate Change*. Cambridge University Press, Cambridge, UK/New York.
Telak, L.J., Pereira, P., Ferreira, C.S.S., Filipovic, V., Filipovic, L., and Bogunovic, I. (2020) Short-term impact of tillage on soil and the hydrological response within a fig (*Ficus carica*) orchard in Croatia. *Water* 12, 3295–3310.
Trad, M., Gaaliche, B., Renard, C.M.G.C., and Mars, M. (2013) Inter- and intra-tree variability in quality of figs: influence of altitude, leaf area and fruit position in the canopy, *Scientia Horticulturae* 162, 49–54.
Wang, L., Jiang, W., Ma, K., Ling, Z., and Wang, Y. (2003) The production and research of fig (*Ficus carica* L.) in China. *Acta Horticulturae* 605, 191–196.
Waring, R.H. (1985) Imbalanced forest ecosystems: assessments and consequences. *Forest Ecology and Management* 12, 93–112.
Yablowitz, Z., Nir, G., and Erez, A. (1998) Breba fig production in Israel: regular and pesticide-free systems. *Acta Horticulturae* 480, 137–141.
Yamakura, T., Hosomi, A., and Hirayama, D. (2008a) Effect of tree spacing on vegetative growth and reproduction in an early growth stage in two cultivars of *Ficus carica* L. *Journal of the Japanese Society for Horticultural Science* 77, 7–16.
Yamakura, T., Hosomi, A., and Hirayama, D. (2008b) Analysis of aboveground vegetative growth by using the logistic theory of the density effect in young monocultures of *Ficus carica* L. *Journal of the Japanese Society for Horticultural Science* 77, 17–23.
Yilmaz, E. and Kaptan, M.A. (2017) Environmental impact of geothermal power plants in Aydın, Turkey. In: *E3S Web of Conferences, 23 October*. Les Ulis, Cedex A, France.

8 Soil Management, Fertilization, and Irrigation

Mehmet Eşref İrget[1]* and Mehmet Kamil Meriç[2]
[1]Department of Soil Science and Plant Nutrition, Faculty of Agriculture, Ege University, İzmir, Turkey; [2]Bergama Vocational School, Ege University, İzmir, Turkey

8.1. Introduction

Until the beginning of the 1950s, fig was assumed to be a crop tolerant to poor soils and one that did not need to be fertilized (Chandler, 1947; Ülkümen *et al.*, 1948). In this respect, Chandler (1947) reported that fig trees preferred well-aerated soils, although they tolerated compacted soils and, unlike peach trees, performed quite well in soils low in available nitrogen (N). It was also stated that N fertilization might result in deterioration of the fruit quality causing splitting and souring and may increase the sensitivity to cold damage (Condit, 1947). Ultimately, the nutrition of figs was neglected, and thus a limited number of publications were released (Condit and Enderud, 1956). Additionally, fig might have not been among the research priorities of the major fig-producing countries located in the Mediterranean Basin. These reasons might explain why fig did not initially occur in major reference textbooks on fruit nutrition (Childers, 1966).

The first studies on fig nutrition and fertilization were conducted by Proebsting and Tate (1952) and Proebsting and Warner (1954) in the USA, followed by Hirai *et al.* (1964), Hirai (1966), and Dekazos and Crane (1966). Research on fig gained momentum in the 1980s and 1990s. During this period, studies on seasonal variation of nutrients (Kabasakal, 1983; Gaşgil, 1993; Brown, 1994; Ersoy *et al.*, 2003) and the nutritional status of fig orchards (Aksoy *et al.*, 1987a,b; Anaç *et al.*, 1987; Eryüce *et al.*, 1987; Anaç *et al.*, 1992; Brown, 1994; Düzbastılar and Güleç, 1995; Aşkın *et al.*, 1998; Moreno *et al.*, 1998; Elmacı *et al.*, 2010) originated mostly from Turkey due to the economic importance of fig culture. Fertilizer trials (Proebsting and Warner, 1954; El-Kassas, 1975; Hernandez *et al.*, 1994a; Eryüce *et al.*, 1995; Hakerlerler *et al.*, 1999; İrget *et al.*, 1999, 2008; Tepecik, 2010; Mordoğan *et al.*, 2013; Ertan *et al.*, 2017; Sotiropoulos *et al.*, 2020) were carried out chiefly on the major fig varieties such as Sarılop (= Calimyrna). The nutritional status and fertilizer requirement of fig can be evaluated by: (i) visual diagnosis; (ii) leaf analysis; (iii) soil analysis; and (iv) a combination of nutrition, vegetative growth, yield, and quality relationships. These are discussed in detail in this chapter.

The basic principles of fig irrigation are not fundamentally different from those of other deciduous fruit trees, and so very few studies were carried out on fig trees. Those that give guidance to fig production and help our understanding of

* Email: esref.irget@ege.edu.tr

 Advances in Fig Research and Sustainable Production (M.A. Flaishman and U. Aksoy)
DOI: 10.1079/9781789242492.0008

the relationship with growth and productivity in fig trees are summarized in this chapter.

8.2 Visual Diagnosis

The nutritional status of fig plants can be assessed by visual assessment of the leaf, shoot, and fruit. In general, leaves are the best indicator tissues, reflecting nutrient deficiency and toxicity of plants. Therefore, leaves are used for visual diagnosis purposes as a first step. However, unlike other fruit trees, fig rarely shows visual symptoms of nutrient deficiencies and toxicities through the leaves, excluding salinity stress and boron (B) toxicity (Fig. 8.1A). Despite the absence of distinctive deficiency symptoms, leaf analysis results revealed the existence of various nutrient deficiencies in a survey study conducted with 272 fig orchards in the Aegean Region of Turkey (Elmacı *et al.*, 2010), which indicated that nutrient deficiencies in fig may particularly occur in the form of hidden hunger (i.e. without visual deficiency symptoms), and are reflected by reduced shoot growth, fruit yield, and/or quality. Visual symptoms observed on fruits such as cracking, sunscald, and abnormal sizes can also provide helpful information about the nutritional status of fig trees (Fig. 8.1B, C). Other visual symptoms may involve changes in the length and diameter of the shoots. Variety, tree architecture, N fertilization, and irrigation are important factors affecting shoot length and diameter. Shoot length is evaluated differently across varieties. In this regard, relatively longer shoots are desired for parthenocarpic varieties compared with the Smyrna type. Excessive N fertilization can result in undesired elongation of shoot length, particularly in young Smyrna-type fig trees. To achieve the ideal number of fruits (i.e. four to six fruits per shoot) in Sarılop varieties, shoot elongation should be in the range of 15–20 cm (Obenauf *et al.*, 1978; Ferguson *et al.*, 1990), and N fertilization is recommended when it is less than 15 cm (Kabasakal, 1983).

Fig. 8.1. Visual symptoms on fig fruits and leaves. A. Symptoms of boron toxicity in leaves and fruits of fig (Sarılop variety). Photo courtesy of M. Aydın. B. Cracking/splitting of the fruit. C. Sunscald in fig fruits (Sarılop variety). Photo courtesy of M.E. İrget.

8.3 Leaf Analysis

Sampling time, the position of the leaf on the shoot, and reference values are known to be the critical points. Like other fruit trees, mineral nutrient concentrations in fig leaves may vary throughout the growing season, i.e. decreases in N, phosphorus (P) and potassium (K), and increases in calcium (Ca) and magnesium (Mg). The phenological stages during which concentrations of the majority of nutrients remain stable are defined as the common stable period (Fig. 8.2A) and are recommended as the appropriate time for leaf sampling. Two common stable periods, 10 days before and after the beginning of fruit-ripening stage, and at the end of harvest, were reported for the Sarılop variety by Kabasakal (1983). In Turkey, the first stable period is recommended for leaf sampling, i.e. during the fruit-ripening period of around July 28–August 18, and the leaves at the axil of first basal fruit on the shoot (usually the third or fourth leaves) as the leaf position to be sampled (Fig. 8.2B).

Although the common stable period is the ideal time to determine the general nutritional status of fig orchards, this period coincides with the fruit-ripening period during which there is no chance to make any additional fertilizer applications. This problem can be overcome by early leaf sampling and using correction factors for each specific nutrient. Examples of correction factor graphs for N and K are given in Fig. 8.3.

Leaf samples can be analyzed as the whole leaf (blade + petiole) or as the blade and petiole, separately. It is reported that analyzing leaf samples separately as the blade and petiole can be advantageous and helps to evaluate the relationship between nutritional status, fruit yield, and quality (Kabasakal, 1983; Aksoy *et al.*, 1987b; Elmacı *et al.*, 2010; Tepecik, 2010).

The leaf nutrient content is generally evaluated by comparison with the reference values given in Table 8.1 (Beutel *et al.*, 1978). These are widely accepted reference values, except for N. Studies conducted in different years and locations in Turkey revealed that most of the leaf N values fall below the adequate range of 2.0–2.5%. Considering the yield and quality, 1.7% is recommended as a reference value for an adequate leaf N content for the Sarılop variety destined for sun-drying in the Aegean Region. The differences between the reference values of N are attributed to the differences in soil structure and management practices, and especially to irrigation, for the Sarılop fig variety in California (Aksoy *et al.*, 1987a,b; Anaç *et al.*, 1987, 1992). Brown (1994) proposed 2.1% (July sampling) as the tentative reference value for adequate N in irrigated fig orchards in California. A more detailed evaluation of the reference values (Table 8.1) shows deficiency to excess levels for primary (N, P, K, Ca, and Mg) and some of the secondary (copper (Cu), zinc (Zn), manganese (Mn), and boron (B) plant nutrients (Beutel *et al.*, 1978).

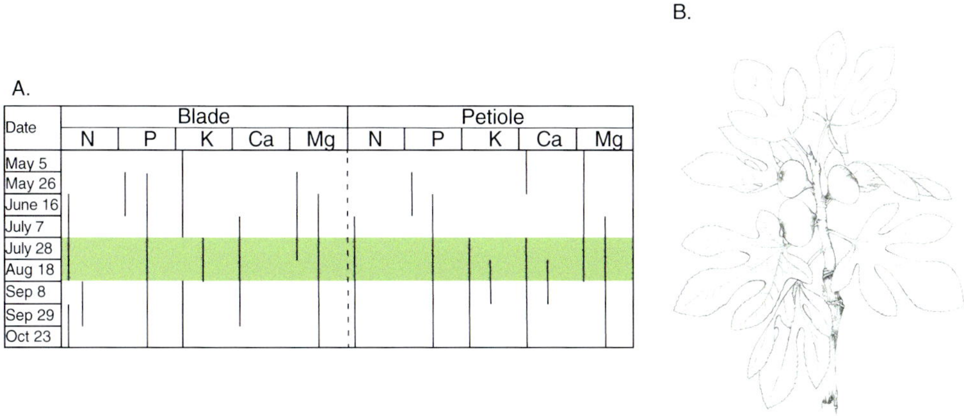

Fig. 8.2. A. Seasonal variation of nutrients in fig leaves during the common stable period. The fruit-ripening period is shaded green. Reproduced from Kabasakal (1983) with permission. B. Leaf sampling position in fig. Photograph courtesy of M.E. İrget.

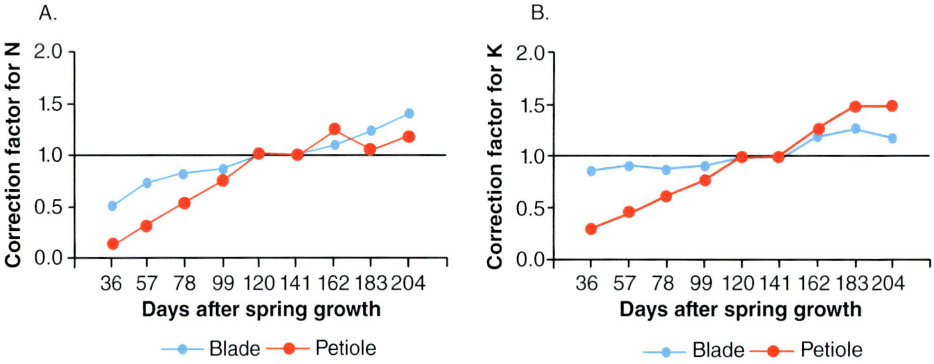

Fig. 8.3. Correction factors for A. nitrogen (N) and B. potassium (K) in different leaf sampling periods. Reproduced from Kabasakal (1983) with permission and courtesy of M.E. İrget.

Table 8.1. Reference values for fig leaf primary and secondary nutrients.

	Definition				
Nutrient	Deficient	Marginal	Adequate	High	Toxic or excessive
Nitrogen (%)	<1.7	1.7–1.9	2.0–2.5	>2.5	
Nitrate nitrogen (%)	<0.4				
Phosphorus (%)			0.1–0.3		
Potassium (%)	<0.7	0.7–1.0	>1.0		
Calcium (%)			>3.0		
Magnesium (%)			>0.75		
Copper (mg kg^{-1})			>4.0		
Zinc (mg kg^{-1})		<15			
Manganese (mg kg^{-1})			>20.0		
Boron (mg kg^{-1})					>300.0

As the parthenocarpic varieties contain higher N than non-parthenocarpic ones (Dekazos and Crane, 1966; Şahin *et al.*, 1999), new variety-specific reference values are required to evaluate fig nutritional status. Table 8.1 does not contain any values for iron (Fe). However, Anaç (1997) stated that 90% of the orchards in the Aegean Region showed a range between 200 and 400 mg kg^{-1}, and thus recommended this range as an optimum level for Fe.

8.4 Mineral Nutrition, Vegetative Growth, Yield, and Quality Relationships

In this section, the effects of nutrients on vegetative growth, yield, and quality of fig are discussed. Basic functions of mineral elements in plants are beyond the scope of this chapter, and readers are referred to Marschner (1995) and Mengel and Kirkby (2001).

8.4.1 Nitrogen

N is one of the most important nutrients affecting vegetative growth. It affects shoot length and number of leaves, yield (fruits per shoot), and quality (fruit size, peel thickness, ostiole size, dry matter, color, sunscald, cracking, and cull ratio) (Proebsting and Warner, 1954; Obenauf *et al.*, 1978; Aksoy *et al.*, 1987b; Kabasakal, 1983, 1990; Ferguson *et al.*, 1990; Anaç *et al.*, 1992; Flaishman *et al.*, 2008; Ertan *et al.*, 2017). Significant differences exist among varieties in terms of their response to N fertilization. For the best management of N in fig, its uptake pattern and seasonal variation should be taken into consideration based on the particular variety.

N concentration in the fig tree, in terms of leaf, fruit, and shoots, is highest prior to the spring growth flush and decreases during the vegetation period. A rapid decrease in N concentration takes place during two different periods: in the spring period of rapid leaf growth, and just before leaf abscission (Fig. 8.4A). The former is due to the increased demand for N because of rapid leaf growth, whereas the latter is due to remobilization of nutrients from leaves to current-season and 1-year-old shoots. In this period, the concentration of N in perennial shoots (≥2 years) does not change significantly (Dekazos and Crane, 1966; Gaşgil, 1993). This specific situation for figs, compared with other deciduous fruit trees, indicates that the current-season and 1-year-old shoots constitute the main N source to support the initial vegetative development period of the following year. The N requirement for the spring growth flush is derived largely from this N reserve. Therefore, the N reserve plays an important role in determining shoot length, and thus the number and size of fruit per shoot.

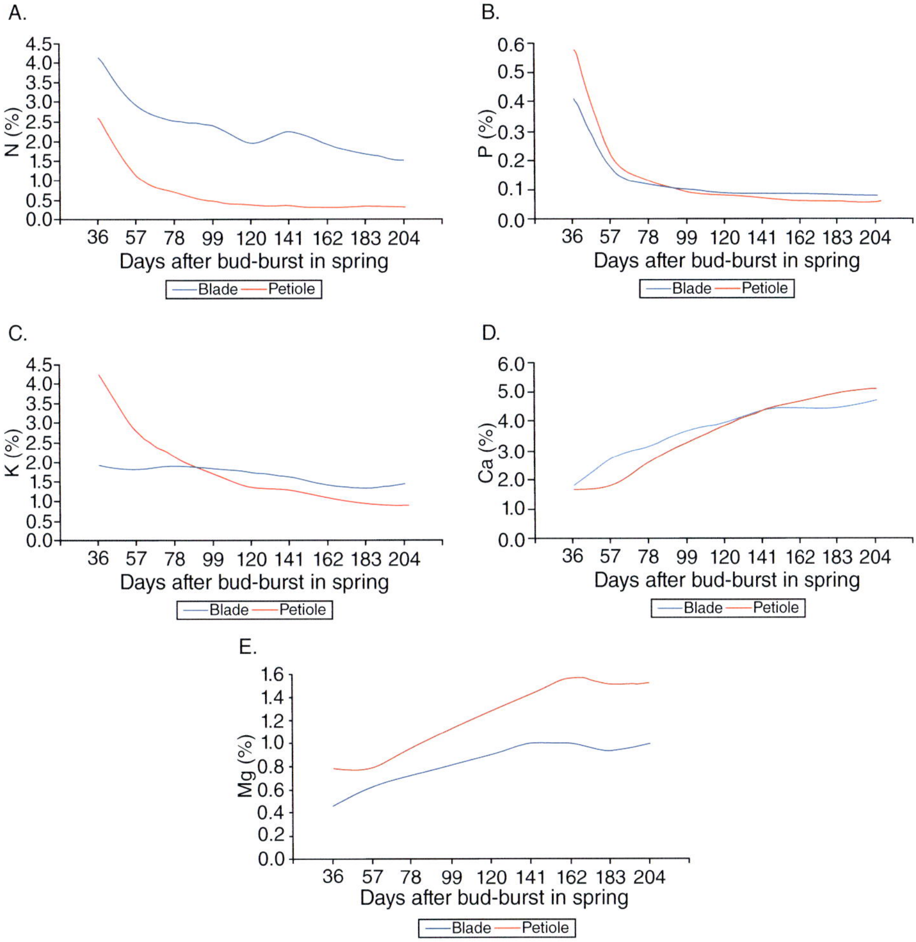

Fig. 8.4. Seasonal variation in nutrient concentrations in a fig leaf. A. Nitrogen (N), B. phosphorus (P), C. potassium (K), D. calcium (Ca), E. magnesium (Mg). Reproduced from Kabasakal (1983) with permission and courtesy of M.E. İrget.

The total N and nitrate nitrogen (NO_3-N) concentrations in fig fruit are highest at the beginning of the vegetation period. N concentrations decrease during the season without depending on the fruit set of the variety, whether parthenocarpic or non-parthenocarpic, except for the increase in the first stage of parthenocarpic varieties (Dekazos and Crane, 1966; Aksoy, 1981). Parthenocarpic varieties contain about three times more nitrate than non-parthenocarpic ones in stages I and II of fruit development. Neither variety contains nitrate in the fruit during the harvest period. High nitrate concentration in parthenocarpic varieties is associated with regulation of indole-3-acetic acid oxidase (Dekazos and Crane, 1966).

Fruit size is an important parameter in terms of both fresh- and dried-fruit quality and is significantly associated with dry matter, cracking, color, culled fruit ratio, titratable acidity, and disease/pest severity (Aksoy *et al.*, 1987b, 1992; Ertan *et al.*, 2017). It is reported that fruit size is related to an increase in water content rather than accumulation of dry matter (Aksoy *et al.*, 1992). High N leads to a decrease in the amount of dry matter in fresh fruit and an increase in water content (U. Aksoy, 2019, personal communication). Following the drying process, due to the low dry-matter content, smaller-sized and darker dried fruits are obtained, which increases the cull ratio. Although fresh fruit size is increased by higher water content, fruit cracking may increase in some varieties due to higher internal pressure exceeding the skin resistance.

The Sarılop variety is sensitive to high N, which negatively affects the dried-fruit quality (Obenauf *et al.*, 1978; Flaishman *et al.*, 2008). Significant negative relationships were also found between leaf N concentrations and dried-fruit quality parameters such as fruit size, color, and sunscald (Aksoy *et al.*, 1987b). It is reported that excessive N enhanced vegetative growth at the expense of fruit yield in the Brown Turkey variety (Flaishman *et al.*, 2008) and resulted in larger but hollow fruits with elongated peduncles in the Bursa Black variety (U. Aksoy, 2019, personal communication).

8.4.2 Phosphorus

The P concentration in the leaves, fruits, and shoots of figs is generally lower than in other deciduous fruit trees. The variation in the P concentration of leaves, fruits, and shoots follows a trend, as for N (Fig. 8.4B) (Aksoy, 1981; Gaşgil, 1993). Due to limited research based solely on P fertilization (Yapar, 1976), the effect of P on vegetative growth, yield, and quality is usually interpreted with observations and in terms of soil–plant relationships (Anaç and Aksoy, 1988; Anaç, 1997). In this regard, leaf P concentration is positively correlated with fruit weight, volume, and size in fresh fruits, and negatively correlated with total soluble solid (TSS) and titratable acidity (TA) content in dried fig fruits. Increased fruit size results in thick receptacle formation in dried fruits. However, this trait may be accepted as a low-quality parameter for table consumption (Aksoy and Anaç, 1994a).

8.4.3 Potassium

A number of studies have been conducted on the relationships between vegetative growth, yield, and quality with respect to K (Kabasakal, 1983; Aksoy *et al.*, 1987b; Anaç *et al.*, 1992; Aksoy and Anaç, 1993; Brown, 1994; İrget *et al.*, 1999, 2008; Tepecik, 2010). These studies demonstrated that K nutrition/fertilization generally had no significant effects on vegetative growth (e.g. shoot length, diameter, number of fruits), but was positively correlated with various fruit-quality parameters such as fruit size, light color, softness, and sunscald-free fruit ratio, and negatively with fruit cracking. It is reported that a negative correlation also exists between leaf K concentration and black neck, which is an important quality parameter of dried fig (Anaç *et al.*, 1992).

Fertilizer trials in this regard revealed that K is highly effective in reducing fruit sunscald; however, a high level of K aggravated fruit cracking, especially in soils with low lime and available Ca content and poor buffering capacity. Therefore, soil properties, and leaf and fruit K/Ca and K/Ca+Mg ratio should be considered in the K fertilization of figs (Table 8.2).

Fig fruit is one of the richest fruits in terms of K. The high K demand by fruits results in a concomitant reduction in leaf K concentration throughout the growing season (Fig. 8.4C) (Kabasakal, 1983; Brown, 1994).

Table 8.2. Effect of different potassium (K) rates (g per tree) on fresh (F) and dried (D) fruit quality of fig (Sarılop). Data from Tepecik (2010).

Treatment	Fruit weight (g) F	Fruit weight (g) D	TSS (g) F	TSS (g) D	Ostiole width (mm) F	Sunscald (%) F	Cracking (%) F	Color (L*) F	Color (L*) D	Cull (%) D
Control	34.65[e]	13.85[c]	19.21[e]	56.58[e]	4.08[d]	15.5[a]	11.4[f]	65.79[b]	62.13[c]	28.55[a]
NP	35.66[de]	13.91[c]	19.23[e]	56.61[e]	4.09[d]	15.6[a]	11.3[f]	66.48[b]	63.59[bc]	28.37[ab]
NP + 150 K_2O	37.11[d]	14.33[c]	19.50[e]	56.79[e]	4.12[cd]	15.3[a]	11.4[f]	69.15[a]	64.37[ab]	27.78[bc]
NP + 300 K_2O	38.77[c]	15.12[b]	20.10[d]	57.61[d]	4.21[cd]	15.0[b]	11.9[e]	69.45[a]	65.59[a]	27.41[c]
NP + 450 K_2O	41.21[b]	15.58[b]	20.60[c]	58.01[cd]	4.30[c]	14.5[c]	12.6[d]	70.60[a]	66.16[a]	26.75[d]
NP + 600 K_2O	42.88[a]	16.31[a]	21.68[b]	58.44[bc]	4.69[b]	13.9[d]	13.0[c]	69.97[a]	65.54[a]	26.66[d]
NP + 750 K_2O	44.18[a]	16.95[a]	22.13[a]	58.75[b]	5.01[a]	13.6[e]	13.6[b]	70.12[a]	64.73[ab]	26.11[d]
NP + 900 K_2O	43.76[a]	16.98[a]	22.16[a]	59.57[a]	5.15[a]	13.6[e]	14.5[a]	70.11[a]	65.33[ab]	26.09[d]

Different superscript letters indicate significant differences between values. NP, nitrogen and phosphorus. Color L* value denotes the lightness (or darkness) of the skin color, and defines black at 0 and white at 100.

In research conducted by Aksoy and Anaç (1993), the soil characteristics and nutritional status of high-quality fruit-bearing orchards were studied, and the average leaf K concentration (1.36%) was found to be higher than the mean value for the Aegean Region (1.19%), as well as the reference value (1.0%). These findings were also confirmed by others, who reported that quality parameters in dried figs such as light color and soft texture are significantly correlated with leaf and soil K contents. In their study, Aksoy and Anaç (1993) highlighted the positive effects of K on healthy and sound (i.e. fruits without sunscald) fruit number, TSS and TA content and concluded that high-quality dried-fig production is associated with high leaf K concentrations.

8.4.4 Calcium

Ca significantly affects the vegetative growth and quality of figs, such as shoot length, fruit dimensions and sunscald, cracking, TSS, TA, fruit size, and color (Kabasakal, 1983; Anaç and Aksoy, 1988; İrget *et al.*, 1999, 2008). Fig tissues (leaf, fruits, and wood) contain more Ca than other deciduous fruit trees. The relatively higher tissue Ca concentration in figs can be associated with its high adaptation to calcareous soils. In contrast to fruit Ca concentration, leaf Ca concentration increases throughout the vegetation period (Fig. 8.4D) (Kabasakal, 1983; Aksoy and Akyüz, 1993; Gaşgil, 1993).

Research has shown that Ca nutrition significantly affects fruit cracking and sunscald. In this context, an optimum Ca application rate (soil and foliar) can help to reduce cracking; however, high Ca application rates are reported to induce sunscald (Aksoy *et al.*, 1987b; İrget *et al.*, 1999, 2008). Therefore, as discussed previously for K, optimization of Ca application to maintain a balance between sunscald and cracking is very important.

Different varieties have different susceptibilities to cracking, and therefore the varietal responses may change in Ca applications for preventing or reducing cracking. In this regard, the results of a study conducted with three different varieties, Bursa Black, Sarılop, and Göklop, indicated Göklop as the most sensitive and most responsive to foliar Ca treatments applied as $CaCl_2$, while the most resistant or least responsive variety was Bursa Black (Aksoy and Anaç, 1994b). In another study, fruit Ca concentrations of these three varieties were examined during the ripening period, and no significant change was found in the more resistant Bursa Black fruits, while a significant decrease was determined in the sensitive Göklop variety (Aksoy and Akyüz, 1993).

8.4.5 Magnesium

Leaf Mg concentrations show significant changes during the vegetation period, with the lowest values during the spring growth flush period, followed by a steady increase with the progress of vegetation. By the time of abscission, leaf Mg concentrations can increase up to two- to threefold in the blade and petiole, respectively (Fig. 8.4E). The increase in leaf Mg concentration can vary among varieties. Proebsting and Warner (1954) observed that the leaf Mg concentration, which was 0.50% at the beginning of the spring vegetation, reached 0.90% and 0.75%, for Adriatic and Calimyrna (Sarılop) varieties, respectively. In contrast to leaves, fruit Mg concentrations are reported to decrease during the vegetation period (Aksoy and Akyüz, 1993).

Magnesium influences several vegetative growth and dried-fruit quality parameters. Significant and positive correlations were found between shoot length and soil Mg (Kabasakal, 1983), and shoot length and leaf Mg (Aksoy *et al.*, 1987b). It is also reported that leaf Mg concentration is positively correlated with the number of fruits per shoot and the fruit-ripening rate, but negatively with light color in the dried fig fruit (Aksoy *et al.*, 1987b). Tepecik (2010) reported a decrease in Ca and Mg concentration in the leaves and fruit with increasing rates of K applications to the soil.

8.4.6 Trace elements

Publications concerning micronutrients in fig are scarce (Hakerlerler *et al.*, 1999; Aydın and Kaptan, 2015). Related studies are based mostly on plant–soil relationships. Hakerlerler *et al.* (1999) reported that foliar and soil application of Zn at different rates affected the yield and quality of the Sarılop fig variety. Survey results revealed that the most effective microelements with respect to fruit composition are Fe and B. These elements are positively associated with TSS and TA content in the fresh fruit. It was also reported that a positive relationship exists between leaf blade and petiole Fe concentrations and the dry matter content of dried fruit (Anaç and Aksoy, 1988).

The role of B needs to be highlighted. Fig is regarded as one of the most sensitive crops to B toxicity. However, fig also appears to be a B accumulator plant (Brown, 1994). For this reason, B toxicity is more common in orchards located close to thermal springs and fault lines. Visual symptoms of B toxicity usually appear above 300 mg kg^{-1} and are intensified between 450 and 800 mg kg^{-1}, depending on the variety. Leaf edge scorch, defoliation, dieback, a decrease in fruit set, cracking, and hollow fruits are the typical visual symptoms of B toxicity in figs (Fig. 8.1A). Therefore, orchard establishment on soils with high B concentrations must be done with caution.

Two consecutive studies were conducted in a 4 ha fig orchard with high soil B concentration (2.01–3.69 mg kg^{-1}) and with different fruit-cracking levels (Aydın and Kaptan, 2015). The first study examined the relationship between cracking, soil properties, and nutrition, while the second studied the effect of foliar application of N, K, Ca, Mg, and B on cracking severity. The soil had lower available K and Ca and higher B concentration in the location with the severe fruit-cracking problem. Fruit cracking was negatively correlated with the soil Ca:B ratio, as well as the leaf and peel Ca concentrations. The most effective treatment to reduce fruit cracking was found to be foliar Ca application.

8.5 Nutrient Requirements

Nutrient requirements vary according to the variety, as well as management practices affecting vegetative growth and yield. Two examples related to the nutrient requirements of the fig varieties Masui Dauphine and Sarılop, are given in Tables 8.3 and 8.4, respectively.

The nutrient requirement for the Smyrna type can be expressed as shown in Equations 8.1 and 8.2:

$$\text{Nutrient requirement} = \text{current-season growth} + \text{expansion of perennial framework} \tag{8.1}$$

$$\text{Nutrient requirement} = (\text{wood} + \text{leaves} + \text{fruits}) + (\geq 1\text{-year-old shoot wood} + \text{trunk} + \text{roots}) \tag{8.2}$$

As some portion of the nutrient requirement (particularly during the onset of spring growth flush) is provided by remobilization from woody storage tissues, the net nutrient requirement can be expressed by Equation 8.3:

$$\text{Net nutrient requirement} = (\text{current-season growth} + \text{expansion of perennial framework}) - \text{remobilized amount}) \tag{8.3}$$

Table 8.3. Annual nutrient uptake by a 6-year-old Masui Dauphine (San Piero) fig tree. Modified from Hirai (1966) with permission.

Plant part		Dry weight	Weight (g per plant)				
			N	P_2O_5	K_2O	CaO	MgO
Fruits	Immature	6.0	0.16	0.03	0.10	0.13	0.04
	Mature	3129.2	29.10	9.42	45.37	20.97	2.50
Leaves		980.8	18.44	2.75	13.34	51.10	10.29
Shoots	1 year old	340.3	4.46	1.40	2.82	5.73	1.77
	2–5 years old	624.9	5.95	1.77	4.51	11.92	2.43
Trunk		105.8	1.11	0.48	1.04	2.90	0.53
Roots		493.1	5.06	2.64	7.87	7.01	3.66
Totals		5680.1	64.28	18.49	75.05	97.76	21.13

Table 8.4. Nutrient requirements of the Sarılop variety for fruit and the current year's vegetative growth. Data from M.E. İrget, 2019, unpublished results.

Plant part	Weight (kg ha^{-1})						Weight (g ha^{-1})				
	N	P_2O_5	K_2O	CaO	MgO	S	Fe	Zn	Mn	Cu	B
Fruit	17.50	5.70	30.0	8.75	3.50	1.75	87.50	15.00	18.75	15.00	55.00
Current-season shoots											
Foliage	9.60	1.00	6.90	38.00	8.50	0.60	128.00	7.70	25.60	3.20	76.80
Wood	3.30	0.50	2.00	10.50	2.40	0.15	10.80	3.10	2.90	0.80	4.50
Total	30.40	7.20	38.90	57.25	14.40	2.50	226.30	25.80	47.25	19.00	136.30

Average plant density: 160 trees ha^{-1}; average yield: 2.0 t ha^{-1}.

As abscized leaves return nutrients to the soil by recycling, the required amount of nutrients for leaves could be neglected unless fed to animals or utilized for any other purpose.

Finally, the requirement can be calculated by Equation 8.4:

$$\text{Nutrient requirement} = \text{net nutrient requirement} - \text{soil supply} \quad (8.4)$$

In calculation of the fertilizer rate, fertilizer-use efficiency (for nitrogenous fertilizers, this is approximately 50%) should also be considered.

8.6 Management

Prior to orchard establishment, soil analysis is a very useful tool for preparing a management plan, especially to maintain soil fertility and health. Some parameters in soil analysis are briefly evaluated below in terms of fig nutrition.

8.6.1 pH

For fig, the optimum soil pH range should be between 6.0 and 7.8. In general, a high pH does not significantly affect development and yield. However, under conditions where the pH falls below 5.5, trace element toxicity (e.g. Mn) might occur. Under such conditions, liming can be necessary, and the lime requirement should be determined by a preliminary soil test.

8.6.2 Lime

Fig trees, as described earlier, are tolerant to a high calcium carbonate content in the soil, and some researchers have defined the fig tree as a "lime lover." When establishing a fig orchard on soils with low available Ca content, lime can be either applied to the entire field or, preferably, in planting holes to meet the further Ca requirement of saplings. For the latter practice, the planting holes should be deeper and wider than usual. Liming material (lime or slaked lime) should be placed at the bottom of the planting hole and covered with a layer of soil prior to planting.

8.6.3 Organic matter

Soil organic matter is vital to fig cultivation. It has an essential role in increasing the water-holding capacity of soils under rainfed conditions and preventing erosion of sloping land. Animal manure (Mordoğan *et al.*, 2013), green manure, and compost are important sources that can be used as organic soil amendments. The rate and frequency of organic fertilizers applied should be adjusted according to the prevailing conditions and expected performance of the tree. Applications can be performed every other year or even more frequently, depending on the soil organic matter content and the plant shoot length. If the orchard is under organic management, there could be limitations imposed by the standard taken as reference in certification regarding the addition of external N, P, or K annually.

8.6.4 Fertilization

Fertilization of fig trees has been evaluated for three different growth stages: at the planting stage in juvenile/non-bearing trees, and in bearing/mature trees.

Planting

At planting, no mineral fertilization is recommended for figs (Obenauf *et al.*, 1978). The accustomed practice is in this direction. However, particularly in soils with low available P, applying a low rate of P fertilizer (e.g. 50 g of triple superphosphate per sapling) at planting can help to maintain a balanced shoot:root ratio. As fig is a mycorrhizal plant (Brown, 1994; Çömlekçioğlu *et al.*, 2008), application of mycorrhizae at planting can be beneficial both for enhancing root development and for tolerance to drought under rainfed conditions. Micronutrients can also be applied at planting according to the results of soil analysis, either to the root zone or below (e.g. 5–10 g zinc sulfate and borax per sapling for Zn and B).

Juvenile/non-bearing trees

This stage varies between 3 and 4 years in figs, based on the variety (Stover *et al.*, 2007). Generally, in the year of planting, N application is not recommended because of the possible cold damage to new growth (Obenauf *et al.*, 1978) in areas where the soil is sufficient in terms of nutrients and the risk of cold damage is high. However, in mild climates (e.g. the Aegean Region) and poor soil conditions, fertilizers can be applied at low rates (e.g. 10–15 g N per sapling) in the planting year to support vegetative growth. In subsequent years, N fertilization can be increased by 25–50% or more every year according to the development of plants. Different sources of N fertilizers (ammonium sulfate, ammonium nitrate, or urea) can be used. At this stage, other nutrients such as P and K can be supplied in appropriate doses when necessary, depending on the results of soil analysis.

Bearing/mature trees

Fertilizer rate, form, and application time are determined by the soil properties, variety, yield, and irrigation in the case of mature trees (Table 8.3 and 8.4). The most important criterion in determining the fertilizer rate is the fruit yield. Shoot length is considered as a secondary criterion for N rate and application frequency. Application time and the form of fertilizers varies according to precipitation, irrigation, and soil properties. Under rainfed conditions, applications are generally performed as basic fertilization once and generally during late winter. Under irrigated conditions (furrow, border, or drip irrigation), however, base fertilization can be practiced as in dried-fig production or divided into two or three parts, as in fresh-fig production.

Stover *et al.* (2007) cited Morton and stated that the applied N should not exceed 0.2–0.5 kg year^{-1} per tree with split application from early spring to July. Avoiding the use of N fertilizers in advance of the fruit-ripening stage is an important practice for N management, especially in dried-fig production. A general fertilizer schedule for figs, which does not specify the soil type or variety, has been proposed for the Near East region (Table 8.5).

Table 8.5. Fertilization schedule for fig. Data from Anon, 2007.

Tree age (years)	Application (g year^{-1} per tree)		
	N	P_2O_5	K_2O
1	–	–	–
2	100	–	–
3	150	30	75
4	200	40	100
5	250	50	125
6	300	60	150
7	350	70	175
8	400	80	200
9	450	90	225
10+	500	100	250

8.6.5 Current fertilizer practices

The rate, form, and application time of fertilizers vary considerably in different regions, and for different soil type and varieties. For example, for Sarılop variety in the Aegean Region, mainly diammonium phosphate (DAP; $N:P_2O_5:K_2O$; ratio 18:46:0), ammonium sulfate (AS: 21% N), ammonium nitrate (AN: 26 or 33% N), potassium sulfate (PS: 50% K_2O), and less frequently 15:15:15 and 20:20:0 fertilizers are used in lowlands, and 15:15:15 and, less frequently DAP, AS, AN, and PS, are used in sloping areas. In Turkey, the fertilizer application rate (as a combination of the cited fertilizers above) ranges from about 0.5 to 1.5 kg per tree with an average of 0.75–1.0 kg per tree. There are also farmers who do not apply mineral fertilizers at all or apply them only every 2 or 3 years. In some varieties destined for fresh consumption such as Bursa Black, higher rates of fertilizers are used compared with dried-fig varieties. Bursa Black is also the most common variety that is fertigated with calcium and potassium nitrate by the drip-irrigation systems used in Turkey. It is reported that only N fertilizers (22–45 kg ha^{-1}) are used for the Sarılop variety cultivated in the USA (Obenauf *et al.*, 1978; Ferguson *et al.*, 1990).

8.7 Irrigation

Fig trees are more tolerant to water stress than many other fruit trees (Dominguez, 1990; Stover *et al.*, 2007; Flaishman *et al.*, 2008). Some genotypes have been reported to tolerate up to –4 MPa leaf water potential (Oukabli *et al.*, 2008). Fig is also known to be moderately sensitive to salinity and can tolerate up to 5.5 dS m^{-1} electrical conductivity of irrigation water (Golombek and Lüdders, 1990; Maas, 1990; Flaishman *et al.*, 2008). Significant losses in yield and quality of fig can occur if rainfall is insufficient and supplemental irrigation is not possible (Hallaç Türk and Aksoy, 2011; Gholami *et al.*, 2012; Soliman and Al-Hady, 2017).

The annual rainfall requirement for fig trees is 625 mm (Aksoy *et al.*, 2001) however in addition to the winter rainfall, supplementary irrigation may be required to reach a total of 750 mm (Ferguson *et al.*, 1990). The distribution as well as the amount of rainfall during the year is extremely important in terms of drought stress and quality. In the Aegean Region, which is one of the most important fig centers in Turkey, the annual rainfall is very close to the water requirement of the fig tree and about 20% of the precipitation falls in December. It is reported that rainfall during the harvest period leads to significant quality losses, especially in dried-fig varieties (Çobanoğlu *et al.*, 2010). Supplemental irrigation is recommended if the precipitation falls below 550 mm. To meet this moisture deficit, supplementary winter and early spring irrigation (once or twice) is generally applied.

8.7.1 Crop water use (evapotranspiration)

The water lost by evapotranspiration is considered as the crop water requirement. Irrigation is required in case the soil moisture cannot meet evapotranspiration. Therefore, net irrigation water to be applied during irrigation in the vegetation period should be determined.

Several methods have been used to determine the irrigation water requirement of plants. Water requirement of most fruits including fig can be estimated by the crop coefficient (K_c). In this method, K_c values for different stages of crop development period are multiplied by reference evapotranspiration (ET_0) values calculated by the United Nations Food and Agriculture Organization (FAO) Penman–Monteith or Class A Pan Evaporation method (Brouwer and Heibloem, 1986; Allen *et al.*, 1998). The basic principles of these two methods are briefly given in the following sections.

FAO Penman–Monteith method

Plant water consumption (ET_c) is determined by correcting the ET_o with K_c (Equation 8.5). ET_o reflects the evapotranspiration from actively growing grass and is calculated using the climate data of a particular region:

$$ET_c = K_c \times ET_o \quad (8.5)$$

where ET_c is evapotranspiration (mm), ET_o is the reference evapotranspiration calculated using the FAO Penman–Monteith method (mm), and K_c is the crop coefficient.

The K_c values of different stages of the development period are found experimentally. The average K_c values for fig in Turkey are presented in Fig. 8.5A (Anon., 2017).

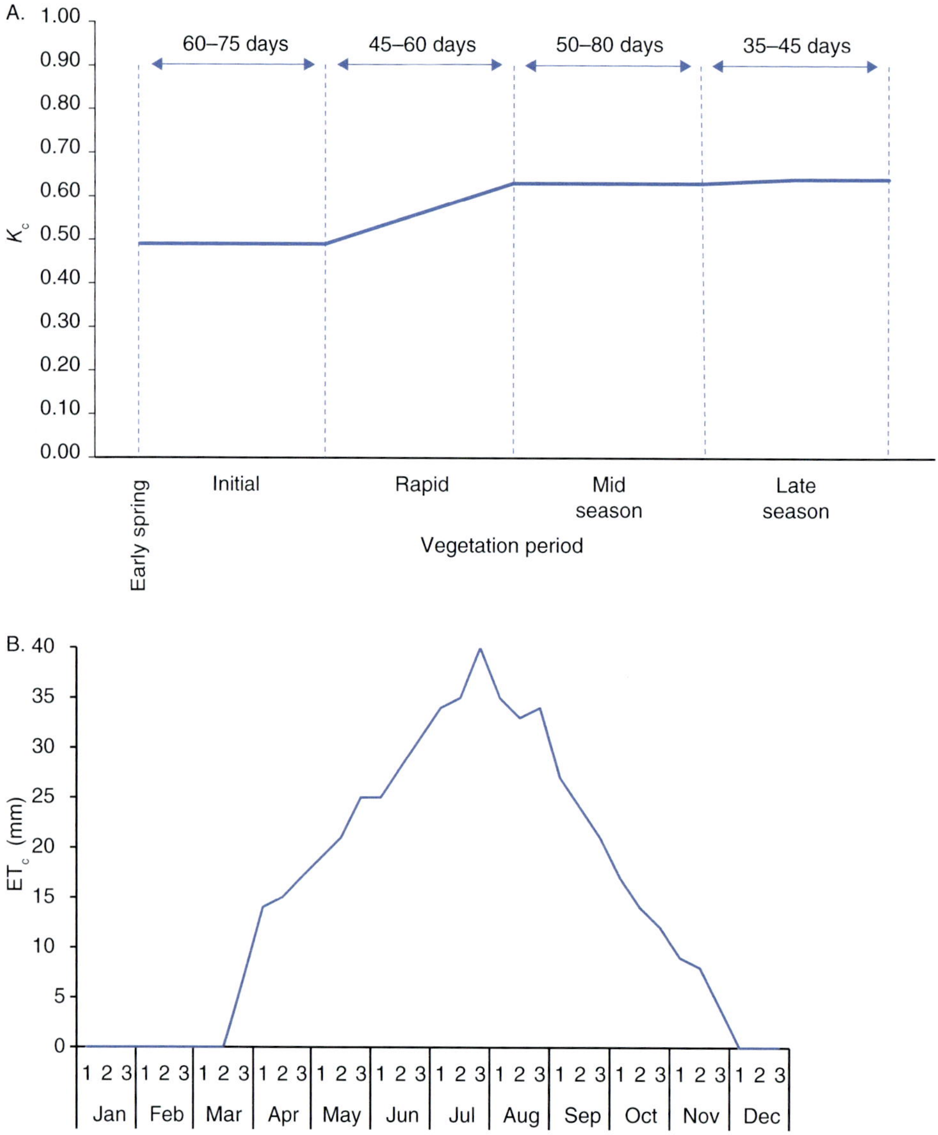

Fig. 8.5. Irrigation values. A. Average K_c values of fig trees in Turkey. B. Ten-day ET_c values for fig trees in the Selçuk township in İzmir, Turkey. Data from Anon. (2017).

The K_c value of 0.47 suggested by Olitta *et al.* (1979) has been used in many studies conducted in Brazil. For 3-year-old drip-irrigated fig trees, the average K_c was determined as 0.71 and 0.51 at 2- and 4-day irrigation intervals, respectively (de Andrade *et al.*, 2014).

Class A Pan Evaporation method

This method can be used to determine ET_o where meteorological data are not available (Equations 8.6 and 8.7). The pan reflects the effect of temperature, humidity, wind speed, and radiation on evaporation from an open water surface (Fig. 8.6A) (Allen *et al.*, 1998).

$$ET_o = K_p \times E_{pan} \quad (8.6)$$

$$ET_c = K_c \times ET_o \quad (8.7)$$

where ET_o is the reference evapotranspiration calculated by the Class A Pan Evaporation method (mm), K_p is the pan coefficient, E_{pan} is the pan evaporation between two consecutive measurements (mm), ET_c is evapotranspiration (mm), and K_c is the crop coefficient.

K_p can vary depending on the wind and surface characteristics of the place where the pan is placed and can also be taken as an average value of 70% (K_p = 0.70) (Brouwer and Heibloem, 1986; Allen *et al.*, 1998). However, it is reported that the use of 50% (d'Andria *et al.*, 1992) and 75% (Hernandez *et al.*, 1994b) pan evaporation at different irrigation intervals has a positive effect on fig tree yield and development. Additionally, Tapia *et al.* (2003) stated that the application of 17% pan evaporation as irrigation water provided sufficient tree development under 2- and 4-day irrigation intervals.

It is recommended that the calculated plant water consumption values should be corrected with the percentage shaded area (P_s) under drip-irrigation conditions (Equation 8.8). For Smyrna varieties with wide planting space (>4 m), the P_s value can be considered as 70% (Yıldırım, 2013).

$$ET_{c,adj} = ET_c \times \left(\frac{P_s}{85}\right) \quad (8.8)$$

where $ET_{c,adj}$ is the corrected evapotranspiration (mm), ET_c is evapotranspiration (mm), and P_s is the percentage shaded area.

In recent years, monthly and total (Table 8.6) and 10-day period (Fig. 8.5B) ET_c values of fig have

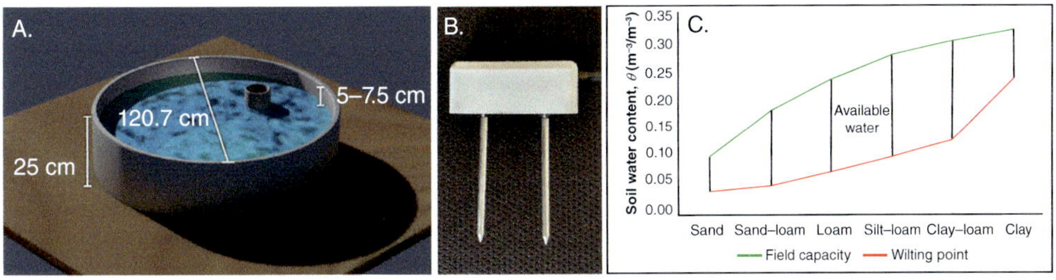

Fig. 8.6. Determination of the amount of irrigation water. A. Class A Pan Evaporation method. Photo courtesy of M.K. Meriç. B. Soil dielectric sensor (Teros 10, Meter Group). C. Determination of the difference in soil water content between field capacity and wilting point in different soil types.

Table 8.6. Monthly and annual ET_c values for the fig tree calculated using the FAO Penman–Monteith method in different districts of the İzmir and Aydın Provinces in Turkey. Data from Anon., 2017, courtesy of M.K. Meriç.

		ET_c (mm)												
Province	Location	Jan	Feb	Mar	Apr	May	Jun	Jul	Aug	Sep	Oct	Nov	Dec	Total
İzmir	Center	–	–	8	54	77	103	138	128	90	58	30	–	686
	Selçuk	–	–	7	46	65	84	109	102	72	43	21	–	549
Aydın	Kuşadası	–	–	7	49	66	86	107	96	72	48	29	–	560
	Sultanhisar	–	–	7	49	72	92	117	109	77	47	21	–	591

been calculated by the FAO Penman–Monteith method for different locations in Turkey (Anon., 2017).

8.7.2 Irrigation water requirements

In addition to ET_c, effective rainfall (P_e) and irrigation efficiencies (conveyance (E_c) and application (E_a)) are taken into consideration when determining the total water to be applied during irrigation.

Effective rainfall (P_e) refers to the part of the rainfall (P) that is stored in the root zone and can be used by the plant (Equations 8.9 and 8.10):

$$P_e = 0.8 \times P - 25\,(P > 75\,\text{mm/month}) \quad (8.9)$$

$$P_e = 0.6 \times P - 10\,(P < 75\,\text{mm/month}) \quad (8.10)$$

where P_e is effective rainfall (mm) and P is rainfall (mm).

Finally, total irrigation water can be calculated using Equation 8.11 (Yıldırım, 2004):

$$I = \frac{ET_{c,adj} - P_e}{E_a \times E_c} \quad (8.11)$$

where I is total irrigation water (mm), $ET_{c,adj}$ is corrected evapotranspiration (mm), P_e is effective rainfall (mm), E_a is application efficiency, and E_c conveyance efficiency.

In addition to the FAO Penman–Monteith and Class A Pan Evaporation methods, tensiometers, dielectric sensors (Fig. 8.6B), and electrical resistance sensors, which can be installed in the plant root zone to monitor the soil water content between the field capacity and the wilting point (Fig. 8.6C), are also used to determine the amount of irrigation water required (Pardossi *et al.*, 2009). The general approach in these methods is to monitor the actual soil water content of the root zone from the sensor data and to apply the amount of water required to bring the soil water content to the specified threshold during irrigation. Although this threshold can be the field capacity in non-deficit conditions, it can also be a value between the field capacity and the wilting point under deficit irrigation conditions.

With the rapid developments in information and communication technologies in recent years, sensors are integrated into wireless sensor networks operating in license-free ISM (industrial, scientific, and medical) frequency bands with different radio modulation technologies (e.g. LoRa, GFSK, FSK) and network communication protocols (e.g. ZigBee, LoRaWan, and other proprietary low-power and low-data rate protocols). Nowadays, these networks are introduced to users and farmers with web and mobile applications by means of various cloud services. For additional information, see Chapter 11 (this volume).

8.7.3 Irrigation methods

The methods used to irrigate fig trees may vary according to the regions, varieties, and management methods. Besides conventional surface irrigation methods such as border and furrow irrigation, modern pressurized irrigation methods such as sprinklers and drip irrigation can be utilized (Sharifzadeh *et al.*, 2012; Abdolahipour *et al.*, 2018). In the case of surface irrigation methods, it is recommended that the land is leveled before planting. However, it has been stated that more than 60 cm excavation or filling causes irregular plant growth (Obenauf *et al.*, 1978).

In the drip-irrigation method, water is applied to each tree in small volumes, at frequent intervals, at a low flow rate and pressure by means of drippers installed on lateral pipes. As the efficiency of the system is closely related to the performance of the drippers, system parameters such as dripper flow rate, dripper spacing, and lateral diameter should be selected carefully. Although the dripper flow rate used in drip-irrigation systems varies from 1 to 70 L h^{-1}, lower flow rates of 1.6, 2, 3 and 4 L h^{-1} are preferred in practice. In heavy soils in particular, low flow rate drippers should be selected to prevent surface flow. Depending on the topographic characteristics of the field, drippers may be preferred with or without pressure compensation. If pressure-compensating drippers are selected, more efficient water and fertilizer distribution can be achieved throughout the orchards. The dripper flow rate and infiltration rate of the soil are determinative in dripper spacing. This distance can be 40, 50, 60 or 75 cm, depending on the dripper flow rate.

To reduce the system cost in orchards, it is preferable to have a single-line lateral layout in each tree row. However, a double-line lateral layout can be arranged in each tree row where adequate moisture distribution in the soil cannot be achieved (Fig. 8.7). Lateral layouts such as a multidripper straight arrangement or a ring are less preferred as they may make soil cultivation difficult.

The application efficiency (E_a) of irrigation methods varies significantly. In surface irrigation, the E_a is calculated as 40–60% for an earth canal and 65–75% in pressurized pipe networks. In sprinkler irrigation systems, this value ranges between 75% and 85%. Compared with other irrigation methods, application efficiency is the highest in drip irrigation and can reach up to 90%. However, the initial investment costs of drip-irrigation systems are high, and dripper clogging is among the most important problems (Phocaides, 2007).

8.8 Water, Plant Growth, Yield, and Quality Relationships

It is stated that fig is very sensitive to drought in the planting year. Drought stress during this period may cause a significant decrease in tree development. Early defoliation and yield and quality decreases are observed in trees that are exposed to water stress during the fruit-bearing stage (Belge *et al.*, 2012). Similarly, because fig roots are susceptible to moisture changes, excessive irrigation during the summer period is not desirable (Dominguez, 1990), as it increases shoot elongation and the possibility of fruit cracking, and decreases the quality of dried figs. Fluctuations or rapid increases in soil moisture, especially during fruit development or the ripening period, may decrease fruit quality and cause fruit splitting (Melgarejo, 1996; Flaishman *et al.*, 2008). Keeping the soil moisture content high with regular irrigation during the summer period increases the fruit size. However, this situation increases the risk of fruit rot/spoilage during drying, resulting in culled figs (Flaishman *et al.*, 2008). As the rainfall between July and September may cause significant reductions in the quality of dried varieties, precipitation between November and June is desired (Çobanoğlu *et al.*, 2010). Bagheri and Sepaskhah (2014) reported that rainfall in spring, especially in April, and in December has a negative effect on fig yield due to

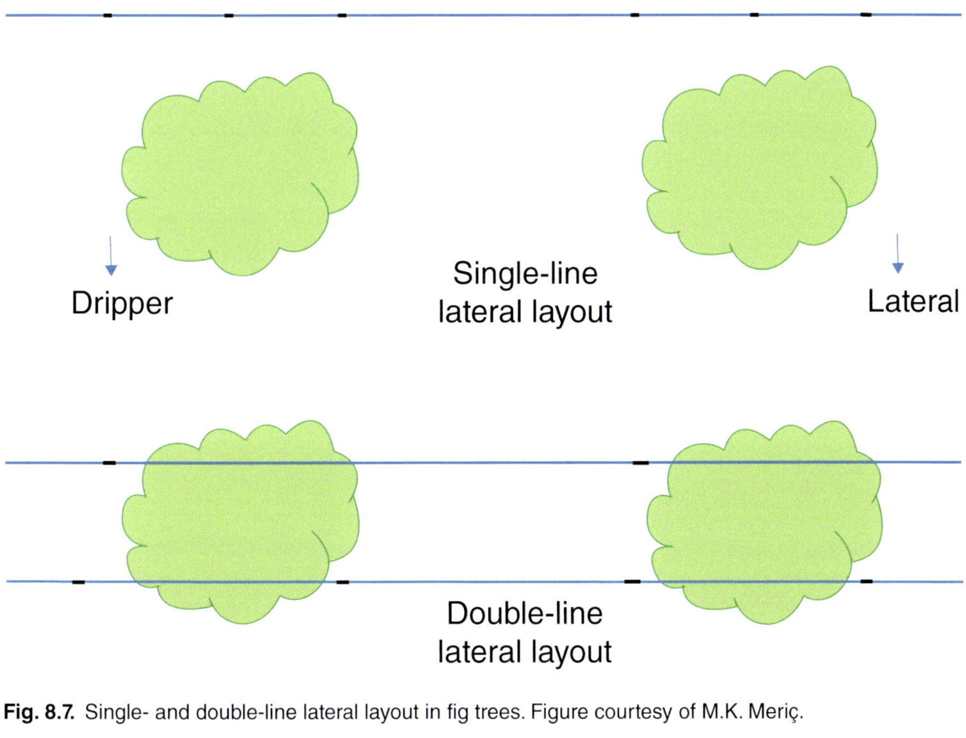

Fig. 8.7. Single- and double-line lateral layout in fig trees. Figure courtesy of M.K. Meriç.

the interference with the life cycle of fig wasps for caprification.

Supplemental irrigation has a positive effect on the yield and morphological properties of fig grown in arid conditions (Abdolahipour *et al.*, 2018). It also reduces the negative effects of water stress on tree development by increasing the soil water content during the critical development period of fig. Irrigation in the early spring period aids in preserving soil moisture content longer than irrigation in summer (Abdolahipour *et al.*, 2018). In a study on Sultani figs in Iran, vegetative development, yield, and fruit quality increased with supplemental irrigation (Al-Desouki *et al.*, 2009) in late winter and/or mid-spring (Kamgar-Haghighi and Sepaskhah, 2015).

In a study conducted by de Andrade *et al.* (2014), two irrigation intervals (2 and 4 days) and non-irrigated conditions were tested on fig trees. Tree yield was 8554 kg ha^{-1} and 7430 kg ha^{-1} for the 2- and 4-day intervals, respectively. In the same study, irrigation at 2-day-interval increased the yield by 27% compared with non-irrigation. While irrigation did not change the shoot length, it increased the number of internodes compared with non-irrigation, which in turn affected the number of fruits and thus the yield.

It is reported that fruit skin damage without affecting other quality parameters can be reduced by applying 25% of the plant water consumption calculated from climate data as irrigation water starting from the second stage of development (weekly fruit diameter change rate <1%) (Kong *et al.*, 2013).

Supplemental irrigation in early spring (first half of April) and mid-summer has a positive effect on leaf gas-exchange parameters (i.e. stomatal resistance, net photosynthesis, and transpiration rate). Application of 2000 L per tree (200 m^3 ha^{-1}) of irrigation water approximately 2 m away from the trunk positively affected the physiological and morphological properties of fig grown under rainfed conditions. However, 1000 L per tree (100 m^3 ha^{-1}) of irrigation water is recommended in terms of conservation of water resources (Abdolahipour *et al.*, 2019).

8.9 Concluding Remarks

Fig is an important crop consumed fresh, sun dried, or processed, and is becoming a worldwide crop. Creating high economic returns is needed to ensure that growers continue farming. Thus, the right agronomic management needs to be used. Nutrient management considering water and soil requirements may positively affect the yield and production capacity of the fig tree. For better fig production, efficient agronomic management needs to be used to eradicate the existing problems caused by internal and external conditions.

Fertigation is gaining acceptance in modern fig production, particularly in young orchards. Increasing soil organic matter to optimize the rhizosphere conditions enable the grower to supply water and nutrients independent of climatic conditions. To further improve yield and quality, more must be learned about the optimal nutritional status of the tree specifically regarding the variety and the dynamic processes and their interactions in plant and soil. While growing fig in a new region, the need to improve fig yields and satisfy the market quality requirements are the most important reasons for switching fig orchards from rainfed to irrigated, even if it is supplementary. Dramatic increases were found in fig yields with improved water and fertilizer management especially with micro-irrigation systems. There are many interactions between the soil type, orchard management, soil fertility, and irrigation that needs to be further studied and evaluated to determine the site-specific conditions.

References

Abdolahipour, M., Kamgar-Haghighi, A.A. and Sepaskhah, A.R. (2018) Time and amount of supplemental irrigation at different distances from tree trunks influence on soil water distribution, evaporation and evapotranspiration in rainfed fig orchards. *Agricultural Water Management* 203, 322–332.

Abdolahipour, M., Kamgar-Haghighi, A.A., Sepaskhah, A.R., Zand-Pars, S., and Honar, T. (2019) Effect of time and amount of supplemental irrigation at different distances from tree trunks on quantity and quality of rain-fed fig production. *Iran Agricultural Research* 38, 35–46.

Aksoy, U. (1981) Akça, Göklop ve Sarılop *İncir* Çeşitlerinde Meyve Gelişmesi, Olgunlaşması ve Depolanması Üzerinde Araştırmalar (Research on fruit growth, ripening and storage of Akça, Göklop and Sarılop fig fruits). PhD thesis, Department of Horticulture, Faculty of Agriculture, Ege University, İzmir, Turkey.

Aksoy, U. and Akyüz, D. (1993) Changes in K, Ca and Mg contents in different part of the fig during development. In: Fragoso, M.A.C., van Beusichem, M.L. and Houwers, A. (eds) *Optimization of Plant Nutrition*. Springer, Dordrecht, The Netherlands, pp. 309–318.

Aksoy, U. and Anaç, A. (1993) Soil properties and mineral content of fig orchards producing high-quality fruits. In: Fragoso, M.A.C., van Beusichem, M.L. and Houwers, A. (eds) *Optimization of Plant Nutrition*. Springer, Dordrecht, The Netherlands, pp. 305–308.

Aksoy, U. and Anaç, D. (1994a) Factors affecting quality of fresh and dried fig fruits. *Acta Horticulturae* 93, 763–766.

Aksoy, U. and Anaç, D. (1994b) The effect of calcium chloride application on fruit quality and mineral content of fig. *Acta Horticulturae* 368, 754–762.

Aksoy, U., Anaç, D., Eryüce, N., and Yoltaş, T. (1987a) Ege bölgesi incir bahçelerinin beslenme durumunun saptanması ve değerlendirilmesi [Determination and evaluation of fig orchard nutrition of Aegean Region]. *Journal of Agriculture Faculty of Ege University* 24, 21–35.

Aksoy, U., Anaç, D., Hakerlerler, H., and Düzbastılar, M. (1987b) Germencik Yöresi Sarılop *İncir* Bahçelerinin Beslenme Durumu ve *İncelenen* Besin Elementleri ile Bazı Verim ve Kalite Özellikleri Arasındaki *İlişkiler* (Nutritional status of fig orchards in Germencik region and relationships between investigated plant nutrients and some yield components and quality). Tariş Ar-Ge Project No. 006, Bornova, İzmir, Turkey.

Aksoy, U., Anaç, D., and Gül, N. (1992) Sarılop incir çeşidinde yaş ve kuru meyve kriterleri arasındaki etkileşimler [Interacation between fresh and dried quality criteria of Sarılop fig variety]. In: *Türkiye I. Ulusal Bahçe Bitkileri Kongresi Cilt I: Meyve*. İzmir, Turkey, pp. 427–434.

Aksoy, U., Can, H.Z., Hepaksoy, S., and Şahin, N. (2001) *İncir Yetiştiriciliği*. TÜBİTAK TARP Türkiye Tarımsal Araştırma Projesi Yayınları, Turkey.

Al-Desouki, M.I., Abd El-Rhman, I.E., and Sahar, A.F. (2009) Effect of some antitranspirants and supplementary irrigation on growth, yield and fruit quality of Sultani fig (*Ficus carica* L.) grown in the Egyptian Western Coastal Zone under rainfed conditions. *Research Journal of Agriculture and Biological Sciences* 5, 899–908.

Allen, R.G., Pereira, L.S., Raes, D., and Smith, M. (1998) *Crop Evapotranspiration – Guidelines for Computing Crop Water Requirements*. FAO Irrigation and Drainage Paper 56, Rome, Italy.

Anaç, D. (1997) Fig nutrition. In: *Advanced Course on Fig Production*. Department of Horticulture, Faculty of Agriculture, Ege University, İzmir, Turkey, 16–28 June.

Anaç, D. and Aksoy, U. (1988) Effect of primary and secondary plant nutrients on vegetative growth and fruit quality of fig (*Ficus carica* L.). In: *Proceedings of VII Colloquium on Optimization of Plant Nutrition*, Nyborg, Denmark.

Anaç, D., Aksoy, U., and Eryüce, N. (1987) Ege bölgesi incir bahçelerinin makro ve bazı mikro besin elementleri içeriklerinin yaprak analizleri ile saptanması üzerinde araştırmalar: I. Büyük Menderes Havzası [Determination of macro and micronutrient contents of fig orchards in the Aegean Region through leaf analysis: I. Big meander basin]. *Journal of Agriculture Faculty of Ege University* 24, 75–90.

Anaç, D., Aksoy, U., Hakerlerler, H., and Düzbastılar, M. (1992) Küçük Menderes *İncir* Bahçelerinin Beslenme Durumu ve *İncelenen* Toprak ve Yaprak Besin Elementleri ile Bazı Verim ve Kalite Özellikleri Arasındaki *İlişkiler* [Nutritional status of fig orchards in small meander basin and relationships between soil and leaf nutrients and some yield and quality characteristics]. Tariş Ar-Ge Project No. 004, Bornova, İzmir, Turkey.

Anon. (2007) *Near East Fertilizer Use Manual*. FAO, Rome.

Anon. (2017) *Türkiye'de Sulanan Bitkilerin Bitki Su Tüketimi Rehberi*. Tarımsal Araştırmalar ve Politikalar Genel Müdürlüğü, Ankara, Turkey.

Aşkın, A., Ceylan, Ş., and Yener, H. (1998) A study on the nutritional status of fig orchards in Birgi-İrimağzı. *Acta Horticulturae* 480, 239–246.

Aydın, M. and Kaptan, M.A. (2015) Effect of nutritional status on fruit cracking of fig (*Ficus carica* L. cv. Sarılop) grown in high level boron contained soils. *Scientific Papers, Series A: Agronomy* 8, 20–25.

Bagheri, E. and Sepaskhah, A.R. (2014) Rain-fed fig yield as affected by rainfall distribution. *Theoretical and Applied Climatology* 117, 433–439.

Belge, A, Özen, M., Çobanoğlu, F., Konak, R., Cocatas, H., *et al.* (2012) *İncir Ağaçlarında Erken Yaprak Dökümünün (Zamansız Yaprak Dökümü) Nedenlerinin Araştırılması*. T.C. Gıda Tarım ve Hayvancılık

Bakanlığı Tarımsal Araştırmalar ve Politikalar Genel Müdürlüğü, TAGEM/BBAD/99/A08/P07/01 No'lu Proje, Aydın, Turkey.

Beutel, J., Uriu, K., and Lilleland, O. (1978) Leaf analysis for California deciduous fruits. In: Reisenauer, H.M. (ed.) *Soil and Plant Tissue Testing in California*. Bulletin No. 1879. University of California, Berkeley, California, pp. 11–14.

Brouwer, C. and Heibloem, M. (1986) *Irrigation Water Management: Irrigation Water Needs*. FAO, Rome.

Brown, P.H. (1994) Seasonal variation in fig (*Ficus carica* L.) leaf nutrient concentrations. *HortScience* 29, 871–873.

Chandler, W.H. (1947) *Decidous Orchards*. Lea & Febiger, Philadelphia, Pennsylvania.

Childers, F.N. (1966) *Temperate to Tropical Fruit Nutrition*. Rutgers – The State University, Somerville, New Jersey.

Çobanoğlu, F., Ertan, B., and Şahin, B. (2010) Farklı azot dozu uygulamaları ile kuru incir verimi arasındaki ilişkinin belirlenmesi. In: *Proceedings of Türkiye IX Tarım Ekonomisi Kongresi*. Şanlıurfa, Turkey, pp. 810–816.

Çömlekçioğlu, S., Akpınar, C., Bayazit, S., Ortaş, İ., and Küden, A.B. (2008) Effect of mycorrhizae applications on the mineral uptake in 'Alkuden' (01-IN-06) fig genotype. *Acta Horticulturae* 772, 513–520.

Condit, I.J. (1947) *The Fig*. Chronica Botanica Co., Waltham, Massachusetts.

Condit, I.J. and Enderud, J. (1956) A bibliography of the fig. *Hilgardia* 25, 1–663.

d'Andria, R., Calandrelli, D., Grassi, G., Limongelli, F., and Maggio, A. (1992) Effetti dell' irrigazione sulla produzione e sull' accrescimento del fico nel Cilento. *Rivista di Frutticoltura e di Ortofloricoltura* 11, 75–81.

de Andrade, I.P.S., de Carvalho, D.F., de Almeida, W.S., Silva, J.B.G., and da Silva, L.D.B. (2014) Water requirement and yield of fig trees under different drip irrigation management. *Engenharia Agricola* 34, 17–27.

Dekazos, E.D. and Crane, J.C. (1966) Fluctuations of total and nitrate nitrogen in parthenocarpic and nonparthenocarpic fig varieties. *Hilgardia* 37, 155–164.

Dominguez, A.F. (1990) *La Higuera, Frutal Mediterraneo Para Climas Calidos*. Mundi-Prensa Libros, Madrid.

Düzbastılar, M. and Güleç, I. (1995) Ege bölgesi incir bahçelerinin beslenmesi açısından toprak özelliklerinin belirlenmesi (Analysis of soil properties in respect to nutrition of fig orchards in the Aegean Region). In: Munsuz, N., Ünver, İ., and Çaycı, G. (eds) *Proceedings of İlhan Akalan Toprak ve Çevre Sempozyumu, Cilt 2*. Soil Science Society of Turkey, Ankara, pp. 73–80.

El-Kassas, S.E. (1975) Effect of moisture levels and nitrogen fertilization on yield, shoot growth and leaf composition of fig trees. *Annals of Agricultural Science of Moshtohor* 4, 155–186.

Elmacı, Ö.L., İrget, M.E., and Aksoy, U. (2010) Büyük Menderes ve Küçük Menderes Havzası İncir (*Ficus carica* L.) Bahçelerinin Beslenme Durumunun *İ*ncelenmesi (Investigation on nutritional status of fig (*Ficus carica* L.) orchards in small and big meander basin). Ege University Faculty of Agriculture Scientific Research Project No. 07-ZRF-014, İzmir, Turkey.

Ersoy, N., Gözlekci, S., and Kaynak, L. (2003) Seasonal variations in the content of nutrient elements in the leaves of fig (*Ficus carica* L. Yeşilgüz). *Acta Horticulturae* 605, 269–275.

Ertan, B., Doğan, Ö., Ertan, E., İrget, M.E., and Tepecik, M. (2017) Sarılop incir çeşidinde farklı azot dozlarının kuru meyve kalite özellikleri ve verim üzerine etkileri (Effect of different nitrogen rates on quality and yield of dried fig (*Ficus carica* L. cv. Sarılop)). In: *5. Uluslararası Katılımlı Toprak ve Su kaynakları Kongresi, Bildiriler Kitabı Cilt 2*. Kırklareli, Turkey, pp. 980–988.

Eryüce, N., Aksoy, U., and Anaç, D. (1987) Ege bölgesi incir bahçelerinin makro ve bazı mikro besin elementleri içeriklerinin yaprak analizleri yolu ile saptanması üzerinde araştırmalar: II. Küçük Menderes Havzası (Determination of macro and micronutrient contents of fig orchards in the Aegean Region through leaf analysis: II. Small meander basin). *Journal of Agriculture Faculty of Ege University* 24, 151–160.

Eryüce, N., Çolakoğlu, H., Aydın, Ş., and Çokuysal, B. (1995) The effect of K and Mg fertilization on some quality characteristics and mineral nutrition of fig. *Acta Horticulturae* 370, 199–204.

Ferguson, L., Michailides, T.J., and Shorey, H.H. (1990) The California fig industry. *Horticultural Reviews* 12, 409–490.

Flaishman, M.A., Rodov, V., and Stover, E. (2008) The fig: botany, horticulture and breeding. In: Janick J. (ed.) *Horticultural Reviews* 34, 113–196.

Gaşgil, N. (1993) İncir bitkisinde yaprak aya, sap ve sürgündeki makro ve mikro besin elementlerinin mevsimsel değişimi ve birbirleriyle ilişkileri üzerinde araştırmalar (Seasonal changes in macro and micro element content of leaf blade, petiole and shoot and their relationships). MSc thesis, Department of Soil Science, Faculty of Agriculture, Ege University, İzmir, Turkey.

Gholami, M., Rahemi, M., and Rastegar, S. (2012) Use of rapid screening methods for detecting drought tolerant cultivars of fig (*Ficus carica* L.). *Scientia Horticulturae* 143, 7–14.

Golombek, S. and Lüdders, P. (1990) Gas exchange of *Ficus carica* L. in response to salinity. In: Beusichem, M.L. (ed.) *Plant Nutrition Physiology and Applications*. Kluwer Academic Publishers, Dordrecht, The Netherlands, pp. 487–493.

Hakerlerler, H., Aydın, Ş., İrget, M.E., Aksoy, U., and Tutam, M. (1999) The effect of soil and foliage applied zinc on yield and quality of fig (*Ficus carica* L.) for drying. In: *Soils with Mediterranean Type of Climate 6th International Meeting. Extended Abstracts*. University of Barcelona, Barcelona, Spain, pp. 256–259.

Hallaç Türk, F. and Aksoy, U. (2011) Comparison of organic, biodynamic and conventional fig farms under rain-fed conditions in Turkey. *Journal of Cell and Plant Sciences* 2, 22–33.

Hernandez, F.B.T., Modesto, J.C., Suzuki, M.A., Corrêa, L.S., and Reichardt, K. (1994a) Effect of irrigation and nitrogen levels on qualitative and nutritional aspects of fig-trees (*Ficus carica* L.). *Scientia Agricola* 51, 292–297.

Hernandez, F.B.T., Suzuki, M.A., Buzetti, S., and Corrêa, L.S. (1994b) Resposta da figueira (*Ficus carica* L.) aouso da irrigação e nitrogênionaregião de Ilha Solteira. *Scientia Agricola* 51, 99–104.

Hirai, J. (1966) Anatomical, physiological and biochemical studies of the fig fruit. *Bulletin of the University of Osaka Prefecture Series B* 18, 169–218.

Hirai, J., Nakagawa, S., Nanjo, Y., and Hirata, N. (1964) Studies on the nutrition of fig trees. I. Effect of nitrogen, phosphoric acid and potassium concentration on growth yield and quality of fruits. *Journal of the Japanese Society for Horticultural Science* 33, 273–279.

İrget, M.E., Aydın, Ş., Oktay, M., Tutam, M., Aksoy, U., and Nalbant, M.A. (1999) Effect of foliar potassium nitrate and calcium nitrate application on nutrient content and fruit quality of fig. In: Anaç, D. and Martin-Prével, P. (eds) *Improved Crop Quality by Nutrient Management*. Kluwer Academic Publishers, Dordrecht, The Netherlands, pp. 81–85.

İrget, M.E., Aksoy, U., Okur, B., Ongun, A.R., and Tepecik, M. (2008) Effect of calcium based fertilization on dried fig (*Ficus carica* L. cv. Sarılop) yield and quality. *Scientia Horticulturae* 118, 308–313.

Kabasakal, A. (1983) Sarılop incir çeşidinde bazı mineral besin maddelerinin mevsimsel değişimi ve toprak-bitki-sürgün ve meyve gelişmesi ilişkileri üzerinde araştırmalar (Research on seasonal changes of some plant nutrients and relationships between soil–plant–shoot and fruit growth). PhD thesis, Department of Soil Science, Faculty of Agriculture, Ege University, İzmir, Turkey.

Kabasakal, A. (1990) *İncir Yetiştiriciliği (Fig Growing)*. TAV Yayın No. 20, Yalova, Turkey.

Kamgar-Haghighi, A.A. and Sepaskhah, A.R. (2015) *Effects of Different Levels of Supplementary Irrigation and Pruning Times on Rainfed Fig Trees in Wet and Dry Years*. National Drought Research Institute, Shiraz, Iran.

Kong, M., Lampinen, B., Shackel, K., and Crisosto, C.H. (2013) Fruit skin side cracking and ostiole-end splitting shorten postharvest life in fresh figs (*Ficus carica* L.) but are reduced by deficit irrigation. *Postharvest Biology and Technology* 85, 154–161.

Maas, E.V. (1990) Crop salt tolerance. In: Wallender, W.W. and Tanji, K.K. (eds) *Agricultural Salinity Assessment and Management*. ASCE Manual Reports on Engineering Practices No. 71. American Society of Civil Engineer, Reston, Virginia, pp. 262–304.

Marschner, H. (1995) *Mineral Nutrition of Higher Plants*, 2nd edn. Academic Press, London.

Melgarejo, P. (1996) La higuera (*Ficus carica* L.). Universidad Politécnica de Valencia, Orihuela, Valencia, Spain.

Mengel, K. and Kirkby, E.A. (2001) *Principles of Plant Nutrition*, 5th edn. Kluwer Academic Publishers, Dordrecht, The Netherlands.

Mordoğan, N., Hakerlerler, H., Ceylan, Ş., Aydın, Ş., Yağmur, B., and Aksoy, U. (2013) Effect of organic fertilization on fig leaf nutrients and fruit quality. *Journal of Plant Nutrition* 36, 1128–1137.

Moreno, D.A., Pulgar, G., Villora, G., and Romero, L. (1998) Nutritional diagnosis of fig tree leaves. *Journal of Plant Nutrition* 21, 2579–2588.

Obenauf, G., Gerdts, M., Leavitt, G., and Crane, J. (1978) *Commercial Dried Fig Production in California*. University of California Agricultural Extension Service Leaflet 21051, Oakland, California.

Olitta, A.F.L., Sampaio, V.R., and Barbin, D. (1979) Estudo da lâmina e frequência de irrigação por gotejo na cultura do figo (Water depth and frequency of drip irrigation on fig crop (*Ficus carica*)). *O Solo* 71, 9–22.

Oukabli, A., Mekaoui, A., Ibnouali-El-Aloui, M., and Bari, A. (2008) Contribution to identification of fig (*Ficus carica* L.) genotypes tolerant to drought. *Acta Horticulturae* 798, 87–93.

Pardossi, A., Incrocci, L., Incrocci, G., Malorgio, F., Battista, P., *et al.* (2009) Root zone sensors for irrigation management in intensive agriculture. *Sensors* 9, 2809–2835.

Phocaides, A. (2007) *Handbook on Pressurized Irrigation Techniques*, 2nd edn. FAO, Rome.
Proebsting, E.L. and Tate, R. (1952) Seasonal change in nitrate content of fig leaves. *Proceedings of the American Society for Horticultural Science* 60, 7–10.
Proebsting, E.L. and Warner, R.M. (1954) The effect of fertilizers on yield quality and leaf composition of figs. *Proceedings of the American Society for Horticultural Science* 63, 10–18.
Şahin, B., Aydın, Ş., and Can, H.Z. (1999) Yellop meyvesi olgunlaşan bazı incir çeşitlerinin bitki besin maddesi içerikleri (Mineral nutrient content of some breba fig varieties). *Journal of Agriculture Faculty of Ege University* 36, 65–71.
Sharifzadeh, M., Kamgar-Haghighi, A.A., Sepaskhah, A.R., Honar, T., Abdolahipour-Haghighi, M., *et al.* (2012) *Factors Contributing to Application of Supplemental Irrigation Technique in Fig Production: Evidence from a Survey in Iran*. National Drought Research Institute, Shiraz, Iran.
Soliman, H.I.A. and Al-Hady, M.R.A.A. (2017) Evaluation of salt tolerance ability in some fig (*Ficus carica* L.) cultivars using tissue culture technique. *Journal of Applied Biology & Biotechnology* 5, 29–39.
Sotiropoulos, S., Paschalidis, C., Kavvadias, V., Chatzissavvidis, C., Koriki, A., and Xirogiannis, G. (2020) Effect of N-P-K fertilization on the yield and nutrient status of fig (*Ficus carica* L. cv. Kalamon) trees grown under Mediterranean conditions. *International Journal of Fruit Science* 20, 1920–1928.
Stover, E., Aradhya, M., Ferguson, L., and Crisosto, C.H. (2007) The fig: overview of an ancient fruit. *HortScience* 42, 1083–1087.
Tapia, R., Botti, C., Carrasco, O., Prat, L., and Franck, N. (2003) Effect of four irrigation rates on grapevine on growth of six fig tree varieties. *Acta Horticulturae* 605, 113–118.
Tepecik, M. (2010) Farklı Potasyum Dozlarının İncirde Kaliteye Etkileri (Effect of different rates of potassium on quality of fig). PhD thesis, Department of Soil Science, Faculty of Agriculture, Ege University, İzmir, Turkey.
Ülkümen, L., Özbek, S., and İleri, M. (1948) *İncir ve Hastalıkları (Fig and Fig Diseases)*. Yüksek Ziraat Enstitüsü Yayınları, Ankara.
Yapar, T. (1976) *Kimyevi gübrelerin mahsüldar incir ağaçlarında kuru mahsül kalite ve kantitesine etkileri üzerinde araştırmalar (Effect of mineral fertilizer on yield and quality of fig)*. Erbeyli İncir Araştırma Enstitüsü, Aydın, Turkey.
Yıldırım, O. (2004) *Sulama (Irrigation)*. Ankara Üniversitesi Ziraat Fakültesi Yayın No. 1540, Ankara, Turkey.
Yıldırım, O. (2013) *Sulama Sistemlerinin Tasarımı (Designing Irrigation Systems)*. Ankara Üniversitesi Ziraat Fakültesi Yayın No.1594, Ankara.

9 Horticultural Practices under Various Climatic Conditions

Moshe A. Flaishman*
Institute of Plant Sciences, Agriculture Research Organization, Volcani Center, Rishon LeZion, Israel

9.1 Introduction

The genus *Ficus* comprising trees, shrubs, and vines occupies a wide variety of ecological niches; most are evergreen, but some deciduous species are endemic to areas outside the tropics, such as fig trees. The fig is an unusual tree in that certain fig types produce two crops of fruit each year and some fig types need pollen. The first crop, called the breba crop, is borne laterally on the growth of the previous season from buds produced in leaf axils. These buds develop in the following spring, and the fruit matures between June and July. The second crop of figs is produced laterally in the axils of leaves on shoots of the concurrent season. Fruit maturation starts in July and may last until the temperature drops, between October and December. At the end of the growth period, the tree defoliates and enters a dormancy period. Reproductive buds that do not produce fruit during the growing season remain dormant over the winter to give rise to the first spring breba crop. Environmental factors such as temperature, photoperiod, humidity, rain, and wind affect fig tree development, quality, and yield. Growing figs under unsuitable conditions may cause crop losses and various types of fruit damage (Flaishman *et al.*, 2008).

Fig trees are widespread in temperate, subtropical, and tropical zones. In general, fig grows best and produces high-quality fruit in Mediterranean and drier warm temperate climates, under intense solar radiation, with summer temperatures ranging between 20 and 40°C following moderate winters, and at low relative humidity (Ferguson *et al.*, 1990; Botti *et al.*, 2003; Flaishman *et al.*, 2008). The tree goes through a period of defoliation followed by dormancy during the winter months. Dormancy is brought about by decreasing day length (photoperiodism), cold temperatures, and sometimes drought conditions. As such, fig growth and production are heavily dependent on climate. Fig production is scattered across different climatic zones worldwide. Reduced temperature in autumn, cold conditions in winter, and growth temperature and rain during the main season will affect tree growth and crop production. Different horticultural practices and growing systems have been developed to increase productivity in these diverse climates. This chapter summarizes the growing conditions that overcome biotic stress and the adaptations of horticultural practices in fig orchard growing systems that enhance fruit quality and productivity in the different climatic zones.

* Email: vhmoshea@agri.gov.il

 Advances in Fig Research and Sustainable Production (M.A. Flaishman and U. Aksoy)
DOI: 10.1079/9781789242492.0009

9.2 Annual Growth and Arrest Under Different Climatic Conditions

9.2.1 Effect of climatic conditions on vegetative growth and tree development

Deciduous fruit trees growing in temperate zones have a dormancy phase of development during the winter that enables the trees to survive unfavorable conditions (Faust *et al.*, 1997). Chilling temperatures or a short photoperiod lead to meristem inactivity. Environmental conditions, temperature, and photoperiod are perceived by the plants. In trees, the photoperiodic signal controlling seasonal growth is perceived in the leaves. Growth cessation, dormancy phases, and bud break are regulated through complex molecular networks involving endogenous signals, plant hormones, oxidative stresses, and energy metabolism (Singh *et al.*, 2017). Changes in temperature or day length induce dormancy in different species; below a certain threshold, they respond by ceasing growth. Each species and variety has a specific chilling requirement for normal bud break, which can be estimated by different chilling accumulation models (Luedeling and Brown, 2011). Bud dormancy has been well studied in the Rosaceae (Lang *et al.*, 1987; Erez *et al.*, 1990; Fadón *et al.*, 2020). In response to short days, symplasmic intracellular communication in the shoot apical meristem is blocked by the deposition of callose (β-1,3-glucan) in the plasmodesmata. It is believed that this blockage restricts the transport of growth regulators, leading to growth arrest and, eventually, dormancy. The plant hormone abscisic acid (ABA) accumulates in the shoot apical meristem in response to short days and contributes to plasmodesmal closure (Tylewicz *et al.*, 2018). Many fruit species are extremely vulnerable to global warming because temperature is the major factor affecting dormancy release and bud break (Cook and Jacobs, 1999; Guo *et al.*, 2014).

In fig, a dropping temperature in autumn will first halt shoot growth, and a typical terminal bud will be produced. This is followed by defoliation, which arrests fruit development and maturation. There is limited information on fig dormancy. The chilling requirement for fig bud break is lower than that for rosaceous fruit trees. Due to its low requirement for chilling units, the fig is well adapted to Mediterranean conditions, with the length of the dormant period depending on the local climatic conditions (Erez and Shulman, 1982). Under hot climatic conditions, in several areas in South America such as Brazil, the tree can grow continuously as an evergreen. In colder weather, however, the tree stops growing, defoliates, develops a typical terminal bud and enters a dormancy period. Kawamata *et al.* (2002) estimated the intensity of bud dormancy in the fig variety Masui Dauphine. Endodormancy of fig buds was classified into three phases: introductory, deepest, and waking. They found that fig shoots can break dormancy in a short period, even when they are in the deepest phase of dormancy. It was concluded that fig shoots can be heated to obtain double cropping or year-round production. Oukabli and Mekaoui (2012) evaluated the pattern of bud break in seven fig varieties collected from Morocco, Italy, Spain, and France. The period from full defoliation to bud break was characterized by small variations (10–20 days) that did not appear to reflect the origin of the varieties. Dormancy was not deep and its period was short. In our fig collection, located in the coastal area of Israel, bud break occurred with less than 200 h of chilling in all varieties between the end of February and the beginning of March, and coincided with the rise in temperature in the spring. Winter pruning treatments stimulated spring growth but did not influence winter dormancy. Hence, successful cultivation of fig trees is possible in a warm growing season following a relatively cold winter to break dormancy.

Cold temperatures cause fruit trees to become hardened to frost damage, and prevent freezing through a decrease in the freezing point. Fig can be grown successfully in cool areas where winters are moderate. However, the common fig is not cold conditioned, and often suffers from cold injury in winter. In cold weather, when temperatures drop to −6°C, shoots and buds will suffer severe damage and will sometimes die. Freezing of the branches and the upper parts of the fig tree in winter limits the cultivation of this tree in temperate regions (Ferguson *et al.*, 2001). Temperatures below −5°C will likely winterkill the branches, but the fig will regrow from its roots and often fruit in the same year. Orchards in China use several techniques to protect figs with mulching. The simplest is to put down a heavy layer of mulch (fallen leaves, salt hay, or wood mulch) around the base of the fig tree to protect the roots. Mulching can also be used to protect the branches. On a young tree, the branches are bent to the ground, pinned, and then mulched. On older trees, the oldest and stiffest branches are pruned and the rest are bent to the ground. In addition, fig trees are wrapped in the winter to protect them from frost damage in cold growing areas. The best strategy is to trim the fig and tie all of the branches together, and then wrap the tree. Wrapping is done in China when the cold weather arrives in November, and removed when the temperature warms up in March.

Karami *et al.* (2018) compared cold hardiness among seven commercial fig varieties at three different developmental stages. All varieties investigated in that study were grown commercially in the Estahban region (central Iran). The fig varieties showed significant differences at the start of cold acclimation. Siah Angir had the highest cold hardiness at the early stage in November, and was more resistant than the other varieties. Field observations also indicated that Siah Angir enters endodormancy earlier in autumn, and is better adapted to cold temperatures. Cold hardiness was estimated by determining the temperature that kills 50% of the samples (LT_{50}). Based on the LT_{50} results, Atabaki showed very good resistance to winter freezing in comparison with the other varieties. In the same growing areas, fig trees are vulnerable to frost damage during spring frosts. The tree tips are injured by the frost, and the whole tree might die under severe frost conditions. As already noted, several practices have been developed to avoid frost damage and allow the fig trees to survive during the cold winter, such as wrapping them in layers of burlap and fallen leaves in late autumn or early winter to keep them from dying back during the cold winter. In other places, during the autumn, trees are pruned, and the shoots are bent and buried for overwintering and to protect the fruiting hardwood from the cold. Long periods of cold weather will kill fig trees. There are some cold-hardy common fig varieties that can grow in winter temperatures of −7°C. Michurinska-10 is a common fig variety in Bulgaria that is suggested to be the most cold-hardy fig in the world. In the coldest places in northern and western Bulgaria, the plant freezes to the ground every year. Damage from the cold starts at temperatures below −16°C to −19°C, and old trees freeze to the ground at temperatures below −22°C. Plants that freeze to the ground will resume growth in spring and produce a main crop that ripens in the summer.

Climate change may influence plants through direct or indirect changes in dormancy. The natural bud break in two different climatic regions in Israel influences fruit production. Due to the low chilling requirements of fig varieties, heat units are the main initiator of vegetative development. In the Israeli desert areas, early bud break and vegetative growth result in enhanced fruit ripening (Fig. 9.1). The earliest ripening variety in the desert is a breba, which is ready to harvest in mid-May, followed by breba varieties growing in the coastal area, which ripen 3–4 weeks later. Common varieties will ripen in the desert and coastal areas by the end of June and mid-July, respectively (Fig. 9.1).

Forcing fig buds during the coldest period allows bud break under a variety of conditions. The use of chemical bud break permits continuous spring growth and may extend fig-growing areas. Spraying chemical bud break growth regulators such as hydrogen cyanamide will induce bud break. The Smyrna-type fig variety Zidi was exposed to 1% and 1.5% hydrogen cyanamide treatment and non-treated control. At 1.5%, shoot length, fruit number per shoot, and yield were significantly increased; bud break was advanced by 10 days and fruit ripening by 7 days compared with the non-treated control. Regarding fruit quality, a dose of 1.5% hydrogen cyanamide significantly increased dry-matter content, while fruit diameter and total soluble solids were unaffected (Gaaliche *et al.*, 2017). However, a regulatory decrease in the use of synthetic hydrogen cyanamide because of its toxic substances has limited this treatment worldwide (Vargas *et al.*, 2008). In the search for new alternatives to break fruit dormancy in temperate climates, Kubota and Miyamuki (1992) conducted a study on the use of garlic paste on grapevines. They found that the most abundant sulfate in garlic is an allyl group, mainly diallyl disulfide, which acts as a dormancy-breaking compound. Botelho (2007) reported successful use of garlic extract for sprouting induction in apple trees. Leonel *et al.* (2015) compared dormancy break in the fig variety Roxo de Valinhos treated with

	JANUARY	FEBRUARY	MARCH	APRIL	MAY	JUNE	JULY	AUGUST
Israeli continental climate: Average temperature	15°C	17°C	18°C	22°C	30°C	32°C	33°C	34°C
Israeli continental climate: Average humidity	64%	64%	65%	50%	35%	42%	42%	44%
Israeli coastal plain: Average temperature	13°C	14°C	15°C	18°C	23°C	26°C	28°C	28°C
Israeli coastal plain: Average humidity	71%	71%	74%	65%	59%	67%	67%	67%

Average temperature Average humidity

Fig. 9.1. Natural bud break in two different climatic regions of Israel. A. Dormancy period. B. Bud break. C. Breba variety ripening. D. Common variety ripening.

hydrogen cyanamide or garlic extract. Application of 1.5–2% hydrogen cyanamide 40 days before bud break or 3% garlic extract were most effective at increasing shoot length and yield.

Fig trees can grow continuously as an evergreen under several climate conditions. Khemira and Mars (2017) described fig production in subtropical south-western Saudi Arabia, an arid region that has a subtropical climate with warm winters on the plains and cool winters in the mountains. They found that under mild winter climates, shoot internodes shorten and the tree takes on a bushy shape, and thus fails to fill its allocated space. To stimulate vegetative growth and induce fruiting, local growers cut back all of the concurrent year's growth in late summer, leaving only the main scaffolds and giving the tree a small goblet form. This practice induces vigorous shoot growth during the autumn and the production of an off-season crop that matures in winter. Similarly, in desert areas in Israel, where the winter temperature is above 6–10°C, defoliation and dormancy are eliminated. Around the Dead Sea area in Israel, where the winter night and day temperatures are between 6°C and 18°C, the fig variety Brown Turkey never loses its leaves or aborts fruitlets, and continues to produce fruit during the winter (Flaishman *et al.*, 2008). In areas where temperatures throughout the entire production cycle range on average between 16°C and 17°C, trees continuously generate basal stems. However, growth slows down, and the tree must be pruned to induce new growth. In tropical climates, such as Malaysia where the temperature ranges between 22°C and 33°C with an average daily temperature of 26.5°C, figs will not enter dormancy. However, biotic and abiotic stresses can slow growth and result in defoliation. A protective construction is needed for year-round production.

Fig-tree architecture, shaping, and pruning

Tree architecture is the result of a balance between the genetic potential of the tree and its environment. The fig cultivation system, plant spacing, and training affect tree growth and fruiting. Architectural analysis is thus a prerequisite for tree management (Lauri and Laurens, 2005). The diversity in architecture among nine Tunisian fig varieties grown in east-central Tunisia was explored during 4-yearly growth cycles (Gaaliche *et al.*, 2016). For all varieties, axis dimensions and internode number decreased with increasing relative branching order. The results showed different patterns of vegetative growth related to the variety and relative branch order. All low-order shoots issued from all varieties had more growth than high-order shoots. A general trend of shoot vigor loss was observed for the different annual growth. To adapt the best training system for a particular variety, the nature of the vegetative annual shoot growth must be evaluated. In Israel, the variety Brown Turkey has moderate growth with a tendency toward a wiping shoot shape. In contrast, the cultivar Autumn Honey has vigorous growth, producing long upright thick shoots.

Interactions between tree architecture and agronomic practices constitute the second step in investigating the branching process of fruit trees (Costes *et al.*, 2006). To adapt the best training system to a particular variety, the nature of the annual vegetative shoot growth must be evaluated. In several Brown Turkey fig orchards in Israel, a trellising training system has been developed (Fig. 9.2). In contrast, Autumn Honey is grown in an open-vase shape, usually

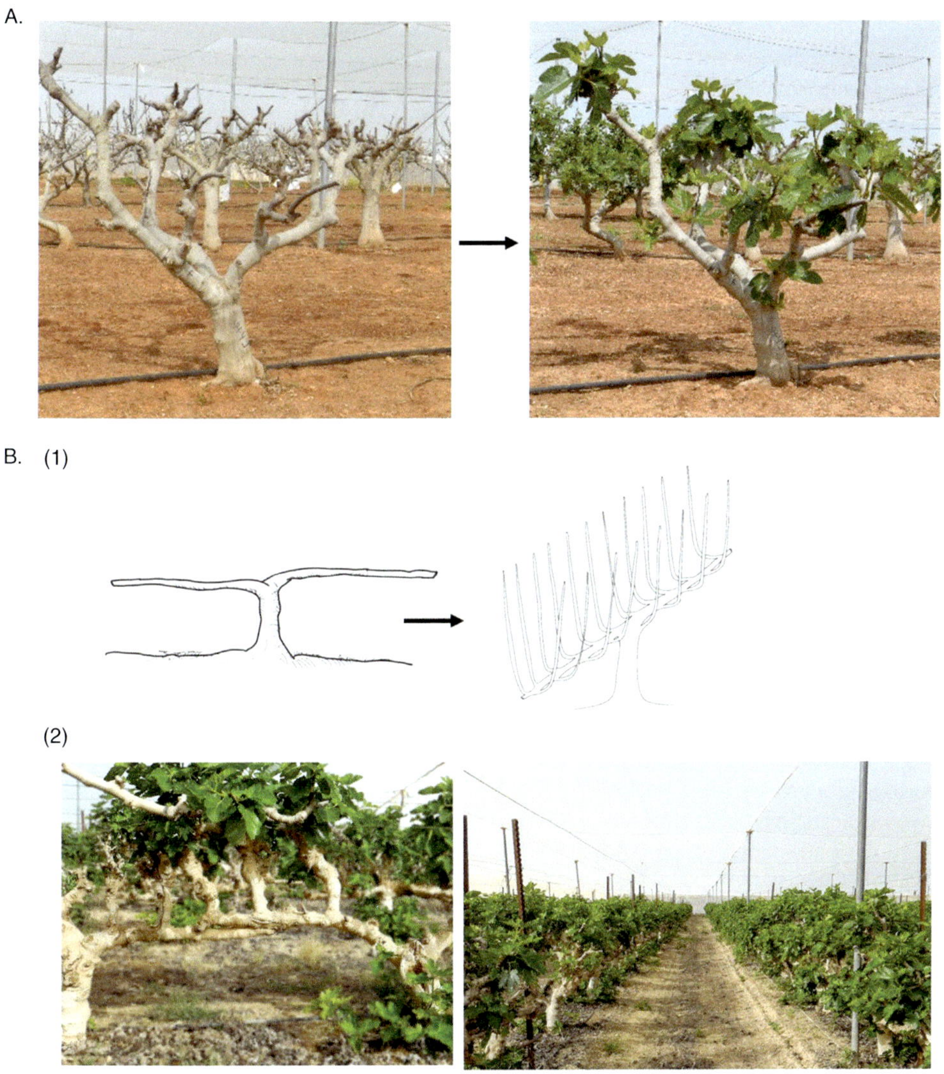

Fig. 9.2. Different training systems in fig orchards. A. Open-vase low fig tree shaping after winter pruning and at bud break. B. Trellis shaping: (1) schematic and (2) trellised fig trees in the orchard.

with four or five main structural limbs (Fig. 9.2); this is the usual practice in most fig orchards worldwide. Although the open-vase shape system is adapted to the wide canopy of the fig tree, it may not be the best system for intensive culture. Erez *et al.* (2003) compared two training systems in a Nazareth variety breba fig orchard, planted in the coastal area of Israel: open vase with four to five scaffolds planted at 5 × 4 m spacing, and a perpendicular V with two scaffolds per tree spaced at 2 × 5 m. The yield was significantly higher in the V-shape system than in the open-vase one. No difference in fruit quality was noted between the two training systems. In Israel and some other countries, spur pruning is used and the young tree is trained to produce four main branches. These are tied down so that they grow almost horizontally. Each year, shoots are allowed to grow vertically from these branches. At the end of the year, they are all cut down to small spurs, similar to the way in which grapevines are managed. When left to grow naturally, the tree canopy can reach 15 m in height. This size is suitable for the production of dry figs that are picked from the ground. In new orchards producing figs for fresh consumption, the trees are kept at 3 m height by pruning to allow easy access during fruit picking (Fig. 9.2).

Pruning in figs is variety dependent. With varieties producing figs for dry fruit, pruning is essential only in the initial years. With fresh-fig-producing varieties, trees should be trained according to fig type. Summer fig fruit varieties can be trained easily in the winter, leaving a few buds to produce the whole tree canopy in spring. With summer main-crop-producing varieties, such as Kadota and Brown Turkey, winter pruning is performed. Winter pruning affects productivity by stimulating new wood growth, which increases the main crop. Gaaliche *et al.* (2011b) studied the interaction between annual vegetative growth and fruiting, and concluded that fruit number in each fig generation is closely related to shoot length and growth during that year. A phenological study was conducted to evaluate the optimum pruning of 1-year-old shoots, comprising four length categories in several fig varieties, and identified the most suitable length category for each variety (Gerber *et al.*, 2010). They found that the three varieties studied required different winter pruning strategies to attain consistently high yields of large fruit. In Israel, with the same fig varieties, two main crops were achieved by applying summer tip pruning. The possibility of producing a second main crop led to the question of when the best time is for tip pruning: on a small branch bearing eight to ten fruit, or a long branch bearing 15–18 fruit? Some compromise has to be adopted, depending on the economics of the two crops and the climatic conditions. If the late second summer crop is the most important economically, pruning after eight to ten fruit will delay the appearance of the autumn fruit but will still be able to regenerate the second autumn crop. In the coastal area of Israel, application of summer tip pruning in July results in resumed shoot growth, giving a second late summer yield. We produce full second crops in areas with the favored condition of high temperature in the autumn.

With breba- and main-crop-producing varieties, the breba crop is formed on the previous season's wood. Therefore, winter pruning will result in loss of this crop. The possibility of producing bearing wood for breba crops of different sizes led to the question of what the optimal bearing branch is. It is always a 1-year-old branch, but it could be a long branch bearing 15 fruit or a small one bearing five fruit. In the San Pedro fig breba variety Nazareth, it was found that small branches bear the best fruit, especially if they are not pushed to grow vigorously by irrigation and high nitrogen fertilization. The reason for this is the ratio between vegetative development and fruit mass. In breba production, the dynamics of vegetative and reproductive development is similar: both buds start to develop at the same time. Thus, the supply of carbohydrates to the fruit is delayed until the developing leaves first reach full size and become exporters of carbohydrates. The time left for the foliage to supply the fruit until ripening is no more than about 6 weeks. Furthermore, in many breba varieties, only the terminal vegetative bud develops. Therefore, in both long and short branches, only a single vegetative bud develops, which has to supply the developing fruit on the branch. Thus, it is clear that fewer fruit drawing carbohydrates from the vegetative sink on the same branch will accumulate more sugars and will have a better chance of developing into quality fruit. With breba varieties, we have to leave all or most of the previous year's wood to obtain early fruit.

This delays pruning to the period after harvest. Especially complicated is the situation in which the tree produces both breba and summer fruit. Under such conditions, a certain compromise has to be adopted, depending on the economics of the crop. If the breba crop is the most important economically, pruning after breba harvest will delay the appearance of the summer fruit, but the whole tree canopy will still be regenerated and carry breba fruit in the following year (Erez *et al.*, 2003). With the Nazareth fig in Israel, we prune the trees immediately after harvest in early June and adopt the limited growth rate of young branches producing short new branches with the summer crop (Erez *et al.*, 2003). Puebla *et al.* (2003) studied pruning dates and intensities in a San Pedro-type fig variety producing mainly an early commercial breba crop. They found significant differences in yield and productivity depending on the dates and type of cut. The highest yields were obtained when pruning was carried out on the earliest date after cropping. Early pruning increased the length of new breba-producing shoots (Puebla *et al.*, 2003; Caetano *et al.*, 2005).

9.2.2 Effect of climatic conditions on fruit development

Fruit growth and development

As with all *Ficus* species, the fig bears a unique closed inflorescence structure—the syconium—that is not present in any other fruit in the human diet. The multiple fig fruit is composed of small individual drupelets that develop from the ovaries enclosed in the succulent receptacle to form a single accessory fruit (Storey *et al.*, 1977). Development of the fig's female fruit is characterized by a double-sigmoid growth curve comprising three phases (Marei and Crane, 1971). Phase I is characterized by a rapid growth in size and a slow rate of increase in fresh and dry weight with no sugar accumulation, and lasts for 5–6 weeks. During phase II, the quiescent stage, the fruit remains nearly the same in size, color, and firmness. There is a great difference among fig varieties in the duration of this second, quiescent period of development. In varieties such as Mission, the second period lasts 3–4 weeks, whereas in the autumn-producing cultivar Autumn Honey, the second period lasts 6–8 weeks. Phase III is the ripening phase, which last 3–5 weeks. During phase III, the fruit exhibit rapid growth, color change, softening, and alteration of the pulp texture to an edible state. During this phase of growth, the fruit accumulates over 70% of its total dry weight and 90% of its total sugar content. Dramatic pigment changes occur during this period in many dark varieties, as the chlorophyll content in the fruit skin decreases rapidly and turns from green to bluish-black (Crane, 1986; Rosianskey *et al.*, 2016). In addition, fruit size increases and tissue softening occurs during the last stage of fig fruit development. The inflorescence buds are borne in the axils of leaves that develop along the same branch at different times along the growing season and therefore fruit maturation is gradual and harvest can last for a long period.

Fig varieties are divided into four types according to their pollination requirement (called caprification in figs) and cropping season (Flaishman *et al.*, 2008). The first type, the caprifig, commonly termed "male fig," is not edible, and is the commercial planting source for pollen. Two edible fig types, Smyrna and San Pedro, require caprification to bear their main crops. The unique San Pedro types yield the early parthenocarpic breba crop, in addition to the pollinated main crop. The fourth type is the common female fig, also called "persistent" fig, which yields one or two crops per year, with or without caprification (Galil and Neeman, 1977; Flaishman *et al.*, 2008). Pollination of the different fig types is a consequence of a complex symbiotic relationship with the fig wasp. The pollen-carrying wasp *Blastophaga psenes* accesses the inner fig cavity through the fruit ostiole and pollinates the female flowers (Galil and Neeman, 1977). Depending on this random process of pollination by *B. psenes* wasps in common-type fig fruit, both caprified and non-caprified fruit may develop on the same branch.

Unlike the fig syconia that develop on the same branch in the same season, breba figs have different patterns of growth and development. Shortly after initiation, the breba syconium enters winter dormancy. In spring, the breba syconium resumes growth, which continues for 9–11 weeks until maturation in June–July. About 2–3 weeks before maturation of the breba fruit, growth rate and sugar accumulation increase significantly. As all breba fig syconia on the same branch are at similar developmental stages, the fruit-harvesting period is shorter, lasting only 2–3 weeks.

Fig is one of the earliest domesticated plants, and human-assisted caprification of its Smyrna- and San Pedro-type varieties is one of the oldest known horticultural practices (Zohary and Spiegel-Roy, 1975; Kislev *et al.*, 2006). Chai *et al.* (2017) performed transcriptomic analyses of young breba and main-crop figs at four stages of developmental phase I, including two young preanthesis fruit stages, and flowers at anthesis and post-anthesis stages. They showed that divergence of the parthenocarpic fates of breba and the main crop in San Pedro-type figs occurs at a very early stage of syconium development, where auxins may play a major role, in the absence of pollination. Main-crop abscission at the end of phase I was modulated by differentially expressed phytohormone metabolism and signaling pathways. Auxin application was found to stimulate fruit set in non-parthenocarpic Smyrna-type figs (Crane and van Overbeek, 1965; Crane, 1986). Various applied growth regulators, including auxins, gibberellins, and cytokinins, could induce parthenocarpy in the Smyrna-type variety Calimyrna (Crane, 1986). The auxin-induced parthenocarpic fruit take somewhat longer to mature than the caprified ones, but their morphology is similar.

The third edible type, the common-type fig, has facultative parthenocarpic fruit, producing both pollinated and parthenocarpic fruit during the summer. Morphologically, the pollinated fig syconium creates true fruit, whereas the non-pollinated fig syconium presents an enlarged inflorescence with multiple long-styled female flowers. This feature makes the common-type fig a good platform to study the effect of fig caprification on fruit development, quality, and post-harvest performance. Development and ripening processes differ in pollinated and parthenocarpic fruit. While the facultative parthenocarpic common-type fig fruit serves as a receptacle for flower development, it becomes fleshy via either pollination or a parthenocarpic process. Rosianski *et al.* (2016a,b) performed advanced analyses of parthenocarpic versus pollinated common-type Brown Turkey figs. The pollinated fruit showed different developmental processes. The ripened pollinated fruit were round, in contrast to the pear-like shape of the parthenocarpic fruit. The pollinated fruit also had a larger diameter and weight, and improved firmness compared with the parthenocarpic fruit. At harvest, the pollinated fruit exhibited more commercially desirable physical and taste characteristics, with advanced fertile nutlets compared with the sterile, undeveloped, non-bearing nutlets of the parthenocarpic fruit.

In developmental phase III of the common-type fig fruit, ripening occurs with fruit growth, a change in peel color, and alteration of the pulp texture to an edible state (Flaishman *et al.*, 2008). Furthermore, fruit-quality attributes, such as fresh weight, texture, firmness, fruit diameter, soluble solids content, total acidity, are affected as the figs mature. Fruit variety and maturity affect consumer preferences. Consumer acceptance is higher for tree-ripened figs than for commercially matured ones, with citric acid, total sugars, flavor, and texture contributing to acceptance. Citric acid content is higher in unripe fruit than in ripe fruit, while total sugars, flavor, and texture are lower in unripe fruit and change dramatically during fruit ripening. Changes in cell-wall polysaccharides occur within the distinct tissues of the receptacle and pulpy drupelets during sequential ripening in fig fruit (Owino *et al.*, 2004).

The distinct physiological green phase II and ripening phase III differ significantly in their gene-transcription patterns in both pulp and inflorescence tissues. Gas chromatographic analysis of whole fruit was performed to determine ripening-related hormone levels. Ethylene, ABA, and auxin increased during fruit ripening. The tissue-specific transcriptome seemed to reveal different metabolic gene patterns for ethylene, auxin, and ABA (Rosianski *et al.*, 2016a; Lama *et al.*, 2019). In mature ripe fig fruit, the receptacle tissue, of vegetative origin, and the reproductive pulp tissue with drupelets are clearly distinct. Fruit ripening is differentially coordinated by the reproductive and vegetative parts of the syconium. Chapter 12 (this volume) details the transcriptional and metabolic changes during fig-fruit ripening in the receptacle and pulp tissue.

The ripening pattern of fig differs from that of other fruit. On most fruit trees, multiple fruit ripen simultaneously, whereas fig fruit ripen individually on the shoot. Therefore, the harvest period can last 8–10 weeks. Increasing the number of harvested fruit, to minimize harvesting costs for the growers, is a challenge. Uniform fig maturation is stimulated by applying olive oil to the ostiole, spraying ethephon, or applying ABA during late phase II of fruit development. Lama *et al.* (2019) demonstrated that on-tree spraying of exogenous ABA stimulates growth and ripening and integrates the ripening process for a cluster of different-sized mature fruit from the same branch. In two consecutive years, application of 2.27 mM ABA to Brown Turkey figs with pinkish ostioles, and 2.84 mM ABA to Autumn Honey figs, in clusters of three to five fruit per branch, resulted in uniform early onset of ripening. ABA application resulted in significantly increased fruit size and weight, and lower firmness compared with controls for both varieties (Fig. 9.3). During storage of ABA-treated fig fruit, weight, diameter, firmness, acidity, and soluble solid contents were not significantly different from their respective controls. Thus, normalized concentrations of exogenous ABA sprayed on clusters of fruit of the selected size increased the yield of fresh figs up to fivefold, reducing harvesting costs and accommodating market demand with no adverse effects during commercial storage (Lama *et al.*, 2019).

Marker assistance to determine timing of horticultural practices

Common fig is a widely cultivated, deciduous tree characterized by remarkable vegetative growth and good adaptation to different environmental factors and climates. The use of horticultural practices is based on understanding the physiological status of the fig tree. For the application of horticultural practices to overcome stress effects, we need an early and simple detection procedure to analyze tree status. Changes in plant photosynthesis rate, stomatal conductance, and stress-related gene expression can be used as markers for optimal timing of efficient horticultural practices.

Studies have shown that fig photosynthetic rate reflects tree status, environmental stresses, and fruit development. A study of seasonal changes in photosynthetic activity and stomatal conductance from April (leaf emergence) to August (fruit maturation) showed significant changes during leaf development (Vemmos *et al.*,

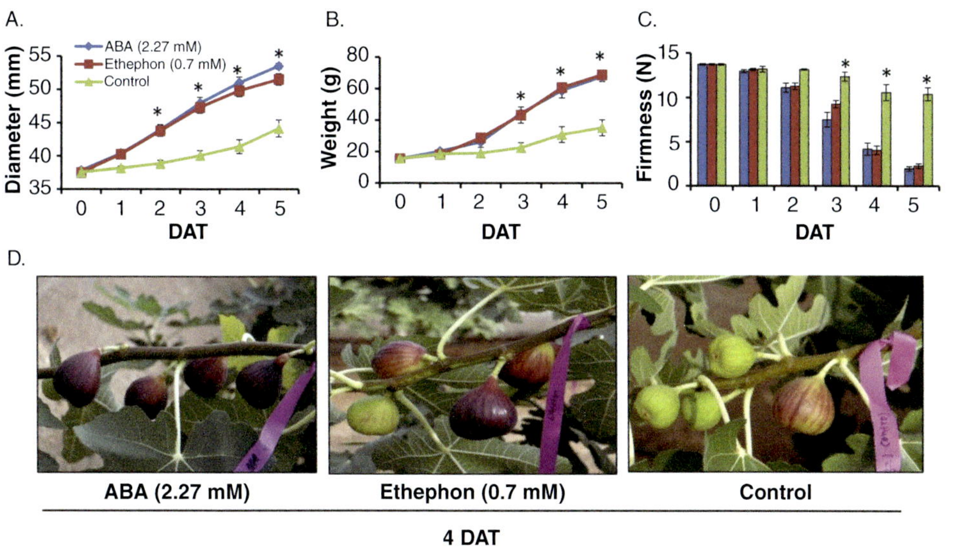

Fig. 9.3. Effect of exogenous abscisic acid (ABA) application on fig fruit variety Brown Turkey (summer crop). A. Fruit diameter. B. Fruit weight. C. Fruit firmness. D. Images showing the colors and ripening stages of fig fruit 4 days after treatment (DAT). Fruit treated with ethephon were used as a positive control. *Indicates significantly different when compared to control. Data in A–C are means ± standard error of 15 fruit from three to four clusters. Modified from Lama *et al.* (2019).

2013). Can and Aksoy (2007) suggested that the photosynthetic rate is high in August due to the increase in carbohydrate demand during fruit growth and maturation. Fig photosynthetic rate was genotype dependent, with some varieties showing better adaptation to environmental stresses (Can *et al.*, 1998; Flaishman *et al.*, 2008). Ammar *et al.* (2020) evaluated the seasonal variation of some physiological traits of two different fig types (Smyrna and San Pedro) grown under conditions of optimal water availability by measuring stomatal conductance and gas-exchange characteristics. They demonstrated the adaptive behavior of fig in coping with stressful situations such as high temperature.

Fig trees and fruit contain typical latex-secreting cells. Latex is the cytoplasmic fluid of laticiferous tissues. The properties of fig latex have been attributed to the presence of several hydrolytic enzymes. It has been suggested that latex secretion is a defense against mechanical wounding and/or herbivores, such as insects, vertebrates, microorganisms, and fungi (John, 1993). Expression of complementary DNA encoding latex, and the stress-related genes peroxidase, trypsin inhibitor and type I chitinase was investigated. The stress-related genes showed differential expression after various abiotic stresses or plant hormone treatments, suggesting crosstalk between the signal-transduction pathways elicited by abiotic stresses and hormones in plants (Kim *et al.*, 2003).

Soil salinity can be a major limiting factor for tree productivity. It is one of the most important environmental factors reducing the growth, development, and production of plants (Sevengor *et al.*, 2011). Fig is a moderately salt-tolerant crop (Golombek and Lüdders, 1990). The response of commercial fig varieties to salinity stress indicates significant variation in morphological characteristics and photosynthetic efficiency. Salinity stress results in reduced leaf area and leaf water content, and decreases in stem and root length, fresh and dry fig productivity, fruit quality, and dry weight (Salimpour *et al.*, 2019). Salt-tolerant fig varieties may be required when using continuous groundwater irrigation that leads to the build-up of mineral salts in the rooting zone. Some fig varieties that show moderate resistance to salinity can be grown on saline land for fig production. A study examined the growth and photosynthetic rate of several commercial Iranian fig varieties exposed to six salt treatments, and showed that Siyah and Sabz were the most salt tolerant, while Shah Anjiri was the most salt sensitive; the latter could not balance transpiration rate and leaf water content under salt treatment higher than 4 dS m^{-1} (Salimpour *et al.*, 2019).

Vangelisti *et al.* (2019) performed a transcriptomic analysis of the *F. carica* variety Dottato in leaf samples after 24 and 48 days of exposure to 100 mM NaCl. Relatively small transcriptome changes were observed after 24 days of salt treatment, showing that fig plants initially tolerate salt stress and qualify as a potential crop for drained, unfertile soils or coastal areas. However, major transcriptomic changes were observed after 48 days of salinity. Seven weeks of 100 mM NaCl dramatically changed the repertoire of expressed genes, leading to activation or reactivation of many cell functions. They identified several salt-regulated genes, providing information for future studies aimed at evaluating the effect of brackish-water irrigation on field-grown fig trees to identify early salt stress and salt-tolerant fig species.

In addition to horticultural practices to overcome saline conditions, transgenic fig plants have been developed (Metwali *et al.*, 2015). The vacuolar Na^+/H^+ antiporter gene (*AgNHX1*) isolated from the halophytic plant species *Atriplex gmelinii* was introduced into fig variety Black Mission via *Agrobacterium*-mediated transformation. The authors demonstrated that expression of *AgNHX1* in fig plantlets conferred tolerance to salt stress in the transgenic plantlets. The physiological mechanisms responsible for enhanced tolerance to salt stress could be accounted for by enhanced accumulation of proline, inhibition of sodium ion (Na^+) accumulation, and retention of potassium ions (K^+) via upregulation of *AgNHX1*. The areas of fig production reveal large climatic variability that severely affects fruit quality. Salt-tolerant fig varieties can be used to expand fig-tree orchards under diverse climatic conditions while maintaining high fruit quality.

9.3 Horticultural Practices to Overcome Fruit Disorders

Growing fig trees in unsuitable conditions (under abiotic or biotic stress) may cause crop loss and fruit damage. Environmental conditions such as high and low temperature, drought, rain, hail, and wind can reduce fruit size, fruit quality, and productivity. In this section, we describe fig fruit cracks, and disorders in fruit size, color, yield, and caprification.

9.3.1 Fruit cracks

Fruit cracking is an important physiological disorder that causes fruit loss and reduces fruit marketability and quality. Cracking has been described as a physical failure of the fruit skin in the form of fractures in the cuticle or skin, which typically do not penetrate the flesh. In contrast, splitting is an extreme form of cracking that does penetrate the flesh. Both physiological disorders are a result of changes in fruit–water relationships and fruit skin properties. Fruit cracking in fig is dependent on the variety. Some varieties have a tendency toward ostiole-end splitting, whereas in others, the cracks cover the whole fruit peel. Splitting and cracking are the result of sudden changes in internal fruit pressure brought on by cool temperatures and/or high humidity as the fruit matures (Crisosto *et al.*, 2011). Environmental conditions and orchard irrigation can increase fruit cracking, which occurs when the fruit takes in water faster than the skin can stretch. Numerous deep cracks may reflect irrigation problems. Watering fig trees consistently may reduce cracking. Kong *et al.* (2013) showed that fig skin side cracking and ostiole-end splitting, which occur during fruit growth and development, may be decreased in some varieties by changes in irrigation management before the cracking occurs. Deficit irrigation (55% of crop evapotranspiration) applied when 25% of the fruit were ripening reduced fruit skin damage and increased the percentage of intact fruit, without negatively influencing other quality attributes. In coastal areas, rain may cause fruit splitting. As the fruit matures, the incidence of peel cracking and ostiole-end splitting increases, leaving marks on the peel tissues of the ripened fruit. The incidence of such physiological disorders decreases fruit quality after harvest and during cold storage. Strong winds in the ripening season whip the foliage and cause scarring of the fruit of, for example, Kadota and Brown Turkey varieties. In Israel, Brown Turkey orchards are covered with netting, which reduces the fruit cracking and scarring and improves postharvest fruit quality. Splitting can also result from too many developing seeds and can be controlled by balancing the number of caprifigs used during caprification.

9.3.2 Fruit size

As already noted, the development of female fig fruit is characterized by a double-sigmoid growth curve, with fruit size increasing two- to threefold during maturation. To facilitate the fruit's rapid growth, an increase in carbohydrates is needed. However, although the final fruit size is dependent on variety, in many varieties, environmental conditions may dramatically affect this parameter. Fruit quality of the main summer crop of the common fig Brown Turkey was evaluated in orchards located in the coastal and desert areas of Israel. During fig maturation in mid-summer, day and night temperatures in the desert area are on average 10°C higher than in the coastal area, with maxima of 40°C and 28°C, respectively. The environmental high-temperature stress reduces the fig's photosynthetic rate and carbohydrate production, resulting in fruit that are about 20% smaller. However, the main benefit of growing figs in desert areas is early maturation. The early-matured fruit are harvested between May and July, and therefore most of the production escapes the high-temperature stress, resulting in the development of high-quality early fruit (Fig. 9.1).

Ferguson *et al.* (2001) reported the use of trunk girdling to increase dried Black Mission fig size and value per tree. A single trunk girdling from early June through mid-July significantly increased fig size, as well as tree yield. For three successive seasons, girdling in early June produced, respectively, 22, 28, and 29 kg per tree of dried figs, compared with 18, 20 and 21 kg per tree on control trees. Similarly, these girdling dates produced higher fig size classifications compared with the non-girdled controls.

9.3.3 Fruit color

Color directly affects fruit quality and consumer preference. Anthocyanin in fig syconia, the female flower tissue, is found in the receptacle and peel tissues; however, anthocyanin pigmentation of the receptacle and peel differs. Anthocyanins in the female flower tissue were identified mainly as pelargonidin 3-glucoside and cyanidin 3-rutinoside; in the peel, the major anthocyanins were cyanidin 3-*O*-glucoside and cyanidin 3-rutinoside (Solomon *et al.*, 2006; Wang *et al.*, 2019). Anthocyanin content was significantly higher in the female flower tissue versus the peel before fig ripening. At ripening, the anthocyanin content in the peel was five times higher than that in the female flower tissue. Pathway expression in the peel was strongly regulated by light. Light deprivation treatment strongly inhibited anthocyanin pigmentation of the peel but not the female flower tissue. Poor coloration of net-house-grown fig fruit due to insufficient sunlight stimulation reduces marketability. Rodov *et al.* (2012) studied color improvement of harvested Brown Turkey figs through the use of pulsed light emitted by a high-energy xenon flash lamp postharvest. This treatment seemed to be a feasible means of compensating for insufficient sunlight stimulation of color development in figs.

9.3.4 Yield

Fruit-tree yield depends strongly on the balance between vegetative growth and fruiting. Environmental factors such as temperature, photoperiod, and humidity affect the development and yield of fig trees. Growing fig trees under unsuitable conditions may cause crop losses. In fig trees, the number and size of fruit buds are closely related to the vegetative growth of the tree and to the size of the crop maturing in summer (Flaishman *et al.*, 2008). This balance is particularly complex in these trees, where two (one breba and one main crop), and sometimes three (one breba and two main crops) generations of figs develop along an annual shoot and compete with consecutive growth seasons. During winter dormancy, fig syconia at phase II will drop, while fig syconia at phase I will form scales that protect the developing fruit from the low winter temperatures. In this case, the fruit remain quiescent and produce a breba crop in the spring. In general, autumn production will reduce the number of dormant buds and therefore, the next year's breba crop production. Breba crop production can be successful in relatively moderate winters. When grown in cooler areas, fig trees are often injured by early or late frosts that kill the younger branches and can damage the syconium buds of breba fruit. When shoots and breba fruit initiate in April, a shortage of available carbohydrates and competition between the new foliage and the breba syconium can cause syconium drop and elimination of the breba crop. Application of anti-gibberellins to San Pedro-type Nazareth figs, or nipping of the terminal bud, will temporarily stop new foliage development and allow breba growth, by reducing competition and syconium drop (Yablowitz *et al.*, 1998).

The number of main crops produced by fig trees directly influences their carbohydrate balance. Smyrna-type figs, producing a single main crop, have maximal starch concentrations in early spring, mid-summer, and late autumn (Crane, 1986). Common-type fig can produce up to two main crops. The first summer crop is in July. Early summer pruning can induce a second shoot growth wave and consequently a second, late-autumn crop. In areas with night temperatures above 8°C, fig trees produce a second fruit crop in November and December and fruit maturation continues until leaf fall (Flaishman *et al.*, 2008).

Gaaliche *et al.* (2011b) studied the interactions between vegetative and crop generations of five fig varieties throughout three growth seasons. In all varieties, annual shoot extension exhibited two consecutive phases of unequal duration, separated by a cessation in growth. The number of reproductive buds and fruit in each generation is closely related to shoot length and, more precisely, to the proportion of shoots corresponding to growth rhythmicity within a year. Thus, the number of breba figs in a concurrent year depends on shoot length in the previous season, whereas the number of main-crop figs is highly correlated with shoot length in the concurrent season.

Optimum tree density along with appropriate canopy management increase the bearing surface per unit land area, leading to higher productivity. An open-field trial study of the fig cultivar Deanna and variety Poona under tree densities of 1000, 800, 666, 571, and 500 plants ha^{-1} indicated that high-density planting (1000 or 800 plants ha^{-1}) was superior to other planting densities (Chithiraichelvan *et al.*, 2017). High-density planting of Poona and Deanna at 1000 plants ha^{-1} with initial training of trees to an open-vase frame and annual pruning of the previous season's shoots to the basal six nodes in September showed promise for maintaining tree vigor and fruit yield. There was no difference in fruit-quality attributes, such as total soluble solids and acidity, among the trees grown at the different planting densities. High-density planting of Poona and Deanna at 1000 plants ha^{-1} enhanced fruit productivity. During the first 6 years in the orchard under the mild tropical, semi-arid conditions, the mean productivity of Poona and Deanna was 14.02 t ha^{-1} and 9.25 t ha^{-1}, respectively.

Fig yield is generally dependent on the fig variety, growing conditions, and harvest. As already noted, they carry two crops in a year, the second main crop in late summer and autumn, and the first or breba crop produced from May to June. Fig yield is usually between 5 and 12 t ha^{-1}. The combination of careful practices under optimal growing conditions and in sophisticated growing facilities could increase, and even double, the regular yield. For Nazareth figs grown in a nethouse in Israel, Yablowitz *et al.* (1998) described 12 t ha^{-1} production of breba crop and an additional 5–7 t ha^{-1} produced for the main summer crop. In Israel, in several orchards growing Brown Turkey, fig production from July to December amounted to up to 25 t ha^{-1} of high-quality crop (M.A. Flaishman, unpublished data). Similarly, in Spain, the mean of 4 years of crops for the Pellejo Duro variety was 25.8 t ha^{-1} (Moreno *et al.*, 1998).

9.3.5 Caprification

Fig varieties are divided into four types according to their pollination requirement (caprification) and cropping season (Flaishman *et al.*, 2008). Pollination of the different fig types relies on a complex symbiotic relationship with the fig wasp. Caprified figs are usually larger and greener than the non-caprified ones (Condit, 1947; Oukabli *et al.*, 2001; Michailides *et al.*, 2008). Studies on the effect of caprification on fruit quality in the common-type fig variety Kadota reported higher fruit weight together with a tendency toward splitting, and changes in several fruit characteristics, including darker skin and red-colored inflorescences, in addition to the sweeter and richer flavor of the pollinated fruit (Gaaliche *et al.*, 2011b). Caprified San Pedro-type fruit showed modifications in their cell-wall components, which moderately affected several textural parameters, as well as stimulation of anthocyanin and total polyphenol biosynthesis (Trad *et al.*, 2013).

Some of the best dried- and fresh-fig varieties, such as Sarılop (= Calimyrna) and Bursa Black, are Smyrna-type figs, which require pollination for fruit set. Commercial Smyrna-type figs are pollinated in early summer with pollen from the profichi caprifig. In general, one caprifig tree is needed for every 15–20 Smyrna-type trees. Planting the caprifig trees within the block is not recommended, as pollination is not even: the trees closest to the caprifigs can be overfertilized and their fruit can split, while more distant trees may not be pollinated. Caprifig trees should be planted in a separate block. The profichi caprifigs (with wasps) are picked and placed in wire baskets around the orchard when the first wasps start to emerge, with six or seven figs per basket. The profichi caprifigs need to be replaced every 3 days for about 3 weeks, as not all of the syconia of Smyrna-type figs are receptive at the same time. It is useful to have more than one variety of caprifig so that the pollination period is extended. The effect of the pollen source on the characteristics of Sarılop fruit was studied, and the caprifig variety had a significant effect on Sarılop fruit size (Aksoy *et al.*, 2003).

Gaaliche *et al.* (2011b) examined the effect of pollination intensity and frequency, and of pollen source on Bouhouli (San Pedro type) and Zidi (Smyrna type) fig productivity and fruit quality, using different numbers of caprifigs and repetitions of caprification. The results indicated that caprification four times at 4-day intervals could be efficient at improving fruit set and fruit quality. Compared with other caprifigs, the pollen of caprifig Djebba 2 resulted in higher productivity and fruit weight, thicker flesh, and a better fruit taste. These results suggested that each variety requires an adequate pollinator caprifig.

Rosianski *et al.* (2016a) examined the influence of pollination on the development of the common-type facultatively parthenocarpic Brown Turkey fig fruit, and on its performance under commercial fruit storage conditions. A comparison of parthenocarpic with pollinated fruit development clearly revealed the superiority of the latter over the former in growth, width, weight, firmness, and taste qualities. It was clear that the pollinated, ripened fruit are larger in both diameter and weight than the parthenocarpic fruit. Similarly, several authors have shown that strongly human-assisted caprified fruit are larger and heavier than non-caprified ones (Gaaliche *et al.*, 2011a,c; Trad *et al.*, 2012). It is likely that in all of the above cases, the larger and heavier fruit are a consequence of inflorescence swelling, seed development, and closure of the inner cavity. The taste of the ripened pollinated fruit was more sour than that of the parthenocarpic fruit, with higher total acidity, whereas the caprified fruit described by Gaaliche *et al.* (2011a) and Trad *et al.* (2012) showed no differences in acidity compared with non-caprified fruit. The soluble solids content was lower in pollinated fruit than in caprified fruit in the study by Gaaliche *et al.* (2011a), which showed no significant difference in total soluble solids compared with non-caprified fruit. The difference in fruit quality could have been due to the use of different varieties. Storage and shelf life of the pollinated common-type figs were improved compared with the parthenocarpic fruit. Stored pollinated fruit retained their commercially favored physiological characteristics, while also maintaining their superior size, higher firmness, and better taste parameters. The differential development of pollinated and parthenocarpic figs resulted from the presence or absence of functional seeds. Comparison of the pollinated and parthenocarpic tissues during fruit ripening showed differential gene expression, especially in the seeds developing in the fruit inflorescence. Tissue-specific transcriptome analysis revealed seemingly different metabolic gene patterns for ethylene, auxin, and ABA in pollinated versus parthenocarpic fruit (Fig. 9.4). Fig-fruit ripening is coordinated by the reproductive part of the syconium, and the differences in ABA production between pollinated and parthenocarpic fig fruit might be key to their different ripening characteristics (Rosianski *et al.*, 2016b).

Many areas worldwide lack caprifigs and the pollinator wasp. Therefore, the symbiotic cycle of pollination is not available and the caprifigs are not grown in Smyrna- or San Pedro-type commercial fig orchards. In countries like China, only common-type figs are currently available.

9.4 Fruit Maturation: Harvest

Fig fruit must be allowed to ripen fully on the tree before they are picked. They will not ripen if picked when immature. Application of ethephon to the dried-fig variety Mission late in the second period of fruit development stimulates growth and ripening within 8 days. Dramatic pigment changes occur during this period as the fruit skin turns from green to black (Puech *et al.*, 1976). The chlorophyll decreases for an additional 4 days, at which time the fruit reach their maximum size and are considered ripe. The ethephon speeds up ripening and thus shortens the harvest period by as much as 10 days. Proper timing of the growth regulator is crucial to fruit quality. Similarly, application of ethrel is known to enhance fruit development, resulting in early and compact ripening. Ethrel applications at the end of the second fruit-growth period on Bursa Black fruit advanced fruit ripening by 5 days compared with the untreated control (Çelikel *et al.*, 1996). In California, where most figs are grown for drying, ground sweepers harvest them mechanically during September and October. After harvest, the dried figs are washed and can be stored for a few days at 0–1°C. The fruit is then dried in the sun or by using an electric dryer at a temperature of 60–70°C before processing as dried figs. Fresh figs are picked when they begin to soften and their color changes, indicating maturity. When picking, gloves should be worn to prevent damaging the fresh fruit and to prevent skin irritation caused by the ficin-containing white sap that exudes from the broken stem. As fresh figs ripen irregularly, picking should be done daily or weekly during the long harvest period (4–6 weeks). Although figs have to be picked fully ripe, this is problematic for storage; therefore, in commercial orchards, immature figs are harvested, and stored and shipped under cold conditions.

9.5 Growing Fig in a Controlled Environment

Fig will produce high-quality fruit under high summer temperatures and moderate winters with low relative humidity. The lack of heat units limits the expansion of cultivation into

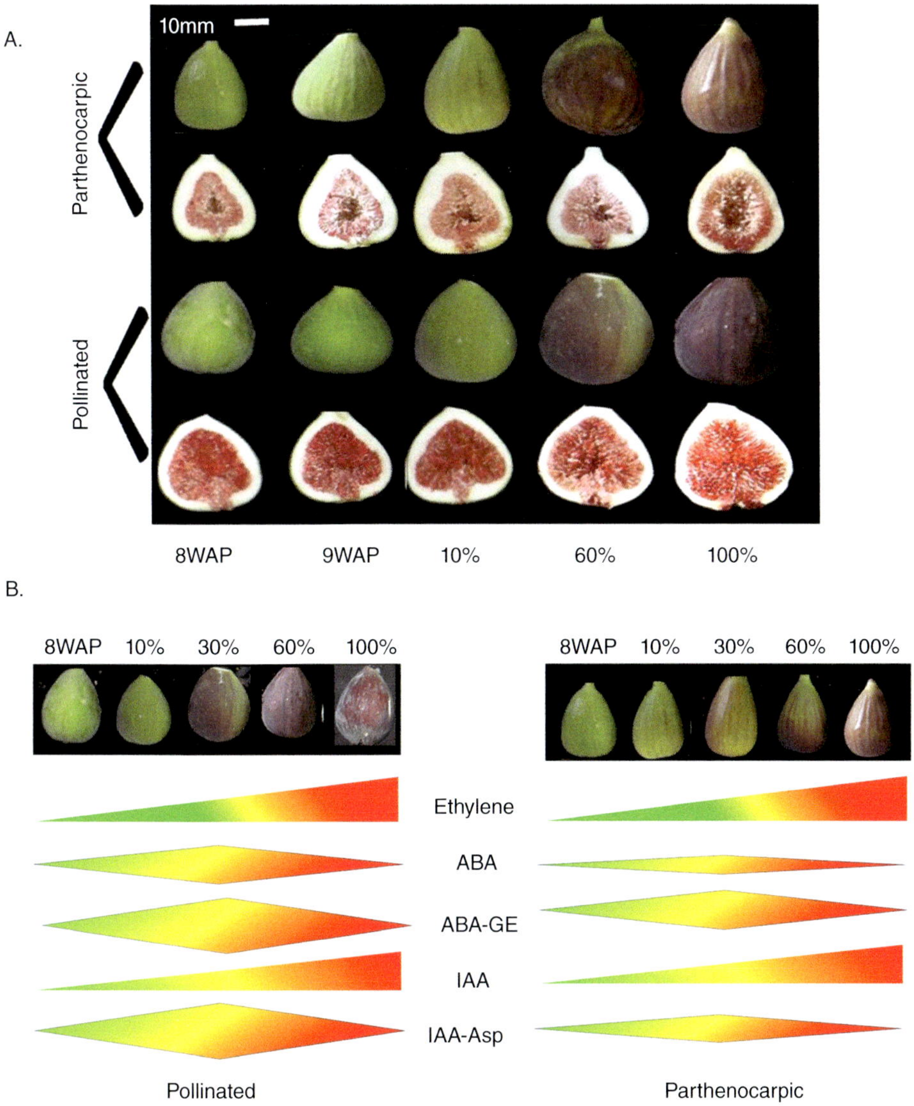

Fig. 9.4. Fig-fruit developmental stages and schematic model of hormone production during fruit ripening. A. Developmental stages of parthenocarpic and pollinated fruit displaying pulp and inflorescence. B. Hormone production level is expressed by schematic width. WAP, weeks after pollination; X%, % of fruit ripening; ABA, abscisic acid; ABA–GE, abscisic acid glucosyl ester; IAA, indole acetic acid; IAA-Asp, indole acetic acid–aspartic acid. Modified from Rosianski *et al.* (2016b).

low-temperature regions. Under unsuitable climatic conditions, open-field fig production has some problems, such as a high incidence of pests and diseases, fruit decay during the rainy season, and severe damage caused by frost, leading to lower yields and eventually to neglected plantations. In these situations, production systems in controlled environments or protected agriculture are good alternatives. Protected environmental conditions, such as in nethouses or greenhouses, increase productivity, fruit quantity, and commercial opportunities.

9.5.1 Nethouses

Nets are widely used in crop production as barriers against abiotic and biotic factors that compromise crop productivity and fruit quality (Sivakumar *et al.*, 2018). Nets also reduce farmers' reliance on agrochemicals, making them an environmentally friendly alternative to chemical pesticides (Briassoulis *et al.*, 2007). The ability of nets to modify the microclimate may enhance crop quality, uniformity, and productivity. Nets may also protect plants against undesirable environmental elements. It is believed that some nets reduce the maximum temperature during the day. This decrease in temperature is beneficial for trees in warm climates, as it may increase the rate of photosynthesis during the hotter hours of the day and protect against sunburn (Kalcsits *et al.*, 2017). Some experiments have shown that nets increase the nightly minimum temperature due to a greenhouse effect and limited night-time radiation (Iglesias and Alegre, 2006). Nets can partially transmit and diffuse solar radiation. This is important for photosynthesis in shaded leaves at low levels in the tree (Tanny, 2013). Moderate shading can decrease leaf temperature and evaporative demand, thereby increasing photosynthesis. This increase may lead to higher carbohydrate production and enhanced water-use efficiency, resulting in better yield and quality (Manja and Aoun, 2019). Studies have shown that covering trees with black nets reduces the transmission of photosynthetically active radiation, thereby reducing fruit coloration. In contrast, the use of a white transparent net is beneficial to many fruit-tree species, including figs. Erez *et al.* (2003) reported growing Nazareth breba fig in a nethouse covered with a 17-mesh UV-treated white net manufactured by Klayman Meteor Ltd., Israel, at a height of 3.5 m above the orchard. In the covered structure, trees grown in the open-vase and V-shaped systems with the two arms perpendicular to the row axis were used. Trees were normally pruned after harvest to keep their height below 3 m. In Israel, Brown Turkey fig orchards, in both the coastal and desert climate regions, are widely covered with nets, most commonly 17- to 25-mesh white nets. The net-covered fig crop has prolonged harvesting windows and improved fruit quality, and the net provides a barrier to various insects that normally compromise crop productivity and fruit quality. The net covering was found to be very efficient at protecting against the fig fly *Silba adipata*, the Mediterranean fruit fly *Ceratitis capitata*, and the fruit fly (*Drosophila* species), all of which cause severe damage in warm growing regions toward fruit maturation (Yablowitz *et al.*, 1998).

9.5.2 Greenhouses

In horticulture, greenhouse production systems are commonly used to produce a wide range of fruit and vegetables. Protected cultivation systems (e.g. greenhouses and high tunnels) provide numerous advantages, such as high yield, premium quality, prolonged harvest windows and production seasons, protection against adverse environmental stresses (including freezing), and controlled microclimatic conditions (Demchak, 2009). The easiest way to overwinter figs in cold climates is to take advantage of the microclimates achieved through the use of protected growing facilities. The greenhouse production system is becoming popular in fig cultivation as a means of extending the harvest window and season of fig fruit. In China, fruit and vegetables are grown out of season solely through the use of solar energy, even when it is freezing outside. The Chinese passive solar greenhouse has three walls of brick or clay. Only the southern side of the building consists of transparent material (usually plastic foil) through which the sun can shine. During the day, the greenhouse captures heat from the sun in the thermal mass of the walls, which is released at night. At sunset, an insulating sheet—made of straw, pressed grass, or canvas—is rolled out over the plastic, increasing the isolating capacity of the structure. The walls also block the cold, northern winds, which would otherwise accelerate heat loss in the greenhouse. The indoor temperature of the structure can be up to 25°C higher than the outdoor temperature. To explore methods that are key to annual production of fig fruit in greenhouses, Masui Dauphine fig was used to produce fruit through the summer and in a solar heating greenhouse during the winter in China. In the solar heating greenhouse, the fig fruit developed for 92 days and ripened from July to December in the following year (Liao *et al.*, 2018).

Byeon and Lee (2020) explored Masui Dauphine fruit quality and the difference in major primary metabolites in open-field versus greenhouse cultivation systems. The fig trees were planted at 2 × 3 m spacing in the open field and 1 × 2 m spacing in the greenhouse. Premium-class fresh fig fruits were harvested in mid-October; the fruits weighed more than 80 g and had more than 90% peel color coverage and a wide opening at the ostiole. In mid-October, the fig fruit from the open field were nearing the end of the harvest season, while those in the greenhouse system were still being actively produced, thereby extending the harvest window. The authors then evaluated the quality parameters of fruit from open fields and from greenhouse cultivation systems during cold storage at 0.5°C for up to 30 days. Weight loss increased with storage duration and was much greater for fruit grown in the greenhouse than for those grown in the open field. Fruit length and diameter at the peduncle end decreased with storage duration and were much shorter in the greenhouse fruit than in the open-field fruit. The fruit respiration rate was higher in the open-field fruit than in the greenhouse fruit during cold storage. The color variables gradually decreased in the peel tissues regardless of the tissue region, and glucose and fructose levels increased with storage duration, regardless of the production system. The organic acid content was not consistent during cold storage—citric acid content was higher but fumaric acid content was lower in the open-field fruit than in the greenhouse fruit. Overall, the results suggested that production systems can affect fig fruit quality and storability.

9.5.3 Growing fig in soil-less hydroponic culture

Today, about 3.5% of the global cultivation area under tunnels and in greenhouses for vegetable production has adopted soil-less agriculture techniques based on hydroponic solutions (Sambo *et al.*, 2019). The huge potential offered by this cultivation approach ranges from productive and qualitative advantages to environmental benefits due to more efficient use of water and nutritional resources. However, soil-less hydroponic cultivation for fruit-tree species is almost non-existent, with only small experimental plots, and it is very difficult to adjust the optimal growing conditions for tree species cultivation in this environment. Kawamata *et al.* (2002) generated double cropping of the fig variety Masui Dauphine in a greenhouse under hydroponic conditions. The trees were kept at over 15°C and supplied with half-strength Enshi solution. The yield of fruit harvested from the primary shoots between June 7 and September 30 was about 1.5 kg or 15.0 fruit per primary shoot with an average weight of 104 g and Brix of 14. For the second cropping, all shoots were pinched at their 30th nodes and half were headed back again to their third nodes. The yield from the secondary shoots harvested from 24 November to 14 February was 1.3 kg or 15.8 fruit per shoot. The percentage of fruit set at the 12th–20th nodes on the secondary shoots of the pinched trees was smaller than that on other nodes. They concluded that fig trees grown hydroponically can be induced to yield double crops of high-quality fruit. Melgarejo *et al.* (2007) tested the growth and production of the Super Fig 1 (SF1) variety, a parthenocarpic variety with a dark color and large fruit size, for both the brebas and main crop, in soil-less hydroponic culture. Two annual crops were obtained. They showed that SF1 can be grown year-round, because it does not require dormancy. Based on growing conditions and planting dates, it was possible to schedule harvesting dates. They concluded that soil-free culture allows irrigated farms to increase their fig-production yield by 18-fold compared with traditional farming. Recently, Mendoza-Castillo *et al.* (2017) evaluated six intensive fig-production systems under hydroponic and greenhouse conditions. Production systems were established by varying the number of productive stems to between three and eight per plant, with densities of 1.25 plants m^{-2}, in five replications with a pot or plant as the experimental unit. The highest fig-fruit yield was achieved with eight productive stems. The handling of productive stems increased leaf area index, harvest index, and fruit yield. These results suggested that intensive production of fig under hydroponic and greenhouse conditions is a valid alternative means of fresh-fig production to plantations in the open field.

Growth of trees in hydroponic solutions is still limited. Thus, the presented results must be considered to be preliminary. To increase the use of hydroponic systems for fig, adequate nutrient concentrations to improve fig quality and yield are needed. New emerging technologies, including the use of nanoparticles and beneficial microorganisms such as plant growth-promoting rhizobacteria, can be tested. Finally, multi-element sensors and analysis using interpretation algorithms based on machine learning logic might be exploited in a smart agriculture approach to monitor tree growth and the hydroponic solution, and to adjust its composition in real time (Sambo *et al.*, 2019).

9.6 Concluding Remarks

Fig is adapted to semi-arid subtropical regions but is also grown in both mild temperate climates and tropical regions. Regular plantations are located between the latitudes of 20° and 40° in both the northern and southern hemispheres. In most fig-producing countries, commercial orchards are at the lower altitudes, but in the tropics, they can be found at high altitudes of 1500–3000 m. Weather has an important effect on fig productivity and quality. Understanding fig-tree physiology is very important in the context of climate change. Global climate change in recent decades has brought about a range of changes in microclimate, which must be considered for successful future fig cultivation.

Fig-production areas present large climate variability that markedly affects fruit size, shape, skin, and pulp. Growing fig at low temperatures holds some risks for fig-fruit development. New growth in the early spring or secondary yield growth in late autumn may be damaged at temperatures below 5°C. The tree requires little chilling to break dormancy. During fruit development and ripening, higher temperatures are required for adequate sugar accumulation and proper maturation. Rain or high humidity during fruit ripening has a negative impact on fruit quality. Growing fig in nethouses or greenhouses may extend the fig-growing area to new environmental conditions, enhancing fruit productivity, fruit quantity, and commercial opportunities. Understanding the environmental factors and their interaction with tree and fruit physiological processes is extremely important for improving fig horticultural practices, with the aim of optimizing photosynthetic carbon assimilation and increasing fruit productivity and crop quality.

Wide genetic variation exists in the developmental responses of figs to environmental factors. Several selection and adaptation studies have been reported on fig varieties. In Turkey, the main genetic origin of the world's domesticated figs, various types of fig varieties can be found in the different climatic conditions of Anatolia (Küden and Tanriver, 1998). Strains used for fresh fruit are cultivated under environmental conditions that differ from those used for dried figs. Similarly, a study was conducted in Chile to evaluate the effect of different climatic conditions on the three fig varieties Kadota, Kennedy, and Larga de Burdeos (Botti *et al.*, 2003). The study demonstrated the strong effect of climatic conditions on fresh-fig production in terms of yield, type of production (breba and main crop), timing of production, and fruit quality. A comparison of the roles of climate in the production of dried and fresh fig revealed that dried-fig production is strongly dependent on climatic conditions and is mostly successful under dry and warm temperate climates. The main fig variety for drying is the Smyrna-type fig Sarılop (Calimyrna), which is widely cultivated in both Turkey and California. Fresh figs, however, can be cultivated under a wider range of ecological conditions (Polat and Caliskan, 2008). Currently, one of the best-adapted and most widely disseminated fresh-fig varieties is Brown Turkey. It goes by several names and grows in diverse climates in many locations worldwide, such as Brazil (known as Roxo de Valinhos), Mexico, Chile, India, Saudi Arabia, Israel, California (known as California Brown Turkey), and Japan (known as Masui Dauphine), among others.

References

Aksoy, U., Balci, B., Can, H.Z., and Hepaksoy, S. (2003) Some significant results of the research-work in Turkey on fig. *Acta Horticulturae* 605, 173–181.

Ammar, A., Ben Aissa, I., Mars, M., and Gouiaa, M. (2020) Seasonal variation of fig tree (*Ficus carica* L.) physiological characteristics reveals its adaptation performance. *South African Journal of Botany* 132, 30–37.

Botelho, R.V. (2007) Effect of the garlic extract on dormancy breaking of apple trees. *Brazilian Journal of Fruticulture* 29, 403–405.

Botti, C., Franck, N., Prat, L., Ionnandis, D., and Morales, B. (2003) The effect of climatic conditions on fresh fig fruit yield, quality and type of crop. *Acta Horticulturae* 605, 37–44.

Briassoulis, D., Mistriotis, A., and Eleftherakis, D. (2007) Mechanical behaviour and properties of agricultural nets—part I: testing methods for agricultural nets. *Polymer Testing* 26, 822–832.

Byeon, S.E. and Lee, J. (2020) Fruit quality and major primary metabolites differ across production systems in cold-stored figs (*Ficus carica* L.). *Scientia Horticulturae* 274: 109669.

Caetano, L.C.S., Cordeiro de Carvalho, A.J., Campostrine, E., Fernandes de Sousa, E., Murakami, K.R.N., and Cereja, B.S. (2005) Effect of the number of productive branches on the leaf area development and fig tree yield. *Revista Brasileira de Fruticultura* 27, 426–429.

Can, H.Z. and Aksoy, U. (2007) Seasonal and diurnal photosynthetic behaviour of fig (*Ficus carica* L.) under semi-arid climatic conditions. *Acta Agriculturae Scandinavica, Section B: Soil & Plant Science* 57, 297–306.

Can, H.Z., Hepaksoy, S., Arkoy, U., and Kutlu, E. (1998) Leaf characteristics and net gas exchange of fig cultigens adapted to different climatic zones. *Acta Horticulturae* 516, 131–137.

Çelikel, F.G., Kaynas, K., Özelkök, S., and Ertan, Ü. (1996) Effects of ethephon on fruit development and ripening of the fig (*Ficus carica* L.) variety "Bursa Siyahi". *Acta Horticulturae* 441, 145–152.

Chai, L., Wang, Z., Chai, P., Chen, S., and Ma, H. (2017) Transcriptome analysis of San Pedro-type fig (*Ficus carica* L.) parthenocarpic breba and non-parthenocarpic main crop reveals divergent phytohormone-related gene expression. *Tree Genetics & Genomes* 13: 83.

Chithiraichelvan, R., Kurian, R.M., Awachare, C.M., and Laxman, R.H. (2017) Performance of fig (*Ficus carica* L.) under different planting densities. *International Journal of Current Microbiology and Applied Sciences* 6, 2603–2610.

Condit, I.J. (1947) *The Fig*. Chronica Botanica Co., Waltham, Massachusetts.

Cook, N.C. and Jacobs, G. (1999) Suboptimal winter chilling impedes development of acrotony in apple shoots. *Hortscience* 34, 1213–1216.

Costes, E., Lauri, P.E., and Regnard, J.L. (2006) Analyzing fruit tree architecture: implications for tree management and fruit production. In: Janick, J. (ed.) *Horticultural Reviews*, Vol. 32, pp. 1–61.

Crane, J.C. (1986) *CRC Handbook of Fruit Set and Development*. CRC Press, Boca Raton, Florida.

Crane, J.C. and van Overbeek, J. (1965) Kinin-induced parthenocarpy in the fig, *Ficus carica* L. *Science* 147, 1468–1469.

Crisosto, H., Ferguson, L., Bremer, V., Stover, E., and Colelli, G. (2011) Fig (*Ficus carica* L.). In: Yahia, E.M. (ed.) *Postharvest Biology and Technology of Tropical and Subtropical Fruits – Fundamental Issues*. Woodhead Publishing, Cambridge, UK, pp. 134–160.

Demchak, K. (2009) Small fruit production in high tunnels. *HortTechnology* 19, 44–49.

Erez, A. and Shulman, Y. (1982) Growing breva figs in a meadow orchard; a possibility for commercial fig growing. *Hassadeh* 63, 969–971.

Erez, A., Fishman, S., Linsley-Noakes, G.C., and Allan, P. (1990) The dynamic model for rest completion in peach buds. *Acta Horticulturae* 276, 165–174.

Erez, A., Flaishman, M., and Yablowitz, Z. (2003) Breba production under net: effect of training system. *Acta Horticulturae* 605, 119–123.

Fadón, E., Fernandez, E., Behn, H., and Luedeling, E. (2020) A conceptual framework for winter dormancy in deciduous trees. *Agronomy* 10: 241.

Faust, M., Erez, A., Rowland, L.J., Wang, S.Y., and Norman, H.A. (1997) Bud dormancy in perennial fruit trees: physiological basis for dormancy induction, maintenance, and release. *HortScience* 32, 623–629.

Ferguson, L., Michalides, T.J., and Shorey, H.H. (1990) The California fig industry. *Horticultural Reviews* 12, 409–490.

Ferguson, L., Mariscal, M., Reyes, H., and Metheney, P. (2001) Using girdling to improve Black Mission fig size. *Acta Horticulturae* 605, 167–169.

Flaishman, M.A., Rodov, V., and Stover, E. (2008) The fig: botany, horticulture, and breeding. *Horticultural Reviews* 34, 132–196.

Gaaliche, B., Hfaiedh, L., Trad, M., and Mars, M. (2011a) Caprification efficiency of some Tunisian local fig (*Ficus carica* L.) cultivars. *Pakistan Journal of Agricultural Sciences* 48, 295–298.

Gaaliche, B., Lauri, P.E., Trad, M., Costes, E., and Mars, M. (2011b) Interactions between vegetative and generative growth and between crop generations in fig tree (*Ficus carica* L.). *Scientia Horticulturae* 131, 22–28.

Gaaliche, B., Trad, M., and Mars, M. (2011c) Effect of pollination intensity, frequency and pollen source on fig (*Ficus carica* L.) productivity and fruit quality. *Scientia Horticulturae* 30, 737–742.

Gaaliche, B., Aïachi-Mezghani, M., Trad, M., Costes, E., Lauri, P.E., and Mars, M. (2016) Shoot architecture and morphology of different branch orders in fig tree (*Ficus carica* L.). *International Journal of Fruit Science* 16, 378–394.

Gaaliche, B., Ghrab, M., and Ben Mimoun, M. (2017) Effect of hydrogen cyanamide on vegetative growth, yield, and fruit quality of fig cv. Zidi in a warm production area. *International Journal of Fruit Science* 17, 63–71.

Galil, J. and Neeman, G. (1977) Pollen transfer and pollination in the common fig (*Ficus carica* L.). *New Phytologist* 79, 163–171.

Gerber, H.J., Steyn, W.J., and Theron, K.I. (2010) The optimum one-year-old shoot length to optimize yield and fruit size of three fig (*Ficus carica* L.) cultivars. *HortScience* 45, 1321–1326.

Golombek, S.D. and Lüdders, P. (1990) Gas exchange of *Ficus carica* in response to salinity. In: Beusichem, M.L. (ed.) *Plant Nutrition – Physiology and Applications*. Kluwer Academic Publishers, Dordrecht, The Netherlands, pp. 487–493.

Guo, L., Dai, J., Ranjitkar, S., Yu, H., Xu, J., and Luedeling, E. (2014) Chilling and heat requirements for flowering in temperate fruit trees. *International Journal of Biometeorology* 58, 1195–1206.

Iglesias, I. and Alegre, S. (2006) The effect of anti-hail nets on fruit protection, radiation, temperature, quality and profitability of 'Mondial Gala' apples. *Journal of Applied Horticulture* 8, 91–100.

John, P. (1993) *Biosynthesis of the Major Crop Products*. Wiley, Chichester, New York.

Kalcsits, L., Musacchi, S., Layne, D.R., Schmidt, T., Mupambi, G., *et al.* (2017) Above and below-ground environmental changes associated with the use of photoselective protective netting to reduce sunburn in apple. *Agricultural and Forest Meteorology* 237, 9–17.

Karami, H., Rezaei, M., Sarkhosh, A., Rahemi, M., and Jafari, M. (2018) Cold hardiness assessment in seven commercial fig cultivars (*Ficus carica* L.). *Gesunde Pflanzen* 70, 195–203.

Kawamata, M., Ohara, H., Ohkawa, K., Marata, Y., Takahashi, E.Y., and Matsui, H. (2002) Double cropping of fig hydroponic culture. *Journal of the Japanese Society for Horticultural Science* 71, 68–73.

Khemira, H. and Mars, M. (2017) Fig production in subtropical south-western Saudi Arabia. *Acta Horticulturae* 1173, 169–172.

Kim, J.S., Kim, Y.O., Ryu, H.J., Kwak, Y.S., Lee, J.Y., and Kang, H. (2003) Isolation of stress-related genes of rubber particles and latex in fig tree (*Ficus carica*) and their expressions by abiotic stress or plant hormone treatments. *Plant and Cell Physiology* 44, 412–414.

Kislev, M.E., Hartmann, A., and Bar-Yosef, O. (2006) Early domesticated fig in the Jordan Valley. *Science* 312, 1372–1374.

Kong, M., Lampinen, B., Shackel, K., and Crisosto, C.H. (2013) Fruit skin side cracking and ostiole-end splitting shorten postharvest life in fresh figs (*Ficus carica* L.), but are reduced by deficit irrigation. *Postharvest Biology and Technology* 85, 154–161.

Kubota, N. and Miyamuki, M. (1992) Breaking bud dormancy in grapevines with garlic paste. *Journal of the American Society for Horticultural Science* 117, 898–901.

Küden, A.B. and Tanriver, E. (1998) Plant genetic resources and selection studies on figs in the East Mediterranean and South East Anatolia Regions. *Acta Horticulturae* 480, 49–54.

Lama, K., Modi, A., Peer, R., Izhaki, Y., and Flaishman, M.A. (2019) On-tree ABA application synchronizes fruit ripening and maintains keeping quality of figs (*Ficus carica* L.). *Scientia Horticulturae* 253, 405–411.

Lang, G.A., Early, J.D., Martin, G.C., and Darnell, R.L. (1987) Endodormancy, paradormancy, and ecodormancy—physiological terminology and classification for dormancy research. *HortScience* 22, 371–377.

Lauri, P.É. and Laurens, F. (2005) Architectural types in apple (*Malus × domestica* Borkh.). In: Dris, R. (ed.) *Crops: Growth, Quality and Biotechnology*. WFL Publisher, Helsinki, pp. 1300–1314.

Leonel, S., Tecchio, M.A., and Cóser, G.M. de A.G. (2015) Dormancy breaking of the fig tree with hydrogen cyanamide and garlic extrate. *Current Journal of Applied Science and Technology* 1–10.

Liao, Y., Zhang, Q., and Shen, Y. (2018) Methods key to annual production of fig fruit in greenhouse. *Acta Horticulturae Sinica* 45, 2437–2441.

Luedeling, E. and Brown, P.H. (2011) A global analysis of the comparability of winter chill models for fruit and nut trees. *International Journal of Biometeorology* 55, 411–421.

Manja, K. and Aoun, M. (2019) The use of nets for tree fruit crops and their impact on the production: a review. *Scientia Horticulturae* 246, 110–122.

Marei, N. and Crane, J.C. (1971) Growth and respiratory response of fig (*Ficus carica* L. cv. Mission) fruits to ethylene. *Plant Physiology* 48, 249–254.

Melgarejo, P., Martínez, J.J., Hernández, F., Salazar, D.M., and Martínez, R. (2007) Preliminary results on fig soil-less culture. *Scientia Horticulturae* 111, 255–259.

Mendoza-Castillo, V.M., Vargas-Canales, J.M., Calderon-Zavala, G., Mendoza-Castillo, M.D.C., and Santacruz-Varela, A. (2017) Intensive production systems of fig (*Ficus carica* L.) under greenhouse conditions. *Experimental Agriculture* 53, 339–350.

Metwali, E.M., Soliman, H.I., Fuller, M.P., Al-Zahrani, H.S., and Howladar, S.M. (2015) Molecular cloning and expression of a vacuolar Na^+/H^+ antiporter gene (*AgNHX1*) in fig (*Ficus carica* L.) under salt stress. *Plant Cell, Tissue and Organ Culture* 123, 377–387.

Michailides, T.J., Morgan D.P., Felts, D., and Doster, M.A. (2008) Control of decay in caprifigs and calimyrna figs with fungicides. *Acta Horticulturae* 798, 269–275.

Moreno, D.A., Pulgar, G., Víllora, G., and Romero, L. (1998) Nutritional diagnosis of fig tree leaves. *Journal of Plant Nutrition* 21, 2579–2588.

Oukabli, A. and Mekaoui, A. (2012) Dormancy of fig cultivated under Moroccan conditions. *American Journal of Plant Sciences* 3: 18699.

Oukabli, A., Mamouni, A., Laghezali, M., Ater, M., Roger, J.P., and Khadari, B. (2001) Local caprifig tree characterization and analysis of interest for pollination. *Acta Horticulturae* 605, 61–64.

Owino, W.O., Nakano, R., Kubo, Y., and Inaba, A. (2004) Alterations in cell wall polysaccharides during ripening in distinct anatomical tissue regions of the fig (*Ficus carica* L.) fruit. *Postharvest Biology and Technology* 32, 67–77.

Polat, A.A. and Caliskan, O. (2008) Fruit characteristics of table fig (*Ficus carica*) cultivars in subtropical climate conditions of the Mediterranean region. *New Zealand Journal of Crop and Horticultural Science* 36, 107–115.

Puebla, M., Toribjo, F., and Montes, P. (2003) Determination of fruit bearing pruning date and cutting intensity in San Pedro (*Ficus carica* L.) type fig cultivars. *Acta Horticulturae* 605, 147–157.

Puech, A.A., Rebeiz, C.A., and Crane, J.C. (1976) Pigment changes associated with application of ethephon ((2-chloroethyl)phosphonic acid) to fig (*Ficus carica* L.) fruits. *Plant Physiology* 57, 504–509.

Rodov, V., Vinokur, Y., and Horev, B. (2012) Brief postharvest exposure to pulsed light stimulates coloration and anthocyanin accumulation in fig fruit (*Ficus carica* L.). *Postharvest Biology and Technology* 68, 43–46.

Rosianskey, Y., Dahan, Y., Yadav, S., Freiman, Z.E., Milo-Cochavi, S., *et al.* (2016) Chlorophyll metabolism in pollinated vs. parthenocarpic fig fruits throughout development and ripening. *Planta* 244, 491–504.

Rosianski, Y., Doron-Faigenboim, A., Freiman, Z.E., Lama, K., Milo-Cochavi, S., *et al.* (2016a) Tissue-specific transcriptome and hormonal regulation of pollinated and parthenocarpic fig (*Ficus carica* L.) fruit suggest that fruit ripening is coordinated by the reproductive part of the syconium. *Frontiers in Plant Science* 7: 1696.

Rosianski, Y., Freiman, Z.E., Cochavi, S.M., Yablovitz, Z., Kerem, Z., and Flaishman, M.A. (2016b) Advanced analysis of developmental and ripening characteristics of pollinated common-type fig (*Ficus carica* L.). *Scientia Horticulturae* 198, 98–106.

Salimpour, A., Shamili, M., Dadkhodaie, A., Zare, H., and Hadadinejad, M. (2019) Evaluating the salt tolerance of seven fig cultivars (*Ficus carica* L.). *Advances in Horticultural Science* 33, 553–565.

Sambo, P., Nicoletto, C., Giro, A., Pii, Y., Valentinuzzi, F., *et al.* (2019) Hydroponic solutions for soilless production systems: issues and opportunities in a smart agriculture perspective. *Frontiers in Plant Science* 10: 923.

Sevengor, S., Yasar, F., Kusvuran, S., and Ellialtioglu, S. (2011) The effect of salt stress on growth, chlorophyll content, lipid peroxidation and antioxidative enzymes of pumpkin seedling. *African Journal of Agricultural Research* 6, 4920–4924.

Singh, R.K., Svystun, T., AlDahmash, B., Jönsson, A.M., and Bhalerao, R.P. (2017) Photoperiod- and temperature-mediated control of phenology in trees – a molecular perspective. *New Phytologist* 213, 511–524.

Sivakumar, D., Jifon, J., and Soundy, P. (2018) Spectral quality of photo-selective shade nettings improves antioxidants and overall quality in selected fresh produce after postharvest storage. *Food Reviews International* 34, 290–307.

Solomon, A., Golubowicz, S., Yablowicz, Z., Grossman, S., Bergman, M., *et al.* (2006) Antioxidant activities and anthocyanin content of fresh fruits of common fig (*Ficus carica* L.). *Journal of Agricultural and Food Chemistry* 54, 7717–7723.

Storey, W.B., Enderud, J.E., Saleeb, W.F., and Nauer, E.M. (1977) *The Fig: Its Biology, History, Culture, and Utilization*. Jurupa Mountains Cultural Center, Riverside, California.

Tanny, J. (2013) Microclimate and evapotranspiration of crops covered by agricultural screens: a review. *Biosystems Engineering* 114, 26–43.

Trad, M., Ginies, C., Gaaliche, B., Renard, C.M.G.C., and Mars, M. (2012) Does pollination affect aroma development in ripened fig (*Ficus carica* L.) fruit? *Scientia Horticulturae* 134, 93–99.

Trad, M., Le Bourvellec, C., Gaaliche, B., Ginies, C., Renard, C.M., and Mars, M. (2013) Caprification modifies polyphenols but not cell wall concentrations in ripe figs. *Scientia Horticulturae* 160, 115–122.

Tylewicz, S., Petterle, A., Marttila, S., Miskolczi, P., Azeez, A., *et al.* (2018) Photoperiodic control of seasonal growth is mediated by ABA acting on cell–cell communication. *Science* 360, 212–215.

Vangelisti, A., Zambrano, L.S., Caruso, G., Macheda, D., Bernardi, R., *et al.* (2019) How an ancient, salt-tolerant fruit crop, *Ficus carica* L., copes with salinity: a transcriptome analysis. *Scientific Reports* 9: 2561.

Vargas, I., Corrales, C., and Martinez, M. (2008) Compounds derived from garlic as bud induction agent in organic farming of table grape. *Chilean Journal Agricultural Research* 68, 94–101.

Vemmos, S.N., Petri, E., and Stournaras, V. (2013) Seasonal changes in photosynthetic activity and carbohydrate content in leaves and fruit of three fig cultivars (*Ficus carica* L.). *Scientia Horticulturae* 160, 198–207.

Wang, Z., Song, M., Li, Y., Chen, S., and Ma, H. (2019) Differential color development and response to light deprivation of fig (*Ficus carica* L.) syconia peel and female flower tissues: transcriptome elucidation. *BMC Plant Biology* 19: 217.

Yablowitz, Z., Nir, G., and Erez, A. (1998) Breba fig production in Israel: regular and pesticide-free systems. *Acta Horticulturae* 480, 137–141.

Zohary, D. and Spiegel-Roy, P. (1975) Beginnings of fruit growing in the Old World. *Science* 187, 319–327.

10 Sustainable Farming Systems and Organic Fig Production

Uygun Aksoy*
Department of Horticulture, Faculty of Agriculture, Ege University, Bornova, İzmir, Turkey

10.1 Introduction

Agricultural practices, which became more input intensive especially during the second half of the 20th century, ignited thorough discussions on sustainability. However, the discussions focusing merely on environmental issues at the initial stage have now turned into a concept comprising social, cultural, political, and economic dimensions and their interactions. Altieri (2015) states that to better understand the agricultural setting, there is a need to study "the global environment and social systems given that agricultural development results from the complex interaction of a multitude of factors." Studying the ecology of agricultural systems may then help to develop a management system that is best suited to the site-specific conditions, resulting in more sustainable agriculture (Altieri, 2015).

The common fig tree is widely grown around the Mediterranean Basin and in regions where a Mediterranean climate prevails; however, it is now extending toward tropical or temperate climate zones. In major fig-producing countries, fig is grown under low-input and rainfed management systems such as perennial monocultures, mostly on marginal land where other species cannot perform well. This brings advantages regarding food security and environmental quality but makes both soil health and production capacity more fragile and affected by the yearly climatic conditions. In order to develop an agroecosystem that is long-lived and productive throughout the lifespan of the fig tree, it is crucial to understand the challenges and opportunities. The Mediterranean region, the main area where figs are grown, is threatened by land degradation, high nitrification rates, low soil organic matter, and water shortages (Li *et al.*, 2020). Furthermore, salinity, heat, or cold waves, or long dry or wet periods are expected to be faced more frequently in the coming years because of climate change.

Agroecosystems are not limited to the physical surroundings; humans, animals, plants and microorganisms share the same environment, above or underground, and impact the performance of the agroecosystem and the production capacity, alone or through their interactions. Current commercial fig production, whether destined for the fresh or dried industry, relies heavily on manual manpower; mechanization remains limited due to the delicate nature of the produce or to existing physical conditions such as the topography. In western Mediterranean countries such as Portugal, Spain, and Italy, high production costs were incurred due to increasing wages, resulting in sharp decreases

* Email: uygun.aksoy@gmail.com

 Advances in Fig Research and Sustainable Production (M.A. Flaishman and U. Aksoy)
DOI: 10.1079/9781789242492.0010

in fig production around the 1970s. Countries in the eastern or southern Mediterranean region gained competitiveness economically in the global markets and became major producers. This chapter focuses on the major threats to fig production that are currently faced and that may prevail in the short and medium term, and evaluates the sustainability of fig production from a multi-dimensional viewpoint with a closer look at environmental issues. However, conditions may change rapidly, and new initiatives or research fields may be required to achieve more sustainable fig production in the coming decades.

10.2 Environmental Adaptability

Rich genetic resources help fig, allowing a wider adaptability to diverse ecological zones. Commercial fig production is concentrated in subtropical or mild temperate climate zones, but it can be grown in tropical regions, as in Brazil, or in some cold temperate zones, such as Uzbekistan. Fig trees can be successfully grown to a height of 1525 m above sea level in tropical regions (Hiwale, 2015), and are found at altitudes of 1450–2800 m on the Andes Mountains in Colombia (Raz and Agudelo Zamora, 2020). In colder climates, fig can be grown in protected backyards or by pruning down the tree, mainly common-type varieties, to ground level. Climates minimizing an early frost risk in late autumn to winter and having dry periods during fruit development and ripening are ideal for commercial fig cultivation. In case of low annual rainfall or sandy soils, additional irrigation may be required for optimum yield and quality. Figs destined for sun drying may have additional climatic requirements, especially during the drying period. Although a few fig varieties dominate the world fig trade, the genetic richness of fig allows high adaptability to different environmental conditions and helps to create new production sites outside its traditional boundaries. In this regard, knowing a particular variety in more depth with regard to its fruiting characteristics and response to biotic and abiotic stress conditions will help to improve its performance under certain conditions or using a particular orchard management system.

10.2.1 Water and heat stress

Turkey and Iran are the two leading countries in world fig production. In these countries, figs destined for sun drying are grown as rainfed monocultures on large surface areas such as the Meander Basin in Turkey or the Estahban region in Iran. One of the major problems that may arise under rainfed conditions is drought, which restricts plant growth and reduces fruit yield and quality. Under Mediterranean conditions, the vegetation cycle of the deciduous fig tree starts with bud burst in early spring, continues with fruit growth and maturation during hot and dry summers, and then terminates with complete leaf fall in late autumn or early winter (Abdel-Razik and El-Darier, 1991). In the case of severe water shortage prior to fruit maturation, rapidly growing fig fruit acts as a powerful sink and draws water from the leaves, causing earlier and excessive leaf fall (Fig. 10.1A). This later prevents fruit from enlarging and reaching full maturity (Darwish *et al.*, 2015). Total yield and marketable-quality fruit decrease due to a higher ratio of low-quality, small-sized, and sun-scalded fruit. In addition to the soil and floor management, the total water requirement alters according to the variety. Annual rainfall of around 550–650 mm distributed between late autumn–winter and early spring is sufficient for optimum vegetative growth and fruit yield in fig trees of the Sarılop (= Calimyrna) variety in western Turkey (Aksoy *et al.*, 2001). Similarly, in Iran, the main variety, Sabz, comprising 90% of plantations, is highly tolerant to drought and performs well under the traditional management and rainfed conditions of Iran with around 300–350 mm annual rainfall (FAO, 2011; Jafari *et al.*, 2017; Roointan *et al.*, 2018). The Sabz variety is grown as a multi-trunk (four to eight trunks) bush. Trials showed that reducing the number of trunks to two to three (trunk thinning) at the end of winter (February) had a significant effect on mitigating the effects of drought and maintaining vegetation and obtaining higher-quality dried figs (Jafari *et al.*, 2017).

Under rainfed conditions, water harvesting to catch runoff and rainfall is a traditional method that is also applied in fig production. Different shapes of micro- or macro-catchment methods collecting surface waters help prevent

A. B.

Fig. 10.1. Effect of dry and hot summer conditions. A. Signs of drought, browning, and early leaf fall. Photograph by U. Aksoy, 2020. B. Air and leaf temperature, and photosynthesis and transpiration measurements of Sarılop leaves. Upper graph: Air and leaf temperatures (°C) measured in Aydın, Turkey, during the fruit development and maturation period under sun or shade at northern (N) or southerly (S) aspects. Lower graph: Midday (12:00–13:00) net photosynthesis (A; μmol m^{-2} s^{-1}) and transpiration (E; mmol m^{-2} s^{-1}) measurements of Sarılop leaves under sun or shade at northern or southerly aspects. Modified from Demir (2005).

erosion and enrich the root zone. In addition to water harvesting, the orchard site and variety selection, use of drought-resistant and deep-rooted rootstocks, supplementary irrigation practices during the critical periods (e.g. rapid shoot growth), and soil management are prominent factors affecting success of rainfed orchards (Roointan *et al.*, 2018).

When selecting the variety and/or the location, it is important to know the morphological and ecophysiological characteristics of fig regarding environmental adaptability. During the fruit development and maturation period, the average photosynthesis and transpiration rates are both enhanced because of the prevailing dry and hot conditions in summer (Fig. 10.1B). Even at the same geographic location, the rates are higher in leaves exposed to the sun, as in southerly slopes, whereas they are much lower in shaded leaves on the northerly slopes as a consequence of the difference in leaf temperatures. Higher leaf area index values were found on trees located on southerly slopes (2.03±0.074) compared with northerly slopes (1.65±0.10) due to higher photosynthetic capacity (Demir, 2005). The aspect impacts substantially, especially during the period when stress builds up, at the end of July and in August (Fig. 10.1B).

Water-use efficiency (WUE), calculated as net photosynthesis/transpiration, decreases throughout the fruit development period during the dry summer period under rainfed conditions (Demir, 2005). During water-deficit conditions, restricted carbon dioxide (CO_2) availability due to stomatal closure may lead to increased susceptibility. Photoinhibition is a slowly reversible decline of maximum quantum yield of photosynthesis (ratio of variable fluorescence:maximum fluorescence, F_v:F_m) associated with loss of photosystem II (PSII) activity. Adverse environmental conditions, such as high temperature and water stress, that strongly limit photosynthetic carbon metabolism can intensify photoinhibition in fig trees (Can and Aksoy, 2007).

In a pot trial designed with rooted stem cuttings of four Iranian fig varieties, Deym, Sabz, Siah, and Shah Anjiri, 4-month-old saplings were subjected to two different water levels: 100% field capacity as optimal irrigation, and limited irrigation as 14-day exposure to water. The results displayed significant differences regarding the response of the varieties to drought and the recovery process. Deym was the most tolerant variety, followed by Sabz, the main Iranian fig variety for sun drying. Shah Anjiri was the most affected. Drought caused increases in pigments and enzymatic and non-enzymatic antioxidants in the tested fig varieties, reacting through different scavenging mechanisms to overcome oxidative stress. Enhanced levels of antioxidants were attributed to resistance of the fig trees to adverse arid conditions (Gholami *et al.*, 2012).

In a similar trial in northern Egypt, fig plants received 75% and 50% of the total water requirement during summer. A reduction in irrigation levels retarded vegetative growth in 1-year-old trees of the Gizy and Aboudi fig varieties. Drought caused a decrease in photosynthesis and induced a reduction in stem diameter, leaf area, and leaf fresh weight. Leaf proline and protein contents varied significantly by variety and irrigation regime, and high levels were measured at 50% irrigation. The Aboudi variety, reported to be drought tolerant, had higher proline content compared with Gizy. Proline is stated to stabilize cell membranes to protect cells from oxidative stress and function as an osmotic regulator.

One-year-old plants of the Dottato variety were tested under three irrigation regimes in a pot trial: 120% of water consumption applied as full irrigation, and 60% and 40% as water-deficit scenarios. Stress conditions were applied for 4 weeks. The effect of stress was relatively low due to a short period of exposure. Leaf dry weight was reduced, and the root:shoot ratio was increased, displaying adaptation to drought. These responses confirmed that fig can tolerate short periods of moderate drought (Caruso *et al.*, 2017).

Water stress reduces leaf size, fresh and dry weight, and water content significantly in fig trees. Leaf area is a varietal characteristic and may vary according to the position on the shoot. However, water availability, temperature, and photosynthetic activity exert more pronounced effects on leaf growth. In the case of water deficit and restrictions in photosynthetic activity, leaf growth slows down, and when rapid fruit growth (enlargement) coincides with water stress, abscission occurs, leading to early leaf fall. Relative water content (RWC, %), determined as the percentage of water in the fresh leaf relative to the water content of the water-saturated leaf, is a parameter used to assess water stress in plants. During the dry season (May–August), leaf RWC in rainfed Sarılop fig trees decreased from 87.9% (June) to 65.7% (August) with the lowest value found at the southerly located leaves in August, showing the level of the stress (Can and Aksoy, 2007). Drought-resistant varieties might adapt better to rapid water loss. Leaves with smaller and thick-walled cells providing a lower fresh:dry weight ratio may survive better (Can *et al.*, 2000; Darwish *et al.*, 2015). High leaf water content, especially during the rapid fruit-growth period starting just before the first harvest and continuing until the peak harvest, is taken as a good sign of the ability to overcome severe drought. Thus, varieties that retain the initial RWC levels throughout the dry summer season can maintain a higher internal water status to tolerate drought under rainfed conditions.

Around the Mediterranean Basin, fig-fruit development and maturation coincide with increasing temperatures during hot and dry summer seasons. In this respect, the air and leaf temperature regimes during fruit development and maturation and their diurnal variation may exert marked effects on the performance of fig varieties. During the summer, air and leaf temperatures were measured in a fig orchard of the variety Sarılop located at 847 m above sea level in the Meander Basin (Aydın, Turkey). The temperatures increased in summer and peaked around 11 August and then started to decrease (Fig. 10.1B). Relative humidity was around 60% at the end of June and then steadily decreased, reaching to 40% at the beginning of September. Leaf temperature is the main factor limiting gas exchange of the fig tree, and leaf temperatures increase especially when exposed to direct sun (Demir, 2005). The highest photosynthetically active radiation (PAR) and air temperature values were determined at 13:00 and 15:00, respectively, and these values were higher at the southern parts of the canopy than the northern parts during the sunny period from May to September. Parallel to the daily changes in PAR and air temperature, stomatal conductance (g_s, mmol m^{-2} s^{-1}) varied and reached its maximum level at 11.00, while leaf temperature values approached 33°C and 27°C in the leaves at the southern and northern parts of the canopy, respectively. Maximum leaf temperature values were measured at 15.00 during all months in leaves at the southern parts of the canopy (Can and Aksoy, 2007). In many Mediterranean countries, fig trees are located on slopes. The aspect of a slope impacts the performance of the orchard by creating a microclimate. Shaded leaves on slopes facing north are less exposed to the sun and thus have the lowest leaf temperatures. Northern slopes can be preferred in cases where high temperatures pose a risk to fig production.

Transpiration and photosynthesis are strongly affected by a decrease in g_s; however, the relationship between g_s and transpiration is marked because photosynthesis is controlled by stomatal, mesophyllic, and boundary layer resistances, whereas g_s is the main resistance mechanism for controlling transpiration. Internal CO_2 may be used by mesophyll cells during photosynthesis after a decrease in g_s. To avoid water loss, stomata closure and transpiration losses lead to improved WUE. Total CO_2 assimilation also decreases, but the plant conserves water. Photosynthesis and transpiration were reduced in the leaves at the southern part at 13.00 and 15.00, depending on the decrease in g_s, especially in August and September when drought stress occurred (Can and Aksoy, 2007). However, photosynthesis decreased more sharply than transpiration, and thus drastic decreases were determined in WUE, particularly due to the lower photosynthesis values. This result shows the importance of non-stomatal effects, such as photoinhibition, as much as stomatal limitations on the photosynthetic capacity of the fig tree.

Studies on the Sarılop fig variety have shown that higher photosynthesis values observed in the morning overlap with the second rapid (or third) growth period phase of the fruit, when most of the sugars accumulate. Lower photosynthetic activity was measured at 13.00 and 15.00, despite high internal CO_2 concentration, showing that CO_2 in the stomatal cavity cannot be used in photosynthesis due to non-stomatal limitations (Can and Aksoy, 2007). If stomata impose a significant limitation on photosynthesis, a decrease in substomatal CO_2 concentration is expected. Significantly high PAR values in the leaves located at the south side of the canopy compared with the north did not exert marked effects on photosynthetic activity. Fig appears to be a somewhat shade-adapted species or possesses tolerance mechanisms. This is not due to a greater intrinsic efficiency of PSII to utilize light in shaded leaves but rather due to shade-adapted leaves allocating proportionally more of their photosynthetic apparatus to light-harvesting chlorophyll than to electron transport itself. Can and Aksoy (2007) proposed PAR values of 1619.6 and 1216.9 mmol m^{-2} s^{-1} in the southern and northern parts of the canopy, respectively, as the threshold values for photosynthesis in terms of irradiance. Photosynthetic activity decreased sharply below the threshold PAR values. High air temperature and irradiance together may cause damage to the photosynthetic system because the light energy cannot be used fully in photosynthesis owing to the inactivation of PSII reaction centers. Many plants are tolerant to high irradiance, and their photosynthetic capacity is inhibited only when high irradiance is accompanied by additional stress factors. Photoinhibition due to high irradiance is a well-known event. Plants of semi-arid regions are exposed to abiotic stresses such as high irradiance combined with high temperature and drought during the summer months, and one of the main factors of photoinhibition is drought stress (Can and Aksoy, 2007).

In an analysis of two Tunisian fig varieties, Zidi and Bither Abiadh, under plastic cover, changes in photosynthetic CO_2 assimilation, stomatal conductance, and maximum photochemical efficiency of PSII (F_v:F_m) were not linked to the varieties. The yearly variation was more pronounced because of the differences in maximum temperatures. The highest stomatal conductance values were recorded at the end of the spring season when daily temperatures were around 28–31°C, with a peak of 35°C in summer. Photosynthesis and stomatal conductance displayed similar variations during the season, confirming that both processes are dependent on temperature (Can and Aksoy, 2007; Ammar *et al.*, 2020). Ammar *et al.* (2020) presented a positive correlation between chlorophyll index and the F_v:F_m ratio and stated that a high level of photosynthetic pigments might enhance light absorption leading to the maximum yield of PSII. The obtained data also suggested the significance of environmental stresses such as high light intensity and high temperatures in fig production.

Leaf gas-exchange properties of nine fig varieties grown in Aydın, Turkey, originally coming from three different climatic regions were compared for their responses to dry and hot summer conditions and water stress prevailing during the fruit-ripening period. The regions and analyzed varieties were as follows: Aegean: Sarılop, Göklop, and Morgüz; Marmara: Bursa Black, Patlıcan, and Sultan Selim; and south-east: Halebi, Sultani, and Morözer. The leaf area varied among varieties, with Sarılop having the lowest leaf area; however, its leaf succulence was highest compared with the others. The varieties showed significant differences in transpiration and photosynthesis values measured in August and September, although the monthly variation was not pronounced. Similarly, when the varieties were grouped according to their origins, WUE was highest in the Aegean Region varieties, followed by the south-east and then the Marmara varieties (Can *et al.*, 2000). When caprifig varieties were tested for drought resistance, significant differences also appeared among the varieties. The caprifig genotypes Khormaei and Shah Anjiri were ranked as drought tolerant as they maintained water in their leaves during the drought period and had a fast recovery after irrigation. Dane Sephid and Pouz Donbali, in contrast, were drought-sensitive caprifig varieties. The RWC and leaf water potential were evaluated in the drought-tolerant caprifig genotypes. Proline accumulating during the stress period contributed to the osmoregulation and fast recovery. Potassium accumulation was not at significant levels and did not play a marked role in drought tolerance; however, sodium is proposed to have a more significant role in osmoregulation in caprifigs (Rostami and Rahemi, 2013). Leaf characteristics and metabolism that can be affected by the environmental conditions play a crucial role in drought tolerance and recovery of fig trees. The genetic variations form the basis of adaptation, and variations may appear within the same fig variety. Kutlu *et al.* (2000) investigated leaf properties, gas-exchange capacity, and physiological responses of clones selected within the Sarılop variety in Turkey. The results showed significant differences among 11 Sarılop clones studied. WUE was negatively correlated with the number of stomata per unit leaf area. Clones with comparatively smaller leaf areas had higher gas-exchange capacity. Leaf parameters were found to be good indices for selection of drought-tolerant fig genotypes.

The ecophysiology of the fig tree may affect not only the vegetative parameters and gas-exchange capacity but also fruit quality. The aspect of the slope through differences in the air and leaf temperatures impacts on dried-fruit quality of the Sarılop variety. Moderate or severely sun-scalded (44.2%) and small-sized substandard fruit (9.5%) were higher on the south-facing slope compared with the northern aspect, where the sun-scalded ratio was 33.3% and small-sized fruit 3.1%, respectively. In contrast, ostiole-end cracks were more frequent (14.2%) on north-facing aspects compared with 4.9% on the southern aspects. The total share of culled figs was higher on the southern aspects (Demir, 2005).

Cracking or splitting, whether at the ostiole end or on the sides, lowers fig fruit quality in both the fresh- and dried-fig market. In many cases, it is linked to the characteristics of the variety; however, environmental conditions and cultural practices may also affect it directly or indirectly (Aksoy and Akyüz, 1993; Aksoy and Anaç, 1994). A regulated deficit irrigation (55% evapotranspiration) trial performed on the Brown Turkey variety and Sierra cultivar from the first week of August until the end of harvest had a marked influence, especially on Brown Turkey, a variety known to have a high culled fruit ratio. Reduced midday stem water potential significantly decreased side cracking and ostiole-end splitting in Brown Turkey. Cracked or split fig fruit are more susceptible to skin damage and decay during postharvest handling and are accepted as defected. Full irrigation followed by a deficit irrigation program starting from the second growth phase of fig-fruit development decreased the fruit size and thus the total yield;

however, it increased the marketable yield by reducing the culled fruit ratio that appeared because of skin damage (Kong *et al.*, 2013).

Within common fig, the summer crops of the San Pedro and Smyrna types require pollination for fruit set and development. This is accomplished by the fig wasp (*Blastophaga psenes*), which has mutually evolved with *F. carica*. Other members of the family Agonidae have similar relationships with more than 750 fig species worldwide. Each *Ficus* species is dependent on one or a few species of fig wasps from the family Agonidae. The coevolution of the fig wasp and *Ficus* species has resulted in obligate mutualism, which has very wide ecological significance, especially in the tropical and subtropical regions. *Ficus* species provide fruit to more than 1200 vertebrate species, and supply the main diets of some birds, bats, and primates. Fig trees are also important for reforestation and supporting wildlife by planting or through spontaneously grown male or female fig trees (Cook and Rasplus, 2003; Jevanandam *et al.*, 2013). In this regard, the existence of the mutual relationship is important not only in commercial fig production of certain varieties but also for sustainability of natural ecosystems. The pollinating fig wasps of the family Agonidae, including *B. psenes*, are small (1–2 mm) and short-lived (1–2-day lifetime for adults) and fly during the day, making them more vulnerable to high temperatures or heavy winds. During transfer from the male profichi crop to the female to produce seeds or to the mammoni crop of the male fig to produce larvae, fig wasps may face a high risk of dehydration. Once they enter the receptive fig fruits, they are surrounded by a more favorable climate providing humidity and ambient temperatures (Cook and Rasplus, 2003; Al-Khalaf *et al.*, 2015). The receptive fig fruit maintain suitable temperatures by evaporative cooling; otherwise, the high temperatures could be lethal to the fig wasp larvae (Jevanandam *et al.*, 2013).

To simulate the impact of climate change, Jevanandam *et al.* (2013) tested increased temperatures on the adult lifespan of four pollinator fig wasps (*Valisia malayana*, pollinator of *Ficus grossularioides*), *Ceratosolen constrictus hewitti* (*Ficus fistulosa*), *Ceratosolen appendiculatus* (*Ficus variegata*) and *Eupristina verticillata* (*Ficus microcarpa*). The scenarios forecast an increase of 2.2–3.0°C for South-east Asia, and the projections suggested that an increase of 3°C in daytime temperatures would reduce the lifespan duration by 33–60%.

10.2.2 Cold tolerance

Fig trees can be found in various regions of the world, sometimes in cold temperate climates either as isolated trees in protected back yards, in pots, or under protection. It can similarly be present in the tropical regions; however, traditionally, commercial production is spread between the regions lying between latitudes of 23–45°N and 23–40°S. This production potential may increase or shift to more northern or southern regions as an aftereffect of climate change. The common fig is among the most cold-hardy species within the genus *Ficus*. However, among the commercial fig varieties known today, very few varieties such as Brown Turkey or Kadota have had the chance to adapt to many other regions of the world. One of the new areas of research is the dormancy and cold tolerance/susceptibility of fig varieties/types. Cold temperatures may have a profound effect on fig production by freezing the young shoots and, if extreme, the frost damage continues until the whole trunk dies.

In mild climates of the Mediterranean, the growth of the current year's shoot stops by the middle or end of June, and is then followed by fruit growth and maturation during the summer months. The susceptibility of fig to cold injury increases in the case of excess or delayed irrigation, excess or late nitrogen application, or in common-type fig varieties, with continued shoot growth throughout the summer and autumn delaying the hardiness of the current year's shoots. Early frosts may cause severe damage, especially on young shoots, but when dormant, fig trees can tolerate temperatures as low as −10°C. The trunk may die when the soil freezes, but the roots remain alive, and new growth helps to regenerate the fig tree. In some regions, an ancient practice to overcome injury from frost in common-type fig varieties is to train the tree low and then cut back shoots every winter leaving one to three nodes and cover the plant with soil.

In some fig varieties, seasonal shoots may continue elongating as long as the temperatures allow, letting many fruits form consecutively in the leaf axils. However, every fig fruit requires a certain amount of heat to mature. If temperatures drop rapidly in late autumn to early winter, the fruit formed on the distant part of the shoot do not carry onward to maturity and remain unripe. Under such conditions, selection of a fig variety with a compact harvest period becomes crucial to obtain adequate marketable yields.

Under critical winter temperatures, selection of not only female but also male (caprifig) varieties plays an important role. Fig wasps (*B. psenes*) may die while overwintering in male fruit due to low temperatures. The eggs of the fig wasp, which overwinter in the mamme crop of the male trees, are susceptible to cold injury as temperatures have an effect on the survival of the wasp. If the wasp eggs freeze in gall flowers, no wasps emerge in the spring to carry pollen to the female fruit. In such cases, site selection for caprifig trees becomes essential. Male figs are recommended to be planted in frost-free regions. In this case, the pollen-ripening period of the male should be compatible with the receptive stage of the female. Otherwise, caprifigs become rare, availability decreases, and profichi crop prices increase. A further effect could be low productivity in commercial fig orchards. The risk increases when winter temperatures fall below −10°C. Klingaman (2008) reported Brown Turkey (also called Texas Everbearing locally) and Blanche (also called White Russian) as cold-hardy fig varieties that could be recommended for southern growers in the USA. The Celeste fig variety is resistant down to −17.8°C and Chicago Hardy down to −12.0°C (Ernst, 2018).

10.2.3 Dormancy and chilling/heat requirements

Weather plays a significant role in the productivity and quality of figs. Under Mediterranean conditions, trees start shedding their leaves in mid-autumn and enter the dormant period. Bud break is in March or early April depending on the site conditions. Figs perform well under mild conditions with high temperatures and dry periods in summer, as characterized by the typical Mediterranean climate. However, fig is currently cultivated commercially beyond this zone in Argentina, Brazil, Peru, Mexico, and Chile, or under protected conditions in Mediterranean climates. In some regions, commercial fig production is realized by the application of dormancy-breaking growth regulators such as hydrogen cyanamide. Special orchard management systems are also developed and/or growth regulator application is coupled with pruning and irrigation. The Masui Dauphine fig variety is grown commercially in Japan, and using hydroponic systems, the productivity can be more than doubled. Bud dormancy can be limiting in year-round production of figs. In 8-year-old Masui Dauphine fig trees, bud dormancy and its physiological basis displayed three-phased endodormancy of the fig bud as: (i) introductory phase (early October to early November); (ii) deepest phase (late November); and (iii) awaking phase (middle December to late January) (Kawamata *et al.*, 2002). The nodal position affects the intensity of dormancy calculated as days for bud break and the percentage of bud break. The intensity of bud dormancy in fig was weaker than grape and weaker in the basal nodes compared with the upper positions. In fig trees, current-season shoots build up starch during the second phase of dormancy and then during the third period prior to bud break convert it into soluble sugars to protect from frost damage. Abscisic acid (ABA) also increases rapidly in current shoots during the first phase of endodormancy and then decreases (Kawamata *et al.*, 2002).

Dormancy studies carried out on node cuttings of seven fig varieties, Filalia, Ournakssi, Ghouddane, Kadota, Lerida, Bourjassote Noire, and Longue d'Août, of Moroccan or Mediterranean origin, showed that dormancy started increasing in mid-November and then decreased sharply in February. The steady-state level that mimicked the winter dormant period revealed that the total amount of chilling hours (<7°C) required for removal of dormancy is 400–450 h. Fig trees did not show any prolonged dormancy, and 400–450 chilling hours fulfilled in winter were sufficient for timely bud break. Varieties displayed differences for the mean time required for bud break irrespective of their origin and varied between 35.6 and 77.3 days. However, no buds stayed dormant under the chilling conditions of 592 and 566 h, in the first and second year of the experiment, respectively, demonstrating that common fig is a species with a low chilling requirement (Oukabli and Mekaoui, 2012).

In all fig varieties bearing the spring crop, the breba fruit generally ripen homogeneously within a very short duration (10–15 days). However, cold, and humid conditions may delay ripening by up to a month. The second-crop fig fruit form on the current season's branches at intervals, and the harvest period may vary between 1 and 3 months. The fig fruit display a double-sigmoid curve consisting of rapid, slow, and rapid growth phases. The length of the second (slow) growth phase varies significantly among fig varieties. The fruit development period is longer in varieties with longer second phases. Especially in common-type fig varieties, fruit continue forming as the shoots continue growing. The number of ripe fig fruit harvested then depends mostly on the climatic conditions and the practices applied. Heat summation is generally adequate in summer or early autumn for fruit to ripen in most of the commercial production sites. However, rarely, farmers plant varieties bearing fruit at longer intervals and/or having a longer fruit-development period without considering the prevailing conditions in autumn, resulting in many unripe fruit. These fruit develop up to a certain point but cannot ripen due to inadequate heat summation. This may occur in cases where irrigation or rainfall continues late in the season, or excess nitrogen is applied at the wrong time triggering shoot growth.

10.2.4 Salt tolerance

Fig trees are widely dispersed in the natural flora of the Mediterranean coastline where salinity is and will continue to be a problem due to over pumping of water sources. Salinity in general restricts gas-exchange capacity, root and vegetative growth, water uptake, and RWC, and reduces yield; however, the responses of fig varieties vary. Several studies have been carried out to test the responses of fig varieties under variable conditions to assess their tolerance. Tissue culture is often used to screen varieties for salinity. The morphological and biochemical responses of five fig varieties to 8-month exposure to salinity ranging from 0 to 12,000 ppm. NaCl under *in vitro* conditions revealed that, as NaCl concentrations increased, growth was reduced both during multiplication and at the rooting stages after 5 weeks. Among the tested varieties, Masone Black and Shami Stihy performed better and were more tolerant to salinity within the tested range than the other three varieties, Achtoy White, Zeiblly Red and Zeiblly Flair, in decreasing order. Under the highest salt stress, the proline content and K^+:Na^+ ratios were higher in the two tolerant varieties compared with the others (Soliman and Abd Alhady, 2017). The Black Mission, Brown Turkey, and Brunswick fig varieties were subjected to NaCl-induced stress under *in vitro* conditions. The salt concentrations applied for 4 weeks were 0, 2, 4, 6, 8, 9, 10, 11, and 12 g L^{-1} NaCl. As salt concentration increased, growth was reduced, and chlorophyll, potassium, iron, and zinc concentrations decreased. The response of the varieties differed, and their tolerance decreased in the order of Black Mission, Brunswick, and Brown Turkey. Proline content is known to be a good marker for selection of tolerant varieties (Metwali *et al.*, 2014).

In figs grown in a saline nutrient solution (up to 100 mM NaCl), net photosynthesis was negatively affected as leaf Na^+ and Cl^- concentrations increased, and the clones and light intensity had an effect on the reduction rate of net photosynthesis. In the case of short exposure to salinity, WUE was increased due to enhanced net assimilation and reduced transpiration rate. The advanced assimilation was due to the increase in non-stomatal conductance. The root mass played a substantial role with respect to rate of photosynthesis after short-term and gradual NaCl treatment. The assimilation rate was not altered or reduced in fig plants with few roots, whereas it was stimulated during the early days of the treatment. Long-term exposure decreased the net assimilation and transpiration rates in figs; however, the decrease in transpiration was more pronounced, leading to higher WUE (Golombek and Lüdders, 1990, 1993).

To examine the effect of the Na^+/H^+ antiporter gene (*AgNHX1*) in salt tolerance, the gene was transferred to Black Mission fig variety using *Agrobacterium tumefaciens* (Metwali *et al.*, 2015). Transgenic Black Mission plantlets were cultured in saline conditions of 200–300 mM NaCl in a greenhouse. The transgenic plantlets continued shoot growth under the highest concentration and proline accumulated fourfold compared with non-transgenic plants. Proline, as in the case of drought, accumulates under salt stress and operates as a reserve source for recovery.

10.2.5 Invasiveness

Fig—whether a female or a male tree—can easily be spread by seeds, suckers, and cuttings. Birds help the dispersal of figs in nature. Suckers can grow rapidly and, if not managed, develop into thick bushes. Currently, fig is considered as a low-risk species regarding invasiveness and its possible threat to other species (Dartiles, 2015).

10.3 Quality Guarantee and Sustainable Farming Systems

All actors of the agri-food system aim to achieve sustainability, and this target pressurizes the actors more, as water crises, food insecurity, global warming, and climate-change data are produced. Furthermore, consumers and the public at large have become more aware of the threats and negative side effects of intensive and unsustainable farming systems. The role of agriculture in integrating knowledge and best practices becomes crucial to reverse these negative effects and continue producing healthy agri-food products without any environmental or social degradation. The initiatives for change may come from farmers, research, industry, or the state; historically, it appeared in two directions, top to bottom or bottom up. After a certain point, ongoing processes create an opportunity and become more of a system innovation (Elzen *et al.*, 2012). These innovative approaches may become a trans-sectoral policy in some countries, as in the European Union (EU) Green Deal, whereas in most cases, these initiatives develop as guarantee systems shaping the international trade to a certain extent. The COVID-19 pandemic and governments' declarations on nationally determined contributions to climate change bring the discussions to a higher level of redesigning the agri-food systems. The quality guarantee systems affecting all the value-chain actors are based on ecological/environmental and/or socio-economic principles, have defined guidelines, and start at the farm, and may include processing and trade phases. The main objective of the guarantee systems, which may lead to a certificate, label and/or logo, is to build trust among consumers who are distant from the producers in a globalized market. Traceability, reference standards, and conformity assessment leading to a certificate are the key tools. The focus of these quality systems may vary and shift from protection of the "terroir" (geographical indications) to severe restriction of synthetic off-farm inputs (www.ifoam.bio, accessed December 1, 2021) to highlighting working conditions (SA 8000 certification) or ethical and fair-trade systems (www.fairtradecertified.org/, accessed December 1, 2021). Such systems do not focus solely on the farm level but cover the whole value chain. Currently, three quality guarantee systems are dominant in worldwide fig trade: geographical indications (GIs), Good Agricultural Practices (GAP), and organic production system. However, third-party multi-certification systems such as organic + GI or organic + fair-trade certifications are also increasing globally. The other trend, spreading especially rapidly in developing countries, is to focus on Participatory Guarantee Systems based on internal quality assurance rather than a third-party conformity assessment (www.ifoam.bio, accessed December 1, 2021).

10.3.1 Geographical indications

In most of the markets, consumers seek "local" or "regional" products for authenticity, appeal, and taste. GI is a sign that shows that the product has a certain quality or reputation linked to a specific geographical origin. GIs are administrated at a national or global level. The World Intellectual Property Organization (WIPO) governs through the Lisbon System based on the Lisbon Agreement for the Protection of Appellations of Origin and their International Registration of 1958 ("the Lisbon Agreement") and its latest revision, the Geneva Act of the Lisbon Agreement on Appellations of Origin and Geographical Indications of 2015 ("the Geneva Act"). The Lisbon System for the International Registration of Appellations of Origin and Geographical Indications provides protection for an "appellation of origin" or a "geographical indication" in the contracting parties through a single registration procedure.

The EU agricultural quality program also includes GIs for agricultural products and foodstuffs as an intellectual property right with the aim of protecting a specific region or a specific quality linked to the specific conditions of the territory. The aim is to value the regional characteristics that affect the final quality and thus contribute to rural development. It is also a cultural issue maintaining future sustainability of the product or the technique. The EU quality scheme comprises three GIs: Protected Designation of Origin (PDO) (for food and wine), Protected Geographical Indication (PGI) (for food and wine) and Traditional Specialty Guaranteed (TGS) (without a link to a specific geographical area) (www.ec.europa.eu/info/food-farming-fisheries/food-safety-and-quality, accessed December 1, 2021). PDO status requires that the product is traditionally produced as raw material, processed, and prepared in a specific region that gives the desired attributes. PGI status is given when traditionally only part of the whole chain of production and processing is in the mentioned region. The EU system not only registers products from the member states but also protects products/regions from non-EU countries either as part of a bilateral agreement or through direct applications. Many countries have their own GI systems with a defined scope to protect a product, processing method, country, or region. The EU process also requires that the approval first comes from the country of origin. This registry empowers the holder to prevent marketing of products that do not conform to the standard laid down during the registry. The registered products are legally protected by the institutions under intellectual property rights within a specific country, group of countries (e.g. EU), or globally. Moreover, labeling and the use of a logo, which are compulsory in some systems, help consumer awareness. eAmbrosia, the EU registry system for GIs, gives the following figs from France, Italy, Greece, and Turkey as having PDO status approved and protected within the EU: Figue de Sollies (PDO-FR-0544), Fichi di Cosenza (PDO-IT-0682), Fico Bianco del Cilento (PDO-IT-0282), Xira Syka Taxiarchi (dried fig; PDO-GR-0790), and Aydın İnciri (dried fig; PDO-TR-1116) (Fig. 10.2).

10.3.2 Good Agricultural Practices

As the markets globalized, management of food safety risks and traceability to identify the critical points became essential. After various scandals of high pesticide residues or non-conformity cases in the fruit and vegetable sector, national

A. B.

Fig. 10.2. Dried-fig varieties with Protected Denomination of Origin (PDO) in the EU. A. Aydın inciri (Sarılop) fig orchards in Aydın, Turkey. Photograph courtesy of Uygun Aksoy. B. Fruit and leaf of the Kalamata fig variety (Xira Syka Taxiarchi) for sun drying in Greece. Photograph courtesy of Vasileios Stournaras.

authorities and companies started to develop standards, check lists, and certification systems to meet the demands of consumers for safe food and to minimize risks. Hazards can be present at all stages of the value chain, starting with the seed, sapling, or soil, and continuing through packaging, storage, processing, and distribution. Thus, farm-to-fork strategies target safety and sustainability throughout the whole value chain. GAP is defined by the Food and Agriculture Organization of the United Nations (FAO) as, a "collection of principles to apply for on-farm production and postproduction processes, resulting in safe and healthy food and non-food agriculture products, while taking into account economic, social and environmental sustainability." A similar approach of identifying the hazards, evaluating the risk, and developing a system where risks are minimized is applied at the storage stage through Good Storage Practices (GSP) and at the processing stage through Good Manufacturing Practices (GMP). The GAP system ensures that the product conforms to the conditions framed in the legislative acts developed by the national authority. The GAP manuals are prepared for a group of crops (FAO, 2016) or target a specific product, as is the case for figs. A GAP book was released in Malta after the die-out of fig trees by the invasive beetle *Hypocryphalus scabricollis*. The aim was to provide guidelines for fig farmers on ways to manage this invasive bark beetle (Cutajar and Mifsud, 2017). The Codex Alimentarius (2008) Code of Practice is developed according to GAP and GSP and aims to "provide uniform guidance for all countries to consider in attempting to control and manage contamination by various mycotoxins, specifically aflatoxins." A similar manual was prepared in Turkey to guide farmers to apply best practices, especially for aflatoxins, and pest and disease management (Şahin *et al.*, 2012).

Among existing national, regional, and international GAP certification systems, the most widespread one is GLOBALG.A.P., which is a private system that first became effective in 1997 as an initiative of European fresh produce retailers and was then introduced at a global level. The earlier name EUREPGAP was changed to GLOBALG.A.P., and the scope of the system enlarged to include various plant- and animal-based products or processes (www.globalgap.org/, accessed December 1, 2021). GLOBALG.A.P. is a trademark and has a set of standards on GAP and a certification scheme to provide safe products, to have less impact on the environment, and to care for the safety and welfare of workers and animals. GLOBALG.A.P. initially covered fresh fruit and vegetables, and fresh figs sold in European retail chains have been included in its certification for decades. The database is accessible to members only, which allows them to search GLOBALG.A.P.-certified products including figs and/or farmers. The final product at the consumer level is not labeled. GLOBALG.A.P. has its own conformity criteria and checklist, and the allowed inputs and maximum limits consider the food safety legislative framework in the producing and the importing country. In some countries, national GAP systems similar to GLOBALG.A.P. exist, and the final product is labeled to provide visibility to the consumer. One of the criticisms brought forward is the high level of paperwork, which is difficult for small farmers, and the high costs incurred for conformity to these quality criteria.

Economic analysis of Turkish dried figs grown conventionally or under the GAP system showed that GAP resulted in higher production costs. The production costs of the two farming systems were \$0.94 (conventional) and \$1.45 (GAP) kg^{-1} of sun-dried figs, yielding a difference of \$0.51 kg^{-1} at the farm level. The main contributors in the conventional system were increased labor and the use of synthetic chemical fertilizers and pesticides. In the case of GAP, the high costs of farmyard manure and the cost of certification created the difference; however, state support for GAP reduced this difference. At the middle-men stage, GSP and purchasing of plastic crates created an additional cost for GAP, making a difference of \$0.22 kg^{-1}. At the processing stage, the additional cost of implementing GSP and GMP was \$0.41 kg^{-1} of dried figs (Çobanoğlu *et al.*, 2010).

10.3.3 Organic fig production

Organic farming is defined in various ways; however, it is a management system that excludes or severely restricts the use of synthetic agrochemicals, helps to build ecological cycles, and relies on on-farm or locally available natural resources. Techniques such as diversification and using green manure, intercropping, cover crops, compost, and animal manure form the basis for maintaining long-term soil fertility. With respect to orchard pest and disease management, only cultural practices such as orchard hygiene and allowed inputs such as traps, microorganisms, beneficials, mineral oils, sulfur, Bordeaux mixture, or copper preparations can be used conditionally. The rules apply to the whole production chain. In dried-fig production, sulfur as a post-harvest treatment is not allowed, and storage-pest control requires a major shift from chemical fumigation to non-chemical alternatives.

In many traditional producer countries, fig production is a low-input system under rainfed conditions. Low reliance on synthetic off-farm inputs has urged transition to organic fig production in many countries. This demand is increasing due to the health benefits of the phytochemical-rich fig fruit, whether fresh, dried, or processed. In the world organic market, every importing country is required to fulfill the requirements of the national regulation unless there is a bilateral agreement or equivalency of the systems. Willer *et al.* (2020) reported that, as of 2018, out of 186 countries where organic agriculture is practiced, 103 have their own legislation, some of which is not yet implemented. These regulations are based on similar principles but may have differences in terms of scope, transition period, and allowed inputs.

The statistical database of the Research Institute of Organic Agriculture (FIBL) shows that in 2018, Argentina, Croatia, Ecuador, Egypt, Israel, Italy, Jordan, Lebanon, Madagascar, Mauritius, Mexico, Montenegro, North Macedonia, Serbia, Spain, Turkey, and the USA had fig orchards that were certified under organic management fully converted or under conversion (https://statistics.fibl.org; accessed 1 December, 2021). As the prime producer, Turkey has the largest certified organic fig area in hectares, far more than the others, followed by Spain, the USA, and Croatia. According to the official figures, approximately 45% of the Turkish fig area is certified as organic, including those in transition. In Turkey, organic figs are mainly sun-dried Sarılop fruit and are destined mainly for the EU and US markets. Fresh figs of the Bursa Black variety are also available for the export market, mainly destined for the EU (www.tarim.gov.tr/, accessed December 1, 2021).

In a research work carried out in Turkey as a preliminary study, the response of fig trees to drought conditions was evaluated in biodynamic, organic, and conventional management systems. All experimental plots were under rainfed and low-input conditions. In fig orchards, the impact of drought appeared as necrotic patches on leaves and early leaf fall. Observations carried out at the end of July showed that necrosis was at the initial stage (~1%) in organic and biodynamic orchards, whereas necrotic areas were more advanced in conventional orchards (~5%) and early leaf fall was already starting. Evaluation of the orchards revealed that organic orchards have higher soil organic matter content and cation exchange capacity. Biodynamic management also led to increased chlorophyll content and maximum fluorescence:minimum fluorescence (F_m:F_o) ratio in the fig leaves, possibly because of the biodynamic preparations used that promote chlorophyll integrity. Other related parameters such as soil respiration or nutrient uptake rates showed variations and did not display a parallel trend. There were no marked differences among the farm management systems with respect to the analyzed fruit quality attributes (Hallaç Türk *et al.*, 2011).

One of the major problems faced in dried-fig production and trade is the storage pests that are generally transferred from the orchard. The dominating pest species vary according to the production area. In Turkey and Greece, the major storage pest is almond moth, *Ephestia cautella* (Walker) (Pyralidae, Lepidoptera) (Erakay and Özar, 1979; Eliopoulos and Athanassiou, 2004), whereas in California, dried-fruit beetle or sap beetles, *Carpophilus* species (Nitidulidae), are more common (Coviello and Bentley, 2009). The traditional method to control storage pests was fumigation with efficient and cost-effective methyl bromide; however, the use and production of methyl bromide has been phased out due to its impact on ozone-layer depletion. In commercial organic fig production, there are two alternative methods utilized widely: CO_2 fumigation and deep freezing. The almond moth is sensitive to temperature, and Aldawood *et al.* (2013) report that 25°C promotes fecundity, egg hatchability, and survival, whereas at 5°C, 100% mortality is attained. In practice, short-term deep freezing of dried figs at –20 to –40°C before processing helps to control all stages of the fig storage pests. Şen *et al.* (2009) compared atmospheric (5 days in gas-tight plastic cubes) and high-pressure (2 h in tanks at 25 bars) CO_2 applications to control *E. cautella* and *Carpoglyphus lactis* L. (Carpoglyphidae, Acari) in dried figs. Both methods effectively control these target pests and have no detrimental effect on dried-fig quality after the treatment or after 2 months of ambient storage. Use of pressurized tanks requires a high investment cost but allows fast and effective control, even at very short exposures. However, CO_2 treatment in gas-tight cubes for 5 days is cost-effective, especially in terms of the required infrastructure.

Organic practices, namely soil fertility management, if applied according to the basic principles, also help carbon sequestration (Babu *et al.*, 2018). The total carbon emissions arising from dried-fig production in Turkey was calculated and the effects of different practices evaluated. When carbon emissions of conventional and organic dried fig were compared, the emissions in conventionally managed orchards were 570 kg CO_2-eq ha^{-1} and were 2.3 times higher than in organic orchards (242 kg CO_2-eq ha^{-1}). The emissions per kilogram of dried fig were 0.315 and 0.192 kg CO_2-eq in conventional and organic management systems. In conventional orchard management, major carbon emissions are due to synthetic fertilizer use, namely nitrogenous fertilizers (32%), followed by irrigation and soil tillage. The organic orchards are mostly located at higher elevations and are under low-input rainfed conditions, and therefore carbon emissions are very low. In organic orchards, 12% of carbon emissions arose from soil tillage. Regarding control of storage pests, chemical and non-chemical fumigation methods showed significant differences with respect to carbon emissions, and therefore all applied treatments were included in the assessment. In conventional fig processing, two fumigation treatments with phosphine gas resulted in 54,891 kg CO_2-eq. In organic dried-fig processing, 50% of the companies deep-freeze fig fruit by keeping them at −24°C for 20 h to kill all stages of storage pests. CO_2 treatment under atmospheric pressure, which is an alternative method to control storage pests, is recommended due to lower carbon emission (7161 kg CO_2-eq) compared with deep freezing (249,716 kg CO_2-eq). Parallel to the extension of the marketing period over the year, more companies are using cold storage (2–4°C and 60–65% relative humidity) to preserve dried-fruit quality throughout the summer months, which contributes to higher carbon emissions. Assessment showed that major practices or operations contributing to and/or elevating the carbon footprint of the dried-fig sector in conventional management systems are use of synthetic fertilizers (chiefly nitrogenous fertilizers), flood irrigation, and phosphine treatments, whereas in organic management, soil tillage, cold storage, and deep freezing for controlling storage pests are the main contributors. Thus, CO_2 treatment under atmospheric pressure is recommended to organic dried-fig processors due to lower carbon emissions compared with deep freezing. Decision makers, farmers and processing companies should aim to optimizing their practices to reduce carbon emissions and develop climate-smart solutions in fig production (Okan Arikan *et al.*, 2020).

10.4 Concluding Remarks

Demand for fresh, dried, and processed figs is increasing as more of the health benefits are disclosed. This creates a stimulus for new producing countries and regions, leading to the spread of fig beyond its origins, the Mediterranean. Climate-change phenomena will play a crucial role in the medium and long-term production figures of current fig-producing countries, as well as those newly entering the global trade. Rich biodiversity among fig germplasm may ensure environmental adaptability, if varietal and site-specific characteristics are well known and risks are manageable. Furthermore, growing figs under protection is becoming more practiced, either to avoid adverse climatic conditions or to produce an earlier harvest so that figs can be on the market for longer periods, bringing the sustainability of the farming system into question. Additionally, enhanced communication skills and capacities raise other issues in sustainability due to the long-distance transfers between the producer and the consumer. The distant consumer wants to know where, how, and by whom the figs are produced. Quality and safety assurance is regulated by official and private standards certified by an independent control body or by participatory systems. During the post-pandemic period, threatened more by climate change, the discussions will focus on agri-food systems that are based on long-distance transfer of inputs and final products in globalized markets, which increase carbon emissions and food security risks. Agri-food policies targeting reduction of external inputs, promoting sustainable farming systems, and favoring self-sufficiency status are expected to trigger organic, local, and GI-labeled products both at the local level and in the global markets. As well as environmental sustainability, cultural, ethical, and socio-economic issues throughout the value chain are currently subject to discussions that will continue in the coming years.

References

Abdel-Razik, M.S. and El-Darier, S. (1991) Functional adaptations of fig trees (*Ficus carica* L.) in agroecosystems of the western Mediterranean Desert of Egypt. *Qatar University Science Journal* 11, 183–199.

Aksoy, U. and Akyüz, D. (1993) Changes in K, Ca and Mg contents in different parts of the fig fruit during development. In: Fragoso, M.A.C., van Beusichem, M.L., and Houwers, A. (eds.) *VIII International Colloquium On Optimization of Plant Nutrition*, 31 August–8 September, Lisbon, Portugal. Kluwer Academic Publishers, Dordrecht, The Netherlands, pp. 309–312.

Aksoy, U. and Anaç, D. (1994) The effect of calcium chloride application on fruit quality and mineral content of fig. *Acta Horticulturae* 368, 754–762.

Aksoy, U., Can, H.Z., Hepaksoy, S., and Şahin, N. (2001) *İncir Yetiştiriciliği (Fig Growing), Turkish Scientific and Technical Council*. Turkish Agricultural Research Publications, Agriculture, Forestry and Food Technologies Research Group, İzmir, Turkey.

Aldawood, A.S., Rasool, K.G., Alrukban, A.H., Soffan, A. Husain, M., *et al.* (2013) Effect of temperature on the development of *Ephestia cautella* (Walker) (Lepidoptera: Pyralidae): a case study for its possible control under storage conditions. *Pakistan Journal of Zoology* 45, 1573–1578.

Al-Khalaf, A., Quinnell, R.J., and Compton, S.G. (2015) Influence of temperature on the reproductive success of a fig wasp and its host plant. *African Journal of Agricultural Research* 10, 1625–1630.

Altieri, M. (2015) *Agroecology: Principles and Strategies for Designing Sustainable Farming Systems*. University of California, Berkeley, California.

Ammar, A., Ben Aissa, I., Mars, M., and Gouiaa, M. (2020) Seasonal variation of fig tree (*Ficus carica* L.) physiological characteristics reveals its adaptation performance. *South African Journal of Botany* 132: 3037.

Babu, S., Das, A., Yadav, S.G., and Ngachan, S.V. (2018) Carbon sequestration in organic farming. In: Dutta, D., Ravishankar, N., Meena, A.L., Ghasal, P.C., Meena, L.K., *et al.* (eds) *Carbon Sequestration and GHG Measurements*. ICAR – Indian Institute of Farming System Research Modipuram, Meerut, India, pp. 26–35.

Can, H.Z. and Aksoy, U. (2007) Seasonal and diurnal photosynthetic behaviour of fig (*Ficus carica* L.) under semi-arid climatic conditions. *Acta Agriculturae Scandinavica, Section B: Soil & Plant Science* 57, 297–306.

Can, H.Z., Hepaksoy, S., Aksoy, U., and Kutlu, E. (2000) Leaf characteristics and net gas exchange of fig cultigens adapted to different climatic conditions. *Acta Horticulturae* 516, 131–138.

Caruso, G., Gennai, C., Ugolini, F., and Gucci, R. (2017) Gas exchange and growth response of young *Ficus carica* L. plants to water deficit and relief. *Acta Horticulturae* 1173, 157–162.

Çobanoğlu, F., Aksoy, U., Tunalioglu, R., and Karaman, A.D. (2010) Differences in the economic costs of implementing conventional practices and good practices intended for aflatoxin management in the dried fig supply chain. *Toxin Reviews* 29, 87–98.

Codex Alimentarius (2008) *Code of Practice for the Prevention and Reduction of Aflatoxin Contamination in Dried Figs*. CAC/RCP 65-2008. Food and Agriculture Organization of the United Nations, Rome/World Health Organization, Geneva.

Cook, J.M. and Rasplus, J.Y. (2003) Mutualists with attitude: coevolving fig wasps and figs. *Trends in Ecology and Evolution* 18, 241–248.

Coviello, R.L. and Bentley, W.J. (2009) Fig: UC IPM Pest Management Guidelines. University of California Agriculture and Natural Resources, Los Angeles, California.

Cutajar, S. and Mifsud, D. (2017) *Good Agricultural Practice (GAP) for Fig Tree Cultivation*. Plant Protection Directorate, Malta University Consulting, Lija, Malta.

Dartiles, M.J. (2015) *Ficus carica* (common fig). In: *Invasive Species Compendium*. CAB International, Wallingford, UK. Available at: www.cabi.org/isc/datasheet/24078 (accessed December 1, 2021).

Darwish, D.A., El-Berry, I.M, Mustafa, N.S, Moursy, F.S., and Hagagg, L.F. (2015) Detecting drought tolerance of fig (*Ficus carica*, L.) cultivars depending on vegetative growth and peroxidase activity. *International Journal of Advanced Research* 2, 830–844.

Demir, Ö. (2005) Effect of aspect and inclination on tree growth, yield and quality in organic fig orchards. MSc thesis, Graduate School of Natural Sciences, Ege University, İzmir, Turkey.

Eliopoulos, P. and Athanassiou, C. (2004) Seasonal occurrence of dried fig pests and their parasitoids in a fig warehouse in Greece. *IOBC/WPRS Bulletin* 27, 159–171.

Elzen, B., Barbier, M., Cerf, M., and Grin, J. (2012) Stimulating transitions towards sustainable farming systems. In: Darnhofer, I., Gibbon, D., and Dedieu, B. (eds) *Farming Systems Research into the 21st Century: The New Dynamic*. Springer, Dordrecht, The Netherlands, pp. 431–455.

Erakay, S. and Özar, A.İ. (1979) Ege Bölgesinde kuru incirlerde zararli böceklerin bulaşma oranlari ve savaş yöntemleri üzerinde ön araştirmalar (Preliminary studies on infestation rates of dried fig pests in the Aegean region and control methods). *Plant Protection Bulletin* 19, 159–173.

Ernst, M. (2018) *Figs*. CCD-CP-135. Center for Crop Diversification, University of Kentucky College of Agriculture, Food and Environment, Lexington, Kentucky. Available at: www.uky.edu/ccd/sites/www.uky.edu.ccd/files/figs.pdf (accessed December 1, 2021).

FAO (2011) *Estahbanat rainfed fig system*. Iranian Agricultural Heritage System. Globally Important Agricultural Heritage Systems (GIAHS). Available at: www.fao.org/fileadmin/templates/giahs/PDF/Fig_traditional_system_Final_.pdf (accessed December 1, 2021).

FAO (2016) A Scheme and Training Manual on Good Agricultural Practices (GAP) for Fruit and Vegetables. Available at: www.fao.org/3/i5739e/i5739e.pdf (accessed December 1, 2021).

Gholami, M., Rahemi, M., Bahman Kholdebarin, B., and Rastegar, S. (2012) Biochemical responses in leaves of four fig cultivars subjected to water stress and recovery. *Scientia Horticulturae* 148, 109–117.

Golombek, S.D. and Lüdders, P. (1990) Gas exchange of *Ficus carica* in response to salinity. In: van Beusichem, M.L. (ed.) *Plant Nutrition – Physiology and Applications*. Springer, Wageningen, The Netherlands.

Golombek, S.D. and Lüdders, P. (1993) Effects of short-term salinity on leaf gas exchange of fig (*Ficus carica* L.). *Plant and Soil* 148, 21–27.

Hallaç Türk, F. and Aksoy, U. (2011) Comparison of organic, biodynamic and conventional fig farms under rain-fed conditions in Turkey. *Cell Plant Science* 2, 22–33.

Hiwale, S. (2015) Fig (*Ficus carica*). In: Hiwale, S. (ed.) *Sustainable Horticulture in Semiarid Dry Lands*. Springer, New Delhi, India, pp. 159–175.

Jafari, M., Rahemi, M., and Zare, H. (2017) Increasing the tolerance of fig tree to drought stress by trunk thinning. *Acta Horticulturae* 1173, 189–194.

Jevanandam, N., Goh, A.G.R., and Corlett, R.T. (2013) Climate warming and the potential extinction of fig wasps, the obligate pollinators of figs. *Biology Letters* 9: 20130041.

Kawamata, M., Nishida, E., Ohara, H., Ohkawa, K., and Matsui, H. (2002) Changes in the intensity of bud dormancy and internal composition of current shoot in fig. *Journal of the Japanese Society for Horticultural Science* 71, 177–182.

Klingaman, G. (2008) *Plant of the week: White Russian Fig. Extension News – 1 August 2008*. Division of Agriculture, University of Arkansas, Fayetteville, Arkansas.

Kong, M., Lampinen, B., Shackel, K., and Crisosto, C.H. (2013) Fruit skin side cracking and ostiole-end splitting shorten postharvest life in fresh figs (*Ficus carica* L.), but are reduced by deficit irrigation. *Postharvest Biology and Technology* 85, 154–161.

Kutlu, E., Can, H.Z., Aksoy, U., and Hepaksoy, S. (2000) Evaluation of gas exchange capacity and physiological responses of selected Sarilop (=Calimyrna) fig clones. *Acta Horticulturae* 517, 59–64.

Li, Z., Zeng, Z., Tian, D., Wang, J., Fu, Z., *et al.* (2020) Global patterns and controlling factors of soil nitrification rate. *Global Change Biology* 26, 4147–4157.

Metwali, E.M.R., Soliman, H.I.A., Al-Zahrani, H.S., Howladar, S.M., and Fuller, M.P. (2014) Influence of different concentrations of salt stress on *in vitro* multiplication of some fig (*Ficus carica* L.) cultivars. *Life Science Journal* 11, 386–397.

Metwali, E.M.R., Soliman, H.I.A., Fuller, M.P., Al-Zahrani, H.S., and Howladar, S.M. (2015) Molecular cloning and expression of a vacuolar Na^+/H^+ antiporter gene (*AgNHX1*) in fig (*Ficus carica* L.) under salt stress. *Plant Cell Tissue and Organ Culture* 123, 377–387.

Okan Arikan, E. and Aksoy, U. (2020) Calculating carbon emissions in Turkish organic and conventional dried fig production. *Acta Horticulturae* 1286, 269–276.

Oukabli, A. and Mekaoui, A. (2012) Dormancy of fig cultivated under Moroccan conditions. *American Journal of Plant Sciences* 3, 473–479.

Raz, L. and Agudelo Zamora, H. (2020) *Ficus carica* L. In: *Catalogo de Plantes y Liquenes de Colombia Version 1.1*. Universidad Nacional de Colombia. Bogotá DC, Colombia.

Roointan, R., Yadollahi, A., Sarikhani Khorami, S., Arab M.M., and Vahdati, K. (2018) Rainfed fruit orchards in sloping lands: soil erosion reduction, water harvesting and fruit production. *Acta Horticulturae* 1190, 107–111.

Rostami, A.A. and Rahemi, M. (2013) Responses of caprifig genotypes to water stress and recovery. *Journal of Biological and Environmental Sciences* 7, 131–139.

Şahin, B., Kösoğlu, İ., Tan, N., Özen, M., Konak, R., *et al.* (2012) Kuru incir yetiştiriciliği ile önemli hastalık ve zararlıları el kitabı. In: Aksoy, U. (ed.) *Manual on Dried Fig Production and Major Pests and Diseases*. Aegean Exporters' Association, İzmir, Turkey.

Şen, F., Meyvacı, K.B., Aksoy, U., Emekçi, M., and Ferizli, A.G. (2009) Effects of the post-harvest application of methyl bromide alternatives on storage pests and quality of dried figs. *Turkish Journal of Agriculture and Forestry* 33, 403–412.

Soliman, H.İ.A. and Abd Alhady, M.R.A. (2017) Evaluation of salt tolerance ability in some fig (*Ficus carica* L.) cultivars using tissue culture technique. *Journal of Applied Biology and Biotechnology* 5, 29–39.

Willer, H., Schlatter, B., Travnicek, J., Kemper, L., and Lernoud, J. (2020) *The World of Organic Agriculture. Statistics and Emerging Trends 2020*. Research Institute of Organic Agriculture (FIBL), Frick and IFOAM – Organics International, Bonn, Germany.

11 Robotics and Remote Sensing in Orchards

Avital Bechar* and Yafit Cohen
Institute of Agricultural Engineering, Agricultural Research Organization, The Volcani Institute, Rishon LeZion, Israel

11.1 Introduction

Labor is the largest single cost contributor in agriculture (Bechar and Eben-Chaime, 2014). It is also the main limiting factor in the development of agricultural industries in both developed and developing countries. Labor force shortages, damage farmer revenue, yield, and durability. In countries where labor wages are low, their transient nature reduces production capability and quality. In addition, some of the manual work can cause musculoskeletal disorder risks and chronic health issues to workers (Perez-Ruiz *et al.*, 2014). Despite the differences and large variability, labor contributes up to 40% of the operational costs in most crops. The enormous labor force required for the various operations causes bottlenecks, downgrading productivity, reducing yield, and increasing costs. Moreover, problems such as aging of the workforce and a shortage of farmers contribute to the lack of manpower (Iida *et al.*, 2013). This high manual labor requirement impedes cost reductions and increases the demand for robotics and automation (Bechar *et al.*, 2007).

In the 20th century, technological progress in developed countries reduced the work force that was traditionally available for farming activities by a factor of 80 (Ceres *et al.*, 1998). Agricultural productivity has significantly and continuously increased over the centuries as a result of mechanization and intensification, and, more recently, with the introduction of automation (Nof, 2009) by increasing efficiency, reliability, and precision, and reducing the need for human involvement. However, agriculture is still in distress from a critical lack of trained workers, especially in orchard crops.

Robots are perceptive machines that can be programmed to perform specific tasks, make decisions, and act in real time. They are required in various fields that normally call for reductions in manpower and workload, and are best suited for applications requiring repeatable accuracy and high yield under stable conditions (Holland and Nof, 2007). However, robots lack the capability to respond to ill-defined, unknown, changing, and unpredictable events (Bechar and Vigneault, 2016).

In agriculture, the environment is unstructured and demands the motion of robots unlike that of machines in a factory or of vehicles in a parking lot (Canning *et al.*, 2004). It is characterized by rapid changes in time and space, such as military, underwater, and space environments (Bechar and Edan, 2003). The terrain, vegetation,

* Email: avital@volcani.agri.gov.il

 Advances in Fig Research and Sustainable Production (M.A. Flaishman and U. Aksoy)
DOI: 10.1079/9781789242492.0011

landscape, visibility, illumination, and other atmospheric conditions are not well defined; they continuously vary, have inherent uncertainty, and generate unpredictable and dynamic situations (Bechar, 2010). Complexity increases with involvement of natural objects, such as those encountered in medical and agricultural environments, because of the high variability in shape, texture, color, size, orientation, and position of such objects (Bechar *et al.*, 2009). In addition, the product being dealt with is of relatively low cost, and therefore the cost of an automated or robotic system must be low in order for it to be economically justified, and the seasonal nature of agriculture makes it difficult to achieve the high utilization found in the manufacturing industries. All these factors are therefore subject to inherent uncertainty, heterogeneity, and unpredictable situations, resulting in lack of information (Oberti and Shapiro, 2016).

The complex agricultural environment, combined with the intensive production, requires robust systems at relatively low cost. The advent of agricultural robots has the potential to raise the quality of fresh produce, lower production costs, reduce the drudgery of manual labor, and, in some parts of the world, compensate for the lack of workers in some agricultural sectors (Bechar, 2010).

Fruit-tree productivity is highly dependent on irrigation (Naor, 2006) and is affected by field heterogeneity (Konopatzki *et al.*, 2016). Irrigated fig orchards obtain increased yield in comparison with rainfed figs (Andrade *et al.*, 2014). The positive effect of the irrigation may be dependent on soil type. Irrigation has also a positive impact on fig fruit size and thus on the profitability (Goldhamer and Salinas, 1999). However, water resources are dwindling, and the cost of water for agricultural use is rising. This necessitates developing more efficient irrigation tools to avert water scarcity. Field-level irrigation management using remote sensing as an aid for water-status monitoring may serve a viable option to meet this challenge.

This chapter focuses on the research effort and the development of agricultural robots and remote sensing to perform different agricultural tasks in orchards, as well as on irrigation management using remote sensing. To the best of our knowledge, not much work has been conducted on robotic remote-sensing irrigation for figs; nevertheless, there is much similarity between agricultural tasks in figs and in other orchard crops that the robotic research can reflect on.

11.2 Agricultural Robot Principles

Robots or automated systems for horticultural field production comprise multiple subsystems to enable them to operate and perform their tasks. These subsystems and devices deal with mobility, steering and control, sensing techniques, path planning, navigation or guidance abilities, manipulators or similar functional devices, and produce contact (via end effectors). It can be divided into two sets of abilities. The first deals with the robot functionalities, such as obstacle avoidance (Belkhouche and Belkhouche, 2005; Bergerman *et al.*, 2015; Gonzalez-de-Soto *et al.*, 2016), self-localization and map building (Gomez-Gil *et al.*, 2013; Bayar *et al.*, 2015; Gimenez *et al.*, 2015; Underwood *et al.*, 2015; Lepej and Rakun, 2016), and navigation and path planning (Bodur *et al.*, 2012; Hameed *et al.*, 2013, 2016; Kraus *et al.*, 2013; Vroegindeweij *et al.*, 2014). The second deals with specific applications, such as yield transportation (Morimoto *et al.*, 2005; Bakker *et al.*, 2011; Thanpattranon *et al.*, 2016; Zhang *et al.*, 2016), harvesting (Foglia and Reina, 2006; Bac *et al.*, 2014; Mann *et al.*, 2014; Mehta and Burks, 2014; Riyaz Ahammed *et al.*, 2015; Vitzrabin and Edan, 2016b; Zhao *et al.*, 2016; Bloch *et al.*, 2018), and pruning and thinning (Rosa *et al.*, 2008; Corbett-Davies *et al.*, 2012; Linker and Neer, 2014; Poni *et al.*, 2016).

Sensors are commonly used in orchard robots for a wide variety of missions, such as mapping, navigation, guidance, plant detection, and recognition, and for performing tasks such as plant protection control, harvesting, thinning, and pruning. They are used to support the performance, operation, and activation of the robot, and in execution of the task. In addition, sensors can be separated into two groups: (i) internal sensors that measure the internal state of the system, e.g. encoders that report the joint or wheel angles, accelerometers that measure linear accelerations, and gyroscopes that measure angular accelerations; these sensors are usually used for dead reckoning; and (ii) external sensors that collect information from the environment and about the state of the system relative to the environment; such sensors include Global Positioning System (GPS), infrared (IR), machine vision, laser radar (lidar), and ultrasonic waves (Rovira-Más *et al.*, 2015; Barnea *et al.*, 2016; Bietresato *et al.*, 2016; Vitzrabin and Edan, 2016a).

11.3 Orchard Tasks

Extensive research over the past four decades has been focused on the application of robots and automated systems to a variety of horticultural and orchard tasks, and technical feasibility has been widely demonstrated (Bechar and Vigneault, 2017). This section reviews the research and development in agricultural robotics, as applied to orchards, and is divided according to application type: navigation and guidance, including transportation tasks; tree pruning and fruit thinning; harvesting; and spraying and other operations.

11.3.1 Navigation, guidance, and transportation

A core feature of every mobile orchard robot is autonomous navigation, which depends on sensor-based robot localization in the orchard environment (Blok *et al.*, 2019). Developing a robot that can navigate autonomously in changing and dynamic outdoor agricultural environments is a challenging and difficult task but is crucial to any orchard autonomous system. The tasks of autonomous navigation, guidance, and transportation in an orchard could be considered the main task of the robotic system, e.g. transporting the yield from the field to the packinghouse, or enabling the system to perform its main mission, such as spraying, fruit or flower thinning, branch pruning, or harvesting.

Automatic guidance has been the most active research area throughout the history of automation of agricultural machinery (Nof, 2009). Commercial guidance systems for tractors are available, but all require supervision by a human operator. There are three main approaches.

In the first approach, the tractors are guided along a predetermined path based primarily on input from a GPS. This approach is technically simpler but suffers from the inability to respond to unexpected changes in the field (Stentz *et al.*, 2002) and disturbances to the GPS signal.

In the second approach, the tractor operates with reference to the crop line by using machine vision (Bak and Jakobsen, 2004). This approach enables a machine to tailor its operation to individual plants as they change over time but is technically more difficult (Stentz *et al.*, 2002). Kim *et al.* (2020) proposed a novel and efficient patch-based approach for autonomous path detection in semi-structured environments such as orchards. A patch-based, four-layer, convolutional neural network was used for image classification to achieve path area segmentation for tree and path classification, path area segmentation, and target path detection by using boundary line determination. The achieved maximum intersection over union is approximately 0.81 for path area localization, the average lateral and angular errors are 0.051 and 7.8°, respectively, and the processing time is 0.111 s. Blok *et al.* (2019) used two-dimensional lidar scanner for in-row robot navigation in orchards with two localization algorithms: a particle filter with a laser beam model, and a Kalman filter with a line-detection algorithm. Experiments were executed in a commercial Dutch apple orchard to assess navigation performance. The results revealed that at 0.50 m s^{-1}, the root mean square error was 0.062–0.091 m. It was found that a particle filter with a laser beam model was better compared with a line-based Kalman filter for the in-row navigation of an autonomous orchard robot. Ye *et al.* (2017) designed and built a robot prototype and a navigation system for bin management in orchards. Tests of straight-line tracking showed successful tracking of the alleyway centerline at speeds of up to 1 m s^{-1} with a root mean square error offset of 0.07 m.

The third approach is a combination of the previous two. Although current navigation systems for agricultural vehicles rely on GPS as the primary sensor for steering control, an alternative backup method is required in case the satellite signals to the GPS receiver are blocked (e.g. in orchards), or in case the accuracy is insufficient (e.g. for in-row navigation) (Subramanian *et al.*, 2006).

For orchard or forest tasks, Subramanian *et al.* (2006) developed a guidance system based on machine vision and lidar, in order to overcome GPS signal losses and to enable a John Deere 6410 tractor to navigate between the rows of a citrus grove at up to 4.4 m s^{-1} with errors lower than 60 mm. Machine vision and lidar were used individually for guidance, a rotary encoder was used to provide feedback of the steering angle, and a proportional–integral–derivative controller was developed to minimize the path error. Khot *et al.* (2006) developed a navigation technique based on sensor fusion for tree nurseries. Real-time kinematic (RTK)-GPS and dynamic measurement sensors were used to determine the position and orientation of the robot, and lidar was used to locate the tree positions within a selected range. The RTK-GPS error was calculated by a second-order autoregressive model, and error states were incorporated into an extended Kalman filter design. The mean error and its standard deviation decreased from 40.5 to 22.1 mm and from 82.7 to 18.9 mm, respectively. The georeferenced tree positions along the navigational paths were recovered by using a *k*-means clustering algorithm, and achieved an average tree-position error of 44 mm. Yekutieli and Garbati Pegna (2002) developed an automatic guidance system for a tractor in a vineyard by using a mechanical sensor. The sensor was positioned on one side of the tractor where a spring continuously pushed the sensor arm toward the guidance line, and the angle between the arm and the tractor frame controlled the steering. The guidance line was a wire stretched along the posts of every second row. At one end of the row, the tractor would turn according to dead reckoning. Yekutieli and Garbati Pegna (2002) revealed that with only a two-stage on/off control and a simple curved sensor arm, or a sensor arm consisting of two hinged parts, it was possible to guide a crawler tractor automatically in a vineyard, with the vines used as a guidance line.

The autonomous control of a tractor-trailer system in orchards and plantations has frequently been observed for the transportation, loading, and unloading of products between plants and trees. Thanpattranon *et al.* (2016) developed a control algorithm for a single-sensor tractor-trailer navigation system for navigating within a row plantation and for traveling between plots. A lidar was used to navigate a full-size autonomous agricultural tractor equipped with a two-wheeled trailer. Compared with the tractor-trailer system driven by humans, the control system could navigate the tractor trailer with root mean square differences of 0.275, 0.373, and 0.518 m for wide curve, tight curve and U-turn experimental paths, respectively. The control system also stopped the tractor trailer at the specified landmarks. The results show that the navigation of the tractor trailer was demonstrated with satisfactory accuracy.

Penizzotto *et al.* (2015) proposed a control strategy for a mobile robot to autonomously navigate through the corridors formed by plants or fruit vines and olive trees. The controller uses an algorithm for estimating the states of the system based on information from a laser sensor, which is enhanced by an extended Kalman filter based on considering the model of the robot–environment interaction. It also incorporates the use of odometer sensors for automation of rotation between the end and beginning of plant corridors. The method has been tested in a typical olive grove using a Pionner 3AT robot.

11.3.2 Tree pruning

Pruning is a labor-intensive task, with labor making up more than 25% of the total costs. The main objectives of this task are to increase exposure to sunlight, control tree shape, and remove protruding branches. In most orchards, this task is conducted once a year, and up to 20% of the branches are selectively removed. Although no autonomous robot or automated systems for flower and fruit thinning or tree pruning have been reported in the literature, several ongoing studies are working on systems for mechanizing and automating these processes.

Since machine introduction in the early 1970s, much work has been expended to adapt pruners to vine trellises and physiological requirements, especially regarding the higher bud load their non-selective cuts leave compared with manual trimming. While units have successfully met the former requirement, efforts to meet the latter have been hampered by a broad range of variables depending on variety, machine type, environmental conditions, and any manual follow-up (Poni *et al.*, 2016).

An electromechanical arm shaker for fruit thinning in nectarine, peach, and prune orchards was developed to reduce manual labor and costs (Rosa *et al.*, 2008). The system deals with individual branches and includes a special mechanism to prevent the clamping of bunches. A system was developed for vineyards enabling the mechanization of dormant and summer pruning, leaf removal, shoot and fruit thinning, canopy management, and harvesting, while maintaining or enhancing fruit quality (Morris, 2007). In a comparison with the standard manual processes, yield and quality characteristics of grapes harvested by the respective methods were not statistically different. Use of the complete vineyard mechanization system would reduce the required manual labor by 45–62% compared with the trellising method.

In ongoing research, selective tree pruning by human–robot systems is being developed (Bechar *et al.*, 2014; Linker and Neer, 2014), and a visual servoing methodology and human–robot systems for selective tree pruning have been tested. The system consists of: a manipulator, a color camera, a single-beam laser distance sensor, a human–robot interface, and a cutting tool based on a circular saw developed specifically for this task. The cutting tool, camera, and laser sensor are mounted on the manipulator's end effector, aligned parallel to each other. The system works in two phases. First, the camera transfers a two-dimensional image of the tree to a human operator, who marks the branches to be removed on a display. Secondly, the system works in autonomous mode. The manipulator maneuvers the laser sensor to measure the branch distance and calculates a trajectory to the cutting point on the screen. It then follows the trajectory to perform the cut. Two types of motion planning are conducted between the tool's initial location and the cutting point: linear motion in global Cartesian coordinates and in the robot joint space. This system is in the testing phase.

Botterill *et al.* (2017) developed a robotic system for the automatic pruning of grape vines. A mobile platform straddles the row of vines, taking images with a stereo camera as it moves. A computer vision system builds a three-dimensional (3D) model of the vines, an artificial intelligence algorithm decides which canes to prune, and a six-degree-of-freedom robot arm makes the required cuts. The robotic system moves at a velocity of 0.25 m s^{-1}, with the error below 1% when modeling a 96 m row with 59 plants. Pruning each vine requires the robot arm to cut an average of 8.4 canes. Field experiments showed that the cycle time to prune one plant is about 2 min, which is similar to worker time. Zahid *et al.* (2020) developed an end-effector prototype for apple-tree pruning considering the maneuvering, spatial, mechanical, and horticultural requirements. The prototype is based on two rotary motors, a pneumatic cylinder, and a pair of bypass shear blades. A three-directional linear manipulator system and a control system were built for moving the end effector to targeted locations. Field tests were conducted and the results indicated that the end effector successfully cut branches up to 12 mm in diameter and was able to cut branches in a wide range of possible orientations in a given 3D space. Most fresh-fig orchards are pruned during the winter. Robotic pruning in figs is not available. The trellis shaping system that was recently developed for fig orchards could be more suitable for robotic pruning than the common open-vase system.

11.3.3 Harvesting

Harvesting or fruit picking is one of the most common tasks in agriculture and also one of the most demanding and challenging areas for agricultural robotics. Research on harvesting and fruit-picking robots started in the late 1970s on crops such as citrus, apple, cotton, industry tomato, melon, and watermelon (Brandt and French, 1983; Grand d'Esnon, 1984; Harrel and Levi, 1988; Chen *et al.*, 1990; Edan *et al.*, 1990, 2000; Harrell *et al.*, 1990; Edan and Miles, 1993; Ceres *et al.*, 1998; Umeda *et al.*, 1999; Riyaz Ahammed *et al.*, 2015; Mehta *et al.*, 2016). Mechanisms, manipulators, end effectors, and tools were developed to perform specific operations (Kawamura *et al.*, 1985, 1987; Harrell *et al.*, 1990; Edan and Miles, 1993; Iida *et al.*, 1996; Ceres *et al.*, 1998; Monta *et al.*, 1998; Kataoka *et al.*, 1999; Ceccarelli *et al.*, 2000). Although some progress was made in performing mechanical and physical operations, robotic harvesters did not match the human worker's ability, accuracy, success, and productivity, and rarely showed minimal economic justification. The current hit rate of robotic pickers/harvesters is around 75–85% detection of the fruits that need to be picked, but these agricultural robots are still often unable to support real-time abilities due to difficulties in avoiding obstacles, and in reaching, grasping, and handling fruit in the agricultural environment.

Electronics research in the last decade has generated continuous increases and improvements in computation power and sensors of various types, such as RTK-GPS, 3D cameras, and lidar. Based on these important advances, the research effort has shifted to modifying and integrating these sensory systems and related algorithms into fruit-picking technology (Jiménez *et al.*, 2000; Bulanon *et al.*, 2002; Sivaraman and Burks, 2006; Hannan *et al.*, 2007; Bulanon *et al.*, 2009; Irie *et al.*, 2009; Arefi *et al.*, 2010; Dong *et al.*, 2011; Yin *et al.*, 2011; Ji *et al.*, 2012; Bac *et al.*, 2013; Li and Wang, 2013; Mehta and Burks, 2014; Barnea *et al.*, 2016; Nguyen *et al.*, 2016; Mehta *et al.*, 2016; Zhao *et al.*, 2016).

Ceres *et al.* (1998) presented the agribot approach, a human–robot integrated system combining human and machine functions for harvesting apples. The human operator detects and identifies each fruit to be harvested using a joystick and an IR laser beam. This signal tells the 3D measurement system to calculate the position of the target. The spherical coordinates, indicating the location of the fruit, include two orthogonal angles and a radius, and are generated within a cycle time for the position measurement of 29 ms. As the IR laser beam attenuates with distance, the measuring range is limited to 2 m, with an average position error less than 5 mm. Their objective was to reach a productivity rate equivalent of 1.25 fruit s^{-1} including fruit transfer to the apple bulk container, which corresponds to 2.5 men of manual harvesting productivity, as a minimum to generate an affordable robot. However, while taking into account simultaneous two arms in operation, their best grasping/detaching operation time was 2 s per fruit, excluding all other operations included in a complete harvesting cycle.

Jiménez *et al.* (2000) developed an IR laser-based range-finder to detect spherical objects in non-structured environments based on the technology of Ceres *et al.* (1998). The image-analysis algorithms associated with their system integrated both range and reflectance information of region primitives respecting sphere characteristics such as crown, contour, convexity, and reflectivity. When the global arrangement of the primitives formed circular arcs, the system concluded evidence of the presence of a spherical object. The output of this system included the radius, the surface reflectivity and the 3D position of each spherical object detected. Their tests resulted in a hit rate (true positive) in the range 80–90% of the visible fruits with 0% false detection when respecting their established rules.

A 3D vision sensor was developed for robotic harvesting of grapes (Luo *et al.*, 2016) and cherries (Tanigaki *et al.*, 2008). In cherries, the 3D vision sensor was equipped with red and IR laser diodes. By processing the images from the 3D vision sensor, the positions of the fruits and obstacles were recognized, and the end-effector trajectory was calculated (Tanigaki *et al.*, 2008).

For a berry-harvesting task, Hayashi *et al.* (2010) assembled a system that included a cylindrical manipulator, end effector, machine-vision unit, suction-cup device, storage unit, and traveling unit. During the evaluation trials, the system obtained a detection performance of 60% for the peduncle of the target strawberry, and a harvesting robot achieved 41.3% of the fruits at the desired maturity level when the fruit were picked by a suction device before cutting the peduncle. The harvesting cycle time was 11.5 s per fruit, which included fruit transfer to the tray. Cui *et al.* (2013) developed a Cartesian-type robot for strawberry in-row hilltop culture. The system consisted of two color cameras, a harvesting hand, an optic-fiber sensor, and a mobile control unit. The system detected the strawberry fruits, evaluated their ripeness, picked the fruits one at a time, and cut their peduncles. The harvesting robot detection rate was 93.6% of the fruit at the desired ripeness level, while the cycle time reached 16.6 s per fruit.

Fig harvesting is a great challenge. Most of the manpower demand in fig is around fruit harvesting. Many traditional fig-growing areas in Europe have been neglected because of manpower limitations during fig harvest. Both berries and figs are small fragile fruits. Therefore, fig growers can adapt the berry-harvesting developments and adjust them to fig harvest in order to facilitate fig-fruit harvesting operations.

Zhao *et al.* (2011) developed a robotic device for harvesting apples consisting of a five-degrees-of-freedom PRRRP structure (one prismatic joint followed by three rotation joints, and ending with another prismatic joint), an end effector, and an image-based vision servo control system. The manipulator was geometrically optimized to provide quasi-linear behavior, simplifying the control strategy. The spoon-shaped end effector, assisted by a pneumatic gripper, harvested apples with a success rate of 77% with a cycle time of 15 s per fruit.

Hohimer *et al.* (2019) developed a pneumatic 3D-printed soft-robotic end effector for apple picking. The end effector was integrated into a robotic system with five degrees of freedom that was designed to simplify the picking sequence and reduce costs. A detachment success rate of 67% was achieved, with an average time of 7.3 s per fruit from separation to the storage bin.

Silwal *et al.* (2017) designed and developed a robotic apple picker. The system was tested in a commercial apple orchard. The machine-vision system had an average localization time of 1.5 s per fruit. A picking robot with seven degrees of freedom successfully picked 84% of the fruit with an average picking time of 6 s per fruit.

To improve robot performance for agricultural tasks, and decrease the cost, the robot can be optimally designed for a specific task in a specific working environment. As the orchard environment defines the robot optimal design, the environment itself should also be optimized for optimal robot performance. This methodology was demonstrated by an example of a tree orchard design for an apple-harvesting robot (Bloch *et al.*, 2018). First, an optimal robot structure for the apple-picking task was found for a number of tree architectures. The results indicated that to minimize the average apple-picking time, the tall spindle architecture was preferable for the robotic harvesting of both a single tree and a tree row. The influence of the robot platform motion time on the chosen robot kinematics and the tree training system was also analyzed. The results showed that, for fast platforms, the tall spindle architecture was advantageous. If the platform movement between positions near the trees was slow, then the central leader architecture was favorable. Additionally, the tilt angle of the Y-trellis training system was analyzed using simulated models created by the L-system simulations. The optimal tilt angle was found to be nearly horizontal (85°), allowing the robot designer to choose the optimal combination of the robot kinematics, number of robot harvesting positions around the tree, and the tree training system.

Arikapudi and Vougioukas (2021) investigated the linear fruit reachability of high-density pear and cling-peach trees, to be used by telescopic robot arms. Simulation results for pears and peaches showed that 85.5% and 83.5% were reachable with a gripper of 11 cm size and an arm extension of 150 and 200 cm, respectively. This suggests that, in some crops, telescoping arms, which are less complex, can be used to harvest certain tree crops.

Fei and Vougioukas (2021) transformed a conventional harvesting platform into a collaborative robot (co-robot) platform. The platform estimates the incoming fruit distribution using a vision system, measures each worker's picking speed using instrumented picking bags, and controls the heights of hydraulic lifts that move workers up and down. Apple-harvesting experiments were performed in a commercial orchard, where 2307 kg of apples were picked: 1045 kg in variable-height zone harvesting mode, and 1262 kg in fixed-zone harvesting mode, with workers at fixed heights that were set by the grower. Variable-height zone harvesting mode throughput was 327.6 versus 298.8 kg h^{-1} for fixed-zone harvesting mode at a human-controlled platform moving speed, resulting in an improvement of 9.5%.

Williams *et al.* (2019) designed and evaluated a multi-arm kiwifruit-harvesting robot designed to operate autonomously in pergola-style orchards. The harvester consists of four robotic arms, each with a dedicated end effector developed to enable safe harvesting of the kiwifruit. The vision system consists of a deep neural network algorithm and stereo matching for reliably detecting and locating kiwifruit in real-world lighting conditions. The performance of the harvester has been measured in a comprehensive and realistic field trial in a commercial orchard. The results showed that the presented harvester was capable of successfully harvesting 51% of the total number of kiwifruit within the orchard with an average cycle time of 5.5 s per fruit.

11.3.4 Spraying and other tasks

The considerable spacing between trees along orchard rows causes a lot of pesticides to be wasted by conventional continuous spraying, which is not targeted to the tree canopy and therefore releases toxic chemicals into the environment. Asaei *et al.* (2019) developed a site-specific orchard sprayer by using machine-vision technology that applies a precise amount of pesticide based on the tree green canopy and improves chemical application efficiency in orchards. The field performance of the sprayer in an olive orchard was evaluated under two conditions of continuous and discrete-targeted spraying at three forward speeds and four replications. The results showed that discrete-targeted spraying resulted in 54% less chemical consumption compared with conventional continuous spraying. Further reduction of pesticide usage is expected in newly established orchards with younger trees and larger intra-row spacing.

Crop segmentation is an important task in precision agriculture, where the use of aerial robots with an on-board camera has contributed to the development of new solution alternatives. Fuentes-Pacheco *et al.* (2019) developed a convolutional neural network with an encoder–decoder architecture that classifies each pixel as crop or non-crop using only raw color images as input for fig plant segmentation in top-view RGB (red, green, blue) images. They achieved an accuracy of 93.85% despite the complexity of the background and a highly variable visual appearance of the leaves.

Remote sensing for irrigation in fig orchards

Irrigation of trees is applied to maximize crop value while maintaining the desired tree water status according to the needs of each fruit growth stage. The FAO-56 model is one of the most commonly used and most practical approaches for the estimation of crop water requirements and thus irrigation management methods (Allen *et al.*, 1998). The irrigation amount is calculated as:

$$\text{Irrigation} = K_s \times K_c \times \text{ET}_0$$

where ET_0 is the reference evapotranspiration, K_c is the crop coefficient, and K_s is the stress coefficient. K_c represents the atmospheric evaporative demand together with the crop water requirements. One of the most reliable sources of K_c estimation is spectral vegetation indices derived from optical remote sensing (see Kaplan *et al.*, 2021, for references) The spectral vegetation indices-based-K_c approach was mostly developed for annual crops but several studies have shown its applicability to orchards (Amri *et al.*, 2014, Beeri *et al.*, 2019). Since 2017, the Sentinel-2 constellation serves as a reliable satellite imagery source for the application of this method with high spatial (10, 20, or 60 m, depending on the band) and temporal (5 days) resolution (Kaplan *et al.*, 2021). K_c values for figs based on multi-year experience are available for a few locations (Goldhamer and Salinas, 1999; Andrade *et al.*, 2014). The crop coefficient aims for 100% return of the canopy evapotranspiration. In one experiment in an orchard of the Black Mission fig variety in the Madera area in the USA, following calendar-based K_c resulted in undesirable declines in stem water potential (SWP) values in two periods along the irrigation season, suggesting that surplus irrigation amounts were applied (Goldhamer and Salinas, 1999). In another irrigation experiment in Chile, two varieties showed that full irrigation may decrease yield (Tapia *et al.*, 2003). K_c calculated based on timely satellite images aims to improve the estimation of real-time water consumption of the canopy in the orchard, thus improving the amount of irrigation applied.

In various orchards, such as vineyards, olives, and peach, regulated deficit irrigation is often implemented during reproductive growth stages, mainly to improve the yield quality with minimal or no tonnage reduction (Naor, 2006). Irrigation experiments in six fig varieties in Chile showed that figs can be treated with deficit irrigation (up to 50%) with no negative effect on yield (Tapia *et al.*, 2003). Imposing regulated water stress, however, necessitates frequent monitoring of water status to minimize the risk of severe water stress. SWP is considered to be a very sensitive and robust indicator for tree water status (Naor, 2000; Naor *et al.*, 2006). SWP was also found to be a sensitive water status measure for fig trees (Goldhamer and Salinas, 1999; Gholami *et al.*, 2012). Nevertheless, this method is laborious, time-consuming, and discrete, and thus measurements for the whole orchard are done using a few representative trees, mostly very close to each other.

Thermal imaging for water status estimation

Evapotranspiration is the process by which water stored in the soil or vegetation is converted from a liquid into the vapor phase and is transferred to the atmosphere. Evapotranspiration decreases plant temperature. Stomatal regulation plays a key role in the plant response to water stress. As plant stomata close, the evapotranspiration rate decreases, the energy heat balance between the vegetation and its environment changes, and leaf temperature rises. Thus, leaf temperature may be used as an indicator of plant water status and plant health. Leaf temperature can be sensed by measuring the thermal infrared (8–14 μm) radiation emitted. First attempts to apply canopy temperature for assessing plant water status and plant health status were made in the 1960s (Gates, 1964). Crop temperature measurements with thermal infrared thermometers are reliable and non-invasive, but they are usually based on a few point measurements and therefore depend on the assumption of uniformity of soil water content and plant density over large areas (Alchanatis *et al.*, 2010). Remote thermal imagery is a viable alternative to point measurements, as the canopy temperature of the whole orchard can be measured at once and a map of the plant water status in the orchard can be produced (Meron *et al.*, 2010). The availability of thermal cameras led to a significant evolution of thermal remote sensing in the 2000s for plant water status estimation (Maes and Steppe, 2012).

Canopy temperature (T_c) is determined not only by the water status of the plant but also by environmental conditions, primarily incoming short-wave radiation, wind speed, air temperature, and humidity. In order to use T_c as a water status indicator, thermal water status indices that calibrate the environmental conditions were developed. One of the commonly used calibration methods is the use of lower and upper reference temperatures in the format of the crop water stress index (CWSI) (Idso *et al.*, 1981; Jackson *et al.*, 1981), defined as:

$$\mathrm{CWSI} = (T_c - T_{wet}) / (T_{dry} - T_{wet})$$

where T_{wet} is the lower reference temperature of a leaf transpiring at the maximum potential rate and T_{dry} is the upper reference temperature of a non-transpiring leaf.

Using aerial thermal images, CWSI and other thermal-based indices have become acceptable indices to map in-field variability in various crops, such as cotton (Cohen *et al.*, 2017) and grain sorghum (O'Shaughnessy *et al.*, 2012), and in various orchards, such as grapevines (Bellvert *et al.*, 2014; Bahat *et al.*, 2021), olives (Egea *et al.*, 2017), apples (Osroosh *et al.*, 2016), and peach (Gonzalez-Dugo *et al.*, 2013; Bellvert *et al.*, 2016). The stomatal response to environmental conditions varies among species and must be taken into consideration when canopy temperature is proposed as a proxy for plant water status in general and for figs in the context of this book. Most studies that used temperature-based indices as a proxy to plant water status used data from midday when net radiation peaks. Aerial thermal images at midday capture sunlit leaves with minimal shadowing interference. Some species, in reaction to high light levels during summer days, display midday depression of photosynthesis and stomatal closure, even under well-watered conditions (Barber and Andersson, 1992). Hu *et al.* (2009) showed a reduction in both net photosynthesis and stomatal conductance from the morning toward midday in Satsuma mandarin (*Citrus unshiu* Marc.) trees. This might explain the limited success of thermal imaging in estimating the water status of citrus and in detecting water shortage stress (Bellvert *et al.*, 2016). Diurnal measurements of stomatal conductance in fig trees that were under different irrigation regimes have indicated that the highest stomatal activity occurred around midday (Goldhamer and Salinas, 1999). Moreover, SWP readings around midday were sensitive to plant water status in figs (Goldhamer and Salinas, 1999) and to drying and wetting cycles (Gholami *et al.*, 2012).

Remote sensing for irrigation management

As described above, multi-spectral remote sensing in the reflected range of the spectrum can contribute to timely crop coefficient (K_c) estimation. Sanchez *et al.* (2017) utilized satellite remote-sensing K_c estimation to apply precision irrigation in a drip-irrigated Cabernet Sauvignon vineyard in California. The vineyard was split into 140 15 × 15 m irrigation zones, which were watered independently during three seasons with weekly schedules based on estimated actual evapotranspiration. Compared with an adjacent, conventionally drip-irrigated section of the vineyard, the variable rate irrigation increased yield and water-use efficiency in all three years. Bellvert *et al.* (2021) used an integrated methodology based on a vine water consumption model and remote-sensing data to optimize precision irrigation of a 100 ha commercial vineyard. In a 2-year trial, they showed that the minimal net benefit was €20 ha^{-1}. Using a biweekly spatial decision support irrigation protocol, which is based on SWP maps extracted from aerial thermal images, Katz *et al.* (2022) showed that, compared with uniform irrigation, precision irrigation decreased water status variability and increased yield and water-use efficiency in a commercial peach orchard.

11.4 Concluding Remarks

To conclude, fruit picking is one of the most complicated agricultural tasks for agricultural robots. This task consists of several different stages and involves physical contact with the fruit with high location and orientation accuracy, decision making in real time, detachment of the fruit from the plant without damaging either, and temporarily storing the fruit under safe conditions. The performance of fruit-picking robots has not improved significantly over the last three decades, yielding on average a detection performance for acceptable localization of 85%, 75% for detachment, 66% for harvesting, and a cycle time of 33 s per fruit (Bac *et al.*, 2014); moreover, 5% of the fruits and 45% of peduncles are damaged. In the harvesting task, the most important gaps to be addressed are: decreasing the cycle time; increasing the production rate and detection performances, and being able to operate in true real time. To solve the difficulties faced by apple harvesters, the development of mechanized apple harvesting and modern orchard structure applications must be accelerated (Bu *et al.*, 2020).

Fig is a fragile fruit that needs an intensive harvesting operation due to selective ripening of the fruit along the branch. The development of efficient harvesting robots for figs will dramatically increase fresh-fig orchard productivity.

In pruning tasks, tree training systems, pruning strategies, 3D structure reconstruction of tree branches, and practice mechanisms or robotics are some of the developments that need to be addressed for an effective tree branch pruning system (He and Schupp, 2018).

Despite the large amount of research, almost no fully automated and cost-efficient solution has been developed. The viability of the prototypes could be increased by modification of the tree structures (Bloch *et al.*, 2017) and, consequently, the orchard architecture, to make it "robot ready." Another direction is in combining several techniques to accomplish individual tasks in one multi-purpose robot platform to increase robot usability and efficiency (Verbiest *et al.*, 2021).

The plant's instantaneous response to water stress can be captured using thermal cameras and potentially narrow-band hyperspectral sensors. Integrated use of thermal and multispectral imagery may be the simplest yet most effective sensor combination to capture the overall water status of the tree. With regard to irrigation scheduling, further developments are required to establish crop-specific thresholds of remotely sensed indices to decide when and how much to irrigate (Gautam and Pagay, 2020).

References

Alchanatis, V., Cohen, Y., Cohen, S., Moller, M., Sprinstin, M., *et al.* (2010) Evaluation of different approaches for estimating and mapping crop water status in cotton with thermal imaging. *Precision Agriculture* 11, 27–41.

Allen, R.G., Pereira, L.S., Dirk, R., and Smith, M. (1998) *Crop Evapotranspiration – Guidelines for Computing Crop Water Requirements*. FAO Irrigation and Drainage Paper 56. United Nations Food and Agriculture Organization, Rome.

Amri, R., Zribi, M., Lili-Chabaane, Z., Szczypta, C., Calvet, J.C., and Boulet, G. (2014) FAO-56 dual model combined with multi-sensor remote sensing for regional evapotranspiration estimations. *Remote Sensing* 6, 5387–5406.

Andrade, I.P. de S., de Carvalho, D.F., de Almeida, W.S., Gonçalves Silva, J.B., and da Silva, L.D.B. (2014) Water requirement and yield of fig trees under different drip irrigation management. *Engenharia Agrícola* 34, 17–27.

Arefi, A., Motlagh, A.M., and Teimourlou, R.F. (2010) A segmentation algorithm for the automatic recognition of tomato at harvest. *Journal of Food Agriculture and Environment* 8, 815–819.

Arikapudi, R. and Vougioukas, S.G. (2021) Robotic tree-fruit harvesting with telescoping arms: a study of linear fruit reachability under geometric constraints. *IEEE Access* 9, 17114–17126.

Asaei, H., Jafari, A., and Loghavi, M. (2019) Site-specific orchard sprayer equipped with machine vision for chemical usage management. *Computers and Electronics in Agriculture* 162, 431–439.

Bac, C.W., Hemming, J., and van Henten, E.J. (2013) Robust pixel-based classification of obstacles for robotic harvesting of sweet-pepper. *Computers and Electronics in Agriculture* 96, 148–162.

Bac, C.W., van Henten, E.J., Hemming, J. and Edan, Y. (2014) Harvesting robots for high-value crops: state-of-the-art review and challenges ahead. *Journal of Field Robotics* 31, 888–911.

Bahat, I., Netzer, Y., Grünzweig, J.M., Alchanatis, V., Peeters, A., *et al.* (2021) In-season interactions between vine vigor, water status and wine quality in terrain-based management-zones in a 'Cabernet Sauvignon' vineyard. *Remote Sensing* 13, 1636–1655.

Bak, T. and Jakobsen, H. (2004) Agricultural robotic platform with four wheel steering for weed detection. *Biosystems Engineering* 87, 125–136.

Bakker, T., van Asselt, K., Bontsema, J., Muller, J., and van Straten, G. (2011) Autonomous navigation using a robot platform in a sugar beet field. *Biosystems Engineering* 109, 357–368.

Barber, J. and Andersson, B. (1992) Too much of a good thing: light can be bad for photosynthesis. *Trends in Biochemical Sciences* 17, 61–66.

Barnea, E., Mairon, R., and Ben-Shahar, O. (2016) Colour-agnostic shape-based 3D fruit detection for crop harvesting robots. *Biosystems Engineering* 146, 57–70.

Bayar, G., Bergerman, M., Koku, A.B., and Konukseven, E.I. (2015) Localization and control of an autonomous orchard vehicle. *Computers and Electronics in Agriculture* 115, 118–128.

Bechar, A. (2010) Robotics in horticultural field production. *Stewart Postharvest Review* 6, 1–11.

Bechar, A. and Eben-Chaime, M. (2014) Hand-held computers to increase accuracy and productivity in agricultural work study. *International Journal of Productivity and Performance Management* 63, 194–208.

Bechar, A. and Edan, Y. (2003) Human–robot collaboration for improved target recognition of agricultural robots. *Industrial Robot* 30, 432–436.

Bechar, A. and Vigneault, C. (2016) Agricultural robots for field operations: concepts and components. *Biosystems Engineering* 149, 94–111.

Bechar, A. and Vigneault, C. (2017) Agricultural robots for field operations. Part 2: operations and systems. *Biosystems Engineering* 153, 110–128.

Bechar, A., Meyer, J., and Edan, Y. (2009) An objective function to evaluate performance of human–robot collaboration in target recognition tasks. *IEEE Transactions on Systems Man and Cybernetics Part C: Applications and Reviews* 39, 611–620.

Bechar, A., Yosef, S., Netanyahu, S., and Edan, Y. (2007) Improvement of work methods in tomato greenhouses using simulation. *Transactions of the ASABE* 50, 331–338.

Bechar, A., Bloch, V., Finkelshtain, R., Levi, S., Hoffman, A., *et al.* (2014) Visual servoing methodology for selective tree pruning by human–robot collaborative system – AgEng 2014. In: *Proceedings International Conference of Agricultural Engineering, 6–10 July 2014, Zurich, Switzerland*.

Beeri, O., Pelta, R., Shilo, T., Mey-Tal, S., and Tanny, J. (2019) Accuracy of crop coefficient estimation methods based on satellite imagery. In: Stafford, J.V. (ed.) *Precision Agriculture '19*. Wageningen Academic Publishers, Wageningen, The Netherlands, pp. 437–444.

Belkhouche, F. and Belkhouche, B. (2005) Modeling and controlling a robotic convoy using guidance laws strategies. *IEEE Transactions on Systems, Man, and Cybernetics, Part B: Cybernetics* 35, 813–825.

Bellvert, J., Marsal, J., Girona, J., and Zarco-Tejada, P.J. (2014) Seasonal evolution of crop water stress index in grapevine varieties determined with high-resolution remote sensing thermal imagery. *Irrigation Science* 33, 81–93.

Bellvert, J., Marsal, J., Girona, J., Gonzalez-Dugo, V., Fereres, E., *et al.* (2016) Airborne thermal imagery to detect the seasonal evolution of crop water status in peach, nectarine and saturn peach orchards. *Remote Sensing* 8: 39.

Bellvert, J., Mata, M., Vallverdú, X., Paris, C., and Marsal, J. (2021) Optimizing precision irrigation of a vineyard to improve water use efficiency and profitability by using a decision-oriented vine water consumption model. *Precision Agriculture* 22, 319–341.

Bergerman, M., Maeta, S.M., Zhang, J., Freitas, G.M., Hamner, B., *et al.* (2015) Robot farmers: autonomous orchard vehicles help tree fruit production. *IEEE Robotics and Automation Magazine* 22, 54–63.

Bietresato, M., Carabin, G., Vidoni, R., Gasparetto, A., and Mazzetto, F. (2016) Evaluation of a LiDAR-based 3D-stereoscopic vision system for crop-monitoring applications. *Computers and Electronics in Agriculture* 124, 1–13.

Bloch, V., Bechar, A., and Degani, A. (2017) Development of an environment characterization methodology for optimal design of an agricultural robot. *Industrial Robot* 44, 94–103.

Bloch, V., Degani, A., and Bechar, A. (2018) A methodology of orchard architecture design for an optimal harvesting robot. *Biosystems Engineering* 166, 126–137.

Blok, P.M., van Boheemen, K., van Evert, F.K., Ijsselmuiden, J., and Kim, G.H. (2019) Robot navigation in orchards with localization based on particle filter and Kalman filter. *Computers and Electronics in Agriculture* 157, 261–269.

Bodur, M., Kiani, E., and Hacisevki, H. (2012) Double look-ahead reference point control for autonomous agricultural vehicles. *Biosystems Engineering* 113, 173–186.

Botterill, T., Paulin, S., Green, R., Williams, S., Lin, J., *et al.* (2017) A robot system for pruning grape vines. *Journal of Field Robotics* 34, 1100–1122.

Brandt, J.A. and French, B.C. (1983) Mechanical harvesting and the California tomato industry – a simulation analysis. *American Journal of Agricultural Economics* 65, 265–272.

Bu, L.X., Chen, C.K., Hu, G.R., Sugirbay, A., and Chen, J. (2020) Technological development of robotic apple harvesters: a review. *Inmateh Agricultural Engineering* 61, 151–164.

Bulanon, D.A., Burks, T.E., and Alchanatis, V. (2009) Fruit visibility analysis for robotic citrus harvesting. *Transactions of the ASABE* 52, 277–283.

Bulanon, D.M., Kataoka, T., Ota, Y., and Hiroma, T. (2002) A segmentation algorithm for the automatic recognition of Fuji apples at harvest. *Biosystems Engineering* 83, 405–412.

Canning, J.R., Edwards, D.B., and Anderson, M.J. (2004) Development of a fuzzy logic controller for autonomous forest path navigation. *Transactions of the ASAE* 47, 301–310.

Ceccarelli, M., Figliolini, G., Ottaviano, E., Mata, A.S., and Criado, E.J. (2000) Designing a robotic gripper for harvesting horticulture products. *Robotica* 18, 105–111.

Ceres, R., Pons, F.L., Jimenez, A.R., Martin, F.M., and Calderon, L. (1998) Design and implementation of an aided fruit-harvesting robot (Agribot). *Industrial Robot* 25, 337–346.

Chen, L.H., Ahmad, T., and Willcutt, M.H. (1990) Cotton harvesting machinery management via simulation. ASAE Paper No. 901566, In: *Winter Meeting of ASAE, 18–21 December. Illinois, Chicago.*

Cohen, Y., Alchanatis, V., Saranga, Y., Rosenberg, O., Sela, E., and Bosak, A. (2017) Mapping water status based on aerial thermal imagery: comparison of methodologies for upscaling from a single leaf to commercial fields. *Precision Agriculture* 18, 801–822.

Corbett-Davies, S., Botterill, T., Green, R., and Saxton, V. (2012) An expert system for automatically pruning vines. In: McCane, B., Mills, S., and Deng, J. (eds) *27th Image and Vision Computing New Zealand Conference, 26–28 November 2012, New York.* Association for Computing Machinery, New York, pp. 55–60.

Cui, Y., Gejima, Y., Kobayashi, T., Hiyoshi, K., and Nagata, M. (2013) Study on Cartesian-type strawberry-harvesting robot. *Sensor Letters* 11, 1223–1228.

Dong, F.H., Heinemann, W., and Kasper, R. (2011) Development of a row guidance system for an autonomous robot for white asparagus harvesting. *Computers and Electronics in Agriculture* 79, 216–225.

Edan, Y. and Miles, G.E. (1993) Design of an agricultural robot for harvesting melons. *Transactions of the ASAE* 36, 593–603.

Edan, Y., Flash, T., Shmulevich, I., Sarig, Y., and Peiper, U.M. (1990) An algorithm defining the motions of a citrus picking robot. *Journal of Agricultural Engineering Research* 46, 259–273.

Edan, Y., Rogozin, D., Flash, T., and Miles, G.E. (2000) Robotic melon harvesting. *IEEE Transactions on Robotics and Automation* 16, 831–835.

Egea, G., Padilla-Díaz, C.M., Martinez-Guanter, J., Fernández, J.E., and Pérez-Ruiz, M. (2017) Assessing a crop water stress index derived from aerial thermal imaging and infrared thermometry in super-high density olive orchards. *Agricultural Water Management* 187, 210–221.

Fei, Z.H. and Vougioukas, S.G. (2021) Co-robotic harvest-aid platforms: real-time control of picker lift heights to maximize harvesting efficiency. *Computers and Electronics in Agriculture* 180: 105894.

Foglia, M.M. and Reina, G. (2006) Agricultural robot for radicchio harvesting. *Journal of Field Robotics* 23, 363–377.

Fuentes-Pacheco, J., Torres-Olivares, J., Roman-Rangel, E., Cervantes, S., Juarez-Lopez, P., *et al.* (2019) Fig plant segmentation from aerial images using a deep convolutional encoder–decoder network. *Remote Sensing* 11: 1157.

Gates, D.M. (1964) Leaf temperature and transpiration. *Agronomy Journal* 56, 273–277.

Gautam, D. and Pagay, V. (2020) A review of current and potential applications of remote sensing to study the water status of horticultural crops. *Agronomy* 10: 140.

Gholami, M., Rahemi, M., and Rastegar, S. (2012) Use of rapid screening methods for detecting drought tolerant cultivars of fig (*Ficus carica* L.). *Scientia Horticulturae* 143, 7–14.

Gimenez, J., Herrera, D., Tosetti, S., and Carelli, R. (2015) Optimization methodology to fruit grove mapping in precision agriculture. *Computers and Electronics in Agriculture* 116, 88–100.

Goldhamer, D.A. and Salinas, M. (1999) Black Mission fig production improved by heavier irrigation. *California Agriculture* 53, 30–34.

Gomez-Gil, J., Ruiz-Gonzalez, R., Alonso-Garcia, S., and Gomez-Gil, F.J. (2013) A Kalman filter implementation for precision improvement in low-cost GPS positioning of tractors. *Sensors* 13, 15307–15323.

Gonzalez-de-Soto, M., Emmi, L., Perez-Ruiz, M., Aguera, J., and Gonzalez-de-Santos, P. (2016) Autonomous systems for precise spraying – evaluation of a robotised patch sprayer. *Biosystems Engineering* 146, 165–182.

Gonzalez-Dugo, V., Zarco-Tejada, P., Nicolas, E., Nortes, P.A., Alarcon, J.J., *et al.* (2013) Using high resolution UAV thermal imagery to assess the variability in the water status of five fruit tree species within a commercial orchard. *Precision Agriculture* 14, 660–678.

Grand d'Esnon, A. (1984) Robotic harvesting of apples. In: *Proceeding of AgrMation I*. ASAE and SME, St. Joseph, Michigan, pp. 210–214.

Hameed, I.A., Bochtis, D., and Sorensen, C.A. (2013) An optimized field coverage planning approach for navigation of agricultural robots in fields involving obstacle areas. *International Journal of Advanced Robotic Systems* 10: 231.

Hameed, I.A., La Cour-Harbo, A., and Osen, O.L. (2016) Side-to-side 3D coverage path planning approach for agricultural robots to minimize skip/overlap areas between swaths. *Robotics and Autonomous Systems* 76, 36–45.

Hannan, M.W., Burks, T.F., and Bulanon, D.M. (2007) A real-time machine vision algorithm for robotic citrus harvesting, Paper No. 073125. In: *2007 ASAE Annual Meeting, 17–20 June 2007, Minneapolis, Minnesota*.

Harrel, R.C. and Levi, P. (1988) Vision controlled robots for automatic harvesting of citrus. Paper No. 88.426. In: *International Conference on Agricultural Engineering, 2–6 March 1988, Paris*.

Harrell, R.C., Adsit, P.D., Munilla, R.D., and Slaughter, D.C. (1990) Robotic picking of citrus. *Robotica* 8, 269–278.

Hayashi, S., Shigematsu, K., Yamamoto, S., Kobayashi, K., Kohno, Y., *et al.* (2010) Evaluation of a strawberry-harvesting robot in a field test. *Biosystems Engineering* 105, 160–171.

He, L. and Schupp, J. (2018) Sensing and automation in pruning of apple trees: a review. *Agronomy* 8: 211.

Hohimer, C.J., Wang, H., Bhusal, S., Miller, J., Mo, C., and Karkee, M. (2019) Design and field evaluation of a robotic apple harvesting system with a 3D-printed soft-robotic end-effector. *Transactions of the ASABE* 62, 405–414.

Holland, S.W. and Nof, S.Y. (2007) Emerging trends and industry needs. In: Nof, S.Y. (ed.) *Handbook of Industrial Robotics*. Wiley, Hoboken, New Jersey, pp. 31–40.

Hu, M.J., Guo, Y.P., Shen, Y.G., Guo, D.P., and Li, D.Y. (2009) Midday depression of photosynthesis and effects of mist spray in citrus. *Annals of Applied Biology* 154, 143–155.

Idso, S.B., Jackson, R.D., Pinter, P.J., Reginato, R.J., and Hatfield, J.L. (1981) Normalizing the stress-degree-day parameter for environmental variability. *Agricultural Meteorology* 24, 45–55.

Iida, M., Furube, K., Namikawa, K., and Umeda, M. (1996) Development of watermelon harvesting gripper. *Journal of JSAM* 58, 19–26.

Iida, M., Suguri, M., Uchida, R., Ishibashi, M., Kurita, H., *et al.* (2013) Advanced harvesting system by using a combine robot. *IFAC Proceedings* 46, 40–44.

Irie, N., Taguchi, N., Horie, T., and Ishimatsu, T. (2009) Asparagus harvesting robot coordinated with 3-D vision sensor. In: *2009 IEEE International Conference on Industrial Technology, 10–13 February, Churchill, Victoria, Australia*.

Jackson, R.D., Idso, S.B., Reginato, R.J., and Pinter, P.J. Jr. (1981) Canopy temperature as a crop water stress indicator. *Water Resources Research* 17, 1133–1138.

Ji, W., Zhao, D., Cheng, F.Y., Xu, B., Zhang, Y., and Wang, J.J. (2012) Automatic recognition vision system guided for apple harvesting robot. *Computers and Electrical Engineering* 38, 1186–1195.

Jiménez, A.R., Ceres, R., and Pons, J.L. (2000) A vision system based on a laser range-finder applied to robotic fruit harvesting. *Machine Vision and Applications* 11, 321–329.
Kaplan, G., Fine, L., Lukyanov, V., Manivasagam, V.S., Malachy, N., *et al.* (2021) Estimating processing tomato water consumption, leaf area index, and height using Sentinel-2 and VENμS imagery. *Remote Sensing* 13: 1046.
Kataoka, T., Hiroma, T., and Ota, Y. (1999) Development of a harvesting hand for apples. *Advanced Robotics* 13, 293–294.
Katz, L., Ben-Gal, A., Litaor, M.I., Naor, A., Peres, M. *et al.* (2021) Spatiotemporal normalized ratio methodology to evaluate the impact of field-scale variable rate application. *Precision Agriculture*. doi: 10.1007/s11119-022-09877-4.
Kawamura, N., Namikawa, K., Fujiura, T., and Ura, M. (1985) Study on agricultural robot. (III) Detection and detaching devices of fruits. (IV) Improvement of manipulator of fruit harvesting robot. In: *Research Report on Agricultural Machinery*. Kyoto University, Kyoto, Japan.
Kawamura, N., Namikawa, K., Fujiura, T., and Ura, M. (1987) Study on agricultural robot. (VII) Hand of fruit harvesting robot. (VIII) Sigling manipulator. In: *Research Report on Agricultural Machinery*. Kyoto University, Kyoto, Japan.
Khot, L.R., Tang, L., Blackmore, S.B., and Norremark, M. (2006) Navigational context recognition for an autonomous robot in a simulated tree plantation. *Transactions of the ASABE* 49, 1579–1588.
Kim, W.S., Lee, D.H., Kim, Y.J., Kim, T., Hwang, R.Y., and Lee, H.J. (2020) Path detection for autonomous traveling in orchards using patch-based CNN. *Computers and Electronics in Agriculture* 175: 105620.
Konopatzki, M.R.S., Souza, E.G., Nóbrega, L.H.P., Bazzi, C.L., and Rocha, D.M. (2016) Spatial variability of chemical attributes of the soil, plant and yield in a pear orchard. *Journal of Plant Nutrition* 39, 323–336.
Kraus, T., Ferreau, H.J., Kayacan, E., Ramon, H., de Baerdemaeker, J., *et al.* (2013) Moving horizon estimation and nonlinear model predictive control for autonomous agricultural vehicles. *Computers and Electronics in Agriculture* 98, 25–33.
Lepej, P. and Rakun, J. (2016) Simultaneous localisation and mapping in a complex field environment. *Biosystems Engineering* 150, 160–169.
Li, B. and Wang, M.H. (2013) In-field recognition and navigation path extraction for pineapple harvesting robots. *Intelligent Automation and Soft Computing* 19, 99–107.
Linker, R. and Neer, O. (2014) Tree skeleton extraction by close-range laser telemetry as first step toward robotic pruning. In: *International Conference of Agricultural Engineering, 6–10 July, Zurich, Switzerland*.
Luo, L., Tang, Y., Zou, X., Ye, M., Feng, W., and Li, G. (2016) Vision-based extraction of spatial information in grape clusters for harvesting robots. *Biosystems Engineering* 151, 90–104.
Maes, W.H. and Steppe, K. (2012) Estimating evapotranspiration and drought stress with ground-based thermal remote sensing in agriculture: a review. *Journal of Experimental Botany* 63, 4671–4712.
Mann, M.P., Rubinstein, D., Shmulevich, I., Linker, R., and Zion, B. (2014) Motion planning of a mobile Cartesian manipulator for optimal harvesting of 2-D crops. *Transactions of the ASABE* 57, 283–295.
Mehta, S.S. and Burks, T.F. (2014) Vision-based control of robotic manipulator for citrus harvesting. *Computers and Electronics in Agriculture* 102, 146–158.
Mehta, S.S., Mackunis, W., and Burks, T.F. (2016) Robust visual servo control in the presence of fruit motion for robotic citrus harvesting. *Computers and Electronics in Agriculture* 123, 362–375.
Meron, M., Tsipris, J., Orlov, V., Alchanatis, V., and Cohen, Y. (2010) Crop water stress mapping for site-specific irrigation by thermal imagery and artificial reference surfaces. *Precision Agriculture* 11, 148–162.
Monta, M., Kondo, N., and Ting, K.C. (1998) End-effectors for tomato harvesting robot. *Artificial Intelligence Review* 12, 11–25.
Morimoto, E., Suguri, M., and Umeda, M. (2005) Vision-based navigation system for autonomous transportation vehicle. *Precision Agriculture* 6, 239–254.
Morris, J.R. (2007) Development and commercialization of a complete vineyard mechanization system. *HortTechnology* 17, 411–420.
Naor, A. (2000) Midday stem water potential as a plant water stress indicator for irrigation scheduling in fruit trees. In: Ferreira, M.I. and Jones, H.G. (eds) *III International Symposium on Irrigation and Horticultural Crops*. International Society for Horticultural Science, Lisbon, pp. 447–454.
Naor, A. (2006) Irrigation scheduling and evaluation of tree water status in deciduous orchards. In: Janick, J. (ed.) *Horticultural Reviews* 32, 111–165.

Naor, A., Gal, Y., and Peres, M. (2006) The inherent variability of water stress indicators in apple, nectarine and pear orchards, and the validity of a leaf-selection procedure for water potential measurements. *Irrigation Science* 24, 129–135.
Nguyen, T.T., Vandevoorde, K., Wouters, N., Kayacan, E., de Baerdemaeker, J.G., and Saeys, W. (2016) Detection of red and bicoloured apples on tree with an RGB-D camera. *Biosystems Engineering* 146, 33–44.
Nof, S.Y. (2009) *Springer Handbook of Automation*. Springer-Verlag, Berlin/Heidelberg.
Oberti, R. and Shapiro, A. (2016) Advances in robotic agriculture for crops. *Biosystems Engineering* 146, 1–2.
O'Shaughnessy, S.A., Evett, S.R., Colaizzi, P.D., and Howell, T.A. (2012) Grain sorghum response to irrigation scheduling with the time-temperature threshold method and deficit irrigation levels. *Transactions of the ASABE* 55, 451–461.
Osroosh, Y., Peters, R.T., and Campbell, C.S. (2016) Daylight crop water stress index for continuous monitoring of water status in apple trees. *Irrigation Science*, 34, 209–219.
Penizzotto, F., Slawinski, E., and Mut, V. (2015) Laser radar based autonomous mobile robot guidance system for olive groves navigation. *IEEE Latin America Transactions* 13, 1303–1312.
Perez-Ruiz, M., Slaughter, D.C., Fathallah, F.A., Gliever, C.J., and Miller, B.J. (2014) Co-robotic intra-row weed control system. *Biosystems Engineering* 126, 45–55.
Poni, S., Tombesi, S., Palliotti, A., Ughini, V., and Gatti, M. (2016) Mechanical winter pruning of grapevine: physiological bases and applications. *Scientia Horticulturae* 204, 88–98.
Riyaz Ahammed, S., Sankar Reddy, P., Vennishmuthu, V., Hushein, R., and Gayathri, P. (2015) Modeling and analysis of cost effective lemon harvesting robot. *International Journal of Applied Engineering Research* 10, 9697–9710.
Rosa, U.A., Cheetancheri, K.G., Gliever, C.J., Lee, S.H., Thompson, J., and Slaughter, D.C. (2008) An electro-mechanical limb shaker for fruit thinning. *Computers and Electronics in Agriculture* 61, 213–221.
Rovira-Más, F., Chatterjee, I., and Sáiz-Rubio, V. (2015) The role of GNSS in the navigation strategies of cost-effective agricultural robots. *Computers and Electronics in Agriculture* 112, 172–183.
Sanchez, L.A., Sams, B., Alsina, M.M., Hinds, N., Klein, L.J., and Dokoozlian, N. (2017) Improving vineyard water use efficiency and yield with variable rate irrigation in California. *Advances in Animal Biosciences* 8, 574–577.
Silwal, A., Davidson, J.R., Karkee, M., Mo, C.K., Zhang, Q., and Lewis, K. (2017) Design, integration, and field evaluation of a robotic apple harvester. *Journal of Field Robotics* 34, 1140–1159.
Sivaraman, B. and Burks, T.F. (2006) Geometric performance indices for analysis and synthesis of manipulators for robotic harvesting. *Transactions of the ASABE* 49, 1589–1597.
Stentz, A., Dima, C., Wellington, C., Herman, H., and Stager, D. (2002) A system for semi-autonomous tractor operations. *Autonomous Robots* 13, 87–104.
Subramanian, V., Burks, T.F., and Arroyo, A.A. (2006) Development of machine vision and laser radar based autonomous vehicle guidance systems for citrus grove navigation. *Computers and Electronics in Agriculture* 53, 130–143.
Tanigaki, K., Fujiura, T., Akase, A., and Imagawa, J. (2008) Cherry-harvesting robot. *Computers and Electronics in Agriculture* 63, 65–72.
Tapia, R., Botti, C., Carrasco, O., Prat, L., and Franck, N. (2003) Effect of four irrigation rates on growth of six fig tree varieties. In: Corrales, M.L. and Bernalte García, M.J. (eds) *II International Symposium on Fig*. International Society for Horticultural Science, Caceres, Spain, pp. 113–118.
Thanpattranon, P., Ahamed, T., and Takigawa, T. (2016) Navigation of autonomous tractor for orchards and plantations using a laser range finder: automatic control of trailer position with tractor. *Biosystems Engineering* 147, 90–103.
Umeda, M., Kubota, S., and Iida, M. (1999) Development of "STORK", a watermelon-harvesting robot. *Artificial Life and Robotics* 3, 143–147.
Underwood, J.P., Jagbrant, G., Nieto, J.I., and Sukkarieh, S. (2015) Lidar-based tree recognition and platform localization in orchards. *Journal of Field Robotics* 32, 1056–1074.
Verbiest, R., Ruysen, K., Vanwalleghem, T., Demeester, E., and Kellens, K. (2021) Automation and robotics in the cultivation of pome fruit: where do we stand today? *Journal of Field Robotics* 38, 513–531.
Vitzrabin, E. and Edan, Y. (2016a) Adaptive thresholding with fusion using a RGBD sensor for red sweet-pepper detection. *Biosystems Engineering* 146, 45–56.

Vitzrabin, E. and Edan, Y. (2016b) Changing task objectives for improved sweet pepper detection for robotic harvesting. *IEEE Robotics and Automation Letters* 1, 578–584.
Vroegindeweij, B.A., van Willigenburg, G.L., Groot Koerkamp, P.W.G., and van Henten, E.J. (2014) Path planning for the autonomous collection of eggs on floors. *Biosystems Engineering* 121, 186–199.
Williams, H.A.M., Jones, M.H., Nejati, M., Seabright, M.J., Bell, J., *et al.* (2019) Robotic kiwifruit harvesting using machine vision, convolutional neural networks, and robotic arms. *Biosystems Engineering* 181, 140–156.
Ye, Y.X., Wang, Z.D., Jones, D., He, L., Taylor, M.E., *et al.* (2017) Bin-dog: a robotic platform for bin management in orchards. *Robotics* 6: 12.
Yekutieli, O. and Garbati Pegna, F. (2002) Automatic guidance of a tractor in a vineyard. In: *Automation Technology for Off-Road Equipment, Proceedings of the July 26–27, 2002 Conference, Chicago, Illinois*. American Society of Agricultural and Biological Engineers, St. Joseph, Michigan, pp. 252–260.
Yin, H., Chai, Y., Yang, S.X., and Mittal, G.S. (2011) Ripe tomato detection for robotic vision harvesting systems in greenhouses. *Transactions of the ASABE* 54, 1539–1546.
Zahid, A., He, L., Zeng, L., Choi, D., Schupp, J., and Heinemann, P. (2020) Development of a robotic end-effector for apple tree pruning. *Transactions of the ASABE* 63, 847–856.
Zhang, C., Noguchi, N., and Yang, L. (2016) Leader–follower system using two robot tractors to improve work efficiency. *Computers and Electronics in Agriculture* 121, 269–281.
Zhao, D.A., Lv, J.D., Ji, W., Zhang, Y., and Chen, Y. (2011) Design and control of an apple harvesting robot. *Biosystems Engineering* 110, 112–122.
Zhao, Y., Gong, L., Huang, Y., and Liu, C. (2016) Robust tomato recognition for robotic harvesting using feature images fusion. *Sensors* 16: 173.

Part III

Fig-fruit Ripening and Postharvest Management

12 Fig-fruit Ripening and Maturation

Kumar Lama[1]* and Moshe A. Flaishman[2]
[1]*Department of Life Sciences, School of Science, Kathmandu University, Dhulikhel, Nepal;* [2]*Institute of Plant Sciences, Agricultural Research Organization, Rishon LeZion, Israel*

12.1 Introduction

Ripening of fig (*Ficus carica* L.) fruit involves the well-orchestrated coordination of several regulatory steps together with strong metabolic and physiological changes, as in ripening of other fleshy fruits. Classically, fruits are categorized into climacteric and non-climacteric groups depending on the two distinct respiration patterns during ripening. Climacteric fruits show an increase in respiration (rise in the production of carbon dioxide) at the beginning of ripening, a rise in endogenous ethylene production during ripening, and further induction of ethylene synthesis following exogenous ethylene application (Paul *et al.*, 2012). In addition, climacteric fruits can ripen fully if they are harvested at the completion of their growth period. In contrast, non-climacteric fruits do not show these characteristics; no rise in respiration or ethylene production rates are detected during fruit ripening, ethylene synthesis is regulated in an auto-inhibitory manner, and fruit ripen fully only if they remain attached to the plant. Despite the difference between climacteric and non-climacteric fruits, recent studies have reported that abscisic acid (ABA) regulates fruit ripening and senescence in both climacteric and non-climacteric fleshy fruit (Zhang *et al.*, 2009; Li *et al.*, 2011; Leng *et al.*, 2014). Surprisingly, some fruits species present both climacteric and non-climacteric ripening patterns. Melons (*Cucumis melo*) and pears (*Pyrus communis*) possess both climacteric and non-climacteric cultivars (Downs *et al.*, 1991; Périn *et al.*, 2002; Obando-Ulloa *et al.*, 2009), while some plum cultivars present a suppressed-climacteric behavior compared with classic climacteric cultivars (Abdi *et al.*, 1998). Guava also presents both climacteric and non-climacteric behavior, depending on the cultivar and fruit developmental stage (Azzolini *et al.*, 2005). Traditionally, fig fruit is categorized as climacteric because its ripening process follows an increase in respiration rate and ethylene production (Chessa *et al.*, 1992). However, unlike climacteric fruit, figs harvested before they are fully ripe never complete their ripening process to reach commercially desirable parameters of size, color, taste, aroma, and texture (Flaishman *et al.*, 2008). Alternatively, if harvested late, fig fruits tend to perish due to over-ripening and high susceptibility to pathogens (Chessa and Mitra, 1997). Recent studies have suggested that climacteric ripening of figs is coordinated mainly by the inflorescence (Freiman *et al.*, 2015; Rosianski *et al.*, 2016a; Lama *et al.*, 2019b). These studies suggest a tissue-specific unique ripening nature of fig fruit, separate from the traditional definition

* Email: kumar.lama@ku.edu.np

DOI: 10.1079/9781789242492.0012

of climacteric fruits. Here, we describe the unique fig fruit growth and maturation and compare them with other fleshy fruits.

12.2 Fruit Growth and Maturation

The fig fruit bears a unique closed inflorescence structure termed the syconium. The syconium is constructed from a portion of the stem that expands into a sac containing an inflorescence that grows internally. This closed inflorescence produces an aggregate fruit, which is composed of small individual drupelets that develop from the receptacle-enclosed ovaries. Growth and development of common fig fruit in terms of fruit diameter can be described by a double-sigmoid curve, consisting of three phases (Fig. 12.1). In phase I, growth is rapid, accompanied by intense cell division and differentiation. Phase II is characterized by a long period where mitotic and fruit activity cease, and the fruit remains nearly the same size, color, and firmness. In phase III, ripening occurs, with a second phase of rapid fruit growth mainly by cell expansion and ethylene production. This is followed by a color change, softening, and alteration of the pulp texture to an edible state (Flaishman *et al.*, 2008).

In fig, the development and ripening processes differ in pollinated and parthenocarpic fruit (Fig. 12.2). Progression of the pollinated inflorescence pistils is more conspicuous than that of the parthenocarpic fruit, with style deterioration and color alterations beginning 1 week after pollination (WAP). Pollinated styles become brown (Fig. 12.2A) and are absent in the parthenocarpic fig. At this stage, the embryo is already developing in the pollinated ovary (Fig. 12.2B3). At 5 WAP, pollinated pistils changed their color to red, 2 weeks earlier than the parthenocarpic pistils (Fig. 12.2A, B6, and B7). As the fruit reaches the final green stage before ripening onset, both parthenocarpic and pollinated fruit display red inflorescences, although the seeds of the pollinated fruit harbor a mature embryo while the styles of the parthenocarpic fruit pistils remained underdeveloped (Fig. 12.2B7 and B8). The facultative parthenocarpic common-type fig fruit serves as a receptacle for flower development while at the same time pollinated figs possessing functional seeds and those with non-bearing seeds present different developmental and ripening features. Signals originating from the seed initiate fruit development, and separate the specific pathways between pollinated and non-pollinated fruit. Normally, the ripened pollinated fruit are round with a larger diameter and weight, while the parthenocarpic fruit continue to a pear-like shape with a smaller diameter and less weight (Rosianski *et al.*, 2016b). In addition, pollinated fruit exhibits more commercially desirable physical and taste characteristics, with advanced fertile nutlets compared with the sterile undeveloped non-bearing nutlets of the parthenocarpic fruit at the time of harvest.

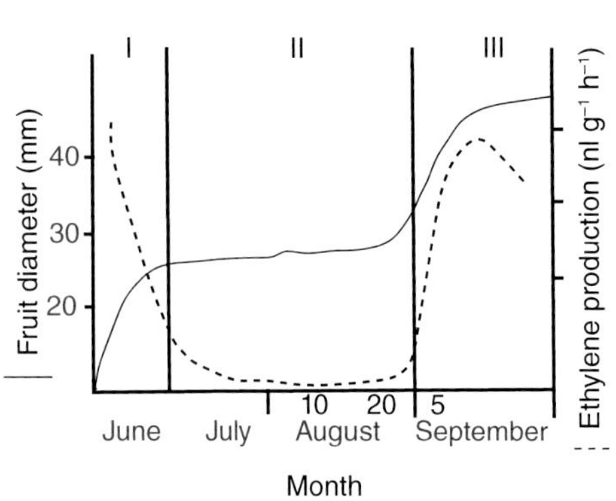

Fig. 12.1. The double-sigmoid growth curve of the fig, showing increases in fruit diameter and ethylene production during phases I, II, and III of fruit growth and development. Adapted and modified from Owino *et al.* (2006).

12.3 Hormone Changes during Ripening of Fig Fruit

Plant hormone content in the fruits changes during fruit development, maturation, and ripening. In particular, the ripening process of fleshy fruits is regulated mainly by the plant hormones ABA and ethylene (Seymour *et al.*, 2013; Kumar *et al.*, 2014). In fig, five plant hormones namely—ethylene, ABA, auxin, gibberellin, cytokinins and their derivatives—have been well studied from the onset of and throughout ripening of the parthenocarpic and pollinated common-type fig-fruit variety Brown Turkey (Rosianski *et al.*, 2016a). Gas chromatographic analysis of whole fig fruits revealed that biologically active

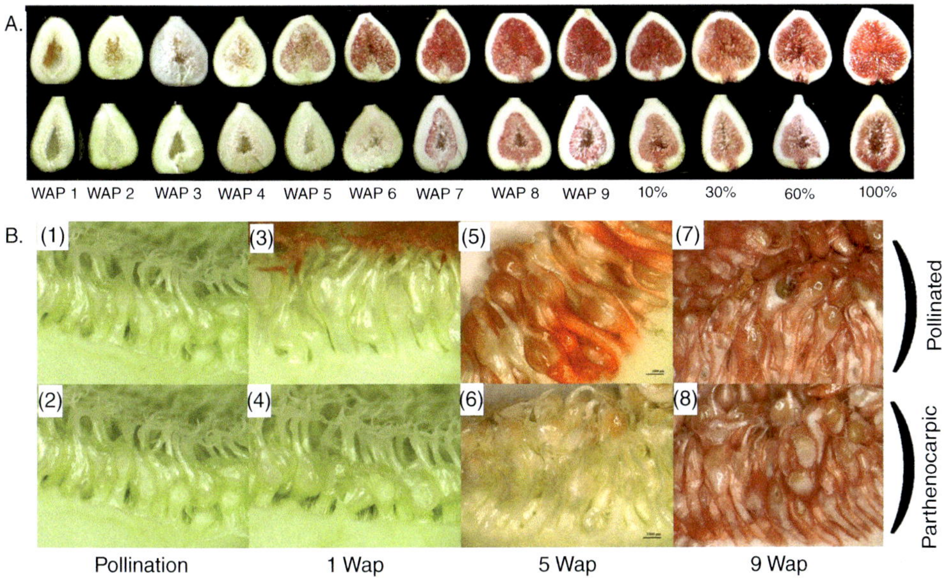

Fig. 12.2. Inflorescence progression in pollinated and parthenocarpic fig. A. Inflorescence development of pollinated (top) and parthenocarpic (bottom) fruits at the indicated weeks after pollination (WAP) or at the indicated percentage ripening stage. B. (1) Pollinated and (2) parthenocarpic fruit at pollination. (3) Pollinated and (4) parthenocarpic fruit at 1 WAP. (5) Pollinated and (6) parthenocarpic fruit at 5 WAP. (7) Pollinated and (8) parthenocarpic fruit at 9 WAP. Modified from Rosianski *et al.* (2016b).

endogenous auxin (indoleacetic acid and its non-active aspartic acid conjugate), ABA and its derivatives (ABA–glucose ester (ABA–GE)), and ethylene increase, whereas no production of active gibberellins and cytokinins occurs during ripening, irrespective of pollination. Interestingly, the prominent ABA degradation pathway seemed to be via the ABA–GE conjugate as its concentration was three to four times higher in both parthenocarpic and pollinated fruit than the concentration of ABA, and no decline in ABA–GE was reported in either case until full ripening (Rosianski *et al.*, 2016a). Moreover, tissue-specific analysis of fig confirmed that ABA increases up to the full ripening stage in both inflorescence and receptacles, as in non-climacteric fruit such as strawberry (Lama *et al.*, 2019b).

12.3.1 Ripening hormones and their biosynthesis genes

To understand the biosynthesis of the fig-fruit major ripening hormones ethylene and ABA, molecular studies have been conducted on some commercial varieties. Ethylene production during fig-fruit ripening comprises three enzymatic reactions: *S*-adenosyl methionine (SAM) synthetase converts methionine to SAM; 1-aminocyclopropane 1-carboxylate (ACC) synthase (ACS) converts SAM to ACC; and ACC oxidase (ACO) degrades ACC to release ethylene, as in other climacteric fruits. All of these enzyme-coding genes have been identified in both inflorescence and receptacle tissues of fig fruit (Owino *et al.*, 2006; Lin *et al.*, 2009; Rosianski *et al.*, 2016a). The general expression patterns of the majority of genes related to the ethylene-biosynthesis and signal-transduction family that were up- or downregulated during the ripening process are shown in Fig. 12.3. In particular, two SAM genes (*FcSAM2* and *FcSAM3*), four ACS genes (*FcACS1*, *FcACS2*, *FcACS3*, and *FcACS4*) and three ACO genes (*FcACO1*, *FcACO2*, and *FcACO3*) were isolated and their expression patterns analyzed during ripening (Owino *et al.*, 2006; Freiman *et al.*, 2015; Rosianski *et al.*, 2016a). It appeared that all ethylene-synthesis genes are active during system 2 ethylene synthesis, with expression increasing in the inflorescence during

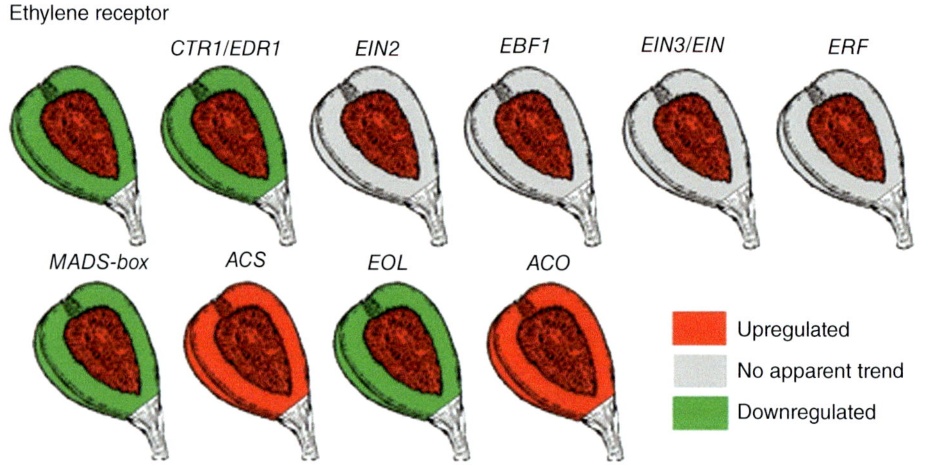

Fig. 12.3. Schematic representation of the activities of different gene families in fig fruit during on-tree ripening. The general expression pattern was concluded from the majority of genes in a family being up- or downregulated during the ripening process. No apparent trend was defined when expression patterns showed combined trends in one gene family or a changing pattern with ripening (Freiman *et al.*, 2015). *CTR1/EDR1*, constitutive triple response 1/enhanced disease resistance 1; *EIN*, ethylene-insensitive; *EBF1*, ethylene-insensitive 3-binding F-box 1; *ERF*, ethylene-responsive factor; *ACS*, 1-aminocyclopropane 1-carboxylate (ACC) synthase; *EOL*, target of ethylene overproducer (ETO1)-like; *ACO*, ACC oxidase.

on-tree ripening. Except for *FcACO3*, all genes also showed increasing expression in the receptacle, although to lower expression levels (Freiman *et al.*, 2015). In addition, several transcription factors have been reported to precede ethylene synthesis and trigger ripening onset, advancing the fig-fruit ripening process. In particular, MADS-box transcription factor and ethylene signal-transduction genes were analyzed during ripening of fig fruit (Freiman *et al.*, 2014, 2015). Transcription of a potential ripening regulator, *FcMADS8*, increased during ripening of fig fruit on the tree. Lately, NAC (standing for NAM, ATAF, and CUC) genes, which are involved in the ripening of several fruit, such as tomato (Ma *et al.*, 2014) and banana (Shan *et al.*, 2012), via interaction with ethylene biosynthesis and ethylene signaling components, have been identified during the ripening of fig fruit (Freiman *et al.*, 2014), and their responses have been further investigated using a number of treatments (Lama *et al.*, 2019b).

The involvement of ABA in the ripening process of fig fruit has been explored, as in other fleshy fruits such as climacteric tomato and non-climacteric strawberry fruits. Recently, genes encoding one zeaxanthin epoxidase, three 9-*cis*-epoxycarotenoid dioxygenases (NCEDs), one short-chain alcohol dehydrogenase (ABA2), and one ABA-8′-hydroxylase enzyme in the ABA biosynthesis and catabolism pathways were identified and isolated from fig fruit (Rosianski *et al.*, 2016a). Their expression patterns during natural ripening were characterized and have provided the first documentation of gene expression with respect to endogenous ABA regulation in fig fruit during ripening.

12.3.2 Hormone treatments to induce fig-fruit maturation and coordinate gene expression

As the tissue-specific ripening nature of fig fruit suggested that the climacteric ripening process is coordinated by the inflorescence (Freiman *et al.*, 2015), the unique nature of fig-fruit ripening has demanded further investigation. The practice of stimulating growth to obtain greater uniformity of maturation and ripening has been the subject of several studies. It has been known since the third century BCE that a drop of olive oil applied to the ostiole (a process termed olerification) of the fig fruit stimulates growth and leads to enhanced fruit ripening (Chessa and Mitra, 1997; Owino *et al.*, 2006). Application of several vegetable oils—rapeseed oil, sesame oil, soybean oil, camellia oil, and linseed oil—was shown to have the same effect as olive oil; animal oil also stimulated fruit ripening but less strongly than the vegetable oils. In addition, ripening of mature fig fruit (late phase II and phase III) is stimulated by exposure to ethylene gas or by spraying the ethylene precursor ethephon (Celikel *et al.*, 1997; Owino *et al.*, 2006). On-tree injection of exogenous ABA, ethephon (which releases ethylene), and the ABA inhibitors fluridone and nordihydroguaiaretic acid (NDGA) inside the mature fruit through the ostiole was carried out to further understand this unique ripening mechanism (Lama *et al.*, 2019b). In this study, exogenous ABA treatment induced earlier ripening of fig fruit, as in other fleshy fruits such as tomato (Zhang *et al.*, 2009; Mou *et al.*, 2016), banana (Jiang *et al.*, 2000), and strawberry (Jia *et al.*, 2011), followed by a rapid increase in fruit diameter, softening, and purple coloration of the outer peel. In contrast, addition of the ABA inhibitors NDGA and fluridone to the mature fig fruit resulted in a delay in onset of ripening, followed by a slower increase in diameter and slower decrease in firmness, as occurs in tomato (Zhang *et al.*, 2009; Mou *et al.*, 2016) and strawberry (Jia *et al.*, 2011). Metabolic analysis has confirmed the strong suppression in endogenous ABA production in both the inflorescence and receptacle tissues of NDGA- and fluridone-treated fig fruits, which ultimately delayed the production of ethylene and the onset of ripening compared with controls (Lama *et al.*, 2019b). As reported in tomato (Zhang *et al.*, 2009; Mou *et al.*, 2016), fig ABA biosynthesis (particularly expression of *FcNCED2* and *FcABA2*), ethylene biosynthesis (particularly *FcACS4*, *FcACOL*, and *FcACO2*), and transcription factor (*FcMADS8*, *-14*, and *-15*, *FcNAC1*, *-2*, and *-5*, and *FcERF9006*) genes were upregulated by exogenous ABA and ethephon treatment but downregulated by NDGA and fluridone treatment of the inflorescence and receptacle of the fruit compared with their controls, as summarized in Fig. 12.4 (Lama *et al.*, 2019b).

The inhibitor of ethylene action 1-methylcyclopropene (1-MCP) has been used successfully in various plant species to arrest the plant's response to elevated internal ethylene level. In figs picked at the commercial-maturity stage, postharvest 1-MCP application was ineffective (d'Aquino *et al.*, 2003) or had only a limited effect expressed as some retardation of softening (Sozzi *et al.*, 2005; Gözlekçi *et al.*, 2008). However, 1-MCP markedly enhanced ethylene production in figs after pre- or postharvest treatment (Sozzi *et al.*, 2005; Owino *et al.*, 2006; Freiman *et al.*, 2012), similar to its effect in non-climacteric fruit such as citrus (McCollum and Maul, 2007). This reaction showed that, at commercial maturity, the harvested figs have auto-inhibitory regulation of ethylene synthesis, rather than the autocatalytic pattern typical of climacteric fruit. When picked at a more advanced stage, the figs showed no response to 1-MCP, in either fruit quality or ethylene production rate (Sozzi *et al.*, 2005). It was suggested that the low efficacy of 1-MCP with figs might be due to the fact that, at harvest, the fruit are already at a postclimacteric stage (Sozzi *et al.*, 2005; Flaishman *et al.*, 2008). Moreover, the ripening process of expression patterns of ethylene-biosynthesis genes (*FcACS1*, *FcACS3*, and *FcACO1*) was inhibited by postharvest 1-MCP treatment of ripe fruits, indicating positive regulation by ethylene, whereas *FcACS2*, *FcACS4*, and *FcACO3* gene expression was induced by 1-MCP. In addition, *FcMADS8* and *FcERF* genes were inhibited following 1-MCP treatment, whereas the expression level of *FcEIL3* was upregulated by the treatment (Freiman *et al.*, 2015).

12.3.3 The different fig syconium parts exhibit climacteric and non-climacteric ripening processes

The fig is an aggregate fruit composed of individual small drupes; each develops from the ovaries in a closed inflorescence, known as a syconium, which exhibits vegetative and reproductive origins. While the individual small drupes of the fruit inflorescence are the reproductive part, the fruit receptacle—the flesh and peel—have a vegetative origin. The fig fruit was categorized as climacteric, but application of 1-MCP was reported to enhance ethylene production (Sozzi *et al.*,

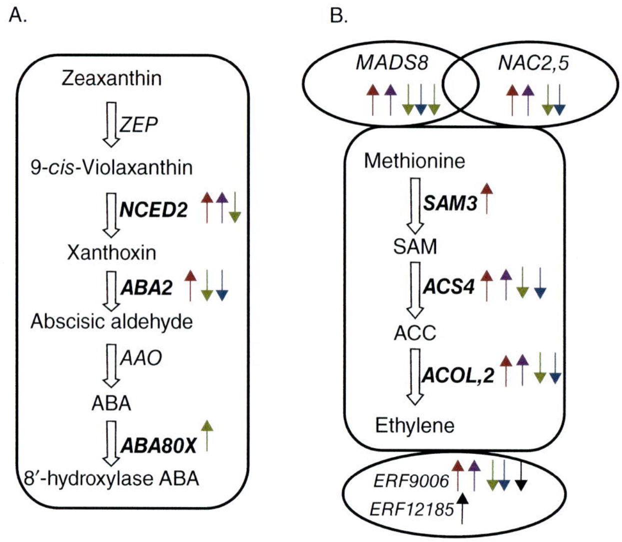

Fig. 12.4. Alterations in gene-expression patterns of abscisic acid (ABA) metabolic pathway genes (A.) and ethylene-biosynthesis, MADS-box, NAC (standing for NAM, ATAF and CUC), and ethylene-responsive factor (ERF) genes (B.) in the fig fruit following on-tree exogenous application of ABA, ethephon, nordihydroguaiaretic acid (NDGA), fluridone, and 1-methylcyclopropene (1-MCP) (Freiman *et al*., 2015). Downward arrows indicate downregulated genes in treated fruit compared with untreated controls; upward arrows indicate upregulated genes. The arrows are color coded as follows: red, ABA; purple, ethephon; green, NDGA; blue, fluridone; black, 1-MCP. *ZEP*, zeaxanthin epoxidase; *NCED2*, 9-*cis*-epoxycarotenoid dioxygenase 2; *ABA2*, short-chain alcohol dehydrogenase; *AAO*, abscisic aldehyde oxidase; *ABA80X*, ABA-8′-hydroxylase *MADS8*, MADS-box gene 8; *SAM3*, *S*-adenosyl methionine 3; *ACS4*, 1-aminocyclopropane 1-carboxylate (ACC) synthase 4; *ACOL,2*, ACC oxidase-like and ACC oxidase 2. Adapted from Lama *et al.* (2019b).

2005; Owino *et al.*, 2006; Freiman *et al.*, 2012), similar to its effect in non-climacteric fruit such as citrus (McCollum and Maul, 2007). Apart from ethylene, the increase in accumulation of endogenous ABA during the onset of fig fruit ripening is similar to other climacteric fruit such as tomato, avocado, and apple (Chernys and Zeevaart, 2000; Lara and Vendrell, 2000; Zhang *et al.*, 2009). However, instead of tapering off after the onset of fruit ripening, as in climacteric fruit, endogenous ABA content continues to rise until the fig fruit is fully ripe, as in non-climacteric strawberry fruit (Jia *et al.*, 2011). Interestingly, this phenomenon is observed in both the inflorescence and receptacle, with high expression of major ABA biosynthesis pathway genes (Lama *et al.*, 2019b). In contrast, expression of ethylene-related genes is dissimilar for the two tissues, with only the inflorescence showing expression similar to that in climacteric fruit (Freiman *et al.*, 2015). Ethylene is produced at high levels at the 30–50% ripening stage, with significantly high expression of the ethylene-biosynthesis gene *FcACS4* in the inflorescence, although expression levels rose in both tissues (Freiman *et al.*, 2012, 2015). Furthermore, increased expression of ethylene-biosynthesis genes in petioles (the vegetative tissue most close to the fruit) of fruit at different ripening stages—from green to fully ripe—was noted (Lama *et al.*, 2019b). Interestingly, the expression levels of these genes in the petioles were very low compared with that in the inflorescences and receptacles, and they played no part in fruit ripening. Similarly, very low expression of ethylene-biosynthesis genes in receptacles compared with inflorescence tissue could lead to the production of minor amounts of ethylene, which might not be enough for the ripening process. The high amount of endogenous ABA production until full ripeness, as in strawberry, could lead to non-climacteric ripening by promoting sugar accumulation, fruit pigmentation, and fruit softening, as described in a non-climacteric fruit-ripening model (Li *et al.*, 2011). This could be the major reason why fig harvested prior to complete ripening never reaches the commercially desirable parameters of size, color, flavor, and texture (Flaishman *et al.*, 2008). This establishes that ABA triggers ripening by boosting ethylene production, as in climacteric fruit, especially in the inflorescence of fig fruit. However, only ABA-dependent ripening occurs in the receptacle of fig fruit, as in non-climacteric fruits (Fig. 12.5).

12.3.4 Uniform ripening simulation of mature fig fruit

Fig ripening is asynchronous: a single fruit ripens per shoot, and the next fruit on the shoot only starts to ripen when the first is fully ripe. Several practices, such as application of vegetable oil, animal oil, and ethylene and its precursors, have been studied to stimulate greater uniformity in fig-fruit growth, maturation, and ripening (Chessa and Mitra, 1997; Owino *et al.*, 2006). Recently, injection of exogenous ABA into the mature fig fruit was noted to cause early onset of ripening (Lama *et al.*, 2019b), but injection into individual fruits will never be a cost-effective approach for commercial fig producers. An alternative method was developed by spraying exogenous ABA or ethephon on to the clusters of matured fruit of both summer and autumn figs, which resulted in uniform early onset of ripening followed by significantly increased fruit size and weight, and lower firmness compared with untreated fruits (Lama *et al.*, 2019a). Similarly, application of ethephon to Mission variety fig fruit late in phase II stimulated early ripening by increasing fruit size and changing the color from green to bluish-black (Crane *et al.*, 1970).

Apart from ABA and ethephon, on-tree application of 1-MCP to mature Brown Turkey variety fig fruit delayed the ripening process but maintained uniform ripening compared with untreated fruit (Freiman *et al.*, 2012). The synchronized ripening of different sizes of mature fig fruits by exogenous ABA or ethephon application during the main-crop seasons could decrease harvesting costs. Growers could harvest three to five fruits per branch at one time instead of singly. Fig growers could also essentially adjust the harvest time of fig fruit to market demand.

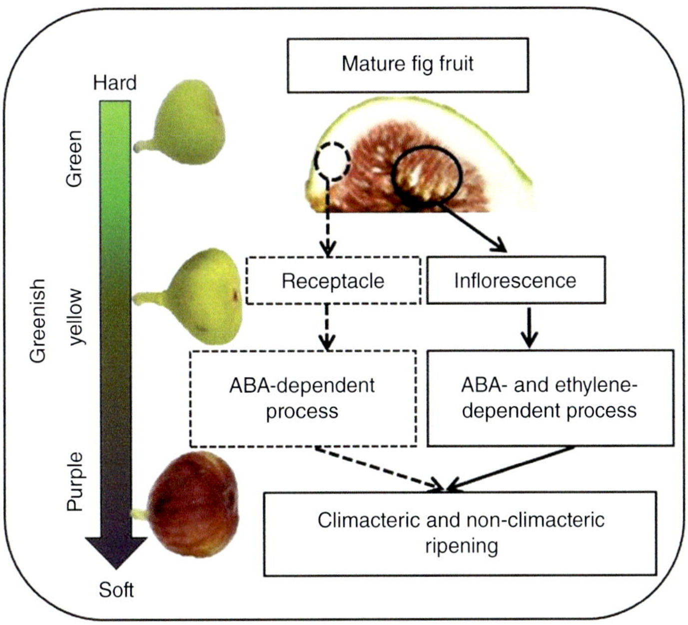

Fig. 12.5. Physiological changes and tissue-specific ripening model of mature fig fruit. ABA, abscisic acid.

On-tree application of exogenous ABA or ethephon has no effect on the fig fruit during cold storage (Lama *et al.*, 2019b). However, the high level of postharvest weight loss, decreased diameter size, and strong softening observed in both treated and untreated fig fruit confirmed that ripening figs are prone to desiccation, a trait that enables their efficient drying (Flaishman *et al.*, 2008).

12.4 Major Metabolic Changes Regulating Fruit Quality during Ripening

The change in fleshy fruit from the unripe to ripe stages occurs via several metabolic activities, which ultimately change the fruit's physiology. These complex metabolic conversions are led by primary and secondary metabolic changes (Carrari *et al.*, 2006; Tohge *et al.*, 2014). In particular, primary metabolites, such as sugars and acids, directly influence fruit taste, while secondary metabolites, such as anthocyanin, carotenoids, volatiles, terpenoids, and polyphenols are responsible for color, aroma, and health-promoting characteristics (Pott *et al.*, 2019). Thus, major metabolic changes during ripening, such as organic acid metabolism for acidic taste, sugar metabolism and transport for a sweet taste, and anthocyanin biosynthesis for color development, are a major field of interest in the ripening of fig, as well as other fleshy fruits, toward improving fruit quality.

12.4.1 Organic acid changes during fig-fruit ripening

Fleshy fruit acidity, which is responsible for the acidic or sour taste, is an important component of the fruit's overall organoleptic quality, along with sugar and aromas, yielding the optimal product for consumers (Etienne *et al.*, 2013). Fruit acidity is due to the presence of organic acids, particularly malic and citric acids, which are the main acids found in most ripe fruits (Seymour *et al.*, 1993). Similar to other fleshy fruits, citric and malic acids are also reported as the dominant organic acids of fig fruit, contributing more than 80% of the total organic acids throughout fruit development, maturation, and ripening (Shiraishi *et al.*, 1996). Citric and malic acids have been detected in different varieties of fig fruit varying from dark to yellow-green skin color (Slatnar *et al.*, 2011; Trad *et al.*, 2014). Due to their nutritional and medicinal values, citric and malic acids have also been studied in the leaves and different fruit parts (peel and pulp) of fig at the optimal ripening stage (Oliveira *et al.*, 2009; Pande and Akoh, 2010). Recently, a higher quantity of acidity (H^+) was reported in pollinated fig fruit at 30% and 60% ripening compared with their parthenocarpic counterparts (Rosianski *et al.*, 2016b). A comparative study of organic acid accumulation and metabolism in the inflorescence and receptacle tissues during the ripening of pollinated and parthenocarpic fig fruit was studied by Lama *et al.* (2020b). In this study, citric and malic acids were found to be the dominant organic acids, as reported by Shiraishi *et al.* (1996), with higher contents among the detected organic acids in the inflorescence and receptacle of both parthenocarpic and pollinated fig fruits. In particular, citric and malic acids were found to be at significantly higher levels in the pollinated inflorescence compared with the parthenocarpic inflorescence at 10% ripening. This difference causes a high quantity of total acidity in the pollinated fruit compared with the parthenocarpic one (Rosianski *et al.*, 2016b). In comparison between the tissues, the inflorescence was seen to have a higher level of organic acids (mainly citric and malic acids) compared with the receptacle, as occurs in the achenes (reproductive part) and receptacles (vegetative part) of strawberry fruit during ripening (Fait *et al.*, 2008).

Variations in fleshy fruit acidity occur due to the metabolism of malate and citrate in the fruit itself (Sweetman *et al.*, 2009; Etienne *et al.*, 2013). Essentially, phosphoenolpyruvate (PEP) carboxylation, the tricarboxylic acid (TCA) cycle, citrate catabolism, and the glyoxylate cycle are involved in malate–citrate metabolism (Etienne *et al.*, 2013). In fig fruit, 14 genes related to PEP carboxylation (*FcPEPC* encoding PEP carboxylase, *FcMDH3* encoding malate dehydrogenase, and *FcNADP-ME1,2* encoding $NADP^+$-malic enzyme), the TCA cycle (*FcCS1* encoding citrate synthase, and *FcMDH1* and *FcFUM* encoding fumarase), citrate catabolism (*FcACO* encoding aconitase, *FcICDH* encoding isocitrate dehydrogenase, and *FcACL* encoding ATP-citrate lyase) and the glyoxylate cycle (*FcCS2*, *FcICL* encoding isocitrate lyase, *FcMLS* encoding malate synthase, and *FcMDH2*) have been reported from the pollinated and parthenocarpic fig transcriptome database (Rosianski *et al.*, 2016a) and further studied by Lama *et al.* (2020b) at different ripening stages to understand the respective gene expression patterns (Fig. 12.6). The expression level of *FcPEPC* was reported to be significantly higher in the pollinated inflorescence compared with the parthenocarpic fruit and was suggested to contribute malic acid accumulation during fig-fruit maturation as does the *PEPC* gene of apple (Yao *et al.*, 2009). Surprisingly, expression of fig-fruit TCA cycle-related genes *FcCS1* and *FcMDH1* was not significantly different between parthenocarpic and pollinated fruit, although their citric acid content differed significantly at 10% and 60% ripening (Lama *et al.*, 2020b). Therefore, the difference in citric acid concentration between the pollinated and parthenocarpic fruit may not be attributable to the difference in expression level of *FcCS1* and *FcMDH1*, as a stronger expression of *PpCS1* was not correlated with organic acid content in ripening peach fruit (Etienne *et al.*, 2002). Most importantly, the expression of mainly *FcMLS* and *FcICL* from the glyxoylate cycle was found to be significantly different at all stages of ripening in both inflorescence and receptacle tissues (Lama *et al.*, 2020b). The possible involvement of malate synthase (MLS) in the synthesis of organic acid/malic acid has been reported in other fleshy fruits such as banana and grape berry (Pua *et al.*, 2003; Terrier *et al.*, 2005). The remarkable increase in expression of *FcMLS* and *FcICL* in the pollinated inflorescence could be governed by the seeds; however, further experiments with different fig varieties are needed to confirm this.

12.4.2 Characterization of sugar accumulation during fig-fruit ripening

Sweetness is perhaps the main indicator of fruit quality in fig and is determined by the soluble sugar concentration. Ripe fig fruit are very rich in sugar, and sugar accumulation is a developmental process. Sugar levels remain low during

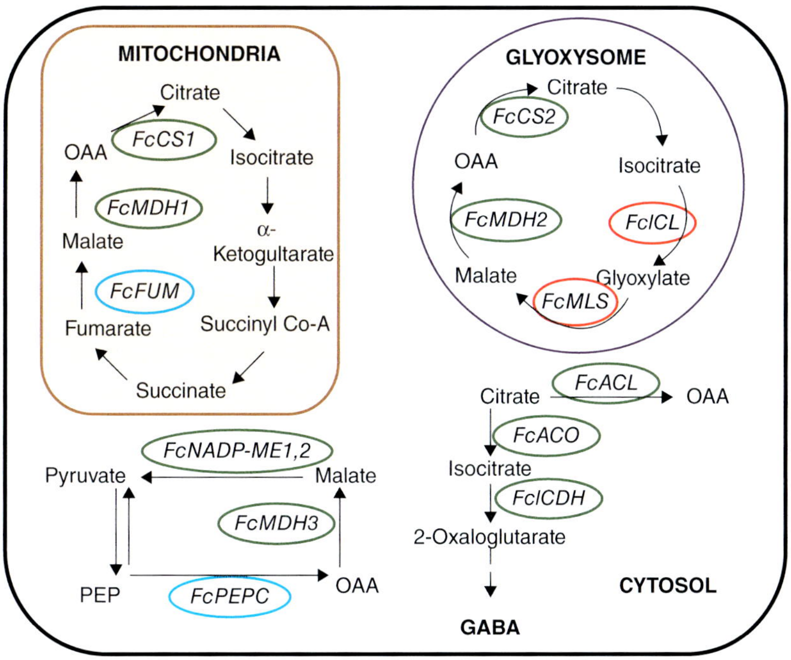

Fig. 12.6. Changes in gene-expression patterns of citrate–malate metabolic pathway genes by pollination in fig fruit. Green circles, no significant difference; blue circles, significantly high expression level in pollinated fruit compared with parthenocarpic fruit; red circle, genes likely to be induced by pollination. OAA, oxaloacetic acid; *CS1*, citrate synthase 1; *MDH1*, malate dehydrogenase isoform 1; *FUM*, fumarase; *ICL*, isocitrate lyase; *MLS*, malate synthase; *NADP-ME1,2*, NADP+-malic enzyme; PEP, phosphoenolpyruvate; *PEPC*, PEP carboxylase; *ACL*, ATP-citrate lyase; *ACO*, 1-aminocyclopropane 1-carboxylate oxidase; *ICDH*, isocitrate dehydrogenase; GABA, γ-aminobutyric acid; Modified from Lama *et al.* (2020b).

phases I and II of development, and then concentrations increase considerably during the final stages of ripening, until harvest (Flaishman *et al.*, 2008). In general, 90% of the total sugar content in female fig fruit is accumulated during the last few days of ripening, and the hexose sugars glucose and fructose are the major sugars that accumulate during this short period (Trad *et al.*, 2012; Vemmos *et al.*, 2013). Detailed developmental analysis in fig-fruit ripening has indicated a sharp increase in sugar content during a short period. The results present fig as an extreme example of sugar import and accumulation, being one order of magnitude higher than other sugar-accumulating sweet fruit species (Lama *et al.*, 2022). Essentially, fig is a monosaccharide-accumulating fruit; sucrose is present as well but at low concentrations (Trad *et al.*, 2012). As sucrose is the predominant sugar in fig leaves (Vemmos *et al.*, 2013), sugar metabolism in fig fruit is likely to be sucrose metabolism, which has been characterized in detail in sucrose-metabolizing plant tissues, including fruit (Hubbard *et al.*, 1990; Li *et al.*, 2012; Osorio *et al.*, 2014; Ruan, 2014). However, for fruits that transport sucrose and accumulate monosaccharides, the maintenance of fruit invertase activity is crucial for glucose and fructose accumulation during ripening (Takayanagi and Yokotsuka, 1997; Miron *et al.*, 2002).

The transport of sugars from source to sink and the inter- and intracellular allocation of sugars involve sugar transporters. There are three main families of transporters—sucrose transporter (SUT/SUC), monosaccharide transporter (MST) and SWEET—participating in the distribution of sugars within most plant cells (Ludewig and Flügge, 2013; Chen *et al.*, 2015). To explore novel sugar-accumulation characteristics of fig, 80 fig sugar metabolism and transporter genes were identified from the transcriptomic data published previously by our laboratory (Freiman *et al.*, 2014; Rosianski *et al.*, 2016a). In these studies, *FcCIN2* and *FcSUS2* showed increased expression along with ripening of the fig fruit. The gene responsible for sucrose synthesis, *FcSPS*, and those of the sucrose synthase (SUS) family (enzymes capable of both synthesis and cleavage of sucrose) were upregulated during sugar accumulation. Although the fig fruit accumulates the breakdown products of sucrose, there is likely to be a role for hydrolysis–resynthesis–hydrolysis of the translocated sucrose during accumulation. Such a cycle has been reported for other sugar-accumulating fruits (Nguyen-Quoc and Foyer, 2001; Miron *et al.*, 2002). In addition, a SWEET transporter—*FcSWEET9* the orthologue of the *Arabidopsis AtSWEET1*, the first reported SWEET, which was also characterized as a glucose efflux transporter—that localized to the plasma membrane (Chen *et al.*, 2010) was upregulated during ripening. Recently, proteomic analysis of commercial-ripe and on-tree fully ripe fig fruit also revealed upregulation of SUS enzymes and sucrose-phosphate synthase (SPS), and suggested the major players in sugar partitioning and accumulation (Cui *et al.*, 2019).

12.4.3 Fruit pigmentation and color development

Color plays an important role in fig-fruit quality in terms of market demand and consumer preference. Fig fruits present a wide diversity of colors, ranging from dark purple to green, depending on the variety. Essentially, the degree of coloration in fig fruit is determined by the content and composition of anthocyanins. The amount of two anthocyanins, cyanidin 3-*O*-rutinoside and cyanidin 3-*O*-glucoside, mainly determine the color of fig fruit during ripening (Solomon *et al.*, 2006; Dueñas *et al.*, 2008). For example, Brown Turkey fig (a commercial variety adopted worldwide) changes its color from green to purple as a result of anthocyanin accumulation during the ripening process (Flaishman *et al.*, 2008). In addition, pollination has been noted to improve the color as well as the taste of fig fruits (Rosianski *et al.*, 2016b; Lama *et al.*, 2020b). Recently, Lama *et al.* (2020a) reported significantly higher quantities of cyanidin 3-O-glucoside and cyanidin 3-O-rutinoside in exogenous ABA- and ethephon-treated fruit with an early dark purple color compared with untreated fruit. Similarly, ABA treatment enhances fruit color in grape (Cantín *et al.*, 2007), strawberry (Kadomura-Ishikawa *et al.*, 2015), red-colored sweet cherry (Shen *et al.*, 2014), and lychee (Wei *et al.*, 2011). In contrast, fig fruit treated with the ABA inhibitors NDGA and fluridone had significantly lower levels of cyanidin 3-*O*-glucoside and cyanidin 3-*O*-rutinoside, with less purple color coverage (Lama *et al.*, 2020b).

Transcriptome analysis has shown high expression levels of anthocyanin biosynthesis pathway genes in a range of varieties during ripening (Ikegami *et al.*, 2013; Wang *et al.*, 2017). Anthocyanin biosynthesis leads to the formation of dihydroflavonols via phenylalanine ammonia lyase (PAL), chalcone synthase (CHS), chalcone isomerase (CHI), flavanone 3-hydroxylase (F3H), flavanone 3′-hydroxylase and flavonoid 3′5′-hydroxylase. Dihydroflavonols are converted into colorless leucoanthocyanidins by dihydroflavonol 4-reductase (DFR) and subsequently to colored anthocyanidins by anthocyanidin synthase (ANS). Anthocyanidins are further glycosylated by UDP-glucose:flavonoid 3-*O*-glycosyl-8 transferase (UFGT) to anthocyanins (Takos *et al.*, 2006; Jaakola, 2013; Liu *et al.*, 2018). In addition, expression of the structural genes involved in anthocyanin biosynthesis has been shown to be regulated by the MBW complex (MYB transcription factor in a complex with basic helix–loop–helix and WD40 proteins) in several species (Allan *et al.*, 2008). Lately, Lama *et al.* (2020a) investigated expression of 15 anthocyanin biosynthesis genes and three MBW complex transcription factors of fig fruit using ABA, ABA inhibitor, and ethephon treatments. Among the reported genes, expression levels of the fig anthocyanin biosynthesis genes (*FcPAL*, *FcCHS2*, *FcCHI*, *FcF3H*, *FcDFR*, *FcANS*, *FcUFGT*, and *Fc3RT*) were upregulated by exogenous ABA and ethephon treatment, as occurs in red sweet cherry (Shen *et al.*, 2014), grape (Ban *et al.*, 2003; Koyama *et al.*, 2018) and strawberry (Kadomura-Ishikawa *et al.*, 2015; Li *et al.*, 2015). In fig fruits treated with the ABA inhibitors (NDGA and fluridone), anthocyanin biosynthesis gene were downregulated, as in red sweet cheery (Shen *et al.*, 2014) and strawberry (Kadomura-Ishikawa *et al.*, 2015; Li *et al.*, 2015). In the MBW complex of fig fruit, the expression levels of *FcMYB113*, *FcGL3*, and *FcWD40* were upregulated by exogenous ABA and ethephon and downregulated by NDGA and fluridone (Lama *et al.*, 2020a). The alterations in expression levels by exogenous ABA and the ABA inhibitors NDGA and fluridone suggest the possibility of interdependent relationships of fig MBW complex genes, as in other fruit such as peach (Rahim *et al.*, 2014), pear (Yao *et al.*, 2017), strawberry (Lin-Wang *et al.*, 2014), apple (Espley *et al.*, 2007), and pomegranate (Ben-Simhon *et al.*, 2011). In addition, a MADS-box transcription factor gene of fig (*FcMADS9*) showed a positive regulatory role in anthocyanin biosynthesis and further showed that ethylene is involved in *FcMADS9*-promoted anthocyanin accumulation (Li *et al.*, 2021). However, the mechanism behind *FcMADS9*-regulated coloration still needs further study in fig.

12.5 Cell-Wall Dynamics and Softening

The fig is a highly perishable fruit subject to rapid physiological breakdown, which shortens the postharvest life. Profound cell-wall modification processes occur within the tissues during maturation (Chessa and Mitra, 1997). The softening process is primarily due to the change in cell-wall metabolism, resulting in a net decrease in certain structural components. As the fruit matures, there is slight decrease in polysaccharides and crude fibers. The cell wall of the fruit generally consists of pectin, hemicelluloses, and cellulose polysaccharides. Quantitative and qualitative changes in cell-wall polysaccharides occur during ripening in both receptacle and inflorescence tissues; qualitative variations between tissues occur only in the pectic polymers, not in the hemicellulosic polymers (Owino *et al.*, 2004a). Several pectin-degrading and modifying enzymes such as exo- and endopolygalacturonases (PGs), pectate lyases (PLs), pectin methylesterases (PMEs), galactosidases, and arabinofuranosidases may act synergistically within the cell wall to reduce the apparent molecular size of pectic polymers, leading to depolymerization and increased solubility of pectic polysaccharides. It is likely that fruits of all species have the same range of enzyme activities but that they are present at different levels (Hadfield and Bennett, 1998).

The effort to understand the molecular basis of softening in fig was started by isolating the complementary DNAs (cDNAs) responsible for cell-wall expansion and disassembly (Owino *et al.*, 2004b). In this study, cDNAs isolated from ripe Masui Dauphine fig variety encoded two divergent endo-1,4-β-glucanases (*FcCel1* and *FcCel2*) and three xyloglucan endotransglycosylases

(*FcXETl*, *FcXET2*, and *FcXET3*). Southern blot analyses indicated that the isolated xyloglucan endotransglycosylases (XETs) and endoglucanases exist as single-copy genes in the fig-fruit genome. Propylene stimulated the accumulation of *FcCel1* mRNA, while 1-MCP inhibited its accumulation, indicating that this gene is upregulated by ethylene. *FcXET1* mRNA accumulation was detected only in the 1-MCP-treated fruit, indicating that this gene is downregulated by ethylene. The *FcCel1* and *FcXET2* mRNAs showed more or less constitutive expression with both treatments, indicating that these genes are ethylene independent and are developmentally regulated. These results suggest that fig-fruit XETs and endoglucanases comprise gene families with divergent members showing differential regulation during fig-fruit ripening. A combination of ethylene and other developmental factors influence the expression of these genes, suggesting that multiple activities are required for the cooperative modification of the hemicellulose network during softening of fig fruit. The authors concluded that, similar to most studied fruit species, the gene products of the isolated 11 cDNAs, putatively encoding cell-wall-related enzymes, are coordinated in both time and amount during fig-fruit development and ripening, and act concertedly to achieve softening (Owino *et al.*, 2004b).

Softening of fig fruit during ripening has been reported to alter by pollination (caprification). In Tunisian figs, namely the variety Bouhouli, which is San Pedro type and Zidi, Thgagli, Bidhi, and Khedri, which are Smyrna types, the fruits were found to be firmer by caprification compared with non-caprified fruits (Trad *et al.*, 2014). However, the author has stressed the rapid softening process in caprified figs, which produces the same firmness in both caprified and non-caprified fruits at the full ripening stage. In another study, pollinated Brown Turkey fig fruits were reported to be firmer at commercial ripening stages with a thicker cell wall and to have an extended shelf life during cold storage compared with parthenocarpic fruits (Rosianski *et al.*, 2016b). At the molecular level, the expression pattern of softening-related enzyme-encoding genes of pollinated and parthenocapric Brown Turkey figs was studied in the tissue-specific transcriptomic data published previously by Rosianski *et al.* (2016a). In total 16 genes encoding different softening enzymes were identified and their expression patterns were characterized in the inflorescence and receptacle part of pollinated and parthenocarpic fig fruit (Fig. 12.7). Generally, high expression levels of the genes were noted at the matured green stages in parthenocarpic inflorescences compared with the pollinated ones. In particular, expansin-encoding genes (*FcPG1*, *FcPL1*, *FcXTH1*, *FcXTH2*, and *FcEXP2*) had relatively lower expression levels in pollinated inflorescences compared with parthenocarpic ones. The lower expression levels of these genes indicated the specific genes that are highly influenced by pollination. In the receptacle, *FcPG1* showed a low expression level in the pollinated fruits compared with the parthenocarpic fruits. Interestingly, at the full ripening stage (100% ripening), all 16 genes showed similar patterns of expression in the receptacle and inflorescence of pollinated and parthenocarpic fig fruits.

12.6 Concluding Remarks

In conclusion, the ripening process of matured fig fruit is tissue specific and mainly governed by the plant hormones ABA and ethylene. The role of ABA and ethylene in the onset of ripening and later in the process is to changes the sugar and organic acid accumulation to alter the fruit taste and anthocyanin biosynthesis to change the skin color during fig-fruit ripening (Fig. 12.8). The increased amount of ABA in the mature fig fruit at the end of phase II triggers ethylene production to allow the onset of ripening and continued increase of ABA production until the fruit is fully ripe. In particular, the ABA biosynthesis genes *FcNCED2* and *FcABA2* play key roles in the increased amount of ABA content, while the ethylene-biosynthesis genes *FcACS4* and *FcACOL2* play major roles in the production of ethylene to cause the onset of ripening of the fig fruit. Other genes from the MADS-box, NAC and ERF gene families are also involved during the onset of fig-fruit ripening.

The organic acid content of the fig fruit is decreased while the sugar content increases

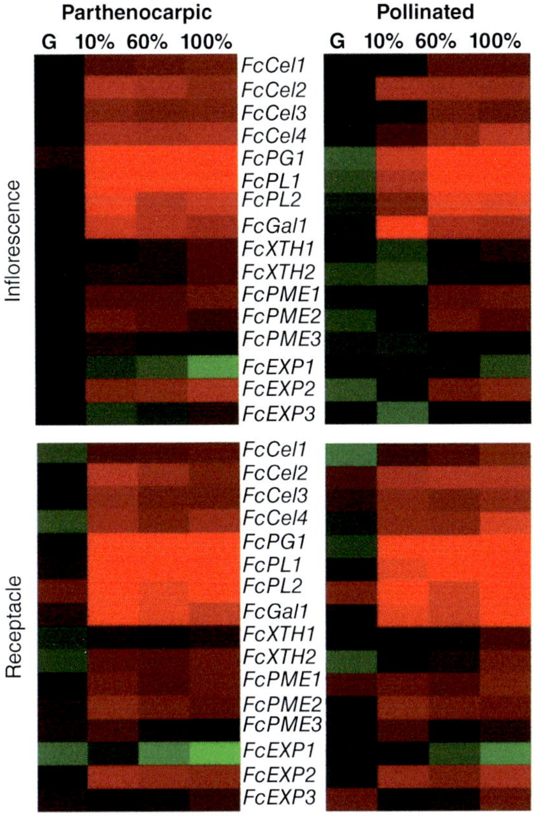

Fig. 12.7. Expression patterns of cell-wall-modifying enzyme-related genes in the inflorescence and receptacle tissues of pollinated and parthenocarpic fig fruits at different stages of maturity. G, mature green fruits; 10%, 60% and 100%, percentage of ripening. Red shading indicates higher expression levels and green shading indicates lower expression levels.

as the ripening process progresses. For a better acidic taste, pollination adds organic acids by increasing the expression of the glyoxylic cycle genes *FcMLS* and *FcICL*, mainly in the inflorescence of the fig fruit. Regarding the sweet taste, global analysis of sugar accumulation-related gene analysis predicted that the sugar metabolic pathway genes *FcCIN2* and *FcSUS2* and the sugar transporter gene *FcSWEET9* were the potentially crucial genes involved in rapid sugar accumulation during fig-fruit ripening. In addition, primary roles for the anthocyanin biosynthesis genes *FcDFR* and *FcANS* and the MBW complex gene *Fc-MYB113* were observed in anthocyanin accumulation and color development. Overall, we suggest that genome editing of the key ABA biosynthesis gene, *FcNCED2*, and the major ethylene-biosynthesis gene, *FcACS4*, in the future to alter the rapid ripening process of fig fruit, which may be helpful to extend the postharvest life. In addition, the relationship between pollination, organic acids, and the aroma of the fig fruit can be explored and breeding projects undertaken to develop early-ripening-stage, sugar-accumulating varieties. For commercial purposes, exogenous ABA might be used in the fig orchard to enhance fruit color and ripening, and to synchronize the ripening process in mature fig fruits after field tests based on varieties and climatic conditions.

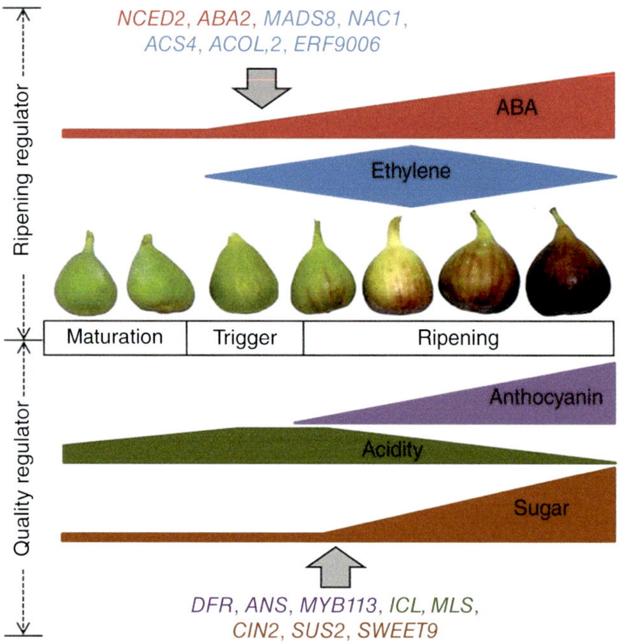

Fig. 12.8. Suggested model of abscisic acid (ABA)- and ethylene-regulated ripening and metabolic changes in fig fruit. Genes related to ABA, ethylene, anthocyanin, acidity, and sugar involved at the onset of ripening to change the fruit from the unripe to the ripe stage are shown in matching colors. The change in size of the frame indicates the amount of metabolic activity. *NCED2*, 9-*cis*-epoxycarotenoid dioxygenase 2; *ABA2*, short-chain alcohol dehydrogenase; *MADS8*, MADS-box gene; *ACS4*, 1-aminocyclopropane 1-carboxylate (ACC) synthase 4; *ACOL,2*, ACC oxidase-like and ACC oxidase 2; *ERF*, ethylene-responsive factor; *DFR*, dihydroflavonol 4-reductase; *ANS*, anthocyanidin synthase; *MYB113*, myeloblastosis 113; *ICL*, isocitrate lyase; *MLS*, malate synthase; *CIN2*, cytoplasmic neutral invertase 2; *SUS2*, sucrose synthases 2; *SWEET9*, Sugars Will Eventually be Exported Transporters 9.

References

Abdi, N., McGlasson, W.B., Holford, P., Williams, M., and Mizrahi, Y. (1998) Responses of climacteric and suppressed-climacteric plums to treatment with propylene and 1-methylcyclopropene. *Postharvest Biology and Technology* 14, 29–39.

Allan, A.C., Hellens, R.P., and Laing, W.A. (2008) MYB transcription factors that colour our fruit. *Trends in Plant Science* 13, 99–102.

Azzolini, M., Jacomino, A.P., Bron, I.U., Kluge, R.A., and Schiavinato, M.A. (2005) Ripening of 'Pedro Sato' guava: study on its climacteric or non-climacteric nature. *Brazilian Journal of Plant Physiology* 17, 299–306.

Ban, T., Ishimaru, M., Kobayashi, S., Goto-Yamamoto, N., and Horiuchi, S. (2003) Abscisic acid and 2,4-dichlorophenoxyacetic acid affect the expression of anthocyanin biosynthetic pathway genes in 'Kyoho' grape berries. *Journal of Horticultural Science and Biotechnology* 78, 586–589.

Ben-Simhon, Z., Judeinstein, S., Nadler-Hassar, T., Trainin, T., Bar-Ya'akov, I., *et al.* (2011) A pomegranate (*Punica granatum L.*) WD40-repeat gene is a functional homologue of *Arabidopsis TTG1* and is involved in the regulation of anthocyanin biosynthesis during pomegranate fruit development. *Planta* 234, 865–881.

Cantín, C.M., Fidelibus, M.W., and Crisosto, C.H. (2007) Application of abscisic acid (ABA) at veraison advanced red color development and maintained postharvest quality of 'Crimson Seedless' grapes. *Postharvest Biology and Technology* 46, 237–241.

Carrari, F., Baxter, C., Usadel, B., Urbanczyk-Wochniak, E., Zanor, M.-I., *et al.* (2006) Integrated analysis of metabolite and transcript levels reveals the metabolic shifts that underlie tomato fruit development and highlight regulatory aspects of metabolic network behavior. *Plant Physiology* 142, 1380–1396.

Celikel, F.G., Kaynas, K., Ozelkok, S., and Ertan, U. (1997) Effects of ethephon on fruit development and ripening of the fig (*Ficus carica* L.) variety Bursa Siyahi. *Acta Horticulturae* 441, 145–151.

Chen, L.-Q., Hou, B.-H., Lalonde, S., Takanaga, H., Hartung, M.L., *et al.* (2010) Sugar transporters for intercellular exchange and nutrition of pathogens. *Nature* 468, 527–532.

Chen, L.-Q., Cheung, L.S., Feng, L., Tanner, W. and Frommer, W.B. (2015) Transport of sugars. *Annual Review of Biochemistry* 84, 865–894.

Chernys, J.T. and Zeevaart, J.A.D. (2000) Characterization of the 9-*cis*-epoxycarotenoid dioxygenase gene family and the regulation of abscisic acid biosynthesis in avocado. *Plant Physiology* 124, 343–353.

Chessa, I. and Mitra, S.K. (1997) *Postharvest Physiology and Storage of Tropical and Subtropical Fruits*. CAB International, Wallingford, UK, pp. 245–268.

Chessa, I., Nieddu, G., and Schirra, M. (1992) Growth and ripening of main-crop fig. *Advances in Horticultural Science* 6, 112–115.

Crane, J.C., Marei, N., and Nelson, M.M. (1970) Growth and maturation of fig fruits stimulated by 2-chloroethylphosphonic acid. *Journal of the American Society of Horticultural Science* 95, 367–70.

Cui, Y., Wang, Z., Chen, S., Vainstein, A., and Ma, H. (2019) Proteome and transcriptome analyses reveal key molecular differences between quality parameters of commercial-ripe and tree-ripe fig (*Ficus carica* L.). *BMC Plant Biology* 19: 146.

d'Aquino, S., Palma, A., Dore, A., and Agabbio, M. (2003) Non conventional treatments to reduce figs decay. In: Tijskens, L.M.M. and Vollebregt, H.M. (eds) *International Conference on Quality in Chains. An Integrated View on Fruit and Vegetable Quality*. International Society for Horticultural Science, Wageningen, The Netherlands, pp. 817–821.

Downs, C.G., Brady, C.J., Campbell, J., and McGlasson, W.B. (1991) Normal ripening cultivars of *Pyrus serotina* are either climacteric or non-climacteric. *Scientia Horticulturae* 48, 213–221.

Dueñas, M., Pérez-Alonso, J.J., Santos-Buelga, C., and Escribano-Bailón, T. (2008) Anthocyanin composition in fig (*Ficus carica* L.). *Journal of Food Composition and Analysis* 21, 107–115.

Espley, R.V., Hellens, R.P., Putterill, J., Stevenson, D.E., Kutty-Amma, S., and Allan, A.C. (2007) Red colouration in apple fruit is due to the activity of the MYB transcription factor, MdMYB10. *Plant Journal* 49, 414–427.

Etienne, A., Génard, M., Lobit, P., Mbeguié-A-Mbéguié, D., and Bugaud, C. (2013) What controls fleshy fruit acidity? A review of malate and citrate accumulation in fruit cells. *Journal of Experimental Botany* 64, 1451–1469.

Etienne, C., Rothan, C., Moing, A., Plomion, C., Bodenes, C., *et al.* (2002) Candidate genes and QTLs for sugar and organic acid content in peach [*Prunus persica* (L.) Batsch]. *Theoretical and Applied Genetics* 105, 145–159.

Fait, A., Hanhineva, K., Beleggia, R., Dai, N., Rogachev, I., *et al.* (2008) Reconfiguration of the achene and receptacle metabolic networks during strawberry fruit development. *Plant Physiology* 148, 730–750.

Flaishman, M.A., Rodov, V., and Stover, E. (2008) *The Fig: Botany, Horticulture, and Breeding*. *Horticultural Reviews* 43, 113–196.

Freiman, Z.E., Rodov, V., Yablovitz, Z., Horev, B., and Flaishman, M.A. (2012) Preharvest application of 1-methylcyclopropene inhibits ripening and improves keeping quality of 'Brown Turkey' figs (*Ficus carica* L.). *Scientia Horticulturae* 138, 266–272.

Freiman, Z.E., Doron-Faigenboim, A., Dasmohapatra, R., Yablovitz, Z., and Flaishman, M.A. (2014) High-throughput sequencing analysis of common fig (*Ficus carica* L.) transcriptome during fruit ripening. *Tree Genetics and Genomes* 10, 923–935.

Freiman, Z.E., Rosianskey, Y., Dasmohapatra, R., Kamara, I., and Flaishman, M.A. (2015) The ambiguous ripening nature of the fig (*Ficus carica* L.) fruit: a gene-expression study of potential ripening regulators and ethylene-related genes. *Journal of Experimental Botany* 66, 3309–3324.

Gözlekçi, S., Erkan, M., Karaşahin, I., and Şahin, G. (2008) Effect of 1-methylcyclopropene (1-MCP) on fig (*Ficus carica* cv. Bardakci) storage. In: Leitao, J. and Neves, M.A. (eds) *Proceedings of the Third International Symposium on Fig, 16–20 May 2005, Vilamoura, Algarve, Portugal*. International Society for Horticultural Science, Leuven, Belgium, pp. 325–330.

Hadfield, K.A. and Bennett, A.B. (1998) Polygalacturonases: many genes in search of a function. *Plant Physiology* 117, 337–343.

Hubbard, N.L., Pharr, D.M., and Huber, S.C. (1990) Role of sucrose phosphate synthase in sucrose biosynthesis in ripening bananas and its relationship to the respiratory climacteric. *Plant Physiology* 94, 201–208.

Ikegami, H., Habu, T., Mori, K., Nogata, H., Hirata, C., *et al.* (2013) *De novo* sequencing and comparative analysis of expressed sequence tags from gynodioecious fig (*Ficus carica* L.) fruits: caprifig and common fig. *Tree Genetics and Genomes* 9, 1075–1088.

Jaakola, L. (2013) New insights into the regulation of anthocyanin biosynthesis in fruits. *Trends in Plant Science* 18, 477–483.

Jia, H.F., Chai, Y.M., Li, C.L., Lu, D., Luo, J.-J., *et al.* (2011) Abscisic acid plays an important role in the regulation of strawberry fruit ripening. *Plant Physiology* 157, 188–199.

Jiang, Y., Joyce, D.C., and Macnish, A.J. (2000) Effect of abscisic acid on banana fruit ripening in relation to the role of ethylene. *Journal of Plant Growth Regulation* 19, 106–111.

Kadomura-Ishikawa, Y., Miyawaki, K., Takahashi, A., Masuda, T., and Noji, S. (2015) Light and abscisic acid independently regulated *FaMYB10* in *Fragaria* × *ananassa* fruit. *Planta* 241, 953–965.

Koyama, R., Roberto, S.R., de Souza, R.T., Borges, W.F.S., Anderson, M., *et al.* (2018) Exogenous abscisic acid promotes anthocyanin biosynthesis and increased expression of flavonoid synthesis genes in *Vitis vinifera* × *Vitis labrusca* table grapes in a subtropical region. *Frontiers in Plant Science* 9: 323.

Kumar, R., Khurana, A., and Sharma, A.K. (2014) Role of plant hormones and their interplay in development and ripening of fleshy fruits *Journal of Experimental Botany* 65, 4561–4575.

Lama, K., Modi, A., Peer, R., Izhaki, Y., and Flaishman, M.A. (2019a) On-tree ABA application synchronizes fruit ripening and maintains keeping quality of figs (*Ficus carica* L.). *Scientia Horticulturae* 253, 405–411.

Lama, K., Yadav, S., Rosianski, Y., Shaya, F., Licheter, A., *et al.* (2019b) The distinct ripening processes in the reproductive and non-reproductive parts of the fig syconium are driven by ABA. *Journal of Experimental Botany* 70, 115–131.

Lama, K., Harlev, G., Shafran, H., Peer, R., and Flaishman, M.A. (2020a) Anthocyanin accumulation is initiated by abscisic acid to enhance fruit color during fig (*Ficus carica* L.) ripening. *Journal of Plant Physiology* 251: 153192.

Lama, K., Peer, R., Shlizerman, L., Meir, S., Doron-Feigenboim, A., *et al.* (2020b) Tissue-specific organic acid metabolism in reproductive and non-reproductive parts of the fig fruit is partially induced by pollination. *Physiologia Plantarum* 168, 133–147.

Lama, K., Chai, L.J., Peer, R., Yeselson, Y., Schaffer, A.A., and Flaishman, M.A. (2022) Extreme sugar accumulation in late fig ripening is accompanied by global changes in sugar metabolism and transporter gene expression. *Physiologia Plantarum* e13648.

Lara, I. and Vendrell, M. (2000) Development of ethylene-synthesizing capacity in preclimacteric apples: interaction between abscisic acid and ethylene. *Journal of the American Society for Horticultural Science* 125, 505–512.

Leng, P., Yuan, B., Guo, Y., and Chen, P. (2014) The role of abscisic acid in fruit ripening and responses to abiotic stress. *Journal of Experimental Botany* 65, 4577–4588.

Li, C., Jia, H., Chai, Y., and Shen, Y. (2011) Abscisic acid perception and signaling transduction in strawberry: a model for non-climacteric fruit ripening. *Plant Signaling and Behavior* 6, 1950–1953.

Li, D., Li, L., Luo, Z., Mou, W., Mao, L., and Ying, T. (2015) Comparative transcriptome analysis reveals the influence of abscisic acid on the metabolism of pigments, ascorbic acid and folic acid during strawberry fruit ripening. *PLoS One* 10: e0130037.

Li, J., Ma, N., An, Y., and Wang, L. (2021) *FcMADS9* of fig regulates anthocyanin biosynthesis. *Scientia Horticulturae* 278: 109820.

Li, M., Feng, F., and Cheng, L. (2012) Expression patterns of genes involved in sugar metabolism and accumulation during apple fruit development. *PLoS One* 7: e33055.

Lin, Z., Zhong, S., and Grierson, D. (2009) Recent advances in ethylene research. *Journal of Experimental Botany* 60, 3311–3336.

Lin-Wang, K., McGhie, T.K., Wang, M., Liu, Y., Warren, B., *et al.* (2014) Engineering the anthocyanin regulatory complex of strawberry (*Fragaria vesca*). *Frontiers in Plant Science* 5: 651.

Liu, Y., Tikunov, Y., Schouten, R.E., Marcelis, L.F.M., Visser, R.G.F., and Bovy, A. (2018) Anthocyanin biosynthesis and degradation mechanisms in solanaceous vegetables: a review. *Frontiers in Chemistry* 6: 52.

Ludewig, F. and Flügge, U.I. (2013) Role of metabolite transporters in source-sink carbon allocation. *Frontiers in Plant Science* 4: 231.

Ma, N., Feng, H., Meng, X., Li, D., Yang, D., *et al.* (2014) Overexpression of tomato *SlNAC1* transcription factor alters fruit pigmentation and softening. *BMC Plant Biology* 14: 531.

McCollum, G. and Maul, P. (2007) 1-Methylcyclopropene inhibits degreening but stimulates respiration and ethylene biosynthesis in grapefruit. *HortScience* 42, 120–124.

Miron, D., Petreikov, M., Carmi, N., Shen, S., Levin, I., *et al.* (2002) Sucrose uptake, invertase localization and gene expression in developing fruit of *Lycopersicon esculentum* and the sucrose-accumulating *Lycopersicon hirsutum*. *Physiologia Plantarum* 115, 35–47.

Mou, W., Li, D., Bu, J., Jiang, Y., Khan, Z.U., *et al.* (2016) Comprehensive analysis of ABA effects on ethylene biosynthesis and signaling during tomato fruit ripening. *PLoS One* 11: e0154072.

Nguyen-Quoc, B. and Foyer, C.H. (2001) A role for 'futile cycles' involving invertase and sucrose synthase in sucrose metabolism of tomato fruit. *Journal of Experimental Botany* 52, 881–889.

Obando-Ulloa, J.M., Jowkar, M.M., Moreno, E., Souri, M.K., Martinez, J.A., *et al.* (2009) Discrimination of climacteric and non-climacteric melon fruit at harvest or at the senescence stage by quality traits. *Journal of the Science of Food and Agriculture* 89, 1743–1753.

Oliveira, A.P., Valentão, P., Pereira, J.A., Silva, B.M., Tavares, F., and Andrade, P.B. (2009) *Ficus carica* L.: metabolic and biological screening. *Food and Chemical Toxicology* 47, 2841–2846.

Osorio, S., Ruan, Y.L., and Fernie, A.R. (2014) An update on source-to-sink carbon partitioning in tomato. *Frontiers in Plant Science* 5: 516.

Owino, W.O., Nakano, R., Kubo, Y., and Inaba, A. (2004a) Alterations in cell wall polysaccharides during ripening in distinct anatomical tissue regions of the fig (*Ficus carica* L.) fruit. *Postharvest Biology and Technology* 32, 67–77.

Owino, W.O., Nakano, R., Kubo, Y., and Inaba, A. (2004b) Coordinated expression patterns of genes encoding cell wall modifying enzymes during ripening in distinct anatomical tissue regions of the fig (*Ficus carica* L.) fruit. *Postharvest Biology and Technology* 32, 253–261.

Owino, W.O., Manabe, Y., Mathooko, F.M., Kubo, Y., and Inaba, A. (2006) Regulatory mechanisms of ethylene biosynthesis in response to various stimuli during maturation and ripening in fig fruit (*Ficus carica* L.). *Plant Physiology and Biochemistry* 44, 335–342.

Pande, G. and Akoh, C.C. (2010) Organic acids, antioxidant capacity, phenolic content and lipid characterisation of Georgia-grown underutilized fruit crops. *Food Chemistry* 120, 1067–1075.

Paul, V., Pandey, R., and Srivastava, G.C. (2012) The fading distinctions between classical patterns of ripening in climacteric and non-climacteric fruit and the ubiquity of ethylene – an overview. *Journal of Food Science and Technology* 49, 1–21.

Périn, C., Gomez-Jimenez, M., Hagen, L., Dogimont, C., Pech, J.-C., *et al.* (2002) Molecular and genetic characterization of a non-climacteric phenotype in melon reveals two loci conferring altered ethylene response in fruit. *Plant Physiology* 129, 300–309.

Pott, D.M., Osorio, S., and Vallarino, J.G. (2019) From central to specialized metabolism: an overview of some secondary compounds derived from the primary metabolism for their role in conferring nutritional and organoleptic characteristics to fruit. *Frontiers in Plant Science* 10: 835.

Pua, E.C., Chandramouli, S., Han, P., and Liu, P. (2003) Malate synthase gene expression during fruit ripening of Cavendish banana (*Musa acuminata* cv. Williams). *Journal of Experimental Botany* 54, 309–316.

Rahim, M.A., Busatto, N., and Trainotti, L. (2014) Regulation of anthocyanin biosynthesis in peach fruits. *Planta* 240, 913–929.

Rosianski, Y., Doron-Faigenboim, A., Freiman, Z.E., Lama, K., Milo-Kochavi, S., *et al.* (2016a) Tissue-specific transcriptome and hormonal regulation of pollinated and parthenocarpic fig (*Ficus carica* L.) fruit suggest that fruit ripening is coordinated by the reproductive part of the syconium. *Frontiers in Plant Science* 7: 1696.

Rosianski, Y., Freiman, Z.E., Cochavi, S.M., Yablovitz, Z., Kerem, Z., and Flaishman, M.A. (2016b) Advanced analysis of developmental and ripening characteristics of pollinated common-type fig (*Ficus carica* L.). *Scientia Horticulturae* 198, 98–106.

Ruan, Y.-L. (2014) Sucrose metabolism: gateway to diverse carbon use and sugar signaling. *Annual Review of Plant Biology* 65, 33–67.

Seymour, G.B., Taylor, J., and Tucker, G.A. (1993) *Biochemistry of Fruit Ripening*. Chapman & Hall, London.

Seymour, G.B., Ostergaard, L., Chapman, N.H., Knapp, S., and Martin, C. (2013) Fruit ripening and development. *Annual Review of Plant Biology* 64, 219–241.

Shan, W., Kuang, J.F., Chen, L., Xie, H., Peng, H.-H., *et al.* (2012) Molecular characterization of banana NAC transcription factors and their interactions with ethylene signalling component EIL during fruit ripening. *Journal of Experimental Botany* 63, 5171–5187.

Shen, X., Zhao, K., Liu, L., Zhang, K., Yuan, H., *et al.* (2014) A role for PacMYBA in ABA-regulated anthocyanin biosynthesis in red-colored sweet cherry cv. Hong Deng (*Prunus avium* L.). *Plant and Cell Physiology* 55, 862–880.

Shiraishi, S.C., Kawakami, K., Widodo, S.E., Shiraishi, M., and Makiko, K. (1996) Organic acid profiles in the juice of fig fruits. *Journal of the Faculty of Agriculture, Kyushu University* 41, 29–33.

Slatnar, A., Klancar, U., Stampar, F., and Veberic, R. (2011) Effect of drying of figs (*Ficus carica* L.) on the contents of sugars, organic acids, and phenolic compounds. *Journal of Agricultural and Food Chemistry* 59, 11696–11702.

Solomon, A., Golubowicz, S., Yablowicz, Z., Grossman, S., Bergman, M., *et al.* (2006) Antioxidant activities and anthocyanin content of fresh fruits of common fig (*Ficus carica* L.). *Journal of Agricultural and Food Chemistry* 54, 7717–7723.

Sozzi, G.O., Abraján-Villaseñor, M.A., Trinchero, G.D., and Fraschina, A.A. (2005) Postharvest response of 'Brown Turkey' figs (*Ficus carica* L.) to the inhibition of ethylene perception. *Journal of the Science of Food and Agriculture* 85, 2503–2508.

Sweetman, C., Deluc, L.G., Cramer, G.R., Ford, C.M., and Soole, K.L. (2009) Regulation of malate metabolism in grape berry and other developing fruits. *Phytochemistry* 70, 1329–1344.

Takayanagi, T. and Yokotsuka, K. (1997) Relationship between sucrose accumulation and sucrose-metabolizing enzymes in developing grapes. *American Journal of Enology and Viticulture* 48, 403–407.

Takos, A.M., Ubi, B.E., Robinson, S.P., and Walker, A.R. (2006) Condensed tannin biosynthesis genes are regulated separately from other flavonoid biosynthesis genes in apple fruit skin. *Plant Science* 170, 487–499.

Terrier, N., Glissant, D., Grimplet, J., Barrieu, F., Abbal, P., *et al.* (2005) Isogene specific oligo arrays reveal multifaceted changes in gene expression during grape berry (*Vitis vinifera* L.) development. *Planta* 222, 832–847.

Tohge, T., Alseekh, S., and Fernie, A.R. (2014) On the regulation and function of secondary metabolism during fruit development and ripening. *Journal of Experimental Botany* 65, 4599–4611.

Trad, M., Gaaliche, B., Renard, C., and Mars, M. (2012) Quality performance of 'Smyrna' Type figs grown under Mediterranean conditions of Tunisia. *Journal of Ornamental and Horticulture Plants* 3, 139–146.

Trad, M., Le Bourvellec, C., Gaaliche, B., Renard, C.M.G.C., and Mars, M. (2014) Nutritional compounds in figs from the southern Mediterranean region. *International Journal of Food Properties* 17, 491–499.

Vemmos, S.N., Petri, E., and Stournaras, V. (2013) Seasonal changes in photosynthetic activity and carbohydrate content in leaves and fruit of three fig cultivars (*Ficus carica* L.). *Scientia Horticulturae* 160, 198–207.

Wang, Z., Cui, Y., Vainstein, A., Chen, S., and Ma, H. (2017) Regulation of fig (*Ficus carica* L.) fruit color: metabolomic and transcriptomic analyses of the flavonoid biosynthetic pathway. *Frontiers in Plant Science* 8: 1990.

Wei, Y.Z., Hu, F.C., Hu, G.B., Li, X.J., Huang, X.-M., and Wang, H.-C. (2011) Differential expression of anthocyanin biosynthetic genes in relation to anthocyanin accumulation in the pericarp of litchi chinensis sonn. *PLoS One* 6: e19455.

Yao, G., Ming, M., Allan, A.C., Gu, C., Li, L., *et al.* (2017) Map-based cloning of the pear gene *MYB114* identifies an interaction with other transcription factors to coordinately regulate fruit anthocyanin biosynthesis. *Plant Journal* 92, 437–451.

Yao, Y.X., Li, M., Liu, Z., You, C.X., Wang, D.-M., *et al.* (2009) Molecular cloning of three malic acid related genes *MdPEPC*, *MdVHA-A*, *MdcyME* and their expression analysis in apple fruits. *Scientia Horticulturae* 122, 404–408.

Zhang, M., Yuan, B., and Leng, P. (2009) The role of ABA in triggering ethylene biosynthesis and ripening of tomato fruit. *Journal of Experimental Botany* 60, 1579–1588.

13 Harvest and Postharvest Physiology and Technology of Fresh Fig Fruit

Eleni Tsantili[1]* and Victor Rodov[2]

[1]*Laboratory of Pomology, Agricultural University of Athens, Athens, Greece;*
[2]*Department of Postharvest Science, Agriculture Research Organization, The Volcani Institute, Rishon LeZion, Israel*

13.1 Introduction

Fig fruit is an important crop due to its attractiveness in appearance, taste, and nutritional value (Flaishman *et al.*, 2008). Figs are rich in monosaccharides, organic and amino acids (Byeon and Lee, 2020a), minerals such as iron, calcium, and potassium (Soni *et al.*, 2014), fiber (Vinson *et al.*, 2005; Soni *et al.*, 2014), phenolic antioxidants (Solomon *et al.*, 2006; Veberic and Mikulic-Petkovsek, 2016; Arvaniti *et al.*, 2019), and phytosterols (Pande and Akoh, 2010). At the same time, they are cholesterol free and low in sodium and fat (Vinson *et al.*, 2005; Solomon *et al.*, 2006). During the last decade, there has been an increasing interest of consumers and researchers in fig fruit due to its health benefits related to prevention of plasma lipoprotein oxidation, neurodegenerative disorders, cardiovascular problems, certain types of cancer, and other diseases. These benefits of figs are attributed primarily to phenolic acids and flavonoids (Vinson *et al.*, 2005; Solomon *et al.*, 2006, 2010).

Fresh figs are among the most perishable commodities, and this is the main reason why the fig fruit is mostly consumed in a dried form worldwide (Aksoy, 1997). However, the fresh fruits are more attractive for consumers than the dried ones and provide a much higher income for farmers. Postharvest technologies (Belay *et al.*, 2016; Beaudry, 2017), innovative packaging materials for modified-atmosphere packaging (MAP) (Beaudry, 2017; Wilson *et al.*, 2019), and treatments to delay ripening (Watkins, 2008; Doerflinger *et al.*, 2019; Opara *et al.*, 2019) are being developed in order to store the fruits in fresh form for a longer time with quality retention.

The keeping quality of fig fruit is determined by its structural and developmental peculiarities. Fig is an aggregate fruit known as a syconium, a composite inflorescence/infructescence comprising pulp made up of numerous parthenocarpic or non-parthenocarpic drupelets, enclosed within a non-climacteric succulent receptacle covered with green, yellow, red, purple, or black skin. The syconium communicates with the external environment through an aperture surrounded by scales called the ostiole (Freiman *et al.*, 2015).

There are three types of fig varieties bearing edible fruit. The first is the Smyrna type producing one crop (main crop) annually, which needs pollination (also called caprification). The second is the San Pedro type, with biannual crops, the first being parthenocarpic (breba), while the second (main crop) needs pollination. In the third type, the common type, the main

* Email: ekotsantili@gmail.com

 Advances in Fig Research and Sustainable Production (M.A. Flaishman and U. Aksoy)
DOI: 10.1079/9781789242492.0013

crop does not necessarily need pollination and usually develops parthenocarpically (Flaishman *et al.*, 2008). Breba figs are set on the previous year's twigs and ripen in late spring, whereas those of the main crop ripen on the current season's shoots and ripen in summer and/or autumn (Crisosto *et al.*, 2011). Certain antagonism exists between breba and the main crop, as well as between the main crop and both the current year's vegetative growth and brebas of the next year's crop (Gaaliche *et al.*, 2011a). Among the best-known varieties, Brown Turkey and Mission belong to the common type, Dauphine and King to the San Pedro type, and Sarılop (= Calimyrna), Black Bursa, and Zidi to the Smyrna type (Condit, 1955; Flaishman *et al.*, 2008). Pollination in figs is a unique process involving the symbiotic fig wasp, *Blastophaga psenes* (Galil and Neeman, 1977).

This chapter gives an overview of factors affecting the postharvest quality and longevity of fresh figs, and good practices during fruit harvest, storage, transportation, and marketing in order to answer the consumer demand for high-quality produce.

13.2 Factors Affecting Fig Quality at Harvest

13.2.1 Fruit growth and ripening physiology

Fig-fruit development follows a double-sigmoidal curve, consisting of three phases, with the size increase rate being rapid during phase I, very slow during phase II, and very rapid during phase III (Marei and Crane, 1971; Puech *et al.*, 1976). In phase II, the visual changes are limited, whereas in phase III the fruit ripens and changes size, color, and firmness; it accumulates sugars and obtains all the characteristics that make it edible and tasty. Here, a concise description of ripening is presented in order to connect the maturity stage with the fruit quality at harvest and during storage, and with its responses to pre- and postharvest treatments.

Fruits can be classified into climacteric and non-climacteric groups. Concisely, climacteric is the final physiological process that marks the end of fruit maturation and the beginning of fruit senescence. Its defining point is a sudden rise in respiration of the fruit, and this normally takes place without any external influences. After the climacteric period, respiration rate (RR) (measured as carbon dioxide (CO_2) production or oxygen (O_2) uptake) returns to the pre-climacteric level or below. By contrast, non-climacteric fruits do not show different rates of respiration between maturation and ripening. Ethylene production exhibits a pattern similar to RR in each fruit group during ripening; the burst of ethylene production in climacteric fruits occurs more or less concomitantly with their peak of RR (Biale *et al.*, 1981; Abeles *et al.*, 1993). Fruits such as apples, bananas, apricots, and tomatoes, among others, are considered climacteric fruits, whereas citrus, grapes, and strawberries are non-climacteric (i.e. they ripen without ethylene and respiration bursts). After the climacteric event, fruits are more susceptible to fungal invasion and start to degrade by cell death (Alexander and Grierson, 2002; Watkins, 2002). Figs are classified as climacteric fruits (Marei and Crane, 1971; Owino *et al.*, 2006; Freiman *et al.*, 2012), although some tissues within the syconium demonstrate non-climacteric behavior (Lama *et al.*, 2019). Ethylene or ethylene-releasing formulation (ethephon) promotes fig growth, as shown in the Mission (Marei and Crane, 1971; Puech *et al.*, 1976) and Bursa Black (Çelikel, *et al.*, 1997) varieties. The peak of ethylene production in Brown Turkey fig occurs before or at the beginning of phase III during fruit development (Freiman *et al.*, 2012). This stands in contrast to other climacteric fruits where ethylene peaks after the fruit has attained its final size, in phase III. Thus, figs are harvested at or after the ethylene peak, in the climacteric or postclimacteric state, respectively (Fig. 13.1) (Freiman *et al.*, 2012). The commercial-maturity or commercial-ripe stage precedes the fully ripe or tree-ripe one, which is characterized by optimum eating quality. However, figs are harvested at the commercial-maturity stage because at this stage they are less susceptible to transport and storage damage, although inferior to the tree-ripe stage in flavor and color. Lately, it has been demonstrated that the fig fruit consists of vegetative and reproductive parts with non-climacteric and climacteric behavior, respectively (Freiman *et al.*, 2012, 2015). Although ripening seemed to be coordinated by the ethylene production of

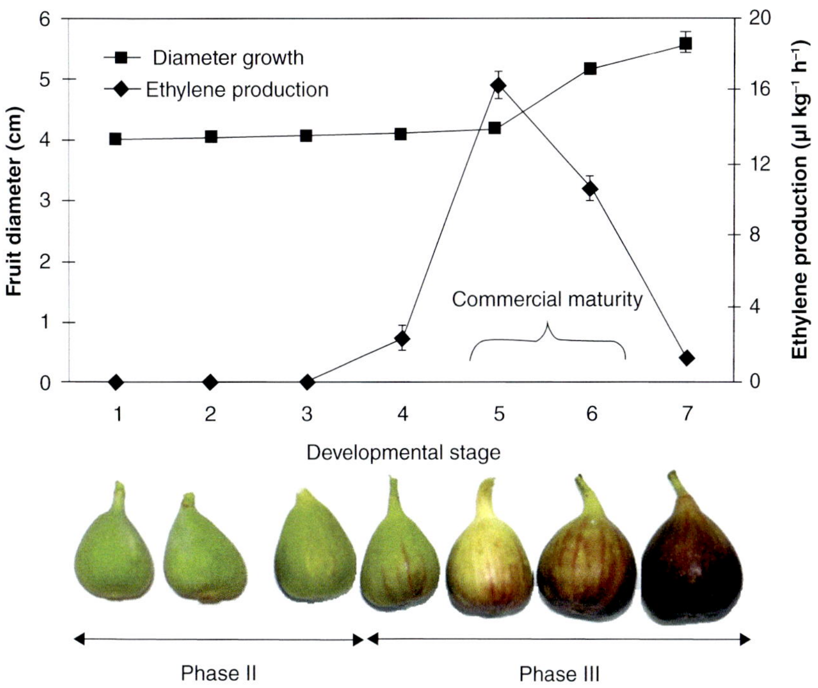

Fig. 13.1. Development, growth and ethylene production of the Brown Turkey fig-fruit variety. The commercial-maturity stages are indicated. Reproduced and modified from Freiman *et al.* (2012) with permission.

the reproductive part of the fruit (Freiman *et al.*, 2015; Rosianski *et al.*, 2016a), abscisic acid (ABA) regulates the ripening of both tissue types (Lama *et al.*, 2019), and can initiate anthocyanin synthesis (Lama *et al.*, 2020). However, preharvest application of an ethylene-releasing compound (ethephon) resulted in upregulation of the structural genes of anthocyanin synthesis in the receptacle but downregulation in the female flowers (Cui *et al.*, 2021). Qiao *et al.* (2021) also proposed an interaction between ethylene and ABA. These fig responses render the fruit particularly interesting for studies on ripening physiology, but at the same time may contribute to certain difficulties in postharvest management.

Rates of the physiological processes of respiration and ethylene production by fruits affect their postharvest behavior. Figs are classified as commodities with moderate ethylene production and RRs, being, at 20°C, within the ranges of 1–10 μl ethylene kg^{-1} h^{-1} (Kader, 2002b) and 40–60 mg CO_2 kg^{-1} h^{-1} (Gross *et al.*, 2016), respectively. For comparison, a highly respiring commodity such as passion fruit has, at the same temperature, an ethylene production rate >100 μl ethylene kg^{-1} h^{-1} (Kader, 2002b) and a RR of 175–349 mg CO_2 kg^{-1} h^{-1} (Gross *et al.*, 2016). Many climacteric fruits, such as apples, apricots, and tomatoes, demonstrate RR ranges similar to or even slightly lower than figs (Gross *et al.*, 2016) but produce much more ethylene, in the range 10–100 μl kg^{-1} h^{-1} (Kader, 2002b). Ethylene production and the RR of fig fruit depend on variety, stage of maturity, and production year, as in other fruits (Crisosto *et al.*, 2017). In particular, at the commercial-ripe stage, both processes exhibit higher values than at the following tree-ripe stage (Fig. 13.1) (Crisosto *et al.*, 2010a; Freiman *et al.*, 2012, 2015; Crisosto *et al.*, 2017). In a 2-year study, tree-ripe Mission figs, at 20°C, showed ethylene production rates of 3.33 and 10.67 μl kg^{-1} h^{-1} in the first and second year, respectively. The corresponding RRs were approximately 65.0 and 142.0 mg CO_2 kg^{-1} h^{-1} (Crisosto *et al.*, 2017) or, upon transformation, according to Gross *et al.* (2016), 36.36 and 78.67 ml kg^{-1} h^{-1}, respectively. Even higher RRs, at 20°C, of 161 mg kg^{-1} h^{-1} (80.0 ml kg^{-1} h^{-1}) were reported by Tsantili *et al.* (2003) for Mavra Markopoulou figs.

Ripening occurs very rapidly at phase III (Tsantili, 1990; Chessa, 1997; Freiman *et al.*, 2014). From a biochemical viewpoint, the ripe figs are characterized by catabolism of most amino acids (Freiman *et al.*, 2014; Byeon and Lee, 2020a), an increase in anthocyanin synthesis and ascorbate oxidase activity, and a decrease in wax synthesis (Freiman *et al.*, 2014). In addition, they demonstrate upregulation of sucrose synthase, sucrose-phosphate synthase, hexokinase, endoglucanase, expansin, β-galactosidase, pectin esterase, and aquaporins at the transcriptional and/or protein level (Cui *et al.*, 2019). Compared with pre-ripe fruit, the figs that reach the fully ripe stage on the tree (tree-ripe) are typically characterized by increased weight, size, and sugar content, advanced color, decreased firmness, and higher sensitivity to mechanical stress (Crisosto *et al.*, 2010a, 2017; Cui *et al.*, 2019; Byeon and Lee, 2021). At the same time, they often demonstrate side cracking, ostiole-end splitting, juice exudation from the ostiole (Kong *et al.*, 2013), and increased disease susceptibility (Chessa, 1997; Villalobos *et al.*, 2016a). Therefore, the choice of optimal harvest maturity is a matter of compromise between the desired eating quality and postharvest lifespan. Harvest season also affects fruit traits. Although at the same ripening stage, late-harvested figs are often softer, and have a higher sugar content, lower acidity, and shorter shelf life than fruit harvested earlier in the season (Rodov *et al.*, 2002; Aksoy *et al.*, 2003a; Crisosto *et al.*, 2017).

13.2.2 Hormonal regulation

Postharvest treatments with the inhibitor of ethylene perception 1-methylcyclopropene (1-MCP) have been proved effective in retarding ripening of many fruits (Blankenship and Dole, 2003; Watkins, 2008) and therefore, 1-MCP has been used commercially for storage of various crops and especially apples (Doerflinger *et al.*, 2019). However, postharvest 1-MCP-treated figs exhibited a somewhat reduced softening rate (Sozzi *et al.*, 2005; Villalobos *et al.*, 2018a; Song *et al.*, 2019). Indeed, these treatments showed limited or no effect on ripening, as the harvested fruit were close to or had already passed the peak of ethylene production (Sozzi *et al.*, 2005; Gözlekçi, *et al.*, 2008). Actually, 1-MCP treatment at the ripening onset, either on detached fruit (Sozzi *et al.* 2005; Owino *et al.* 2006) or when still on the tree (Freiman *et al.* 2012), resulted in a sharp rise in ethylene production as a response of the non-climacteric tissue (McCollum and Maul, 2007). In one reported case, Brown Turkey figs were treated with 5 μl l^{-1} 1-MCP at the preharvest state after the onset of chlorophyll degradation (with fruits enclosed in bags for 1-MCP application). The fruits harvested at the commercial-ripe stage and stored at 1–2°C exhibited a slower ripening rate and maintained a better quality, resulting in 7-day extension of storage compared with the control (Fig. 13.2) (Freiman *et al.*, 2012). According to the preliminary results reported by the same authors, applying the preharvest 1-MCP treatment to a whole branch with fruits at various stages of maturity had no adverse effect on their development and tended to shorten the harvest period. This change might enable decreased harvesting frequency and reduced labor cost. Recently, a formulation for preharvest 1-MCP application has been released under the commercial name Harvista (Doerflinger *et al.*, 2019) but has not been tried on figs so far.

The peculiar hormonal mechanism of fig ripening has been deciphered in a series of intensive studies (Freiman *et al.*, 2014, 2015). Ethylene-synthesis patterns of expression and signal-transduction genes were in accordance with other climacteric fruits during ripening, at least in the flesh, suggesting the climacteric pattern of these tissues. However, at the onset of climacteric period, there was one gene in the receptacle (*FcERF*, involved in the ethylene response) that was upregulated after preharvest 1-MCP treatment, and its transcripts remained at higher levels than untreated fruit for several days. This gene was considered responsible for ethylene synthesis in the non-climacteric tissue of the receptacle in response to 1-MCP (Freiman *et al.*, 2015). Further studies have demonstrated that growth in both pollinated and parthenocarpic figs is parallel to increases in auxin and ethylene synthesis with no gibberellin or cytokinin activity (Rosianski *et al.*, 2016a).

At the same time, ripening is regulated by the reproductive part (flesh) of the syconium through hormonal development of seeds, and in

Fig. 13.2. Effect of preharvest 1-MCP treatment on the appearance of the Brown Turkey fig variety after 12 days of storage at 1–2°C and 2 days of shelf life at 1–2°C. The fruit were harvested at commercial maturity. Left: 1-MCP-treated fruit, right: untreated fruit. From Freiman *et al.* (2012), with permission.

particular by ABA, which showed an increase shortly before the ethylene peak. Indeed, proteomic and transcriptomic analyses by Cui *et al.* (2019) on the receptacle of Brown Turkey figs showed a decreased ability of ethylene synthesis at the fully ripe stage. At the same time, ABA synthesis was also increased in parthenocarpic and pollinated figs but reached much higher levels in the latter (Rosianski *et al.*, 2016a). By comparison, pollination of the same variety, which is a common-type fig, retarded ripening and senescence and extended the postharvest life. Therefore, due to hormonal differences, ripening is slower in pollinated than in non-pollinated figs (Rosianski *et al.*, 2016b). It was further confirmed that ABA is responsible for ethylene production and regulation of ripening (Lama *et al.*, 2019). However, in the receptacle, ABA levels increased simultaneously with the low expression of ethylene-synthesis genes, and therefore a primary role of ABA was suggested to be a restraining factor in ripening, as in non-climacteric tissues (Lama *et al.*, 2019). The coexistence of both climacteric and non-climacteric tissues in the fruit has also been observed in banana (Inaba, 2007). Nevertheless, the receptacle role for fruit set and abscission in fig is not negligible, as found in a study on the main crop of San Pedro-type figs (Chai *et al.*, 2019). Moreover, comparison between the first (parthenocarpic) and the second (pollinated) crop of the same variety by transcriptome analysis showed diverged expression of hormone-related genes (Ikegami *et al.*, 2013; Chai *et al.*, 2017).

13.2.3 Genotype effects on fruit morphology

Undoubtedly, all fruit characteristics are affected by genotype. Researchers make efforts to select genotypes that combine intense flavor at the commercial harvest stage with good postharvest sustainability (King *et al.*, 2012; Crisosto *et al.*, 2017). Figs of large size and high weight are preferred, unless they become less firm and more susceptible to handling damage. On average, fig-fruit weight varies from 30 to 70 g. In a selection of 31 varieties for the fresh market from western Turkey the weight ranged from 34 to 75 g (Aksoy *et al.*, 2003a). The weight ranges were 20–60 g in ten Californian varieties (Crisosto *et al.*, 2011), 37–90 g in seven varieties from Greece (Pisimisi *et al.*, 2012), and 34–96 g in 17 varieties from Tunisia (Gaaliche *et al.*, 2012). For wild figs found in Iran, the fruit weight ranges were 0.6–28.7 g in 92 genotypes of edible figs, and 0.12 to 8.25 g in 68 genotypes of caprifigs (Mirheidari *et al.*, 2020).

The fruit length may vary from 25 to 100 mm (Crisosto *et al.*, 2011). Among both Turkish and Tunisian selections, fruit length varied between 35 and 64 mm, and width (transverse diameter) between 36 and 61 mm (Aksoy *et al.*, 2003a; Gaaliche *et al.*, 2012). In some varieties, a sort of "neck" exists, averaging usually less than 10 mm, but there are exceptions, with some neck lengths reaching about 21 mm (Aksoy *et al.*, 2003a). Pear-shaped fruits are considered more attractive visually than oblate (Gaaliche *et al.*, 2012). The ratio of width to length may vary between 0.47 and 1.30 (Çalişkan and Polat, 2008). The weight and size also depend on environmental conditions and cultivation practices. For example, the weight of Bursa Black figs ranged from 50 g (Çalişkan and Polat, 2008) to 88 g (Ersoy *et al.*, 2017), while the width was from 43 to 58 mm, respectively, and the length from 46 to 58 mm, respectively. Trad *et al.* (2013a) observed that high temperature promotes fig weight and size but reduces firmness and titratable acidity (TA). According to Crisosto *et al.* (2017), there was no significant difference in weight and size between commercial and tree-ripe figs. However, these results are not consistent with other reports, which have stated that width and weight increased significantly from the commercial to the tree-ripe stage in Brown Turkey cultivated in China (Cui *et al.*, 2019).

13.2.4 Pollination (caprification)

Pollinated figs are characterized by a slower ripening rate than non-pollinated figs of the same variety and have different quality traits, as observed for Brown Turkey figs, which belong to the common type. Pollination usually results in relatively larger fruits (Rodov *et al.*, 2005; Flaishman *et al.*, 2008; Gaaliche *et al.*, 2011b; Rosianski *et al.*, 2016b), as mentioned earlier. However, seedless Smyrna-type Calimyrna figs sprayed with indole 3-butyric acid were similar in size and in Brix or soluble solids content (SSC) to pollinated ones (Crane and Blondeau, 1949). SSC, which includes soluble sugars, acids, and many other soluble compounds in the fruit juice, increases as the maturity stage advances and therefore is used as a maturity index. In pollinated figs, seed development and the inflorescence swelling fill the internal cavity of the syconium, exhibit slow ripening (Rosianski *et al.*, 2016b), and promote cell thickening (Trad *et al.*, 2014; Rosianski *et al.*, 2016b). In Brown Turkey figs, these changes resulted in increased firmness, better storability, a lower tendency to split, and reduced mold growth compared with the non-pollinated fruit (Rosianski *et al.*, 2016b). These observations are in contrast to the moderate decrease in firmness after pollination of the Tunisian varieties Zidi and Thgagli (Smyrna type) and Bouhouli (San Pedro type), as reported by Trad *et al.* (2013b, 2014).

Pollinated figs have higher polyphenol concentrations and darker flesh and skin color, with more distinct stripes in the skin, a superior flavor (Flaishman *et al.*, 2008; Gaaliche *et al.*, 2011a; Trad *et al.*, 2013b; Rosianski *et al.*, 2016b), and a more intense aroma (Trad *et al.*, 2012). In addition, pollinated figs exhibit reduced apocarotenoid volatile production than non-pollinated figs, in contrast to carotenoid pigments, which are in higher amounts in pollinated figs, by comparison (Nawade *et al.*, 2020).

Pollination increased TA and slightly decreased SSC in Brown Turkey figs (Rosianski *et al.*, 2016b) but had no effect regarding these parameters in Tunisian varieties (Gaaliche *et al.*, 2011b; Trad *et al.*, 2012). Frequent hand pollination (with 20 profici caprifigs per tree, each time for up to four repeated times at 4-day intervals) improved fruit weight, which reached 112 g in Bouhouli and 102 g in Zidi, and increased the number of fertile seeds per fruit from 220 in the control to 804 seeds in the pollinated fruit (Gaaliche *et al.*, 2011c). However, increasing the number of seeds can result in ostiole splitting (Ferguson *et al.*, 1990; Crisosto *et al.*, 2010a) or a fully open ostiole (Zare, 2008). Splitting propensity depends on variety (Baskaya and Crane, 1950), and can be reduced by auxin treatment (Crane and Blondeau, 1949). Thus, a combination of optimal pollination frequency with a sufficient quantity of pollen from good-quality profichi is recommended (Gaaliche *et al.*, 2011c).

13.3 Quality Parameters of Fig Fruit

13.3.1 Quality evaluation methodology

Fruit quality, according to Kader (2002a), can be defined as the combination of all characteristics determining their value to the consumer. A number of attributes comprise the overall quality of a fig fruit: (i) organoleptic properties evaluated by sensory analyses, such as taste, aroma, and appearance (Aksoy *et al.*, 2003a; King *et al.*, 2012); (ii) objective characteristics, such as firmness, color, SSC, TA, phytonutrient content, and antioxidant capacity (Crisosto *et al.*, 2010a, 2017); these objective characteristics are typically measured by instrumental tests; and (iii) undesirable traits that reduce fruit value, in severe cases rendering it unmarketable or non-usable, including microbial contamination and decay indices (Villalobos *et al.*, 2017a), as well as fermentation (Colelli *et al.*, 1991).

Flavor, sweetness, skin color, and texture were found to be the main sensory attributes for quality evaluation of Dottato and Melanzana (Sortino *et al.*, 2017). King *et al.* (2012) developed a detailed sensory system, using the answers of a trained group of panelists, and described 12 varieties harvested at both the commercial and tree-ripe stages. Color is described by hue angle and C* (chroma), which have a good correlation with visual estimation. In particular, L^* (lightness) varies from 0 (black) up to +100 (white), a^* from −100 (green) to 100 (red), and b^* from −100 (blue) to 100 (yellow). After computation of a^* and b^*, the hue angle (arctangent b^*/a^*) denotes the color (yellow, bluish-green, blue, and reddish purple, at 90°, 180°, 270°, and 0° or 360°, respectively) and C^* $[(a^{*2} + b^{*2})^{½}]$ the intensity of the color (McGuire, 1992). The fruit skin color darkens after harvest, but storage conditions may retard the darkening. The dark red- or purple-skinned figs are recommended for consumption, as the dark color is attributed to anthocyanins, which are beneficial to human health (Pereira *et al.*, 2017). Preferable skin parameters are also the moderate thickness and reduced cracking (Kong *et al.*, 2013). In some varieties, the skin thickness is reduced at the tree-ripe stage, such as Orphan (Crisosto *et al.*, 2017). Ease of peeling is another good characteristic of fresh figs (Aksoy *et al.*, 2003a). No disorders or defects should be observed on the skin, such as sun burning or damage caused by latex exudation from the pedicel of under-ripe figs (Flaishman *et al.*, 2008; Crisosto *et al.*, 2011).

13.3.2 Physicochemical characteristics

Regarding the physical characteristics, the ostiole will preferably display a small aperture, without splitting and juice exudation (Aksoy *et al.*, 2003a; Crisosto *et al.*, 2010a, 2011). The flesh is preferred without a large cavity, and should be juicy, sweet, and tasty, and with an attractive red color, with only a few small achenes that are not very hard when chewed (Aksoy *et al.*, 2003a).

Concerning the chemical characteristics, taste comprises the ultimate factor of consumer acceptance and is defined based on SSC and TA levels. Indicatively, the values of SSC ranged between 13.27 and 19.02%, in a Tunisian selection (Gaaliche *et al.*, 2012) and 16.0–27.1% in Turkish varieties (Çalişkan and Polat, 2012). Values of TA (w/w expressed in citrate) were 0.17–0.25% (Gaaliche *et al.*, 2012) and 0.10–0.35% (Çalişkan and Polat, 2012), with the ratio of SSC:TA in the last study being 64–231. However, there are conflicting reports on whether citric or malic acid is the predominant acid of figs, and this may be ascribed to different varieties. Malic acid prevailed in the genotypes of a Moroccan selection, ranging between approximately 2.5 and 5 g kg^{-1} dry weight (DW) (Hssaini *et al.*, 2020).

The amounts of fructose and glucose in 76 local Mediterranean accessions averaged 8.2 g 100 g^{-1} fresh weight (FW) and 6.2 g 100 g^{-1} FW, respectively, while the accession Kırmızı 2 had the highest fructose (11.9 g 100 g^{-1} FW) and glucose (9.7 g 100 g^{-1} FW) levels. Sucrose was detected in small amounts in some accessions, averaging in these accessions 0.14 g 100 g^{-1} FW (Çalişkan and Polat, 2011). In Masui Dauphine, Bongresi, and Banane, soluble sugar concentrations were similar among the three varieties at harvest, averaging approximately 6.0 g 100 g^{-1} FW for fructose and 6.0 g 100 g^{-1} FW for glucose (Byeon and Lee, 2020a). In the Kalamata or Tsapela fig varieties harvested at the tree-ripe stage, the average concentrations of fructose, glucose, and sucrose in two successive years fluctuated between 8.0 and 13.0 g 100 g^{-1} FW, 8.0 and 8.2 g 100 g^{-1} FW and 10.7 and 17.9 g 100 g^{-1} FW, respectively, indicating the year effect (Tsantili, 1990). Among 11 varieties grown in Morocco, glucose varied from approximately 5.5 to 30.0 g kg^{-1} DW and fructose from 6.0 to 28.0 g kg^{-1} DW (Hssaini *et al.*, 2020).

Figs at the tree-ripe stage have higher levels of soluble sugars and SSC but lower TA than at the commercial stage (Tsantili, 1990; Byeon and Lee, 2021), and therefore a relatively higher SSC:TA ratio. A taste panel showed their preference for figs at the tree-ripe stage against the commercial-stage figs (Crisosto *et al.*, 2010a; King *et al.*, 2012; Crisosto *et al.*, 2017). As fig fruits do not ripen normally after harvest (Aksoy, 1997; Rodov *et al.*, 2002), this difference cannot be overcome.

Regarding health benefits, flesh with a high antioxidant capacity is obviously preferable. There has been an increasing number of studies determining the phytochemicals in various genotypes and evaluating their benefits for human health (Solomon *et al.*, 2006; Pereira *et al.*, 2017; Arvaniti *et al.*, 2019). The antioxidant capacity in fig flesh is attributed primarily to phenolic compounds. Total content of phenolic compounds in fig flesh ranged from 23.3 to 34.3 mg of gallic acid equivalents (GAE) 100 g^{-1} FW in nine fig varieties grown in Spain (Pereira *et al.*, 2017), and from 28.6 to 211.9 mg GAE 100 g^{-1} FW in 76 accessions grown in Turkey (Çalişkan and Polat, 2011). The content of phenolics in whole figs of 11 varieties grown in Morocco ranged between 131.2 and 478.6 mg GAE 100 g^{-1} DW (Hssaini *et al.*, 2020). Total antioxidant capacity in fig flesh ranged between 20.8 and 357.5 μmol Trolox equivalents 100 g^{-1} FW in five varieties grown in Israel (Solomon *et al.*, 2006). The dark skin contributes largely to the antioxidant capacity of whole fruits (Pereira *et al.*, 2017). In addition to phenolics, the antioxidant capacity in fig flesh is also contributed by carotenoids, ascorbic acid, and other compounds. Total content of carotenoids in fresh figs is within the range of ca. 5 to 8 μg g^{-1} FW, lycopene being the prevalent compound (Su *et al.*, 2002; Veberic and Mikulic-Petkovsek, 2016; Nawade *et al.*, 2020). The total carotenoid levels decreased during growth and ripening. The amino acids aspartic acid, arginine, γ-aminobutyric acid, methionine, tryptophan, phenylalanine, leucine, lysine, and glutamic acid have been identified in fig fruit (Byeon and Lee, 2020a, 2020b, 2021).

Studies on aroma volatiles are relatively limited in fig fruits (Gibernau *et al.*, 1997; Grison-Pigé *et al.*, 2002; Oliveira *et al.*, 2010a,b; Gözlekçi *et al.*, 2011; Trad *et al.*, 2012; Villalobos *et al.*, 2018b). It is suggested that variation in the composition of aroma volatile compounds in figs is largely determined by variety (Pereira *et al.*, 2020), whereas variation in their quantity is ascribed to the maturity stage (Villalobos *et al.*, 2018b). For example, in Tunisian varieties Bouhouli, Zidi, and Thgagli at the mature stage, the major volatiles were acetoin, butyl acetate, isoamyl acetate, and hexyl acetate (Trad *et al.*, 2012), whereas pyranones, hexanal, furaneol, and 3-heptanol were among the most abundant volatiles in Cuello Dama Negro (Villalobos *et al.*, 2018b). In addition, it was found that pollination slightly affects the aroma volatiles in Tunisian varieties (Trad *et al.*, 2012).

13.3.3 Microbial decay

The most common losses of fig fruits in their postharvest life are due to the development of molds such as *Botrytis cinerea*, *Monilinia laxa*, *Alternaria alternata*, *Fusarium verticillioides* (formely known as *Fusarium moniliforme*), *Rhizopus stolonifer*, *Aspergillus niger*, *Cladosporium herbarum*, and *Phytophthora palmivora*. No microorganism development on the skin is permitted at harvest, and thus all the proper cultivation practices should be applied in order to minimize external or internal contamination (Villalobos *et al.*, 2017a).

Fig fruit attached to the tree can be attacked by various fungi (*Fusarium*, *Alternata*, *Penicillium*, and *Aspergillus* species), bacteria, and yeasts. Among the fungal-related decay, there are two main diseases, called endosepsis and smut. Endosepsis is attributed to infection by *F. verticillioides*. This sort of contamination is invisible from the outside, and renders the fruit soft and the flesh watery with an off-odor (Michailides, 2003; Crisosto *et al.*, 2011). Smut is caused by *A. niger*, which produces black spores in the cavity, while the visual symptoms start from the ostiole and appear as a developing watery-soaked spot. Bacteria or yeast can also result in a third type of decay called souring, led by fermentation (Michailides *et al.*, 1996; Michailides, 2003; Doster and Michailides, 2007).

Decay reduction may be achieved by the removal of the profichi and breba crop from the orchard, as well as by fungicide treatments of the profichi and winter crop of the wild tree, as they are the main sources of contamination (Michailides *et al.*, 2008). The three main diseases mentioned (endosepsis, smut, and souring) can cause a loss of up to 50% of the annual production, as was seen in California (Crisosto *et al.*, 2011). In cases when breba is not a commercially important crop, it can be a contamination source after falling on the ground and therefore must be removed. A simple and economical way of reducing contamination from breba fruits is to treat the trees with ethephon. Treatments with ethephon at 300 μl l^{-1} in the autumn or at 500 μl l^{-1} in spring before the full expansion of leaves resulted in approximately 30% and 92% reduction of the breba crop, respectively, without negative effects on the tree and following main crop (Crisosto *et al.*, 2010b).

As the fig wasp usually transfers fungi, yeast, or bacteria, decay is much more likely to occur in pollinated figs than in parthenocarpic ones. Fig, as a succulent and sweet fruit, becomes a good substrate for microbial development, particularly at the mature and ripe stages. Figs with a larger aperture in the ostiole are more affected by many fig wasps or even other insects, and therefore their cultivation should be avoided. Note that wasps prefer to visit trees close to the wild tree. Thus, the wild fig-tree cultivation has to be located further than 50 m from the orchard borders (Michailides, 2003). Preharvest contamination will not disappear after harvest but will continue to develop at a rate depending on the conditions of transit or storage.

Excessive pollination may result in skin cracking or ostiole-end splitting, depending on the skin characteristics of the variety. Excessive irrigation during fig ripening on the tree and careless handling can also promote/accelerate fruit decay. Therefore, all skin damage should be avoided (Colelli and Amodio, 2020). In the USA, there are restrictions that render these damaged figs unmarketable (Kong *et al.*, 2013). In Europe, however, slight or moderate cracking and splitting damage of the skin is permitted in figs classified as Class I or Class II fresh figs under the respective regulations of the United Nations Economic Commission for Europe (2017). To avoid side cracking or/and ostiole splitting, it is recommended to apply regulated deficit irrigation is applied (Kong *et al.*, 2013), as well as soil fertigation with potassium and calcium (Holstein *et al.*, 2017), a foliar potassium spray (Mimoun *et al.*, 2017; Gaaliche *et al.*, 2019), or postharvest treatment with calcium chloride ($CaCl_2$) (Irfan *et al.*, 2013). In addition, juice exudation from the ostiole at the advanced ripening stage is a good substrate for microbial growth and results in consumer rejection, and thus these figs should be removed.

13.4 Harvesting

In contrast to many other climacteric fruits, figs do not ripen after harvest. Although they can soften and darken during storage, fruit harvested before reaching optimal maturity never attain the desirable size and flavor (Aksoy, 1997; Rodov *et al.*, 2002). In contrast, figs harvested too late tend to over-ripen, resulting in excessively low firmness and increased susceptibility to decay (Chessa, 1997; Flaishman *et al.*, 2008). In general, the more advanced the fruit is in terms of maturation by harvest point, the better its flavor, but this will shorten its quality sustainability and storability. Thus, due to rapid ripening, the harvest timing for commercial purposes is very limited and difficult to define, being a compromise between storability and consumer preferences.

There are no universal recommendations to determine the optimal harvest stage as it depends on the variety and horticultural practices (Flaishman *et al.*, 2008), as well as on supply chain demands. No more than 3 days should pass before selling the fruit at a local market (Özer and Sen, 2003), and even less for a roadside vendor, so in these cases harvesting at a more advanced stage can be afforded. In contrast, figs intended for distant shipment or delayed marketing need a much longer lifespan. Obviously, crops destined for processed products such as jams or dried figs will require different sets of considerations and might justify a more flexible harvest window.

A combination of fruit firmness and skin color usually serves as a practical marker of harvest maturity. For commercial harvest, fruit firmness should slightly resist a gentle finger pressure. In terms of skin color, figs should change their initial green shade and gain about 70–90% of the final color (King *et al.*, 2012). Reaching the desired skin color is a more subtle harvest indicator. The exact harvest criteria are affected by the variety, the desired fruit lifespan, and the growing conditions. For example, nethouse-grown figs tend to be paler at harvest maturity than figs from uncovered trees (Rodov *et al.*, 2012). Harvesters need to be trained to recognize this stage (Crisosto *et al.*, 2017), and instructed accordingly prior to harvest.

Figs on the shoot ripen in succession, usually starting from the basal node and continuing toward the upper end. This means that figs should be collected on multiple harvest dates. The main crop has a longer harvesting period than the breba crop. The fact that figs should be harvested at the commercial-ripe stage in order to maintain proper postharvest handling, along with the occurrence of their rapid ripening, forces farmers to harvest every second day in most cases. It is well known that figs for fresh consumption have to be hand harvested from short trees to enable easy detachment and avoid fruit blemishes. Tree design trimming should also result in shoots that are well aerated and exposed to sunshine not only for photosynthesis products but also to minimize microorganism development and prevent a bruised appearance on the skin caused by the fruit being in contact with the leaves and shoots.

Harvesters should harvest figs only from the tree and should never collect fruits that have fallen on the ground. Harvesting should be done with very careful handling and without damaging the fruit. Fruits with any defect (e.g. open ostiole, splitting or cracking, attacks from insects on the fruit surface, discolorations, shriveling) should be discarded and removed from the field. The harvesters may wear clean gloves to protect themselves from fruit latex. They should detach the fruit from the tree with a clean cut across the stalk, and place the fruit in a clean and sanitized harvest basket, without letting the milky latex drip on the other figs. It is important that harvest baskets are relatively small to avoid compression damage and that they are sanitized frequently before they are reused, particularly in cases when the varieties show cracks or splits on their surface (Crisosto *et al.*, 2011). Harvesting should be carried out early in the morning to minimize the field heat (Flaishman *et al.*, 2008; Crisosto *et al.*, 2011).

13.5 Postharvest Technology

13.5.1 Sorting, packaging, and transportation

Figs in harvest baskets have to be transferred to the packing house, protected by a cover from sun, rain, and dust. They should be sorted and precooled as soon as possible upon harvest. The cold chain should be maintained throughout the postharvest life of the fruit. In the packing house, a trained group, wearing gloves, performs the sorting, grading, and final packaging (Crisosto *et al.*, 2011; Ait-Oubahou *et al.*, 2019). Nowadays, in addition to manual/visual sorting, some industrial packing houses employ sophisticated optical grading equipment and packing lines, capable of handling up to 100 t of fresh figs a day (Alara Agri, 2018; MAF RODA AGROBOTICS, 2020).

Packaging has several aims. Primarily, appropriate packaging protects the fruit from mechanical injuries (e.g. abrasion, puncturing), bruises, vibration stress, and contamination during transport and storage. In addition, it should maintain an optimal microenvironment around the fresh produce, including relative humidity (RH) and atmospheric composition, and should allow fast cooling and cold storage of the fruit. Finally, it has to provide the consumer with all the required information according to the market regulations.

As fig is a highly perishable fruit, it requires very careful handling to ensure minimal bruising. It is imperative to make sure that figs are protected from impact, vibration, and compression bruises during harvest, transport, and storage. Bruising is the result of cell membrane breakage of individual cells, leading cytoplasmic enzymes to react with vacuolar compounds and filling the intercellular spaces. It is accompanied by increased respiration, ethylene production rates, and water loss, induction of microbial spoilage, and sometimes by a decrease in nutritional values (Li and Thomas, 2014). The impact bruising may occur if some fruit hits any surface or another fruit during the filling of packages or during transport of partially filled packages. Compression is observed with overfilled packages, while vibration occurs during transport on rough roads without good vehicle suspension (Ait-Oubahou *et al.*, 2019).

Sensitivity to bruising depends on the fig variety. Mao *et al.* (1995) observed that Masui Dauphine was more sensitive to mechanical stress than Celeste and Brunswick, as manifested by their physiological responses to vibration. In particular, at 26°C and 6 × g vibration, Masui Dauphine, Brunswick, and Celeste exhibited ethylene production rates of approximately 18,

9, and 5 μl kg^{-1} h^{-1}, respectively. The respective values of RR were approximately 186, 146, and 128 mg kg^{-1} h^{-1}, and cellular leakage was approximately 20, 18 and 13 mg kg^{-1} min^{-1} (Mao *et al.*, 1995). Comparisons made among the varieties Yediveren, Black Bursa, and Sarılop showed that Yediveren figs were the most resistant to transportation stress, followed by Black Bursa, whereas Sarılop was the most sensitive variety (Aksoy *et al.*, 2003b). Further work on these varieties reconfirmed the off-road high sensitivity of Sarılop, but the most sensitive for long distances on a highway road was Black Bursa (Çakmak *et al.*, 2010). The same authors tested transportation stress in three different boxes—a cardboard box, extruded polystyrene, and expanded polystyrene—and found that the best-suited package for transportation was the extruded polystyrene, while the worst was the cardboard box, which did not secure or immobilize the fruit (Çakmak *et al.*, 2010). Yenge and Nidoni (2014) studied the Poona variety packaged in corrugated fiberboard boxes, transported up to 500 km, and then exposed to ambient temperature for 5 days. They compared the reduction in quality traits such as firmness, weight loss, and color darkening when using polyurethane foam sheets, corrugated fiberboard with paper shavings, a polyethylene foam sheet, and an air bubble polyethylene film. They concluded that polyurethane foam sheets showed the best performance in reducing quality deterioration (Yenge and Nidoni, 2014). Decay incidence was also lowest in corrugated fiberboard boxes with polyurethane in the boxes, as evaluated after 5 days of storage (Yenge *et al.*, 2014).

To enable efficient precooling, the fruit can be placed in suitable ventilated packages, such as open trays (Fig. 13.3A), perforated clamshells (Fig. 13.3B) and vented carton boxes (Fig. 13.3C). When figs are stored for relatively long periods (a few weeks), transported by vehicles, or destined for distant markets, they are usually placed on nested inserts located within laminated or

Fig. 13.3. Ventilated fig packages. A. A stack of open cardboard trays with figs placed on protective nested inserts. Note the four corner protrusions to interlock the trays in order to fix their position in the stack. B. Figs in clamshell containers. Note the slot perforations on the container lids to enable efficient precooling. C. Palletizing the fig carton boxes with the use of rigid L-shaped edge corner protectors. In the foreground: an enlarged carton image. Note the macroperforated plastic film on top of each box and side vents for efficient precooling.

corrugated fiberboard trays or boxes in order to absorb the exerted impact energy transferred to the fruit (Hussein *et al.*, 2020). In this type of packaging, only one layer of fiberboard (paperboard carton) package is used (Fig. 13.3A). The package can be covered by a non-transparent carton lid, or with a transparent macroperforated film (Fig. 13.3C). This type of packaging is relatively expensive, but the cost is justified by longer quality retention. Both clamshells and tray carton packages restrict water loss to some extent but do not change the atmospheric composition inside the package (Rodov *et al.*, 1998a; Ben-Yehoshua and Rodov, 2003). The venting area is among the important characteristics of packaging when figs will be transported or stored at low temperature in air for dissipation of heat, air cooling, and excess water vapor. The venting area can be approximately 5% as a compromise between heat transfer and package strength (Kader, 2002b).

The packages are usually arranged in a pallet. To avoid compression, the bottom packages should withstand the upper load of the pallet without collapsing. Therefore, reinforcing the pallet with rigid L-shaped edge corner protectors is a good practice (Fig. 13.3C). Moreover, protrusions (Figs. 13.3A and 13.3C) can be used to stack the packages and nets or wrapping material used to immobilize the pallet. The package dimensions should be appropriate for palletization. Special attention should be given to the total venting area, the shape of the vents, and the alignment of package vents horizontally in a pallet (Elansari *et al.*, 2019). All packaging materials and the packaging area should be clean and safe, all operations should be applied according to Good Manufacturing Practice and the personnel must follow good hygienic practices. In cases when packing houses require a certification of Good Agricultural Practices, they must operate according to ISO 9001, ISO 22000, or other packaging standards (Ait-Oubahou *et al.*, 2019). Labels on any kind of package are necessary to inform about the product weight, variety, area/country of cultivation, orchard, harvest date, and harvesting group so that all produce can be traced from the field to the consumer (Ait-Oubahou *et al.*, 2019).

When the fruit is destined for travel over long distances and/or for relatively prolonged storage periods, figs can be packaged and stored in a reduced O_2 and/or elevated CO_2 atmosphere, achieved using the methods of controlled atmosphere (CA) or MAP. However, the fruit must be properly precooled before applying these methods. The CA and MAP approaches are presented in separate sections of this chapter.

13.5.2 Precooling

Precooling is the initial cooling that takes place soon after harvest to remove field heat and vital heat, including the harvest stress. Therefore, precooling suppresses the respiratory loss of energy and substrates that cannot be recovered after being detached from the tree (Saltveit, 2018). Figs are chilling resistant. Generally, it is recommended to store figs at −1 to 0°C. Under these conditions, the RH should remain at 90–95%. When the temperature is rapidly reduced from 20 to 0°C, the RR and ethylene production rate decrease drastically (Gross *et al.*, 2016), resulting in inhibited ripening. For fresh produce, there is an inverse relation between RR and duration of postharvest life. Therefore, fast precooling and maintenance of the cold chain until reaching the consumer are important for fruit preservation (Gross *et al.*, 2016; Saltveit, 2018).

Precooling is carried out mainly by keeping the produce in refrigerated rooms ("room cooling") or by "forced-air cooling" (Elansari *et al.*, 2019). Precooling is performed before the final packaging to reduce the temperature of the fruit and of its surrounding air. As cold air holds less water vapor than warm air, it is less prone to water condensation. Therefore, precooling is imperative when MAP storage is used in order to avoid condensation of water vapor and respiratory depletion of O_2. Examples of precooling and MAP use are presented by Villalobos *et al.* (2014, 2016a, 2018a). Forced-air cooling is performed by circulation of cold air through the produce packages; it is a faster and much more effective method of temperature reduction than room cooling, although there is always a risk of increased water loss. However, a decrease in the airflow by 50% before the completion of the forced-air cooling process minimizes the water loss of fruit placed in open or vented packages (Elansari *et al.*, 2019). Other precooling methods seem to be less effective for figs. For example, excessive water loss was observed in figs precooled by vacuum cooling and therefore this method is not recommended (Ito *et al.*, 1987).

Concerning the lowering temperature effect, a decrease of 10°C, between 0 and 30°C, results in a reduction of the velocity of a physiological/biochemical reaction in fruit by approximately two- to threefold, on average, and vice versa. This also applies to the RR. Therefore, at two temperatures, t and T, where $T = (t + 10°C)$, the ratio of $RR^T:RR_t$ is called the temperature coefficient or quotient of the two RRs (Q_{10}) (Kader and Saltveit, 2003). This factor can be calculated and used to predict the RR at a given temperature. It is noteworthy that the higher the Q_{10} value, the greater the requirement for fast cooling to reduce quality deterioration and extend the lifespan, particularly when the produce exhibits high RR values. Waghmare *et al.* (2014) found that Q_{10} values for RR in fig varies between 4 and 5. These values denote the high significance of low temperature throughout the postharvest life of the studied (unspecified) variety. The reported RRs for the Mission variety at 0, 5, 10, and 20°C were 4–8, 10–16, 18–24 and 40–60 mg CO_2 kg^{-1} h^{-1}, respectively (Gross *et al.*, 2016). Lowering the temperatures from 25–30°C to 0–2°C, the RR in Cuello Dama Blanco, Cuello Dama Negro and San Antonio decreased from approximately 181.3, 163.4, and 229.5 mg CO_2 kg^{-1} h^{-1} to 49.0, 55.6, and 33.5 mg CO_2 kg^{-1} h^{-1}, respectively (Villalobos *et al.*, 2016a). However, temperature may also affect fruit susceptibility to bruising. In cherries, low temperature increased the susceptibility to impact bruising but decreased the vibration bruising (Crisosto *et al.*, 1993). There is no such information regarding figs. However, it is worth noting that cold figs are still very susceptible to bruising.

Water loss is controlled by fast precooling after harvest due to the reduced transpiration and respiration processes. Therefore, a delay in precooling enhances the water loss of fruit. For example, Brown Turkey figs cooled to 0°C immediately after harvest or after a 3- or 6-h delay, exhibited water loss of 0.6%, 4.8%, and 5.9%, respectively. At the same time, the water loss rate is genotype-dependent: under similar conditions the Kadota figs showed a higher water loss rate than the Brown Turkey (Crisosto *et al.*, 2011).

Precooling extends the market life of figs as it also retards softening and decay. For example, Bursa Black variety figs could be stored at 1°C at 90% RH for 6 weeks without excessive softening only if the harvested fruit had undergone rapid (3.5 h) precooling from 25°C to 4°C (Türk, 1989). It was also observed that fast forced-air cooling of Bursa Black from 20°C to 0°C within 4 h resulted in reduced weight loss and decay of fruit, and doubled the storage period to 4 weeks, by comparison (Çelikel and Karaçali, 1998). However, the variety is also an important factor for firmness loss. In San Pietro figs, softening was advanced after an 8 h delay in cooling, although no difference was observed in softening in Melanzano figs (Eccher Zerbini *et al.*, 1986).

The precooling efficiency is determined by technological progress. A temperature decrease from 20 to 1°C was achieved after 110 min of forced-air cooling of 48 cartons of Roxo de Valinhos figs (24 fruit per carton), arranged in a tunnel, with an air flow of 2.8 l s^{-1} kg^{-1} and an average air velocity of 2.0 m s^{-1} (Dussán-Sarria *et al.*, 2006). In wooden packages with 1.5 m s^{-1} air velocity, the same temperature decrease was achieved in 100 min (Dussán Sarria *et al.*, 2010). These studies emphasize the significance of the packaging and operating conditions during precooling. To minimize water loss, humidifying techniques may be applied during the operation (Elansari *et al.*, 2019). Recently, a portable forced-air cooling unit was used to treat small product batches, but this has not yet been implemented for figs (Elansari *et al.*, 2019).

13.5.3 Cold storage

Low-temperature storage is the basic principle of postharvest technology. It remains the primary and unreplaceable method to reduce quality deterioration and extend storage periods of fresh fruits (Kays, 1997; Cantín *et al.*, 2020). The development of storage technology and modification of atmospheres augment the low-temperature performance (Beaudry, 2017). The importance of low-temperature maintenance was confirmed for figs stored under CA or modified-atmosphere (MA) conditions (Tsantili *et al.*, 2003; Villalobos *et al.*, 2014, 2016a,b, 2017a,b, 2018b; Gross *et al.*, 2016; Hong *et al.*, 2016; Bahar and Lichter, 2018).

Upon cooling from 5°C to 0°C, the Mission fig variety exhibited a twofold lower RR and a fivefold lower ethylene production rate (Colelli *et al.*, 1991), indicating the slower rate of ripening. Comparing the storage temperatures of 0, 2.2 and 5°C, the same authors observed that the lower temperature resulted in the greatest RR reduction. In commercially mature figs of the Mavra Markopoulou variety, during air storage at −1°C, both RR and ethylene production showed abrupt decreases during the first 7 days with small fluctuations afterwards (Tsantili *et al.*, 2003).

In most cases, it is recommended that figs are stored at 0°C in 90–95% RH. The quality of cold-stored figs depends mainly on the combination of storage temperature and duration, variety, and stage of maturity at harvest. Temperature largely affects RR, water loss, and fruit softening. Generally, the lower the storage temperature and the shorter the storage duration, the less quality deterioration of the fresh produce occurs.

The storage duration for figs depends largely on variety, maturity stage, and temperature (Morton and Dowling, 1987; Park *et al.*, 1998; Yahata and Nogata, 2000). In particular, Black Mission and Sarılop figs can be stored at low temperature (from −1°C to 0°C) in air for 7–14 days. Figs that are properly precooled immediately after harvest can subsequently be stored for up to 20 days at 1°C, for 7 days at 10°C, but only for 2–3 days at 20°C (Ito *et al.*, 1987). Similarly, Ipoh Blue Giant, a popular variety in Malaysia, could be stored for 5 weeks at 5°C, but for 28 days at 10°C and for 14 days at 15°C, based on visual estimation (Jusoh *et al.*, 2019). Postharvest temperature manipulation treatments, such as a delay of cooling for 6 h or intermittent cooling, were not beneficial to fig quality preservation (Cantín *et al.*, 2020). It has been reported that fresh-cut figs at 0°C or 5°C could be kept in air for 6 days with excellent quality and the figs were still marketable for 12 days at 0°C but only for 9 days at 5°C (Hong *et al.*, 2016). Hong *et al.* (2016) claimed that cutting of figs was not detrimental for their storage and did not affect RR at any storage temperature, for any variety studied. However, more experiments are required to investigate the behavior of other varieties after cutting.

Ripening processes

Cultivation practices usually affect the quality parameters not only at harvest but also during storage. Byeon and Lee (2020b) also demonstrated that during air storage at 0.5°C for 30 days, figs from greenhouses exhibited higher weight loss (%) but lower RR and concentrations of citric acid and most free amino acids determined than figs harvested from the open field. However, the maturity stage at harvest plays an important role in the physiological and biochemical changes occurring during storage, even at very low temperatures, as well as after removal from storage. A recent work of Byeon and Lee (2021) investigated in detail the effect of low temperature on many quality variables in Masui Dauphine figs harvested at different maturity stages for 30 days storage at 0.5°C in air and after transfer to 20°C in air for 1 day. In particular, they found that the levels of SSC, glucose, fructose, and TA increased during storage in figs of both maturity stages, but glucose and fructose concentrations were consistently higher, whereas concentrations of sucrose, citric acid, shikimic acid, and fumaric acid were lower in more mature figs than in less mature ones throughout storage. In the same study, the initial levels of ascorbate and dihydroascorbate were lower, and of anthocyanins were higher, in the advanced mature fruit than in less mature fruit, but those of total phenolics were similar in all harvested fruit. Byeon and Lee (2021) also observed differences in the content of amino acids, with glutamine, serine, and alanine being at high levels during storage of advanced mature fruit. The content of other amino acids decreased in all fruits during storage.

Fruit firmness, a major quality attribute for figs, decreases after cold storage (Tsantili *et al.*, 2003; Gözlekçi *et al.*, 2008; Freiman *et al.*, 2012; Bahar and Lichter, 2018). Softening rates are reduced at lower temperatures, in comparison with higher temperatures (Türk *et al.*, 1994; Tsantili *et al.*, 2003). Houraishi variety figs could be kept for a shorter time at 15°C than at 5°C due to rapid softening associated with increased water-insoluble pectic substances in the cell wall at 15°C (Yahata and Nogata, 2000). Concerning other changes, SSC increased in most studies of figs (Çelikel and Karaçali, 1998; Bahar and Lichter, 2018; Byeon and Lee, 2020a,b), while TA levels remained rather constant in most studies (Türk *et al.*, 1994; Çelikel and Karaçali, 1998; Kaynak *et al.*, 1998; Gözlekçi *et al.*, 2008; Villalobos *et al.*, 2016a) or increased (Byeon and Lee, 2020a). In the skin of figs, color darkening was demonstrated in studies during storage, but the final color depended on the initial color of the mature or ripe fruit variety at harvest (brown or greenish-yellow) (Alturki, 2013; Allegra *et al.*, 2017). It is worth noting that flavonoids and particularly anthocyanin levels increased at low temperature. This was demonstrated for Ottomanit figs by fluorescence measurements using a Multiplex III system (Bahar and Lichter, 2018).

Postharvest treatments with 1-MCP were not beneficial to figs compared with other fruits. Treatment of Bursa Black with 0.5 and 1 $\mu l\ l^{-1}$ 1-MCP, after precooling at 1°C, resulted in a limited effect on ripening delay during 10-day storage at the same temperature (Özkaya *et al.*, 2014). Tofanelli *et al.* (2018) found similar results with 1-MCP treatment of Roxo de Valinhos at 20°C and stored thereafter at 4°C. Figs of the Cuello Dama Negro variety treated with 1-MCP at 20°C and then stored at 1°C in 95% RH also exhibited no beneficial effect in comparison with untreated. Instead, a higher incidence of side cracking occurred in the 1-MCP-treated figs (Cantín *et al.*, 2020).

Shelf life after removal from stores

Generally, after storage, a lower RR, higher total SSC, and lower acidity are expected to be found in figs harvested at a more mature stage compared with less mature figs. However, changes in biochemical processes during storage are not always predictable, because at high fruit weight losses during or after storage, the determined variables may result in higher concentrations when the results are presented on a FW basis. Moreover, some variables can be estimated in cold fruit or after equilibration to ambient temperature. Additionally, upon removal from stores, a change in temperature might result in stress responses of the fresh fruits, which in turn may affect quality variables. Thus, experimental results are not always directly comparable among studies. Nevertheless, the transfer to ambient temperature occurs inevitably in some cases in local markets and provides an estimate of fruit quality before figs reach the consumers. An exception to this is the consumer demand for cold figs in Mediterranean countries during hot days in summer.

The transfer of cold figs to ambient temperature results in considerable weight loss and microbial decay, as demonstrated in Ottomani (Bahar and Lichter, 2018) and Cuello Dama Negro (Cantín *et al.*, 2020) figs. This can be explained by the increased respiration and transpiration rates. In Mavra Markopoulou figs, the RR at 20°C after storage did not change consistently; RR decreased after 7 days, increased after 14 days, and remained almost stable for up to 28 days of storage (Tsantili *et al.*, 2003). By contrast, the RR of Masui Dauphine and Banane, determined at 20°C after storage at 0.5°C, increased after the first 10 days, decreased after 20 days, and thereafter remained at stable levels (Byeon and Lee, 2020a). These differences may be attributed to different genotypes, maturity stages, and experimental conditions.

Although the percentage of deterioration of some quality parameters at ambient temperature may be greater in figs stored previously at a lower temperature than at a higher one (Baccaunaud *et al.*, 1995), the overall quality is superior after the lower temperature storage (Flaishman *et al.*, 2008). Figs stored at a higher storage temperature, and for a longer duration, became more susceptible to cracking and decay when exposed to 20°C (Cantín *et al.*, 2020). The response of cold-stored figs to ambient temperature is also greatly affected by the maturity stage at harvest, the variety, and the storage conditions, as confirmed by most studies (Cantín *et al.*, 2011; Kong *et al.*, 2013; Villalobos *et al.*, 2018b; Byeon and Lee, 2020a, 2021). Generally, figs exposed to 20°C after low-temperature air storage have a postharvest life of 1–2 days (Chessa, 1997; Crisosto *et al.*, 2011).

13.5.4 Effects of relative humidity

Humidity in the atmosphere around the fresh produce is another important factor for quality maintenance during or after cold storage. Transpiration is reduced at elevated RH, resulting in lower weight loss. Losses in fruit weight are always observed during storage (Bahar and Lichter, 2018). Water loss increases at the advanced maturity stage of the fruit (Byeon and Lee, 2021) and with extended storage duration (Aksoy, 1981; Çelikel and Karaçali, 1998; Tsantili *et al.*, 2003; Gözlekçi *et al.*, 2008). Moreover, the water loss rate is variety dependent (Aksoy, 1981; Kaynak *et al.*, 1998; Villalobos *et al.*, 2016a; Byeon and Lee, 2020a). In particular, Bahar and Lichter (2018) found that water loss increased gradually in Ottomanit figs stored in air at 0°C under 95% RH and reached 14.4% after 30 days of storage. The water loss in the three studied varieties, Masui Dauphine, Bongresi, and Banane increased progressively with advanced storage time and ranged between 12% and 20% after 30 days of storage at 0.5°C with 90% RH and 1 day at room conditions, indicating the effect of the variety (Byeon and Lee, 2020a). Weight loss of less than 3% was observed in Bursa Black (Türk *et al.*, 1994) and Mavra Markopoulou (Tsantili *et al.*, 2003) figs during low-temperature storage. However, the low water loss in the latter study was recorded only during cold storage, which had followed the initial transport from field where the figs might already have lost a large part of their initial weight.

Water loss affects many quality variables, either directly or indirectly, such as firmness and sugar content (Aharoni *et al.*, 2007a,b). Efforts have been made to reduce water loss using humidifiers. Hung *et al.* (2011) reduced the water loss and softening in figs more effectively with a nano mist humidifier (nano mists) than with an ultrasonic one (ultrasonic mists). Humidity should always be under consideration in conjunction with storage temperature. There were only slight differences in firmness, color, SSC, and TA in Cuello de Dama figs stored at 0°C, 3°C, and 6°C in fruits covered with a semi-permeable film, but the differences between covered and uncovered figs at 6°C increased during storage (García *et al.*, 2003). The last result stresses the positive effect of proper packaging on water loss, allowing additional energy savings by storing at a relatively high storage temperature. At the same time, air storage under high humidity may result in decay advancement after transfer of figs to an elevated temperature (Türk *et al.*, 1994; Bahar and Lichter, 2018). Efforts to alleviate this problem are discussed in the following sections.

13.5.5 Controlled-atmosphere storage

CA principles and types

CA technology allows the fruit to be kept under continuously regulated temperature, humidity, and concentrations of O_2, CO_2, and other gases, which are actively maintained at the desired levels throughout storage. A dynamic CA is a novel CA technology allowing storage of the fruits under the lowest O_2 concentration tolerable without passing the anaerobic compensation point (ACP), i.e. without switching the fruit metabolism to anaerobic fermentation, which causes off-flavor generation and quality decline. The anaerobiosis onset is monitored in the DCA systems either by ethanol vapor accumulation, or by the respiratory quotient (the ratio of CO_2 emission to O_2 uptake), or by chlorophyll fluorescence (Beaudry, 2017).

It is well documented that keeping fruit in reduced O_2 and/or elevated CO_2 atmospheres decreases the RR, biosynthesis and perception of ethylene, and production of aroma volatiles. It also delays color changes, decay, and cell-wall metabolism, and preserves the fruit firmness. In fact, decreasing O_2 and increasing CO_2 has synergistic effects on retardation of the ripening processes and senescence. The combination of low O_2 and elevated CO_2 also suppresses pathogens, either indirectly or directly (Beaudry, 1999). In fact, high levels of CO_2 (above 10%) and very low levels of O_2 (typically below 1%) are required for fungal growth control. However, these levels of O_2 and CO_2 may result in fermentation and the production of acetaldehyde, ethanol, and ethyl acetate that render the fruit inedible due to off-flavor and off-odor (Beaudry, 1999; Ben-Yehoshua *et al.*, 2005). In addition, the excessive CO_2 concentrations can be injurious, as in Ottomanit figs stored under 15% CO_2 (Bahar and Lichter, 2018). Therefore, to avoid hypoxic fermentation and CO_2 injury, the CO_2 levels should not exceed critical points and O_2 levels should not decrease below the anaerobic compensation point. Hazard points are known for some commodities (Beaudry, 1999), but safe limits depend on maturity stage and environmental conditions.

CA and DCA technologies require specific and expensive facilities and equipment and incur high operation costs. They are usually implemented for large quantities of produce that are stored for a long time, such as apples. Therefore, CA is not usually employed for commercial storage of figs, because of the short storage duration (up to 6 weeks), small areas of cultivated land in most countries, and extended harvesting period, which do not allow filling of the CA room.

Figs stored under CA

Optimal CA conditions for fig storage include a temperature of 0°C with 90–95% RH and an atmospheric composition of 15–20% CO_2 and 5–10% O_2. These conditions result in a reduction in respiration and ethylene production, firmness retention, and decay control (Gross *et al.*, 2016). Keeping Black Mission and Calimyrna figs grown in California under the above conditions extended their storage life from 1–2 to 3–4 weeks (Gross *et al.*, 2016). However, the applicability of these recommendations to all varieties under CA or MAP is questionable. Concerning ethylene removal, which usually accompanies the CA storage of other climacteric fruits, it is known that prolonging fig storage did not rely on ethylene removal because of the already climacteric or postclimacteric stage of the harvested fruit. Experiments with ethylene absorption showed limited efficacy (Rodov *et al.*, 1998b; Bouzo *et al.*, 2012).

At the same time, CA experiments may help to determine the optimal atmospheric compositions for fig-fruit preservation, which may be further reproduced in more affordable MAP systems. The experimental investigation of fig storage under a controlled environment started with increasing CO_2 levels. As early as 1952, Claypool and Ozbek (1952) observed that the storage life of Mission figs at 20°C could be extended slightly in atmospheres containing up to 60% CO_2. Colelli *et al.* (1991) studied the quality of Mission figs stored at 0, 2.2, or 5°C in atmospheres enriched with 15% and 20% CO_2. The authors observed firmness and appearance preservation, inhibition of ethylene production, and reduction of decay, but there was induction of ethanol and acetaldehyde production and off-flavors with a risk of not recovering after storage due to the high CO_2. Similar results have also been obtained by others (Mathooko *et al.*, 1993; Park *et al.*, 1998; Park and Jung, 2000). However, Colelli (1995) recommended the transportation of figs in relatively high CO_2 atmospheres, as is usually done for strawberries.

Reductions in O_2 levels with/without increased CO_2 levels were investigated. Türk *et al.* (1994) investigated the effect of CO_2 and O_2 at different ratios on the decay of Bursa Black figs over 30 and 40 days of CA storage at 0°C and 90–95% RH. Different quality aspects were studied, such as weight loss (WL), RR, firmness, flavor, TA, SSC, pH, and concentrations of vitamin C, chlorophyll, and carotene. The authors observed that 3% CO_2 + 3% O_2 or 5% CO_2 + 5% O_2 levels were beneficial for storage up to 4 weeks, whereas figs could not be stored for 40 days under any of the storage conditions tested. However, CA storage with 1% CO_2 + 1% O_2 were more beneficial over CO_2:O_2 ratios equal to 2:1, 3:1 or 0:20 (2% CO_2 + 1% O_2, 3% CO_2 + 1% O_2 or 0% CO_2 + 20% O_2, respectively). In Ottomanit figs, higher firmness retention, lower decay, lower weight loss, and the highest fruit integrity were achieved when figs were stored at 0°C in 95% RH in 5% CO_2 + 5% O_2 for 30 days followed by a 2-day shelf life at 20°C in 85% RH compared with the control and those stored in higher CO_2 concentrations (Bahar and Lichter, 2018). In Mavra Markopoulou figs, 2% O_2 at −1°C reduced softening during storage periods longer than 1 week. In addition, after 14 days of storage at 4°C in 2% O_2, the fruit firmness was similar to that stored at −1°C in air (Tsantili *et al.*, 2003). The new approach of cut figs was also studied under CA conditions. Brown Turkey and Sierra figs that had been sanitized (100 mg L^{-1} sodium hypochlorite) before cutting were stored at 5°C under humidified flows of 3% O_2 and 6, 12, or 18% CO_2 and were maintained at excellent quality for 6 days. However, after 6 days, the high levels of CO_2 increased the ethanol and acetaldehyde levels (Hong *et al.*, 2016).

The results of CA experiments on figs are not conclusive so far, and there is no recommendation for optimal O_2 and CO_2 concentrations for all figs. The particular composition of CA for storing figs depends on the variety, maturity stage, storage temperature, and duration. Therefore, more studies on experimental and pilot scales have to be carried out for CA storage of any fig variety in order to confirm the best atmospheric composition before installing CA rooms.

13.5.6 Modified-atmosphere packaging

MAP principles

The basic idea of MAP is to enclose the respiring produce in semi-permeable packaging material (typically a plastic film) where the respiratory activity of the produce results in the generation of a MA with reduced O_2 and elevated CO_2 levels inside the package. A desirable atmospheric composition, which extends the product shelf life, is reached by selecting a packaging material with proper barrier properties, matching the produce RR under given storage conditions. This technology is less expensive than CA storage and can be commercially implemented with fig fruits.

The MAP can be passive or active, depending on the initial in-package atmospheric composition. In passive MAP, air is used as the initial atmosphere inside the package and MA generation is purely the result of the interaction between the product respiration and gases diffusion through the packaging material. After a certain transient period, steady-state O_2 and CO_2 concentrations are established in such packages due to the equilibrium between the respiratory and diffusion processes. Therefore, passive MAP is also called "equilibrium MAP." Desirable barrier properties of the packaging material can be achieved by choosing an appropriate plastic polymer, film thickness, and perforation. A relatively new type of passive MAP uses microperforated films bearing a certain number and size of laser-pierced or mechanically pierced holes of approximately 50–150 μm in diameter, preventing the risk of the generation of anaerobic conditions in the package.

In active MAP, the package is initially filled with a gas mixture containing predetermined reduced O_2 and/or elevated CO_2 concentrations. This approach allows a transient period in MA generation, when the produce is exposed to non-optimal atmosphere composition, to be omitted. Further evolution of the MA is determined by respiration–diffusion interaction. The active MAP may include additional components that release or scavenge gases or other bioactive compounds in the package atmosphere. In Europe, this should be done according to European Commission Regulation 450/2009 (EFSA, 2009; Beaudry, 2017; Siracusa and Lotti, 2019).

Characteristics of common packaging films and their uses have been presented by Opara *et al.* (2019). Besides barrier properties, other film attributes (e.g. strength, transparency) are also important. In MAP, gas concentrations and humidity levels are affected by the combination of film properties, respiration and transpiration rates of the product, and environmental conditions outside the package (Belay *et al.*, 2016). Respiration and transpiration, in turn, are influenced by the produce genotype and maturity stage at harvest. Steady-state atmospheric composition and duration of the transient period depend on the produce weight and surface:volume ratio, the package void volume, and the film material, thickness, and surface area (Belay *et al.*, 2016). Different gases have different permeation rates. For example, the rate of CO_2 transmission through common plastic packaging materials is two- to eightfold higher than that of O_2; for low-density polyethylene this ratio (permselectivity) is close to 4 (Beaudry, 2017). Note that permeation is affected by the temperature and gas pressure gradient. Additionally, at elevated CO_2 levels, fruits become less tolerant to low O_2, which may lead to anaerobic metabolism. This situation is often observed in fruit packaged in MAP and exposed to elevated temperatures after cold storage. For the same reason, fruit should be packaged in MAP after cooling, and a safety limit for gas concentrations at abuse temperatures has to be taken into consideration (Beaudry, 1999; Belay *et al.*, 2016).

MAP effects on fig preservation: water loss versus decay

The early MAP investigations with figs revealed the opposing effect of plastic packaging on the fruit weight loss and microbial decay. To reduce weight loss, Niedda Longa figs harvested at the ripe stage were placed on polystyrene disks, wrapped with 15 μm thick extensible polyvinyl chloride, or 15 or 19 μm thick polyolefin heat-shrinkable films and stored at 0°C and 95% RH with subsequent transfer to 20°C and 75% RH. The results showed that all plastic films reduced weight loss similarly, compared with unwrapped figs. However, the films increased the percentage of rotten figs, particularly at the higher temperature (Piga *et al.*, 1995). In another trial, Craxiou de Porcu main-crop figs were wrapped in 19 μm polyethylene shrinkable film, filled with a MA of either 100% N_2 or 100% CO_2 and exposed to simulated market conditions at 15°C and 80% RH (d'Aquino *et al.*, 1998). This study showed that MAP reduced the weight loss, inhibited fruit aging and allegedly extended the shelf life from 2 to 5 days. The N_2 atmosphere was relatively effective in reducing spoilage, but the decay incidence after 5 days was still too high (30%). Similarly, wrapping with 12 μm cling film decreased weight loss, but increased decay in comparison with unwrapped Yediveren figs (Özer and Sen, 2003).

All fig packages with high humidity retention exhibited weight loss reduction (Eccher Zerbini *et al.*, 1986; Piga *et al.*, 1995; Hernández Méndez *et al.*, 2003; Lima *et al.*, 2005). However, in many cases, the high in-package humidity tended to increase fungal decay of the figs and to cause excessive softening, degradation, and exudation of juice from the ostiole, providing a good substrate for microbial growth (Flaishman *et al.*, 2008). Allegra and Colelli (2017) showed that Dottato breba figs stored at 4°C in 90% RH exhibited an excellent quality retention for up to 7 days when left in the open air but started rotting within 3 days in rigid bioriented polystyrene bags (under passive MAP). In contrast, no enhanced fruit deterioration was reported in a recent study with Black Mission figs stored in expanded polystyrene trays covered with pliofilm, compared with clamshell containers (Martínez-Damián *et al.*, 2020). When combined with cold storage at 1°C, the passive MAP of expanded polystyrene trays covered with pliofilm was beneficial for reducing fruit weight loss and preserving its appearance and antioxidant capacity. These results might be related to the special properties of the packaging material used.

One of the ways to reduce the negative consequences of excessive in-package humidity is to use a hydrophilic packaging material with a high water-vapor transmission rate, or to include in the package an insert containing a hygroscopic moisture absorber. Keeping Brown Turkey figs for 21 days at 2°C in Xtend packages (StePac L.A. Ltd, Tefen, Israel) characterized by enhanced water-vapor permeability resulted in lower weight loss and better external and internal appearance, with no fungal decay observed, as opposed to unpackaged fruit. Moreover, there was no indication of fermentation in the packaging, as the film also had increased permeability to CO_2 (Bouzo *et al.*, 2012). The Xtend MAP was found to be more beneficial for preservation of Cuello Dama Negro figs than all the other techniques tested, including LifePack passive MAP, air storage under various temperature and RH conditions, 1-MCP application, and sulfur dioxide (SO_2)-releasing pads (Cantín *et al.*, 2020).

The potential for extending the life of Bianco del Cilento figs by inclusion of a moisture absorbent in plastic packages was reported by Matteo *et al.* (1999). In a more recent study, moisture absorbers containing calcium sulfate were used to prevent the excessive in-package humidity and control microbial growth on Domenico Tauro breba figs kept in active MAP with 20% O_2 and 20% CO_2. This treatment proved efficient in firmness retention and microbial reduction without an appreciable effect on fruit sensory quality (Villalobos *et al.*, 2017b). Cantín *et al.* (2020) tested an alternative way for reducing both weight loss and decay of Cuello Dama Negro figs by including SO_2-releasing pads in the polyethylene-wrapped packages. The use of pads reduced fig decay but detrimentally affected the fruit quality due to enhanced side cracking, bleaching, and discoloration. Ethylene-absorbing package additives were also of limited benefit for fig preservation (Alique and de la Plaza, 1990; Rodov *et al.*, 1998b; Bouzo *et al.*, 2012; Zidi *et al.*, 2020), presumably because the harvested fruits were already at the climacteric or postclimacteric stage.

Optimizing O_2 and CO_2 concentrations in active and passive MAP

Several studies have been carried out to define the beneficial CO_2 and O_2 concentrations for fresh-fig preservation. In Nazareth figs, relatively high levels of CO_2 (above 10%) resulted in abnormal ripening, with off-flavors and advanced softening. However, CO_2 levels of 4% and 6% inhibited fig softening (Rodov *et al.*, 1998b). In agreement with these results, the equilibrium concentrations of 4.3% CO_2 and 10.7% O_2 were reported as the optimal MA composition for Collar figs, extending the fruit life by 7 days compared with the control (Guillén *et al.*, 2015). Using CO_2-releasing agents or absorbers of CO_2 could be of help for modulating the in-package atmospheric composition (Rodov *et al.*, 1998b).

Active MAP of 20% CO_2 and 10% O_2 (balanced with N_2) preserved the quality of Bursa Black figs for 15 days; however, the same lifespan was reached in passive MAP. A further increase in O_2 concentration to the super-atmospheric level of 70% enhanced fruit softening and affected its sensorial quality (Ayhan and Cara Cay, 2011). Active MAP comprising 10% O_2, 20% CO_2, and 70% N_2 extended the life of common figs grown in Saudi Arabia compared with both passive MAP and unsealed controls (Alturki, 2013). Storage duration was doubled from 14 days in the control to 28 days in active MAP based on weight loss, SSC, firmness, skin color, and, notably, decay incidence (Alturki, 2013).

An original approach to active MAP preservation of figs was used by Kim *et al.* (2012) who enclosed Masui Dauphine fruit in microperforated polypropylene packages flushed with gas mix containing 70% CO_2 (balanced by N_2). The performance of these packages, realizing the principle of microperforated active modified-atmosphere (MAMA) packaging (Rodov *et al.*, 2007; Shinde *et al.*, 2018), was compared with non-perforated active MAP of the same initial composition, and with microperforated and non-perforated passive MAP. After the initial fruit exposure to 70% CO_2, within 3–4 days of storage at 5°C, the CO_2 level in the MAMA packages stabilized at the level of 10% with a steady-state O_2 level of around 7%. Perforated passive MAP reached this composition only after a 9-day transient period, while the non-perforated packages developed the detrimental conditions of hypoxia and excessive CO_2 levels, causing damage to the fig fruit. The MAMA-type packaging was found to be highly effective in maintaining firmness and reducing weight loss and decay of figs.

Optimizing package perforation is an efficient way to obtain a desirable MA composition for produce preservation. The effect of microperforated biaxially oriented polypropylene films on the storage of several fig varieties at 0°C was investigated in a series of studies performed by Villalobos *et al.* (2014, 2016a, 2018b). Three types of experimental packages termed M10, M30, and M50 were tested, bearing respectively 16, five, or three microperforations (100 μm in diameter) per package of approximately 400 g of fruit, in comparison with the control bearing five 9 mm perforations per package. The samples with three microperforations per package (M50) consistently showed the best performance in terms of reducing weight loss, maintaining fruit quality, and delaying decay. The steady-state CO_2/O_2 concentrations (in kPa) established in such packages were 13.6/8.7, 11.9/10.0, and 17.25/4.95 for Cuello Dama Blanco, Cuello Dama Negro, and San Antonio figs, respectively (Villalobos *et al.*, 2016a). In a separate trial, the M50 packages of Banane breba figs accumulated approximately 19 kPa CO_2 and 3 kPa O_2 (Villalobos *et al.*, 2014). Aroma is one the most storage-sensitive quality parameters of fresh figs (Tsantili and Knee, 1991). Keeping the fruit in optimal M50 microperforated MAP packages, among other advantages, also allowed maintenance of the composition of aroma volatiles, minimizing the storage-associated changes, especially in the San Antonio variety (Villalobos *et al.*, 2018b). Microperforated MAP was also beneficial for the storage of rare Algerian figs for up to 10 days at 4°C, achieving a delay of ripening and retention of bioactive compounds (Zidi *et al.*, 2020).

New trends in MAP of figs

MAP technology has undergone considerable progress associated with the introduction of cutting-edge approaches in order to address specific requirements of crops (Belay *et al.*, 2016). An incomplete list of such innovations includes interactive packaging materials with differential thermoresponsive permeability, nanotechnology applications, integrated sensors, non-traditional atmospheric ingredients (e.g. noble gases), functional films, and edible coatings with special (e.g. antimicrobial or antioxidant) properties (Wilson *et al.*, 2019). The design of novel packaging solutions such as MAMA packaging is based on modeling respiration and gas-diffusion processes involved in MAP generation (Shinde *et al.*, 2018). One of the most acute issues in food packaging today is replacing the conventional environment-polluting plastic films with sustainable biodegradable/compostable materials (Mistriotis *et al.*, 2016).

Many of these innovations have been implemented with figs. In particular, impressive results have been achieved with fresh-fig preservation using edible coatings, as presented in the following section. Mathematical modeling of fig-fruit respiration serving as a basis for MAP optimization was performed by Waghmare *et al.* (2014), and Ghosh and Dash (2020). The efficacy of the MAMA packaging approach for the preservation of Masui Dauphine figs was demonstrated by Kim *et al.* (2012). An original approach to MAP design was used by Ma *et al.* (2019) who assembled a package comprising pieces of different functional films responsible for water-vapor exchange, O_2 supply, and 1-MCP release, so that the relative area of these films determined the package performance. Packaging Blanrick figs in such multiple-piece MAP extended their life under non-refrigerated (25°C) conditions from 2 to 6 days while all determined variables (water loss, firmness, decay, SSC, TA, vitamin C content, and sensory evaluation) agreed with efficient quality maintenance. A similar approach was adopted by Liu *et al.* (2020) using a tray of polyethylene terephthalate covered with two films, one water-vapor permeable and the other O_2 permeable. Storage of figs at 25°C was doubled to 4 days, with levels of O_2 and CO_2 being approximately 4% and 17%, respectively.

Sanjay (2018) reported that packaging fresh figs in biodegradable chitosan/polyvinyl alcohol composite film with the addition of silver nanoparticles extended their shelf life from 3 to 7 days by maintaining fruit color and firmness, reducing weight loss, controlling microbial growth, and delaying decay development. Another formulation of biodegradable composite film, comprising starch, polyvinyl alcohol, glycerol, and citric acid, exerted strong antibacterial activity toward both Gram-positive (*Listeria monocytogenes*) and Gram-negative (*Escherichia coli*) foodborne pathogens and inhibited the deterioration of the film-wrapped fresh figs (Wu *et al.*, 2017). Interestingly, citric acid was the film ingredient responsible for its antimicrobial effect.

13.5.7 Edible coatings

In contrast to other fruits (e.g. citrus), ripening of figs is accompanied by degradation of their protective epicuticular wax layer (Chessa *et al.*, 1992). Therefore, fully mature figs are especially prone to desiccation and shriveling. This trait is beneficial for fruit drying but detrimental for preservation of fresh products. Covering the fruit with an artificial edible layer may compensate for the loss of its natural surface protection and extend the postharvest shelf life. Edible coatings are typically prepared from biopolymers (proteins, polysaccharides, lipids, or combinations of these components) in order to protect the product from physical, chemical, and biological deterioration during storage and distribution (Galus, 2019). In addition to desiccation control, the coatings modulate gas exchange of the fruit, delay its aging, and may carry additives helpful for maintenance of the product quality (Saberi and Golding, 2019).

Very few studies on edible coatings for fig fruits were performed in the 1990s and early 2000s. Coating figs with the sucrose-ester formulation Semperfresh reduced the incidence of microbial spoilage (decay) of spring-season fruit but had little effect on other aspects of the fruit quality (Baccaunaud *et al.*, 1995). Applying 2% antitranspirant pinolene (Vapor Gard) was effective in controlling weight loss of Bursa Black figs. However, a higher concentration (5%) of the same material enhanced fruit decay incidence (Çelikel *et al.*, 1998).

During the last decade, however, edible coatings have become a popular theme, suggested by many authors as a preservation means for fresh figs. Most studies use polysaccharides as the base for the fig coating, typically one of two types: chitosan or natural plant-derived gel (mucilage). Chitosan is a deacetylated derivative of chitin, an abundant naturally occurring polysaccharide, a biopolymer found in the exoskeleton of crustacean and fungal cell walls. A good film-forming capacity and intrinsic antibacterial and antifungal activity make chitosan a promising material for food edible coatings (Elsabee and Abdou, 2013). Chitosan coating was shown to reduce the weight loss of stored figs and to improve their antioxidant status, preventing the fruit from browning (Meng *et al.*, 2014; Adiletta *et al.*, 2019). Furthermore, a chitosan coating reduced decay incidence and quantitative losses of Sultani figs, extending their life from below 1 week up to 3 weeks at 0°C, with a subsequent 2 days of shelf life at room temperature (Mostafa *et al.*, 2019).

Further improvement of the chitosan-based coating performance may be reached by amending the formulation with additional ingredients, particularly antimicrobial materials such as thymol essential oil (Saki *et al.*, 2019) and zinc oxide nanoparticles (Lakshmi *et al.*, 2018). Aparicio-García *et al.* (2020) studied the preservation of Black Mission figs as affected by coating formulations comprising chitosan with or without the addition of propolis and/or nanoparticles fabricated from the same ingredients (i.e. chitosan and propolis). A pure chitosan coating was most efficient for weight loss reduction during 12 days of storage at 4°C, while adding propolis and chitosan nanoparticles enhanced the antifungal effect against *Aspergillus flavus*, without changing the sensory characteristics of the fruit. The most remarkable finding of this study was an approximate 10-fold reduction in the aflatoxin level in the fruit by all composite coatings. Another antifungal coating formulation comprising chitosan, cinnamon essential oil, and aqueous extract of Roselle calyces (*Hibiscus sabdariffa* L.) was tested by Saavedra *et al.* (2020) with Black Mission figs inoculated with *Alternaria alternata*. When the coating was applied to figs before inoculation (protective mode), it prevented the physiological changes associated with pathogenesis (enhanced respiration and weight loss) and reduced the rot incidence during cold storage at 5°C. However, the efficacy of the coating was insufficient to prevent disease after transfer to 20°C, or in cases when the coating was applied after inoculation (curative mode).

Reyes-Avalos *et al.* (2016) used a composite alginate–chitosan bilayer coating on Mission figs. In such a coating, the inner alginate layer ensures good surface adhesion due to the calcium-mediated binding to fruit compounds (e.g. pectin), while the chitosan layer is tightly bound to the alginate by electrostatic forces, providing antimicrobial properties to the coating (Poverenov *et al.*, 2014). The alginate–chitosan coating improved firmness and color retention by Mission figs and reduced their weight loss and decay indices, extending the fruit shelf life from 6 to 15 days, when stored at 6°C (Reyes-Avalos *et al.*, 2016). Furthermore, the same team showed that the alginate–chitosan bilayer coating also preserved a stable level of bioactive compounds such as anthocyanins and other polyphenols, and a high antioxidant capacity of the fruit (Reyes-Avalos *et al.*, 2019). The alginate–chitosan coating delayed fig aging via modification of the internal fruit atmosphere (reducing O_2 and enhancing CO_2 levels) without deterioration of their sensory characteristics throughout the 15 days of storage (Reyes-Avalos *et al.*, 2019). In contrast to many other studies, the polysaccharide-based coating tested by Paolucci *et al.* (2020) contained no chitosan but a blend of alginic acid and agar. This coating inhibited fruit weight loss, softening, and microbial contamination, especially when the formulation was amended with pomegranate peel extract.

Mucilages are natural viscous water-dispersible plant hydrocolloids, typically of a polysaccharide nature (Lee *et al.*, 2014). For example, *Aloe vera* mucilage consists of 99.5% water with mannose-rich polysaccharides (e.g. acetylated glucomannan) making up most of the remaining dry matter (Hamman, 2008). Similarly, linear galactan and highly branched xyloarabinan were identified as major high-molecular-weight mucilage components in the cladodes of the prickly pear cactus, *Opuntia ficus-indica* (di Lorenzo *et al.*, 2017). Coating Dottato breba figs with mucilage solution from *Opuntia* cladodes maintained the fruit FW, firmness, appearance, and total carotenoid content for 14 days, at 4°C. Furthermore, it reduced the *Enterobacteriaceae* load on the fruit throughout storage (Allegra *et al.*, 2017), increased the amount of carbohydrates, and mitigated the decline of amino acid content (Allegra *et al.*, 2018). Significant improvement of keeping quality was reported for fresh Poona figs coated with *A. vera* gel. For example, the coating reduced decay incidence from 50% to less than 10% in fruit stored for 4 days under non-refrigerated conditions, with a concomitant decrease of fruit shriveling and browning (Marpudi *et al.*, 2013). Siyah figs coated with *A. vera* gel maintained higher firmness and nutritional quality (contents of anthocyanins, total phenolic compounds, flavonoids, ascorbic acid, and total antioxidant capacity) compared with the non-coated control (Mirshekari *et al.*, 2020).

The potential of materials other than polysaccharides to serve as a fig-fruit coating has attracted much less attention. Pinto *et al.* (2020) compared the efficacy of coating formulations based on polysaccharides (starch), proteins (whey protein), and lipids (coconut oil) for preservation of figs stored for 6 days at 25°C. While all coating types slowed down fruit degradation compared with the non-treated control, the lipid-based formulation was most efficient in maintaining the fruit appearance and reducing weight loss and desiccation due to its higher water-vapor barrier properties. Preservation of fresh figs with defatted soybean meal extract, and in particular controlling weight loss, was partially attributed to its coating effect (Villalobos *et al.*, 2016b). Using a composite polysaccharide-protein coating based on chitosan/gelatin combination allowed successful preservation of fresh Brown Turkey and Autumn Honey figs for 4 weeks at 1–2°C cold storage with a subsequent 3 days of shelf life at 10–12°C (M.A. Flaishman, Agriculture Research Organization, Israel, 2019, personal communication). Such storage duration can enable marine transportation of fresh figs to distant markets.

13.5.8 Chemical and biological treatments

Due to the high perishability of fresh fig fruits, many attempts have been made to extend their postharvest life by applying various chemical, physical, and biological factors, mostly with the aim of controlling the causative spoilage agents. In this and subsequent sections, we present some of the treatments used for this purpose.

Fungicide treatments can control many postharvest fig pathogens. However, no postharvest fungicides have been registered for this commodity (Molinu *et al.*, 2007). In the case of pollinated Smyrna-type varieties, an alternative strategy involves a preharvest fungicide treatment of inedible caprifigs—a major source of infection transmitted by fig wasps to the edible fruit (Michailides *et al.*, 1996). Preharvest application of fungicides on edible common figs is practiced in some countries, provided the appropriate preharvest interval is maintained. For example, preharvest application of the fungicide mancozeb was found to be more effective for quality preservation of Roxo de Valinhos figs than the copper sulfate spray usually used in Brazil, especially when followed by a 4% $CaCl_2$ dip (de Paula *et al.*, 2007). Kurubar *et al.* (2013) reported extension of non-refrigerated storage of Poona figs from 2 to 4 days by combining preharvest $CaCl_2$ sprays with postharvest carbendazim application. Postharvest application of the reduced-risk fungicide azoxystrobin was tested in low doses (d'Aquino *et al.*, 2015). The treatment reduced the incidence of external rots on the fruit but had no effect on internal decay. It was recommended to combine the fungicide application with efficient preharvest pest control, as insects are often the carriers of the pathogen conidia (d'Aquino *et al.*, 2015). However, the regulatory considerations of postharvest fungicide applications are debated.

Another approach to postharvest disease control in figs is to apply non-specific disinfectants. For example, chlorine dioxide fogging disinfected the storage atmosphere, controlled the epiphytic microbial populations on the fruit surface, and significantly reduced the decay indices of Bursa Black figs, predominantly caused by gray mold (Karabulut *et al.*, 2009). A similar effect was reached by fogging Brown Turkey and Autumn Honey figs with an active chlorine-releasing sanitizer, sodium troclosene, resulting in 5–6 days of shelf-life extension at 10°C and 2–4 days at 20°C (Horev *et al.*, 2015).

Several inorganic compounds showed good potential to improve the keeping quality of fresh figs. Dipping the fruit in 1% solutions of sodium carbonate (Venditti *et al.*, 2005) and sodium bicarbonate (Antunes *et al.*, 2008) significantly reduced the decay incidence during storage of different fig varieties. Postharvest treatment with 4% $CaCl_2$ extended the storage and shelf life of Poona figs, primarily by improving the fruit texture, controlling the microbial population on their surface, and inhibiting aging (Irfan *et al.*, 2013). Combined treatment with 2% sodium chloride (NaCl) and $CaCl_2$ extended the shelf life of figs at room temperature by 1 day, while a 2-day extension was reached when the fruits were dipped in 1.5% potassium metabisulfite ($K_2S_2O_5$) solution (Adiyaman *et al.*, 2013). The latter effect was most probably related to the generation of SO_2, which is known to control fungal decay on grapes and other fruits, including figs (de la Plaza, 2003). A detailed study of the effect of SO_2 on the storage of dark and green figs was performed by Cantín *et al.* (2011). The treatment controlled phytopathogenic fungi on the fig surface, such as *Alternaria*, *Rhizopus*, *Botrytis* and *Penicillium* species, and reduced the fruit decay indices. Applying 25 $\mu l\ l^{-1}$ of SO_2 as a single fumigation agent before entering cold storage alleviated the fruit discoloration caused by other SO_2 continuous treatments, such as the use of SO_2-generating pads.

Various organic compounds were also tested as potential natural preservatives of fresh figs. Fumigation with 100 ppm. of acetic acid reduced the decay of fresh figs after 14 days of storage at 8°C from 39.3% to 2.4% with no visible phytotoxic damage (Venditti *et al.*, 2005). Applying acetic acid vapor after curing at 36°C improved the efficacy of fumigation (Molinu *et al.*, 2007). Dipping the fruits in 96% ethanol, alone or with the addition of 1% cinnamic acid, reduced the decay of Bianca figs, especially when applied after the first 3 days of storage at 15°C (d'Aquino *et al.*, 2003). However, treatments with acetic acid caused browning of the epicarp of whitish varieties (Molinu *et al.*, 2007), and cinnamic acid further affected the fruit flavor (d'Aquino *et al.*, 2003).

Dipping the figs in thymol solutions of 0.3 mM, and especially 0.6 mM, markedly improved the storage capacity of Sultani figs. Thymol solutions of 0.6 mM reduced the decay indices from 75–85% to 9%, and prolonged storage at 0°C from 1 to 3 weeks (Mostafa *et al.*, 2019). Similar trends were reported by Saki *et al.* (2019) for Siah figs and a thymol concentration of 1.3 mM. The sensory characteristics of the thymol-treated fruit were above or insignificantly different from those of the control. In both studies, fruit preservation was further improved when thymol application was combined with a chitosan coating.

Villalobos *et al.* (2016b) investigated the effect of soybean extract on the storage of Cuello Dama green (Blanco) and dark (Negro) figs. The extract was rich in phenolic compounds and prepared by 80% ethanol extraction of defatted soybean meal with further removal of ethanol and redispersion of the extract in water. Application of the extract significantly reduced the decay incidence after 14 days of cold storage and controlled the mold population on the fruit surface, in particular in the dark variety (Villalobos *et al.*, 2016b). The preservation effect was greatly enhanced by combining the soybean extract with MAP. A further in-depth study confirmed the inhibitory effect of the soybean extract on the mold and yeast populations on the surface of several fig varieties (Villalobos *et al.*, 2017a). Another antifungal compound extracted from legume seeds, the chickpea glycoprotein vicilin, showed promising potential for controlling fig spoilage. Treating with vicilin reduced the disease severity in *Alternaria*-inoculated figs from 55% to 15% after 7 days of incubation (Abbas *et al.*, 2020).

Biological control is the natural alternative to the use of synthetic pesticides. It is based on the application of natural microbial antagonists for the control of postharvest pathogens (Droby *et al.*, 2016). The microbial communities of fig fruit and orchards have served as a source for isolation of many microbial species used as biocontrol agents, often for controlling diseases of other crops. Fifty-five bacterial species producing compounds related to phytopathogen biocontrol were isolated from figs and fig orchard soils (Mota *et al.*, 2017). In particular, six strains of *Bacillus thuringiensis* and *Bacillus mojavensis* were isolated from fig orchard soils based on their high chitosanase activity and showed potential to act as biocontrol agents against mycotoxigenic fungi of fig orchards (Öztopuz *et al.*, 2018). *Paecilomyces lilacinus*, a fungus naturally occurring in the cavities of figs, was evaluated by Subbarao *et al.* (1993) as a potential antagonist for the biocontrol of the fig pathogens *A. niger* and *F. moniliforme*, the causative agents of smut and endosepsis, respectively. Spraying a spore suspension of *P. lilacinus* on caprifigs reduced endosepsis incidence by 50% compared with untreated trees. However, the same approach was ineffective with Calimyrna figs, presumably due to the drier conditions during the fruit development period.

Yeast samples isolated from fig fruit were presented as promising biocontrol agents against postharvest fruit diseases (Ruiz-Moyano *et al.*, 2016), in particular against pathogens of citrus (Arras *et al.*, 1999), apples (Scherm *et al.*, 2003), nectarines (Ruiz-Moyano *et al.*, 2016), and cherries (de Paiva *et al.*, 2017). However, the efficacy of yeast isolates for biocontrol of fig diseases has not yet been reported. Overall, the application of biocontrol approaches for the preservation of fresh figs is understudied and still at the initial steps, but is a promising approach.

13.5.9 Physical treatments

Heat treatments and cold plasma

Postharvest heat treatments are known to modulate host–pathogen interactions (Schirra *et al.*, 2000). Various heat applications were tested with fig fruit. For example, curing at 36°C for 24 h just slightly reduced fruit decay when applied alone but showed much better efficacy if followed by acetic acid fumigation. For example, with Niedda Longa figs, the decay incidence was reduced by this combined treatment from 83% in non-treated fruit to 18% after 2 weeks of storage at 5°C (Molinu *et al.*, 2007). Similar heat treatment of 38°C for 24 h had a negligible effect on the decay of Petrelli breba figs stored at 25°C but caused severe fruit browning at 0°C (Piazzolla *et al.*, 2018). Keeping the fruit at 50°C for 60 min was also detrimental, almost doubling the decay incidence from 35.1% in the non-treated control to 65.3% (both after 3 days at 25°C). At the same time, the 30 min exposure to 50°C showed a positive effect, reducing the decay to 8.1%.

Dipping Lampa Preta figs for 2 min in water at 45°C resulted in skin damage that was aggravated in the case of heated 1% $CaCl_2$ solution (Antunes *et al.*, 2003). However, treatment with hot sodium carbonate solutions (1 min in 0.5% solution at 60°C) had a positive effect on the keeping quality of Niedda Longa figs, reducing the decay from 26% in the control to 0% after 1 week of storage at 5°C and from 50% to 14% after 2 weeks (Molinu *et al.*, 2006). This apparent discrepancy in the effects of heated salt solutions was most probably related to variety peculiarities and different treatment conditions.

In addition to heat applications, other postharvest physical treatments were also tested with fig fruit. Abbaszadeh *et al.* (2018) investigated the effect of cold plasma on the preservation of Siyah figs. During the treatment, the fruit were either kept inside plastic bags or left open with subsequent plastic packaging. After 11 days of storage at 3°C, the plasma-treated fruit maintained better quality (color, texture, odor, appearance, and general acceptability) than the non-treated control. Optimal exposure duration was 30 s for the fruit treated within plastic bags, and 90 s for the fruit treated out of bags.

Electromagnetic irradiation

Infrared radiation was used for fig heat treatment by Trivittayasil *et al.* (2014) with further fruit storage for 6 days at 7°C. Heating the fig-fruit (Toyomitsuhime variety) surface on the ostiole side to 50°C took 30 s and resulted in a reduction of decay incidence from 50% to 5.6% without significant heat injury. Further heating up to 70°C completely prevented fruit decay but took 1.8 min and resulted in fruit softening and discoloration. The efficacy of infrared decontamination of the fig-fruit surface was enhanced by combination with ultraviolet C (UVC) irradiation (Hamanaka *et al.*, 2011). The potential of 4 s of steam application for extending the life of Yediveren breba figs was tested by Özer and Sen (2003). The treatment resulted in significant enhancement of weight loss and decay incidence, presumably due to the heat damage.

A 20 min exposure to short-wave UVC irradiation reduced the decay and cracking of Bursa Black figs and kept their quality above the marketability limit for 28 days of storage at 0–1°C, compared with only 14 days in the non-treated control (Bal, 2012). Treatment of Roxo de Valinhos figs with UVC at doses of 0.71, 1.32, 2.64, and 4.01 kJ m^{-2} (corresponding to exposure durations of 16, 30, 61, and 91 s) had no effect on their ripening but reduced the rot incidence in fruit stored for 7 days at 10°C or 20°C (Usberti and Ferraz, 2021). However, the highest UVC dose applied demonstrated some phytotoxic damage, expressed as enhanced wilting and stains.

As mentioned above, UVC enhances the efficacy of infrared irradiation for the preservation of fresh figs (Hamanaka *et al.*, 2011). The improvement in fruit color (Fig. 13.4), anthocyanin, and total phenolic compounds content was reached by brief exposure to irradiation from a xenon flash lamp emitting short pulses of high-intensity, broad-spectrum light comprising ultraviolet, visible, and near-infrared wavelengths (Rodov *et al.*, 2012).

Gamma-irradiation in doses 1 or 2 kGy did not significantly improve the keeping quality of green figs harvested unripe for processing into traditional Brazilian confectionary (Silva *et al.*, 2012). However, irradiation of Poona figs at doses of 0.5 or 1 kGy allowed maintenance of their quality for up to 15 days of storage at 5°C (Waghmare and Annapure, 2018). At the end of storage, the irradiated figs had lower decay and microbial load, and better aroma, visual appearance, and general acceptance, compared with the non-treated control. Fruit preservation was further improved when irradiation was combined with MAP by treating the fruit enclosed in polypropylene bags. At the doses mentioned above, the fruit firmness was close to that of the control, but higher irradiation doses (3–4 kGy) resulted in considerable fig softening.

Fig. 13.4. Effect of 90 s exposure to pulsed light on the appearance of Brown Turkey figs. The fruit were exposed to pulsed light on the day of harvest and subsequently kept for 5 days in the dark at 20°C. During the pulsed light treatment, part of the fruit surface was covered with light-impermeable rectangular adhesive patches, which were removed immediately after the treatment. Note the pale color of the fruit surface protected from the pulsed light by the patches. Reproduced from Rodov *et al.* (2012), with permission.

13.6 Concluding Remarks

Fresh figs are one of the most challenging produce for postharvest handling due to their high perishability and the dilemma between fruit sensory quality and longevity. Thus, tree-ripe fruit of the best eating quality is too perishable to be delivered successfully to consumers, especially to the distant ones, while fruit of commercial maturity can endure the supply chain conditions but it has inferior sensory quality. Due to this situation, most consumers have little chances to enjoy the "genuine" fig flavor. Obviously, such a dilemma exists with some other delicate fruits as well, but fresh figs present one of the extreme examples. The traditional value chain of fresh figs uses the basic means of produce preservation, such as careful selective manual harvest, cautious handling, protective packaging, fast precooling, optimal storage temperature, and prompt delivery under strict cool chain conditions in order to reach a satisfactory compromise between fruit quality and longevity (Fig. 13.5). This process depends on extensive employment of skilled manual labor and costly, environmentally unfriendly airfreight transportation, and is still associated with an undeniable risk of produce losses.

The last decade was marked by a peak in fresh-fig research that resulted in a vast body of new knowledge and numerous new publications on this important and challenging subject, including postharvest aspects. These new developments have provided additional chances to extend the postharvest life of fresh figs without compromising their sensory quality and nutritional value. Much progress has been made in understanding the biological mechanisms underlying fig-fruit development and ripening, and in particular the superior quality and storability of pollinated fruit compared with parthenocarpic fruit. Insights into the hormonal and molecular regulation of this phenomenon opens up avenues for the improvement of maintenance of fruit quality, for example by imitating the hormonal status of pollinated fruit in parthenocarpic common figs. Regulating ethylene sensitivity by preharvest 1-MCP treatment appears to be more efficient for common fig preservation than traditional postharvest application because the developing fruit passes the most competent 1-MCP-sensitive stage while still on the tree. However, the efficacy of applying 1-MCP to figs as orchard sprays demands further investigation.

Currently, manual harvest remains the only feasible option with fresh figs. However, further technical developments may allow the use of robotic picking in fig harvests, saving many labor costs. The basic feasibility of this approach has been confirmed by successful testing of picking robots with other delicate fruits such as strawberries. Other innovations facilitating the selective fig harvest might be minute sensors for instant non-destructive evaluation of fruit maturity and/or detection of disorders. An additional area of fig industry where sensor technologies and robotics are already implemented, or will be available soon, are packinghouse operations, such as grading, sorting, and packaging.

Encouraging progress has been achieved in extending the fruit life by postharvest technologies such as MAP, edible coatings, and modern chemical and physical treatments. In recent years, the approach of edible coatings has attracted much attention as a potential means for fresh-fig preservation. Several promising formulations have been developed, typically comprising a biopolymer matrix (e.g. chitosan) amended with natural antimicrobials, often in a nanoparticle form. Such formulations have proven efficient for alleviating fig-fruit weight loss and physiological deterioration, while at the same time controlling microbial spoilage.

In MAP technology, novel packaging solutions allow overcoming the dilemma between weight loss control caused by high in-package humidity and the risk of enhanced fruit decay that was often encountered in early studies. These solutions include modified-humidity packaging using materials with enhanced water-vapor permeability such as Xtend, active MAP based on flushing packages with optimal gas mixes and/or modifying the atmospheric composition by gas-releasing or absorbing inserts, and applying microperforated materials in the form of MAMA packaging. One of the advantages of MAP is its applicability for fruit transportation, for example in the form of active MA pallets (e.g. the TransFresh system). The challenge of the food-packaging industry today is switching from conventional environment-polluting plastic films to sustainable (e.g. compostable) materials.

CA storage has been known for decades as an efficient method for fresh-fig preservation. However, the traditional CA facilities designed for long-term storage of bulky fruit loads such as apples are normally not suitable for short-life commodities such as figs. At the same time, the more flexible smaller-scale CA technologies such as Palliflex that are successfully used with berries and other soft fruits can provide a good solution for fresh figs as well. Modern chemical and physical treatments such as disinfectant fogging or UVC irradiation have also shown promising results for fresh-fig preservation. It should be kept in mind, however, that while some approaches, such as disinfectant fogging, may be combined with the existing packinghouse practices relatively easily, others, such as UVC treatment, would demand significant changes in the workflow. From this viewpoint, methods allowing fruit treatment within the package, such as cold plasma application, are of special interest.

Improving the storage potential of figs by innovative approaches may have several positive outcomes for the industry and for the consumers (Fig. 13.5). First, efficient postharvest preservation can control spoilage and allow marketing of fruit harvested at a relatively advanced (tree-ripe) maturity stage that has the best edible quality. Thus, the new technologies can alleviate the dilemma between edible quality and storage potential of the fresh figs determined by their harvesting stage. Such technologies may extend the postharvest life of figs to the duration enabling their marine shipment to distant markets without compromising on fruit quality, instead of the expensive airfreight transportation. Shifting from air to marine transportation would greatly improve the cost efficiency of the industry and the growers' rewards. In addition, without disputing the general importance of postharvest temperature management, extending the life of non-refrigerated figs by 2–4 days as reported in some studies cited in this chapter, may open up additional possibilities for their short-term delivery and marketing, for the benefit of local growers and consumers. Another modern trend that may be important for reducing fresh-fig losses is the expansion of direct marketing schemes such as e-commerce that shorten the route between the grower and the consumer and exclude the most problematic supply chain link of "shelf life" associated with produce wastage.

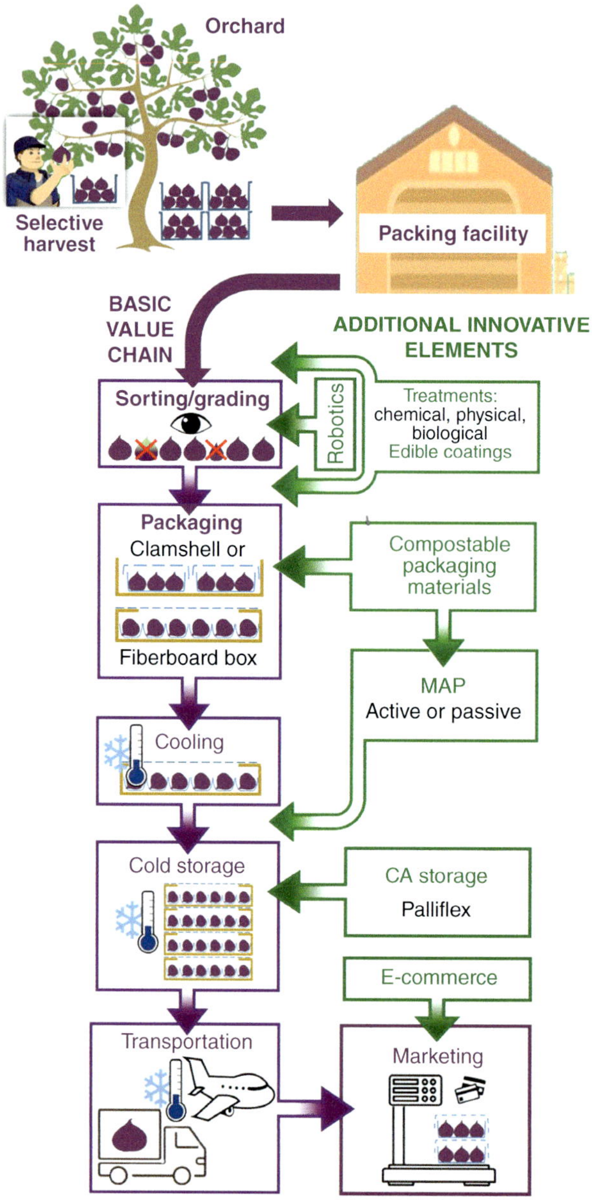

Fig. 13.5. Fresh figs value chain: a schematic diagram. The basic traditional path is shown in purple, while innovative elements are shown in green. CA, controlled atmosphere; MAP, modified-atmosphere packaging.

References

Abbas, E., Osman, A., and Sitohy, M. (2020) Biochemical control of *Alternaria tenuissima* infecting post-harvest fig fruit by chickpea vicilin. *Journal of the Science of Food and Agriculture* 100, 2889–2897.

Abeles, F.B., Morgan, P.W., and Saltveit, M.E. (1993) *Ethylene in Plant Biology*, 2nd edn. Academic Press, San Diego, California.

Abbaszadeh, R., Alimohammad, K., and Zarrabi Ekbatani, R. (2018) Application of cold plasma technology in quality preservation of fresh fig fruit (Siyah): a feasibility study. *International Journal of Horticultural Science and Technology* 5, 165–173.

Adiletta, G., Zampella, L., Coletta, C., and Petriccione, M. (2019) Chitosan coating to preserve the qualitative traits and improve antioxidant system in fresh figs (*Ficus carica* L.). *Agriculture* 9, 84–95.

Adiyaman, P., Kanchana, S., Hemalatha, G., Ananthan, M., Gopal, N.O., *et al.* (2013) Influence of pretreatments and prepackaging on post harvest shelf life of underutilized fruits (star fruit, egg fruit and fig). *Indian Journal of Science and Technology* 6, 5263–5270.

Aharoni, N., Rodov, V., Fallik, E., Afek, U., Chalupowicz, D., *et al.* (2007a) Modified atmosphere packaging for vegetable crops using high-water vapor-permeable films. In: Wilson, C.L. (ed.) *Intelligent and Active Packaging for Fruits and Vegetables*. CRC Press, New York, pp. 73–112.

Aharoni, N., Rodov, V., Fallik, E., Porat, R., Pesis, E., and Lurie, S. (2007b) Controlling humidity improves efficacy of modified atmosphere packaging of fruits and vegetables. *Acta Horticulturae* 804, 121–128.

Ait-Oubahou, A., Brecht, J.K., and Yahia, E.M. (2019) Packing operations. In: Yahia, E. (ed.) *Postharvest Technology of Perishable Horticultural Commodities*. Woodhead Publishing, Cambridge, UK, pp. 311–351.

Aksoy, U. (1981) Studies on fruit development, maturation and cold storage of the fig (*Ficus carica* L.) fruits cvs Akça, Göklop and Sarilop. PhD thesis, Faculty of Agriculture, Ege University, Bornova, Izmir, Turkey (in Turkish).

Aksoy, U. (1997) Harvest and drying of figs. *Advanced Course on Fig Production*. Ege University, Izmir, Turkey.

Aksoy, U., Can, H.Z., Misirli, A., and Kara, S. (2003a) Fig (*Ficus carica* L.) selection study for fresh market in western Turkey. *Acta Horticulturae* 605, 197–203.

Aksoy, U., Balci, B., Can, H.Z., and Hepaksoy, S. (2003b) Some significant results of the research-work in Turkey on fig. *Acta Horticulturae* 605, 173–181.

Alara Agri (2018) Fig. Available at: http://alaraagri.com/fig/ (accessed December 7, 2021).

Alexander, L. and Grierson, D. (2002) Ethylene biosynthesis and action in tomato: a model for climacteric fruit ripening. *Journal of Experimental Botany* 53, 2039–2055.

Alique, R. and de la Plaza, J. (1990) Effects of high CO_2 pretreatments and ethylene removal on the ripening of refrigerated figs. In: *Proceedings of XXIII International Horticulture Congress*, Firenze, Italy.

Allegra, A. and Colelli, G. (2017) The effect of passive atmosphere on quality of 'Dottato' breba fig stored at low temperature. *Acta Horticulturae* 1173, 315–318.

Allegra, A., Sortino, G., Inglese, P., Settanni, L., Todaro, A., and Gallota, A. (2017) The effectiveness of *Opuntia ficus-indica* mucilage edible coating on post-harvest maintenance of 'Dottato' fig (*Ficus carica* L.) fruit. *Food Packaging and Shelf Life* 12, 135–141.

Allegra, A., Gallotta, A., Carimi, F., Mercati, F., Inglese, P., and Martinelli, F. (2018) Metabolic profiling and post-harvest behavior of 'Dottato' fig (*Ficus carica* L.) fruit covered with an edible coating from *O. ficus-indica*. *Frontiers in Plant Science* 9: 1321.

Alturki, S. (2013) Utilization of modified atmosphere packaging to extend the shelf-life of fresh figs. *Biotechnology* 12, 81–86.

Antunes, M.D.C., Oliveira, R., Miguel, M.G., Martins, M.A., Pica, C., and Neves, M.A. (2003) The effect of calcium chloride and thermal postharvest treatments on storage ability and quality of fig fruit. *Acta Horticulturae* 604, 715–719.

Antunes, M.D.C., Costa, P., Miguel, M.G., Martins, D., Neves, M.A., *et al.* (2008) The effect of postharvest treatments with sodium bicarbonate or acetic acid on storage ability and quality of fig fruit. *Acta Horticulturae* 798, 279–284.

Aparicio-García, P.F., Ventura-Aguilar, R.I., del Río-García, J.C., Hernández-López, M., Guillén-Sánchez, D., *et al.* (2020) Edible chitosan/propolis coatings and their effect on ripening, development of *Aspergillus flavus*, and sensory quality in fig fruit, during controlled atmosphere. *Plants* 10: 112.

Arras, G., Dessi, R., Sanna, P., and Arru, S. (1999) Inhibitory activity of yeasts isolated from fig fruits against *Penicillium digitatum*. *Acta Horticulturae* 485, 37–46.

Arvaniti, O.S., Samaras, Y., Gatidou, G., Thomaidis, N.S., and Stasinakis, A.S. (2019) Review on fresh and dried figs: chemical analysis and occurrence of phytochemical compounds, antioxidant capacity and health effects. *Food Research International* 119, 244–267.

Ayhan, Z. and Kara Cay, E. (2011) Preservation of the 'Bursa Siyahi' fresh fig under modified atmosphere packaging (MAP) and cold storage. *International Journal of AgriScience* 1: e27.

Baccaunaud, M., Raynal, J., and Babotilene, F. (1995) Figs: maintaining quality after harvest. *Arboriculture Fruitière* 489, 37–42.

Bahar, A. and Lichter, A. (2018) Effect of controlled atmosphere on the storage potential of Ottomanit fig fruit. *Scientia Horticulturae* 227, 196–201.

Bal, E. (2012) Effect of postharvest UV-C treatments on quality attributes of fresh fig. *Bulgarian Journal of Agricultural Science* 18, 191–196.

Baskaya, M. and Crane, J.C. (1950) Comparative histology of naturally parthenocarpic, hormone-induced parthenocarpic, and caprified fig syconia. *Botanical Gazette* 111, 395–413.

Beaudry, R. (1999) Effect of O_2 and CO_2 partial pressure on selected phenomena affecting fruit and vegetable quality. *Postharvest Biology and Technology* 15, 293–303.

Beaudry, R. (2017) Atmosphere control in packages and CA rooms: biology, technology, and future opportunities. *Acta Horticulturae* 1256, 541–548.

Belay, Z.A., Caleb, O.J., and Opara, U.L. (2016) Modelling approaches for designing and evaluating the performance of modified atmosphere packaging (MAP) systems for fresh produce: a review. *Food Packaging and Shelf Life* 10, 1–15.

Ben-Yehoshua, S. and Rodov, V. (2003) Transpiration and water stress. In: Bartz, J.A. and Brecht, J.K. (eds) *Postharvest Physiology and Pathology of Vegetables*. Marcel Dekker, New York, pp. 111–159.

Ben-Yehoshua, S., Beaudry, R.M., Fishman, S., Jayanty, S., and Mir, N. (2005) Modified atmosphere packaging and controlled atmosphere storage. In: Ben-Yehoshua, S. (ed.) *Environmentally Friendly Technologies for Agricultural Produce Quality*. Taylor & Francis, Boca Raton, Florida, pp. 61–112.

Biale, J.B. and Young, R.E. (1981) Respiration and ripening in fruits–retrospect and prospect. In: Friend, J., Rhodes, M.J.C. (eds) *Recent Advances in the Biochemistry of Fruit and Vegetables*. Academic Press, New York, pp. 1–39.

Blankenship, S.M. and Dole, J.M. (2003) 1-Methylcyclopropene: a review. *Postharvest Biology and Technology* 28, 1–25.

Bouzo, C.A., Travadelo, M., and Gariglio, N.F. (2012) Effect of different packaging materials on postharvest quality of fresh fig fruit. *International Journal of Agriculture and Biology* 14, 821–825.

Byeon, S.E. and Lee, J. (2020a) Differential responses of fruit quality and major targeted metabolites in three different cultivars of cold-stored figs (*Ficus carica* L.). *Scientia Horticulturae* 260: 108877.

Byeon, S.E. and Lee, J. (2020b) Fruit quality and major primary metabolites differ across production systems in cold-stored figs (*Ficus carica* L.). *Scientia Horticulturae* 274: 109669.

Byeon, S.E. and Lee, J. (2021) Fruit maturity differentially affects fruit quality and responses of targeted metabolites in cold-stored figs (*Ficus carica* L.). *Journal of the Science of Food and Agriculture* 101, 673–683.

Çakmak, B., Alayunt, F., Akdeniz, C., Can, Z., and Aksoy, U. (2010) The assessment of the quality losses of fresh fig fruits during transportation. *Journal of Agricultural Sciences* 16, 180–193.

Çalişkan, O. and Polat, A.A. (2008) Fruit characteristics of fig cultivars and genotypes grown in Turkey. *Scientia Horticulturae* 115, 360–367.

Çalişkan, O. and Polat, A.A. (2011) Phytochemical and antioxidant properties of selected fig (*Ficus carica* L.) accessions from the eastern Mediterranean region of Turkey. *Scientia Horticulturae* 128, 473–478.

Çalişkan, O. and Polat, A.A. (2012) Morphological diversity among fig (*Ficus carica* L.) accessions sampled from the Eastern Mediterranean Region of Turkey. *Turkish Journal of Agriculture and Forestry* 36, 179–193.

Cantín, C.M., Palou, L., Bremer, V., Michailides, T.J., and Crisosto, C.H. (2011) Evaluation of the use of sulfur dioxide to reduce postharvest losses on dark and green figs. *Postharvest Biology and Technology* 59, 150–158.

Cantín, C.M., Giné-Bordonaba, J., and Echeverría, G. (2020) Optimal handling and postharvest strategies to reduce losses of 'Cuello Dama Negro' dark figs (*Ficus carica* L.). *International Journal of Fruit Science* 20, S414–S431.

Çelikel, F.G. and Karaçali, I. (1998) Effects of harvest maturity and precooling on fruit quality and longevity of 'Bursa Siyahi' figs (*Ficus carica* L.). *Acta Horticulturae* 480, 283–288.

Çelikel, F.G., Kaynas, K., Özelkök, S., and Ertan, Ü. (1997) Effects of ethephon on fruit development and ripening of the fig (*Ficus carica* L.) variety 'Bursa Siyahi'. *Acta Horticulturae* 441, 145–152.

Çelikel, F.G., Kaynas, K., Özelkök, S., and Ertan, Ü. (1998) Effect of antitranspirant treatment on water loss and keeping quality of the fig cultivar Bursa Siyah. *Journal of Ataturk Central Horticultural Research Institute (Turkey)* 25, 21–28 (in Turkish).

Chai, L., Wang, Z., Chai, P., Chen, S., and Ma, H. (2017) Transcriptome analysis of San Pedro type fig (*Ficus carica* L.) parthenocarpic breba and non-parthenocarpic main crop reveals divergent phytohormone-related gene expression. *Tree Genetics and Genomes* 13: 83.

Chai, P., Dong, S., Chai, L., Chen, S., Flaishman, M., and Ma, H. (2019) Cytokinin-induced parthenocarpy of San Pedro type fig (*Ficus carica* L.) main crop: explained by phytohormone assay and transcriptomic network comparison. *Plant Molecular Biology* 99, 329–346.
Chessa, I. (1997) Fig. In: Mitra, S. (ed.) *Postharvest Physiology and Storage of Tropical and Subtropical Fruits*. CAB International, Wallingford, UK, pp. 245–268.
Chessa, I., d'Aquino, S., Cossu, A., and Mannu, L. (1992) Epicuticular wax structure and accumulation on main-crop fig fruit (*Ficus carica* L.) during development. *Advances in Horticultural Science* 6, 47–54.
Claypool, L. and Ozbek, S. (1952) Some influences of temperature and carbon dioxide on the respiration and storage life of the Mission fig. In: *Proceedings of the American Society for Horticultural Science*, Vol. 60, pp. 226–230.
Colelli, G. (1995) Physiological aspects of ripening and postharvest technology of fig (*Ficus carica* L.) fruits. *Rivista di Frutticoltura e di Ortofloricoltura* 57, 71–77 (in Italian).
Colelli, G. and Amodio, M.L. (2020) Subtropical fruits: fig. In: Gil, M. and Beaudry, R. (eds) *Controlled and Modified Atmospheres for Fresh and Fresh-Cut Produce*. Academic Press, New York, pp. 427–434.
Colelli, G., Mitchell, F.G., and Kader, A.A. (1991) Extension of postharvest life of mission figs by CO_2-enriched atmospheres. *HortScience* 26, 1193–1195.
Condit, I. (1955) Fig varieties: a monograph. *Hilgardia* 23, 323–538.
Crane, J.C. and Blondeau, R. (1949) The use of growth-regulating chemicals to induce parthenocarpic fruit in the 'Calimyrna' fig. *Plant Physiology* 24, 44–54.
Crisosto, C.H., Garner, D., Doyle, J., and Day, K.R. (1993) Relationship between fruit respiration, bruising susceptibility, and temperature in sweet cherries. *HortScience* 28, 132–135.
Crisosto, C.H., Bremer, V., Ferguson, L., and Crisosto, G.M. (2010a) Evaluating quality attributes of four fresh fig (*Ficus carica* L.) cultivars harvested at two maturity stages. *HortScience* 45, 707–710.
Crisosto, C.H., Bremer, V., Norton, M., Ferguson, L., and Einhorn, T. (2010b) Preharvest ethephon eliminates first crop figs. *HortTechnology* 20, 173–178.
Crisosto, C.H., Ferguson, L., and Bremer, V. (2011) Fig. In: Yahia, E.M. (ed.) *Postharvest Biology and Technology of Tropical and Subtropical Fruits: Fundamental Issues*. Woodhead Publishing, Cambridge, UK, pp. 134–158.
Crisosto, C.H., Ferguson, L., Preece, J.E., Michailides, T.J., Haug, M.T., *et al.* (2017) Developing the California fresh fig industry. *Acta Horticulturae* 1173, 285–292.
Cui, Y., Wang, Z., Chen, S., Vainstein, A., and Ma, H. (2019) Proteome and transcriptome analyses reveal key molecular differences between quality parameters of commercial-ripe and tree-ripe fig (*Ficus carica* L.). *BMC Plant Biology* 19: 146.
Cui, Y., Zhai, Y., Flaishman, M., Li, J., Chen, S., *et al.* (2021) Ethephon induces coordinated ripening acceleration and divergent coloration responses in fig (*Ficus carica* L.) flowers and receptacles. *Plant Molecular Biology* 105, 347–364.
d'Aquino, S., Palma, A., Dore, A., and Agabbio, M. (2003) Non conventional treatments to reduce figs decay. *Acta Horticulturae* 604, 817–821.
d'Aquino, S., Palma, A., Satta, D., de Pau, L., and Schirra, M. (2015) Influence of azoxystrobin dip treatments on postharvest decay of second-crop fig (*Ficus carica*) fruits from Sardinian germoplasm. *Advances in Horticultural Science* 29, 141–144.
d'Aquino, S., Piga, A., Molinu, M.G., Agabbio, M., and Papoff, C.M. (1998) Maintaining quality attributes of 'Craxiou de porcu' fresh fig fruit in simulated marketing conditions by modified atmosphere. *Acta Horticulturae* 480, 289–294.
de la Plaza, J.L. (2003) Sulfur dioxide achieves the long term storage of fresh figs. *Acta Horticulturae* 605, 225–228.
de Paiva, E., Serradilla, M.J., Ruiz-Moyano, S., Córdoba, M.G., Villalobos, M.C., *et al.* (2017) Combined effect of antagonistic yeast and modified atmosphere to control *Penicillium expansum* infection in sweet cherries cv. Ambrunés. *International Journal of Food Microbiology* 241, 276–282.
de Paula, L.A., Isepon, J.D.S., and Corrêa, L.D.S. (2007) Postharvest quality of figs of the cv. Roxo-de-Valinhos with application of calcium chloride and fungicides. *Acta Scientiarum: Agronomy* 29, 41–46.
di Lorenzo, F., Silipo, A., Molinaro, A., Parrilli, M., Schiraldi, C., *et al.* (2017) The polysaccharide and low molecular weight components of *Opuntia ficus indica* cladodes: structure and skin repairing properties. *Carbohydrate Polymers* 157, 128–136.
Doerflinger, F.C., Nock, J.F., Miller, W.B., and Watkins, C.B. (2019) Preharvest aminoethoxyvinylglycine (AVG) and 1-methylcyclopropene (1-MCP) effects on ethylene and starch concentrations of 'Empire' and 'McIntosh' apples. *Scientia Horticulturae* 244, 134–140.

Doster, M.A. and Michailides, T.J. (2007) Fungal decay of first-crop and main-crop figs. *Plant Disease* 91, 1657–1662.
Droby, S., Wisniewski, M., Teixidó, N., Spadaro, D., and Jijakli, M.H. (2016) The science, development, and commercialization of postharvest biocontrol products. *Postharvest Biology and Technology* 122, 22–29.
Dussán-Sarria, S., Honório, S.L., and Nogueira, D.H. (2006) Precooling parameters of 'Roxo de Valinhos' figs (*Ficus carica* L.) packed in a carton box. *Fruits* 61, 401–405.
Dussán Sarria, S., Honório, S.L., and Teruel, B.J. (2010) Precooling characteristics of the fig fruits 'Roxo de Valinhos' packed in wooden box. *Acta Horticulturae* 864, 283–287.
Eccher Zerbini, P., Grassi, M., and Vaccari, S. (1986) The influence of packaging and of delayed cooling on the quality of fresh figs after refrigerated transport. In: *Annali dell'Istituto Sperimentale per la Valorizzazione Tecnologica dei Prodotti Agricoli*. Istituto Sperimentale per la Valorizzazione, Milan, Italy.
Elansari, M.A., Fenton, D.L., and Callahan, C.W. (2019) Precooling. In: Yahia, E.M. (ed.) *Postharvest Technology of Perishable Horticultural Commodities*. Woodhead Publishing, Cambridge, UK, pp. 161–207.
Elsabee, M.Z. and Abdou, E.S. (2013) Chitosan based edible films and coatings: a review. *Materials Science and Engineering: C* 33, 1819–1841.
Ersoy, N., Gözlekçi, S., Gok, V., and Yilmaz, S. (2017) Fig (*Ficus carica* L.) fruit: some physical and chemical properties. *Acta Horticulturae* 1173, 329–334.
EFSA (European Food Safety Authority) (2009) Guidelines on submission of a dossier for safety evaluation by the EFSA of active or intelligent substances present in active and intelligent materials and articles intended to come into contact with food. *European Food Safety Authority Journal* 7, 1208–1217.
Ferguson, L., Michailides, T.J., and Shorey, H.H. (1990) The California fig industry. *Horticultural Reviews* 12, 409–490.
Flaishman, M.A., Rodov, V., and Stover, E. (2008) The fig: botany, horticulture, and breeding. *Horticultural Reviews* 34, 113–197.
Freiman, Z.E., Rodov, V., Yablovitz, Z., Horev, B., and Flaishman, M.A. (2012) Preharvest application of 1-methylcyclopropene inhibits ripening and improves keeping quality of 'Brown Turkey' figs (*Ficus carica* L.). *Scientia Horticulturae* 138, 266–272.
Freiman, Z.E., Doron-Faigenboim, A., Dasmohapatra, R., Yablovitz, Z., and Flaishman, M.A. (2014) High-throughput sequencing analysis of common fig (*Ficus carica* L.) transcriptome during fruit ripening. *Tree Genetics and Genomes* 10, 923–935.
Freiman, Z.E., Rosianskey, Y., Dasmohapatra, R., Kamara, I., and Flaishman, M.A. (2015) The ambiguous ripening nature of the fig (*Ficus carica* L.) fruit: a gene-expression study of potential ripening regulators and ethylene-related genes. *Journal of Experimental Botany* 66, 3309–3324.
Gaaliche, B., Lauri, P.E., Trad, M., Costes, E., and Mars, M. (2011a) Interactions between vegetative and generative growth and between crop generations in fig tree (*Ficus carica* L.). *Scientia Horticulturae* 131, 22–28.
Gaaliche, B., Trad, M., and Mars, M. (2011b) Effect of pollination intensity, frequency and pollen source on fig (*Ficus carica* L.) productivity and fruit quality. *Scientia Horticulturae* 130, 737–742.
Gaaliche, B., Hfaiedh, L., Trad, M., and Mars, M. (2011c) Caprification efficiency of three Tunisian fig (*Ficus carica* L.) cultivars. *Journal of Natural Products and Plant Resources* 1, 20–25.
Gaaliche, B., Saddoud, O., and Mars, M. (2012) Morphological and pomological diversity of fig (*Ficus carica* L.) cultivars in northwest of Tunisia. *International Scholarly Research Notices* 2021: 326461.
Gaaliche, B., Ladhari, A., Zarrelli, A., and Mimoun, M.B. (2019) Impact of foliar potassium fertilization on biochemical composition and antioxidant activity of fig (*Ficus carica* L.). *Scientia Horticulturae* 253, 111–119.
Galil, J. and Neeman, G. (1977) Pollen transfer and pollination in the common fig (*Ficus carica* L.). *New Phytologist* 79, 163–171.
Galus, S. (2019) Development of edible coatings in the preservation of fruits and vegetables. In: Gutierrez, T.J. (ed.) *Polymers for Agri-Food Applications*. Springer International Publishing, Cham, Switzerland, pp. 377–390.
García, B., Hernández Méndez, M.T., and Lozano Ruiz, M. (2003) Effect of temperature on the shelf-life of fig fruit during refrigerated storage. *Acta Horticulturae* 605, 229–232.
Ghosh, T. and Dash, K.K. (2020) Modeling on respiration kinetics and modified atmospheric packaging of fig fruit. *Journal of Food Measurement and Characterization* 14, 1092–1104.
Gibernau, M., Buser, H.R., Frey, J.E., and Hossaert-McKey, M. (1997) Volatile compounds from extracts of figs of *Ficus carica*. *Phytochemistry* 46, 241–244.

Gözlekçi, S., Erkan, M., Karaşahin, I., and Şahin, G. (2008) Effect of 1-methylcyclopropene (1-MCP) on fig (*Ficus carica* cv. Bardakci) storage. *Acta Horticulturae* 798, 325–330.
Gözlekçi, S., Kafkas, E., and Ercisli, S. (2011) Volatile compounds determined by HS/GC-MS technique in peel and pulp of fig (*Ficus carica* L.) cultivars grown in Mediterranean region of Turkey. *Notulae Botanicae Horti Agrobotanici Cluj-Napoca* 39, 105–108.
Grison-Pigé, L., Hossaert-McKey, M., Greeff, J.M., and Bessière, J.M. (2002) Fig volatile compounds – a first comparative study. *Phytochemistry* 61, 61–71.
Gross, K.C., Wang, C.Y., and Saltveit, M.E. (2016) *The Commercial Storage of Fruits, Vegetables, and Florist and Nursery Stocks*. US Department of Agriculture, Agricultural Research Service, Washington, DC.
Guillén, F., Castillo, S., Valero, D., Zapata, P.J., Martínez-Romero, D., *et al.* (2015) Use of modified atmosphere packaging improves antioxidant activity and bioactive compounds during postharvest storage of 'Collar' figs. *Acta Horticulturae* 1071, 263–268.
Hamanaka, D., Norimura, N., Baba, N., Mano, K., Kakiuchi, M., *et al.* (2011) Surface decontamination of fig fruit by combination of infrared radiation heating with ultraviolet irradiation. *Food Control* 22, 375–380.
Hamman, J.H. (2008) Composition and applications of *Aloe vera* leaf gel. *Molecules* 13, 1599–1616.
Hernández Méndez, T., Gervasini Rodriguez, C., and Bernalte García, J. (2003) Sensory quality of cv. *Tiberio breba. Acta Horticulturae* 605, 233–235.
Holstein, H., Crisosto, G., and Crisosto, C. (2017) Fertigation with potassium increases size and yield in fresh figs growing in California. *Acta Horticulturae* 1173, 177–181.
Hong, G., Crisosto, C., and Cantwell, M.I. (2016) Quality and physiology of two cultivars of fresh-cut figs in relation to ripeness, storage temperature and controlled atmosphere. *Acta Horticulturae* 1141, 213–220.
Horev, B., Daus, A., Yablowicz, Z., Kiesel, H., Freedman, Y., *et al.* (2015) Postharvest disinfectant fogging can extend the shelf life of fresh figs. In: *Proceedings of the 5th International Symposium on Fig, 31 August–3 September 2015, Naples, Italy*. Book of Abstracts, International Society for Horticultural Science, Leuven, Belgium, p. 99.
Hssaini, L., Charafi, J., Razouk, R., Hernández, F., Fauconnier, M.L., *et al.* (2020) Assessment of morphological traits and fruit metabolites in eleven fig varieties (*Ficus carica* L.). *International Journal of Fruit Science* 20, 8–28.
Hung, V.H., Tanara, F., Uchino, T., and Hiruma, N. (2011) Using nanomist humidifier to maintain postharvest quality of fig (*Ficus carica* L.) fruit in high humidity storage environment. *Journal of the Faculty of Agriculture* 56, 361–365.
Hussein, Z., Fawole, O.A., and Opara, U.L. (2020) Harvest and postharvest factors affecting bruise damage of fresh fruits. *Horticultural Plant Journal* 6, 1–13.
Ikegami, H., Habu, T., Mori, K., Nogata, H., Hirata, C., *et al.* (2013) *De novo* sequencing and comparative analysis of expressed sequence tags from gynodioecious fig (*Ficus carica* L.) fruits: caprifig and common fig. *Tree Genetics and Genomes* 9, 1075–1088.
Inaba, A. (2007) Studies on the internal feedback regulation of ethylene biosynthesis and signal transduction during fruit ripening, and the improvement of fruit quality. *Journal of the Japanese Society for Horticultural Science* 76, 1–12.
Irfan, P.K., Vanjakshi, V., Prakash, M.K., Ravi, R., and Kudachikar, V.B. (2013) Calcium chloride extends the keeping quality of fig fruit (*Ficus carica* L.) during storage and shelf-life. *Postharvest Biology and Technology* 82, 70–75.
Ito, H., Iwamoto, S., and Sato, E. (1987) The effect of low temperature on quality maintenance in fig fruits. *Research Bulletin of the Aichi-ken Agricultural Research Center (Japan)* 19, 286–295 (in Japanese).
Jusoh, N.A.B.M., Ding, P., and Yeat, C.S. (2019) Performance of fig (*Ficus carica* L. cv. Ipoh blue giant) fruit stored at various storage temperatures. *International Journal of Agriculture, Forestry and Plantation* 8, 9–11.
Kader, A.A. (2002a) Quality parameters of fresh-cut fruit and vegetable products. In: Lamicanra, O. (ed.) *Fresh-cut Fruits and Vegetables*. CRC Press, Boca Raton, Florida, pp. 20–29.
Kader, A.A. (2002b) *Postharvest Technology of Horticultural Crops*. Agriculture and Natural Resources, University of California, Oakland, California.
Kader, A.A., and Saltveit, M.E. (2003) Respiration and gas exchange. *Postharvest Physiology and Pathology of Vegetables* 2, 7–29.
Karabulut, O.A., Ilhan, K., Arslan, U., and Vardar, C. (2009) Evaluation of the use of chlorine dioxide by fogging for decreasing postharvest decay of fig. *Postharvest Biology and Technology* 52, 313–315.
Kaynak, L., Gözlekçi, S., and Ersoy, N. (1998) A research on storage and pomological properties of some fig (*Ficus carica* L.) cultivars grown in Antalya conditions. *Acta Horticulturae* 480, 277–282.

Kays, S.J. (1997) *Postharvest Physiology of Perishable Plant Products*. Van Nostrand Reinhold, New York.

Kim, J.S., Chung, D.S., and Lee, Y.S. (2012) Effect of packaging systems with high CO_2 treatment on the quality changes of fig (*Ficus carica* L.) during storage. *Korean Journal of Food Preservation* 19, 799–806.

King, E.S., Hopfer, H., Haug, M.T., Orsi, J.D., Heymann, H., *et al.* (2012) Describing the appearance and flavor profiles of fresh fig (*Ficus carica* L.) cultivars. *Journal of Food Science* 77, S419–S429.

Kong, M., Lampinen, B., Shackel, K., and Crisosto, C.H. (2013) Fruit skin side cracking and ostiole-end splitting shorten postharvest life in fresh figs (*Ficus carica* L.), but are reduced by deficit irrigation. *Postharvest Biology and Technology* 85, 154–161.

Kurubar, A.R., Alloli, T.B., Naik, M.K., and Pujari, B.T. (2013) Effect of pre-and postharvest application of different chemicals on quality traits of 'Poona Fig' fig (*Ficus carica* L.). *Acta Horticulturae* 1012, 91–95.

Lakshmi, S.J., Roopa Bai, R.S., Sharanagouda, H., Ramachandra, C.T., Nadagouda, S., and Nidoni, U. (2018) Effect of biosynthesized zinc oxide nanoparticles coating on quality parameters of fig (*Ficus carica* L.) fruit. *Journal of Pharmacognosy and Phytochemistry* 7, 10–14.

Lama, K., Harlev, G., Shafran, H., Peer, R., and Flaishman, M.A. (2020) Anthocyanin accumulation is initiated by abscisic acid to enhance fruit color during fig (*Ficus carica* L.) ripening, *Journal of Plant Physiology* 251: 153192.

Lama, K., Yadav, S., Rosianski, Y., Shaya, F., Lichter, A., *et al.* (2019) The distinct ripening processes in the reproductive and non-reproductive parts of the fig syconium are driven by ABA. *Journal of Experimental Botany* 70, 115–131.

Lee, C.S., Chong, M.F., Robinson, J., and Binner, E. (2014) A review on development and application of plant-based bioflocculants and grafted bioflocculants. *Industrial & Engineering Chemistry Research* 53, 18357–18369.

Li, Z. and Thomas, C. (2014) Quantitative evaluation of mechanical damage to fresh fruits. *Trends in Food Science and Technology* 35, 138–150.

Lima, L.C., Dias, M.S.C., Castro, M.V.D., Martins, R.N., Ribeiro Junior, P.M., *et al.* (2005) Post-harvest conservation of unripe figs (*Ficus carica* L.) cv. 'Roxo de Valinhos' treated with sodium hypochlorite and stored under refrigeration in passive modified atmosphere. *Ciência e Agrotecnologia* 29, 810–816.

Liu, H., Li, D., Xu, W., Fu, Y., Liao, R., *et al.* (2020) Effect of packaging design with two functional films on the quality of figs stored at ambient temperature. *Journal of Food Processing and Preservation* 44: e14836.

Ma, J., Li, D., Yang, D., Xu, W., Fu, Y., *et al.* (2019) Effects of packaging designs with multiple pieces of function films on the quality of figs stored at ambient temperature. *Scientia Horticulturae* 251, 32–38.

MAF RODA AGROBOTICS (2020) Figs. Available at: www.maf-roda.com/en/page/figs.php (accessed December 7, 2021).

Mao, L., Ying, T., Xi, Y., and Zhen, Y. (1995) Respiration rate, ethylene production, and cellular leakage of fig fruit following vibrational stress. *HortScience* 30, 145.

Marei, N. and Crane, J.C. (1971) Growth and respiratory response of fig (*Ficus carica* L. cv. Mission) fruits to ethylene. *Plant Physiology* 48, 249–254.

Marpudi, S.L., Pushkala, R., and Srividya, N. (2013) *Aloe vera* gel coating for post harvest quality maintenance of fresh fig fruits. *Research Journal of Pharmaceutical, Biological and Chemical Sciences* 4, 878–887.

Martínez-Damián, M.T., Omegar, C.A., and Oscar, C.A. (2020) Effect of modified atmosphere packaging on nutraceutical quality and overall appearance of figs stored at 1°C. *Notulae Botanicae Horti Agrobotanici Cluj-Napoca* 48, 2292–2305.

Mathooko, F.M., Sotokawa, T., Kubo, Y., Inaba, A., and Nakamura, R. (1993) Retention of freshness in fig fruit by CO_2-enriched atmosphere treatment or modified atmosphere packaging under ambient temperature. *Journal of the Japanese Society for Horticultural Science* 62, 661–667.

Matteo, M., Rapacciuolo, M., and Cinquanta, L. (1999) Cold storage of figs using semipermeable plastic film. *Informatore Agrario* 55, 35–36 (in Italian).

McCollum, G. and Maul, P. (2007) 1-Methylcyclopropene inhibits degreening but stimulates respiration and ethylene biosynthesis in grapefruit. *HortScience* 42, 120–124.

McGuire, R.G. (1992) Reporting of objective color measurements. *HortScience* 27, 1254–1255.

Meng, X., Wu, Z., Zhang, Q., and Liu, B. (2014) Effects of chitosan-coating on cold preservation of *Ficus carica* Linn. *Food Research and Development* 35, 114–117.

Michailides, T.J. (2003) Diseases of fig. In: Ploetz, R.C. (ed.) *Diseases of Tropical Fruit Crops*. CAB International, Wallingford, UK, pp. 253–273.

Michailides, T.J., Morgan, D.P., and Subbarao, K.V. (1996) An old disease still a dilemma for California growers. *Plant Disease* 80, 828–841.

Michailides, T.J., Morgan, D.P., Felts, D., and Doster, M.A. (2008) Control of decay in caprifigs and calimyrna figs with fungicides. *Acta Horticulturae* 798, 269–275.
Mimoun, M.B., Yahmed, J.B., and Gaaliche, B. (2017) Effect of foliar potassium application on fruit quality of fig cv. Bouhouli in the North West of Tunisia. In: *Proceedings of the 18th International Plant Nutrition Colloquium, 19–24 August 2017, Copenhegan, Denmark*, pp. 777–778.
Mirheidari, F., Khadivi, A., Moradi, Y., and Paryan, S. (2020) Phenotypic variability of naturally grown edible fig (*Ficus carica* L.) and caprifig (*Ficus carica* var. *caprificus* Risso) accessions. *Scientia Horticulturae* 267: 109320.
Mirshekari, A., Madani, B., Wall, M., and Biggs, A.R. (2020) *Aloe vera* coatings maintain antioxidants of fig (*Ficus carica* L.) fruit during storage. *Advances in Horticultural Science* 34, 205–212.
Mistriotis, A., Briassoulis, D., Giannoulis, A., and d'Aquino, S. (2016) Design of biodegradable bio-based equilibrium modified atmosphere packaging (EMAP) for fresh fruits and vegetables by using micro-perforated poly-lactic acid (PLA) films. *Postharvest Biology and Technology* 111, 380–389.
Molinu, M.G., Venditti, T., Dore, A., d'Hallewin, G., Serusi, A., *et al.* (2006) Postharvest decay reduction of fig fruit (*Ficus carica*) by hot water sodium carbonate solutions dip. *Communications in Agricultural and Applied Biological Sciences* 71, 923–928.
Molinu, M.G., Venditti, T., Dore, A., d'Hallewin, G., Serusi, A., and Agabbio, M. (2007) Reduction of fig fruit (*Ficus carica*) postharvest decay by curing and acetic acid treatments. In: *Novel Approaches for the Control of Postharvest Diseases and Disorders. Proceedings of the International Congress, 3–5 May 2007, Bologna, Italy*.
Morton, J.F. and Dowling, C.F. (1987) *Fruits of Warm Climates*. J.F. Morton, Miami, Florida.
Mostafa, Y.S., El-Berry, I.M., and Mustafa, N.S. (2019) Reducing 'Sultani' figs losses during marketing by thymol and chitosan dipping. *Alexandria Journal of Agricultural Sciences* 64, 33–40.
Mota, M.S., Gomes, C.B., Souza Júnior, I.T., and Moura, A.B. (2017) Bacterial selection for biological control of plant disease: criterion determination and validation. *Brazilian Journal of Microbiology* 48, 62–70.
Nawade, B., Shaltiel-Harpaz, L., Yahyaa, M., Bosamia, T.C., Kabaha, A., *et al.* (2020) Analysis of apocarotenoid volatiles during the development of *Ficus carica* fruits and characterization of *carotenoid cleavage dioxygenase* genes. *Plant Science* 290: 110292.
Oliveira, A.P., Silva, L.R., Andrade, P.B., Valentão, P., Silva, B.M., Pereira, J.A., and de Pinho, P.G. (2010a) Determination of low molecular weight volatiles in *Ficus carica* using HS-SPME and GC/FID. *Food Chemistry* 121, 1289–1295.
Oliveira, A.P., Silva, L.R., de Pinho, P.G., Gil-Izquierdo, A., Valentão, P., Silva, B.M., *et al.* (2010b) Volatile profiling of *Ficus carica* varieties by HS-SPME and GC-IT-MS. *Food Chemistry* 123, 548–557.
Opara, U.L., Caleb, O.J., and Belay, Z.A. (2019) Modified atmosphere packaging for food preservation. In: Galanakis, C. (ed.) *Food Quality and Shelf Life*. Academic Press, New York, pp. 235–259.
Owino, W.O., Manabe, Y., Mathooko, F.M., Kubo, Y., and Inaba, A. (2006) Regulatory mechanisms of ethylene biosynthesis in response to various stimuli during maturation and ripening in fig fruit (*Ficus carica* L.). *Plant Physiology and Biochemistry* 44, 335–342.
Özer, K. and Sen, F. (2003) Effects of some postharvest treatments on preserving quality of Yediveren fig fruits. *Acta Horticulturae* 605, 211–218.
Özkaya, O., Çömlekcioglu, S., and Demircioglu, H. (2014) Assessment of the potential of 1-methylcyclopropene treatments to maintain fruit quality of the common fig (*Ficus carica* L. cv. 'Bursa Siyahi') during refrigerated storage. *Notulae Botanicae Horti Agrobotanici Cluj-Napoca* 42, 516–522.
Öztopuz, Ö., Pekin, G., Park, R.D., and Eltem, R. (2018) Isolation and evaluation of new antagonist bacillus strains for the control of pathogenic and mycotoxigenic fungi of fig orchards. *Applied Biochemistry and Biotechnology* 186, 692–711.
Pande, G. and Akoh, C.C. (2010) Organic acids, antioxidant capacity, phenolic content and lipid characterisation of Georgia-grown underutilized fruit crops. *Food Chemistry* 120, 1067–1075.
Paolucci, M., di Stasio, M., Sorrentino, A., La Cara, F., and Volpe, M.G. (2020) Active edible polysaccharide-based coating for preservation of fresh figs (*Ficus carica* L.). *Foods* 9: 1793.
Park, Y. and Jung, S. (2000) Effects of CO_2 treatments within polyethylene film bags on fruit quality of fig fruits during storage. *Journal of the Korean Society for Horticultural Science* 41, 618–622.
Park, Y.S., Kim, B.W., and Jung, S.T. (1998) Effects of storage temperature and CO_2-enriched atmospheres on the shelf life of fig fruits during storage. *Journal of the Korean Society for Horticultural Science* 39, 184–188.
Pereira, C., López-Corrales, M., Serradilla, M.J., del Carmen Villalobos, M., Ruiz-Moyano, S., and Martin, A. (2017) Influence of ripening stage on bioactive compounds and antioxidant activity in nine fig (*Ficus carica* L.) varieties grown in Extremadura, Spain. *Journal of Food Composition and Analysis* 64, 203–212.

Pereira, C., Martín, A., López-Corrales, M., Córdoba, M.D.G., Galván, A.I., and Serradilla, M.J. (2020) Evaluation of the physicochemical and sensory characteristics of different fig cultivars for the fresh fruit market. *Foods* 9: 619.

Piazzolla, F., Amodio, M.L., and Colelli, G. (2018) Effects of thermal treatments on quality of 'Petrelli' figs during storage. *Acta Horticulturae* 1194, 879–888.

Piga, A., d'Aquino, S., Agabbio, M.C.S., and Papoff, C. (1995) Influenza del confezionamento con film plastici sulla conservazione del fico. *Italus Hortus* 5, 3–7.

Pinto, A.F.J., Santos, K.C., Santos, L.P., Lubenow, L.G.B., Coutinho, W.B.G., *et al.* (2020) Postharvest quality of fig fruits using different edible coatings. *Scientific Electronic Archives* 13, 24–29.

Pisimisi, E., Vemmos, S.N., and Mili, E. (2012) The photosynthetic activity and evaluation of fruit quality in seven fig cultivars (*Ficus carica* L.). *Acta Horticulturae* 940, 341–348.

Poverenov, E., Danino, S., Horev, B., Granit, R., Vinokur, Y., and Rodov, V. (2014) Layer-by-layer electrostatic deposition of edible coating on fresh cut melon model: anticipated and unexpected effects of alginate–chitosan combination. *Food and Bioprocess Technology* 7, 1424–1432.

Puech, A.A., Rebeiz, C.A., and Crane, J.C. (1976) Pigment changes associated with application of ethephon ((2-chloroethyl)phosphonic acid) to fig (*Ficus carica* L.) fruits. *Plant Physiology* 57, 504–509.

Qiao, H., Zhang, H., Wang, Z., and Shen, Y. (2021) Fig fruit ripening is regulated by the interaction between ethylene and abscisic acid. *Journal of Integrative Plant Biology* 63, 553–569.

Reyes-Avalos, M.C., Femenia, A., Minjares-Fuentes, R., Contreras-Esquivel, J.C., Aguilar-González, C.N., *et al.* (2016) Improvement of the quality and the shelf life of figs (*Ficus carica*) using an alginate–chitosan edible film. *Food and Bioprocess Technology* 9, 2114–2124.

Reyes-Avalos, M.C., Minjares-Fuentes, R., Femenia, A., Contreras-Esquivel, J.C., Quintero-Ramos, A., *et al.* (2019) Application of an alginate–chitosan edible film on figs (*Ficus carica*): effect on bioactive compounds and antioxidant capacity. *Food and Bioprocess Technology* 12, 499–511.

Rodov, V., Ben-Yehoshua, S., Fishman, S., Peretz, J., and de la Asuncion, R. (1998a) Modified-humidity and modified-atmosphere packaging of bell-pepper: mathematical modeling and experimental trials. In: *The Post-harvest Treatment of Fruit and Vegetables – Current Status and Future Prospects. Proceedings of the 6th International Symposium of the European Concerted Action Program COST 94, 19–22 October 1994, Oosterbeek, The Netherlands*, pp. 61–66.

Rodov, V., Pesis, E., Marinianski, R., Peretz, J., and Ben-Yehoshua, S. (1998b) Effect of modified-atmosphere packaging on the quality of breba figs harvested at two different firmness levels. *Alon Hanotea* 52, 258–262 (in Hebrew).

Rodov, V., Percelan, J., Horev, B., Vinokur, Y., Ben-Yehoshua, S., *et al.* (2002) Development of dark figs for export: I. Optimal picking criteria for the 'Brazilian' variety. *Alon Hanotea* 56, 372–376 (in Hebrew).

Rodov, V., Horev, B., Goldman, G., Vinokur, Y., Yablowich, Z., *et al.* (2005) Purple fig: Pollination effects on fruit quality and storage potential. *Alon Hanotea* 60, 110–112 (in Hebrew).

Rodov, V., Horev, B., Goldman, G., Vinokur, Y., and Fishman, S. (2007). Model-driven development of microperforated active modified-atmosphere packaging for fresh-cut produce. *Acta Horticulturae* 746, 83–88.

Rodov, V., Vinokur, Y., and Horev, B. (2012) Brief postharvest exposure to pulsed light stimulates coloration and anthocyanin accumulation in fig fruit (*Ficus carica* L.). *Postharvest Biology and Technology* 68, 43–46.

Rosianski, Y., Doron-Faigenboim, A., Freiman, Z.E., Lama, K., Milo-Cochavi, S., *et al.* (2016a) Tissue-specific transcriptome and hormonal regulation of pollinated and parthenocarpic fig (*Ficus carica* L.) fruit suggest that fruit ripening is coordinated by the reproductive part of the syconium. *Frontiers in Plant Science* 7, 1696–1709.

Rosianski, Y., Freiman, Z.E., Cochavi, S.M., Yablovitz, Z., Kerem, Z., *et al.* (2016b) Advanced analysis of developmental and ripening characteristics of pollinated common-type fig (*Ficus carica* L.). *Scientia Horticulturae* 198, 98–106.

Ruiz-Moyano, S., Martín, A., Villalobos, M.C., Calle, A., Serradilla, M.J., *et al.* (2016) Yeasts isolated from figs (*Ficus carica* L.) as biocontrol agents of postharvest fruit diseases. *Food Microbiology* 57, 45–53.

Saavedra, S.C., Ventura-Aguilar, R.I., Bautista-Baños, S., and Barrera-Necha, L.L. (2020) Biodegradable chitosan coating for improving quality and controlling *Alternaria alternata* growth in figs. *World Journal of Advanced Research and Reviews* 7, 115–125.

Saberi, B. and Golding, J.B. (2019) Postharvest application of biopolymer-based edible coatings to improve the quality of fresh horticultural produce. In: Gutierrez, T.J. (ed.) *Polymers for Food Applications*. Springer International Publishing, Cham, Switzerland, pp. 211–250.

Saki, M., ValizadehKaji, B., Abbasifar, A., and Shahrjerdi, I. (2019) Effect of chitosan coating combined with thymol essential oil on physicochemical and qualitative properties of fresh fig (*Ficus carica* L.) fruit during cold storage. *Journal of Food Measurement and Characterization* 13, 1147–1158.

Saltveit, M.E. (2018) Postharvest biology and handling of tomatoes. In: Heuvelink, E. (ed.) *Tomatoes*, 2nd edn. CAB International, Wallingford, UK, pp. 314–336.

Sanjay, P.R. (2018) *Development of chitosan based nanocomposite film for packaging of perishable foods*. PhD thesis, University of Agricultural Sciences, Raichur, India.

Scherm, B., Ortu, G., Muzzu, A., Budroni, M., Arras, G., *et al.* (2003) Biocontrol activity of antagonistic yeasts against *Penicillium expansum* on apple. *Journal of Plant Pathology* 85, 205–213.

Schirra, M., d'Hallewin, G., Ben-Yehoshua, S., and Fallik, E. (2000) Host–pathogen interactions modulated by heat treatment. *Postharvest Biology and Technology* 21, 71–85.

Shinde, R., Rodov, V., Krishnakumar, S., and Subramanian, J. (2018). Active and intelligent packaging for reducing postharvest losses of fruits and vegetables. In: Paliyath, G., Subramarian, J., Lim, L.-T., Subramarian, K.S., Handa, A.K., and Matoo, A.K. (eds) *Postharvest Biology and Nanotechnology*. Wiley, Hoboken, New Jersey, pp. 171–189.

Silva, L.C., Harder, M.N., Arthur, P.B., Lima, R.B., Modolo, D.M., *et al.* (2012) Gamma radiation effects on figs ready-to-eat. *Revista Ecossistema* 36/37, 39–46.

Siracusa, V. and Lotti, N. (2019) Intelligent packaging to improve shelf life. In: Galanakis, C.M. (ed.) *Food Quality and Shelf Life*. Academic Press, Amsterdam, pp. 261–279.

Solomon, A., Golubowicz, S., Yablowicz, Z., Grossman, S., Bergman, M., *et al.* (2006) Antioxidant activities and anthocyanin content of fresh fruits of common fig (*Ficus carica* L.). *Journal of Agricultural and Food Chemistry* 54, 7717–7723.

Solomon, A., Golubowicz, S., Yablowicz, Z., Bergman, M., Grossman, S., *et al.* (2010) EPR studies of $O_2^{\bullet -}$, OH, and $^1O_2$ scavenging and prevention of glutathione depletion in fibroblast cells by cyanidin-3-rhamnoglucoside isolated from fig (*Ficus carica* L.) fruits. *Journal of Agricultural and Food Chemistry* 58, 7158–7165.

Song, C., Li, A., Chai, Y., Li, Q., Lin, Q., *et al.* (2019) Effects of 1-methylcyclopropene combined with modified atmosphere on quality of fig (*Ficus carica* L.) during postharvest storage. *Journal of Food Quality* 2019: 2134924.

Soni, N., Mehta, S., Satpathy, G., and Gupta, R.K. (2014) Estimation of nutritional, phytochemical, antioxidant and antibacterial activity of dried fig (*Ficus carica*). *Journal of Pharmacognosy and Phytochemistry* 3, 158–165.

Sortino, G., Barone, E., Tinella, S., and Gallotta, A. (2017) Postharvest quality and sensory attributes of *Ficus carica* L. *Acta Horticulturae* 1173, 353–357.

Sozzi, G.O., Abraján Villaseñor, M.A., Trinchero, G.D., and Fraschina, A.A. (2005) Postharvest response of 'Brown Turkey' figs (*Ficus carica* L.) to the inhibition of ethylene perception. *Journal of the Science of Food and Agriculture* 85, 2503–2508.

Su, Q., Rowley, K.G., Itsiopoulus, C., and O'Dea, K. (2002). Identification and quantitation of major carotenoids in selected components of the Mediterranean diet: green leafy vegetables, figs and olive oil. *European Journal of Clinical Nutrition* 56, 1149–1154.

Subbarao, K.V., Michailides, T.J., and Morgan, D.P. (1993) Effects of osmotic potential and temperature on growth of two pathogens of figs and a biocontrol agent. *Phytopathology* 83, 1454–1459.

Tofanelli, M.B.D., Cuquel, F.L., Medeiros, J.G.S., and de Oliveira d'Angelo, J.W. (2018) Effects of 1-methylcyclopropene on postharvest quality of 'Roxo-de-Valinhos' fresh ripe figs. *Revista Colombiana de Ciencias Hortícolas* 12, 75–81.

Trad, M., Ginies, C., Gaaliche, B., Renard, C.M., and Mars, M. (2012) Does pollination affect aroma development in ripened fig (*Ficus carica* L.) fruit? *Scientia Horticulturae* 134, 93–99.

Trad, M., Gaaliche, B., Renard, C.M.G.C., and Mars, M. (2013a) Plant natural resources and fruit characteristics of fig (*Ficus carica* L.) change from coastal to continental areas of Tunisia. *Journal of Agricultural Research and Development* 3, 22–25.

Trad, M., Le Bourvellec, C., Gaaliche, B., Ginies, C., Renard, C.M., *et al.* (2013b) Caprification modifies polyphenols but not cell wall concentrations in ripe figs. *Scientia Horticulturae* 160, 115–122.

Trad, M., Ginies, C., Gaaliche, B., Renard, C.M., and Mars, M. (2014) Relationship between pollination and cell wall properties in common fig fruit. *Phytochemistry* 98, 78–84.

Trivittayasil, V., Tanaka, F., Hamanaka, D., and Uchino, T. (2014) Prediction of surface temperature of figs during infrared heating and its effect on the quality. *Biosystems Engineering* 122, 16–22.

Tsantili, E. (1990) Changes during development of 'Tsapela' fig fruits. *Scientia Horticulturae* 44, 227–234.

Tsantili, E. and Knee, M. (1991) Ripening of 'Gloster 69' apples after low ethylene storage. *Journal of the Science of Food and Agriculture* 54, 579–586.
Tsantili, E., Karaiskos, G., and Pontikis, C. (2003) Storage of fresh figs in low oxygen atmosphere. *Journal of Horticultural Science and Biotechnology* 78, 56–60.
Türk, R. (1989) Effects of harvest time and precooling on fruit quality and cold storage of figs (*F. carica* L. cv. 'Bursa Siyahi'). *Acta Horticulturae* 258, 31–37.
Türk, R., Eris, A., Özer, M.H., Tuncelli, E., and Henze, J. (1994) Research on the CA storage of fig cv. Bursa Siyahi. In: Saas, P. (ed.) *Proceedings of International Symposium on Postharvest Treatment of Horticultural Crops*. International Society for Horticultural Science, Kecskemét, Hungary, pp. 830–839.
United Nations Economic Commission for Europe (UNECE) (2017) UNECE Standard FFV-17 concerning the marketing and commercial quality control of fresh figs. Available at: https://unece.org/trade/wp7/FFV-Standards (accessed December 7, 2021).
Usberti, F.C.S. and Ferraz, A.C.O. (2021) UV-C radiation on fresh fig quality. *Scientia Agricola* 78, 3–7.
Veberic, R. and Mikulic-Petkovsek, M. (2016) Phytochemical composition of common fig (*Ficus carica* L.) cultivars. In: Simmonds, M.S.J. and Preedy, V.R. (eds) *Nutritional Composition of Fruit Cultivars*. Academic Press, New York, pp. 235–255.
Venditti, T., Molinu, M.G., Dore, A., d'Hallewin, G., Fiori, P., *et al.* (2005) Treatments with gras compounds to keep fig fruit (*Ficus carica* L.) quality during cold storage. *Communications in Agricultural and Applied Biological Sciences* 70, 339–343.
Villalobos, M., Serradilla, M.J., Martín, A., Ruiz-Moyano, S., Pereira, C., and Cordoba, M.G. (2014) Use of equilibrium modified atmosphere packaging for preservation of 'San Antonio' and 'Banane' breba crops (*Ficus carica* L.). *Postharvest Biology and Technology* 98, 14–22.
Villalobos, M.D.C., Serradilla, M.J., Martín, A., López Corrales, M., Pereira, C., and Cordoba, M.D.G. (2016a) Preservation of different fig cultivars (*Ficus carica* L.) under modified atmosphere packaging during cold storage. *Journal of the Science of Food and Agriculture* 96, 2103–2115.
Villalobos, M.C., Serradilla, M.J., Martín, A., Ruiz-Moyano, S., Pereira, C., and Cordoba, M.G. (2016b) Synergism of defatted soybean meal extract and modified atmosphere packaging to preserve the quality of figs (*Ficus carica* L.). *Postharvest Biology and Technology* 111, 264–273.
Villalobos, M.C., Serradilla, M.J., Martín, A., Hernández-León, A., Ruíz-Moyano, S., and Cordoba, M.G. (2017a) Characterization of microbial population of breba and main crops (*Ficus carica*) during cold storage: influence of passive modified atmospheres (MAP) and antimicrobial extract application. *Food Microbiology* 63, 35–46.
Villalobos, M.C., Ansah, F., Amodio, M.L., Serradilla, M.J., and Colelli, G. (2017b) Application of modified atmosphere packaging with moisture absorber to extend the shelf life of 'Domenico Tauro' breba fruit. *Acta Horticulturae* 1173, 365–370.
Villalobos, M., Serradilla, M., Martín, A., Aranda, E., López-Corrales, M., and Cordoba, M.G. (2018a) Influence of modified atmosphere packaging (MAP) on aroma quality of figs (*Ficus carica* L.). *Postharvest Biology and Technology* 136, 145–151.
Villalobos, M.C., Serradilla, M.J., Ruiz-Moyano, S., Martín, A., López-Corrales, M., and Cordoba, M.G. (2018b). Postharvest application of 1-methylcyclopropene (1-MCP) for preservation of 'Albacor' figs (*Ficus carica* L.). *Acta Horticulturae* 1194, 853–860.
Vinson, J.A., Zubik, L., Bose, P., Samman, N., and Proch, J. (2005) Dried fruits: excellent *in vitro* and *in vivo* antioxidants. *Journal of the American College of Nutrition* 24, 44–50.
Waghmare, R.B. and Annapure, U.S. (2018) Integrated effect of radiation processing and modified atmosphere packaging (MAP) on shelf life of fresh fig. *Journal of Food Science and Technology* 55, 1993–2002.
Waghmare, R.B., Mahajan, P.V., and Annapure, U.S. (2014) Modelling the influence of time and temperature on respiration rate of fresh fig and diced papaya. *International Journal of Food Engineering* 10, 89–96.
Watkins, C.B. (2002) Ethylene synthesis, mode of action, consequences and control. In: Knee, M, (ed.) *Fruit Quality and its Biological Basis*. Sheffield Academic Press, Boca Raton, Florida, pp. 180–224.
Watkins, C.B. (2008) Overview of 1-methylcyclopropene trials and uses for edible horticultural crops. *HortScience* 43, 86–94.
Wilson, M.D., Stanley, R.A., Eyles, A., and Ross, T. (2019) Innovative processes and technologies for modified atmosphere packaging of fresh and fresh-cut fruits and vegetables. *Critical Reviews in Food Science and Nutrition* 59, 411–422.
Wu, Z., Wu, J., Peng, T., Li, Y., Lin, D., *et al.* (2017) Preparation and application of starch/polyvinyl alcohol/citric acid ternary blend antimicrobial functional food packaging films. *Polymers* 9: 102.

Yahata, D. and Nogata, H. (2000) Changes in sugar and polyuronide contents in fig syconia during storage. *Journal of the Japanese Society for Food Science and Technology* 47, 333–337.

Yenge, G.B. and Nidoni, U. (2014) Transportation losses in fresh fig (*Ficus carica* L) fruits. *International Journal of Farm Sciences* 4, 100–109.

Yenge, G.B., Shelar, S.D., and Shinde, K.S. (2014) Effect of packaging materials on quality losses in fresh fig (*Ficus carica* L.) fruits during transportation. *International Journal of Progressing Research* 13, 424–430.

Zare, H. (2008) Comparison of fig caprification vessels, period and caprifig cultivar usable in Iran. *Acta Horticulturae* 798, 233–239.

Zidi, K., Kati, D.E., Benchikh, Y., Bey, M.B., Ouandjeli, D., *et al.* (2020) The use of modified atmosphere packaging as mean of bioactive compounds and antioxidant activities preservation of fresh figs (*Ficus carica* L.) from rare cultivars. *Annals of the University Dunarea de Jos of Galati. Fascicle VI: Food Technology* 44, 149–164.

14 Postharvest Handling of Dried Fig Fruit

Fatih Şen*
Department of Horticulture, Faculty of Agriculture, Ege University, Bornova, İzmir, Turkey

14.1 Introduction

Figs are mostly consumed locally or in domestic markets worldwide, as their fresh fruits are susceptible to decay and their shelf life is short. For this reason, the vast majority of fig fruits that are subject to trade are processed by sun drying. World dried-fig production totals 158,000 t annually according to data from the International Nut and Dried Fruit Council (INC, 2020). Turkey alone undertakes 56% of world production and 55–60% of international dried-fig trade, with yearly exports ranging between 70,000 and 89,000 t (INC, 2020).

Dried figs are preferred because they can withstand deterioration and are a high-energy source and a functional food (Ferguson *et al.*, 1990). In recent years, the share of dried figs in the market has also been growing due to the increasing demand of consumers for healthy nutrition, minimally processed, vegan, and organic products. In line with the health benefits of dried fruits and consumer demands, it is of great importance that the nutritional content should be protected until the product reaches the consumer.

Sun drying is one of the methods of food storage that has been practiced for thousands of years, which humanity has learned from nature. Because the figs can easily be dried outdoors under the sun at low cost, fig is successfully dried in most regions with a suitable climate. A very large amount of fig production is still destined for sun drying, but in recent years, hot-air blowing systems or freeze-drying methods have also been put into practice. In the future, artificial drying systems may be used for drying figs alone or combined with sun drying.

As dried figs are given durability through moisture reduction, the care required during postharvest stages or during storage is often ignored. However, as the storage period extends, deterioration of the product becomes increasingly important. For this reason, the storage of dried figs in controlled cold conditions has become a reasonable alternative.

The large size, light color, and soft texture of the dried figs are indicators of high quality (Aksoy *et al.*, 2008). Additional quality parameters associated with dried figs are the shape of product, visual appeal, flavor, retention of nutrients, microbial load, texture, rehydration properties, water activity and chemical stability, use of preservatives, and freedom from insects, pests, and contaminants, as well as freedom from colorants and off-odors (Perera, 2005). To produce high-quality dried figs, it is necessary to pay attention to the quality changes at every stage of the value chain such as harvest, drying, storage, and transportation, starting from cultivation.

* Email: fatih.sen@ege.edu.tr

 Advances in Fig Research and Sustainable Production (M.A. Flaishman and U. Aksoy)
DOI: 10.1079/9781789242492.0014

This chapter gives an overview on how quality may be maintained by implementing good practices during harvest, drying, storage, and processing, and what the major factors are affecting quality and safety of the produce through the postharvest stages in commercial dried-fig production.

14.2 Harvest

Fig fruit begins to ripen rapidly at the third fruit-development period and become ready for harvest. Although the harvest of breba figs (spring crop) usually begins from the second half of May and continues until the first week of July, the second crop or main crop usually ripens in late July, and the harvest continues until the end of October (Aksoy, 1997a). The length of the harvest period usually varies depending on the variety. The first crop (breba) is harvested within 10–12 days, while this period is extended to 30–90 days for the main crop (Aksoy, 1981). The length of the harvest period depends on the number of fruit setting on the shoot and the time difference between each successive fruit. High temperature differences between day and night or low temperatures that occur during the harvest period can elongate the harvest duration. This elongation provides an advantage in terms of extending the marketing period for fresh varieties cultivated on large areas, while rains observed in the last period of the harvest may create risks for fig fruits dried under the sun. Moreover, temperature drops in autumn may allow extended harvest periods. The Sarılop (= Calimyrna) fig variety destined for commercial sun drying has a harvest period of 85–90 days under conditions prevailing in the Menderes Basin, western Turkey, the main dried-fig region in Turkey (Aksoy, 1981). The harvest period may vary according to the variety, ecology, practices, handling, and yearly conditions (Aksoy, 1997a; Karaçalı, 2016).

At harvest maturity, the fruit water content is high in fresh figs; however, in fig varieties suitable for drying, the moisture content is relatively lower due to higher accumulation of dry matter. Fully ripe figs remain on the shoot during senescence, continue losing water and shrivel. Unlike fig varieties harvested for fresh consumption, figs destined for sun drying in general are not harvested as fresh ripe but continue over-ripening on the tree. Due to intervals between each successive fruit, unripe, ripe, partially dried, and shriveled fruit can be seen on the same shoot (Fig. 14.1A). The moisture content of fruits of the Sarılop variety at the over-ripe or initial stage of shriveling, while on the tree, varies on average from around 69% to 73% (Şen *et al.*, 2017). Fig fruits, which form a layer of abscission after shriveling on the tree, fall to the ground naturally or require shaking of the branches or applying force of any other form. At this stage, the fruit has not yet completed its drying up to a safe water activity level and still contains a significant amount of water. When partially dried fruit contact the soil, risk of contamination with pests, fungi, and foreign substances increases. Although this varies according to the variety and ecology, figs usually start to shrivel 5–6 days after the full fresh maturation stage (Aksoy, 1997a). With the progression of maturation and the onset of shriveling in the fruit, many physicochemical events such as color and hardness begin to gain importance, and the fruit loses its physical and physiological resistance. The

Fig. 14.1. Fig fruits of the Sarılop variety. A. Figs at the unripe, over-ripe, and shriveled stages on the tree. B. Figs with side cracks and severe ostiole-end splitting. C. Figs with black neck.

amount of water in the fruit, which falls to the ground as shriveled fruit, is usually between 30% and 50% (Aksoy 1997a; Şen *et al.*, 2017). As a result of the high fruit water content, shriveled fruits falling to the ground can also suffer mechanical damage, related to the moisture content, the height of the fall, and the condition of the ground. For this reason, to avoid damage, it is necessary to remove weeds before harvest and make the surface smooth and soft. Soil cultivation carried out just before harvest may increase the risk of aflatoxin and ochratoxin A formation, especially by spreading toxigenic molds just below the soil surface. Moreover, soil cultivation, which is easily carried out on flat land, is difficult especially on sloping land, and the soil cannot be processed as desired. Most fig orchards cultivated for sun drying in Turkey, Iran, and Morocco are in mountainous sloping areas. If the soil surface is not smooth enough, it also increases the risks of puncturing, or stones, sand, or stem sinking in the shriveled figs that fall to the rough surface. Sometimes soil and stone particles may remain in the fruit and create problems during marketing by exceeding the tolerated level of foreign material. If the conditions do not enable preharvest soil surface management, spreading covers or sheets over the soil surface is recommended.

In commercial sun-dried fig production, the fruit harvest is carried out in the form of a selective harvest, because the development and ripening processes of the fruit on the tree show significant differences. The main-crop fruits start forming on the axil of the bottom leaves on the current season's shoot. Between the first fruit borne on the shoot and the second one, there is a time difference in harvest correlated with the time the fruit is set and the rate of fruit development. The first or second fruit at the bottom of the shoot begins to develop first, followed by the others in turn. Fruit ripening, shriveling, and falling to the ground follow a similar path. For this reason, shriveled figs gradually fall to the ground, requiring frequent harvest. The number of figs falling to the ground develops in the form of a time-dependent maximum curve; the rate of fruits that shrivel and fall to the ground gradually increases and reaches a maximum, and then decreases again and ends.

Harvesting of dried figs is done mostly by hand by collecting the shriveled figs that fall to the ground spontaneously and by shaking the branches (Aksoy, 1981; Aksoy, 1997b). There are also various harvesting machines used to collect figs that fall to the ground, as practiced in California dried-fig production (Morton, 1987). Shaking the tree to release the remaining shriveled figs can be done by mechanical means. Any fruit that is of advanced maturity or has just begun to shrivel can also be harvested by hand directly from the tree. For late-ripening figs, some farmers twist the fruit while on the tree to increase the rate of abscission layer formation and drying. However, if these fruits start drying under the sun with a high initial moisture content, the risk of aflatoxin formation increases, the color darkens, and quality losses are observed because of the longer drying time (Aksoy, 2017). In mountainous areas with steep slopes, harvesting is a difficult and laborious task. Here, it is difficult to use semi-mechanical and mechanical harvesting systems. However, especially in flat and low-sloping lands where labor availability and high labor cost become a problem, semi-mechanical and mechanical harvesting systems are inevitably used in fig harvest. The labor force used in harvest–drying–storage–classification operations in dried-fig cultivation constitutes a significant part of the total work required. Therefore, dried-fig production requiring high manual labor has decreased in countries like Portugal, Italy, and Greece (Cobanoglu *et al.*, 2010).

From the moment the shriveled figs fall to the ground until they are collected and dried, the fruits are exposed to fungal diseases, pests, mechanical damage, and quality losses. The entry of contamination such as soil, stones, or stalks into the fruit limit its value in international trade (UNECE, 2016). Difficulties can be experienced in removing these impurities from dried products during processing and packaging, and quality problems can be encountered in the trade.

The fig harvest season for drying includes a long period, and in the event of rain during this process, a break is given to dry the soil. When the soil is wet, shaking the tree should be avoided, and those figs that fall naturally must be collected every day and dried separately to prevent the risk of cross-contamination. Shriveled figs falling on wet soil may absorb external moisture, which delays drying and causes significant quality losses. Rainfall during the harvest period or humid or extremely hot and dry winds may also increase quantity and quality losses.

Harvest covers should be laid on the crown projection of the tree so that the problems experienced by falling on to the soil can be prevented. The soil surface under the trees is first smoothened, and mesh, tarpaulin, canvas, or plastic covers are laid on the surface, allowing figs to fall on clean and smooth surfaces. The moisture (29.8 %) of fig fruits taken from mesh covers was found to be lower than that of those collected directly from the soil (32.2%), and the drying time was shortened by 10% (Özen *et al.*, 2008). Shriveled figs falling on the ground are easier to collect, increasing worker productivity and reducing the harvest cost in total production costs. Due to the positive effects of covers on quality and drying, expansion of this method may have a positive economic impact with an increase in overall quality and quantity. Because of the positive effects of covers, it is important to optimize their use and to disseminate this as good practice. Optimization needs to be done according to the site-specific conditions.

As good practice, it is recommended to regularly collect figs that fall to the ground, if possible, twice a day (Aksoy, 1997b). The longer the shriveled fruits stay on the soil surface, the more likely it is that there may be increased pest incidence such as from dried-fruit beetles, birds, and fig moths. Insect damage is lower in figs that fall outside the crown of the fig tree and remain in the direct sun than in those that are in the shade (Shorey *et al.*, 1989). Collecting shriveled figs every day prevents damage from fig moths (*Ephestia cautella* and *Plodia interpunctella*), which is a problem in trade, especially after the ban of methyl bromide, as they lay eggs on shriveled figs under the tree (Aksoy, 2017). However, figs that fall to the ground are not collected every day, but within an interval of a few days or at longer intervals. As the length of time on the soil surface increases, the development of toxigenic fungi in particular increases, causing decreases in safety and quality. The nutrient composition of the shriveled fruit left on the soil surface, water content (>30%) and water activity (>0.70) values, night/day temperature differences, and prolonged contact with the soil constitute favorable conditions for proliferation of fungi (Jay, 2000). Some of the fungi prevalent in fig orchards such as *Aspergillus flavus* and *Aspergillus parasiticus* may produce aflatoxins, secondary metabolites that are harmful to human health (Betina, 1984; Aksoy, 1990). As soil is a suitable environment for pathogenic soil fungi and pests, this type of damage to fruits falling from trees is expected to increase, especially with the extension of the contact period. For these reasons, it is important that figs should not be left on the soil and should be collected daily to produce high-quality dried figs. As harvest covers avoid the the problem of shriveled figs contacting the soil, their use can reduce the dimension of the problems.

The shriveled figs that remain on the soil surface continue to lose water under the influence of heat and radiation emitted from the soil until they reach an equilibrium moisture content. Some fruits that fall to the ground in orchards on sloping lands can end up outside the tree crown, and because they are then under direct sunlight, they lose water faster than the ones within the crown projection and dry faster. If the fruit waits for long periods under high temperatures, as the period is extended there is a risk that the fruit texture will harden due to extreme moisture loss. In orchards located on steep slopes, a barrier is made with shrubs and similar plant parts to prevent figs from rolling and going out of the orchard boundary.

The shriveled fruit is collected from the ground by hand into containers such as baskets or buckets. At the end of the harvest period in particular, in some cases the abscission layer is not fully formed, so application of an external force that will cause the remaining fruits to fall from the tree is mandatory. Hitting or shaking the branches to let shriveled figs fall may cause damage to trees, as well as causing increased time and labor (Aksoy, 1997b; Aksoy *et al.*, 2001). On large farms, tools, equipment, and systems that provide strong air movement can be used, such as helicopters or shakers. When collecting figs, very low-quality, culled figs such as those with severe cracks or sunburn, black neck, or bird damage, and those that cannot be consumed directly as human food, should be collected separately. In addition, black moldy figs, where ochratoxin A incidence is more common, should be collected and destroyed in separate containers, as they will infect harvest and transport containers and may lead to cross-contamination. Figs that are culled should be kept away from clean figs not only at harvest but also during the drying and storage processes (Aksoy, 1997b; Aksoy *et al.*, 2003). Culled figs can be used as feed or as raw material in the food industry if proven safe for animal and human health.

Cracking or splitting in the fruit occurs when the internal pressure overcomes the resistance of the skin and the fig cracks or splits starting from the ostiole end (Fig. 14.1B). Sudden changes in soil and air humidity are effective at causing fig-fruit cracks. Irrigation and rainfall close to or during the harvest period may also cause cracking of the fig fruit, especially in sensitive varieties. Cracked fig fruits dry, losing water faster than other figs, but their texture hardens. It is more difficult for such figs to fall from the tree. In Smyrna-type figs, abundant seed formation and rapid drupelet development due to excessive caprification may lead to side cracks on the skin. Some varieties are more sensitive to such side cracks or splits at the ostiole end. The sun-dried fruit of the Sabz variety is usually cracked and wide open at the ostiole as a varietal characteristic and under conditions prevailing in the Estahban region of Fars Province in Iran. Calcium uptake and partitioning in the fig tree and fruit parts plays a role in ostiole-end cracking, and varieties may show differences in this respect (Aksoy and Anaç, 1994). Black neck is also observed as a defect causing darkening and browning of the neck part of the fruit and appears during the period of shriveling on the tree and drying (Fig. 14.1C). This disorder is attributed to potassium deficiency (Aksoy *et al.*, 1987; Anaç *et al.*, 1987).

To avoid mechanical damage in partially dried shriveled figs, such as tearing or scratching, the containers for carrying or transporting the figs from the orchard to the drying area must be of appropriate size and texture. Collecting bags are not recommended. Boxes or baskets are the routinely used containers. Fig containers should be capable of taking a maximum of 10–15 kg of product to provide ease of collection and transportation and prevent crushing from pressure or cross-contamination (Fig. 14.2A). Containers with a larger capacity make it difficult to work, especially on sloping land, reduce work efficiency,

Fig. 14.2. Sun drying of fig fruits. A. Shriveled figs in traditional baskets collected from the ground with culled figs removed, in Aydın, Turkey. B. Plastic and wooden trays used for sun drying the fruit. C. Sun drying of fig fruits on wooden trays in an open yard. D. Sun drying on plastic trays within a plastic tunnel.

and can cause damage such as putting pressure on the products underneath. Carrying out any required repair of the containers at the beginning of each season is necessary to prevent damage to the fruit. In addition, these containers should be cleaned at regular intervals to prevent secondary contamination. Baskets used for collecting should not be dragged on the soil, and small stone particles should be prevented from entering the basket. The appearance of shriveled figs deteriorates because of mechanical damage and then oxidation may occur, and these wound areas become suitable for fungal growth (Aksoy, 2017; EFSA, 2020). All these events decrease dried-fig quality. At the end of the harvest period, the containers should be cleaned and stored in hygienic and appropriate conditions isolated from any chemicals such as pesticides and fertilizers.

14.3 Drying

Drying fruit to provide durability is the oldest, most common, effective, and inexpensive preservation method, learned by humans from nature, and has been applied since the early ages (Cemeroğlu and Özkan, 2009). The drying process, which is widely preferred in preservation of products, can be defined in a basic sense as the removal of moisture from the product. As fresh fruits and vegetables with high moisture content continue to be metabolically active after harvest, their postharvest life is short and thus deterioration occurs quickly (Kader, 2002). Drying reduces the moisture content and water activity in the product, reducing microbial activity and limiting physical and chemical changes, thus maintaining product quality. This is why drying is the most widely used method in preserving the quality of fig fruits (Farahnaky *et al.*, 2009; Mujić *et al.*, 2014; Kutlu *et al.*, 2015).

Pre-drying treatments in figs accelerate drying, prevent or reduce browning in dried figs, improve texture, and reduce insect damage (Piga *et al.*, 2004; Waskar and Gawade, 2004; Chimi *et al.*, 2008; Mujić *et al.*, 2012; Okşar *et al.*, 2017a). For this purpose, different periods and concentrations of sodium metabisulfite (Chimi *et al.*, 2008; Haug *et al.*, 2013), potassium metabisulfite (Piga *et al.*, 2004; Okşar *et al.*, 2017a), potassium carbonate (Okşar *et al.*, 2017a), ascorbic acid (Mujić *et al.*, 2012; Okşar *et al.*, 2017a), citric acid (Mujić *et al.*, 2012; Okşar *et al.*, 2017a), and sodium chlorate (Chimi *et al.*, 2008), dipping in boiled water (Piga *et al.*, 2004), and sulfur treatments (Mujić *et al.*, 2012) have been practiced since the 19th century for figs collected as fresh or partially dried. The applicability of these and similar pretreatments in sun drying of figs can often be difficult, but they can be applied prior to drying in artificial dryers. When the shriveled figs are treated with 5% citric acid for 2 min, discoloration or browning that appears during drying can be prevented in short-term storage under normal conditions (Okşar *et al.*, 20017a). Pretreatments may also significantly shorten the drying time of figs during artificial drying (Piga *et al.*, 2004; Mujić *et al.*, 2012).

Treating fresh figs with sulfur before drying was practiced for many decades in some regions for varieties that have lighter dried-fruit color, and to prevent storage-pest damage (Aksoy, 2017). In the USA, white fig varieties such as Kadota and Adriatic are sulfur treated to inhibit the Maillard reaction, which causes browning and darkening of color (Haug *et al.*, 2013). Sulfur treatments also reduce shrinkage, improve rehydration, and may facilitate drying (Lewicki, 2006). Treating dried figs with sulfur was widespread as an old fumigation method to prevent storage pests (Aday, 1938). The marketing standard of dried figs (UNECE, 2016) accepts a slight smell of sulfur; however, if the concentrations or the exposure periods are exceeded, an abnormal smell and taste appears, and the texture of the dried fruit may be damaged.

14.4 Drying Methods

Fig fruit is dried using two basic methods: drying under the sun in natural conditions using solar heat, or drying artificially using other heat sources. The amount of primary (e.g. sugar and organic acids) and secondary (e.g. phenolic substances) metabolites in dried fig fruit varies according to the variety, harvest time, and drying method (Slatnar *et al.*, 2011). Therefore, the drying process may exert significant effects.

14.4.1 Sun drying

In most of the dried-fig-producing regions, the harvest period coincides with hot and arid conditions, so drying can be carried out under the sun in natural conditions. Drying conditions are not controlled fully because they depend on weather conditions and usually take longer. In Mediterranean regions, fig drying begins during the last days of July and continues until the end of October. Climatic conditions during the drying period directly affect the quality of the dried figs, in addition to the varietal characteristics. Optimal drying conditions are 40°C as the maximum temperature and 40–50% relative humidity (RH). Dew formation, which may begin in mid- and late August because of differences in night and day temperatures, reduces the quality of dried figs by browning the skin color, promoting fungal growth on the skin, and triggering aflatoxin formation (Aksoy, 1991; Göçmez and Seferoğlu, 2014).

When sun drying figs, the water content is usually reduced to 18–20%, and to an equilibrium where no further moisture reduction is possible (Şen *et al.*, 2017). In sun drying, hygiene conditions cannot be completely controlled. The fruit is open to microorganisms and various insects, birds, or other pathogens during open-air sun drying, even if shriveling begins on the tree and metabolic activity continues (Belessiotis and Delyannis, 2011; Chand *et al.*, 2015). As dry-matter losses may also occur because of slower drying under the sun, the dry-matter content and nutritional properties may decrease slightly compared with artificial drying (Ayensu, 1997; Karaçalı, 2002; Madhlopa *et al.*, 2002; Cemeroğlu and Özkan, 2009). Additionally, there is an obligation to allocate an area within the orchard for drying. Despite such negativities, coloration is more homogeneous (Aksoy, 2017) and the product quality (Singh *et al.*, 2018) is better in fig fruits dried under the sun. The sugar:acid and fructose:glucose ratios, which determine the taste, are higher in figs dried under the sun (Slatnar *et al.*, 2011). The energy costs for drying and carbon emission of sun-dried figs are much lower than for artificial drying. Thus, sun drying is an environmentally friendly technique compared with artificial drying where the energy source, whether renewable or not, creates a major difference (Rahman, 2007).

Preparing the drying yard

Sun drying of shriveled fig fruits is usually completed in drying yards created in or near the production areas. If the spacing between rows allows, drying can be performed between the rows. The drying area should receive direct sunlight, and for this purpose, areas with a light tilt facing south or west, where there is a continuous airflow, should be selected (Şen, 2009). The soil cover or drying trays should not result in contamination with foreign material such as dust, stalks, or straw. The drying area must be clean, and away and protected from contamination risks such as dusty roads, industrial facilities, railways and roads, railway stations, landfills, toilets, and animal shelters (Aksoy, 1997b; Şen, 2009).

To complete drying, the shriveled figs are spread under the sun in an open area by laying them on trays, polypropylene sheets, common reed, or directly on the soil. Placing the figs directly on the soil surface is not recommended due to the risk of contamination, even though the drying time may be shortened. The use of trays, which makes a positive contribution to drying and dried-fig quality, is common in most fig-producing countries. Wooden trays are also used in addition to plastic trays in sizes of 60 × 80 × 12 cm, which can take 6–7 kg of shriveled figs (Fig. 14.2B). Wooden trays can have different sizes (e.g. 60 × 100 cm, 100 × 200 cm, 100 ×300 cm) with legs (at a height of 10–15 cm) (Fig. 14.2C) or without legs. There is a plastic mesh or trellis fence in the middle parts (Aksoy, 1997b; Aksoy *et al.*, 2001). In sun drying, trays with legs of medium height are preferred, as the fruit can benefit from the heat on the soil surface and are well ventilated. On such trays, the figs dry faster and are prevented from contamination. Plastic trays are becoming increasingly widespread due to their ease of use, washability, and no repair or maintenance requirements. Fig fruits laid on the trays do not have contact with the soil, thus preventing the transmission of foreign material and reducing losses caused by diseases and pests. The trays can collect the product easily and can be covered within a short time in adverse situations such as precipitation or dew. Trays used in drying should be cleaned at the beginning of each season, and wooden trays should be checked and repaired. Plastic trays should be washed and disinfected, if necessary, to ensure cleanliness. Ensuring that the trays are clean and will not damage the product is important for maintenance of fig quality. The trays are lined up in single rows at the drying sites, and walking areas are left for the workers to control the drying process and then collect the figs when dried (Fig. 14.2D). Laying black polypropylene covers in all drying areas and/or in the areas between the trays allows clean products to be obtained by preventing dust.

Infested and moldy culled figs must be separated throughout the whole handling process at every possible stage, and the sources of contamination must be minimized. Shriveled figs should be spread on the drying surface as a single layer. Overloading the trays slows down drying, increasing the risk of mycotoxin formation. Moreover, prolonged drying times will increase quality losses (Aksoy, 1997b; Aksoy *et al.*, 2003; Özer Meyvacı, 2009). Culled figs are dried in a separate place. These substandard-quality figs are used sometimes as raw matter in the food industry for preparing molasses or vinegar, or as an additive after analyzing for safety (Aksoy *et al.*, 2008). Leaving low-quality figs on trays or separating and accumulating them near the drying place increases insect infestation. When drying under the sun, it is recommended that the top of the trays is closed at night, especially when there is a risk of rain or dew formation, and then opened immediately in the early morning to prevent moisture accumulation. In addition, covering the trays helps prevent fig moths from laying eggs on the fruit (Aksoy, 1997b; Özen *et al.*, 2007).

Drying in plastic tunnels

Plastic tunnels are widely used to facilitate sun drying of fig fruits. These tunnels may differ in their technical characteristics and size. Drying tunnels are recommended to be 4–6 m wide, 10–15 m long, and 2 m high, because of their ease of use and cost (Aksoy *et al.*, 2013) (Fig. 14.2D). One end of the plastic tunnel must be closed or covered with a mosquito net and the other side must have an entrance door. Along the sides of the tunnel, there should be ventilation within a minimum width of 80 cm on both sides covered with mosquito net (pore width of net <5 mm). The net prevents infestation with pests. Drying tunnels are also effective in preventing contamination with impurities such as dust or sand. As an east–west direction will get better light and indoor temperatures will be higher, the tunnels must be installed in this direction whenever possible (Sevgican, 1999). If the tunnels are long (>20 m) and only the end parts are open, this can cause ventilation problems by slowing down moisture loss and prolonging the drying time. The polyethylene cover material must be ultraviolet and infrared resistant, and have antifog properties (Emekli and Büyüktaş, 2006). If covers without antifog properties are used, the dense moisture coming out of the fruit, especially at the beginning of drying, concentrates inside the cover material and falls on the fruit as droplets. Infrared-resistant covering material prevents heat losses, traps heat inside the tunnel that decreases the difference in diurnal temperature, and increases the indoor temperature by 2–4°C and thus shortens the drying time.

Picking dried figs from the trays

Trays are handled daily, preferably in the morning, before the temperature of the fruit increases and the texture softens. Handling twice per day to collect any fruit that are completely dried is recommended after the dew has gone in the early morning and during the cool hours in late afternoon or evening. When collecting dried fruit, the figs that have not yet dried are turned upside down and left in the drying area. Depending on the weather conditions and the amount of moisture of the figs, drying usually takes 3–5 days (Aksoy *et al.*, 2001). The water content of dried fig fruits reaching a certain degree of dryness (approximately 18–24% for the Sarılop variety) is checked manually and the figs are picked from the trays (Aksoy, 1981; Anon., 2008; Şen *et al.*, 2017). If figs are collected at a higher moisture content before they dry enough, they are at higher risk for yeast and mold growth, and mycotoxin formation. Drying conditions and the water activity level of the dried fruits play a significant role in aflatoxin production (Özay *et al.*, 1995; Özer Meyvacı, 2009).

During the drying period, when the labor demand is high, farmers work at the drying yard in the evening or at night, because the weather is cooler. Because fig moths are attracted to light, farmers should not be working under light during the night hours unless appropriate traps are used.

Shriveled fig fruits laid in trays dry at least 1 day earlier under a plastic tunnel than those dried in the open air without adversely affecting the quality of the fruit. The shriveled fig fruits in plastic tunnels reach an equilibrium moisture content (moisture content ≤22%, water activity ≤0.65) on the second day during the mid-harvest (15–30 August) period and on the third day in the late (15–30 September) harvest period. A temperature increase of 5.2°C and 5.5°C was provided inside the plastic tunnels in the mid- and late seasons compared with outdoor temperatures in Aydın, Turkey (Okşar *et al.*, 2017b) (Fig. 14.3). The temperature increase accelerates drying by increasing the vapor pressure of the water of the fruit and the saturation deficit of the air (Pixton and Warburton, 1976; Şen, 2009), that is, plastic tunnels increase the difference between the wet and dry thermometer degree, enhancing the drying rate (Cemeroğlu and Özkan, 2009). Integrating fans and systems that increase night temperatures in plastic tunnels further reduces the drying time (Green and Schwarz, 2001).

Drying in a plastic tunnel reduces labor costs, as the trays do not need to be stacked together at night in case of rainfall or dew. It also reduces the required total drying area because of shortened drying times. Tunnel drying allows better-quality, cleaner dried figs than outdoor drying, protects them from storage product pests, and reduces the risk of contamination with aflatoxin-forming fungi (Özen *et al.*, 2008; Okşar *et al.*, 2017b). It also improves the microbiological quality of dried figs.

Dried figs are usually put in perforated plastic boxes that can take about 30 kg of dried figs and transported to producer warehouses. In the USA, fig fruits are put in large boxes (bins) after drying. If the fruit is taken from the drying yard to be stored in the producer's warehouses for a while, it must be fumigated to prevent infestation by stored-product pests. The boxes should be clean, and broken or cracked boxes should not be used to prevent mechanical damage. Farmers usually grade their products in their warehouses into different quality classes prior to selling to the processor.

The specific gravity of dried fig fruit varies according to variety and product quality. The specific gravity of dried figs of the Sarılop variety was determined as 1.205 g cm^{-3} on average (Demirayak, 2008). Sacks or bags should not be used for transporting or storing dried figs. Because of the pressure caused by overlapping sacks, irreversible problems such as crushing, splitting, and fragmentation may occur in dried figs, the integrity of the product deteriorates, and the visual quality decreases. Improper means of transport and storage also increase the risk of mold growth and mycotoxin development. Fermentation may begin in hot spots when ventilation is insufficient, and thus rotting can occur (Jay, 2000). Yarns and fibers in jute or cotton sacks can stick to dried fig fruits, especially when the texture is soft.

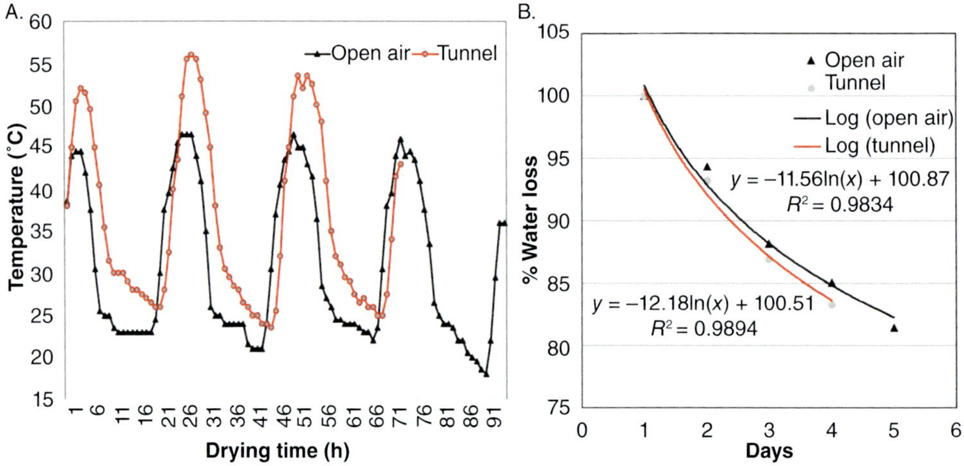

Fig. 14.3. A. Outdoor (open air) and indoor (inside tunnel) diurnal temperatures. B. The drying period for fig fruits. Reproduced and modified from Okşar *et al.* (2017b), with permission.

14.4.2 Artificial drying: drying in mechanical systems

In mechanical drying systems, the drying process is completed under controlled conditions in a shorter time, the quality and flavor are well maintained, the labor demand is lower, and hygiene conditions are ensured. The total sugar and organic acid content of figs dried in an artificial dryer are higher than those dried under the sun. In artificial dryers where controlled drying is ensured, secondary metabolites are better protected (Slatnar *et al.*, 2011; Abul-Fadl *et al.*, 2015). However, these systems demand high investment and operating costs and, depending on the energy source and the system, carbon emissions may be quite high (Karaçalı, 2002; Rahman, 2007; Cemeroğlu and Özkan, 2009; Şen, 2009). Artificial drying has eliminated many of the disadvantages of sun drying (Ayensu, 1997; Madhlopa *et al.*, 2002), and industrial drying technologies are being used in fig drying as they allow better-quality products to be obtained (Babalis and Belessiotis, 2004; Piga *et al.*, 2004; Silva *et al.*, 2013; Chand *et al.*, 2015; Villalobos *et al.*, 2016). However, discoloration or browning may be observed depending on the drying temperature (Ertekin *et al.*, 2003; Manolopoulou *et al.*, 2017).

Many factors affect the drying system to be utilized, such as the fruit characteristics (whole fruit or cut, skin and internal color, sensitivity to oxygen, and temperature), the infrastructure required, the costs of investment and operation, availability and cost of renewable energy sources, and the value of the final product (Desa *et al.*, 2019). In artificial drying, the figs are collected as fresh and/or shriveled and then dried. Where fresh figs are to be dried, the fruits are harvested in full mature period as a whole or after some pre-processes such as peeling, splitting, and slicing.

Although there are many different drying systems, the working principle of hot-air blow dryers is the same for all. Ventilation systems are used for drying the product. Optimization of temperature and airflow parameters in fig drying directly affects product quality (Ertekin *et al.*, 2003; Desa *et al.*, 2019). Babalis *et al.* (2006) conducted two studies on artificial drying of figs using standard industrial drying practice at air temperature and velocity ranges of 55–85°C and 0.5–3 m s^{-1}, respectively. Mechanical dryers performed the best and were cost-effective at a velocity of 1–2 m s^{-1} (Babalis and Belessiotis, 2004; Babalis *et al.*, 2006). Fig fruits acquire the highest quality when dried at 50°C air temperature, while higher temperatures result in discoloration and bad taste and smell (Ertekin *et al.*, 2003). The chosen temperature for drying figs affects the drying time. Drying whole figs (Tsapela variety) without peeling the fruits took an average of 48, 36, and 22 h at air temperatures of 45, 55, and 65°C, respectively (Xanthopoulos *et al.*, 2009). At a drying temperature of 45–65°C, 12–20 h and 22–37 h were required for ground and halved figs to dry, respectively (Martínez-García *et al.*, 2013). However, high temperatures chosen for drying caramelize the sugars in the fruit, causing discoloration (Cemeroğlu and Özkan, 2009). Fruits of black and yellow fig varieties collected in the shriveled (partially dried) stage dry in 2–3 days under the sun, while in an oven dryer adjusted to 60°C, drying was completed in 12 h (Konak *et al.*, 2017). Drying fresh figs in an oven at 60°C took 40 h to dry as a whole fruit and 26 h for quartered fruits. The total phenol content and antioxidant activity are higher, and the skin colors are lighter in fast-dried split figs than in whole figs (Tan, 2017).

In general, the majority of conventional dryers operate at high air temperatures (>65°C), which contribute to the product's visual and quality degradation (Xanthopoulos *et al.*, 2009). Several studies have reported the influence of the fruit color components and the changes that occur during drying (Kamiloglu and Capanoglu, 2015; Ersoy *et al.*, 2017; Konak *et al.*, 2017). Kamiloglu and Capanoglu (2015) used different colored figs—yellow and purple skins—to demonstrate the effect of sun drying on polyphenol content. For both varieties, analyses of anthocyanins, total phenolics, and antioxidant activity revealed lower levels after sun drying (Kamiloglu and Capanoglu, 2015).

To dry whole figs using artificial drying systems that blow hot air, appropriate drying conditions must be selected. If the drying conditions are incorrect and a high temperature is applied during the first stage of drying, a "crust" formation occurs. This rapidly formed dry layer on the surface of sugar-rich fig fruit has a glassy appearance that prevents further diffusion of water. As the water cannot go through this layer in the outer skin, drying stops; the outside of the product becomes hard and dry, while the inside remains soft and wet. For this purpose, when determining the drying conditions, it is necessary to prevent crusting, which affects the quality and shelf life of the dried produce (Karaçalı, 2002; Cemeroğlu and Özkan, 2009).

Cabin dryers

Although there are many types of cabin dryers, the operating principle is similar. The product to be dried is placed on trays and the heated air is circulated through the product with the help of a fan (Yağcıoğlu, 1999). In cabin dryers with a large capacity, these trays are stacked by placing them on top of each other on a trailer, or a wagon, so they can be taken into the drying cabin one after the other, facilitating the operation. The trays are motionless during drying, and hot air enters through the sidewalls, passes over the products, and again passes through the sidewall outside the cabin, reaching the heater (Heldman and Singh, 1981). In cabin dryers, the most important problem is that the same drying rate cannot be achieved at every point on the tray and drying of the fruit on the side where hot air enters is faster than on the opposite side. To avoid this, arrangements should be made in the fan system. Cabin dryers, which are usually made with a capacity to dry several tons of product, have low investment costs and are easy to operate.

Fresh figs are peeled with a knife, lined up on trays, and dried in cabin dryers. Figs dried in this way are white in color (Fig. 14.4). Fig varieties with small fruits are more suitable for this process. The figs may also be sliced and dried in cabin dryers.

Fig. 14.4. Peeled and dried figs of the Dottato variety, from Calabria, Italy.

Tunnel dryers

In tunnel dryers, which are more advanced than cabin dryers, the wagons consisting of stacked trays move on a rail track along a tunnel. Fresh product enters the tunnel from one end, goes along the tunnel, and comes out as dry at the other end. The drying takes place inside the tunnel (Yağcıoğlu, 1999). Tunnel dryers differ according to the direction of air movement and the product. If the hot air moves in the same direction as the products, they are called "parallel-flow tunnels," and if they move in opposite direction, they are called "opposite-flow tunnels" (van Arsdel, 1951). In parallel-flow tunnel dryers, hot dry air first encounters the fresh product, gets colder, and the moisture increases; the drying power then stops toward the end of the tunnel. In this system, the drying rate is very high at the beginning, while the final stage of drying takes place very slowly. This causes very little wrinkling of the product, but air pockets and cracks may form, and it takes a little more time for the product to dry. In opposite-flow tunnel dryers, hot dry air first meets the most dried product, and finally meets the wettest product. For this reason, at the beginning of drying, as the product encounters colder and moist air, water loss is slow, there is full and unobstructed wrinkling, skin binding is prevented, and there is no water leakage in the product. However, overheating of the product toward the end of drying has a negative effect on the color (Cemeroğlu and Özkan, 2009).

Conveyor dryers

Conveyor dryers have a continuous running belt instead of trays and wagons. Hot air is blown from the bottom or top on to the product moving on a stainless steel, sieve-shaped band (Yağcıoğlu, 1999). Although conveyor dryers can be multi-stage, two-stage driers are usually preferred. Two-stage conveyor dryers consist of two separate bands: the first band covers the first stage where the wet product dries quickly, while the second band carries relatively dried product until the final moisture content is achieved (Brown *et al.*, 1973).

Freeze-drying

The freeze-drying method is based on completely different principles from other drying methods. The fresh product is first frozen (–20°C to –30°C), the water in the product turns to ice in its location, and the ice is then removed directly within the system by turning the ice into steam in vacuum dryers. The installation and operating costs of freeze-drying are higher than other methods. In freeze-dried products, the humidity level drops below 5%, usually to around 2–4%. The drying time of figs using this method varies according to the size of the product (whole, halved, quartered, sliced, or cubed), the method applied in heat conduction, and the physical and chemical structure of the fruit (Cemeroğlu and Özkan, 2009; Desa *et al.*, 2019). The size, shape, color, texture, taste, flavor, and nutritional value of the figs to be dried, and the measures of change of these properties after drying vary greatly according to the variety, so appropriate varieties should be selected. İzmir Bardacık, Bursa Black, and Black Mission fig varieties are very suitable for freezing due to their appropriate color, texture, and flavor, while the Sarılop variety is not suitable for freeze-drying due to its peel characteristics. Brown Turkey is also not recommended for freeze-drying due to its light flavor and the occurrence of discoloration (Karaçalı, 2002).

Figs are harvested by hand at full maturity and after washing are frozen using individual quick freeze (IQF) technology. They are cut according to the determined shape (usually transversally), and those of the appropriate size are separated and dried in vacuum dryers. Freeze-dried figs have an extremely porous structure, so they quickly bind moisture and oxygen from the environment. As dried figs are conducive to oxidation, nitrogen gas is bound by the pores and additional nitrogen is inserted during packaging (Satomi, 1990). Because nitrogen gas is inert, it prevents many oxidative reactions in dried figs.

In freeze-drying, because the drying temperature is kept very low, color and structural changes are minimized and quality is maintained. The same quality of freeze-dried figs is not accessible in any other drying system (Hammami *et al.*, 1999). For this reason, although it is an expensive method, it is used commercially for drying figs (e.g. www.freshbak.com; accessed December 9, 2021). The nutritional value of freeze-dried fruits, which also keep their taste, flavor, and color, is very high (Krokida *et al.*, 1999). Nutrients inside the cell do not leave the cell as in other methods, and as water is bound as ice where it is located, liquid movement is not observed during drying. Dried figs have no significant changes in size and shape and almost keep their pre-drying state. The porous structure of freeze-dried figs gives them the ability to rehydrate almost to the same state as fresh figs. However, the porous structure makes it extremely brittle and reduces mass density. Thus, this low mass density can cause some economic disadvantages in terms of packaging, storage, transportation, and the need for wider spaces on the shelf of freeze-dried products (Hammami *et al.*, 1999; Ratti, 2008).

Microwave drying

Microwave drying is one of the alternative methods that minimizes the quality loss of dried products and allows the product to be dried in a short time (Sharifian *et al.*, 2012). Industrial microwave drying technologies have long been studied. However, the use of microwaves in combination with the existing hot-air systems is seen as a good alternative, as microwave energy is not distributed homogeneously and causes high mass transfers, which may have a negative effect on the textural properties of the product (Nijhuis *et al.*, 1998; Sumnu *et al.*, 2005). There are microwave drying systems supported by hot air (Patel and Kar, 2012). In traditional drying, the removal of free moisture initially occurs at a constant drying rate, and diffusion forces are then activated in the region of a reduced drying rate, resulting in mass transfer mechanisms. The pressure difference in the system plays a role in the realization of the microwave drying event. Therefore, the use of microwave energy is generally expected to contribute to the final stage of traditional drying (Datta and Anantheswaran, 2001).

Microwave air convection and microwave vacuum drying methods significantly shorten the drying time and improve antioxidant activity and sensory properties compared with the air convection and vacuum drying methods. Figs dried with microwave convection and a microwave vacuum have better physicochemical quality properties such as color, rehydration ratio, and shrinkage ratio than air convection drying, and contain higher amounts of total phenolic compounds and total flavonoids (Abul-Fadl *et al.*, 2015).

Other methods

In addition to these artificial drying systems, studies have been conducted on osmotic dehydration (Piga *et al.*, 2005), pulsed vacuum osmotic dehydration (Şahin and Öztürk, 2016), spray drying, and infrared drying methods for drying of fig fruits. However, the results have shown that some of these technologies are not very suitable for figs, probably due to the morphological and/or anatomical features of the fig fruit or the variety tested.

14.5 Storage of Dried Figs

Dried figs can be stored for a period by the farmers, traders, processors, or retailers, throughout different stages of the value chain. The main storage conditions for dried figs are usually ambient, uncontrolled storage at farm or trader levels. Processors utilize both uncontrolled and cold storage facilities.

14.5.1 Farmer storages

Most figs dried under the sun ripen gradually, and the period when they become shriveled and fall to the ground is relatively long. Dried figs are marketed at intervals, as the farmer reaches a certain volume or after completion of the drying process. Some farmers store dried figs in their own conditions for some time after drying, usually using the time to find a good price. They then either sell their figs after sorting them into different quality fruits and sizes, or sell them as bulk (Fig. 14.5). Delivery of dried figs to facilities

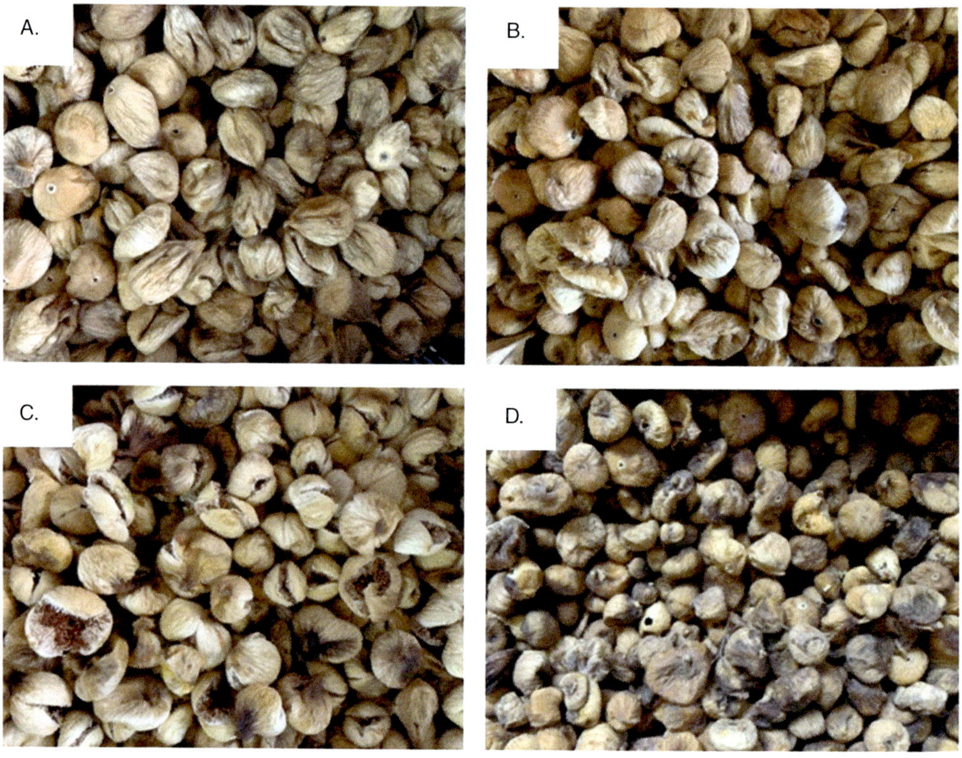

Fig. 14.5. Different quality and size classes of Sarılop variety figs separated in the farmer's storage room. A. High-quality and larger fruits (Extra). B. Class 1 mixed sizes. C. Defective figs that are split or have black neck, D. Defective culled figs.

where they can be handled properly as soon as possible is important as it reduces the risk of quality losses and infestation caused by storage pests.

To maintain the safety and quality of dried figs, the ambient conditions and cleanliness of the storage areas in or outside the orchard are of great importance (Şen, 2009). Storage conditions should be dry, with ventilation, and free from sunlight and external odors. If farmers do not sell the dried figs immediately and store them for a while, on-farm storage conditions become even more important. Fig moths and other dried-fig pests can attack the product in the orchard, drying areas, transportation vehicles, or farmers' storage rooms. It is known that fig moths and dried-fruit beetles start infesting the fruit while on the tree and in the drying area, and then multiply in warehouses (Turanlı, 2003). Light attracts fig moths and triggers infestation in storage areas. In addition, culled figs (Fig. 14.5D) and healthy sound figs (Fig. 14.5A) should not be stored together to prevent cross-contamination (Aksoy, 1997b). In cases where humidity and temperatures are high in farmers' storage areas, the risk of secondary aflatoxin formation in the dried fruit will increase (Aksoy *et al.*, 2003).

In farmer storage areas, all surfaces such as floors, walls, and ceilings should be smooth, and there should be no recesses, cavities, or similar areas where pests and fungal pathogens can nest or grow. A net should be attached to windows and doors closing all types of openings to prevent contamination from outside the storage area (Şen, 2009). Dried figs are taken into storage after they dry and reach the required moisture equilibrium. During this period, the water content should be below 24% and the water activity should preferably be below 0.65 (Okşar *et al.*, 2017b; Şen *et al.*, 2017). Poison-free mouse traps must be placed inside the storage areas to prevent rodent damage.

Dried figs must be kept in clean plastic boxes. The boxes should not rest on the side wall, or be placed directly on the floor; a distance should be left to provide moisture insulation. The boxes should be placed on plastic pallets or similar materials on the floor as good storage practice. If boxes are not available and storing dried figs as piles is inevitable, the pile height should not exceed 50 cm, leaving a gap of 1 m between the stacks, so that the product does not heat up and is ventilated adequately (Karaçalı, 2002; Şen, 2009).

Dried figs are highly susceptible to storage pests, and therefore if stored for a long time, they should first be fumigated with permitted preparations or methods to prevent storage pests. Otherwise, storage pests can multiply rapidly during storage starting from the farmer's warehouses, causing significant problems in the marketing of the product. According to the dried-fig standard, insect damage or damaged fruit must be within the limits allowed. In the United Nations Economic Commission for Europe (UNECE) standard, it can be at most 9.0%, 12.0%, and 16.0% in Extra, Class I, and Class II, respectively, and there is no tolerance for live insects, worms, and other zoological parasites (UNECE, 2016).

14.5.2 Storage at the trader and processing stages

Producers may sell their dried figs directly or after storing them in their own storage facilities. Dried figs are transported from the farmer to the trader's (mid-men) or the processor's storage facilities in plastic boxes. Plastic boxes used in storage must be clean and not damage the product. The bags that were traditionally used cause clumping and quality losses.

Dried-fig marketing begins toward the end of summer or beginning of autumn and continues intensively until the Christmas period. Due to diversification of the markets, nowadays marketing continues for the whole year. This makes appropriate storage mandatory for dried figs. Keeping dried figs in ideal conditions allows the processors and retailers to offer high-quality products to consumers throughout the year. As dried figs become durable by evaporating the water, as in other dry products, the care during storage is generally ignored. However, with the prolongation of storage time, significant product quality losses can be seen. These losses are closely linked to the storage conditions, namely temperature and RH (Karaçalı, 2002; Meyvacı *et al.*, 2003; Cemeroğlu and Özkan, 2009; Şen *et al.*, 2010).

Dried figs must be fumigated against stored-product pests (e.g. *E. cautella*, *Carpophilus hemipterus*, *P. interpunctella*, *Oryzaephilus surinamensis*, *Carpoglyphus lactis*) as soon as possible after drying (Turanlı, 2003) (see Chapter 18, this volume). Methyl bromide (40 g m^{-3} for 24 h) was commonly applied in gas-tight chambers for fumigation of dried figs. However, due to the prohibition of methyl bromide because of its ozone-layer depletion, the use of alternative methods has been investigated. For the fumigation process, phosphine is widely used in conventionally produced figs. Phosphine application under vacuum conditions is also preferred because it shortens the application time (Şen *et al.*, 2015). In organic dried-fig production and trade, carbon dioxide (CO_2) application is widely used either under normal atmospheric conditions (5–7 days exposure at >90% CO_2) or at high pressures (2 h exposure at 2000–3000 kPa, >95% CO_2). Deep freezing at −35°C to −40°C is also used to control storage pests, especially in organic certified dried figs without any negative impact on the quality (Şen *et al.*, 2009; Meyvacı *et al.*, 2010; Aksoy *et al.*, 2012).

Storage conditions and periods may vary for sun-dried, artificially dried, or freeze-dried figs, which are also categorized differently as intermediate or low-moisture dried fruit. In global trade, the majority come from intermediate-moisture sun-dried figs, which are stored in uncontrolled or cold storage conditions. The amount of moisture in dried figs to be stored is ideally around 18–22% and preferably less than 24% (Aksoy, 1997b). If the fruit water content is high, when kept under unsuitable conditions, fermentation and deterioration occur and the risk for toxin formation increases; therefore, the amount of water in dried figs is of the utmost importance in storage (Aksoy and Dokuzoğuz, 1984; Aksoy *et al.*, 2003).

As in farmer's storage areas, good storage principles should be implemented at all stages. At the beginning of each season, warehouses must be cleaned, repaired, well maintained, disinfected, and fumigated with chemicals or methods allowed in the producer country and, in the case of export, also in the importing country. During storage, the figs should be protected from dust and microbial contamination. Dried figs should not be stored together with products that release odors (Aksoy, 1997b; Şen, 2009). In storage, old stocks and new-season products, healthy and culled products, and products whose characteristics are very different from each other (well dried and partially dried, sulfur treated or not treated) should not be put together. Boxes should not rest against the wall or be placed directly on the floor, but should be placed to ensure air circulation and achieve control of the product (Şen, 2009; Karaçalı, 2016).

Under ambient conditions, due to changes in humidity and temperature during the storage period, some deterioration in quality may occur. Storage conditions can also alter the water content of dried figs, leading to undesirable physical properties that are of great importance to the structure, quality, and palatability of the product. In dried figs, the most important negativities that arise due to drying/storage conditions and duration are discoloration (browning), sugaring, and texture hardening (Aksoy and Dokuzoğuz, 1984; Meyvacı *et al.*, 2003; Şen *et al.*, 2010; Okşar *et al.*, 2017a,b). Moreover, microbial load and storage pests can also pose significant problems in inappropriate storage conditions (Jay, 2000; Şen *et al.*, 2009; Meyvacı *et al.*, 2010).

The final product color is affected by many preharvest factors (e.g. variety characteristics, excessive nitrogen fertilization, nutrient deficiencies, irrigation). However, conditions during different stages of the chain such as harvest (humidity, harvesting, and storage conditions), drying, storage (e.g. moisture content, drying conditions, storage conditions), processing (e.g. washing water temperature, use of sorbate or aluminum phosphide) and marketing (e.g. shelf life, market conditions) also exert marked effects (Aksoy, 2017; Desa *et al.*, 2019). Long-term high humidity during the drying season, especially in the morning, negatively affects the quality of dried figs, causing browning and darker colors (Aksoy, 1991). The color of the fig fruits harvested during the later part of the harvest period is darker than the early-ripening fruit as climatic conditions prevailing during the late period prolong the drying time. In the yellow Sarılop and Sarızeybek dried-fig varieties, about 80% of carotenoid compounds are broken down at the end of the first 7 days of sun drying (Yemiş *et al.*, 2012). Therefore, ecological factors, especially prolonging the drying time, can directly affect dried-fig quality (Aksoy, 1991; Göçmez and Seferoğlu, 2014).

Light-colored fig varieties show brown discoloration during drying. This browning can become more pronounced during the storage period, depending on the environmental conditions and the length of the storage. Decreases in the L^* (lightness), C^* (chroma) and $h°$ (hue angle) color values of dried figs during storage under normal conditions are consistent with browning in color (Meyvacı *et al.*, 2003; Şen *et al.*, 2010). Browning of light-colored dried fruits becomes more pronounced with the progression of time under normal storage conditions, while cold storage conditions limit these changes (Şen *et al.*, 2010).

Browning observed in dried figs can be the result of enzymatic or non-enzymatic reactions (Fig. 14.6A). As in other dried fruits, in dried figs browning occurs more often in a non-enzymatic way. In this browning reaction, also called the Maillard reaction, aldehyde groups of sugars and amino groups of proteins play a major role. Hundreds of compounds have been reported to occur because of non-enzymatic browning reactions, and these products are known to affect not only the product color but also the aroma, nutritional content, texture, and stability (Gasparetti, 2012). For the Maillard reaction, there must be a certain level of water in the substrate, and the amount of water in dried figs is not low enough to prevent this reaction. No browning reaction occurs below 2% moisture content (Fennema, 1976). As Maillard reactions occur most rapidly when the fruit moisture level is between 15% and 20%, figs provide a suitable substrate. Apart from the amount of fruit moisture content, the water activity, ambient temperature, and pH value of the product are the main factors affecting the level of browning. The optimum range of water activity is 0.40–0.65 (Maltini *et al.*, 2003). As the browning reaction depends on the temperature, the storage temperature should be kept as low as possible. Every 10°C temperature increase in storage causes a six- to eightfold increase in the browning reaction rate depending on the moisture content of the product (Cemeroğlu and Özkan, 2009). According to Ngammongkolrat *et al.* (1985), non-enzymatic browning is minimal at 4°C but increases rapidly at storage above 20°C.

It is quite difficult to break down or remove the melanoidins that form dark colors in figs after drying. In recent years, it has been shown that melanoidins can have positive effects on human health, participating in antioxidant, antimicrobial, high blood pressure decreasing, and prebiotic activities (Wang *et al.*, 2011). However, studies on compounds formed in dry products related to these effects remain limited.

The surface of the stored dried fig may appear white, as if sprinkled with flour or starch. This physical event is called "sugaring" (Fig. 14.6B). Sugaring is seen in dried figs that are stored for a long time under uncontrolled conditions, especially if the temperatures are high and RH levels are low. Sugaring is indicated as a defect according to the US Dry Fig standard, although it is not considered a defect according to the TS-541 dried-fig standard that meets UNECE standards.

Sugars, which are soluble in water in dried figs, are carried to the fruit surface during dehydration, and as the water evaporates, these sugars remain on the surface of the skin forming a thin crystal layer (Karaçalı, 2002). Similar crystals are also seen in other dried fruits, and consist of glucose monohydrate and anhydrous glucose crystals (Bolin, 1976). The Sarılop dried-fig

Fig. 14.6. Dried-fig (Sarılop variety) storage defects. (A) Original color and discoloration (browning) of dried figs. (B) Sugaring in figs stored in inappropriate conditions. (C) Fig fruits processed as "lerida" and stored in cold (bottom) and uncontrolled ambient (top, with sugar crystals formed) conditions for 12 months.

variety has similar proportions of glucose and fructose in the skin and pulp parts (Elmagülü *et al.*, 2020). Low amounts of sucrose have been found in some Sarılop dried figs (Şen *et al.*, 2017; Elmagülü *et al.*, 2020). Sugaring in dried fruits is directly related to the conditions and duration of storage. Factors creating sudden changes in dried-fig water content encourage sugaring (Aksoy and Dokuzoğuz, 1984). Therefore, storage conditions (proportional changes in humidity and temperature increases) that promote water movements in dried figs cause sugaring, while low temperatures at stable RH limit sugaring (Şen *et al.*, 2010; Badii *et al.*, 2014). No sugaring and browning have been seen in dried figs stored for 12 months at 4°C temperature and 55–65% RH (Fig. 14.6C) (Şen *et al.*, 2010). In dried figs, sugaring was observed from January onwards, followed by texture degradation under uncontrolled storage conditions in western Turkey (Aksoy and Dokuzoğuz, 1984). Some treatments that cause changes in texture and in the skin cells of figs promote sugaring as they facilitate the diffusion of water in or out of the fruit. The application of potassium metabisulfite as a pre-treatment before drying increased the sugar content of dried figs at the end of storage (Okşar *et al.*, 2017a).

If the fig fruits are dried to moisture and water activity levels below the equilibrium values, their texture hardens and they lose palatability (Pala and Saygı, 1983; Perera, 2005). High temperatures and dry winds, which can speed up the drying process, may cause excessive drying of the fruit, resulting in hardening and reduced palatability (Aksoy, 1991). Shriveled figs that fall outside the projection of the crown dry faster and may harden if exposed to direct sun for longer periods. This situation is encountered particularly in fig cultivation on sloping lands. In figs that are left in the drying site for a long time or stored under normal storage conditions at very low humidity levels for a long time, the texture hardens due to significant moisture loss (Şen *et al.*, 2010; Okşar *et al.*, 2017a,b). It is noted that there is an important correlation between fruit moisture content and softness or firmness of the dried fig fruit (Şen *et al.*, 2009). Based on demand, semi-dried or rehydrated dried figs possessing high moisture content and that are soft are prepared and marketed for specific uses.

Storage in cold conditions is recommended to prevent quality losses due to browning, sugaring, and hardening of texture. Dried figs in cold storage maintain their initial quality for 12 months or longer at 2–4°C and 55–65% RH. Cold storage conditions preserve the physicochemical properties of dried figs and allow them to be stored for longer, ensuring higher organoleptic scores (Waskar and Gawade, 2004; Şen *et al.*, 2010). In storage, the temperature and RH should be recorded continuously, and humidity in particular should not increase. Short-term increases in RH in storage rooms can cause increases in the water content and water activity of the dried figs if they are not packed in moisture-tight packaging material. The moisture content of dried fig should be <24% and/or the water activity ≤0.65. Osmophilic yeasts belonging to the genera *Zygosaccharomyces* and *Hanseniaspora* and some xerophilic fungi have been reported to cause significant problems in dried figs as these microorganisms can survive at low water activity levels (Jay, 2000). If the humidity in dried figs is at optimal levels (18–20%) and the water activity is within safe limits (≤0.65), microbial growth is not expected during storage. However, if ambient conditions become favorable, microorganisms found on dried figs that remain alive for a long time can start causing degradation. To prevent rodent damage and contamination of the product, appropriate traps are placed in the appropriate places of dried-fig storage areas.

14.6 Processing of Dried Figs

When dried figs come to the processing facility, they go through a number of stages comprising raw material control, fumigation for control of stored product pests, separation of contaminated (bright greenish-yellow fluorescence, BGYF) fruits under a long-wave ultraviolet (UV) lamp, sizing, washing, draining of excess water, drying, processing, packaging, passing through a metal detector, final inspection, and shipment, respectively.

Representative samples are taken before the dried figs go into the processing facility, and the general quality characteristics, especially the ratio of BGYF fruit, size, and defects, are determined. Major chemical hazards to be considered for prevention or reduction during processing are mycotoxins such as aflatoxin, ochratoxin A, and fumonisins, and the chlorine level of the washing water. Physical hazards are foreign materials such as wood, sand, soil, stones, and plant particles. Biological hazards are microbial load, storage pests, and mites. For traceability from the moment the product enters the facility, labels containing information describing the date of arrival, from which farmer it has been purchased, and the characteristics of the lot are attached to the boxes, and thus a registration system is created for traceability.

Dried figs are fumigated to control live storage pests as soon as they arrive at the processing facility. Fumigation applications vary depending on whether the product is certified as organic or conventional. Dried figs are either processed and packed directly or stored in normal and cold conditions throughout the year and processed according to demand.

In dried figs, aflatoxin contamination is reduced through detection of kojic acid, as aflatoxin-producing fungi also produce kojic acid, which produces BGYF mostly on the outer skin under a 365 nm wavelength UV lamp. As an incoming raw material control, the ratio of the fruit showing fluorescence under the UV lamp is determined in processing plants, and if this ratio is high, the product is rejected. If the fluorescence is chiefly in the inner cavity, this screening is not efficient to decrease aflatoxin contamination of the lots. For aflatoxin control or decreasing the aflatoxin contamination level in the lot, BGYF dried figs are removed under UV lamps in dark rooms whose walls are painted black or in chambers surrounded by black covers. The system of selection varies; the dried figs are poured on to tables placed under the UV lamps and selected by the workers manually and/or separated on a conveyor belt that moves slowly under the UV lamps. The use of optical sorting machines for removal of aflatoxin-contaminated figs is also becoming increasingly common. Dried figs removed and separated from healthy fruit due to aflatoxins are labeled and removed in different colored boxes from the processing area to prevent any intermingling.

The sizing process in dried figs is done by means of sieves with different diameters. The UNECE standard makes sizing optional either as the number of fruits per kilogram or according to the diameter. In the case of the number of fruits per kilogram, there are 11 sizes, ranging from size 1 having less than 40 to size 11 having more than 120 dried figs per kilogram (UNECE, 2016). If dried figs are sized by diameter, the minimum diameter should be 18 mm for black fig varieties and 22 mm for light-colored varieties. In the US Standards for Grades of Dried Figs (USDA, 2001), separate sizes are used for whole and loose black or light-colored dried figs. The sizes are stated as Jumbo size, Extra fancy, Fancy, Extra choice, Choice, and Standard, and the diameter ranges are specified. According to the Turkish dried-fig standard (TS-541), dried figs are divided into ten size classes, with the numbers of fruit per kilogram in each size as in the UNECE standard, excluding size 11 (≥120). The sizes and the range of number of fruits per kg in each size are given in Table 14.1.

Dried figs are washed at the processing unit if demanded by the buyer. In all cases, 3–5% salt is added to the washing water, which is of drinking quality. During washing, the amount of moisture in dried figs increases by 2–3%, especially in the outer peel. To decrease absorbed moisture, the products are kept in boxes to drain this excess water or passed through drying ovens. Drying is carried out by blowing air (45–50°C) with the help of fans on a conveyor belt moving at a constant speed or in stationary ovens. After drying, it is checked whether the moisture content of the product has fallen to the appropriate levels. The marketing standards have different threshold limits. In the USA, grade A and B moisture content limit is set as 24% but if a "safe and suitable mold inhibitor" is used, this can go up to 30%. In the UNECE (2016) standard 26% is allowed as the threshold limit

Table 14.1. The size classes of dried fig fruits and the number of fruits per kilogram in each size according to the Turkish standard TS-541.

Size 1	≤40	**Size 3**	46–50	**Size 5**	56–60	**Size 7**	66–70	**Size 9**	81–100
Size 2	41–45	**Size 4**	51–55	**Size 6**	61–65	**Size 8**	71–80	**Size 10**	≥101

for the moisture content of untreated dried figs and "above 26.0 and not exceeding 40.0% for high-moisture dried figs provided they are labeled as high moisture or equivalent denomination and treated with preservatives or preserved by other means (e.g. pasteurization)."

Dried figs are either packed loose without changing their natural form or shaped manually into different forms and styles such as chocolate, Turkish delight (locum), layer, pulled, protoben, lerida, macaroni, baglama, and garland. Dried figs are generally traded at international markets in 12.5 kg cardboard boxes in a loose form without manipulation. Shaped figs are usually packed with different materials such as cardboard boxes, stand-up pouches, or compressed foam, which can take consumer packages ranging mostly from 250 g to 2500 g. After the products are packed, they pass through a metal detector to detect any metal presence in them. The sensitivity setting of the metal detector is extremely important and this setting needs to be checked often for efficiency.

In the final product, the date and the total quantity of the lot are recorded, and then representative samples are taken randomly according to a sampling plan. Aflatoxins and total microorganism, mold, and yeast counts are analyzed routinely, and the results are recorded. These products are kept in storage or shipped to the destined market. Additional fumigation is performed on the final product before or during shipment against storage pests.

Every producing country has a national quality standard for marketing dried figs such as the US (USDA, 2001) or Turkish (TS-541) standard. For international trade, the UNECE standard (DDP-14) prepared according to the Dry and Dried Product Standard framework is used as the reference in most cases. Before marketing at the retail level, the requirements stated in the reference standard must be met. Although there are some differences among dried-fig standards, the main characteristics include "physical appearance, hygienic conditions, overall texture i.e. free of abnormal external moisture, foreign smell and/or taste and free from living insects." The international trade depends largely on sun drying of fresh or semi-dried figs in the open air or under simple tunnels (Aksoy, 2017). However, the moisture levels can be elevated through different methods to prepare ready-to-eat high-moisture dried figs. As stated in the standards, high moisture levels can be attained safely only if they pass through an antimicrobial treatment. The figs may then be packaged under modified-atmosphere conditions (Meyvacı *et al.*, 2003; Şen *et al.*, 2005).

14.7 Water Activity and its Relationship with Quality and Safety

From past to present, people have developed various processing methods such as drying, salting, and sugar addition, especially for off-season use. The common goal of these methods is to control the water activity value of the product and thus prevent microbial growth, deterioration, and quality losses. Water activity is the ratio of the vapor pressure of pure water at the same temperature to the vapor pressure of water in the product. In other words, it is the point where the moisture content of the sample comes into equilibrium with the RH of the air (Fontana and Campbell, 2004). For this reason, the water activity of a product is the measure of the availability of water in that product for microorganisms (Cemeroğlu and Özkan, 2009). The water activity of figs is reduced by drying them to a limit where microorganisms cannot grow (Koç *et al.*, 2012; Şen *et al.*, 2017). The decrease in water activity is associated with the moisture content of the fruit and is affected by the temperature and RH of the environment (Maltini *et al.*, 2003; Piga *et al.*, 2004; Koç *et al.*, 2012). The water activity value of Sarılop figs is measured as 0.91 when fresh ripe on the tree (water content 71.2%), 0.79 when shriveled (water content 34.9%), and 0.62 at the drying site (water content 21.6%) (Şen *et al.*, 2017).

Water activity is more effective on the physical, chemical, and biological properties of products than the amount of water (Rockland and Nishi, 1980). Water activity is the most important factor limiting chemical, biochemical, and microbiological changes in foods. As water activity decreases, food durability increases. It is generally accepted that all microorganism activity stops at a water activity level below 0.60. Most bacteria in foods are considered to have ceased their activities at a value of <0.90, some halophyte bacteria at <0.75, yeasts at <0.80, and molds at 0.60–0.75 (Labuza, 1970; Karel, 1975; Bone, 1987; Jay, 2000). Mold growth stops completely when the water activity falls below 0.65. Osmophilic yeasts and xerophilic molds, which are dangerous in dried fruits, can grow at water activity values as low as 0.60–0.65 (Troller, 1980; Jay, 2000). If the water activity in dried fruit is above 0.70 (moisture content 24%), some yeasts (e.g. *Saccharomyces rouxii*) cause rancidity, and some mold fungi (e.g. *Aspergillus*, *Penicillium*, and *Xeromyces* species) may cause rotting, souring, and deterioration (Jay, 2000). In general, the target for water activity in dried figs is below 0.70 and preferably below 0.65. The amount of water equivalent to this varies depending on the product but is accepted as 18–24% (Okşar *et al.*, 2017b; Şen *et al.*, 2017). In addition to microbial growth, some toxigenic organisms produce secondary metabolites that can be harmful to humans (Hussein and Brasel, 2001). The values of water activity that fungi need to develop and produce mycotoxins vary (Farkas, 2007). Generally, the water activity required to form mycotoxins is higher than the water activity needed for their growth. Water activity in the product also affects the rates of various enzymatic and non-enzymatic reactions. Lowering the water activity limits or prevents enzymatic changes (Fontana and Campbell, 2004.). Enzymes such as amylase, peroxidase, and phenol oxidase, which are common in fruit products and cause significant degradation, become inactive at water activity values below 0.85 (Labuza, 1975; Fennema, 1976). In conditions where the water activity is low, the substrate–enzyme contact required for an enzymatic reaction is not possible (Labuza, 1975).

Non-enzymatic browning in dry products, the Maillard reaction, is closely related to the water activity of the product as it is a water-dependent reaction. At very low water activities, the Maillard reaction slows down, while at higher water activity values, this reaction accelerates, resulting in rapid browning (Cemeroğlu and Özkan, 2009).

14.8 Concluding Remarks

Dried figs are becoming increasingly important because they can endure ambient conditions without spoilage and are a high-energy and nutritive food. In addition to preharvest practices, harvesting, drying, and storage have substantial importance for marketing high-quality dried figs. For this purpose, it is necessary to implement good practices at every stage regarding quality and safety. Dissemination of these good practices to all production regions such as the harvest nets or drying tunnels will contribute significantly to improving quality and preventing or reducing the formation of mycotoxins that may pose health hazards. As manual harvesting of dried figs is the chief factor contributing to the production cost, innovative low-cost mechanical harvesting techniques with or without training and pruning need to be developed.

In recent years, the use of practical plastic trays both in open-air sun drying or under plastic tunnels has become widespread. Artificial drying is also enhanced as it eliminates many drawbacks of sun drying and allows higher-quality or innovative low-moisture products to be obtained. In recent years, numerous studies related to different artificial drying systems have been concluded. However, in artificial drying, more research is needed on systems that will reduce the cost of energy while still maintaining high quality. Due to bulk drying, low investment and energy costs, and low carbon emissions, sun drying of figs will continue to be widely used as an environmentally friendly technique. Therefore, more research work is essential on hybrid systems where artificial dryers are integrated with sun drying or where mechanized systems take over the required manual labor.

The current yearly dried-fig production is almost totally consumed within the year without any significant stock turnover. However, cold storage is recommended to prevent quality losses such as discoloration or browning in light-colored varieties, sugaring, and hardening of the texture, decreasing palatability. New studies are needed to prepare high-moisture, ready-to-eat dried figs with extended storage and shelf life. Additionally, due to significant varietal differences, studies on sun-dried or artificially dried new fig varieties are needed to introduce alternative tastes to the consumers. Processing and storage conditions of these varieties may need to be optimized.

References

Abul-Fadl, M.M., Ghanem, T.H., El-Badry, N., and Nasr, A. (2015) Effect of some different drying methods on quality criteria of dried fig fruits. *Current Science International* 4, 548–566.

Aday, N. (1938) Yunan Kuru İncirleri (Greek dried figs). In: *Publication of the First Rural and Agricultural Development Congress*. Ministry of Agriculture of the Turkish Republic, Ankara, p. 59.

Aksoy, U. (1981) Akça, Göklop ve Sarılop İncir Çeşitlerinde Meyve Gelişmesi, Olgunlaşması ve Depolanması Üzerinde Araştırmalar (Research on fruit growth, ripening and storage of Akça, Göklop and Sarılop fig fruits). PhD thesis, Ege University, İzmir, Turkey.

Aksoy, U. (1990) *Aflatoksin (Aflatoxin)*. Tarımsal Uygulama ve Araştırma Merkezi Yayım Bülteni No. 2, Bornova, İzmir.

Aksoy, U. (1991) Kuru incir kalitesine etki eden faktörler (Factors on quality of dried figs). In: *İncirde Verimlilik Paneli (Proceedings of the Panel on Productivity of Fig)*, Aydın, Turkey.

Aksoy, U. (1997a) Fruit formation and development. In: *Advanced Course on Fig Production*. Ege University, Faculty of Agriculture Department of Horticulture, İzmir, Turkey, 16–28 June.

Aksoy, U. (1997b) Harvest and drying of figs. In: *Advanced Course on Fig Production, 16–28 June 1997*. Department of Horticulture, Faculty of Agriculture, Ege University, İzmir, Turkey.

Aksoy, U. (2017) The dried fig management and the potential for new products. *Acta Horticulturae* 1173, 377–382.

Aksoy, U. and Anaç, D. (1994) Factors affecting quality of fresh and dried fig fruits. *Acta Horticulturae* 93, 763–766.

Aksoy, U. and Dokuzoğuz, M. (1984) Kuru incirlerde saklama koşullarının meyve kalitesine etkileri (Effect of storage conditions on fruit quality of dried figs). In: TUBİTAK Yayınları No. 578, Ankara, pp. 250–264.

Aksoy, U., Anaç, D., Eryüce, N., and Yoltaş, T. (1987) Ege bölgesi incir bahçelerinin beslenme durumunun saptanması ve değerlendirilmesi (Determination and evaluation of fig orchard nutrition of Aegean Region). *Ege Üniversitesi Ziraat Fakultesi Dergisi* 24, 21–35.

Aksoy, U., Can, H.Z., Hepaksoy, S., and Şahin, N. (2001) *İncir Yetiştiriciliği (Fig Production)*. TÜBİTAK TARP Türkiye Tarımsal Araştırma Projesi Yayınları, Turkey.

Aksoy, U., Sabır, E., Eltem, R., Kıraç, S., Sarıgül, N., *et al.* (2003) Kuru incirlerde okratoksin A'nın potansiyel kontaminasyon riskinin araştırılması (Investigation of potential contamination risk of ochratoxin A in dried figs). In: *Ulusal Mikotoksin Sempozyumu, 18–19 September 2003, İstanbul, Turkey*, pp. 41–47.

Aksoy, U., Can, H.Z., Meyvacı, K.B. and Şen, F. (2008) Kuru incir (Dried fig). In: Aksoy, U. (ed.) *Türk Sultanları (Turkish Sultans)*. Can Dijital Baskı Merkezi, İzmir, Turkey, pp. 51–85.

Aksoy, U., Sen, F., and Meyvacı, K.B. (2012) Effect of post-harvest carbon dioxide application on storage pests and fruit quality of dried figs. In: *Proceeding of the 9th International Conference on Controlled Atmosphere and Fumigation in Stored Products, 15–19 October, Antalya, Turkey*. ARBER Professional Congress Services, Ankara, pp. 166–171.

Aksoy, U., Can H.Z., Meyvacı, K.B., Şen, F., İrget, E., *et al.* (2013) *Kuru İncirde İyi Uygulamaların Aflatoksinlerin Önlenmesi/Azaltılması Üzerindeki Etkilerinin Belirlenmesi (Determining the Effects of Good Practices on Dried Figs on Prevention/Reduction of Aflatoxins)*. Ege İhracatçı Birlikleri Proje Sonuç Raporu, İzmir, Turkey.

Anaç, D., Aksoy, U., and Eryüce, N. (1987) Ege bölgesi incir bahçelerinin makro ve bazı mikro besin elementleri içeriklerinin yaprak analizleri ile saptanması üzerinde araştırmalar: I. Büyük Menderes Havzası (Determination of macro and micro nutrient contents of fig orchards in the Aegean Region through leaf analysis: I. Big Meander Basin). *Ege Üniversitesi Ziraat Fakultesi Dergisi* 24, 75–90.

Ayensu, A. (1997) Dehydration of food crops using a solar dryer with convective heat flow. *Solar Energy* 59, 121–126.

Babalis, S.J. and Belessiotis, V.G. (2004) Influence of the drying conditions on the drying constants and moisture diffusivity during the thin-layer drying of figs. *Journal of Food Engineering* 65, 449–458.

Babalis, S.J., Papanicolaou, E., Kyriakis, N., and Belessiotis, V.G. (2006) Evaluation of thin-layer drying models for describing drying kinetics of figs (*Ficus carica*). *Journal of Food Engineering* 75, 205–214.

Badii, F., Farahnaky, A., and Behmadi, H. (2014) Effect of storage relative humidity on physical stability of dried fig. *Journal of Food Processing and Preservation* 38, 477–483.

Belessiotis, V. and Delyannis, E. (2011) Solar drying. *Solar Energy* 85, 1665–1691.

Betina, V. (1984) *Mycotoxins: Production, Isolation, Separation and Purification*. Elsevier, Amsterdam.

Bolin, H. (1976) Texture and crystallization control in raisins. *Journal of Food Science* 41, 1316–1319.

Bone, D.P. (1987) Practical applications of water activity and moisture relations in foods. In: Rockland, L.B. and Beuchat, L.R. (eds) *Water Activity: Theory and Applications to Food*. Marcel Dekker, New York, pp. 369–395.

Brown, A.H., van Arsdel, W.B., Lowe, E., and Morgan, A.E. (1973) Air drying and drum drying. In: van Arsdel, W.B., Copley, M.J., and Morgan, A.I. (eds) *Food Dehydration*, Vol. 1. AVI Publication Co., Westport, Connecticut, pp. 82–160.

Cemeroğlu, B. and Özkan, M. (2009) Kurutma teknolojisi (Drying technology). In: Cemeroğlu, B. (ed.) *Meyve ve Sebze İşleme Teknolojisi (Fruit and Vegetable Processing Technology)*, Vol. II. Bizim Grup Publisher, Ankara, pp. 479–620.

Chand, T.K., Mohanty, M.K., and Mohanty, R.C. (2015) An overview of solar energy and its application in solar dryers with brief concept of energy and energy analysis. *International Journal of Research* 2, 870–877.

Chimi, H., Ouaouich, A., Semmar, M., and Tayebi, S. (2008) Industrial processing of figs by solar drying in Morocco. *Acta Horticulturae* 798, 331–334.

Codex Alimentarius (2008) Code of practice for the prevention and reduction of aflatoxin contamination in dried figs (CAC/RCP 65-2008). In: *Prevention and Reduction of Food and Feed Contamination*, 1st edn. Food and Agriculture Organization of the United Nations, Rome/World Health Organization, Geneva, pp. 44–54.

Cobanoglu, F., Aksoy, U., Tunalioglu, R., and Karaman, A.D. (2010) Differences in the economic costs of implementing conventional practices and good practices intended for aflatoxin management in the dried fig supply chain. *Toxin Reviews* 29, 87–98.

Datta, A.K. and Anantheswaran, R.C. (2001) *Handbook of Microwave Technology for Food Applications*. Marcel Dekker, New York.

Demirayak, V. (2008) Kuru İncirlerin Bazı Fiziko-Mekanik Özelliklerinin Belirlenmesi (Determination of some physico-mechanical properties of dried figs). MSc thesis, Adnan Menderes University Graduate School of Natural and Applied Sciences, Aydın, Turkey.

Desa, W.N.M., Mohammad, M., and Fudholi, A. (2019) Review of drying technology of fig. *Trends in Food Science and Technology* 88, 93–103.

EFSA (European Food Safety Agency) (2020) Risk assessment of aflatoxins in food. *EFSA Journal* 18: 6040.

Elmagülü, G., Aksoy, U., and Şen, F. (2020) Sugars in 'Sarılop' fig fruits. *Acta Horticulturae* 1280, 115–122.

Emekli, N.Y. and Büyüktaş, K. (2006) Sera örtü malzemelerinin mekanik özellikleri (Mechanical properties of greenhouse covering materials). *Derim* 23, 24–35.

Ersoy, N., Gozlekci, S., Gok, V., and Yilmaz, S. (2017) Fig (*Ficus carica* L.) fruit: some physical and chemical properties. *Acta Horticulturae* 1173, 329–334.

Ertekin, C., Yaldiz, O., and Miihlbauer, W. (2003) Thin layer drying of fig as affected by different drying conditions. *Acta Horticulture* 605, 283–291.

Farahnaky, A., Ansari, S., and Majzoobi, M. (2009) Effect of glycerol on the moisture sorption isotherm of figs. *Journal of Food Engineering* 93, 468–473.

Farkas, J. (2007) Physical methods of food preservation. In: Doyle, M. and Beuchat, L. (eds) *Food Microbiology: Fundamentals and Frontiers*, 3rd edn. ASM Press, Washington, DC, pp. 685–712.

Fennema, O.R. (1976) *Principles of Food Science, Part I: Food Chemistry*. Marcel Dekker, New York.

Ferguson, L., Michailides, T.J., and Shorey, H.H. (1990) The California fig industry. *Horticultural Reviews* 12, 409–490.

Fontana, A.J. and Campbell, C.S. (2004) Water activity. In: Nollet, L.M.L. (ed.) *Handbook of Food Analysis, Physical Characterization and Nutrient Analysis*. Marcel Dekker, New York, pp. 39–54.

Gasparetti, C. (2012) Biochemical and structural characterisation of the copper containing oxidoreductases catechol oxidase, tyrosinase, and laccase from ascomycete fungi. VTT Technical Research Centre of Finland, Espo, Finland. Available at: https://www.vttresearch.com/sites/default/files/pdf/science/2012/S16.pdf (accessed December 9, 2021).

Göçmez, A. and Seferoğlu, H.G. (2014) Fresh fig and dry fig quality parameters and the effective factors on quality. *Turkish Journal of Agriculture Research* 1, 98–108.

Green, M.G. and Schwarz, D. (2001) *Solar Drying Equipment: Notes on Three Driers*. Infogate Information Service/gtz, Eschborn, Germany. Available at: www.researchgate.net/profile/Matthew-Green-12/publication/237503256_Solar_Drying_Technology_for_Food_Preservation/links/546b6e7c0cf20dedafd52fff/Solar-Drying-Technology-for-Food-Preservation.pdf (accessed December 9, 2021).

Hammami, C., René, F., and Martin, M. (1999) Process-quality optimization of the vacuum freeze-drying of apple slice by the response surface method. *International Journal of Food Science* 34, 145–160.

Haug, M.T., King, E.S., Heymann, H., and Crisosto, C.H. (2013) Sensory profiles for dried fig (*Ficus carica* L.) cultivars commercially grown and processed in California. *Journal of Food Science* 78, 1273–1281.

Heldman, D.R. and Singh, R.P. (1981) *Food Process Engineering*, 2nd edn. AVI Publishing, Westport, Connecticut.

Hussein, H.S. and Brasel, J.M. (2001) Toxicity, metabolism and impact of mycotoxins on humans and animals. *Toxicology* 167, 101–134.

INC (International Nut and Dried Fruit Council) (2020) Statistics. Available at: www.nutfruit.org/industry/statistics (accessed December 9, 2021).

Jay, J.M. (2000) Fermentation and fermented dairy products. In: Jay, J.M. (ed.) *Modern Food Microbiology*, 6th edn. Aspen Publishers, Gaithersburg, Maryland, pp. 113–130.

Kader, A.A. (2002) Postharvest biology and technology: an overview. In: Kader, A. (ed.) *Postharvest Technology of Horticultural Crops*. University of California Agricultural and Natural Resources, Publication 3311, USA, pp. 39–47.

Kamiloglu, S. and Capanoglu, E. (2015) Polyphenol content in figs (*Ficus carica* L.): effect of sun-drying. *International Journal of Food Properties* 18, 521–535.

Karaçalı, İ. (2002) *Meyve ve Sebze Değerlendirme* (*Fruit and Vegetable Processing*). Faculty of Agriculture Ege University, İzmir, Turkey.

Karaçalı, İ. (2016) *Bahçe Ürünlerinin Muhafazası ve Pazarlanması* (*Storage and Marketing of Horticultural Product*). Publication No. 494, Faculty of Agriculture, Ege University, İzmir, Turkey.

Karel, M. (1975) Water activity and food preservation. In: Karel, M., Fennema, O.R., and Lund, D.B. (eds) *Principles of Food Science. Part II. Physical Principles of Food Preservation*. Marcel Deckker, New York, pp. 237–263.

Koç, S., Şen, F., and Aksoy, U. (2012) Farklı işlem görmüş kuru meyvelerde su aktivitesinin değişimi üzerine araştırmalar (Research on changes of water activity of different processed dried fruits). In: *5. Bahçe Ürünlerinde Muhafaza ve Pazarlama Sempozyumu, 18–21 September, İzmir, Turkey*, pp. 267–273.

Konak, R., Kösoğlu, İ., and Yemenicioğlu, A. (2017) Effects of different drying methods on phenolic content, antioxidant capacity and general characteristics of selected dark colored Turkish fig cultivars. *Acta Horticulturae* 1173, 335–340.

Krokida, M.K., Kiranoudis, C.T., and Maroulis, Z.B. (1999) Viscoelastic behaviour of dehydrated products during rehydration. *Journal of Food Engineering* 40, 269–277.

Kutlu, N., İşçi, A., and Şakıyan Demirkol, Ö. (2015) Gıdalarda ince tabaka kurutma modelleri (Thin layer drying models in foods). *Gıda* 40, 39–46.

Labuza, T.P. (1975) Oxidative changes in foods at low and intermediate moisture levels, In: Duckworth, R.B. (ed.) *Water Relations of Foods*. Academic Press, New York, pp. 455–474.

Labuza, T.P., Tannenbaum, S.R., and Karel, M. (1970) Water content and stability of low moisture and intermediate moisture foods. *Food Technology* 24, 543–550.

Lewicki, P.P. (2006) Design of hot air drying for better foods. *Trends in Food Science and Technology* 17, 153–163.

Madhlopa, A., Jones, S.A., and Saka, J.D.K. (2002) A solar air heater with composite-absorber systems for food dehydration. *Renewable Energy* 27, 27–37.

Maltini, E., Torreggiani, D., Venir, E., and Bertolo, G. (2003) Water activity and the preservation of plant foods. *Food Chemistry* 82, 79–86.

Manolopoulou, E., Tsiavtari, E., and Demopoulos, V. (2017) Sun drying and hot air drying of figs: performance evaluation. *Acta Horticulturae* 1173, 341–346.

Martínez-García, J.J., Gallegos-Infante, J.A., Rocha-Guzmán, N.E., Ramírez-Baca, P., Candelas-Cadillo, M.G., and González-Laredo, R.F. (2013) Drying parameters of half-cut and ground figs (*Ficus carica* L.) var. Mission and the effect on their functional properties. *Journal of Engineering* 2013: 710830.

Meyvacı, K.B., Sen, F., Aksoy, U., Özdamar, F., and Çakır, M. (2003) Research on prolonging the marketing period of dried and ready-to-eat type figs (*Ficus carica* L.). *Acta Horticulturae* 628, 439–445.

Meyvacı, K.B., Sen, F., and Aksoy, U. (2010) Optimization of magnesium phosphide treatment used to control major dried fig storage pests. *Horticulture, Environment, and Biotechnology* 51, 33–38.

Morton, J.F. (1987) *Fruits of Warm Climates*. J.F. Morton, Miami, Florida.

Mujić, I., Kralj, M., Jokić, S., Jarni, K., Jug, T., and Prgomet, Z. (2012) Changes in aromatic profile of fresh and dried fig – the role of pre-treatments in drying process. *International Journal of Food Science and Technology* 47, 2282–2288.

Mujic, I., Kralj, M.B., Jokic, S., Jug, T., Subaric, D., *et al.* (2014) Characterization of volatiles in dried white varieties figs (*Ficus carica* L.). *Journal of Food Science and Technology* 51, 1837–1846.

Ngammongkolrat, A., Moutounet, M., and Pech, J.C. (1985) Pulp browning of French prunes: methods of measurement and types of reactions involved. *Sciences des Aliments* 5, 393–405.

Nijhuis, H.H., Torringa, H.M., Muresan, S., Yuksel, D., Leguijt, C., and Kloek, W. (1998) Approaches to improving the quality of dried fruit and vegetables. *Trends in Food Science Technology* 9, 13–20.

Okşar, R.E., Şen, F., and Aksoy, U. (2017a) The effects of pre-treatments on color, composition and storability of dried 'Sarılop' fig fruit. *Acta Horticulturae* 1173, 347–352.

Okşar, R.E., Şen, F., and Aksoy, U. (2017b) Effect of drying under plastic tunnels on drying rate and quality of fig (*Ficus carica* L. 'Sarılop'). *Acta Horticulturae* 1173, 297–302.

Özay, G., Aran, N., and Pala, M. (1995) Influence of harvesting and drying techniques on micoflora and mycotoxin contamination of figs. *Nahrung* 2, 156–165.

Özen, M., Çobanoğlu, F., Kocataş, H., Tan, N., Ertan, B., *et al.* (2007) *İncir Yetiştiriciliği* (*Fig Production*). TC Tarım ve Köyişleri Bakanlığı, TAGEM, Erbeyli İncir Araştırma Enstitüsü Müdürlüğü, Aydın, Turkey.

Özen, M., Çobanoğlu, F., Özkan, R., Konak, R., and Tutmuş, E. (2008) Kuru incirde farklı hasat yöntemlerini ve plastik tünel altında kurutmanın aflatoksin oluşumu ve bazı kalite kriterlerine etkisi (Effects of different harvest method and drying products under plastic tunnel on occurring aflatoxin and some quality parameters in dried figs). In: *Bahçe Ürünlerinde IV. Muhafaza ve Pazarlama Sempozyumu, 8–11 October, Antalya, Turkey*, pp. 456–463.

Özer Meyvacı, K.B. (2009) Bahçe ürünlerinde mikotoksin problemi (Mycotoxin problem in horticultural products). In: Şen, F. (ed.) *Hasat Sonrası İyi Tarım Uygulamaları*. Basım Yayım Matbaacılık San. Tic. Ltd. Şti., İzmir, Turkey, pp. 207–221.

Pala, M., and Saygı, Y.B. (1983) Su aktivitesi ve gıda işlemedeki önemi (Water activity and its importance in food processing). *Gıda* 1, 33–39.

Patel, K.K. and Kar, A. (2012) Heat pump assisted drying of agricultural produce—an overview. *International Journal of Food Science and Technology* 49, 142–160.

Perera, C.O. (2005) Selected quality attributes of dried foods. *Drying Technology* 23, 717–730.

Piga, A., Del Caro, A., and Pinna, I. (2005) Osmo-dehydration of fig fruits with five sugars. *Acta Horticulturae* 798, 313–318.

Piga, A., Pinna, I., Özer, K.B., Agabbio, M., and Aksoy, U. (2004) Hot air dehydration of figs (*Ficus carica* L.): drying kinetics and quality loss. *International Journal of Food Science and Technology* 39, 793–799.

Pixton, S.W. and Warburton, S. (1976) The relationship between moisture content and equilibrium relative humidity of dried figs. *Journal of Stored Products Research* 12, 87–92.

Rahman, M.S. (2007) Food preservation: overview. In: Rahman, M.S. (ed.) *Handbook of Food Preservation*. CRC Press, Boca Raton, Florida, pp. 3–18.

Ratti, C. (2008) Freeze and vacuum drying of foods. In: Chen, X.D. and Mujumdar, A.S. (eds) *Drying Technologies in Food Processing*. Blackwell Publishing, Chennai, India, pp. 225–251.

Rockland, L.B. and Nishi, S.K. (1980) Influence of water activity on food product quality and stability. *Food Technology* 34, 42–51.

Şahin, U. and Öztürk, H.K. (2016) Effects of pulsed vacuum osmotic dehydration (PVOD) on drying kinetics of figs (*Ficus carica* L.). *Innovative Food Science and Emerging Technologies* 36, 104–111.

Satomi, K. (1990) Gas-exchange packaging. In: Kadoya, T. (ed.) *Food Packaging*. Academic Press, San Diego, California, pp. 269–277.

Şen, F. (2009) Meyve ve sebzelerin kurutulması (Drying fruits and vegetables). In: Şen, F. (ed.) *Hasat Sonrası İyi Tarım Uygulamaları*. Basım Yayım Matbaacılık San. Tic. Ltd. Şti., İzmir, Turkey, pp. 89–114.

Şen, F., Meyvacı, K.B., Aksoy U., and Cakir, M. (2005) Studies to optimize pre-treatments for high moisture dried figs. *Acta Horticulturae* 798, 293–297.

Şen, F., Meyvacı, K.B., Aksoy, U., Emekçi, M., and Ferizli, A.G. (2009) Effects of the post-harvest application of methyl bromide alternatives on storage pests and quality of dried fig. *Turkish Journal of Agriculture and Forestry* 33, 403–412.

Şen, F., Meyvacı, K.B., Aksoy, U., Koç, S., Sarılar, S., and Kocatürk, R. (2010) Changes in quality of dried fig fruit and processed fig products under cold and ambient storage conditions. *Acta Horticulturae* 877, 853–860.

Şen, F., Aksoy, U., Emekci, M., and Ferizli, A.G. (2015) Determining effect of phosphine (ECO$_2$FUME®) fumigation under atmospheric and vacuum conditions on dried fig quality. *Fresenius Environmental Bulletin* 24, 2046–2051.

Şen, F., Aksoy, U., Özer, K.B., Can, H.Z., Köseoğlu, İ., and Konak, R. (2017) Impact of yearly conditions on major physical and chemical properties of fresh, semi-dried and sun dried fig (*Ficus carica* L. 'Sarılop') fruit. *Acta Horticulturae* 1173, 309–314.

Sevgican (1999) *Örtüaltı Sebzeciliği* (*Greenhouse Vegetables*), Vol. 1. Publication No. 528, Faculty of Agriculture, Ege University, İzmir, Turkey.
Sharifian, F., Motlagh, A.M., and Nikbakht, A.M. (2012) Pulsed microwave drying kinetics of fig fruit (*Ficus carica* L.). *Australian Journal of Crop Science* 6, 1441–1447.
Shorey, H.H., Ferguson, L., and Wood, D.L. (1989) Solar heating reduces insect infestations in ripening and drying figs. *HortScience* 14, 443–445.
Silva, J.M., Cantu, M.G., Rodrigues, V., and Mazutti, M.A. (2013) Influence of osmotic pre-treatment on convective drying kinetics of figs. *International Journal of Food Engineering* 9, 187–196.
Singh, P., Shrivastava, V., and Kumar, A. (2018) Recent developments in greenhouse solar drying: a review. *Renewable and Sustainable Energy Reviews* 82, 3250–3262.
Slatnar, A., Klancar, U., Stampar, F., and Veberic, R. (2011) Effect of drying of figs (*Ficus carica* L.) on the contents of sugars, organic acids, and phenolic compounds. *Journal of Agricultural and Food Chemistry* 59, 11696–11702.
Sumnu, G., Turabi, E., and Oztop, M. (2005) Drying of carrots in microwave and halogen lamp–microwave combination ovens. *LWT – Food Science and Technology* 38, 549–553.
Tan, N. (2017) Effect of cabinet drying method on dried fruit quality and functional properties of 'Sarilop' (*Ficus carica* L.) fig cultivar. *Acta Horticulturae* 1173, 359–364.
Troller, J.A. (1980) Influence of water activity on microorganisms in foods. *Food Technology* 5, 76–82.
Turanlı, F. (2003) Studies on infestation levels of pests on dried fig in Aydın and Izmir provinces. *Turkish Journal of Entomology* 27, 171–180.
UNECE (United Nation Economic Commission for Europe) (2016) *UNECE Standard DDP-14: Dried Figs*. United Nations, New York/Geneva. Available at: https://unece.org/fileadmin/DAM/trade/agr/standard/dry/dry_e/14DriedFigs_E2016.pdf (accessed December 9, 2021).
USDA (2001) *Grades of Dried Figs*. Agricultural Marketing Service, US Department of Agriculture, Washington, DC. Available at: https://www.ams.usda.gov/grades-standards/dried-figs-grades-and-standards (accessed December 9, 2021).
van Arsdel, W.B. (1951) *Tunnel-and-Truck Dehydrators as Used in Vegetable Dehydration*. US Department of Agriculture, Bureau of Agricultural and Industrial Chemistry, Albany, California.
Villalobos, M.C., Serradilla, M.J., Martín, A., Pereira, C., López-Corrales, M., and Córdoba, M.G. (2016) Evaluation of different drying systems as an alternative to sun drying for figs (*Ficus carica* L.). *Innovative Food Science and Emerging Technologies* 36, 156–165.
Wang, H., Qian, H., and Yao, W. (2011) Melanoidins produced by the Maillard reaction: structure and biological activity. *Food Chemistry* 128, 573–584.
Waskar, D.P. and Gawade, M.H. (2004) Storage of fig fruits – a review. *Agricultural Reviews* 25, 76–78.
Xanthopoulos, G., Yanniotis, S., and Lambrinos, G. (2009) Water diffusivity and drying kinetics of air drying of figs. *Drying Technology* 27, 502–512.
Yağcıoğlu, A. (1999) *Tarım Ürünleri Kurutma Tekniği (Agricultural Products Drying Technique)*. Publication No. 536, Faculty of Agrculture, Ege University, İzmir, Turkey.
Yemiş, O., Bakkalbaşı, E., and Artık, N. (2012) Changes in pigment profile and surface colour of fig (*Ficus carica* L.) during drying. *International Journal of Food Science and Technology* 47, 1710–1719.

Part IV

Fig Pests and Diseases

15 Pests of Fig Trees

David Ben-Yakir[1]* and Alain Costa[2]
[1]*Department of Entomology, Agricultural Research Organization, Volcani Center, Rishon LeZion, Israel;* [2]*Technical Agricultural Engineer, Albatera, Spain*

15.1 Introduction

Fig trees are infested by about 100 species of insects and mites. Fig stem borers, leaf rollers, fruit flies, gall midges, mealybugs, and scales are considered the most important pests of fig (Lakra, 2004). Several books and web sites list and describe the local pests of fig. Records of fig pests can be found in the following references and websites: international (Alford, 2007; Hill, 2008; Reddy, 2010), India (Lakra, 2004; Atwal and Dhaliwal, 2013; Singh and Kaur, 2017; Anon., 2021), South Asia (Atwal and Dhaliwal, 2013), Maltese Islands (Mifsud *et al.*, 2012), Europe (Ellis, 2021), and Africa (Wohlfarter *et al.*, 2011; Dunoven, 2014). In Pakistan, lectures were given to train fig growers to identify fig pests and their typical damage (Khan, 2015).

In this chapter, fig pests are presented in four categories according to the plant part they infest and the nature of their damage: (i) trunk and branch borers; (ii) leaf and shoot feeders; (iii) developing fruits; and (iv) ripe and dry fruits. The information given in this chapter does not cover pests controlled by specific chemical pesticides.

Table 15.1 lists arthropod fig pests by their taxonomic attributes, the type of damage they cause, and their geographical distribution.

15.2 Trunk and Branch Borers

15.2.1 Tree-boring cerambycid beetles

Tree-boring cerambycid beetles are important pests of fruit trees in most parts of the world, several species of which are pests of fig trees. Among these species are: the fig-tree borers *Batocera rufomaculata* (Fig. 15.1A), *Trichoferus griseus*, and *Phryneta spinator* (Fig. 15.1B), the mango borer *Batocera rubus*, and the tree borers *Olenecamptus bilobus* (Khan and Maiti, 1982) and *Apriona rugicollis*. The adults are strong, robust, large beetles (body length 3–6 cm) with exceedingly long antennae that resemble long horns (Figs. 1A, B). They are dark colored with distinct colored spots. They have prominent spines on both sides of the thorax. The males are smaller than the females. The adults live for 2–4 months; they chew on green growing tips and on the bark of twigs, and may feed on the skin of unripe figs. Females are active during the summer and lay their eggs near the base of branches. After the larvae hatch, they feed on the bark and then tunnel into the core of the tree trunk or branches. The larvae are flat, cream-colored, and legless with a small dark head bearing powerful mandibles. The large larvae (maximum length 4–10 cm;

* Email: benyak@volcani.agri.gov.il

DOI: 10.1079/9781789242492.0015

Table 15.1. List of insect and mite pests infesting fig trees.

Order	Family	Scientific name	Common name	Damaged parts	Continents
Coleoptera	Cerambycidae	*Batocera rufomaculata*	Fig borer	Trunk, branches	Asia, Africa, America
		Batocera rubus	Mango borer	Trunk, branches	Asia
		Olenecamptus bilobus	–	Trunk, branches	Asia, Australia
		Trichoferus griseus	Fig-tree borer	Trunk, branches	Europe, Asia, Africa
		Apriona rugicollis	–	Trunk, branches	Asia
		Eutrichillus comus	–	Trunk, branches	America
		Neoptychodes trilineatus	–	Trunk, branches	America
		Phryneta spinator	Fig borer	Trunk, branches	Africa
	Curculionidae	*Aclees* spp.	–	Trunk, branches	Europe (Asia)
		Hypoborus ficus	Fig bark beetle	Trunk, branches	Europe, Asia, Africa
		Cryphalus fici	–	Trunk, branches	Solomon Islands
		Cryphalus dilutus	–	Trunk, branches	Asia, Africa, Europe
		Cryphalus discretus	–	Trunk, branches	Asia
		Hypothenemus leprieuri	–	Trunk, branches	Europe, Asia
	Scarabaeidae	*Holotrichea consanguinea*	Sugarcane beetle	Leaves	Asia
		Adoretus spp.	Cockchafer beetle	Leaves	Asia
		Cotinis texana	Fig beetle	Fruits	North America
		Cotinis mutabilis	Figeater beetle	Fruits	North America
		Cotinis nitida	June beetle	Fruits	North America
	Chrysomelidae	*Epitrix* spp.	Flea beetle	Leaves, shoots	South America
		Colaspis spp.	Leaf beetle	Leaves, shoots	America
		Poneridia spp.	Fig leaf beetle	Leaves, shoots	Australia and Pacific Islands
	Tenebrionidae	*Blapstinus fuliginosus*	Darkling ground beetle	Fruits	America
	Nitidulidae	*Carpophilus hemipterus*	Dry fruit beetle	Dry fruits	Cosmopolitan
Lepidoptera	Crambidae	*Azochis gripusalis*	Fig shoot borer	Shoots, branches	America
		Glyphodes cosmarcha	–	Leaves	Asia, Australia
		Cirrhochrista brizoalis	–	Fruits	Asia, Australia
	Cossidae	*Prionoxystus robiniae*	Carpenter worm	Trunk, branches	America
		Zeuzera coffeae	Coffee carpenter	Trunk, branches	Asia
		Zeuzera pyrina	Wood leopard moth	Trunk, branches	Asia, Africa, Europe, America
		Paropta paradoxa	Vine carpenter moth	Trunk, branches	Asia, Africa, Europe
	Choreutidae	*Choreutis nemorana*	Fig leaf roller	Leaves, shoots	Asia, Africa, America, Europe
	Brachodidae	*Phycodes minor*	–	Leaves	Asia
		Phycodes radiata	–	Leaves	Asia
	Erebidae	*Euproctis lunata*	Castor hairy caterpillar	Leaves	Asia
		Spilarctia obliqua	Jute hairy caterpillar	Leaves	Asia

		Perina nuda	Banyan tussock moth	Leaves	Asia
	Hypsidae	*Hypsa ficus*	–	Leaves	Asia
	Pyralidae	*Ephestia cautella*	Almond moth	Ripe and dry fruits	Cosmopolitan
		Ectomyelois ceratoniae	Carob moth	Dry fruits	Asia, Africa, Europe, Australia, America
		Amyelois transitella	Navel-orange worm	Fruits	America
	Noctuidae	*Chrysodeixis chalcites*	Golden twin-spot moth	Leaves, shoots, fruits	Asia, Africa, Europe, America
	Nymphalidae	*Marpesia petreus*	Ruddy daggerwing	Leaves	America
Hemiptera	Miridae	*Zanchius breviceps*	–	Leaves	Asia, Africa, Europe
	Cercopidae	*Cosmoscarta niteara*	Spittlebug	Leaves	Asia
	Cicadellidae	*Velu caricae*	Fig jassid	Leaves	Asia
		Empoasca alsiosa	Green leafhopper	Leaves	Europe
		Ficocyba ficaria	–	Leaves	Asia, Europe
		Tartessus ferrugineus	–	Leaves	Asia
	Issidae	*Agalmatium flavescens*	Issid planthopper	Leaves	Europe, Asia, Africa
		Agalmatium grylloides	–	Leaves	Europe, Asia
	Triozidae	*Pauropsylla buxtoni*	Fig psylla	Leaves (galls)	Middle East
		Pauropsylla depressa	Fig psylla	Leaves (galls)	Asia
	Homotomidae	*Homotoma ficus*	Fig psylla	Leaves	Asia, Europe, America
	Aphididae	*Aphis gossypii*	Cotton aphid	Leaves	Cosmopolitan
		Greenidea ficicola	Weeping fig aphid	Leaves	Cosmopolitan
		Toxoptera aurantii	Black citrus aphid	Leaves	Cosmopolitan
	Monophlebidae	*Drosicha stebbingi*	Giant mealybug	Leaves, twigs, fruits	Asia
		Icerya purchasi	Cottony cushion scale	Leaves, twigs, fruits	Cosmopolitan
	Diaspididae	*Lepidosaphes conchiformis*	Mediterranean fig scale	Trunk, branches, leaves, fruits	Cosmopolitan
		Aspidiotus destructor	Coconut scale	Trunk, branches, leaves, fruits	Cosmopolitan
		Chrysomphalus aonidum	Circular scale	Trunk, branches, leaves, fruits	Cosmopolitan
		Parlatoria oleae	Olive scale	Trunk, branches, leaves, fruits	Asia, Africa, Europe
	Coccidae	*Ceroplastes rusci*	Fig wax scale	Leaves, twigs, fruits	Cosmopolitan
		Ceroplastes rubens	Pink wax scale	Leaves, twigs, fruits	Asia, America, Australia
		Ceroplastes sinensis	Chinese wax scale	Leaves, twigs, fruits	Asia, Europe
		Coccus hesperidum	Brown soft scale	Leaves, twigs, fruits	Cosmopolitan
		Saissetia oleae	Olive black scale	Leaves, twigs, fruits	Cosmopolitan

Continued

Table 15.1. Continued.

Order	Family	Scientific name	Common name	Damaged parts	Continents
	Kerriidae	*Kerria fici*	Fig lac scale	Leaves, twigs, fruits	Asia
	Asterolecaniidae	*Russellaspis pustulans*	Oleander pit scale	Trunk, branches, leaves	Asia, Africa, America
	Pseudococcidae	*Planococcus ficus*	Vine mealybug	Leaves, twigs, fruits	Cosmopolitan
		Planococcus lilacinus	Coffee mealybug	Leaves, twigs, fruits	Asia, Africa, America
		Drosicha mangiferae	Mango mealybug	Leaves, twigs, fruits	Asia
		Planococcus citri	Citrus mealybug	Roots, leaves	Cosmopolitan
		Pseudococcus longispinus	Long-tailed mealybug	Leaves	Cosmopolitan
Diptera	Tephritidae	*Ceratitis capitata*	Mediterranean fruit fly	Fruits (syconia)	Cosmopolitan
		Bactrocera dorsalis	Oriental fruit fly	Fruits (syconia)	Cosmopolitan
		Bactrocera zonata	Peach fruit fly	Fruits (syconia)	Cosmopolitan
	Drosophilidae	*Zaprionus tuberculatus*	Fruit fly	Fruits	Asia, Africa, Europe
		Zaprionus indianus	African fig fly	Fruits	Cosmopolitan
		Drosophila suzukii	Spotted wing drosophila	Fruits	Cosmopolitan
	Lonchaeidae	*Silba adipata*	Black fig fly	Fruits	Asia, Africa, Europe
	Cecidomyiidae	*Anjeerodiplosis peshawarensis*	Fig midge	Fruits	Asia
		Udumbaria nainiensis	Fig midge	Fruits	Asia
Thysanoptera	Phlaeothripidae	*Gigantothrips elegans*	Giant fig thrips	Leaves	Asia
	Thripidae	*Thrips tabaci*	Onion thrips	Fruits	Cosmopolitan
		Frankliniella occidentalis	Western flower thrips		
Trombidiformes	Tetranychidae	*Tetranychus urticae*	Two-spotted spider mite	Leaves	Cosmopolitan
		Tetranychus neocaledonicus	Red cucurbit mite	Leaves	Africa, Asia, Europe, South America, Australia
		Eutetranychus orientalis	Citrus brown mite	Leaves	Africa, Asia, Europe, Australia
		Eutetranychus hirsti		Leaves	Asia
		Panonychus ulmi	European red mite	Leaves	Cosmopolitan
	Eriophyidae	*Aceria fici*	Fig mite	Leaves, fruits	Asia, Africa, Europe, America, Australia
		Rhyncaphytoptus ficifoliae	Fig leaf mite	Leaves	Africa, Asia, Australia
		Anothopoda fici	–	Leaves	Asia, Europe
		Eriophyes ficivorus	Fig gall mite	Leaves	Asia
	Diptilomiopidae	*Asetadiptacus emiliae*	–	Leaves	Europe, Central America
Hymenoptera	Agaonidae	*Blastophaga psenes*	Fig wasp	Pollinator (beneficial)	Cosmopolitan

Fig. 15.1C1) tunnel in the trunk and branches (Fig. 15.1C2), causing the main damage. Tunneling weakens the tree, causes breakage of branches, and may kill the tree after a heavy or prolonged infestation. Larval development takes about 10 months. Pupation occurs in the tunnels, just beneath the bark, in a self-constructed cell. Tunneling activities are often detected by the sawdust and frass that come out of holes in the tree. When entry holes are found in the trunk or branches, iron wire can be inserted into the tunnel to clean it or to kill the larvae. If the larvae are not killed mechanically, the holes can be plugged with cotton wool or rubber septum soaked with volatile toxic chemicals. Injecting toxic vegetal extracts such as *Antholyza ringens* into the tunnels significantly suppresses the larvae. Trunks and branches may be covered with oil-coated aluminum foil or with sleeves of metal net to interfere with adult emergence and egg laying. Chemical insecticides can be applied to the trees against adults and young larvae. Insecticides can be injected into the tunnels or into the trunk as a systemic treatment. Fig-tree borers have only a few biocontrol agents such as entomopathogenic fungi and several species of parasitic wasps of the larvae. At high infestation rates, affected branches or whole trees should be cut off and removed from the orchard.

In a survey of beetles whose larvae bore in fig trees (xylophagous) conducted in Turkey, it was found that many species of the families Bostrichidae, Buprestidae, Cerambycidae, Curculionidae, and Scolytidae occupied the same area (Akşit *et al.*, 2005). Similar results were obtained in a survey conducted in a fig orchard in Mexico, where the cerambycids *Eutrichillus comus* Bates and *Neoptychodes trilineatus* (L.) (Fig. 15.1D) were the dominant species (López-Martínez *et al.*, 2015).

15.2.2 Weevils of the genus *Aclees*

Weevils of the genus *Aclees* also cause damage to the trunk and branches of fig trees (EPPO, 2009; Benelli *et al.*, 2014) (Fig. 15.2A). Females lay eggs at the base of the trunk and the larvae feed at the base of the tree, mostly under the soil surface. Larval feeding causes yellowing, drying of branches, and eventually death of the branch. *Aclees* sp. cf. *foveatus* is an important pest of fig in France and Italy and can cause tree death. Adults feed on the fig inflorescence and on young twigs. During heavy infestation, the adults can also feed on leaves. This pest has caused damage mainly in fig nurseries, private gardens, and abandoned orchards.

15.2.3 Fig bark weevils

Fig bark weevils usually attack weak trees and hasten their death (www.barkbeetles.info/index.php;

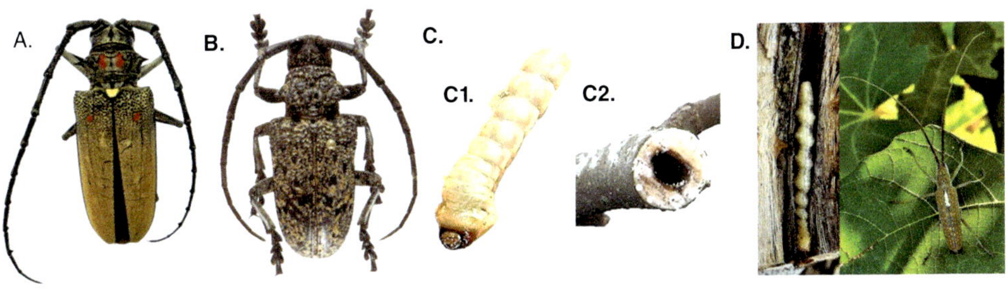

Fig. 15.1. Fig-tree-boring cerambycid beetles: A. *Batocera rufomaculata*. B. *Phryneta spinator*. C. Larva of a fig-tree-boring cerambycid beetle (1) and a tunnel caused by the larvae (2). D. *Neoptychodes trilineatus*, larvae and adult. A. Photo by David Ben-Yakir. B. From: www.coleoptera-atlas.com/IMG/jpg/phryneta-spinator zambia28m_sm01.jpg (accessed December 9, 2021). C(1). Modified from: https://candidegardening.com/GB/insects/eaabf98455d9bdfef9c454fdb2e74581 (accessed December 9, 2021). C(2). Modified from: www.youtube.com/watch?v=6c6OUmBARos (accessed December 9, 2021) by Gregorynts. D. Photos by Juan Gutierrez.

accessed January 10, 2022). These beetles are small (1.0–1.5 mm long) and dark brown. The females drill holes in the bark to create galleries where they lay their eggs. The developing larvae create more galleries as they feed on the bark. Boring dust (frass) in the bark crevices and at the tree base is one of the first signs of bark beetle attack. Often, numerous small holes with a diameter of about 1 mm appear on the trunk or branches of infested trees. Adults are active throughout the year but spend the winter in the galleries. These pests can have three or four generations per year. Removal of injured branches from the tree and of weak trees from the orchard may reduce these pests. Biological enemies of the larvae include parasitic wasps of the family Proctotrupidae and *Pyemotes* mites.

The common bark beetles of figs in the Mediterranean are *Hypoborus ficus* (Erichson) (Fig. 15.2B), *Hypothenemus leprieurii* (Perris), and *Cryphalus (Hypocryphalus) dilutus* Eichhoff (Mifsud *et al.*, 2012; Johnson *et al.*, 2017; Yang, 2020). *Cryphalus (Hypocryphalus) fici* Browne has also been listed in some publications as a pest of fig. The bark beetle *Cryphalus discretus* (syn. *Hypocryphalus scabricollis*) (Eichhoff) was first recorded in Malta in 1991 where it caused severe damage to fig trees in the Maltese Islands (Fig. 15.2C; Mifsud *et al.*, 2012). Results of taxonomic and genetic analyses of fig pests of the genus *Hypocryphalus* were published recently (Johnson *et al.*, 2017). A list of species in the genus *Cryphalus* has been published by Yang (2020). Web-based automatic traps for early detection of bark beetles have been developed (Rassati *et al.*, 2016).

15.2.4 Wood-boring moths

The larvae of wood-boring moths bore into fig branches and may kill the tree when infestation levels are high. Active galleries are recognized by copious amounts of sap and sawdust from the larval burrows. *Bacillus thuringiensis*, *Beauveria bassiana* and *Steinernema feltiae* can be used as biocontrol agents against these larvae.

The larvae of the fig shoot borer moth *Azochis gripusalis* (Fig. 15.3A) tunnel in new shoots and may bore down through the trees to the roots. Carpenterworm moths such as *Prionoxystus robiniae*, *Paropta paradoxa* (Mesbah *et al.*, 1993), and *Zeuzera coffeae* and the wood leopard moth, *Zeuzera pyrina*, occasionally attack fig trees.

15.3 Leaf and Shoot Feeders

15.3.1 Chrysomelid beetles

Several small chrysomelid beetles feed and develop on fig leaves and shoots. Among them are flea beetles of the genus *Epitrix* and leaf beetles of the genera *Colaspis* and *Poneridia* (Fig. 15.3B). These beetles usually do not cause economic damage except at very high infestation levels.

15.3.2 Leaf-eating moths

Larvae of several moths feed and develop on the leaves of figs and are collectively called fig-tree skeletonizers or fig leaf rollers. The moth *Choreutis nemorana* is reddish-brown with whitish

Fig. 15.2. A. Fig-tree-boring weevil of the genus *Aclees*. B. The fig bark beetle *Hypoborus ficus*: (1) larva, (2) pupa, (3) adult. C. Damage caused by the fig bark beetle *Cryphalus discretus* (syn. *Hypocryphalus scabricollis*). A. From Benelli *et al.* (2014), courtesy of Springer Publishing. B. Photos by Alain Costa. C. From: https://agrikoltura.gov.mt/en/phd/Documents/guidelines/figTreeBookletOnlineEN.pdf (accessed January 10, 2022).

Fig. 15.3. A. Fig shoot borer: (1) larva, (2) adult. B. Fig leaf beetles: (1) larvae, (2) adults. A1. Jonas Janner Hamann, Universidade Federal de Santa Maria (UFSM), Bugwood.org image no. 5553608. A2. From: www.agrolink.com.br/problemas/broca-da-figueira_382.html (accessed December 9, 2021). B1. From: http://lepidoptera.butterflyhouse.com.au/none/fig.html (accessed December 9, 2021). Photograph courtesy of Chris Quinn, Frenchs Forest, New South Wales. B2. From: www.pestnet.org/fact_sheets/ficus_sandpaper_cabbage_beetle_238.htm (accessed December 9, 2021).

markings and its body length is about 13 mm (Fig. 15.4A1). It develops two generations per year, in the spring and at the end of summer. Larvae of the second generation enter diapause in the autumn. Eggs are laid on the leaves and the emerging larvae feed on the leaves (Fig. 15.4A2). The larvae roll the leaves and use silk to create a shelter during their feeding and pupation (Fig. 15.4A3). Usually this pest does not cause economic damage except at high infestation levels. If management is required, insecticides based on *B. thuringiensis* can provide good control.

Larvae of the moths *Phycodes minor* and *Phycodes radiata* also develop on leaves of fig trees (Fig. 15.4B). These larvae also roll the leaves using silk. Their peak activity is in the late summer. Up to 80% defoliation can occur during high infestation (Kumar and Ramamurthy, 2010).

Moth larvae named castor hairy caterpillar and jute hairy caterpillar (*Euproctis lunata* and *Spilarctia obliqua*, respectively) feed on many plants including fig (Fig. 15.4C; Islam *et al.*, 1988). In India, these larvae feed on leaves, tender shoots, and young buds of fig and can cause great damage, especially to young plants. These pests are active from July to September. Up to 15% defoliation can occur during high infestation.

The ruddy daggerwing, *Marpesia petreus* (Cramer), is a butterfly (Nymphalidae) whose larvae feed on the leaves of fig (Fig. 15.4D).

Other moth larvae that feed on the leaves of various *Ficus* species, and also can damage fig trees, are *Glyphodes cosmarcha* Meyrick, the banyan tussock moth *Perina nuda* (F.), and the golden twin-spot moth, *Chrysodeixis chalcites* (Esper).

15.3.3 Plant-sucking hemipterans

Plant-sucking hemipterans are small insects that feed on plant sap and can develop on leaves and branches of plants. Most of the hemipteran pests feed on many host plants and only a few are specific to fig. The feeding of these pests usually causes stippling of leaves, loss of green coloration (chlorosis), dropping of leaves, and stunting of shoots. A few of these small pests enter the fruits through the ostiole.

Among these pests are the mirid bug *Zanchius breviceps* (Fig. 15.5A), the spittlebug *Cosmoscarta niteara*, and the leafhoppers *Velu caricae* (fig jassid), *Empoasca alsiosa*, *Ficocyba ficaria* and *Tartessus ferrugineus* (Walker).

The issid planthopper *Agalmatium flavescens* (Olivier) and *Agalmatium grylloides* (F.) are about 6 mm long with body and wings colored light brown (Fig. 15.5B). These insects overwinter as eggs, which require high moisture for hatching in the spring.

The fig psylla *Pauropsylla buxtoni* infests leaves of fig and induces conspicuous galls (Fig. 15.5C1). These galls are found on the upper part of the leaf surface and usually are clustered. Each gall contains one developing nymph and the gall size increases as the nymph grows inside. High gall density can reduce the yield of cultivated figs. The fig psylla produces one generation per year. Leaves bearing galls remain attached to the tree in the winter until the adults emerge from them in the spring (Fig. 15.5C2). Thus, removal and burning of the overwintering galls in the autumn can reduce the psyllid

Fig. 15.4. A. The fig leaf roller *Choreutis nemorana*: (1) adult, (2) larva, (3) leaf rolling and feeding damage. B. Fig leaf rollers of the genus *Phycodes*: (1) *P. minor* adult, (2) *P. radiata* adult, (3) *P. radiata* larva. C. Hairy caterpillars: (1) adult of the castor hairy caterpillar, *Euproctis lunata*, (2) adult of the jute hairy caterpillar, *Spilarctia obliqua*, (3) the jute hairy caterpillar larva. D. *Marpesia petreus*: (1) larva, (2) adult. A1. From: https://upload.wikimedia.org/wikipedia/commons/4/41/Choreutidae_-_Choreutis_nemorana.JPG (accessed December 9, 2021). A2 and A3. Photographs by A. Georis in De Prins *et al.* (2014) *Choreutis nemorana* (Lepidoptera: Choreutidae) well established in Belgium. *Phegea* 42: 29. B1. Photograph by Native HK, via iNaturalist, from: https://api.gbif.org/v1/image/unsafe/https%3A%2F%2Fstatic.inaturalist.org%2Fphotos%2F5134776%2Foriginal.jpeg%3F1475766940 (accessed December 9, 2021). B2 and B3. From: Kumar and Ramamurthy (2010) with permission. C1. From: www.wikidata.org/wiki/Q13476738#/media/File:Euproctis_lunata_(5046736742).jpg (accessed December 9, 2021). CC BY-SA 2.0. Photo by Dinesh Valke from Thane, India. C2. From: https://en.wikipedia.org/wiki/Spilosoma_obliqua#/media/File:Jute_hairy_caterpillar_moth_@_Kanjirappally.jpg (accessed December 9, 2021). CC BY-SA 4.0. C3. From: https://en.wikipedia.org/wiki/Spilosoma_obliqua#/media/File:Spilosoma_obliqua_at_Kadavoor.jpg (accessed December 9, 2021). Photogrpah by Jeevan Jose. D1. Photograph by Juan Gutierrez. CC BY-SA 4.0. D2. From: https://upload.wikimedia.org/wikipedia/commons/b/bc/DSCN7048W.JPG (accessed December 9, 2021).

population. Susceptibility of fig trees to this pest is highly dependent on variety (Burckhardt and Batta, 2018).

The psylla *Pauropsylla depressa* Crawford infests several plants of the genus *Ficus*. This pest also induces galls on the dorsal surface of the leaf. Maximum infestation occurs from August to October. A newly formed gall is green, while mature gall is pink. Economic damage occurs only at high infestation levels.

Another fig psylla, *Homotoma ficus*, develops on fig leaves. The nymphs usually feed along the major veins on the underside of the leaves (Fig. 15.6A) (Jerinić-Prodanović, 2011). This psylla produces one generation per year and overwinters as an egg. Nymphs develop in the spring and adults are active during the summer. Feeding by this pest can distort the leaf tissue, but economic damage is rare.

Several species of aphids can develop on fig trees, mainly during the spring and autumn. Nymphs and adults suck sap from leaves, young twigs, and, in severe cases, young fruits, but usually cause little economic damage. Aphids excrete honeydew that attracts ants and serves as substrate for sooty mold. *Greenidea ficicola* usually develops on *Ficus benjamina* but also feeds on fig trees (Fig. 15.6B). The cotton aphid, *Aphis gossypii* Glover, and the black citrus aphid, *Toxoptera aurantii*, feed on many hosts, including fig trees.

15.3.4 Scale insects

Several species of scale insects infest fig trees. Most of these scales are not specific to fig and

Fig. 15.5. A. The mirid bug *Zanchius breviceps*. B. The issid planthopper *Agalmatium* species. C. The fig psylla *Pauropsylla buxtoni*: (1) nymph-induced galls; (2) overwintering galls containing adults. A. Photograph by Barış Çerçi, from: http://ukrbin.com/show_image.php?imageid=86866 (accessed December 9, 2021). B. Photograph by Hectonichus, from: https://commons.wikimedia.org/w/index.php?curid=73223915 (accessed December 9, 2021). CC BY-SA 4.0. C. Photographs by David Ben-Yakir.

also develop on other hosts. They feed on plant sap and develop on leaves, branches, and the trunk of fig trees. Most scales inject toxic saliva during feeding, which results in the appearance of colored spots on the infested plant tissues. Eggs are usually protected under the shell of females. Heavy infestations cause yellowing of the leaves, followed by partial or complete defoliation. Economic damage usually occurs when scales settle on the fig fruits.

The cottony cushion scale, *Icerya purchasi*, is a minor pest of fig trees. This pest mainly feeds on the underside of leaves, young shoots, and near the base of the fruit (Fig. 15.6C). The peak activity of this pest is in the summer.

The females of the Mediterranean fig scale, *Lepidosaphes conchiformis*, are elongate (about 2.4 mm long) and shaped like oyster shells. This pest has two to four generations per year. Overwintering females are dark brown with a greasy-looking wax coat. Summer broods and

Fig. 15.6. A. The fig psylla *Homotoma ficus*: (1) female, (2) nymphs feeding along the major veins. B. An adult *Greenidea ficicola* aphid. C. A colony of the cottony cushion scale, *Icerya purchasi*. A1. Photograph courtesy of Joyce Gross, from: https://bugguide.net/node/view/785381 (accessed December 9, 2021). A2. Photograph modified from Figure 7 in Dušanka, J.-P. (2011) The first finding of the fig psylla *Homotoma ficus* L. (Hemiptera, Psylloidea, Homotomidae) in Serbia. *Pesticidi i Fitomedicina* 26, 205–212. B. From: www.inaturalist.org/places/dove-myer-robinson-park-parnell-auckland-nz (accessed December 9, 2021). CC0 1.0. C. Photograph by Clemson University, USDA Cooperative Extension Slide Series, Bugwood.org, image No. 1435060. From: https://commons.wikimedia.org/w/index.php?curid=3173765 (accessed December 9, 2021). CC BY 3.0.

younger scales are light brown. First-generation scales settle on leaves, but later generations may settle on twigs and fruits. Their feeding on the fruit causes dark green spots and gives the fruit a rough appearance. A parasitic wasp (*Aphytis* sp.) generally provides good control of this scale.

Similar damage may be caused to fig by the coconut scale, *Aspidiotus destructor* (Sign.) (Colvée), the circular scale, *Chrysomphalus aonidum* (L.), and the olive scale, *Parlatoria oleae* (Colvée).

The fig wax scale, *Ceroplastes rusci* (L.) infests many hosts and is usually a minor pest of fig. However, at high infestation levels, especially on the fruits, damage can be serious. The body of the female is dark red, 3–6 mm long, and covered by a pinkish wax (Fig. 15.7A). The wax consists of a large median block and eight surrounding small blocks that are clearly separated in young females. These scales have one to two generations per year. Adult females of the autumn generation are overwintering. Damage to the figs is due to colored spots on the fruits and contamination of the tree by honeydew. Parasitic wasps such as *Aprostocetus ceroplastae* and *Scutellista caerulea* infest this scale. Treatment of the first generation with paraffin oil was found to give good control (Morsi and Mousa, 2004).

Similar damage may be caused to figs by the pink wax scale, *Ceroplastes rubens* Mask (Williams and Watson, 1990), the Chinese wax scale, *Ceroplastes sinensis*, the brown soft scale, *Coccus hesperidum* L., the olive black scale, *Saissetia oleae*, and the fig lac scale, *Kerria fici* (Green).

The oleander pit scale, *Russellaspis pustulans* (Cockerell) (formerly known as *Asterolecanium pustulans*), infests many hosts including fig. The body of the female is oval, about 1 mm long, greenish, and its margins bear whitish filaments (Fig. 15.7B). This scale lies in a pit surrounded by plant tissue that has been induced to grow around it. This scale has two to three generations per year.

Mealybug females are 2–4 mm long and their bodies are covered with meal-like wax showing clear segmental lines. Immatures and females are mobile using their legs. Males are dark colored, about 1.5 mm long, and have one pair of wings. Males locate their mates during flight by using the pheromone that the females emit (Zada *et al.*, 2003). Traps baited with the species-specific pheromone are used for monitoring the males. Damage is mainly due to contamination of the leaves and fruits with honeydew, which leads to the development of sooty mold. Mealybugs may enter the fig fruit through the ostiole. These pests are usually controlled by natural enemies.

Several species of mealybugs infest fig trees: the vine mealybug, *Planococcus ficus* (Signoret), the coffee mealybug, *Planococcus lilacinus* (Cockerell), the citrus mealybug, *Planococcus citri*, and the long-tailed mealybug, *Pseudococcus longispinus* (Targioni Tozzetti).

The giant mealybug, *Drosicha stebbingi* (Green), is a pest of mango in Asia that occasionally infests fig trees. Eggs are laid in the soil

Fig. 15.7. A. The female fig wax scale, *Ceroplastes rusci* (L.): (1) on fig twigs, (2) on a fruit. B. Oleander pit scale, adult. A1. Photograph by Doug Caldwell, University of Florida. A2. Slide no. 68 in www.slideshare.net/NicolasValanides1/pests-and-diseases-71251140 (accessed December 9, 2021). B. Photograph from the US National Collection of Scale Insects, USDA Agricultural Research Service, Bugwood.org.

around infected trees during the late spring. Nymphs emerge when the temperature rises during January and climb up the trees via the stem to feed on shoots and the base of fruits. The lifespan for males and females are 7 and 15–35 days, respectively. Damage is caused by the nymphs and wingless females. At high infestation levels, they can cause leaf drop, twig dieback, and fruit drop. They secrete honeydew, which serves as substrate for sooty mold. Damage occurs primarily in older, well-shaded groves planted on heavy soils. Physical methods to reduce this pest include plowing to expose the eggs to natural enemies, heat, and creating a barrier for nymphs climbing up the trunk. Lacewings and predatory flies are natural enemies of this mealybug.

15.3.5 Thrips

Thrips are minute (about 1 mm long), slender insects with fringed wings. They feed on plants by puncturing and sucking up the contents of plants' epidermal and mesophyll cells. Feeding by thrips results in discoloration, deformities, and reduced marketability of the crop. Some thrips enter fig fruits, causing feeding scars that hinder the development of those fruits (see later).

In India, the giant fig thrips, *Gigantothrips elegans* Zimm. (Fig. 15.8A), feed on young leaves, tender twigs, and fruits of fig. During severe infestation, the leaves dry out and curl up.

15.3.6 Mites

Mites are small arthropods belonging to the class Arachnida. Their body has two major sections and adults have four pairs of legs. They feed by puncturing plant cells and sucking the cell sap. Spider mites (Tetranychidae) usually live on the undersides of leaves of plants and often spin protective silk webs. The generation time is 2–3 weeks at optimal conditions of 25°C and 40% relative humidity. They usually spend the winter in diapause. Females disperse between plants and sites by ballooning on winds. Infested leaves have blotches and may dry up and fall off as a result of a severe infestation. Predatory phytoseiid mites often control them.

Fig. 15.8. A. The adult and nymphs of the giant fig thrip, *Gigantothrips elegans*. B. The spider mite, *Tetranychus urticae*. C. The fig blister mite, *Aceria fici*: (1) a mite, (2) fruit russet, (3) a mite (arrow) inside a fruit. A. From: www.gbif.org/occurrence/2573812793 (accessed December 9, 2021). B. From: https://biologicalservices.com.au/pests/two-spotted-mite-90.html (accessed December 13, 2021). C1. Photo by Eric Erbe; digital colorization by Chris Pooley (USDA, ARS, EMU). From: https://commons.wikimedia.org/w/index.php?curid=515635 (accessed December 13, 2021). C2 and C3. Photographs by Alain Costa.

Typically, spider mites are not host specific. Several species of spider mites are found on fig trees. The two-spotted spider mite, *Tetranychus urticae* Koch, is small (up to 0.5 mm long), green–yellow in color, with two darker internal spots (Fig. 15.8B). Adults of the European red mite, *Panonychus ulmi* (Koch), are red and bear prominent dorsal setae rising from whitish circles. The citrus brown mite, *Eutetranychus orientalis* (Klein), *Eutetranychus hirsti* Pritchard & Baker, and the red cucurbit mite, *Tetranychus neocaledonicus* Andre, also affect figs.

Gall mites (Eriophyidae) are small (about 0.2 mm long) worm-like, have two pairs of legs and are colored yellow to pink. They are usually host specific and often confined to one organ of their host plant. Their generation time is about 14 days. Their primary mode of spread is by wind. They usually infest buds and young leaves, often inducing the formation of galls. Some serve as vectors of viral diseases. Predatory phytoseiid mites often feed on them.

The fig blister mite, *Aceria fici* (Ewin) (Fig. 15.8C1), cause blisters on the surface of leaves and fruits until they become russeted (Fig. 15.8C2). This mite can enter developing fruits and cause some damage inside (Fig. 15.8C3). The mites feed mainly in the interior portion of the canopy and may cause leaf drop and stunting of twigs. This mite transmits the fig mosaic virus. Similar damage is caused by the fig leaf mite *Rhyncaphytoptus ficifoliae* Keifer, the fig gall mite, *Eriophyes ficivorus* Channabasavanna, and *Anothopoda fici*. *Asetadiptacus emiliae* Carmona is another similar mite found on fig trees.

15.4 Developing Fruits

15.4.1 Fruit flies

Fruit flies are attracted to ripe fruits and vegetables and lay their eggs in them. Their larvae (also known as maggots) feed on and develop in fruits and vegetables, causing great economic damage. Fruit fly larvae are whitish, legless, and worm-like with a maximum body length of

7–9 mm. When the larvae have completed their development, they leave the fruit and pupate in the soil. There are two main families of economically important fruit flies, the Tephritidae (2.5–10 mm, colorful, with pictured wings) and the Drosophilidae (2–4 mm, pale yellow to reddish-brown to black). Fruit flies are monitored using yellow sticky traps and attractive baits (e.g. ammonia). Removal of ripe and fallen fruits can reduce the populations of these flies. Control methods include mass trapping, establishing natural enemies (mainly hymenopterous parasitoids), and releasing sterile males. The major pest species are not host specific and often infest fig fruits. The Mediterranean fruit fly, *Ceratitis capitata* (Wiedemann), is a highly invasive pest that is widely spread (Gonçalves *et al.*, 2008). The life cycle of the Mediterranean fruit fly spans 2–5 weeks at 24°C, depending on the host. Adults can live for several months but do not disperse over long distances. In warm regions, they are active all year round and develop six to eight generations per year. Similar important pests are the oriental fruit fly, *Bactrocera dorsalis* (Hendel), and the peach fruit fly, *Bactrocera zonata* (Saunders).

Fruit flies (Drosophilidae) in the genus *Zaprionus* have white longitudinal stripes on their head and thorax. Usually, these flies are secondary pests developing in over-ripe, fallen, or rotting fruit. However, the African fig fly, *Zaprionus indianus* Gupta, has been reported as a primary pest of fig in Brazil and Jordan (Pfeiffer *et al.*, 2019). The females oviposit near the fig ostiole and the larvae enter the fruit and feed, causing the fruit to become soft and unmarketable (https://gd.eppo.int/taxon/ZAPRIN; accessed December 13, 2021). Spotted wing drosophila, *Drosophila suzukii* Matsumura, flies have red eyes and black transverse stripes on the abdomen (Fig. 15.9). Females infest thin-skinned fruits before harvest and feeding of the larvae make the fruits soft and unmarketable.

15.4.2 Other flies

The black fig fly, *Silba adipata* (McAlpine), develops only in the fruits of fig. The fly is glossy black with a bluish hue, 3.5–4.5 mm long, with large red eyes and brown legs. Females deposit the eggs under the bracts of the ostiole of unripe fruits (Fig. 15.10A1). The larvae are whitish, up to 8 mm long, feed in groups, and bore galleries in the fruit (Fig. 15.10A2). When the larvae have completed their development, they exit to pupate in the soil, leaving the fruits with noticeable exit holes (Fig. 15.10A3). These flies have three to six generations per year. Peak populations are during late spring and early summer. Their economic damage is due to premature

Fig. 15.9. *Drosophila suzukii*: male (left) and female (right). Photo by Shane F. McEvey, Australian Museum. From: https://doi.org/10.6084/m9.figshare.4644793.v1 (accessed December 9, 2021). CC BY 4.0.

fruit drop and reduced fruit quality at harvest. The flies are attracted to ammonium sulfate, protein hydrolysate, and hexanol (Katsoyannos and Guerin, 1984). They are also attracted to volatiles of fig trees, especially from unfertilized fruits (Nawade *et al.*, 2020). These volatiles can be used for monitoring and mass trapping of the flies. Infested fruits change color from green to brown. The parasitoid *Pachycrepoideus vindemmiae* Rondani (Pteromalidae) parasitize the pupae of this pest.

In India, fig fruits are infested by two species of fig midge, *Udumbaria nainiensis* Grover and *Anjeerodiplosis peshawarensis* Mani. The adults are small (2–4 mm) and are a light brown color. The larvae are creamy white, and feed and bore in the fruit. Infested fruits become hard, hollow, deformed, remain yellowish green and fail to ripen. These pests are active throughout the year except in the summer. The life cycle takes 35–50 days and they develop seven generations per year. Cultural management methods include the collection and destruction of fallen infested fruits and raking the soil around the trees to expose and kill the pupae.

15.4.3 Fruit-feeding beetles

The figeater beetle, *Cotinis mutabilis* (Gory & Percheron), the fig beetle, *Cotinis texana* Casey, and the green June beetle, *Cotinis nitida* (L.), are North American scarabs that can feed on fig fruits. These beetles are 3–4 cm long and green colored with a brownish yellow band around the edge of the wings (Fig. 15.10B). They are active during the day in the summer and feed on ripening soft-skinned fruits (www.youtube.com/watch?v=XWYd8Tgal8U; accessed December 13, 2021). They scrape the skin, drill holes in the fruits, and contaminate the fruit surface with excrement. At high densities, they can cause economic damage. Their larvae (grubs) develop in the soil, usually feeding on decomposing organic matter. The larva of the figeater beetle can crawl on its back. Larvae pupate in soil cells. These beetles usually produce one generation per year. Reducing organic matter in the soil, keeping the soil surface dry and packed, or flooding can reduce the population of these pests.

The darkling ground beetle, *Blapstinus fuliginosus* Casey, causes damage to ripe fig fruits in California (www2.ipm.ucanr.edu/agriculture/fig/Darkling-Ground-Beetle/; accessed December 13, 2021). The beetle is oval shaped, dull black in color, and about 0.6 cm long. The larvae, called false wireworms, are up to 1.2 cm long and cream to pale tan in color. The larvae feed on decaying organic matter in and on the soil.

Larvae of moths belonging to the superfamily Pyraloidea can develop in fig fruits, for example, the navel orangeworm, *Amyelois transitella* (Walker), in America and *Cirrhochrista brizoalis* (Walker) in Asia. Infested fruits have brown spots on their surface or drop from the tree before harvest time.

15.4.4 Fruit-feeding thrips

Thrips often enter and develop in fig fruits, causing damage to the flesh of the fruit. In Korea, the

A. B.

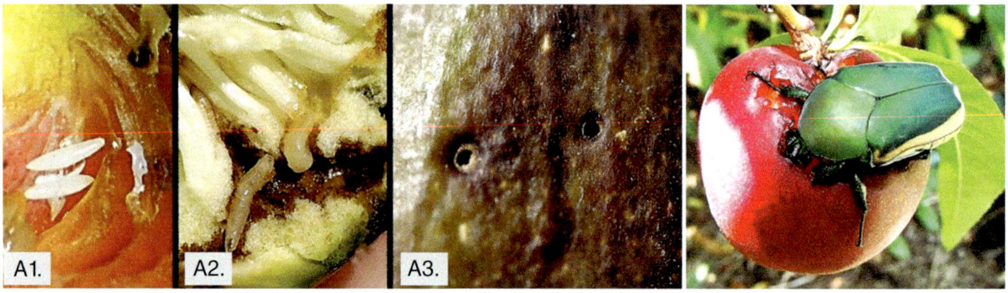

Fig. 15.10. A. The black fig fly, *Silba adipata*: (1) eggs, (2) larvae, (3) larval exit holes. B. The figeater beetle, *Cotinis mutabilis*: an adult feeding on a nectarine. A1–A3. Photos by Alain Costa. B. Modified from photograph by Davefoc, https://commons.wikimedia.org/wiki/Cotinis_mutabilis#/media/File:Cotinis Mutabilis_7864.JPG (accessed January 10, 2022). CC BY-SA 3.0.

species *Thrips tabaci* Lindeman, *Frankliniella occidentalis* Pergande, and *Frankliniella intonsa* Trybom cause internal damage to fig fruits grown in greenhouses. Thrips-damaged fruits have discoloration of the pericarp and a rough surface. The inside of the fruit contained dead bodies and exuviae of the thrips, and their feeding causes brown scars that prompt the growth of molds (Kim *et al.*, 2014). Similar fruit damage was observed in Israel caused by *T. tabaci* and *F. occidentalis* (Fig. 15.11A). Highly infested fruits were found to have about 30 thrips per fruit at all life stages. The infestation rates of fruits by thrips ranged from 5% to 50%, and one-third of the infested fruits were unmarketable (Ben-Yakir *et al.*, 2009). Fruit damage caused by thrips was also recorded in South Africa (Wohlfarter and Venter, 2012). In Japan, *T. tabaci* and *F. occidentalis* cause damage to fig fruits. Covering the soil of the orchard with reflective plastic mulch reduces thrips damage by one-third (Morishita, 2002). Insect-proof netting and diffuse-reflection mulch was found to protect figs from thrips (Shirotsuka *et al.*, 2018). Covering the fruits with non-woven surgical tape reduced the rate of fruit damage from 61% to 4% (Ichikawa and Naganawa, 2004). Surrounding the fig orchard with sorghum plants served as a thrips barrier and enhanced the activity of natural enemies (Horikawa *et al.*, 2016; Ito *et al.*, 2017).

15.4.5 Fig wasp

Another insect that may be found in fig fruits is the fig wasp, *Blastophaga psenes*, which pollinates the common fig. A fertilized female, loaded with pollen, penetrates an unripe fig via a narrow opening at the top. It oviposits in the female

A.

Fig. 15.11. A. The onion thrips, *Thrips tabaci*, inside a fig fruit: (1) Adults (yellow arrows), (2) feeding damage. B. Dried-fruit beetle of the genus *Carpophilus*: (1) larva, (2) adult. A1 and A2. Photos by David Ben-Yakir. B1. From: www.agric.wa.gov.au/citrus/dried-fruit-beetle-carpophilus-pest-stone-fruit (accessed December 13, 2021). B2. Photograph by Udo Schmidt. https://commons.wikimedia.org/wiki/File:Carpophilus_hemipterus_(Linn%C3%A9,1758)_(9720396701).png (accessed December 13, 2021). CC BY-SA 2.0.

florets within the fig, which develop into a small, globular shining gall; crawling around within the fig, she fertilizes many more florets than are galled. From the galls emerge females and apterous males, which, after having fertilized a female, remain in the fig. The young females leave the fig covered in pollen (www.youtube.com/watch?v=kctIFkVaZQs; accessed December 13, 2021).

15.5 Ripe and Dry Fruits

15.5.1 Flies

Fig fruits have soft skin and a relatively short shelf life. Ripe figs emit volatiles, fermentation by-products that are very attractive to fruit flies. The common fruit fly, *Drosophila melanogaster* (Meigen), can infest and cause decay and fermentation in ripe and dry figs. Late-ripening varieties are more susceptible to damage. In orchards, removal of decaying fruits will reduce the fly population. Traps with lures may be used to attract and kill these flies. Protecting the fruits with nets or boxes can prevent infestation.

15.5.2 Beetles

Beetles of the genus *Carpophilus*, such as *Carpophilus hemipterus* (L.), infest ripe and dry fruits. Both larvae and adults feed on fig fruits.

The larvae are white to straw colored, 2–5 mm long, and have two horn-like structures on the anal end (Fig. 15.11B1). The adults are brown, about 4 mm long, and have yellowish spots on their elytra (Fig. 15.11B2). These beetles can damage figs by reducing their quality, encouraging contamination by microorganisms, and increasing the attractiveness of the fruit to other pests such as fruit flies and the navel orangeworm, *Amyelois transitella*. Baited traps can be used to monitor the beetles before the fruits begin to ripen. Early harvest and orchard sanitation can help reduce the damages caused by these pests. Large containers with lures may be used to attract and kill the beetles. Protecting the harvested and dry fruits with nets or boxes can prevent the infestation.

15.5.3 Moths

Larvae of pyralid fruit and cereal moths can infest ripe and dry figs. Infested fruits and dry foods become contaminated with silk webs, feces, and cast skins. Their sex pheromones can be used for monitoring and mating disruption (Trematerra and Savoldelli, 2013). A modified atmosphere and fumigation are often used to control these pests in dry foods. Larvae of the almond moth, *Ephestia cautella* (Walker), feed on various foods, especially cereals. Under optimum conditions (30–32°C and 75% RH), their generation time is 4–5 weeks. The carob moth, *Ectomyelois ceratoniae* (Zeller), is dark gray, 8–10 mm long, and the larvae are pink and reach 16–18 mm in length. In the field, this pest develops mainly in *Acacia farnesiana* L., carobs, dates and figs. In storage it infests almonds and other nuts. The pest completes four to five generations per year. Acacia pods and left-over almonds serve as larval overwintering sites. Infested figs turn yellow and some drop before harvest. There are several hymenopterous parasitoids that attack this pest.

15.6 Nematodes

Plant-parasitic nematodes are pests of fig trees roots (El-Borai and Duncan, 2005; de O. Abrantes *et al.*, 2008; Wohlfarter *et al.*, 2011). The nematodes cause damage that reduces the developmental rate and the fruit yield of fig trees. Secondary damage often results from infection by other pathogenic microorganisms that exploit the root injuries caused by the nematodes (El-Borai and Duncan, 2005). Using drip irrigation in fig orchards often leads to an increases in nematode densities.

15.6.1 Root-knot nematodes

Root-knot nematodes (Heteroderidae) induce root galls and are the most damaging pests in fig cultivation. Young fig trees planted in nematode-infested soil grew poorly and some died (Wohlfarter *et al.*, 2011). The main species infesting fig roots are *Meloidogyne incognita* (Kofoid and White), *Meloidogyne javanica* (Treub), *Meloidogyne arenaria* Chitwood, *Meloidogyne acrita* Chitwood and *Meloidogyne enterolobii* Yang & Eisenback (Fig 15.12A).

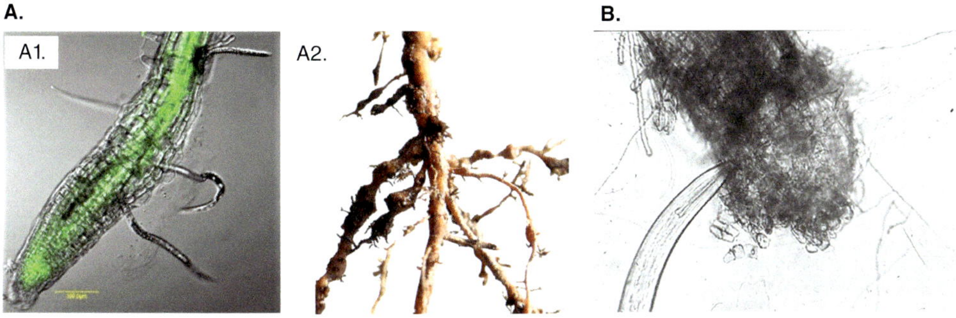

Fig. 15.12. A. Root-knot nematodes of the genus *Meloidogyne*: (1) larvae penetrating the root, (2) root damage. B. A larva of the dagger nematode, *Xiphinema index*, penetrating the root. A1. Photograph by Mishael Mor. A2. Photograph by Alain Costa. B. Photograph by Mishael Mor.

15.6.2 Cyst nematodes

The cyst nematode *Heterodera fici* Kirjanova (Heteroderidae) is a common pest of fig trees (Fanelli *et al.*, 2019). Generally, populations of *H. fici* do not reach damaging levels in mature fig orchards but can cause significant damage in fig nurseries. In addition, when fig trees were infested with both *H. fici* and *M. incognita*, there was complete crop loss (Wohlfarter *et al.*, 2011).

15.6.3 Sheath nematodes

Sheath nematodes (Tylenchulidae, Pratylenchidae) have been reported as pests of fig trees (El-Borai and Duncan, 2005). The main species of this group found in fig roots are *Paratylenchus hamatus* Thorne & Allen, *Pratylenchus vulnus* Allen & Jensen, and *Pratylenchus coffeae* Goodey.

15.6.4 Ring nematodes

The ring nematode *Criconemoides xenoplax* Raski (Criconematidae) has been found in fig roots.

15.6.5 Dagger nematodes

The dagger nematode, *Xiphinema index* Thorne & Allen (Longidoridae), attains extremely high populations near the roots of fig trees, especially in the Mediterranean region (Fig. 15.12B) (El-Borai and Duncan, 2005). Usually, these nematodes do not cause significant damage and do not transmit viruses to fig trees. However, high crop losses due to this nematode have occurred in South Africa (Wohlfarter *et al.*, 2011).

For reducing nematode damages, it is best to combine several measures (Dawabah *et al.*, 2019). Common measures used are: (i) selection of nematode-free planting plots; (ii) weed control (Lima-Medina *et al.*, 2013); (iii) biological and chemical nematicides; and (iv) development of nematode-resistant rootstocks (Hosomi *et al.*, 2002; Costa *et al.*, 2015).

15.7 Concluding Remarks

Fig trees have a good capacity for regeneration and thus can usually overcome damage caused by pests. However, a few pests pose a significant economic risk to fig cultivation. Tree-boring cerambycid beetles are probably the major pests that can kill fig trees of all ages. Nematodes can either kill or reduce the development of young fig trees. Pests that infest the fig fruits can also inflict grave economic losses.

Methods to reduce pest damage include: (i) developing rootstocks resistant to tree borers and nematodes; (ii) growing fruits with small ostioles and dense flesh to reduce pests such as thrips and vinegar flies; (iii) physical protection of trees with insect-proof screens, which can block beetle pests and fruit flies; and (iv) developing cultivation methods (irrigation, fertilization, pruning, weeding) that reduce pests and encourage the activity of natural control agents.

More studies on the biology, control and natural enemies of fig pests in the major growing regions are needed. This is especially important for exotic pests that have recently invaded fig-growing regions.

References

Akşit, T., Çakmak, İ., and Özsemerci, F. (2005) Some new xylophagous species on fig trees (*Ficus carica* cv. Calymirna L.) in Aydın, Turkey. *Turkish Journal of Zoology* 29, 211–215.

Alford, D.V. (2007) *Pests of Fruit Crops: A Colour Handbook*, 2nd edn. CRC Press, Boca Raton, Florida.

Anon. (2021) Fig pests. Available at: https://vikaspedia.in/agriculture/crop-production/integrated-pest-managment/ipm-for-fruit-crops/ipm-strategies-for-fig/fig-pests (accessed December 13, 2021).

Atwal, A.S. and Dhaliwal, G.S. (2013) *Agricultural Pests of South Asia and their Management*. Kalyani Publishers, New Delhi.

Benelli, G., Meregalli, M., and Canale, A. (2014) Field observations on the mating behavior of *Aclees* sp. cf. *foveatus* Voss (Coleoptera: Curculionidae), an exotic pest noxious to fig orchards. *Journal of Insect Behavior* 27, 419–427.

Ben-Yakir, D., Flaishman, M., Chen, M., and Yablowitz, Z. (2009) Thrips damage to fig fruit and developing means to reduce it. In: *Proceedings of the 4th International Symposium on Fig, September 29–October 3, 2009, Meknès, Morocco*. International Society for Horticultural Science, Leuven, Belgium.

Burckhardt, D. and Batta, Y. (2018) Taxonomy and biology of *Pauropsylla buxtoni* comb. *nov*. (Hemiptera: Psylloidea) on *Ficus carica* (Moraceae). *Journal of the Entomological Research Society* 20, 39–52.

Costa, M.G.S., Correia, E., Dos Reis, L.L., and Wilcken, S.R.S. (2015) Reaction of fig tree to three species of root-knot nematodes. *Revista Brasileira de Fruticultura* 37, 617–622.

Dawabah, A.A.M., Al-Yahya, F.A., and Lafi, H.A. (2019) Integrated management of plant-parasitic nematodes on guava and fig trees under tropical field conditions. *Egyptian Journal of Biological Pest Control* 29, 29–37.

de O. Abrantes, I.M., Vieira dos Santos, M.C., da Conceicao, I.L.P.M., de A. Santos, M.S.N., and Vovlas, N. (2008) Root-knot and other plant-parasitic nematodes associated with fig trees in Portugal. *Nematologia Mediterranea* 36, 131–136.

Dunoven, P. (2014) Dealing with fig diseases and pests. Available at: www.farmersweekly.co.za/farm-basics/how-to-crop/dealing-with-fig-pests-and-diseases/ (accessed December 13, 2021).

El-Borai, F.E. and Duncan, L.W. (2005) Nematode parasites of subtropical and tropical fruit tree crops. In: Luc, M., Sikora, R.A., and Bridge, J. (eds) *Plant Parasitic Nematodes in Subtropical and Tropical Agriculture*. CABI Publishing, Wallingford, UK, pp. 467–492.

Ellis, W.N. (2021) *Ficus carica*: fig. Available at: https://bladmineerders.nl/host-plants/plantae/spermatopsida/angiosperma/eudicots/superrosids/rosids/fabids/rosales/moraceae/ficus/ficus-carica/ (accessed December 13, 2021).

EPPO (2009) Insect pests recorded for the first time in Italy. EPPO Reporting Service No. 04-2008. Article No. 2009/071. Available at: https://gd.eppo.int/reporting/article-175 (accessed December 13, 2021).

Fanelli, E., Vovlas, A., Santoro, S., Troccoli, A., Lucarelli, G., *et al.* (2019) Integrative diagnosis, biological observations, and histopathology of the fig cyst nematode *Heterodera fici* Kirjanova (1954) associated with *Ficus carica* L. in southern Italy. *Zookeys* 824, 1–19.

Gonçalves, M.A., Andrade, L., Almeida, L., and Pica, M.C. (2008) Study of *Ceratitis capitata* and *Lonchaea aristella* on fig trees. *Acta Horticulturae* 798, 263–267.

Hill, D.S. (2008) Major tropical crops and their pest spectra. In: Hill, D.S. (ed.) *Pests of Crops in Warmer Climates and Their Control*. Springer, Dordrecht, The Netherlands, pp. 511–658.

Horikawa, H., Harata, D., and Kato, S. (2016) The efficacy of preventing thrips from invading a fig field and generating indigenous natural enemies through the use of the sorghum and petunia. *Annual Report of the Kansai Plant Protection Society* 58, 139–141.

Hosomi, A., Dan, M., and Kato, A. (2002) Screening of fig varieties for rootstocks resistant to soil sickness. *Journal of the Japanese Society for Horticultural Science* 71, 171–176.

Ichikawa, K. and Naganawa, M. (2004) Control of thrips by tape affixation to fig fruit apex. *Research Bulletin of the Aichi-ken Agricultural Research Center* 36, 35–39.

Islam, W., Nargis, A., and Joarder, O.I. (1988) Biology, seasonal occurrence, host range and damage potential of the castor hairy caterpillar, *Euproctis lunata* Walk. (Lymentriidae: Lepidoptera). *Crop Protection* 7, 332–335.

Ito, R., Horikawa, H., and Sakai, S. (2017) The efficacy of using sorghum in preventing injury to fig fruit caused by thrips. *Annual Report of the Kansai Plant Protection Society* 59, 81–83.

Jerinić-Prodanović, D. (2011) The first finding of the fig psylla *Homotoma ficus* L. (Hemiptera, Psylloidea, Homotomidae) in Serbia. *Pesticidi i fitomedicina* 26, 205–212.

Johnson, A.J., Knížek, M., Atkinson, T.H., Jordal, B.H., Ploetz, R.C., and Hulcr, J. (2017) Resolution of a global mango and fig pest identity crisis. *Insect Systematics and Diversity* 1: ixx010.

Katsoyannos, B.I. and Guerin, P.M. (1984) Hexanol – a potent attractant for the black fig fly, *Silba adipata*. *Entomologia Experimentalis et Applicata* 35, 71–74.

Khan, A.D. (2015) Fig insects: A Lecture To ToT trainees (FFS). Available at: www.slideshare.net/AllahDadKhan/fig-insects-a-lecture-to-tot-trainees-ffs-by-mr-allah-dad-khan-provincial-coordinator-ipm-kpk-minfal-pakistan (accessed December 9, 2021).

Khan, T.N. and Maiti, P.K. (1982) Life and fecundity tables for the longicorn beetle borer, *Olenecamptus bilobus* (Fabricius) *(Coleoptera: Cerambycidae). Proceedings: Animal Sciences* 91, 249–257.

Kim, D., Cho, M., Yang, C., Kang, T., Kim, H., and Jeon, S. (2014) Occurrence and damage by thrips on greenhouse-cultivated fig. *Korean Journal of Applied Entomology* 53, 485–490.

Kumar, R. and Ramamurthy, V.V. (2010) Morphology and bionomics of *Phycodes radiata* Ochsenheimer (Lepidoptera: Brachodidae) from New Delhi, India. In: Freitag, P.H. (ed.) *Tijdschrift voor Entomologie*. Brill, Leiden, The Netherlands, pp. 15–24.

Lakra, R.K. (2004) Insect pests of fruits of arid and semiarid regions of India. In: Verma, L.R., Verma, A.K., and Gautam, D.C. (eds) *Pest Management in Horticulture Crops Principles and Practices*. Asiatech Publishers, New Delhi, pp. 364–367.

Lima-Medina, I., Somavilla, L., Carneiro, R., and Gomes, C.B. (2013) Species of *Meloidogyne* associated with fig (*Ficus carica*) and host weeds. *Nematropica* 43, 56–62.

López-Martínez, V., Vargas, O.R., Alia-Tejacal, I., Toledo-Hernández, V.H., Corona-López, A.M., *et al.* (2015) Xylophagous beetles (Coleoptera: Buprestidae and Cerambycidae) from *Ficus carica* L. (Moraceae) in Morelos, Mexico. *Coleopterists Bulletin* 69, 780–788.

Mesbah, H.A., Tadros, A.W., and Shehata, W.A. (1993) The life cycle of the solitary carpenterworm *Parapta paradoxa* H. Schaclf (Lepidoptera: Cossidae) on fig trees. *Egyptian Journal of Agricultural Research* 71, 921–927.

Mifsud, D., Falzon, A., Malumphy, C., de Lillo, E., Vovlas, N., and Porcelli, F. (2012) On some arthropods associated with *Ficus* species (Moraceae) in the Maltese Islands. *Bulletin of the Entomological Society of Malta* 5, 5–34.

Morishita, M. (2002) Reflective plastic mulch as control measure for thrips in fig plant (*Ficus carica* L.). *Annual Report of the Kansai Plant Protection Society* 44, 21–26.

Morsi, G. and Mousa, F. (2004) Seasonal abundance of the fig wax scale insect, *Ceroplastes rusci* Linneus (Homoptera: Coccidae) and its parasitoids in Middle Egypt. *Egyptian Journal of Biological Pest Control* 14, 59–64.

Nawade, B., Shaltiel-Harpaz, L., Yahyaa, M., Bosamia, T.C., Kabaha, A., *et al.* (2020) Analysis of apocarotenoid volatiles during the development of *Ficus carica* fruits and characterization of carotenoid cleavage dioxygenase genes. *Plant Science* 290: 110292.

Pfeiffer, D.G., Shrader, M.E., Wahls, J.C.E., Willbrand, B.N., Sandum, I., *et al.* (2019) African fig fly (Diptera: Drosophilidae): biology, expansion of geographic range, and its potential status as a soft fruit pest. *Journal of Integrated Pest Management* 10: 20.

Rassati, D., Faccoli, M., Chinellato, F., Hardwick, S., Suckling, D.M., and Battisti, A. (2016) Web-based automatic traps for early detection of alien wood-boring beetles. *Entomologia Experimentalis et Applicata* 160, 91–95.

Reddy, P.P. (2010) *Insect Mite and Vertebrate Pests and Their Management in Horticultural Crops*. Scientific Publishers, Jodhpur, Rajasthan.

Shirotsuka, K., Mizuno, A., and Shibao, M. (2018) Control of thrips on fig crops using insect-proof netting and diffuse-reflection mulch sheets. *Annual Report of the Kansai Plant Protection Society* 60, 157–160.

Singh, S. and Kaur, G. (2017) Biodiversity of insect and mite pests infesting fig in the Indian Punjab. *Acta Horticulturae* 1173, 257–261.

Trematerra, P. and Savoldelli, S. (2013) The use of water traps and presence of spermatophores to evaluate mating disruption in the almond moth, *Ephestia cautella*, during exposure to synthetic sex pheromone. *Journal of Pest Science* 86, 227–233.

Williams, D.J. and Watson, G.W. (1990) *The Scale Insects of the Tropical South Pacific Region. Part 3: The Soft Scales (Coccidae) and Other Families*. CAB International, Wallingford, UK.

Wohlfarter, M. and Venter, E. (2012) Thrips spp. recorded from commercial figs, *Ficus carica* (Moraceae), in the Western Cape province of South Africa. *Acta Horticulturae* 940, 393–395.

Wohlfarter, M., Giliomee, J.H., Venter, E., and Storey, S. (2011) A survey of the arthropod pests and plant parasitic nematodes associated with commercial figs, *Ficus carica* (Moraceae), in South Africa. *African Entomology* 19, 165–172.

Yang, L. (2020) Checklist of the genus *Cryphalus* Erichson, 1836 (Coleoptera: Curculionidae: Scolytinae) with special attention to Chinese species. *Zoological Systematics* 45, 176–205.

Zada, A., Dunkelblum, E., Assael, F., Harel, M., Cojocaru, M., and Mendel, Z. (2003) Sex pheromone of the vine mealybug, *Planococcus ficus* in Israel: occurrence of a second component in a mass-reared population. *Journal of Chemical Ecology* 29, 977–988.

16 Fig Pathogens: Viruses, Viroids, and Phytoplasmas

Toufic Elbeaino*
Istituto Agronomico Mediterraneo di Bari (CIHEAM-IAMB), Valenzano, Bari, Italy

16.1 Introduction

Fig (*Ficus carica*) diseases were recorded as early as the time of Theophrastus, in the third century BCE (Blodgett and Gömeç, 1967). Before 2006, a turning point in the identification of fig viral diseases (Elbeaino *et al.*, 2006), all information on pests and pathogens affecting this crop in nature was based exclusively on bacteria, fungi, insects, and nematodes, whereas viruses, viroids, and phytoplasmas were completely out of the infectious scenario. In brief, the most common and important pathogens and pests of figs are: (i) bacterial diseases, such as bacterial canker (*Phytomonas fici*) and crown gall (*Agrobacterium tumefaciens*); (ii) fungal diseases, including leaf rust (*Physopella fici*), leaf spot (*Cylindrocladium scoparium* or *Cercospora fici*), smut (*Aspergillus niger*), blast of branch terminals (*Botrytis* species), web blight (*Thanatephorus cucumeris*/*Rhizoctonia solani*), anthracnose (*Glomerella cingulata*/*Colletotrichum gloeosporoides*), *Nectriella* stem gall (*Nectriella pironii*/*Kutilakesa pironii*), mushroom root rot (*Armillaria tabescens*), and pink limb blight (*Erythricium salmonicolor*); (iii) nematodes, such as root-knot nematodes (*Meloidogyne* species.) and lesion nematodes (*Pratylenchus musicola*); and (iv) insects, including stem-borer (*Bactrocera rufomaculata*), fig wax scale (*Ceroplastes rusci*), fig psylla (*Homotoma ficus*), fig weevil (*Otiorhynchus davricus*), and fig bark beetle (*Hypoborus ficus*). All the above-mentioned disease pathogens, bacteria, and fungi are discussed in a separate chapter (Chapter 17, this volume). Pests are discussed also elsewhere (Chapter 15, this volume). This complementary chapter will thus be dedicated to the newly studied phenomenon of viruses, viroids, and phytoplasmas.

16.2 The Old World of Fig-infecting Viruses

Until recently, all experiments and attempts to isolate viral entities from fig plants showing mosaic-like symptoms were only partially successful. There has always been difficulty in recovering viruses from diseased fig trees by the customary method of mechanical inoculation, used routinely in the isolation of known and unknown viruses. This difficulty resides, in contrast to other crops, in the components of the fig, which are rich in polyphenols, polysaccharides, resins, and tannins, which mechanically inhibit the transmission of viruses on to indicator herbaceous plants, and in the toxic nature of the milky latex, which induces burns and necrosis on the inoculated herbaceous leaves; all of these

* Email: elbeaino@iamb.it

 Advances in Fig Research and Sustainable Production (M.A. Flaishman and U. Aksoy)
DOI: 10.1079/9781789242492.0016

are factors that create confusion about what may be the symptoms of herbaceous hosts believed to be transmission of a virus and simple chemical damage due to phytotoxicity. This challenge has hampered the virological studies of this crop. The first attempt to identify the virus responsible for mosaic-affected fig trees was carried out on a strain of sowbane mosaic virus (genus *Sobemovirus*). Sowbane mosaic virus was inoculated on to herbaceous hosts, after it had been partially purified from mosaic-infected leaves. However, its characterization was not achieved and its etiology in fig mosaic disease (FMD) was not confirmed (Quacquarelli, 1971). Another attempt to characterize a putative species in the genus *Poytvirus* was executed in Herzegovina, where *Myzus persicae* has been used to mechanically transmit the virus on to herbaceous host plants, in a non-persistent manner. The researchers' attempts to isolate the virus from infected *Nicotiana megalosiphon* plants and transfer it to *N. megalosiphon* healthy plants also failed (Grbelja and Eric, 1983).

Another challenge in virus detection is their comprehensive and accurate identification, or rather the lack of it. Parenchymal cells from symptomatic fig accessions grown in the UK were thin sectioned and analyzed using electron microscopy. Virion particles lying within tubules and vacuolate-vesiculate inclusions typical of infections by members of the family *Secoviridae* were detected. This was rather a general identification, and its genome has never been sequenced (Martelli *et al.*, 1993). Numerous viral entities were reported from different origins on fig without providing concrete data on their real nature, for example, the cases of cripavirus-like viruses and potyviruses from Spain (Serrano *et al.*, 2004, 2008), potyviruses from Israel (Salomon *et al.*, 2003) and Japan (Chen *et al.*, 2001), and carlaviruses from Japan (Namba, 1983; Doi, 1989). All of these viruses were identified or described based on microscopic observations or, very rarely, after being transmitted to herbaceous hosts. None of them, even when they were mechanically transmissible, was ever characterized molecularly or serologically.

Of all the recorded diseases associated with figs, FMD is certainly the most ancient, serious, and critical pathological constraint that fig production and fig germplasm exchange are facing (Martelli *et al.*, 1993). The first known reports on FMD were made by Swingle in 1928 (cited in Alfieri, 1967), followed by Condit and Horne (1933). The etiology of this disease has always been considered a dilemma for plant pathologists, who were unable to determine its causal agent. The ultrastructural observations conducted for decades on mosaic-diseased fig trees always showed the occurrence of intracytoplasmic enveloped spherical bodies in infected fig cells, originally denoted "double-membraned bodies" (DMBs) (Bradfute *et al.*, 1970; Plavšic and Milicic, 1980; Appiano, 1982) and predicted to be the viral particles of this disease (Plavšic and Milicic, 1980; Martelli *et al.*, 1993; Appiano *et al.*, 1995; Serrano *et al.*, 2004). A critical study conducted in 2009 on mosaic-diseased plants showed that the causal agent was a virus with a multipartite genome belonging to the family *Bunyaviridae* (Elbeaino *et al.*, 2009a), with a particle shape typical of this virus, and it was named fig mosaic virus (Elbeaino *et al.*, 2009a).

16.3 The New World of Virus and Virus-like Diseases of Fig

In recent years, the number of fig viruses that have been accurately identified has sharply increased, thanks to technological advances made in the field of molecular research. Thus, virus species of the families *Betaflexiviridae* (genus *Trichovirus*), *Fimoviridae*, *Caulimoviridae* (genus *Badnavirus*), *Closteroviridae* (genera *Closterovirus* and *Ampelovirus*), *Partitiviridae* (genus *Deltapartitivirus*), *Tymoviridae*, and *Secoviridae* have been detected in mosaic-affected fig trees. These viruses have been partially or completely characterized, along with a score of other viruses whose identification is based primarily, or only, on distant homology with known representatives of the genera *Luteovirus*, *Cripavirus*, *Potyvirus*, and *Umbravirus*. In the rest of this section, we report a chronological review of newly discovered viruses infecting figs in nature.

16.3.1 Fig viruses

Fig leaf mottle-associated virus 1

A turning point in the identification of fig viruses occurred in 2006, when the first molecular genomic information on a virus infecting fig in nature was deposited in GenBank (www.ncbi.nlm.nih.gov/genbank/; accessed December 13, 2021); this was fig leaf mottle-associated virus 1 (FLMaV-1) (Elbeaino *et al.*, 2006). FLMaV-1 is a putative member of the genus *Closterovirus* identified in Apulia, southern Italy, starting from ds-RNA extracted (~17.5 kb) from a mosaic-diseased fig tree. The virus has filamentous closterovirus-like particles approximately 1500 nm long that accumulate in companion cells and sieve elements of the plant. The level of nucleotide variability found among the isolates of FLMaV-1 in the heat-shock protein 70 kDa homologue protein (HSP70h) can reach 18% (Elbeaino *et al.*, 2009b). In nature, FLMaV-1 is vectored by the wax scale insect *Ceroplastes rusci* (L.) (Yorganci and Açikgöz, 2019). This insect and virus have a wide geographical distribution and have been reported in Albania, Bosnia and Herzegovina, California, Canary Islands, Chile, China, Egypt, Hungary, Iran, Italy, Lebanon, Mexico, Montenegro, Portugal, Saudi Arabia, South Africa, Syria, Tunisia, and Turkey (Elbeaino *et al.*, 2006, 2007a, 2009b, 2011a, 2012; Nahdi *et al.*, 2007; de Stradis *et al.*, 2008; Martelli, 2009; Minafra *et al.*, 2009; Elbeshehy and Elbeaino, 2011; Alhudaib, 2012; Shahmirzaie *et al.*, 2012; Perović *et al.*, 2016; Delić *et al.*, 2017; Mijit *et al.*, 2017).

Fig leaf mottle-associated virus 2

Fig leaf mottle-associated virus 2 (FLMaV-2) is a putative member of the genus *Ampelovirus*, with particles exceeding 2000 nm in length, and with a dsRNA of about 18 kb in length. It induces a cytopathology conforming to that of the family *Closteroviridae* (Elbeaino *et al.*, 2007a). Only five open reading frames (ORFs), spanning 5868 nucleotides (nt), constitute the sequence information available in its genome (GenBank accession number FJ473383). The remaining sequences are yet to be discovered. Similarly to FLMaV-1, FLMaV-2 has a wide geographical distribution and has been identified in mosaic-diseased fig accessions from a variety of Mediterranean countries (Albania, Algeria, Egypt, Greece, Iran, Italy, Lebanon, Syria, Tunisia, and Turkey) but also from the Canary Islands, Bosnia and Herzegovina, and Montenegro (Elbeaino *et al.*, 2007a, 2009b, 2011a, 2012; Minafra *et al.*, 2009; Caglar *et al.*, 2011; Elbeshehy and Elbeaino 2011; Danesh-ahmuz *et al.*, 2013; Delić *et al.*, 2017).

Fig mild mottle-associated virus

Fig mild mottle-associated virus (FMMaV) is a putative member of the genus *Closterovirus*, which was identified in a fig accession showing mild mottling on the leaves (Elbeaino *et al.*, 2009c). The filamentous virion particles are about 2000 nm long, with a positive-sense single-stranded RNA ((+)ssRNA) genome of 6290 nt (GenBank accession number FJ611959) that encompass seven ORFs, with three to five predicted additional ORFs that have not yet been sequenced (Elbeaino *et al.*, 2009c). FMMaV and FLMaV-1 share around 30% divergence in the HSP70h gene at the amino acid level (Elbeaino *et al.*, 2009c). FMMaV was detected in Bosnia and Herzegovina, Canary Islands, China, Egypt, Iran, Montenegro, Syria, and Tunisia (Elbeshehy and Elbeaino 2011; Elbeaino *et al.*, 2012; El-Air *et al.*, 2013; Delić *et al.*, 2016; Mijit *et al.*, 2017; Alishiri *et al.*, 2018a).

Closterovirus-like, umbravirus-like, and luteovirus-like sequences in fig trees

Occasional occurrences of closterovirus sequences in fig trees affected by mosaic have also been reported from different origins such as the USA (Tzanetakis *et al.*, 2010) and Iran (Norozian *et al.*, 2014). Three closteroviral sequences from Iran (469 base pairs (bp)) (GenBank accession numbers MG407557, KM516761, and KM516760) were attributed the name of fig leaf mottle-associated virus 3 (FLMaV-3), whereas the sequences from the USA were denoted Arkansas fig closterovirus-1 (AFCV-1, GenBank accession number JN882590) and Arkansas fig closterovirus-2 (AFCV-2, GenBank accession number EF654103) (Tzanetakis *et al.*, 2010). The isolates of AFCV-1 have remained unverified in GenBank and there are currently no sequences available for AFCV-2, although diagnostic primers are available (Tzanetakis *et al.*, 2010). However, it is not clear yet whether all these sequences derive from the same species or from different viruses/isolates with nucleotide variance of the known closteroviruses already reported in the literature. The first hypothesis is corroborated by BLAST analysis of the AFCV-1 isolate YN4 sequence (GenBank accession number JN882590), which showed 97%, 95%, 92% and 90.01% sequence identity to four FLMaV-3 isolates reported in GenBank (accession numbers EF654103, MG407557, KM516761 and KM516760). More information is needed before a conclusion can be drawn regarding the nature and origin of these closterovirus sequences.

Recent studies based on molecular approaches have also reported the presence of short nucleotide sequences, distantly related to those of luteoviruses and umbraviruses, identified in fig trees from a Californian germplasm collection (Walia *et al.*, 2009).

Fig latent virus 1

Fig latent virus 1 (FLV-1) has a (+)ssRNA genome, with particles 700 nm in length, and an estimated genome size of approximately 8000 nt, 6620 nt of which have been sequenced (Gattoni *et al.*, 2009). Its genomic structural arrangement resembles that of members of the genus *Trichovirus* (family *Betaflexiviridae*). FLV-1 is mechanically transmissible to a very limited number of herbaceous host species in which it multiplies without inducing visible symptoms. This virus is transmitted by seed at a very high rate (90–100%) (Gattoni *et al.*, 2009); however, several experiments conducted on seedlings obtained from FLV-1-infected fig trees showed mixed results, and it was found to be present in only 4% (two out of 50 polymerase chain reaction (PCR)-assayed seedlings) (T. Elbeaino, personal communication). Compared with other fig-infecting viruses, FLV-1 seems to be distributed worldwide, as it was reported in fig accessions from Corsica, Iran, Italy, Mexico, Portugal, Saudi Arabia, South Africa, Syria, Turkey, and the USA (Castellano *et al.*, 2007; Minafra *et al.*, 2009; Elbeaino *et al.*, 2012; Shahmirzaie *et al.*, 2012; El-Air *et al.*, 2013; Elbeshehy *et al.*, 2017).

Fig mosaic emaravirus

Undoubtedly, one of the most intriguing discoveries in the world of fig-tree viruses is the so-called DMBs (Fig. 16.1A). The nature of DMBs remained for decades a subject of astonishment for researchers due to their doubtful interpretation either as virus-induced cytopathological artifacts or virus particles. This doubt arose because of the FMD viral agents and their transmissibility by *Aceria ficus* (Fig. 16.1B) to healthy fig seedlings on which they induced mosaic symptoms (Elbeaino *et al.*, 2009a,c). Fig mosaic emaravirus, also known as fig mosaic virus (FMV), is a negative-sense, single-stranded

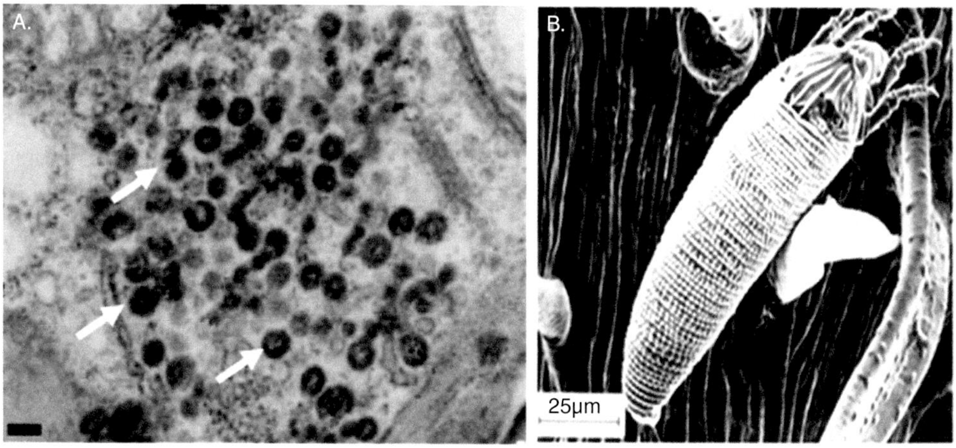

Fig. 16.1. Fig-tree viruses, originally denoted “double-membraned bodies” or DMBs. A. Groups of DMBs in the cytoplasm of cells from a seedling exposed to FMV. Arrows indicate aggregates of convoluted electron-dense filamentous elements that usually accompany the DMBs. Bar, 200 nm. B. Adult of an eriophyid mite (*Aceria ficus*), the vector of FMV. Courtesy of Professor Enrico De Lillo, Dipartimento di Biologia e Chimica Agroforestale e Ambientale (DiBCA) Università degli Studi di Bari, Bari, Italy.

multipartite RNA virus (Elbeaino *et al.*, 2009a), whose genome consists of six viral RNA segments, each encompassing a single ORF of 7093, 2252, 1490, 1472, 1752, and 1212 nt, named RNA-1 to RNA-6, respectively (Elbeaino *et al.*, 2009a,c, 2012). A virus with similar features to FMV is the European mountain ash ringspot-associated virus (Mielke and Muehlbach, 2007). The molecular information gained from both of these viruses launched the molecular characterization of other similar viruses known from the literature to induce diseases on several crops, such as pigeonpea sterility mosaic (Lava Kumar *et al.*, 2003) and High Plains disease of cereals (Skare *et al.*, 2006). The discovery of these viruses enabled their taxonomical classification and the establishment of a new genus (*Emaravirus*) and later of a new family (*Fimoviridae*) (Elbeaino *et al.*, 2018a), now encompassing 20 classified viruses and three putative fimoviral species (Elbeaino and Digiaro, 2021; ICTV Master species list: https://talk.ictvonline.org/files/master-species-lists/m/msl/9601).

Fimoviruses are distantly related to orthotospoviruses and orthobunyaviruses of the families *Tospoviridae* and *Peribunyaviridae*, respectively, and they share several common features: (i) a multipartite, (–)ssRNA genome, which in the case of fimoviruses consists of four to eight segments; (ii) a high sequence similarity with orthologous proteins of members of the order *Bunyavirales* at equivalent genome positions in the first three RNAs (corresponding to the L, M, and S RNA segments); (iii) the presence of five conserved motifs (A–E) in the amino acid sequence of their RNA-dependent RNA polymerase (RdRP), similar to those in the L segment-encoded protein of members of the order *Bunyavirales*; (iv) the virus particles are typically enveloped in membranes; and (v) the presence of conserved nucleotide sequences at both the 5′- and 3′-termini of all genomic RNA segments of fimoviruses, although not identical to those of other members of the order *Bunyavirales*.

In nature, transmission of FMV occurs through the eriophyid mite *A. ficus* (Fig. 16.1B) (Flock and Wallace, 1955; Elbeaino *et al.*, 2009a), but no seed transmission has been recorded (Martelli *et al.*, 1993; Elbeaino *et al.*, 2009a). Among all viruses of fig, FMV is the only virus identified as an etiological agent of FMD (Elbeaino *et al.* 2009a). Affected trees show a wide range of symptoms, appearing mainly on their leaves, in the form of mosaic-like discolorations, various patterns of chlorotic mottling and blotching, vein banding, vein clearing, chlorotic or necrotic ringspot, and line patterns (Fig. 16.2A–F). Some plants may display apical rosetting of the leaves and reduced vitality. In some instances, the fruits show yellow mottling, longitudinal stripes and ringspots, and may drop prematurely (Fig. 16.2A–F). FMD is transmitted experimentally by grafting to fig and other members of the family *Moraceae*, primarily in the genus *Ficus* (16 different species), plus *Cudrania tricuspidata* and *Morus indica*, which are the only two known artificial hosts beside *Ficus* (Burnett, 1960, 1962). FMV can also infect ornamental plant species, such as cyclamen (*Cyclamen persicum* Mill) (Elbeaino *et al.*, 2018b). The cyclamen accession Cic, from Ercolano (Campania region, south of Italy), found infected by FMV, showed symptoms of leaf mottling, light yellowing, deformation, flower discoloration, and mosaic discolorations similar to those demonstrated for figs in nature (Fig. 16.2G–H).

Fig mosaic is a cosmopolitan disease and is present wherever fig is grown; however, its presence was mostly investigated in European and Mediterranean regions such as in Albania, Bosnia and Herzegovina, Bulgaria, Croatia, Cyprus, the UK, France, Germany, Greece, Hungary, Italy, Portugal, Romania, Spain, Algeria, Egypt, Iran, Israel, Jordan, Lebanon, Libya, Morocco, Palestine, Saudi Arabia, Syria, Tunisia, and Turkey. It was also reported in the USA, Mexico, Puerto Rico, Argentina, Chile, South Africa, Australia, Japan, China, New Zealand, and India (Blodgett and Gömeç, 1967; Plavšic and Milicic, 1980; Martelli *et al.*, 1993; Castellano *et al.*, 2007; de Stradis *et al.*, 2007, 2008; Elbeaino *et al.*, 2009c; Caglar *et al.* 2011; Elbeshehy and Elbeaino, 2011; Alhudaib, 2012; El-Air *et al.*, 2012; Ishikawa *et al.*, 2012; Shahmirzaie *et al.*, 2012; Ale-Agha and Rakhshandehroo, 2014; Mijit *et al.*, 2015; Perović *et al.*, 2016; Mijit *et al.* 2017).

Fig cryptic virus

Fig cryptic virus (FCV) was detected in fig for the first time in Italy (Elbeaino *et al.*, 2011b), and is one of the few viruses infecting woody crops in nature. It is a member of the genus *Deltapartitivirus*, in the family *Partitiviridae*. FCV has a dsRNA

Fig. 16.2. Fig mosaic-diseased plants infected with FMV, showing different chromatic and cytopathological alterations on leaves and fruits. A. Sparse ringspots on leaves. B. Accentuated ringspot, vein clearing and banding, mottling, and blotching leading to a general mosaic aspect. C. Complete mosaic aspect and deformation of leaves. D. Line stripes and discoloration on fruits. E. Chlorotic yellowish spots and rings on fruits. F. Mosaic-affected fruits showing small size, and necrotic and chlorotic ringspots. G–I. Cyclamen plants, infected with FMV, showing mosaic light discolorations of leaves (G), diffused mosaic discoloration and deformation (H) and leaf deformation and rosetting, with flower abortion (I).

bipartite genome; the RNA-1 segment is 1696 bp and contains a single ORF (1419 nt) potentially encoding a 54 kDa protein (RdRP). The RNA-2 segment is 1415 bp and has also a single ORF (1014 nt) encoding a polypeptide with a predicted molecular mass of 38 kDa (coat protein). Although FCV was found in mosaic-diseased figs, it does not induce symptoms in fig trees and is not mechanically transmitted (Elbeaino *et al.*, 2011b). It has been detected in fig orchards of many Mediterranean and Asian countries, such as Albania, Algeria, Iran, Italy, Lebanon, Syria, Turkey, and Tunisia (Elbeaino *et al.*, 2011b; Ale-Agha and Rakhshandehroo, 2014; Elçi *et al.*, 2017).

Fig fleck-associated virus

Fig fleck-associated virus (FFkaV) was first reported in Apulia, southern Italy, in 2011 (Elbeaino *et al.*, 2011c). It has a (+)ssRNA genome and belongs to the genus *Maculavirus* in the family *Tymoviridae* (Elbeaino *et al.*, 2011c). The presence of this virus has been reported in many Mediterranean countries, including Albania, Algeria, Italy, Lebanon, Syria, and Tunisia, with different levels of infection (Elbeaino *et al.*, 2011c, 2012). However, its presence has barely spread in Turkey (Elçi *et al.*, 2017) and Iran (Nouri-Ale *et al.*, 2013; Ale-Agha and Rakhshandehroo, 2014). In contrast, relatively high levels of infection were found in China (Mijit *et al.*, 2017). Although FFkaV has been found in mosaic-affected fig trees across several countries, it is still unclear whether it is involved in FMD of figs.

Fig badnavirus 1

Fig badnavirus 1 (FBV-1) was first reported in 2008 in the USA (Tzanetakis *et al.*, 2010), where it was found in both asymptomatic and symptomatic fig trees (Laney *et al.*, 2012). This virus is prevalent in fig, and has been found in various countries such as the USA (Laney *et al.*, 2012), Croatia (Vončina *et al.*, 2015), Iran (Alimoradian *et al.*, 2014; Alishiri *et al.*, 2018b), and China (Mijit *et al.*, 2017). It belongs to the genus *Badnavirus*, family *Caulimoviridae* and has a monopartite, open circular dsDNA genome, with three ORFs and a total length of 7141 bp. Similar to other caulimovirids, FBV-1 can also integrate into the fig genome. Therefore, it is difficult to unequivocally determine whether an FBV-1-positive plant is the result of some genomic DNA residues of FBV-1 that have remained integrated in the genome of the plant or is the physical entity incorporated into the plant. The intuitive question arising is whether and to what extent the FBV-1 could be involved in FMD of figs. Its widespread distribution, compared with FMV, and its finding in most symptomatic fig plants makes it a competitor of FMV in the etiology of FMD. More research is needed to address this point and to develop ways for sanitizing fig plant material from FBV-1.

Strawberry latent ringspot virus

Strawberry latent ringspot virus (SLRSV) is a cosmopolitan virus, capable of infecting several plant species. Its presence in fig was investigated in 140 samples collected from the main fig-growing areas of Lebanon and Turkey, where it was found in 18% and 24.4% of the samples, respectively (Elbeaino *et al.*, 2015). Nucleotide sequence analysis of all isolates showed 93–98% similarity with SLRSV isolates present in GenBank, in particular with the isolate X75165. The possible etiological role of SLRSV in the fig symptoms observed (leaf mottling and vein clearing) is not clear due to the multiple viral infections found within the same infected fig tree analyzed.

Fig viruses A and B

Two novel closteroviruses, fig virus A (FiVA) and fig virus B (FiVB, were recently identified by high-throughput RNA-sequencing technology applied to fig trees from the Republic of Korea (Park *et al.*, 2021). The FiVA and FiVB genomes consist of 19,333 bp (GenBank accession number MN817232) and 18,741 bp (MN817233), respectively. Phylogenetic analysis grouped FiVA and FiVB in one subclade with FMMaV, all as species of the genus *Closterovirus*. There is no biological information on their possible involvement with FMD of figs.

Ficus Tepovirus

A new virus belonging to the genus *Tepovirus* of the family *Betaflexiviridae* has recently been identified in fig in the Republic of Korea as *Ficus* tepovirus A (FiTVA). The total genomic sequence of this virus is 6849 nt (GenBank accession number

MH898491). The genome of FiTVA comprises three ORFs, which encode for replicase, movement, and coat protein genes. There is no biological information on its impact on figs in nature.

16.3.2 Fig viroids: hop stunt viroid, *Citrus* exocortis viroid, apple dimple fruit viroid (ADFVd), and fig hammerhead viroid-like RNA

Viroids of the family *Pospiviroidae*, including hop stunt viroid (HSVd; genus *Hostuviroid*), *Citrus* exocortis viroid (CEVd; genus *Pospiviroid*), and a viroid resembling apple dimple fruit viroid (ADFVd; genus *Apscaviroid*) have been reported in fig trees from Tunisia (Chiumenti *et al.*, 2014) and Italy (Yakoubi *et al.*, 2007). ADFVd was initially identified in figs by high-throughput sequencing, whereas HSVd and CEVd, were identified using specific primers in reverse transcriptase (RT)-PCR assays. The high-throughput sequencing technology has also contributed to a recent discovery of viroid-like RNAs associated with fig trees grown in the Hawaiian Islands (Olmedo-Velarde *et al.*, 2020). One of these RNAs has been characterized and found to be a circular RNA ranging in size from 357 to 360 nt, tentatively named fig hammerhead viroid-like RNA (FHVd-LR) (Olmedo-Velarde *et al.*, 2020). None of these viroids has been associated with any particular fig disease, and it is unlikely that they could be involved in FMD, but this has yet to be determined through further studies.

16.3.3 Fig phytoplasmas: "*Candidatus* Phytoplasma asteris"- and "*Candidatus* Phytoplasma solani"-related strains

A recent study conducted in a fig collection plot in Locorotondo, southern Italy, showed the first occurrence of phytoplasma in different varieties. Two phylogenetically distinct phytoplasmas were found, "*Candidatus* Phytoplasma asteris" and "*Candidatus* Phytoplasma solani" affiliated to the 16SrI and 16SrXII ribosomal groups, respectively (Alsaheli *et al.*, 2020). The 16SrI (GenBank accession numbers LR584982 and LR584983)- and 16SrXII (LR584984)-related strains of fig phytoplasmas were found in trees showing yellowing, deformed leaves, short internodes, mottling, and mosaic symptoms. However, their involvement in the symptoms observed in the diseased fig plants could not be established due to the prevalence of FMV in the infected trees. Note that three fig plants with leaf mottling and deformation symptoms were found to be infected with these phytoplasma strains and free of FMV; thus, this may indicate the involvement of these phytoplasmas with specific disease symptoms in fig.

16.4 Transmission

Figs are propagated by grafting and/or self-rooted cuttings, two methods that disseminate various pests and diseases including the viruses and viroids associated with FMD. However, they may be spread by other means such as aphids, mealybugs, and leafhoppers. The presence of putative members of the genera *Ampelovirus*, *Badnavirus*, *Carlavirus*, *Closterovirus*, *Fimoviruses*, *Potyvirus*, and *Secovirus*, together with phytoplasma agents in diseased figs (Grbelja and Eric, 1983; Namba, 1983; Doi, 1989; Elbeaino *et al.*, 2006, 2007a,b, 2009a, 2015; Tzanetakis *et al.*, 2010) suggests that some aerial vectors may be involved in the natural spread of fig viruses and phytoplasmas in the field. Viruses may also be spread through the soil. The nematode *Xiphinema diversicaudatum*, for example, is known to transmit SLRSV in nature. Vectoring experiments were carried out, mimicking the transmission of FLMaV-1 and FLMaV-2 by mealybugs and scales of the species *Planococcus ficus* and *C. rusci* to healthy fig plants. The results showed that FLMaV-2, could be transmitted by *P. ficus* to healthy figs under experimental conditions (T. Elbeaino, unpublished data), while FLMaV-1 was vectored by *Ceroplastes rusci* (Yorganci and Açikgöz, 2019). Transmission of FMV (DMBs) by *A. ficus* was demonstrated, as *A. ficus* was unequivocally associated with the development of symptoms in mite-inoculated seedlings (Elbeaino *et al.*, 2009a; Caglayan *et al.*, 2012).

16.5 Detection and Control

The detection of fig viruses in infected plant material relies mainly on RT-PCR assays (Table 16.1), or

Table 16.1. A list of specific primers used in RT-PCR assays for the detection of fig viruses.

Virus	Primers	Sequences (5′–3′)	Gene	PCR amplicon (bp)	Reference
FLMaV-1	N17-s	CGTGGCTGATGCAAAGTTTA	HSP70h	352	Elbeaino *et al*. (2006)
	N17-a	GTTAACGCATGCTTCCATGA			
FLMaV-2	F3-s	GAACAGTGCCTATCAGTTTGATTTG	HSP70h	360	Elbeaino *et al*. (2007)
	F3-a	TCCCACCTCCTGCGAAGCTAGAGAA			
FMMaV	LM3-s	AAGGGGAATCTACAAGGGTCG	HSP70h	311	Elbeaino *et al*. (2010)
	LM3-a	TATTACGCGCTTGAGGATTGC			
AFCV-1	AFCV1F	CTGTATCTGTCATTACCTCTTCGGG	HSP70h	375	Tzanetakis *et al*. (2010)
	AFCV1R	ATGCTTCCTCGGCTGC			
AFCV-2	AFCV1F	GTTCGGAATTAGTTAATAVTA	HSP70h	1671	Tzanetakis *et al*. (2010)
	AFCV1R	ACCCGCTAGAGTAATCAGTCAAAGTT			
FMV	E5-s	GGGTACATATGCGTCATTCTTG	RdRP	302	Elbeaino *et al*. (2009b)
	E5-a	CGTTTGTCTTGGATCACAGCAA			
FLV-1	Ffup	CCATCTTCACCACACAAATGTC	RdRP	193	Gattoni *et al*. (2012)
	Ffdown	CAATCTTCTTGGCCTCCATAAG			
FCV-1		TCGGATTGTCTTTGGAGAGG	RdRP	353	Elbeaino *et al*. (2011b)
		CGCATCCACAGTATCCCATT			
FFkaV	D8-s	TCAATCCCAAGGAGGTGAAG	CP	270	Elbeaino *et al*. (2012)
	D8-a	ACACGGTCAATGAGGGAGTC			
FBV-1	1094F	ACCAGACGGAGGGAAGAAAT	Polyprotein	473	Laney *et al*. (2012)
	1567R	TCCTTGCCATCGGTTATCTC			
SLRSV	SLRSV-s	TCAAGGAGAATATCCCTGGCCC	CP	525	Elbeaino *et al*. (2014)
	SLRSV-a	CTAAGTGCCAGAACTAAACC			
ADFVd	ADFVd-F	CCCCCCTGCGCTACTGACTAAAAG		262	Chiumenti *et al*. (2014)
	ADFVd-R	GTGTTTTACCCTGGAGGCTCCACTC			
CEVd	CEVd-F	GGAAACCTGGAGGAAGTCGAGG		370	Yakoubi *et al*. (2007)
	CEVd-R	CCGCCTCTTTTTTCTTTTCCTGCCTGC			
HSVd	78P	AACCCGGGGCAACTCTTCTC		312	Sano *et al*. (2001)
	83M	AACCCGGGGCTCCTTTCTCA			
FHVd-LR	FHVd-5F	CTGATGAGAACAAAAGTTCGAAAC			Olmedo-Velarde *et al*. (2020)
	FHVd-6R	TTGGATCACACAATCCAATACCTT			

See text for full virus names; s, sense; a, antisense; F, forward; R, reverse; HSP70h, heat-shock protein 70-like protein; RdRp, RNA-dependent RNA polymerase; CP, coat protein.

on loop-mediated isothermal amplification in the case of FMV (Ishikawa *et al.*, 2012). However, the detection of FMV, and in general all fig viruses, is hampered by the low virus concentration and sequence variations of isolates in infected plants. Accordingly, a recent study was conducted to evaluate the sensitivity, specificity, and flexibility (multiplex) of a real-time (TaqMan RT-PCR) assay for the detection of FLMaV-1, FLMaV-2, FMMaV, FMV, FLV-1, FCV-1, and FFkaV in infected fig plants (Alsaheli *et al.*, 2021). TaqMan RT-PCR showed a higher sensitivity and flexibility than RT-PCR when used as singleplex and multiplex assays in the detection of single and multiple virus infections in fig plant material.

One approach toward eliminating viruses/FMD from infected trees showed that fig plants respond very well to sanitation through heat therapy and in vitro meristem tip culture (Muriithi *et al.*, 1982; Demiralay *et al.*, 1998). These two approaches were successful in eliminating FMD (Muriithi *et al.*, 1982; G. Bottalico, personal communication), FLMaV-1, and FLMaV-2 from Lebanese fig varieties (Chalak *et al.*, 2009). Maintaining a healthy sanitary status of fig trees in the field when dealing with highly active and viruliferous insect populations (e.g. eriophyid mites, aphids, and mealybugs) is a very daunting task. The use of insecticides following specific calendars linked to the life cycles of these insects can reduce or eliminate the means of transmission of fig viruses in the orchard.

16.6 Concluding Remarks

The detection of FMD in the field is an easy task because of the expression of symptoms that can be observed. Usually, the symptoms are clearly visible in spring and remain throughout the growing season (Fig. 16.3).

However, ambiguity remains in determining the correct causal agent inducing these symptoms. Bioassays are of no help in identifying the etiology of fig diseases because there are no differential fig indicator plants as yet. Mechanical transmission of infected plant sap on to herbaceous hosts remains the easiest and simplest technique for the detection of viruses (known and unknown) in general; however,

Fig. 16.3. Diseased fig leaves showing different types of symptoms. A. Vein feathering. B. Line pattern. C. Mosaic discoloration. D. Clearing of secondary and tertiary veins. E. Vein banding. F. Deformation.

many constraints inhibit the success of this technique due to phenolic compounds and inhibitors naturally present in fig. The difficulty in purifying viruses directly from the extremely milky fig tissues and unsuccessful experimental transmission on hosts other than fig have hindered their identification and thus also impeded the production of specific antisera for their serological detection (Elbeaino *et al.*, 2006, 2007, 2009a, 2010). Therefore, researchers have found it necessary to produce a polyclonal antiserum to FMV, the most important fig virus, using different recombinant protein technologies (Shahmirzaie *et al.*, 2019, 2020).

References

Ale-Agha, G.N. and Rakhshandehroo, F. (2014) Detection and molecular variability of Fig fleck associated virus and Fig cryptic virus in Iran. *Journal of Phytopathology* 162, 417–425.

Alfieri, S.A. (1967) *Fig Mosaic*. Plant Pathology Circular No. 64. Florida Department of Agriculture, Tallahassee, Florida.

Alhudaib, K. (2012) Incidence of fig leaf mottle-associated virus and fig mosaic virus in Eastern Province of Saudi Arabia. *International Journal of Virology* 8, 128–132.

Alimoradian, P., Rakhshandehroo, F., and Shams-Bakhsh, M. (2014) First record of fig badnavirus-1 in fig trees in Iran. *Journal of Plant Pathology* 96, 4–124.

Alishiri, A., Rakhshandehroo, F., Jouzani, G.S., and Shams-bakhsh, M. (2018a) First report of fig mild mottle-associated virus in Iran. *Journal of Plant Pathology* 100, 135–135.

Alishiri, A., Rakhshandehroo, F., Jouzani, G.S., and Shams-bakhsh, M. (2018b) Exploring the genetic diversity and molecular evolution of fig badnavirus-1 from Iran. *Journal of Plant Pathology* 100, 287–299.

Alsaheli, Z., Contaldo, N., Mehle, N., Dermastia, M., Elbeaino, T., and Bertaccini, A. (2020) First detection of "*Candidatus* Phytoplasma asteris"- and "*Candidatus* Phytoplasma solani"-related strains in fig trees. *Journal of Phytopathology* 168, 63–71.

Alsaheli, Z., Abdallah, A., Incerti, O., Shalaby, A., Youssef, S., *et al.* (2021) Development of singleplex and multiplex real-time (TaqMan) RT-PCR assays for the detection of viruses associated with the fig mosaic disease. *Journal of Virological Methods* 293: 114145.

Appiano, A. (1982) Cytological observations on leaves infected with fig mosaic. *Caryologia* 35: 152.

Appiano, A., Conti, M., and Zini, N. (1995) Cytopathological study of the double-membrane bodies occurring in fig plants affected by fig mosaic disease. *Acta Horticulturae* 386, 585–592.

Blodgett, E.C. and Gömeç, B. (1967) Fig mosaic. *Plant Disease Reporter* 51, 893–896.

Bradfute, O.R., Whitmoyear, R.E., and Nault, R.L. (1970) Ultrastructure of plant leaf tissues infected with mite borne viral-like particles. *Proceeding of the Electron Microscopy Society of America* 258, 178–179.

Burnett, H.C. (1960) Species of *Ficus* susceptible to the fig mosaic virus. *Proceedings of the Florida State Horticultural Society* 73, 316–320.

Burnett, H.C. (1962) Additional hosts of the fig mosaic virus. *Plant Disease Reporter* 46, 693–693.

Caglar, B.K., Fidan, H., Guldur, M.E., and Elbeaino, T. (2011) The prevalence of three viruses infecting fig in southern Turkey. *Journal of Phytopathology* 159, 181–183.

Caglayan, K., Elçi, E., Serce, C.U., Kaya, K., Gazel, M., and Medina, V. (2012) Detection of Fig mosaic virus in viruliferous eriophyid mite *Aceria ficus*. *Journal of Plant Pathology* 94, 629–634.

Castellano, M.A., Gattoni, G., Minafra, A., Conti, M., and Martelli, G.P. (2007) Fig mosaic in Mexico and South Africa. *Journal of Plant Pathology* 89, 441–443.

Chalak, L., Elbeaino, T., Elbitar, A., Chehade, A., Fattal, T., and Choueiri, E. (2009) Attempts to virus-free production of Lebanese fig varieties by tissue culture. In: *Proceeding of IV International Symposium on Fig, September 29–October 3, Meknes, Morocco*. International Society for Horticultural Science, Leuven, Belgium.

Chen, J., Chen, J., and Adams, M.J. (2001) A universal PCR primer to detect members of the *Potyviridae* and its use to examine the taxonomic status of several members of the family. *Archives of Virology* 146, 757–766.

Chiumenti, M., Torchetti, E.M., Di Serio, F., and Minafra, A. (2014) Identification and characterization of a viroid resembling apple dimple fruit viroid in fig (*Ficus carica* L.) by next generation sequencing of small RNAs. *Virus Research* 188, 54–59.

Condit, I.J. and Horne, W.T. (1933) A mosaic of the fig in California. *Phytopathology* 23, 887–897.

Danesh-ahmuz, S., Rakhshandehroo, F., Rezaee, S., and Elbeaino, T. (2013) Occurrence of Fig leaf mottle-associated virus 2 in Iran (Disease note). *Journal of Plant Pathology* 95, 663.

de Stradis, A., Gattoni, G., Minafra, A., and Martelli, G.P. (2007) Fig mosaic in Corsica. *Journal of Plant Pathology* 89, S76.

de Stradis, A., Pantaleo, A., Salamon, P., Minafra, A., and Martelli, G.P. (2008) Viruses associated with fig mosaic in Hungary. *Journal of Plant Pathology* 90, 585–585.

Delić, D., Elbeaino, T., Lolić, B., and Đurić, G. (2016) Detection of Fig mosaic virus, Fig leaf mottle-associated virus 1 and Fig mild mottle-associated virus in Bosnia and Herzegovina (Disease note). *Journal of Plant Pathology* 98, 371.

Delić, D., Perović, T., Hrnčić, S., Lolić, B., Đurić, G., and Elbeaino, T. (2017) Detection and phylogenetic analyses of fig-infecting viruses in Bosnia and Herzegovina and Montenegro. *Phytopathologia Mediterranea* 56, 470–478.

Demiralay, A., Yal çin-Mendi, Y., Aka-Kaçar, Y., and Çetiner, S. (1998) *In vitro* propagation of *Ficus carica* L. var Bursa Siyahi through meristem culture. *Acta Horticulturae* 480, 165–168.

Doi, Y. (1989) Fig virus S. In: Hull, R., Brown, F., and Payne, C. (eds) *Directory and Dictionary of Animal, Bacterial and Plant Viruses*. McMillan, London, p. 76.

El-Air, M., Mahfoudhi, N., Elbeaino, T., Dhouibi, M.H., and Digiaro, M. (2012) Occurrence of *Fig mosaic virus* in Tunisian fig orchards (Disease note). *Journal of Plant Pathology* 94 (Suppl. 4), 89.

El-Air, M., Mahfoudhi, N., Elbeaino, T., Dhouibi, M.H., and Digiaro, M. (2013) Presence of *Fig mild mottle-associated virus* and *Fig latent virus 1* in Tunisia. *Journal of Plant Pathology* 95 (Suppl. 4), 72.

Elbeaino, T. and Digiaro, M. (2021) *Fimoviridae*. In: Bamford, D. and Zuckerman, M. (eds) *Encyclopedia of Virology*, 4th edition. Academic Press, Cambridge, Massachusetts, pp. 396–404.

Elbeaino, T., Digiaro, M., de Stradis, A., and Martelli, G.P. (2006) Partial characterization of a closterovirus associated with a chlorotic mottling of fig. *Journal of Plant Pathology* 88, 187–192.

Elbeaino, T., Choueiri, E., Hobeika, C., and Digiaro, M. (2007a) Presence of Fig leaf mottle-associated virus 1 and -2 in Lebanese fig orchards. *Journal of Plant Pathology* 89, 409–411.

Elbeaino, T., Digiaro, M., de Stradis, A., and Martelli, G.P. (2007b) Identification of a second member of the family *Closteroviridae* in mosaic-diseased figs. *Journal of Plant Pathology* 89, 119–124.

Elbeaino, T., Digiaro, M., Alabdullah, A., de Stradis, A., Minafra, A., *et al.* (2009a) A multipartite single-stranded negative-sense RNA virus is the putative agent of fig mosaic disease. *Journal of General Virology* 90, 1281–1288.

Elbeaino, T., Nahdi, S., Digiaro, M., Alabdullah, A., and Martelli, G.P. (2009b) Detection of FLMaV-1 AND FLMaV-2 in the Mediterranean region and sequence variation of the HSP70 gene. *Journal of Plant Pathology* 91, 425–431.

Elbeaino, T., Digiaro, M., and Martelli, G.P. (2009c) Complete nucleotide sequence of four viral RNA segments of Fig mosaic virus. *Archives of Virology* 154, 1719–1727.

Elbeaino, T., Heinoun, K., Digiaro, M., and Martelli, G.P. (2010) Fig mild mottle-associated virus, a novel closterovirus infecting fig. *Journal of Plant Pathology* 92, 165–172.

Elbeaino, T., González-Rodríguez, Á.M., Grajal-Martín, M.J., and Digiaro, M. (2011a) Survey of fig viruses in the Canary Islands. *Journal of Plant Pathology* 93, 737–739.

Elbeaino, T., Kubaa, R.A., Digiaro, M., Minafra, A., and Martelli, G.P. (2011b) The complete nucleotide sequence and genome organization of Fig cryptic virus, a novel bipartite dsRNA virus infecting fig, widely distributed in the Mediterranean basin. *Virus Genes* 42, 415–421.

Elbeaino, T., Digiaro, M., and Martelli, G.P. (2011c) Complete sequence of Fig fleck-associated virus, a novel member of the family *Tymoviridae*. *Virus Research* 161, 198–202.

Elbeaino, T., Digiaro, M., and Martelli, G.P. (2012) RNA-5 and -6, two additional negative-sense RNA segments associated with Fig mosaic virus. *Journal of Plant Pathology* 94, 421–425.

Elbeaino, T., Turgut, F., Gumus, M., and Digiaro, M. (2015) Evidence of Strawberry latent ringspot virus infections in fig (*Ficus carica*). *Phytopathologia Mediterranea* 54, 164–165.

Elbeaino, T., Digiaro, M., Mielke-Ehret, N., Muehlbach, H.P., Martelli, G.P., and ICTV Report Consortium (2018a) ICTV Virus Taxonomy Profile: *Fimoviridae*. *Journal of General Virology* 99, 1478–1479.

Elbeaino, T., Marais, A., Faure, C., Trioano, E., Candresse, T., and Parrella, G. (2018b) High-throughput sequencing reveals *Cyclamen persicum* Mill. as natural host for fig mosaic virus. *Viruses* 10, 684–691.

Elbeshehy, E.K.F. and Elbeaino, T. (2011) Viruses infecting figs in Egypt. *Phytopathologia Mediterranea* 50, 327–332.

Elbeshehy, E.K.F., Aldhebiani, A.Y., and Hassan, W.M. (2017) Partial molecular characterization of the Fig latent virus 1 (FLV-1) infecting figs in Western Saudi Arabia. *Research Journal of Biotechnology* 12, 91–98.

Elçi, E., Hançer, T., and Çağlayan, K. (2017) Molecular identification of Fig cryptic virus and Fig fleck-associated virus in Turkey. *Romanian Biotechnological Letters* 22, 12383–12390.

Flock, R.A. and Wallace, J.M. (1955) Transmission of fig mosaic by the eriophyid mite *Aceria ficus*. *Phytopathology* 45, 52–54.

Gattoni, G., Minafra, A., Castellano, M.A., de Stradis, A., Boscia, D., *et al.* (2009) Some properties of Fig latent virus 1, a new member of the family *Betaflexiviridae*. *Journal of Plant Pathology* 91, 543–552.

Grbelja, J. and Eric, Z. (1983) Isolation of a potyvirus from *Ficus carica* L. *Acta Botanica Croatica* 42, 11–14.

Ishikawa, K., Maejima, K., Nagashima, S., Sawamura, N., Takinami, Y., *et al.* (2012) First report of fig mosaic virus infecting common fig (*Ficus carica*) in Japan. *Journal of General Plant Pathology* 78, 136–139.

Laney, A.G., Hassan, M., and Tzanetakis, I.E. (2012) An integrated badnavirus is prevalent in fig germplasm. *Phytopathology* 102, 1182–1189.

Lava Kumar, P.L., Jones, A.T., and Reddy, D.V.R. (2003) A novel mite-transmitted virus with a divided RNA genome closely associated with pigeonpea sterility mosaic disease. *Phytopathology* 93, 71–81.

Martelli, G.P. (2009) Fig mosaic disease and associated viruses. In: Hadidi, A., Barba, M., Candresse, T. and Jelkmann, W. (eds) *Virus and Virus-like Diseases of Pome and Stone Fruits*. APS Press, St. Paul, Minnesota.

Martelli, G.P., Castellano, M.A., and Lafortezza, R. (1993) An ultrastructural study of fig mosaic. *Phytopathologia Mediterranea* 32, 33–43.

Mielke, N. and Muehlbach, H.-P. (2007) A novel, multipartite, negative-strand RNA virus is associated with the ringspot disease of European mountain ash (*Sorbus aucuparia* L.). *Journal of General Virology* 88, 1337–1346.

Mijit, M., Li, S.F., Zhang, S., and Zhang, Z.X. (2015) First report of *Fig mosaic virus* infecting common fig (*Ficus carica*) in China. *Plant Disease* 99, 422.

Mijit, M., Zhen, H.E., Jian, H., Lu, M., Li, S., and Zhang, Z. (2017) Analysis of fig tree virus type and distribution in China. *Journal of Integrative Agriculture* 16, 1417–1421.

Minafra, A., de Stradis, A., Clara, M.I., and Martelli, G.P. (2009) Viruses in mosaic-affected figs in Portugal and the island of Crete. *Journal of Plant Pathology* 91, 238.

Muriithi, L.M., Rangan, T.S., and Waite, B.H. (1982) *In vitro* propagation of fig through shoot tip culture. *Horticultural Science* 17, 86–87.

Nahdi, S., Elbeaino, T., Digiaro, M., and Martelli, G.P. (2007) First record of Fig leaf mottle virus 1 in Tunisia. *Journal of Plant Pathology* 89, S70.

Namba, S. (1983) Fig virus S. In: Yora, K., Saito, Y., Doi, Y., Inoue, T., and Tomaru, K. (eds) *Handbook of Plant Viruses*. Asakura Shoten, Tokyo, pp. 326–327.

Norozian, E., Rakhshandehroo, F., and Shams-bakhsh, M. (2014) Presence of fig leaf mottle-associated virus 3 in an Iranian fig orchard. *Journal of Plant Pathology* 96 (Suppl. 4), 131.

Nouri-Ale, A., Rakhshandehroo, F., Zamanizadeh, H.R., and Elbeaino, T. (2013) Presence of Fig fleck-associated virus in fig trees in Iran. *Journal of Plant Pathology* 95, 664.

Olmedo-Velarde, A., Navarro, B., Hu, J.S., Melzer, M.J., and di Serio, F. (2020) Novel fig-associated viroid-like RNAs containing hammerhead ribozymes in both polarity strands identified by high-throughput sequencing. *Frontiers in Microbiology* 11: 1903.

Park, D., Goh, C.J., and Hahn, Y. (2021) Two novel closteroviruses, fig virus A and fig virus B, identified by the analysis of the high-throughput RNA-sequencing data of fig (*Ficus carica*) latex. *Acta Virologica* 65, 42–48.

Perović, T., Hrnčić, S., Čizmovic´, M., Lolić, B., Đurić, G., and Delić, D. (2016) First report of Fig mosaic virus, fig leaf mottle-associated virus 1 and fig mild mottle-associated virus infecting fig (*Ficus carica*) in Montenegro. *Journal of Plant Pathology* 98, 678.

Plavšic, B. and Milicic, D. (1980) Intracellular changes in trees infected with fig mosaic. *Acta Horticulturae* 110, 281–286.

Quacquarelli, A. (1971) Il mosaico del fico e il virus latente del *Chenopodium*. *Phytopathologia Mediterranea* 10, 283–286.

Salomon, R., Voronov, N., and Flaishman, M. (2003) The identification of fig mosaic rymovirus, the causative agent of fig mosaic disease, and its elimination from fig. In: *8th International Congress of Plant Pathology, 2–7 February 2003, Christchurch, New Zealand*. International Congress of Plant Pathology, Christchurch, New Zealand, pp. 23–37.

Serrano, L., Ramon, J., Segarra, J., Medina, V., Achon, M.A., *et al.* (2004) New approach in the identification of the causal agent of fig mosaic disease. *Acta Horticulturae* 657, 559–566.

Serrano, L., Benito, A., Medina, V., and Achón, M.A. (2008) Fig mosaic disease: associated protein and cDNA. *Acta Horticulturae* 798, 227–232.

Shahmirzaie, M., Rakhshandehroo, F., Zamani Zadeh, H.R., and Elbeaino, T. (2012) Current status of fig mosaic disease in Iran. *Journal of Phytopathology* 160, 324–330.

Shahmirzaie, M., Safarnejad, M.R., Rakhshandehroo, F., Safarpour, H., Rabbani, H., *et al.* (2019) Production of a polyclonal antiserum against recombinant nucleocapsid protein and its application for the detection of fig mosaic virus. *Journal of Virological Methods* 265, 22–25.

Shahmirzaie, M., Safarnejad, M.R., Rakhshandehroo, F., Safarpourc, H., Shirazid, F.H., *et al.* (2020) Generation and molecular docking analysis of specific single-chain variable fragments selected by phage display against the recombinant nucleocapsid protein of fig mosaic virus. *Journal of Virological Methods* 276: 113796.

Skare, J.M., Wijkamp, I., Denham, I., Rezende, J.A.M., Kitajima, E.W., *et al.* (2006) A new eriophyid mite-borne membrane-enveloped virus-like complex isolated from plants. *Virology* 347, 343–353.

Tzanetakis, I.E., Laney, A.G., Keller, K.E., and Martin, R.R. (2010) New viruses found in fig exhibiting mosaic symptoms. *Julius-Kühn-Archiv* 427, 79–82.

Vončina, D., Pilipović, P., Škorić, D., Krapac, M., Prgomet, Ž., and Miličević, T. (2015) First report of Fig mosaic virus and Fig badnavirus 1 on common fig trees in Croatia. *Journal of Plant Pathology* 97, S71–S74.

Walia, J.J., Salem, N.M., and Falk, B.W. (2009) Partial sequence and survey analysis identify a multipartite, negative-sense RNA virus associated with fig mosaic. *Plant Disease* 93, 4–10.

Yakoubi, S., Elleuch, A., Besaies, N., Marrakchi, M., and Fakhfakh, H. (2007) First report of *Hop stunt viroid* and *Citrus exocortis viroid* on fig with symptoms of fig mosaic disease. *Journal of Phytopathology* 155, 125–128.

Yorganci, S. and Açikgöz, S. (2019) Transmission of fig leaf mottle-associated virus 1 by *Ceroplastes rusci*. *Journal of Plant Pathology* 101, 1199–1201.

17 Fig Diseases

Moshe A. Flaishman[1]*, Reut Peer[1], Dvora Namdar[1], and Uygun Aksoy[2]
[1]Institute of Plant Sciences, Agriculture Research Organization, Volcani Center, Rishon LeZion, Israel; [2]Department of Horticulture, Faculty of Agriculture, Ege University, Bornova, İzmir

17.1 Introduction

As the world population grows and with the increase in attention to healthy and higher-quality products, pest control and disease reduction have become of major interest. Fig trees, in general, are tenacious and rather resistant to diseases. There are varying responses of varieties or *Ficus* species to some diseases. In some cases, the prevailing conditions may be the leading factor promoting disease incidence. Neglecting fig orchards will rarely lead to the death of the complete orchard but rather to drying out of some fig trees. In fact, many abandoned fig orchards are still thriving, despite being neglected. Fig trees usually die because of adverse soil conditions or root decay, rather than from a disease attacking its aerial organs (Condit, 1947). Sunburn can also cause decay of fig trees, mostly during winter when the tree sheds its leaves and the sunlight shines directly on the trunk and branches. In orchards where the main scaffolds are not established properly or not well pruned, there may also be severe damage of the trunk. In well-managed orchards, this will be prevented traditionally by painting the exposed parts white. Moss is another phenomenon that is prevented by proper orchard treatment. In neglected orchards, lichens may serve as the habitat for insects or fruit-decay organisms. Guglielmi (1908) described Californian figs with yellowing and falling leaves in mid-summer as suffering from "apoplexy." This disease is yet not explained, but is probably a physiological process.

Some phenomena causing malformation of fig fruits are not disease related, their causal microorganism has not yet been identified, or they may be triggered only under specific climatic conditions. Fig-fruit dropping and splitting, for example, are problems traditionally not related to any known pathogen. In most cases, fig dropping is related to an unfavorable climate for the given variety, especially if it affects the pollination process in Smyrna and San Pedro types of fig. In this case, the tree should be replaced by a variety known to fruit well in the given conditions. Fig splitting has received numerous explanations and has been attributed to: abundant sugar secretion; abnormal water relationships in the fruit and leaves; physical difficulties in withstanding sap pressure; and low temperature and increased humidity. However, as in many other fruits, splitting occurs when the internal pressure of the fruit is greater than the resistance (linked also to the elasticity) of the skin. Under high temperatures and dry conditions, the calcium that is transported through the

* Email: vhmoshea@agri.gov.il

DOI: 10.1079/9781789242492.0017

xylem goes preferably to the leaves because of the greater evapotranspiration rate of the leaf, changing the ratio of potassium (and magnesium) to calcium. Thus, factors elevating the internal pressure, excess irrigation, high humidity, high nitrogen fertilization, or overpollination resulting in too many seeds become important. Factors affecting skin resistance such as the translocation and presence of adequate calcium in the fruit prior to the third and rapid growth phase also have a marked influence on splitting.

Although rather rare and not severe, multiple diseases have been identified in figs. In general, first-crop figs tend to have more fungal decay than main-crop figs (Doster and Michailides, 2007). Here, we will suggest a general guide for fig diseases based on their source—either fungal or bacterial (Table 17.1); the role of viruses is discussed elsewhere (Chapter 16, this volume).

Generally, the most common and important pathogens and pests of figs are: (i) bacterial diseases, such as bacterial canker (*Phytomonas fici*) and crown gall (*Agrobacterium tumefaciens*); (ii) fungal diseases, including leaf rust (*Physopella fici*), leaf spot (*Cylindrocladium scoparium* or *Cercospora fici*), smut (*Aspergillus niger*), blast of branch terminals (*Botrytis* species), web blight (*Thanatephorus cucumeris*/*Rhizoctonia solani*), anthracnose (*Glomerella cingulata*/*Colletotrichum gloeosporoides*), *Nectriella* stem gall (*Nectriella pironii*/*Kutilakesa pironii*), root rot (*Armillaria tabescens* and *Armillaria mellea*), and pink limb blight (*Erythricium salmonicolor*); (iii) nematodes, such as root-knot nematodes (*Meloidogyne* species), and lesion nematodes (*Pratylenchus musicola*); and (iv) insects, including stem borers (*Bactrocera rufomaculata*), fig wax scale (*Ceroplastes rusci*), fig psylla (*Homotoma ficus*), fig weevil (*Otiorhynchus davricus*), and fig bark beetle (*Hypoborus ficus*).

17.2 Fungal Diseases

Fungi cause most plant diseases. The taxonomy of these pleomorphic organisms remained unclear for many years, which led to a confusing dual-nomenclature system by plant pathologists. The emergence of DNA sequencing revolutionized the taxonomy of this field and clarified relationships between fungi. DNA sequence comparisons also helped to resolve the dual-nomenclature problem and resulted in a more natural classification for fungi (Wingfield *et al.*, 2012). Nevertheless, plant pathogenic fungi identification relies on microscopic observations.

Important fig diseases include *Armillaria* root rot, endosepis, *Phomopsis* canker, *Rosellinia* root rot, and fruit rots caused by agents such as *Aspergillus* and *Phytophthora*. These major diseases and some other minor ones are described in this chapter.

17.2.1 Leaf rust

Fig rust is mostly caused by the fungus *Physopella fici* (McKenzie, 2013). The rust spores are wind-borne and spread by rain splashing on to other leaves and trees (Parthasarathy *et al.*, 2019). Rust spores land on the leaf and fruit surface and germinate within at least 14 h of continuous wetness (McKenzie, 2013). A few days later, they penetrate the leaf tissue, and the first symptoms of the disease in the shape of small yellowish spots start to develop on the upper leaf surface, followed by reddish-brown pustules, which become visible on the underside of the leaves (Latinovic *et al.*, 2015). These blister-like pustules also appear on the fruit skin. Pustules are typically clustered on the leaf edges or in areas where dew collects (McKenzie, 2013). Later, the rust-infected parts form angular brown spots on the lower leaf surface and form necrotic areas, followed by premature shedding of the leaves. Fruits growing on such trees, suffering from premature massive defoliation, will not be able to fully ripen. Windy and rainy weather promote spread and increase germination of spores (Fig. 17.1) (Parthasarathy *et al.*, 2019). Rust spores survive well on fallen leaves and thus become an inoculum source for new infections.

Cerotelium fici fungus also causes infected leaves to twist upwards with rusty and necrotic areas, and finally to dry out, leaving bare branches and unripe fruits (Latinovic *et al.*, 2015). When more abundant, it is characterized by the appearance of yellow- to brown-colored spots on the upper side and reddish-brown pustules on the lower side of the leaves. This disease favors humid and warm conditions (Latinovic

Table 17.1. A field guide to fig diseases.

Causal microorganism	Common name	Targeted part	Commonness	Result	Geographical spread	Season	Control measures
Fungi							
Armillaria mellea	Oak-root fungus	Root	Rare; more common in areas recovered from forests		California (USA), France, Algeria, Malta, Turkey		Avoid establishing orchard where there has been previous root rot
Rosellinia necatrix Prill. (*Dematophora necatrix* Hart.)	*Dematophora* root rot	Root	Widespread; attacks a wide range species	Sudden leaf loss, death of a single tree, and then spreads toward neighboring trees	Turkey (commercial dried-fig region), Greece	Noticeable at the beginning of vegetation as retarded growth; then sudden yellowing of leaves and early leaf fall	*Trichoderma harzianum* is an antagonist
Ganoderma sessile	Root rot	Root			Argentina		
Phymatotrichum omnivorum	Cotton root rot	Root	Common	Root rot, loss of cuttings	Arizona, Texas (USA)		
Corticium laetum	Limb blight	Branches		Leaf death	Louisiana (USA)		
Corticium salmonicolor	Limb blight	Branches		Limb death	Florida (USA)		Pruning
Sclerotinia (S. sclerotiorum?)	Limb blight	Branches	Minor to serious	Branch and trunk death, soft branches	California, Texas (USA)		
Botrytis cinerea	Limb blight	Branches		Mycelial growth into the wood, branch dieback	California (USA), France, UK		
Sclerotium rolfsii	Blight	Trunk	Occasional		Florida (USA)		
Botrytis species	Blight	Branches	Saprophytic	Fruit decay, twig death	UK	Winter	
Botryosphaeria ribis	Twig blight	Branches and twigs		Defoliation, crop loss	Texas (USA)		
Corticium stevensii	Thread blight	Leaves and branches		Defoliation, crop loss	Louisiana (USA)	June	Spray with Bordeaux mixture, pruning
Libertella ulcerata	Canker	Branches	Frequent		UK		

Continued

Table 17.1. Contiuned.

Causal microorganism	Common name	Targeted part	Commonness	Result	Geographical spread	Season	Control measures
Phoma cinerescens	Canker	Branches			UK		
Phomopsis cinerascens	Canker	Branches	Frequent, especially in mountainous areas; triggered by open wounds caused while pruning and by frost damage	Drying out of branches, fruit yield reduction	Turkey, Italy, Brazil, California (USA)		Avoid pruning wounds, remove infected branches, spray with Bordeaux mixture during dormant season
Macrophoma fici	Canker	Branches		Crop loss, bark rupture	Texas, Louisiana (USA)		Pruning
Tubercularia fici	Canker	Branches	Rare	Open wounds, yield reduction	Louisiana, Georgia (USA)		
Septobasidium pedicellatum	Canker	Branches	Common		Mississippi (USA)		
Stilbum cinnabarinum	Stilbum	Branches	Severe to mild, variety dependent	Dead twigs, tree death	Louisiana (USA)		Pruning, infected old trees should be destroyed
Cercospora bolleana	Blotch	Leaves	Rare	Spots 5–15 mm diameter	France		
Cephalosporium acremonium	Blotch	Leaves	Rare	Spots	Louisiana (USA)		
Cercospora fici	Blotch	Leaves	Common	Causes the leaves to fall	Texas (USA)	Late summer	
Fusarium roseum	Blotch	Leaves		Yellow spots			
Phyllosticta fici-carici	Blotch	Leaves	Common	Dark brown spots	Caucasus countries		
Mycosphaerella bolleana	Leaf blotch	Leaves	Common	Yellowing leaves, early defoliation	Mediterranean, Australia, Argentina, UK, Texas (USA)		

Rhizoctonia microsclerotia	Leaf blight	Leaves		Leaf upper parts turn silvery white, lower surface turns brown; fruits are overgrown	Florida (USA)		
Physopella fici	Rust	Leaves	Common	Complete defoliation, fruit sunburn	Worldwide	July–September, November–January	Spray Bordeaux mixture (5:5:50)
Aspergillus niger and other black-spored *Aspergillus* species	Smut	Fruit	Carried by wind, dust, dried-fruit beetle, fruit flies, thrips, and *Blastophaga* species	Internal black mold; some toxigenic strains may cause ochratoxins	Worldwide	Mid-May and mid-June	Spray with 1% lime sulfur
Aspergillus flavus, *A. parasiticus* and other *Aspergillus* ssp.		Fruit	Saprophytic; carried by wind, dried-fruit beetle, fruit flies, and thrips	Some toxigenic strains may cause aflatoxins	Worldwide	Mid-May and mid-June	
Fusarium moniliforme var. *fici*	Endosepsis	Caprifig	Carried by *Blastophaga* species	Internal rot: first brown color inside fruit cavity, then pinkish coloration around the eye and droppings of exudate from the ostiole; formation of fumonisins	Turkey	Infection through caprification, mold growth only after sugar accumulation, fumonisin formation starts at fresh ripe stage	Use of clean and healthy caprifigs, removal of caprifigs with brown or darker coloration in the fruit
Alternaria tenuis	Black spot	Fruit	Common in high humidity conditions	Spots on fresh fruit	Worldwide		Good aeration in areas with relatively high humidity: larger tree distances, row alignment according to wind dominance, pruning; lowering humidity in storage

Continued

Table 17.1. Contiuned.

Causal microorganism	Common name	Targeted part	Commonness	Result	Geographical spread	Season	Control measures
Alternaria alternata	Fruit rot	Fruit		Fruit rot	Montenegro		
Cladosporium herbarum	Black spot	Fruit	Common	Spots on fresh fruit	Worldwide		Good aeration in relatively high humidity areas; lower humidity in storage
Botrytis cinerea	Soft rot	Fruit			UK, Turkey		Good aeration in relatively high humidity areas
Choanephora cucurbitarum	Fruit rot	Fruit		Decay	Texas (USA)		
Glomerella fructigena	Anthracnose	Fruit		Premature fruit fall, rot, whole crop destruction	Louisiana (USA)		
Glomerella cingulata	Bitter rot	Fruit		Premature fruit fall, rot, whole crop destruction	Florida (USA)		
Rhizopus nigricans	"Common bread mold"	Fruit	Common	Fruit is sour, soft, and rotten	California (USA)	Summer with rainy spells during ripening	Good aeration in relatively high humidity areas
Phytophthora palmivora	Rot	Fruit		Fruit spoilage	Japan		
Phytophthora fici	Rot	Fruit			India		
Phytophthora carica	Rot	Fruit			Hawaii (USA)		
Bacteria and yeasts							
Phytomonas tumefaciens	Crown gall	Branches	Rare		Fresno, California (USA)		
Bacterium fici	Bacteriosis/ gummosis	Root and branches	Common	Leaf and fruit fall	Italy	July–September	
Saccharomyces or *Candida* species	Insect infestation	Fruit	Common; carried by insects: dried-fruit beetle, fruit flies, thrips, and mites	Souring	Worldwide		Control of dried-fruit beetle and other carriers

Fig. 17.1. A. Fig rust on leaf, B. stem, and C. fruit caused by *Physopella fici* (original).

et al., 2015). *C. fici* may also appear on fig fruits in tropical and subtropical regions, causing yield loss (Verga and Nelson, 2014).

17.2.2 Leaf spot

C. fici is the most common of the four *Cercospora* leaf spots on fig and is one of a range of other fig leaf-spotting microorganisms (Farr *et al.*, 1989; Alfieri *et al.*, 1994). *C. fici* has been widespread for decades and occurs wherever figs are grown (Chupp, 1953; Farr *et al.*, 1989). Initially, the leaf spots are reddish-brown and angular to irregular in shape, with a slight, diffused yellow halo. Individual leaf spots are up to 8 mm across and may unite, creating larger spots (Schubert *et al.*, 1999). Under favorable hot and humid conditions, like summer rains, the fungus spreads rapidly within leaves. Severe infection can result in leaf drop (Schubert *et al.*, 1999; Simone and Mullin, 1999).

17.2.3 Shoot blight

Shoot blight has been detected in Turkey, caused by *Lasiodiplodia theobromae* (Pat.) Griffon & Maubl. (synonym *Botryodiplodia theobromae*), the anamorph of *Botryosphaeria rhodina* Berk. & Curt. Arx. The fungus creates white colonies that turn grey to black with age and form black pycnidia (Çeliker and Michailides, 2012). The pycnidiospores created are oval and greenish brown, with a septum in the middle, with the dimensions 20.9–27.5 × 11.0–15.4 µm. This disease is rarely observed and was restricted to occurrences in one or two varieties of fig in Turkey. However, this pathogen is also known to cause cankers and dieback on grapevine in Italy (Khanzada *et al.*, 2004), on mango in Sindh Pakistan (Ko *et al.*, 2004), on kumquat in Taiwan (Burruano *et al.*, 2008), and on cacao in Cameroon (Mbenoun *et al.*, 2008).

Another fungus causing twig blight is *Neofusicoccum parvum*. Symptomatic fig cuttings infected by *N. parvum* usually demonstrate twig blight, subcortical discoloration, and apical dieback (Aiello *et al.*, 2020).

17.2.4 Root rot

Armillaria

Armillaria (Basidiomycota, Physalacriaceae) is a genus of ecologically and economically important wood and root-rot fungi that is widely distributed across a range of climates worldwide and impacts more than 500 host species, including fig (Hood *et al.*, 1991; Devkota and Hammerschmidt, 2020). *Armillaria* species include important causal agents of tree root rots. The species boundaries of the *Armillaria* root-rot pathogens and their distribution are not yet clear, and genetics-based taxonomy is much needed (Coetzee *et al.*, 2018).

Species from the family *Physalacriaceae* (Basidiomycota) cause the root white rot disease that weakens and often kills woody plants (Akulova *et al.*, 2020). *Armillaria borealis* Marxm. & Korhonen is a fungus belonging to this family, widely distributed in Eurasia, including Siberia and the Far East. Climatic changes can weaken trees and make fungi and bacteria more destructive (Akulova *et al.*, 2020). *A. borealis* is very aggressive but is less pathogenic than other

Armillaria fungi such as *Armillaria ostoyae*. *A. borealis* acts as a secondary pathogen, killing trees already weakened by other factors. Climatic changes can weaken trees and make them more prone to fungal and bacterial attack causing more destruction (Akulova *et al*., 2020).

Armillaria mellea and *Armillaria solidipes* are the two primary causal agents of *Armillaria* root rot (Devkota and Hammerschmidt, 2020). However, despite extensive research on this disease over the past few decades, there remains a lack of knowledge of the fungal infection and host response. A mucilaginous substance produced at the rhizomorph tip of *A. mellea* and *A. solidipes* is thought to aid attachment of the pathogen to the host. Toxins and cell-wall-degrading enzymes produced by the fungus together with mechanical pressure generated by the rhizomorph are speculated to support fungal penetration. However, a detailed description of the *Armillaria* species infection process is still lacking (Devkota and Hammerschmidt, 2020).

A. mellea is traditionally used as a Chinese medicine and has been proven to possess many biological activities, including anti-oxidation, anti-vertigo and anti-aging biological effects. *A. mellea* has also an anti-inflammatory activity (Geng *et al*., 2017).

Dematophora *root rot*

Dematophora root rot is caused by *Rosellinia necatrix*, a fungus in the family Xylariaceae. It damages a wide range of species in addition to figs and is widespread across various parts of the world. It is reported to be present in rose, apple, tea, almond, cassava, avocado and others (Anon., 2021). It causes die-out of fig trees completely. It may start with a single fig tree in the orchard and then spread from that single point to other neighboring trees. On slopes, it can be carried downward with the water flow. Signs appear with wilting and yellowing of the leaves, which then turn brown and dry out. The cortex becomes soft and the root wood rots. Papachatzis *et al*. (2008) reported the presence of both *A. mellea* and *R. necatrix* in fig orchards of the Smyrna type of fig; the symptoms were very similar and, in addition to the leaf symptoms, splitting was seen in the bark and white mycelia were visible under the bark. The incidence of both root rots is higher in fig orchards near forests. Şahin *et al*. (2013) stated that white mycelia were also found in the soil surrounding the root crown and that high soil moisture content and poor drainage advanced root-rot damage. The fungus may remain in the soil for many years, and therefore it is recommended not to replant fig trees in soils where fig trees have dried out due to root rot (Fig. 17.2).

17.2.5 Fig endosepsis

Fig endosepsis, also called "pink rot," "brown rot," "soft rot," and "eye-end rot," is named after the internal rot characteristics of this disease (Subbarao and Michailides, 1992). Almost 100 years ago, Caldis (1927) recognized *Fusarium moniliforme* var. *fici* as the main cause of fig endosepsis, based on some similarities to the *F. moniliforme* detected in maize (Michailides and Ferguson, 2009), but other species of *Fusarium* can also cause endosepsis. The fig inflorescence, called the syconium, requires pollination to produce fruits. This required pollination is assisted by the fig wasp (*Blastophaga psenes*) (Subbarao and Michailides, 1992). Naturally, fig wasps transport and spread fungal propagules from fig to fig (Michailides and Morgan, 1998), mainly during spring, the time when they pass from one crop to the other in males and pollinate the female fig fruit. Wasps carrying pollen and the fungus enter the green fruits to lay eggs. Infection of the pulp occurs later, when the fruit starts ripening and accumulates sugars. Therefore, parthenocarpic varieties (non-pollinating figs) only occasionally develop fig endosepsis (Michailides and Ferguson, 2009).

Three different species of *Fusarium* (*Fusarium dimerum*, *Fusarium solani*, and *F. moniliforme*) were identified in cultivated and wild caprifigs with endosepsis symptoms in California (Subbarao and Michailides, 1993) and Turkey (Yildiz *et al*., 2008). Identification of the different species is facilitated by the illustrated manual of Nelson *et al*. (1983) for identification of *Fusarium* species.

Infected fruits show discolored areas on the base of the flowers in the shape of internal strips of pink or brown. As the figs ripen, rusty-colored spots appear on many flowers, mostly near the

Fig. 17.2. A. A fig tree typical die-out caused by *Rosellinia* infection. B. Brown lesions on the root. Photographs by Ozlem Dogan, Erbeyli Fig Research Institute, with permission.

eye of the fig, but can also develop on any part of the pulp. As the fruit matures and softens, the pulp starts to look water soaked, mainly around the ostiole. The water-soaked area eventually extends up the sides to the neck and turns purple in color (Michailides and Ferguson, 2009). Even partial infection of the pulp causes off-flavor of the fruit.

Studies of three different *Fusarium* species (*Fusarium ramigenum*, *F. solani*, and *Fusarium proliferatum*) showed that only strains of *F. proliferatum* were fertile (Moretti *et al.*, 2010). *Fusarium* species produce different toxic compounds such as fusaric acid, beauvericin, fumonisin (B_1 and B_2), and fusaproliferin. All three of these *Fusarium* species are pathogenic, with *F. ramigenum* the most pathogenic (Moretti *et al.*, 2010).

17.2.6 Fig-fruit rots

Fungal agents such as *Aspergillus* and *Phytophthora* commonly cause fig-fruit rots. At least 16 different *Aspergillus* species have been found in decaying first-crop figs (Doster and Michailides, 2007). The symptom first to appear in *Phytophthora palmivora* infection is a water-soaked lesion covered with a white, fluffy mycelium, followed by a soft, mushy rot of the infected area on the fruit (Zhang *et al.*, 2013). *P. palmivora* is one of the most severe pathogens of edible figs (El-Gholl and Alfieri, 1985).

17.2.7 Fig smut

Fig smut is a serious disease caused by *Aspergillus niger* and related fungi such as *Aspergillus japonicus*, *Aspergillus carbonarius*, and other black-spored *Aspergillus* species. Fig smut is usually present at approximately the same level in first- and main-crop figs (Doster and Michailides, 2007). Saprophytic *Aspergillus* species are known to produce the mycotoxins aflatoxin and ochratoxin A, which are also found in decaying first-crop figs. At least 16 different *Aspergillus* species have been found to decay first-crop figs. *A. niger* and other related fungi are usually detected in both first- and main-crop figs. However, as spores produced on the first-crop figs can infect main-crop figs, the fungal decay of first-crop figs can result in higher levels of decay on main-crop figs (Doster and Michailides, 2007). When carried with the fig wasp during caprification,

A. niger spores may develop just inside the fruit cavity and are invisible from the outside. As these are mostly resistant to drought, they especially develop in hot and dry conditions and continue developing during the drying phase. Black spores inside the dried fig fruit may cause problems in trade.

17.2.8 Anthracnose

Species of *Colletotrichum* are associated with anthracnose of a wide range of host plants including fig fruits. *Colletotrichum siamense sensu lato* was identified on figs (*Ficus racemosa*) as well as on other tropical fruits and was the dominant group of *Colletotrichum* species among most wild fruits studied (Udayanga *et al.*, 2013). There is a need to re-assess the evolutionary relationships of *Colletotrichum* species occurring on cultivated and wild fruits with emphasis on their ecology and cryptic diversification including sampling at regional and global scales (Udayanga *et al.*, 2013).

Colletotrichum gloeosporioides (Penz.) Penz. & Sacc. was detected in Korea (Choi *et al.*, 2013), where figs were only introduced during the 1990s (Ma *et al.*, 2008). Early symptoms of infection appeared as small, slightly sunken, water-soaked, and circular spots. Later, injuries increased in size and became more depressed, with a darkened central area. A semi-soft decay, sometimes causing premature fruit drop, was the result of this process. The symptoms were restricted to the fruits, as no symptoms were found on leaves. The infected parts were circular to ellipsoid in shape and 50–400 μm in diameter (Weir *et al.*, 2012; Choi *et al.*, 2013). Anthracnose fruit rot of common figs may cause losses not only for fruit production but also for marketing (Choi *et al.*, 2013).

17.2.9 *Alternaria* rot

Edible fig varieties potentially have two crops. The first crop, known as breba and generally less important commercially, ripens in late spring or early summer. The second crop, which is the main one, ripens in late summer. The first-crop figs typically are unharvested and are left in the orchard if fig varieties are commercially grown for their second crop. Thus, relatively, first-crop figs tend to have more fungal decay than main-crop figs, especially *Alternaria* rot (caused by *Alternaria alternata* and *Ulocladium atrum*) (Doster and Michailides, 2007) (Fig. 17.3).

17.3 Bacterial and Yeast Diseases

17.3.1 Canker

Canker, identified as a cause of mortality in *Ficus* trees, is a non-specific symptom caused by a wide range of plant pathogens. Canker may lead to yellowing of leaves, leaf loss, crown thinning, and dieback of branches. Eventually, the whole tree dies within a couple of years of infection (Al-Bedak *et al.*, 2018). In 2018, *Neofusicoccum parvum* was reported to cause canker on figs showing twig blight, subcortical discoloration, and apical dieback (Aiello *et al.*, 2020). Out of the 1500 fig cuttings tested in Catania Province,

Fig. 17.3. *Alternaria* rot caused by *Alternaria alternata* and *Ulocladium atrum* (original).

Sicily (Italy), up to 20% of the plants showed canker and twig blight. *N. parvum* leads to the death of all infected trees (Aiello *et al.*, 2020).

Fig limb dieback is a cosmopolitan disease caused by *Neoscytalidium dimidiatum* (Botryosphaeriaceae), characterized by branch and shoot cankers, discoloration of woody tissues, and dieback. *N. dimidiatum* induces cankers on fig, mainly on wounded shoots. In the last several years, the disease has become prevalent among fig orchards in California (Gusella *et al.*, 2020). Mechanically stressed shoots are more susceptible to infection, and thus summer infections are more severe than winter infections. Figs that require pollination for fruit development are less susceptible to canker than varieties with a persistent fruiting habit. A simple solution is to remove canker-affected branches from the orchard, which may stop the spread of the disease (Gusella *et al.*, 2020).

Different fig species show different resistance to canker. *Ficus carica* is highly susceptible to *Ceratocystis* canker, caused by *Ceratocystis fimbriata* Ellis & Halsted, while the wild fig species *Ficus erecta* Thunb. is resistant (Yakushiji *et al.*, 2012). Interspecific hybridization was performed between the common fig (seed parent) and *F. erecta* (pollen parent) using artificial pollination. When cuttings of all three fig varieties—common fig, *F. erecta*, and their hybrid—were inoculated with *C. fimbriata*, the results showed that common fig varieties died within 10 weeks, while *F. erecta* and hybrids prospered even after 14 weeks (Yakushiji *et al.*, 2012).

Phomopsis *canker* (Phomopsis cinerascens)

The genus *Phomopsis* (teleomorph *Diaporthe*) is distributed worldwide with diverse host associations (Udayanga *et al.*, 2011). Accurate *Phomopsis* species identification is critical for plant disease epidemiology studies and control. Rapid progress in molecular identification has revolutionized disease taxonomic studies and has provided definite species boundaries (Udayanga *et al.*, 2011). Udayanga *et al.* (2011) published an account of the names of phytopathogens currently in use, along with their taxonomic entry, teleomorph, associated hosts, and disease symptoms, including brief summaries of taxonomic and phylogenetic research, which can be very helpful in *Phomopsis* species recognition in infected trees.

Fig branch canker is a major disease in most parts of Iran, especially in rainfed fig plantations throughout Fars Province (Javadi and Banihashemi, 2005; Banihashemi and Javadi, 2009). Infection by the causal agent results in canker formation and killing of young and old branches. Banihashemi and Javadi (2009) isolated *Phomopsis cinerascens* from the cankers. *P. cinerascens* fungus produced pycnidia containing α-conidia on active cankers from autumn to mid-spring. No β-conidia were found under natural conditions. The optimal temperature for fungal infection of branches was 15–25°C, while no infection was detected at lower (5°C) or higher (30°C) temperatures. Thus, infection occurred on branches mostly from autumn to mid-spring. The fungus was pathogenic only on detached or intact fig branches. The pruning wounds remain receptive to the pathogen from autumn to mid-spring. When exposed to humidity, new occurrences of canker were observed on the cut branches (Banihashemi and Javadi, 2009). The fungus overwintered under natural conditions only as dormant mycelium on infected branches, which produced new pycnidia bearing spores that were exuded in white to creamy cirri (Javadi and Banihashemi, 2005). Overwintered pycnidiospores produced in the previous season did not survive during the dormant period. Pruning cuts made in early autumn remained receptive to the pathogen until mid-spring, and the pathogen could not infect plants in late spring or summer during the active growth of the host plants (Javadi and Banihashemi, 2005).

In Turkey, *Phomopsis* canker affects Sarılop (= Calimyrna) variety fig trees located at higher elevations where vegetative growth is relatively less and damage risk due to frost is high. Wounds resulting from frost, hail, sunburn, or pruning interfaces allow the entrance of the fungus. Application of excess nitrogen may trigger frost damage and thus canker formation. The first signs appear as swelling and darkening of the shoots, followed by a dark-colored exudate, and the leaves turn yellow and then dry out. Precautions must be taken to prevent spread of the fungus. Damaged branches should be removed and in cases where the whole tree is affected, the tree has to be uprooted. Taking propagation material must be avoided. Pruning agent should be applied to all wounds as a preventive measure, and in the case of canker these branches must be treated with a fungicide (Şahin *et al.*, 2013).

17.3.2 Crown gall

Crown gall is a plant tumor disease caused by the specific action of the bacterium *Agrobacterium tumefaciens* (de Cleene and de Ley, 1976). Crown gall tumors result from overproduction of the phytohormones auxin and cytokinin specified by *A. tumefaciens* T-DNA genes (Gelvin, 1990). In their extensive study, de Cleene and de Ley (1976) reviewed the susceptibility of 1193 plant species, belonging to 588 genera and 138 families, to crown gall and found that almost 60% of the gymnosperms and the dicotyledonous angiosperms examined were sensitive for diseases caused by crown gall. Their findings undermine the common belief that crown gall susceptibility is limited to members of the orders Liliales and Arales (de Cleene and de Ley, 1976). *A. tumefaciens* was found to cause crown gall in figs grown in the Jordan Valley and uplands of Jordan, where it was widespread (Khlaif *et al.*, 2018). In Isfahan and Fars Provinces of Iran, a total of 50 strains of *A. tumefaciens* were isolated from diseased *Ficus* plants and their morphological, molecular, and biochemical characteristics were published (Aeini *et al.*, 2014).

17.3.3 Bacterial leaf spot

Pseudomonas ficuserectae is a bacterium producing dark brown, water-soaked spots on the leaves and stems of *F. erecta* Thunb., resulting either in defoliation or shoot blight. It is similar to *Pseudomonas amygdali* in many properties (Goto, 1983). However, the main differences between the two *Pseudomonas* species include size of colonies, production of hydrogen sulfide, utilization of saccharides such as ribose, raffinose, mannitol, sorbitol, and malonate, and their pathogenicity. The DNA base composition of *P. ficuserectae* is mostly guanine plus cytosine.

17.4 Concluding Remarks

Compared with many other fruit species, *F. carica* is relatively resistant to diseases, partially due to the toxic compounds in the latex, in both the fruit and leaves. Large amounts of ficin isoforms and cysteine protease are detected in fig latex, as well as isoforms of trypsin inhibitor (Kitajima *et al.*, 2018). These compounds are lethal to *Lepidoptera* caterpillars (Konno *et al.*, 2004), fungi (Huynh *et al.* 1992; Terras *et al.*, 1993; Karnchanatat *et al.*, 2011; López-García *et al.*, 2012), and insects (Hilder *et al.* 1987). The relative amounts of ficins to trypsin inhibitors differ between immature fruit latexes and young petioles (Kitajima *et al.*, 2018), suggesting that these organs have adopted different defense strategies. This diversity might relate to different pests threatening different parts of the plant. In this way, the hardened latex of the trunk protects the plant mainly from microbes, and that of the leaf protects it mainly from herbivores, while the fruit latex has dual interests—it needs not only to protect the fruit from threats but also to attract pollinators (insects) at the early stages, and distributors (animals) after ripening. This is reflected in the chemical composition of the toxins produced by each part (Kitajima *et al.*, 2018).

Climatic changes can weaken trees and make fungi and bacteria more destructive (Akulova *et al.*, 2020). Pathogenic fungi are currently one of the main concerns in forest conservation.

There is also a fundamental problem with the identification and classification of different fungal and bacterial species (Aiello *et al.*, 2020). Defining species boundaries based on DNA sequence-inferred phylogenies has been a central focus of contemporary mycology. It seems that genetic study of repetitive sequences and their comparative analysis will help accurate taxonomy of the different species and reveal their evolutionary relationships. The results of such studies have already in many cases resolved the biogeographic history of species, the mechanisms involved in their dispersal, the taxonomy of species, and how certain phenotypic characteristics have evolved throughout lineage diversification (Coetzee *et al.*, 2018, and references therein).

References

Aeini, M., Mirzaee, H., Taghavi, S.M., Khodakaramian, G.R., and Mazhar, M.A. (2014) Occurrence of crown gall disease on *Ficus benjamina* in Fars and Isfahan provinces of Iran. *Archives of Phytopathology and Plant Protection* 47, 2257–2262.

Aiello, D., Gusella, G., Fiorenza, A., Guarnaccia, V., and Polizzi, G. (2020) Identification of *Neofusicoccum parvum* causing canker and twig blight on *Ficus carica* in Italy. *Phytopathologia Mediterranea* 59, 213–218.

Akulova, V.S., Sharov, V.V., Aksyonova, A.I., Putintseva, Y.A., Oreshkova, N.V., *et al.* (2020) *De novo* sequencing, assembly and functional annotation of *Armillaria borealis* genome. *BMC Genomics* 21: 534.

Al-Bedak, O.A., Mohamed, R.A., and Seddek, N.H. (2018) First detection of *Neoscytalidium dimidiatum* associated with canker disease in Egyptian *Ficus* trees. *Forest Pathology* 48: e12411.

Alfieri, S.A. Jr., Langdon, K.R., Kimbrough, J.W., El-Gholl, N.E., and Wehlburg, C. (1994) *Diseases and Disorders of Plants in Florida*. Bulletin 14. Florida Department of Agriculture and Consumer Services, Division of Plant Industry, Gainesville, Florida.

Anon. (2021) *Rosellinia necatrix* (dematophora root rot). In: *Invasive Species Compendium*. CAB International, Wallingford, UK. Available at: www.cabi.org/isc/datasheet/47860#REF-DDB-102949 (accessed December 16, 2021).

Banihashemi, Z. and Javadi, A.R. (2009) Further investigations on the biology of *Phomopsis cinerascens*, the cause of fig canker in Iran. *Phytopathologia Mediterranea* 48, 454–460.

Burruano, S., Mondello, V., Conigliaro, G., Alfonzo, A., Spagnolo, A., and Mugnai, L. (2008) Grapevine decline in Italy caused by *Lasiodiplodia theobromae*. *Phytopathologica Mediterranea* 47, 132–136.

Caldis, P.D. (1927) Etiology and transmission of endosepsis (internal rot) of the fruit of the fig. *Hilgardia* 2, 287–328.

Çeliker, N.M. and Michailides, T.J. (2012) First report of *Lasiodiplodia theobromae* causing canker and shoot blight of fig in Turkey. *New Disease Reports* 25, 12–13.

Choi, I.Y., Park, J.H., Cho, S.E., and Shin, H.D. (2013). First confirmed report of anthracnose fruit rot caused by *Colletotrichum gloeosporioides* on common fig in Korea. *Plant Disease* 97, 1119.

Chupp, C. (1953). *A Monograph of the Fungus Genus* Cercospora. Ithaca, New York, 1964 pp.

Coetzee, M., Wingfield, B.D., and Wingfield, M.J. (2018) *Armillaria* root-rot pathogens: species boundaries and global distribution. *Pathogens* 7:83.

Condit, I.J. (1947) *The Fig*. Chronica Botanica Co., Waltham, Massachusetts.

de Cleene, M. and de Ley, J. (1976) The host range of crown gall. *Botanical Reviews* 42, 389–466.

Devkota, P. and Hammerschmidt, R. (2020) The infection process of *Armillaria mellea* and *Armillaria solidipes*. *Physiological and Molecular Plant Pathology* 112: 101543.

Doster, M.A. and Michailides, T.J. (2007) Fungal decay of first-crop and main-crop figs. *Plant Disease* 91, 1657–1662.

El-Gholl, N.E., and Alfieri, S.A. Jr. (1985) Fruit rot of fig caused by *Phytophthora palmivora*. In: *Proceedings of the Annual Meeting of the Florida State Horticulture Society*, Vol. 97, pp. 327–328.

Farr, D.F., Bills, G.F., Chamuris, G.P., and Rossman, A.Y. (1989) *Fungi on Plants and Plant Products in the United States*. American Phytopathological Society Press, St. Paul, Minnesota.

Gelvin, S.B. (1990) Crown gall disease and hairy root disease: a sledgehammer and a tackhammer. *Plant Physiology* 92, 281–285.

Geng, Y., Zhu, S., Cheng, P., Lu, Z.M., Xu, H.Y., *et al.* (2017) Bioassay-guided fractionation of ethyl acetate extract from *Armillaria mellea* attenuates inflammatory response in lipopolysaccharide (LPS) stimulated BV-2 microglia. *Phytomedicine* 26, 55–61.

Goto, M. (1983) *Pseudomonas ficuserectae* sp. nov., the causal agent of bacterial leaf spot of *Ficus erecta* Thunb. *International Journal of Systematic and Evolutionary Microbiology* 33, 546–550.

Guglielmi, G. (1908) Coltivazione industriale del fico nel Leccese. In: *Boll. Arbor. Italiana*, Giannini and Figli, Naples.

Gusella, G., Morgan, D.P., and Michailides, T.J. (2020) Further investigation on limb dieback of fig (*Ficus carica*) caused by *Neoscytalidium dimidiatum* in California. *Plant Disease* 105, 324–330.

Hilder, V.A., Gatehouse, A.M., Sheerman, S.E., Barker, R.F., and Boulter, D. (1987) A novel mechanism of insect resistance engineered into tobacco. *Nature* 330, 160–163.

Hood, I.A., Redfern, D.B., and Kile, G.A. (1991) *Armillaria* in planted hosts. In: Shaw, C.G. and Kile, G.A. (eds) *Armillaria Root Disease*. USDA Forest Service, Washington, DC, pp. 122–149.

Huynh, Q.K., Borgmeyer, J.R., and Zobel, J.F. (1992) Isolation and characterization of a 22 kDa protein with antifungal properties from maize seeds. *Biochemistry and Biophysics Research Communications* 182, 1–5.

Javadi, A.R. and Banihashemi, Z. (2005) Biology and pathogenicity of *Phomopsis cinerascens*, the causal agent of fig canker in Fars Province of Iran. *Acta Horticulturae* 798, 219–222.

Karnchanatat, A., Tiengburanatam, N., Boonmee, A., Puthong, S., and Sangvanich, P. (2011) Zingipain, a cysteine protease from *Zingiber ottensii* Valeton rhizomes with antiproliferative activities against fungi and human malignant cell lines. *Preparative Biochemistry and Biotechnology* 41, 138–153.

Khanzada, M.A., Lodhi, A.M., and Shahzad, S. (2004) Mango dieback and gummosis in Sindh Pakistan caused by *Lasiodiplodia theobromae*. *Plant Health Progress* 5, 13–18.

Khlaif, H., Abu-Obeid, I., and Werikat, B. (2018) Occurrence of plant bacterial diseases in Jordan. *African Journal of Agricultural Research* 13, 2104–2117.

Kitajima, S., Aoki, W., Shibata, D., Nakajima, D., Sakurai, N., *et al.* (2018) Comparative multi-omics analysis reveals diverse latex-based defense strategies against pests among latex-producing organs of the fig tree (*Ficus carica*). *Planta* 247, 1423–1438.

Ko, W.H., Wang, I.T., and Ann, P.J. (2004) *Lasiodiplodia theobromae* as a causal agent of kumquat dieback in Taiwan. *Plant Disease* 88, 1383–1383.

Konno, K., Hirayama, C., Nakamura, M., Tateishi, K., Tamura, Y., *et al.* (2004) Papain protects papaya trees from herbivorous insects: role of cysteine proteases in latex. *Plant Journal* 37, 370–378.

Latinovic, J., Radisek, S., and Latinovic, N. (2015) Severe infection of figs by fig rust pathogen *Cerotelium fici* in Montenegro. *Poljoprivreda i Sumarstvo* 61, 101–107.

López-García, B., Hernández, M., and Segundo, B.S. (2012) Bromelain, a cysteine protease from pineapple (*Ananas comosus*) stem, is an inhibitor of fungal plant pathogens. *Letters in Applied Microbiology* 55, 62–67.

Ma, K.C., Byeon, M.H., Bang, G.P., Ko, S.J., and Lee, Y.H. (2008) Incidences of *Phytophthora* fruit rot on fig according to the cultural practice and its eco-friendly control. *Research in Plant Disease* 14, 107–111.

Mbenoun, M., Momo Zeutsa, E.H., Samuels, G., Nsouga Amougou, F., and Nyasse, S. (2008) Dieback due to *Lasiodiplodia theobromae*, a new constraint to cocoa production in Cameroon. *Plant Pathology* 57, 381.

McKenzie, E. (2013) *Cerotelium fici*. In: *PaDIL Species Fact Sheet*. Available at: www.padil.gov.au/maf-border/Pest/Main/143006 (accessed December 13, 2021).

Michailides, T.J. and Ferguson, L. (2009) *Fig Endosepsis*. University of California, Davis, California.

Michailides, T.J. and Morgan, D.P. (1998) Spread of endosepsis in Calimyrna fig orchards. *Phytopathology* 88, 637–647.

Moretti, A., Ferracane, L., Somma, S., Ricci, V., Mule, G., *et al.* (2010) Identification, mycotoxin risk and pathogenicity of *Fusarium* species associated with fig endosepsis in Apulia, Italy. *Food Additives and Contaminants* 27, 718–728.

Nelson, P.E., Toussoun, T.A., and Marasas, F.O. (1983) *Fusarium Species: An Illustrated Manual for Identification*. Pennsylvania State University Press, University Park, Pennsylvania.

Papachatzis, A., Eliopoulos, P., Statharas, G., and Vagelas, I. (2008) *Ficus carica* root rot disease caused by *Armillaria mellea* and *Rosellinia necatrix* in Greece. *Universitătcedilla˜ii din Craiova – Biologie, Horticultura* 13, 143–148.

Parthasarathy, S., Thiribhuvanamala, G., Ramalingam, K., Jebaraj, M.D., and Prabakar, K. (2019) Incidence and ultramicroscopic characterization of *Cerotelium fici* (castagne) rust in *Ficus carica* L. *Journal of Applied Horticulture* 22, 24–26.

Şahin, B., Köseoğlu, İ., Özen, M., Doğan, Ö., and Tutmuş, E. (2013) Dried fig production. In: Aksoy, U. (ed.) *Kuru İncir Yetiştiriciliği ve Aflatoksin Yönetimi El Kitabi*. Basim, İzmir, Turkey.

Schubert, T.S., El-Gholl, N.E., and Alfieri, S.A. (1999) *Cercospora leaf spot of fig. Circular No. 394*. Florida Department of Agriculture & Consumer Services, Division of Plant Industry, Tallahassee, Florida.

Simone, G.W. and Mullin, R.S. (1999) 1999–2000 Florida plant disease management guide. In: *Fruit and Vegetables*. Institute of Food and Agricultural Sciences, Florida Cooperative Extension Service, University of Florida, Gainesville, Florida, pp. 89–90.

Subbarao, K.V. and Michailides, T.J. (1992) A re-evaluation of *Fusarium moniliforme* var. *fici*, the causal agent of fig endosepsis. *Mycological Research* 96, 766–768.

Subbarao, K.V. and Michailides, T.J. (1993) Virulence of *Fusarium* species causing fig endosepsis in cultivated and wild caprifigs. *Phytopathology* 83, 527–533.
Terras, F.R., Schoofs, H.M., Thevissen, K., Osborn, R.W., Vanderleyden, J., *et al.* (1993) Synergistic enhancement of the antifungal activity of wheat and barley thionins by radish and oilseed rape 2S albumins and by barley trypsin inhibitors. *Plant Physiology* 103, 1311–1319.
Udayanga, D., Liu, X., McKenzie, E.H., Chukeatirote, E., Bahkali, A.H., and Hyde, K.D. (2011) The genus *Phomopsis*: biology, applications, species concepts and names of common phytopathogens. *Fungal Diversity* 50, 189–225.
Udayanga, D., Manamgoda, D.S., Liu, X., Chukeatirote, E., and Hyde, K.D. (2013) What are the common anthracnose pathogens of tropical fruits? *Fungal Diversity* 61, 165–179.
Verga, A. and Nelson, S. (2014) Fig rust in Hawai'i. *Plant Disease* 100, 1–5.
Weir, B.S., Johnston, P.R., and Damm, U. (2012) The *Colletotrichum gloeosporioides* species complex. *Studies in Mycology* 73, 115–180.
Wingfield, M.J., de Beer, Z.W., Slippers, B., Wingfield, B.D., Groenewald, J.Z., *et al.* (2012) One fungus, one name promotes progressive plant pathology. *Molecular Plant Pathology* 13, 604–613.
Yakushiji, H., Morita, T., Jikumaru, S., Ikegami, H., Azuma, A., and Koshita, Y. (2012) Interspecific hybridization of fig (*Ficus carica* L.) and *Ficus erecta* Thunb., a source of *Ceratocystis* canker resistance. *Euphytica* 183, 39–47.
Yildiz, A., Benlioğlu, S., and Saribiyik, D. (2008) Fig endosepsis in some cultivated varieties. *Journal of Phytopathology* 156, 573–575.
Zhang, C., Zhang, W., Ma, H.Q., and Zhang, G.Z. (2013). First report of *Phytophthora palmivora* causing fruit rot of fig (*Ficus carica* L.) in China. *Plant Disease* 97, 1252.

18 Postharvest Pests of Dried Figs

Ferit Turanlı*
Department of Plant Protection, Faculty of Agriculture, Ege University, Bornova, İzmir, Turkey

18.1 Introduction

In regions under the influence of a Mediterranean climate, fig is an economically important agri-food product, and its demand is increasing due to its high nutritional value. The fig fruit can be consumed fresh, dried, or processed in other forms. Sun drying is the most widely utilized form as it extends the shelf life of a very delicate fruit, allowing global trading of over 50% of the world's annual fig production. One of the major threats for sun-dried figs in worldwide trade is infestation with storage pests. Storage-pest infestation starts in the orchard and thus prevention of contamination at preharvest, harvest, and postharvest stages means eliminating or minimizing many of the problems that may rise in the following processing and storage stages.

Dried figs are marketed globally as free from any live pests, and with regard to the damage caused by storage pests, all lots must fulfill the requirements in the standards. The marketing quality standard of the United Nations European Economic Commission (UNECE, 2016) requires that dried figs be "free from damage caused by pests, including the presence of dead insects and/or mites, their debris or excreta subject to the special provisions for each class and tolerances allowed." The tolerances for fruit damaged by pests or notified as defective fruit by weight or number are 9, 12, and 15% in Extra, Class I and II, respectively. A dried fig that may have just one hole, a fragment or some webbing is counted as a defective fruit and is subject to tolerance. There is a maximum limit for the total number of defective fruits, so heavy infestation alone may lead to rejection of the dried-fig lot (UNECE, 2016). The tolerance limits allowed may vary in different dried-fig standards. In the USA, the marketing standard on dried figs (USDA, 2001) does not have an article on pests; however, there are limitations imposed through the Food Defect Levels Handbook of the US Food and Drug Administration (FDA). A defect of the dried fig is described as "insect filth and/or dirty fruit or pieces of fruit" due to insect infestation at pre- or postharvest stages. The tolerance that applies to all quality classes of dried figs is stated as a maximum of 10% by count. Separate requirements apply for imports of fig paste (FDA, 2018). Moreover, the major postharvest pests may vary in different dried-fig-producing regions. In Turkey and Greece, the fig moths *Cadra cautella* (Fig. 18.1A, F–H) and *Plodia interpunctella* (Fig. 18.1B, C) are the major pests (Eliopoulos and Athanassiou, 2004; Şen *et al.*, 2009), whereas in California, dried-fruit beetles or sap beetles (*Carpophilus hemipterus* (most common),

* Email: ferit.turanli@ege.edu.tr

 Advances in Fig Research and Sustainable Production (M.A. Flaishman and U. Aksoy)
DOI: 10.1079/9781789242492.0018

Fig. 18.1. A. *Cadra cautella*, adult. B. *Plodia interpunctella*, adult. C. *Plodia interpunctella*, larva. D. *Oryzaephilus surinamensis*, adult. E. *Carpophilus hemipterus*, adult. F. Skin damage, webbing, and excreta of *C. cautella* on the outer surface. G. Webbing inside the dried fig. H. Damage and larva of *C. cautella*.

Carpophilus mutilatus, and *Carpophilus freemani*) cause the main problems for the dried-fig industry (Shorey *et al.*, 1989; Coviello and Bentley, 2009). These pests infest dried figs in the orchard and are carried to the storage and processing facilities, and wholesale and retail stages, if necessary precautions are not taken or if the ambient conditions permit. Methyl bromide was the only fumigant used at the postharvest stage against storage pests until its phase-out according to the Montreal Protocol. The phase-out was January 1, 2005, in developed countries, 2007 in Turkey, and 2015 in developing countries. Since then, methyl bromide has been banned as a Class I ozone-layer-depleting substance, except for critical uses for exemption or pre-shipment applications that allow application within 21 days of export in some countries (www.epa.gov/ods-phaseout/methyl-bromide; accessed December 16, 2021). Methyl bromide alternatives were developed during the phase-out and applied to control storage pests in the dried-fig sector (Meyvacı and Şen, 2007; Şen *et al.*, 2009, 2010, 2015, 2016).

Storing an agri-food product without quality deterioration is an important advantage for keeping the supply–demand balance and thus providing a stable price at market. Storage of dried products such as figs requires special attention, as pests and mycotoxins may create significant quality and safety problems in global trade. The purpose of dried-fig storage is to manage timely product flow to market without any loss in palatability, flavor, texture, color, or marketable quantity. The more ideal the storage conditions, the fewer quantity and quality losses there will be. In many producer countries, however, dried figs are stored under uncontrolled conditions until the following year's crop. This storage environment—whether at the fig farm, with the middle men, or later at the processing plant for raw material or the final product—should provide complete isolation from pests. This isolation will prevent secondary infestation. However, prior to storage, living pests coming from the orchard should be controlled by chemical or non-chemical means allowed in the producing and importing countries. In organic dried-fig trade, storage-pest control utilizes either carbon dioxide as a controlled atmosphere means or extremely low freezing temperatures. As the number of damaged or infested dried figs should be below the tolerance levels, precautions need to be taken starting from the preharvest period. These pests are mostly polyphagous, so many dried commodities can host these harmful species. Therefore, monitoring and identifying the pest and knowing the favorable ecological conditions for their growth and reproduction, their means of infestation, and their distribution areas are extremely important for developing an effective control plan. In this chapter, the major postharvest pests damaging dried figs are introduced and key issues regarding their management are discussed.

18.2 Description: Biology, Significance, and Management of Dried-fig Pests

Stored-product pests may infest fig fruit at pre- and postharvest stages. Monitoring during the ripening, drying, and storage stages are crucial for detecting an infestation problem. In addition to the sanitation recommendations, disinfestation of the dried fruits after their delivery to the processing plant is needed (Şen *et al.*, 2009, 2010).

Climatic conditions prevailing in fig-producing regions provide a suitable environment for many species of dried-fig pests. The most widely found stored-product pests in the dried-fig sector are *Cadra cautella* (Lepidoptera, Pyralidae), *Plodia interpunctella* (Lepidoptera, Pyralidae), *Oryzaephilus surinamensis* (Coleoptera, Silvanidae), and *Carpophilus hemipterus* (Coleoptera, Nitidulidae). *Carpoglyphus lactis* (Acari, Carpoglyphidae) may also cause significant problems especially at long-term storage and/or under humid storage conditions (Turanlı, 2003). Among these pests, the population of fig moth (*C. cautella*) is quite high in fig production areas in Turkey and for many decades has played a major role in reducing the quality and quantity of dried figs by damaging the fruit through feeding and contaminating with other residues (Damarlı *et al.*, 1998; Emekçi *et al.*, 2004). Common dried-fig pests are shown in Table 18.1 and Fig. 18.1, and described below, while control strategies are explained in the later sections.

18.2.1 Tropical warehouse moth or fig moth, *Cadra cautella* (Walker, 1863)

Description

The newly laid eggs are oval shaped and white. Close to hatching, the color of the egg turns to yellowish orange with an irregular surface. The last-instar larvae is about 10–15 mm. The dorsal surface of the larvae is pinkish in color and the rest is light brown with dark brown spots. The larval skin is covered with sparse hairs. The pupae are dark brown and enclosed within a relatively light pupal case. The pupae are about 6–8 mm long. The wingspan of the adult is about 16–20 mm and the body length is 9–10 mm. The adult forewings are grayish-brown and have two dark brown zigzag bands (Fig. 18.1A). The hind wings are whitish gray, and there are short fringes of hairs (Anon., 2008; Akol *et al.*, 2020).

Distribution

The fig moth is found throughout the tropics and subtropics and is more common in non-arid regions. It can be found in temperate zone climates but can survive in winter only in heated, closed areas where temperatures are elevated. It feeds on a range of stored foods, notably cereals and dried fruits. It is the most widely found stored-product pest in fig storage in Turkey (Turanlı, 2003; Anon., 2008; Akol *et al.*, 2020).

Host range

The fig moth is a major pest of a range of stored foods, especially feeding on cereal products, groundnuts and other pods, dried fig, dried mango, dates, nutmeg, and other dried stored products (Anon., 2008; Akol *et al.*, 2020).

Life cycle

Adults live for 7–15 days. The females lay their eggs on stored food. Up to 230–270 eggs are laid in their lifespan. Eggs hatch after approximately 3–5 days. There are five larval instars before pupa and larva development, which is completed in about 42–52 days under optimum conditions. The mature larvae leave the produce to pupate in a cocoon on surfaces such as walls of

Table 18.1. Major stored-product pests of dried figs.

Common name	Species name	Family	Order
Fig moth	*Cadra cautella*	Pyralidae	Lepidoptera
Indian meal moth	*Plodia interpunctella*	Pyralidae	Lepidoptera
Saw-toothed grain beetle	*Oryzaephilus surinamensis*	Silvanidae	Coleoptera
Dried-fruit beetle (sap beetle)	*Carpophilus* species	Nitidulidae	Coleoptera
Dried-fruit mite	*Carpoglyphus lactis*	Carpoglyphidae	Acarina

the store or in between dried fruits. The pupal stage is completed in about 7 days. It produces three generations under cold climates, but in the Mediterranean region, it may reach four generations per year (Anon., 2008; Akol *et al.*, 2020).

Damage

The larvae feed inside the dried fig fruit and then exit, damaging the peel. However, as in most stored products, the main damage is through contamination with silk, fecal pellets, cast skins, and egg or head fragments, as well as product loss through larval feeding (Anon., 2008; Akol *et al.*, 2020).

18.2.2 Indian meal moth, *Plodia interpunctella* (Hübner, 1813)

Description

The white, spherical eggs are about 0.5–1 mm in diameter and are deposited singly or in groups on the larval food. The young larvae of *P. interpunctella* are white, while older larvae are usually yellowish white with pink tinges with no dark dot rows (Fig. 18.1C). The last-instar larvae grow to about 10–12 mm in length. The larvae pupate in a silk cocoon. The brown-colored pupae are about 6–8 mm in length. The adult Indian meal moth can be distinguished from other stored-product moths by its uniformly colored wings (Fig. 18.1B). The outer tip of the forewings is coppery brown with darker markings, while the base of the forewings is gray. The wings are yellowish gray. Adult moths are about 9 mm long, with a wingspan of about 18 mm (Şen *et al.*, 2010; Johnson, 2013; Akol *et al.*, 2020).

Distribution

P. interpunctella is found in stored products in a wide range of climates around the world. It is the second most widely found stored pest in fig storage in Turkey and the major one in Greece (Turanlı, 2003; Eliopoulos and Athanassiou, 2004; Akol *et al.*, 2020).

Host range

P. interpunctella may infect all types of dried food including dried figs and its products (Anon., 2008; Akol *et al.*, 2020).

Life cycle

The entire life cycle of this species lasts between 30–90 days. Females lay between 300 and 400 eggs on food surfaces throughout their 1–2-week lifespan. The larvae begin feeding in the dried fig fruit after hatching. The larval stage lasts from 30 to 52 days encompassing five instars. Adults emerge from the pupal cocoon in about 7 days. In temperate climates, *P. interpunctella* overwinters either as fully grown larvae or as pupae in cocoons (Bell *et al.*, 1979; Johnson, 2013). *P. interpunctella* has two to five generations per year, depending on the climate (Anon., 2008; Akol *et al.*, 2020).

Damage

Larval feeding results in important damage of dried fig fruits, but most of the damage is through contamination of the produce with larval residues such as webbing, fecal pellets, and cast skins as in other moths (Anon., 2008; Akol *et al.*, 2020).

18.2.3 Saw-toothed grain beetle, *Oryzaephilus surinamensis* (Linnaeus, 1758)

Description

The saw-toothed grain beetle *Oryzaephilus surinamensis* (Fig. 18.1D) has a thin, long (2.4–3.0 mm) body. The color of the adult changes from red to dark brown. There are six saw teeth at each side of the prothorax and three ridges on the prothorax. The eggs are white, thin, and slender, and about 0.7 mm long. The larvae are white with a brown head and 3 mm in length. The pupae are 3.5–5 mm long and cream colored (Anon., 2008; Akol *et al.*, 2020).

Distribution

O. surinamensis is found all over the world and is transported in or on dried fruits (Akol *et al.*, 2020).

Host range

O. surinamensis is a pest of a wide range of stored foods including cereals and dried fruits (Akol *et al.*, 2020).

Life cycle

The lifespan of adults is around 6–10 months. Usually, *O. surinamensis* overwinters at the adult stage. The female lays 300–400 eggs singly or in small batches on the dried food product. The incubation period of the eggs is 3–16 days. The larvae have four instars. The life cycle takes around 20–80 days depending on the temperature and substrate. *O. surinamensis* has five to six generations per year (Anon., 2008; Akol *et al.*, 2020).

Damage

Both adults and larvae feed externally on dried fruits and are especially harmful in long-term storage (Anon., 2008; Akol *et al.*, 2020).

18.2.4 Dried-fruit beetle, *Carpophilus* species

Description

Different species of *Carpophilus* have been identified as dried-fig pests. The dried-fruit beetle, *Carpophilus hemipterus* (Fig 18.1E), is the most common species in California fig orchards. The Freeman sap beetle, *Carpophilus freemani*, and the confused sap beetle, *Carpophilus mutilatus*, are also found and their incidence may vary among orchards. Additionally, to a lesser extent, *Carpophilus marginellus*, *Haptoncus luteolus*, and *Urophorus humeralis* may also be present (Coviello and Bentley, 2009).

The adults of dried-fruit beetles are 3–3.5 mm in length. The general color of the adults varies from dark brown to brown according to the species, with yellowish gray bands and dots on the elytra. The wings do not completely cover the abdomen. The eggs are thin and long, and transparent white. The larvae are cylindrical; the head and the last segment of the body are brown, while the other parts are white. The larvae are 6–7 mm in length. The pupae are 3–4 mm long and bright white (Anon., 2008; Coviello and Bentley, 2009).

Distribution

This pest is widespread in tropical, subtropical, and temperate regions. It is the major pest in fig orchards during the harvest and postharvest periods and is also found in dried-fig warehouses (Turanlı, 2003).

Host range

This pest has a large number of hosts, especially fresh and dried fruits. It can cause damage on grapes, figs, pears, and plums, or any fermenting fruit, both in the orchard and in storage. Other hosts are walnuts, hazelnuts, and cocoa (Anon., 2008).

Life cycle

Carpophilus species overwinter at the adult, pupal, or larval stages in the soil, in fruits that have fallen on to the ground, or in stored fruits. Adults fly from the storage rooms to the orchards and lay their eggs on ripe fruits. Under suitable conditions, larvae complete their development in 4–5 weeks. Female can lay about 1000 eggs. The larvae become pupae after three larval stages. *Carpophilus* species can produce five generations per year.

Damage

Dried-fruit beetles damage dried fig fruits in many ways. They can carry decaying or toxigenic fungi such as *Fusarium*, *Aspergillus*, *Alternaria*, and *Penicillum* species, which may start rotting and souring the fruit during the fresh ripe stage. For this reason, this beetle is locally called “souring beetle.” Dried-fruit beetle acts a vector carrying aflatoxigenic fungal spores into the fruit cavity, which may then cause the formation of aflatoxin. In such cases, the characteristic bright greenish-yellow fluorescence is formed inside the dried fig fruit, making removal of contaminated figs difficult. Dried-fruit beetles also attract other pests as fruit flies or navel-orange worms, which also facilitate aflatoxin formation (Coviello and Bentley, 2009). The various types of damage caused by *Carpophilus* species may eventually lead to a decrease in the quality and safety of the product. An increased rate of infestation with this pest as well as the “presence of dead insects and/or mites, their debris or excreta” will also create problems in trade (UNECE, 2016; FDA, 2018).

18.2.5 Dried-fruit mite, *Carpoglyphus lactis* (Linnaeus, 1767)

Description

The eggs are transparent and shiny. The larvae are like the adults but have three pairs of legs. In the nymphal period, they have four pairs of legs. The adults are transparent, 0.4 mm long, and have an oval and flattened shape. Long hairs at the end of the body are typical. Adults have four pairs of legs that are thin and long, and the leg hairs are short with thick ends (Anon., 2008).

Distribution

C. lactis is a cosmopolitan species and has been reported from the Holarctic, Neotropical, Oriental, and Australian regions (Chmielewski, 1972; Klimov *et al.*, 2016).

Host range

Beside agricultural products like dried fruit, *C. lactis* is reported from environments ranging from beehives to different food and seeds (Klimov *et al.*, 2016).

Life cycle

Lifespan of *C. lactis* is 40–50 days and females lay 25–30 eggs during this period. Under favorable conditions, the development period is 15–20 days. The dried-fruit mite prefers environments with high relative humidity and temperature, and thus unsuitable storage conditions favor mite infestation. Fluctuating temperatures in storage may enhance sugaring on dried figs, while also providing suitable conditions for mites. *C. lactis* produces many generations per year under such favorable conditions.

Damage

C. lactis prefers products rich in sugar. It causes problems during storage of dried figs, raisins, and similar dried fruits, as well as in flour and fermented foods. *C. lactis* feeds on sugar secretions on the surface of dried figs. Damage may occur as a result of feeding and contamination by body fragments. Its population increases when conditions are favorable, leading to quality and quantity losses.

18.3 Control Strategies for Postharvest Fig Pests

Among fig varieties, Smyrna types or the second crop of San Pedro types require pollination for fruit set. Caprifigs used for pollination host many decaying fungi and therefore fungal spores are carried to the female fig with caprification, causing decay/internal rot, which then attracts dried-fruit beetles. Thus, varieties that require caprification are more prone to dried-fruit beetle infestation. Additionally, varieties that have larger ostiole openings such as the Sarılop (= Calimyrna) variety are more susceptible to decay and dried-fruit beetle problems. Therefore, persistent-type varieties and/or those with small and closed ostioles should be preferred whenever possible (Codex Alimentarius, 2008; Coviello and Bentley, 2009). Infestation, especially with fig moths, occurs during sun drying, and therefore artificial drying practices or drying under tunnels help to shorten the time required for drying and reduce infestation rates by limiting exposure (Okşar *et al.*, 2017).

In the orchard or in storage rooms, processing facilities or wherever dried figs are present, stored-product pests should be monitored either by visual inspection or using sex pheromone traps. Larvae can be seen on the fruits, on the walls of the storage facility, or around the packages. Mixed webbing and excreta are characteristic symptoms of heavy infestation. Pupae can be detected in their silk cocoons. An integrated pest management strategy must be applied within a Good Agricultural Practices protocol designed according to the stage and the site-specific conditions. Thus, prevention of damage by the pest should start from the early days of its emergence until it reaches the consumer. The first precautions should be taken in the orchard during preharvest, harvest, drying, transportation, and on-farm storage. The moisture level of dried figs should not exceed 24–25%. The storage temperature and humidity should be monitored and kept below 20°C and 65%, respectively. Conditions at the farm and during transportation from the farm to the middle men, commercial stores, or processing plants are also important regarding stored-product pest infestation. Sanitation at the farm and good storage conditions may prevent most of the infections from the start (see Chapter 14, this volume).

18.3.1 Cultural practices and sanitation at the farm

Monitoring pests is important to apply the precise corrective measures. In this respect, a monitoring plan must be developed for visual inspection or using traps, starting before the pests emerge and start causing damage. Further to this step, important sanitation and care measures to prevent infestation at the farm, are detailed below.

Storage rooms

Storage rooms should follow the following rules:

- All doors, windows, and ventilation holes should be closed properly; if open, they should be covered with insect nets.
- All cracks and crevices in the storage room or on the walls should be filled.
- The inner wall surfaces of the storage room should be whitewashed.
- Necessary precautions should be taken to make the storage room cool with adequate ventilation.
- The storage room and all materials used should be clean and, if necessary, fumigated.
- Infested material or any other relevant food source should be removed completely prior to the new crop.

Care at harvest

The following precautions should be taken during harvest (FAO, 1985; Codex Alimentarius 2008; Şen *et al.*, 2009, 2010; Akol *et al.*, 2020):

- Partially-dried figs falling on the ground should be collected frequently. It is recommended that they are picked at least twice a day, in the morning and evening. After harvest, all remaining cull, mummified, and caprifigs must be collected and removed outside the orchard.
- Prior to placing the harvested fruit on trays, infested dried figs must be removed and dried and stored in an isolated place separately to prevent cross-contamination.
- Figs that have dried completely should be selected and taken from the drying tray every day during the daylight, preferably when the air temperature is still cool, and stored in an infestation-free room.
- During sun drying of the fruit, trays should be covered at night and reopened early in the morning if there is a risk of fig moths laying eggs.
- No light should be turned on for any purpose at night, especially around the drying trays or in storage rooms, as fig moths are attracted to light.
- Infestations from low-quality fruits to high-quality ones by storage pests must be limited by grading and keeping low-quality fruits in a separate room.
- At harvest, dried figs should not be kept for a long time in the orchard or in the farmers' stores if an infestation risk is present. They should be delivered quickly to the next step where fumigation or any other efficient control measures can be applied.
- Product transportation should be done during the day. If it must be done at night, the fig fruits should be covered with a tarpaulin.
- Dried figs should not be stored in bulk as piles but stored in boxes that have adequate ventilation.
- If the dried figs are stored in sacks or bags, they should be stacked on a clean pallet and excess pressure on the lower sacks should be avoided.
- At least 0.5 m space should be left between stacks/piles and the wall.

Controlled atmosphere and cold treatments

In suitable infrastructures (gas-tight chambers, pressurized tanks, or flexible and airtight PVC storage), low-oxygen and carbon dioxide-enriched atmospheres with or without pressure can be used to control stored-product pests. A controlled atmosphere plus heat treatment is recommended to control storage pests in the dried-fig industry, but the temperatures applied should not exceed 45°C to prevent further darkening of color (Anon., 2008; Akol *et al.*, 2020; Şen *et al.*, 2009, 2010).

Freezing dried figs in a commercial-scale deep freezer for several hours has proved to be an effective control method for stored-product pests. According to the target pest and the stage, keeping the inner core of the produce at −15°C or −18°C for around 2–3 h could be enough for disinfestation of dried figs from all stages of common pests. Due to slow heat transfer in low-moisture produce, reaching these temperatures at the core of the dried fig requires keeping the fruit at −35°C to −40°C for around 30–36 h. The most resistant stage of the fig moth life cycle is the eggs. This freezing method is widely used for disinfection of organic dried fig in processing plants (Şen *et al.*, 2009; Meyvacı *et al.*, 2010; Aksoy *et al.*, 2012).

18.3.2 Biological pest control

For lepidopterous pests, the following natural enemies have proved to be successful: *Apanteles merula* Reinhard, *Bracon hebetor* Say, *Bracon brevicornis*, and *Phanerotoma flavitestacea* Fisher belonging to the family Braconidae, *Trichogramma* species belonging to the family Trichogrammatidae, and *Diadegma chrysosticta* Gmel., *Mesostenus transfuga* Grav., *Venturia canescens* Grav., and *Pachycrepoideus vindemiae* (Rond.) belonging to the family Ichneumonidae (Hymenoptera). They can be used to control fig moths in fig storage as part of an integrated pest management program.

Mass rearing and release studies were carried out with *B. hebetor* in the 1970–1980s in fig-processing units and warehouses in Aydın Province, Turkey. The results showed that the natural enemy is highly effective against last-instar larvae but could not completely control the pest (Anon., 2008). The effectiveness of biological control strategies of stored pests in storage facilities needs further research work.

18.3.3 Biotechnical methods

Biotechnical control methods can be used to manage insect pests of stored products. They can be monitored, confused, and mass trapped using synthetic sex pheromones. Sex pheromones and other attractive semiochemicals can be used to attract insects to the trap and determine infestation and the density of pests. Mass trapping using pheromone traps or sticky traps combined with pheromones is an effective method to reduce the population density of stored-product pests. Besides the direct use of pheromones for mass trapping, they may also be used for monitoring to enhance the efficacy and precision of chemical controls. This method is especially useful in monitoring the biology and behavior of pests in stored products (Anon., 2008). For dried-fruit beetles, bait traps with culled figs, water, and yeast can be used for monitoring. If the bait traps are placed prior to emergence and replaced regularly, they may be effective in reducing dried fruit beetle populations (Coviello and Bentley, 2009).

18.3.4 Chemical control

Pesticides are used to kill a certain group of insects or mites. They can be plant derived or synthetic, and either broad spectrum or with a specific toxicity. In particular, insecticides show some degree of toxicity to non-target organisms including humans, domestic animals, bees, and poultry, and may pollute water and other natural resources. Because of their impact, chemical applications should be based on correct identification of the pests and expert recommendation of the pesticides registered for use. In the case of international trade, the relevant legislation in the producer and importer countries must be considered prior to application. Some pesticides are registered or chemically approved for use in empty storage facilities or against pests of the stored product. There are three kinds of pesticide application in storage: (i) empty-store spraying; (ii) filled-store fumigation; and (iii) fumigation of dried figs before storage. Spraying of the empty storage room is important before a new crop comes in, and the appropriate time period required before the product is placed in the storage room must be closely followed. Fumigation of dried figs in storage facilities of farmers, dealers, or processing plants with a registered fumigant or other method may control existing infestations but will not protect from reinfestation, so all necessary precautions need to be taken (Akol *et al.*, 2020). Fumigation of dried figs infested with pests can be done before storage in airtight structures, under vacuum or controlled-atmosphere conditions. In such applications, leakage is important, and the concentration of the fumigant needs to be monitored for efficacy (Meyvacı and Şen, 2007). Fumigation of the produce and disinfestation of stores should be done simultaneously to prevent reinfestation. One of the major issues in addition to achieving complete mortality is the effect on dried-fig quality, which may only become apparent after long-term storage (Şen *et al.*, 2016).

18.4 Concluding Remarks

Storage pests create significant problems in the dried-fig sector worldwide. On the one hand, they damage the crop, resulting in quantity and quality losses, while on the other hand, as in the case of dried-fruit beetle, they create safety problems by carrying spores of decay or mycotoxin-producing fungi. Storage-pest problems have existed in the dried-fig trade since the early days of sun drying and storage. Traditionally, fig farmers use simple methods to preserve their home supplies of dried figs such as inserting aromatic herbs like bay leaves, known to delay spoilage, between fig layers, or treating them with hot water or vapor. However, to transfer this knowledge to a commercial scale requires research.

As the dried-fig trade became more important, chemical fumigation of dried figs became recommended or compulsory with methyl bromide, as in Turkey (Erakay and Özar, 1978) or with sulfur as in Greece (Aday, 1938), due to the significance of the pest problem. However, after the phase-out of methyl bromide and labeling requirements for sulfur treatment, as well as enhanced concerns of consumers about pesticide residues, the need to take precautions at the orchard level to prevent/reduce infestations became more pronounced. Additionally, non-chemical commercial alternatives to control storage pests became available and received high acceptance, as in the case of organic products. Ongoing discussions on reducing the tolerances allowed in standards are forcing the actors of the value chain to seek efficient preventive measures. As major fig varieties for sun drying, prevailing cultural practices, and ecological conditions vary among fig-producing regions, major pests may differ as well. Studies on identifying the major pests causing damage at the postharvest stages must be carried out in all newly emerging exporting countries. Even if details are not known, the transportation and retail stages are also among the critical control points for risk analysis. Based on the outcomes of risk analysis carried out locally, an effective set of recommendations and control measures must be developed and disseminated among farmers, storekeepers, processors, and other actors in the value chain.

References

Aday, N. (1938) *Yunan Kuru İncirleri (Greek Dried Figs)*. Publication of the First Rural and Agricultural Development Congress, Ministry of Agriculture of the Turkish Republic, Ankara.

Akol, A.M. Chidege, M.Y., Talwana, H.A.L., and Mauremootoo, J.R. (2020) *Keys and Fact Sheets. BioNET-EAFRINET UVIMA Project.* Available at: https://keys.lucidcentral.org/keys/v3/eafrinet/ (accessed December 16, 2021).

Aksoy, U., Şen, F., and Meyvacı, K.B. (2012) Effect of post-harvest carbon dioxide application on storage pests and fruit quality of dried figs. In: Navarro, S., Banks, H.J., Jayas, D.S., Bell, C.H., Noyes, R.T., *et al.* (eds) *Proceeding of the 9th International Conference on Controlled Atmosphere and Fumigation in Stored Products, October 15–19, 2012, Antalya, Turkey*, pp. 166–171.

Anon. (2008) Stored product pests. In: Aydemir, M. (ed.) *Plant Protection Technical Instructions*. Ankara, pp. 201–248.

Bell, C.H., Bowley, C.R., Cogan, P.M., and Sharma, S. (1979) Diapause in twenty-three populations of *Plodia interpunctella* (Hübner) (Lep., Pyralidae) from different parts of the World. *Ecological Entomology* 4, 193–197.

Chmielewski, W. (1972) Mites occurring on food products. The morphology, biology and ecology of *Carpoglyphus lactis* (L., 1758) (Glycyphagidae, Acarina). *Prace Naukowe-Instytutu-Ochrony-Roslin* 13, 167–186.

Codex Alimentarius (2008) *Code of Practice for the Prevention and Reduction of Aflatoxin Contamination in Dried Figs*. CAC/RCP 65-2008. Food and Agriculture Organization of the United Nations, Rome/World Health Organization, Geneva.

Coviello, R.L. and Bentley, W.J. (2009) *UC IPM Pest Management Guidelines: Fig*. University of California Division of Agriculture and Natural Resources Publication, Los Angeles, California.

Damarlı, E., Gün, H., Özay, G., Bülbül, S., and Oechsle, P. (1998) An alternative method instead of methyl bromide for insect disinfestation of dried figs: controlled atmosphere. *Acta Horticulturae* 480, 209–214.

Eliopoulos, P.A. and Athanassiou, C.G. (2004) Seasonal occurrence of dried fig pests and their parasitoids in a fig warehouse in Greece. In: Navarro, S., Adler, C., Scholler, M., Emekci, M., Ferizli, A.G., and Hansen, L.S. (eds) *Integrated Protection of Stored Products. Proceedings of the Meeting. IOBC-WPRS, September 16–19, 2003, Kusadasi, Turkey*, pp. 159–171.

Emekçi, M., Ferizli, A.G., Tütüncü, Ş. and Navarro, S. (2004) The efficacy of modified atmosphere applications against dried fruit pests in Turkey. In: Navarro, S., Adler, C., Scholler, M., Emekci, M., Ferizli, A.G. and Hansen, L.S. (eds) *Integrated Protection of Stored Products, Proceedings of the Meeting, 16–19 September 2003, Kusadasi, Turkey*. IOBC-WPRS Bulletin, Vol. 27, pp. 227–231.

Erakay, S. and Özar, A.İ. (1978) Ege Bölgesinde bazı ambar zararlılarına karşı methyl bromide fumigasyonunun değişik koşullarda etkisi üzerinde araştırmalar. *Bitki Koruma Bülteni* 18, 32–38.

FAO (Food and Agriculture Organization of the United Nations) (1985) Pest control in stored produce. In: *Prevention of Post-Harvest Food Losses: A Training Manual*. FAO, Rome. Available at: www.fao.org/3/x0039e/X0039E05.htm (accessed December 16, 2021).

FDA (2018) *Food Defect Levels Handbook. US Food and Drug Administration*, Silver Spring, Maryland. Available at: www.fda.gov/food/ingredients-additives-gras-packaging-guidance-documents-regulatory-information/food-defect-levels-handbook#commodities (accessed December 16, 2021).

Johnson, J. (2013) Pest control in postharvest nuts. In: Harris, L.J. (ed.) *Improving the Safety and Quality of Nuts*. Woodhead Publishing, Philadelphia, Pennsylvania, pp. 56–87.

Klimov, P., O'Connor, B., Ochoa, R., Bauchan, G., Redford, A., and Scher, J. (2016) *Carpoglyphus Fact Sheet*. US Department of Agriculture, University of Michigan, Ann Arbor, Michigan. Available at: http://idtools.org/id/mites/beemites/factsheet.php?name=15327 (accessed December 16, 2021).

Meyvacı, K.B. and Şen, F. (2007) Magnezyum fosfit uygulamalarının kuru incir meyve kalitesine etkileri, *Journal of Agricultural Faculty of Ege University* 44, 29–40.

Meyvacı, K.B., Sen, F. and Aksoy, U. (2010) Optimization of magnesium phosphide treatment used to control major dried fig storage pests. *Horticulture, Environment, and Biotechnology* 51, 33–38.

Okşar, R.E., Şen, F., and Aksoy, U. (2017) Effect of drying under plastic tunnels on drying rate and quality of fig (*Ficus carica* L. 'Sarılop'). *Acta Horticulturae* 1173, 297–302.

Şen, F., Meyvacı, K.B., Aksoy, U., Emekçi, M., and Ferizli, A.G. (2009) Effects of the post-harvest application of methyl bromide alternatives on storage pests and quality of dried fig. *Turkish Journal of Agriculture and Forestry* 33, 403–412.

Şen, F., Meyvaci, K.B., Turanlı, F., and Aksoy, U. (2010) Effects of short-term controlled atmosphere treatment at elevated temperature on dried fig fruit. *Journal of Stored Products Research* 46, 28–33.

Şen, F., Aksoy, U., Emekci, M., and Ferizli A.G. (2015) Determining effect of phosphine ($ECO_2FUME^{®}$) fumigation under atmospheric and vacuum conditions on dried fig quality. *Fresenius Environmental Bulletin* 24, 2046–2051.

Şen, F., Meyvacı, K.B., Aksoy, U., and Tan, G. (2016) Effect of sulfuryl fluoride fumigation on dried fig quality. *Acta Horticulturae* 1120, 241–244.

Shorey, H.H., Ferguson, L., and Wood, D.L. (1989) Solar heating reduces insect infestations in ripening and drying figs. *HortScience* 14, 443–445.

Turanlı, F. (2003) Studies on infestation levels of pests on dried fig in Aydın and İzmir provinces. *Turkish Journal of Entomology* 27, 171–180.

UNECE (2016) *UNECE Standard DDP-14 Concerning the Marketing and Commercial Quality Control of Dried Figs*. United Nations, New York and Geneva. Available at: https://unece.org/fileadmin/DAM/trade/agr/standard/dry/dry_e/14DriedFigs_E2016.pdf (accessed December 16, 2021).

USDA (2001) *United States Standards for Grades of Dried Figs*. Processed Products Branch, Fruit and Vegetable Programs, Agricultural Marketing Service, US Department of Agriculture, Washington, DC.

19 Mycotoxins in Fig

Kamer Betül Özer*
Department of Horticulture, Faculty of Agriculture, Ege University, Bornova, İzmir, Turkey

19.1 Introduction

Mycotoxins are low-weight secondary metabolites of fungi that cause hazards for livestock, animals, and humans. However, secondary metabolites may also have antibiotic, cholesterol-lowering, or antitumor properties. They are therefore important for public health for both reasons (Moreau, 1979; Ashiq, 2015; Jeswal and Kumar, 2015). In the 1960s, with the death of 10,000 turkey poults on a British farm, research on this issue was enhanced after the cause was found to be feeding animals with *Aspergillus flavus*- and *Aspergillus parasiticus*-contaminated groundnuts. After this significant initial alert, a large number of studies were conducted, and with the advances in analytical procedures, new mycotoxins are being revealed. More than 400 mycotoxins have been isolated and chemically identified so far, researchers have focused on mycotoxins that have significant harm to humans, livestock, and pets. These include aflatoxins, ochratoxins, trichothecenes, zearalenone, cyclopiazonic acid, fumonisins, patulin, and ergot alkaloids (Bennett and Klich, 2003; Bayman and Baker, 2006; Desjardins and Proctor, 2007; Schaafsma and Hooker, 2007; Zinedine *et al.*, 2007a; Moss, 2008; Trucksess and Scott, 2008; Adejumo and Adejoro, 2014; Jeswal and Kumar, 2015). In addition to these mycotoxins, citrinin and sterigmatocystin have recently gained importance, and in the future, risks created by many more are likely to unfold. Mycotoxins (except fumonisins) are generally lipophilic and therefore tend to accumulate in the fatty parts of plants and animals (Hussein and Brasel, 2001). *Aspergillus*, *Penicillium* and *Fusarium* molds are the leading species responsible for mycotoxin formation in foodstuffs (Bhatnagar *et al.*, 2004), but *Trichoderma*, *Trichothecium* and *Alternaria* are also key food contaminants or plant pathogens (Richard, 2007; Ashiq, 2015). Mycotoxins are a common problem of all countries in the world, although the prevailing molds and/or mycotoxins may differ. Aflatoxins cause more damage especially in tropical and subtropical zones due to the higher temperature and humidity levels, which are ideal for the growth of aflatoxigenic fungi and aflatoxin formation. *Penicillium* molds and toxins are found mostly in temperate regions, while *Fusarium* molds and trichothecenes prevail in colder regions (Özay and Eke, 1988). Most of the toxin-producing fungi can produce more than one mycotoxin in the same product (Bottalico, 1998; Atalla *et al.*, 2003). Products may also be contaminated with various fungal species, some of which may compete, with some becoming dominant as the toxin producer or as non-toxigenic

* Email: kamer.betul.ozer@ege.edu.tr

 Advances in Fig Research and Sustainable Production (M.A. Flaishman and U. Aksoy)

DOI: 10.1079/9781789242492.0019

(Sweeney and Dobson, 1998; Grenier and Oswald, 2011). After molds complete their vegetative development, they enter the idiophase stage during which they produce mycotoxins as their secondary metabolites. These toxins are also produced as defensive mechanisms to suppress other microorganisms sharing the same growth environment. For a toxigenic strain, in forming mycotoxins, humidity, temperature, morphology, and chemical composition of the product and fungal population in the environment play important roles (Heseltine, 1976; Njumbe Ediage *et al.*, 2015).

Mycotoxins are synthesized by various enzymatic routes from polypeptides, amino acids, phenols, and terpenoids, which are the constituents of the fungi. Mycotoxins have varying chemical structures from simple structures (e.g. patulin) to more complex ones (e.g. T-2 toxin). Not every mold can produce toxin, and the presence of mold does not mean the presence of a toxin; similarly, no mold growth does not mean that the product is free of toxins. After mycotoxins are synthesized by molds, they are stable and resistant to any changes that may occur in the physical state of the food such as moisture or pH. Mycotoxins can survive in the absence of molds and preserve their durability (Topal, 1993; Njumbe Ediage *et al.*, 2015).

Mycotoxins may have acute or chronic effects. The disease caused by mycotoxins in humans and animals is called mycotoxicosis. Mycotoxicosis is a complete poisoning and occurs because of consumption of feed and foods contaminated with toxic metabolic substances. The first case of mycotoxicosis (ergotism) was recorded historically as St. Anthony's Fire. Known also as St. Anthony's fever, it is one of the oldest known human foodborne illnesses and caused hallucinatory symptoms and many deaths in France and other European countries during the Middle Ages (Betina, 1989). The effects of mycotoxins on living organisms are remarkably diverse. In addition to toxicities resulting in death, mycotoxins may have significant health effects such as carcinogenic, mutagenic, DNA damaging, hindering RNA and protein synthesis, causing malformations and skin lesions, and suppressing the immune system (Aksoy, 1990). The World Health Organization International Agency for Research on Cancer (WHO-IARC) reported the carcinogenic potential of aflatoxins, ochratoxins, trichothecenes, zearalenones, and fumonisins in 1993. The Joint Food and Agriculture Organization of the United Nations (FAO)/WHO (JECFA) Food Additives Committee has been evaluating mycotoxins as potential human carcinogens since 2008 and emphasizes that reducing dietary exposure to the lowest possible levels is necessary to reduce the potential risk (JECFA, 2008). The degree of toxicity in humans and animals depends on the levels of ingested mycotoxins through contaminated food or feed, the type of toxin in the food, the mechanism of action, metabolism, and the defense mechanism (Hussein and Brasel, 2001). Some mycotoxins have a higher chronic dietary risk factor than synthetic contaminants, food additives, or pesticide residues (van Egmond *et al.*, 2007). In this respect, both country-specific and international regulations set forth special maximum limits for major mycotoxins according to the exposure risks. To protect consumers against food safety risks, countries such as the USA, Japan, and Australia have systems open to the public where they report non-compliant cases found in market controls or at the stage of importing. The European Union (EU) established the Rapid Alert System for Food and Feed (RASFF), in 1979 to allow rapid exchange of information on hazards (chemical, biological, physical, and allergens) in food and feed among EU countries. All food and feedstuffs imported into the EU undergo a control system applied by the competent authorities of the member states. When risks to public health are identified during these checks, the information is communicated through the RASFF to all EU member states, as well as to the country of origin and the exporting countries (Taylor *et al.*, 2013; Kowalska and Manning, 2020). RASFF is a system established between the member states of the EU to share the detected food risks and to inform the consumers at large. Pathogenic microorganisms, mycotoxins, and pesticide residues are the safety hazards listed at the top ranks of the notifications. As well as contaminant levels exceeding the maximum allowed limits, the absence of labeling/insufficient labeling and improper health certificates/absence of health certificates are also reported as notifications. Dried figs are among the products that receive notifications mainly for the presence of aflatoxins and to a lesser extent for the presence of ochratoxin A (OTA).

With its fruit morphology and composition, the fig fruit is highly susceptible to growth of toxigenic fungi and formation of mycotoxins. Additionally, the climatic conditions where common fig is grown are favorable for both fungal growth and aflatoxin formation. Toxin-producing fungi can germinate easily on the juicy and fleshy skin, which also contains sugars making the substrate favorable. The internal cavity of the fruit may also host toxigenic fungi if the fungus is transmitted to the core part of the fruit through vectors. This chapter gives a wide overview of mycotoxin occurrence in figs, chiefly aflatoxins, OTA, and fumonisins, by reporting the results of scientific papers on incidence, levels of contamination, major microbial agents, and conditions favoring the formation of mycotoxins. As dried fruits have a higher risk for mycotoxin formation, the major focus is on dried fruit, although the risks of mycotoxin formation in fresh fig fruit are also discussed.

19.2 Mycotoxins in Dried Figs

Studies on mycotoxin occurrence in dried figs from different origins and of different quality classes, ranging from sound to showing bright greenish-yellow fluorescence (BGYF) or rotten and moldy fruit, have revealed that various mycotoxins and their metabolites may form. These include aflatoxins, OTA, fumonisins, kojic acid, zearalenone, HT-2 toxin, tenuazonic acid, alternariol, cyclopiazonic acid, patulin, fusaric acid, beauvericin, fusaproliferin, and ergosterol (Karaca and Nas, 2008; Gilbert and Şenyuva, 2008; di Sanzo *et al.*, 2018; Galvan *et al.*, 2021).

In fig production, fungal infection and mycotoxins occur primarily in the orchard, and the high-risk stage is while the fig is on the tree (Heperkan, 2006; Aksoy *et al.*, 2013). Dried fig fruits provide a suitable substrate for different fungi due to their high carbohydrate content and water activity levels (a_w, defined as the ratio of vapor pressure of the product to that of pure water; Desjardins *et al.*, 2003) during overripening and drying (Özay and Alperden, 1991). Based on the requirements of the toxigenic species, fresh figs may also be subject to formation of some mycotoxins, as contamination of the fig fruit with the toxigenic fungi mostly occurs before ripening or drying. However, for growth of the fungus and toxin formation, climatic conditions present during ripening, shriveling, and sun drying, and the a_w level of the fruit play major roles, and therefore occurrence of some mycotoxins is greatly affected by yearly conditions (Özer Meyvacı, 2009). In sun drying, the contamination risk increases when partially shriveled fig fruits that still have comparatively high moisture content fall to the ground and have contact with soil-borne toxigenic fungi (Marin *et al.*, 2013).

Surveys carried out on dried Sarılop (= Calimyrna) variety fig fruits in İzmir and Aydın Provinces of Turkey between 1981 and 1984 showed that *Aspergillus niger* was the most prevalent species with an occurrence rate of 98%. The others in decreasing order of prevalence were *Mucor* and *Rhizopus* species (90.4%), *Penicillium* species (26.9%), *Fusarium moniliforme* (23.8%), *Alternaria* species (22.2%), and other *Aspergillus* species (20.6%) (Özar *et al.*, 1986). The presence of different species in the microbial population may exert a marked effect on mycotoxin formation. *A. niger*, *Rhizopus oligosporus*, and *Neurospora* species that are found abundantly in dried figs inhibit aflatoxin formation, even if aflatoxigenic strains are present (Gilbert and Şenyuva, 2008). For some strains of mycotoxic fungi such as *Aspergillus*, *Penicillium*, and *Stachybotrys*, the optimum a_w level for toxin formation is 0.93–0.98, whereas the overall range is 0.71–0.94 (Gaddi, 1973). There is a direct relationship between a_w and mycotoxin formation. The a_w value required for molds to form toxins is generally slightly higher than that of the levels required for mold to grow. For example, *A. flavus* can reproduce at 0.78 a_w, but needs 0.83–0.87 a_w to form aflatoxins. Molds that have optimum temperature requirements of 25–27°C stop aflatoxin formation at temperatures below 10°C (Özay, 1989). It has been stated that fungal growth can be prevented by reducing a_w below 0.65 in agricultural products (Northolt and Soentoro, 1988). Similarly, formation of mycotoxins in figs can be prevented by reducing the a_w below 0.70 rapidly during drying of figs (Codex Alimentarius, 2008; Aksoy *et al.*, 2013). In this respect, unless a preservative is used to prevent microbial growth, the UNECE Standard on Figs (DDP-14) states that the maximum water content should be below 26% (UNECE, 2016), although 24% can be proposed as the maximum humidity level for dried figs in order to prevent risks (Oksar *et al.*, 2017). During sun drying of Sarılop fig fruit, the mean a_w was measured as 0.91 in fresh mature figs, 0.79 in semi-dried and partially shriveled figs, and 0.62 in fully sun-dried figs. Fig with its fruit characteristics and composition is a susceptible substrate for mold growth and mycotoxin formation, and the risk is minimized only after the fruit is fully dried (Şen *et al.*, 2017).

Being a susceptible product, natural occurrence of fungal contamination in dried fruit has been subject to many studies in both producer and importer countries (Zohri and Abdel-Gawad, 1993; Özay *et al.*, 1995; Özer and Derici, 1998; MacDonald *et al.*, 1999; Altındişli *et al.*, 2002; Bayman *et al.*, 2002; Meyvacı, 2003; Moller and Nyberg, 2003; Aksoy *et al.*, 2007; Zinedine *et al.*, 2007b; Özay and Özer, 2008; Şenyuva *et al.*, 2008; Bircan, 2009; Heperkan *et al.*, 2012a; di Stefano *et al.*, 2015; Hedawoo *et al.*, 2017). The fungal flora found most often in dried figs is *Aspergillus* section *Nigri*, *Fusarium* species, *Aspergillus* section *Flavi*, and *Penicillium* species, which can produce mycotoxins, including aflatoxins, citrinin, cyclopiazonic acid, fumonisins, patulin, and OTA (Heperkan *et al.*, 2012b). Mold genera isolated from figs also include *Acremonium*, *Byssochlamys*, *Cladosporium*, *Trichoderma*, *Mucor*, and *Scopulariopsis* (Zorlugenc *et al.*, 2008; Isman and Biyik, 2009). In screening dried figs of the Sarılop variety, the fungal flora and contamination rates were 0.0–29.9% for *A. niger*, 0.0–4.5% for *A. flavus*/*A. parasiticus*, 0.0–10.0% for *Fusarium* spp, 0.0–12.0% for *Penicillium* spp, 0.0–6.1% for *Cladosporium* spp, and 0.0–4.0% for *Alternaria* species (Benlioğlu *et al.*, 2008). Hedawoo *et al.* (2017) isolated 15 fungal species from figs: *Alternaria alternata*, *Aspergillus candidus*, *A. flavus*, *A. fumigatus*, *A. niger*, *A. parasiticus*, *A. versicolor*, *Eurotium chevalieri*, *Cladosporium cladosporioides*, *C. herbarum*, *Fusarium solani*, *Penicillium granulatum*, *P. nigricans*, *P. oxalicum*, and *Rhizopus stolonifer*. Among the fungi, members of the genus *Aspergillus* were the most dominant isolates with six different species, including *A. parasiticus* and *A. flavus*, which produce aflatoxins and pose a threat for animal and human health. In Yemen, *Aspergillus* was again the most frequently isolated genus from dried figs and *A. niger* was the predominant species. The genus *Penicillium* took second place (Alghalibi and Shater, 2004). The environmental conditions play a crucial role in microbial development. Zohri and Abdel-Gawad (1993) found that *Penicillium* was the most predominant genus isolated from dried apricots, figs, prunes, and raisins. Among the isolated *Aspergillus* species, *A. niger* and *A. flavus* were the most common, followed by *A. fumigatus* and *Aspergillus ochraceus*.

Mycotoxins are naturally occurring contaminants, and therefore complete elimination is generally not possible. Most mycotoxins are chemically invariant, stable, and difficult to eliminate once formed, as detoxification is not allowed for human food. Only physical methods that help removal of contamination or that reduce the levels can be applied (Stoev, 2015). Mold and mycotoxin formation can be prevented completely or partially by applying various preventive Good Agricultural Practices (GAP) before and/or after harvest, such as appropriate and timely harvest, cleaning, appropriate drying and storage conditions, selection of fungus-resistant species and varieties, pest control, orchard floor management, and care at harvest (Doko *et al.*, 1995; Bata *et al.*, 2001). Prevention or reduction of mycotoxins first requires systematic identification of hazards and their risk assessment in the food production system, determination of the causal fungi, defining the critical control points, and significantly improving the food safety management based on this assessment. Caprifigs, which are important for some fig varieties that require pollination and fertilization for fruit set, are one of the sources of fungal infections. However, they do not cause a similar threat in persistent fig varieties. If caprification is required, the caprifigs should be healthy and free from fungi, in addition to having abundant and vibrant pollen grains and fig wasps (*Blastophaga psenes* L.). Male fig fruits may host *Fusarium* and *Aspergillus* species and other fungi, which can be transported from the male to the female fig fruits by the fig wasps during pollination. Therefore, it is important to use only clean and healthy caprifigs. Caprifigs that are rotten or showing brown discoloration inside the fruit cavity should be removed and destroyed. After the pollen has been shed and the wasps have stopped emerging, it is recommended that caprifigs placed on female trees are collected and destroyed outside the orchard, as they can host fungal diseases and pests. The only way to achieve orchard hygiene is the application of GAP starting from cultivation until the last stage of production (Reddy *et al.*, 2001; Aksoy *et al.*, 2013; Stoev, 2015). Therefore, whenever possible, selection of resistant species and varieties, appropriate methods in soil cultivation (e.g. no soil tillage prior to fruit ripening), and appropriate drying and storage conditions are recommended. GAP as given in the *Code of Practice for Prevention or Reduction of Aflatoxins in Dried Figs* (Codex Alimentarius, 2008) can provide guidance in prevention and reduction of aflatoxins.

19.3 Major Mycotoxins in Figs

19.3.1 Aflatoxins

Aflatoxins are accepted as the most potent mycotoxins because of their genotoxic and teratogenic effects on animals and humans. Aflatoxins are a group of nearly 20 similar metabolites. However, only aflatoxins B_1 (AFB_1), B_2 (AFB_2), G_1 (AFG_1), and G_2 (AFG_2) are found in nature and cause contamination in feed and food, while M_1 is found in milk resulting from ingestion of aflatoxin-contaminated feed. High levels of aflatoxin intake can have an acute poisoning effect (aflatoxicosis) and may have a significant health impact, especially on the liver. Toxicological data obtained from animal studies show that the relative mutagenic and hepatocarcinogenic effects are in the order of $AFB_1 > AFG_1 >> AFB_2$ and AFG_2 (WHO/FAO, 2018). Due to the impact of AFB_1 on health, specific maximum limits are set for AFB_1, in addition to the total aflatoxins.

The major aflatoxin-producing fungi are *A. flavus*, *A. parasiticus*, and *A. nomius*. Many other species closely related to *A. flavus* (*A. minisclerotigenes*, *A. korhogoensis*, *Aspergillus aflatoxiformans*, and *A. texensis*) or to *A. parasiticus* (*Aspergillus novoparasiticus* and *Aspergillus arachidicola*) produce AFB and AFG. However, different aflatoxigenic strains and the classification of aflatoxigenic *Aspergillus* are still under thorough discussion. In addition to aflatoxins, these fungi may form other mycotoxins such as aflatoxicol, versicolorin, and sterigmatocystin. The aflatoxin-producing *Aspergillus* species are found especially in areas with hot, humid climates, and contamination may occur at the pre- or postharvest stages. The aflatoxigenic fungi are saprophytic and can be found in the fig orchard dispersed in the soil, on the tree, branches, fruit, and leaves, or grow on decaying debris. In general, *A. flavus* is known to produce AFB_1 and AFB_2 and is hosted more by the above-ground parts of the vegetation in the orchard, while *A. parasiticus* produces all four aflatoxins, AFB_1, AFB_2, AFG_1, and AFG_2, is better adapted to the soil, and is located mainly in the top 10–15 cm soil layer (di Sanzo *et al.*, 2018; Schrenk *et al.*, 2020). Site-specific factors play a significant role in prevalence of aflatoxigenic fungi and their virulence. In soil, leaf, and male and female fig fruit samples taken from Fars Province, the major dried-fig-producing region in Iran, the soil *Aspergillus* population was found to be more diverse, and soil was noted as the main source of *Aspergillus* species in fig orchards (Farjood and Banihashemi, 2013). When all analyzed samples were considered, the prevalences of identified *Aspergillus* species were: *A. niger* var. *niger* (39.6%), *A. flavus* (35.1%), *Aspergillus carbonarius* (7.44%), *A. fumigatus* (6.38%), *A. japonicus* (4.25%), *A. alliceus* (3.13%), *A. ochraceus* (2.13%), and *A. terreus* (2.13%). The last four species were present chiefly in the soil samples. The results showed high genetic diversity among the isolated aspergilli. Toxin formation was not linked to the ability to form sclerotia. The toxigenic *A. flavus* isolates from soil samples had the highest aflatoxin-producing potential, signifying the role of soil management in fig orchards. Caprifigs remaining in the female fig orchard from the previous year were also a source of aspergilli. Previous studies carried out in Iran and California revealed the presence of *A. parasiticus*, but the aflatoxin formation potential was lower (Farjood and Banihashemi, 2013). In analysis of dried-fig microflora in Iraq, *A. niger* (73.4%), *A. flavus* (66.0), *A. carbonarius* (31.3%), and *A. parasiticus* (31.3%) were the species displaying the highest frequencies among 31 species from 14 genera isolated. Three isolates selected from *A. flavus* and *A. parasiticus* were tested *in vitro* for aflatoxin formation, and two of the isolates of *A. flavus* and all three *A. parasiticus* isolates proved to be aflatoxigenic (Saadullah and Abdullah, 2015).

The diversity within *A. flavus*, especially in terms of higher intra-strain differences, regarding lag phase and growth, was demonstrated in Spanish figs (Galvan *et al.*, 2021). The lag phase ranged between 0.11 and 1 day, showing that *A. flavus* may start growing readily on a dried-fig substrate. Selected strains of *A. flavus* were tested over a temperature range of 5–37°C and showed no growth at 5°C. The fastest growth occurred at 30°C and 37°C, and the intra-strain differences were minimal. As the temperature dropped toward 25°C and 16°C, the growth rates slowed down. The optimum conditions for aflatoxin formation of AFB_1 and AFB_2 were found to be 4 days at 25°C. As the temperature decreased, the incubation period required for aflatoxin formation was longer. The aflatoxin concentration increased as the incubation period extended. At 16°C, AFB_1 and AFB_2 formed after 12 days of incubation. Expression of the aflatoxin gene (*aflR*) by the two strains of *A. flavus* selected in the study exposed to different temperatures and incubation periods displayed differences. The optimal conditions for *A. flavus* were reported as 25°C and 0.96 a_w (Galvan *et al.*, 2021). Dried fig, as a monosaccharide (glucose and fructose)-rich substrate, supplies the carbon source for the fungus after sugars start to accumulate rapidly during the last week prior to ripening. The fig fruit then starts to dry on the tree and reaches the ideal a_w levels necessary for fungal growth and toxin formation. In areas where figs are sun dried, the ambient temperatures during fruit ripening and the drying period favor growth and toxin formation of aflatoxigenic fungi. Stress conditions may further promote aflatoxin formation. Dried figs grown traditionally under rainfed conditions in primary producing regions are subject to frequent drought and heat stress that promote the production of proline, which favors aflatoxin formation (Bircan *et al.*, 2008b; Gilbert and Şenyuva, 2008).

There are many factors affecting mold growth and toxin production in agricultural products and foods. These include the prevalent fungal species and strain, microbial populations, plant susceptibility, species, sex, age, plant health, the orchard (i.e. soil-management techniques), soil properties, irrigation, stress conditions such as drought, delayed harvest, mechanical damage, crop composition, temperature, relative humidity, oxygen, improper storage, and the presence of other toxins (Ciegler, 1978; Jonathan and Esho, 2010). Factors affecting fungal contamination on the fruit are mainly the location and abundance of spores, the efficiency of spore distribution, the ability of spores to penetrate the fruit, and suitability of the fruit as a suitable substrate for fungal growth and aflatoxin production (Buchanan *et al.*, 1975). Growth of toxigenic fungi on dry products increases if stress conditions such as drought, insect damage, or frost are present. The fig fruit becomes susceptible after fresh maturity although fungal infection may occur when the fruit is still green and unripe. The rapid accumulation of sugars just prior to ripening and the softening of the peel may make the fruit more disposed to fungal growth and aflatoxin formation. In traditional sun drying of figs, fresh fig fruits were not harvested at fresh maturity but remain on the tree, lose water, shrivel, and then fall to the ground as partially-dried fruit. The harvest is done by picking partially-dried figs, which are further sun dried on trays. The fig fruit is prone to fungal growth and aflatoxin formation during the period that starts with full maturity when a_w is theoretically 1.0 and ends when a_w drops to below 0.65 when fully dried. Incubation of a mixture of *A. flavus* and *A. parasiticus* on sterilized dried-fig slurry at 28°C showed that aflatoxin formation peaked around day 7 and then decreased (Gilbert and Şenyuva, 2008). These results confirmed that the most critical stage for aflatoxin formation is while the fruit continues drying on the tree. Drying causes many changes in the product as the cells lose their water and die. In addition, the physiological protection mechanism provided by living cells against saprophytes is also lost (Phillips *et al.*, 1980).

Aflatoxin incidence in dried figs was reported first in 1973, almost a decade after aflatoxin was identified, through a notification from Denmark. After the notification, studies on aflatoxin presence in dried figs were initiated in the 1980s and still continue. The correlation between BGYF under long-wave ultraviolet (UV) light and aflatoxin was confirmed and recommended as a physical method to be utilized for reduction of aflatoxins in maize lots (Dickens and Whitaker, 1981). The first study suggesting a close link between external BGYF and aflatoxin presence in dried figs paved the way toward reduction of aflatoxins in contaminated dried-fig lots (Steiner *et al.*, 1988). Aflatoxigenic fungi also produce kojic acid and BGYF on some substrates, including fig. The incidence of external and internal fluorescence varied between Turkey (Aydın) and USA (California), although the samples were collected from the same variety, Sarılop (= Calimyrna). In the USA, the internal BGYF was in 20–62% of samples analyzed, while external colonization as well as the BGYF was found in less than 10%. However, 82% of figs with external colonization had external fluorescence. In a comparison of the Conadria cultivar and Calimyrna variety, BGYF incidence was nearly 50% less in Conadria, with Calimyrna being more prone to internal infections due to widely spread dried-fruit beetle infestations (Doster and Michailides, 1998). As external fluorescence is more prominent in Turkey, all dried-fig processing plants utilize the selection of BYGF as a part of the Hazard Analysis and Critical Control Points (HACCP) and Good Manufacturing Practices. BGYF dried figs are removed manually under long-wave UV light in dark chambers or tunnels, or there are automatic sorting machines removing fluorescent dried fig fruit. The intensity of the fluorescence or the area of BGYF has no correlation with the aflatoxin concentration but shows the presence of aflatoxins to a certain extent. All BGYF are not 100% contaminated with aflatoxins (Demir *et al.*, 1990). The toxin formation on the individual fruits is limited to the area of the BGYF (Steiner *et al.*, 1988). The efficacy of automatic sorters is not so high as to completely clean the lot, so after automatic removal, all dried-fig lots in Turkey undergo visual inspection under long-wave UV, and BGYF is further checked manually (Fig. 19.1). The routine selection operation is based on external BGYF, whereas in dried figs fluorescence may also develop in the internal cavity when spores are carried into the cavity by vector insects. In the USA, BGYF can be found more inside the fruit, so the selection and removal according to external BGYF cannot be put into practice. The percentage of external and internal BGYF also varies according to site-specific factors, and is especially linked to the presence of vector insects, namely dried-fruit beetle (*Carpophilus* species) and fruit flies (*Drosophila* species) that carry the toxigenic spores inside the fruit. In processing units where dried figs are cut into slices or diced, these cut figs are additionally screened under UV light to identify and remove any with existing internal fluorescence.

Due to the health hazard posed by aflatoxins, some countries such as Sweden and Switzerland established low maximum aflatoxin limits after reportings. Later, the EU set maximum levels for some mycotoxins in certain foodstuffs, which included aflatoxins in dried figs (EC, 2002). The maximum limits could be subject to revision in time, either lowered or increased, as scientific studies reveal supporting results (EC, 2010). The EU limits for ready-to-eat dried figs or those destined directly for the consumer were set as 2 μg kg^{-1} for AFB_1 and 4 μg kg^{-1} for total aflatoxins. For dried figs channeled to industry for further processing,

Fig. 19.1. Sorting dried fig fruits to remove those contaminated with aflatoxins showing bright greenish-yellow fluorescence under long-wave ultraviolet light.

the limits are 5 μg kg^{-1} for AFB_1 and 10 μg kg^{-1} for total aflatoxins. In addition, sampling and analytical methods were introduced for official controls (EC, 2006a,b). The distribution of aflatoxin-contaminated dried figs shows a negative binomial distribution, meaning that only very few individual figs are contaminated whereas the majority have no contamination. This skewed distribution requires the development of sampling plans specific to the toxin and the product, based on operational curves to balance the "seller" and "buyer" risks (Codex Alimentarius, 2009). The Codex Alimentarius Commission based on the exposure assessment of JECFA agreed to set one maximum level in ready-to-eat dried figs for total aflatoxins at 10 μg kg^{-1} (Codex Alimentarius, 2009). Following this decision, the EU revised the maximum limits for aflatoxins and, considering the contribution of AFB_1 to the total aflatoxin levels, these limits are set as 6 μg kg^{-1} for AFB_1 and 10 μg kg^{-1} for total aflatoxins for ready-to-eat dried figs. For products going to further processing, no separate maximum limits were announced (EC, 2012).

In studies carried out on aflatoxin presence in figs, the size, origin, number of samples (4–4917), toxins, and parameters analyzed (AFB_1, total aflatoxins, range, mean or frequency, or percentage below a maximum limit) and methods of analyses have varied significantly (Table 19.1). Some authors have provided more information on the analyzed samples. The analytical methods utilized have also advanced, starting with thin-layer chromatography in the 1980s, later moving to high-performance liquid chromatography, and more recently to multi-toxin analysis by liquid chromatography with tandem mass spectrometry (LC-MS/MS) as the widely accepted method. Similar to the prevalence of aflatoxigenic *Aspergillus* species, the frequency of AFB_1 and total aflatoxins varied within a wide range. For fresh figs, aflatoxin contamination was not reported except for the study by Demir *et al.* (1990), which reported the presence of quantifiable amounts of total aflatoxins but did not specify the a_w of the analyzed figs, which could have been at senescence (Table 19.1). These studies were initially aimed at screening dried figs of interest for the presence of aflatoxins and in some cases microbial populations, and later for exposure assessment to calculate the consumer's risk arising from dried-fig consumption.

19.3.2 Ochratoxin A

OTA is a mycotoxin produced by a few species, mainly *Aspergillus* and *Penicillium* molds, which are quite common in nature (Duarte *et al.*, 2010). OTA presence is generally reported in cocoa, coffee beans, cereals, dried fruits, wine, beer, grapes, nuts, and spices (Marin *et al.*, 2013; Şen and Nas, 2010). It was first described by van der Merwe *et al.* (1965) from a culture of *A. ochraceus* (Zinedine *et al.*, 2007b). However, later studies suggested that OTA is produced by many fungal organisms belonging to genus *Aspergillus* and *Penicillium*. Fungi of the *A. ochraceus* group are generally found in soil and decaying plant residues. *A. ochraceus* can also be found in wheat, maize, cottonseed, legumes, dried figs, hazelnuts, onions, peaches, peppers, and apples with damaged surfaces (Rodricks and Lovett, 1984). The conidia producing OTA are on the grain or the fruit, and when the ambient conditions are suitable, they germinate and cause the outer surface of the product to become contaminated (Frank, 1991).

In the late 1990s, based on research work, OTA was reported as a mycotoxin with carcinogenic, nephrotoxic, teratogenic, immunotoxic, and possibly neurotoxic effects. It is also associated with Balkan endemic nephropathy in humans and has a long half-life in the human body (Barkai-Golan, 2008; Khoury and Atoui, 2010; Raiola *et al.*, 2012). The kidney is the target organ, as evidenced by functional and morphological changes in animal studies. as a result of these effects, interest in OTA contamination has increased (Hassan *et al.*, 2010). The IARC considers OTA a potential human carcinogen (class 2B) (Aish *et al.*, 2004). *In vivo* and *in vitro* studies have shown that one of the most striking effects of OTA is to inhibit protein synthesis (Pohland *et al.*, 1992). OTA has also been associated with the development of urinary tract tumors and kidney disease in different parts of the world (Esteban *et al.*, 2006). An exposure assessment of OTA in Europe revealed that the most important contributors to chronic dietary exposure were preserved meat, cheese, and grains

Table 19.1. Results of studies on frequency and concentration of aflatoxins present in dried-fig samples.

Number of fig samples analyzed	Aflatoxin-positive samples: Frequency (%)		Range (μg kg^{-1})		Mean (μg kg^{-1})		Additional remarks	Reference
	AFB_1	Total aflatoxins	AFB_1	Total aflatoxins	AFB_1	Total aflatoxins		
30	–	0	0	–	–	–	Green figs	Demir *et al.* (1990)
30	–	20	–	–	–	–	Fresh figs	Demir *et al.* (1990)
30	–	33.3	–	>10	–	–	Shriveled	Demir *et al.* (1990)
60	–	30	–	>10	–	–	Sun dried on drying tray	Demir *et al.* (1990)
30	–	50	–	<4–>10	–	–	Dried in storage	Demir *et al.* (1990)
39	–	33.3	–	<4–>10	–	–	At processing plant	Demir *et al.* (1990)
54	11.1	22.2	–	–	–	–		Sanlı *et al.* (1990)
112	–	24.0	–	–	–	–	Fig paste	Sharman *et al.* (1991)
93	–	9.0	–	<4	–	–	Dried figs	Sharman *et al.* (1991)
20	–	20.0	–	–	–	–	Dried figs	Anon. (1996)
17	0	0	0	0	0	0	Fresh figs	Anon. (2002)
4	0	0	0	0	0	0	Dried figs	Anon. (2002)
162	–	19.2	–	–	–	–	–	Özkaya *et al.* (2002)
20	–	10	–	–	–	–	–	Alghalibi and Shater (2004)
19	58	–	<4	–	–	–	–	Iamanaka *et al.* (2007)
20	5	30	–	–	–	–	–	Juan *et al.* (2008)
2643	–	11.8	–	–	–	–	–	Bircan *et al.* (2008a)
4917	–	32.0	–	–	–	–	–	Bircan *et al.* (2008b)
22	–	14.3	–	0–8.82	–	–	–	Kaya and Tosun (2013)
10	–	–	–	–	–	–	–	Asghar *et al.* (2017)
26	–	–	–	–	–	–	–	Mimoune *et al.* (2018)
473	–	–	0.83–12.15	0.14–13.54	3.17	4.03	2014	Bakırcı (2020)
390	–	–	0.61–8.63	0.20–9.37	1.73	2.87	2015	Bakırcı (2020)
387	–	–	0.59–32.30	0.30–56.07	6.08	9.15	2016	Bakırcı (2020)
467	–	–	0.63–60.76	0.22–87.75	5.95	7.47	2017	Bakırcı (2020)
292	–	–	0.61–69.92	0.28–132.37	7.24	13.14	2018	Bakırcı (2020)

AFB_1, aflatoxin B_1.

and grain-based products. Dried and fresh fruit such as grapes, figs, and dates, as well as fruit juices and nectars, were also contributing to the exposure, especially in "toddlers" and "other children" groups, but to a lesser extent than the three major categories (Anon., 2020). The EU has set common maximum limits for OTA in coffee and cereals as 5 μg kg^{-1}, 3 μg kg^{-1} for cereal products, 2 μg kg^{-1} for wine, 0.5 μg kg^{-1} for baby food, and 10 μg kg^{-1} for dried vine fruit (EC, 2006a). Although there is no common maximum limit for dried figs as yet, an 8 μg kg^{-1} limit is under discussion, as countries such as the Netherlands and Germany currently have national limits of 10 and 8 μg kg^{-1} for OTA in dried figs, respectively. Thus, many European countries are already demanding OTA analysis of dried figs from companies exporting to Europe (Pavon *et al.*, 2012a).

Ochratoxin derivatives other than OTA are ochratoxin B, ochratoxin C, methyl OTA, and ochratoxin B methyl and ethyl esters. Among these, OTA is the most common in products posing the highest health risk (Remiro *et al.*, 2012). Mold growth and subsequent OTA production are related to many factors such as temperature, a_w, pH, substrate composition, variety, cultural management, and the presence of stress conditions such as drought (Abrunhosa *et al.*, 2001; Serra *et al.*, 2003; Ilic *et al.*, 2007). In addition, the ability of *Aspergillus* species to thrive in different climates and vegetation has gained worldwide importance. Therefore, it is important to determine the conditions that affect mold formation and toxin production specific for the crop and the agroecosystem in terms of food safety (Kapetanakou *et al.*, 2009). Abarca *et al.* (1994) reported for the first time that *Aspergillus* section *Nigri* members are OTA producers. The black-spored *Aspergillus* section *Nigri* members are resistant to sun and UV light, and they form the dominant flora in sun-dried fruits and grow in a wide temperature range (Aish *et al.*, 2004). In most studies examining OTA production in *A. niger* and *A. carbonarius*, the main OTA producer is reported as *A. carbonarius* (Pitt, 2000; Battilani *et al.*, 2006; Belli *et al.*, 2006; Magnoli *et al.*, 2006; Noonim *et al.*, 2008; Frisvad *et al.*, 2011; Palumbo *et al.*, 2011; Techarat and Dachoupakan, 2012). Although OTA is produced in grapes and grape products by many species belonging to the genus *Aspergillus*, the most common species are *A. carbonarius* and *A. niger*. Unlike these molds seen in regions with hot climates, *Penicillium verrucosum* and *A. ochraceus* are the most common OTA-producing molds in cereal products grown in colder climates (Hocking *et al.*, 2002). *Aspergillus westerdijkiae* and *Aspergillus steynii*, belonging to the genus *Aspergillus* section *Circumdati*, are two new ochratoxigenic species reported in recent years, but the most virulent OTA producer is still *A. ochraceus* (Hayat *et al.*, 2012). *Aspergillus alliaceus*, *Aspergillus melleus*, and *Aspergillus sulphureus* also contribute to ochratoxin contaminations (Anlı and Alkış, 2010).

Aspergillus section *Nigri* is frequently reported as the cause of OTA contamination in dried figs (Iamanaka *et al.*, 2005; Zinedine *et al.*, 2007b; Karbancıoğlu-Güler and Heperkan, 2008; Bircan, 2009). In commercial sun-dried fig-production regions in western Turkey, *A. niger* is the most frequently isolated fungus in dried figs, followed by *A. ochraceus* (Şenyuva *et al.*, 2008). *A. carbonarius* is considered the major OTA producer in Turkish dried figs (Heperkan *et al.*, 2012b; Mutlu-İngök and Karbancıoğlu-Güler, 2014). In contrast, the main source of OTA contamination in dried figs in California was identified as *A. alliaceus* (Bayman *et al.*, 2002). In California, *A. niger* Teigh is also reported as the most common *Aspergillus* species in Calimyrna figs (Doster *et al.*, 1996). Although *A. niger* is widespread in dried-fig regions and causes smut, it may not be a virulent OTA producer. More studies are required on ochratoxigenic fungal species and OTA incidence in dried figs, as the site-specific and yearly conditions play crucial roles.

Studies on OTA contamination in Turkish dried figs (Sarılop) showed that within isolated fungal populations, *A. niger* aggregate (65.2%) was more predominant than that of *A. carbonarius* (21.7%). However, regarding virulence, 45–47% of *A. niger* aggregate isolates produced OTA whereas the rate was 88–90% for *A. carbonarius* isolates (Heperkan *et al.*, 2012a; Mutlu-İngök *et al.*, 2020). These studies confirmed that *A. carbonarius*, which is more virulent, and *A. niger* aggregate, which is widespread, are responsible for OTA contamination in Turkish dried figs. When screening ochratoxigenic molds in dried fruits, *A. carbonarius* and *A. niger* were found in figs at a rate of 4.5%, and *A. ochraceus* was not determined in dried figs (Iamanaka *et al.*, 2006).

The challenge in OTA formation arises because it can be found in fresh fruits, and there are different ochratoxin-producing species. Thus, the conditions for fungal growth and toxin formation vary greatly according to the virulent species, increasing the risk compared with aflatoxins where only a few *Aspergillus* species are responsible. Among major OTA producers in dried figs, *A. carbonarius* produces OTA within a very wide temperature range (5–45°C), but the risk is highest at 15–30°C when optimum OTA production occurs. *A. ochraceus* can produce OTA between 5°C and 40°C, but the optimum temperature is 25–30°C. *A. ochraceus* can grow and produce OTA at a temperature of 10°C within an a_w range of between 0.85 and 0.87 (Desjardins *et al.*, 2003). The optimum temperature for OTA production by *A. niger* is 13–35°C, and for *P. verrucosum* is 24–25°C. At optimum a_w, *Penicillium viridicatum* and *Penicillium cyclopium* can produce OTA at temperatures varying between 4°C and 31°C, while *P. verrucosum* can produce OTA starting from 0°C (Amezqueta *et al.*, 2012). This feature makes these *Penicillium* species the main producers in cooler temperate climates, whereas *Aspergillus* species remain the major cause of OTA in milder climates.

In the main fig-producing Aegean region of Turkey, the dominant flora in sun-dried fig fruits are *Aspergillus* section *Nigri*. Black *Aspergillus* members, which can grow in a wide temperature range (6–47°C), can also form OTA between 10°C and 45°C (Karbancıoğlu-Güler and Heperkan, 2008). *A. carbonarius* isolated from Turkish dried figs grew at temperatures of 15–35°C and produced OTA. At temperatures of 5°C and 10°C, *A. carbonarius* isolates ceased growth. Although the temperature requirement for OTA production is lower than the optimum temperature required for the growth of *A. carbonarius*, both growth and OTA formation require high a_w levels. The minimum a_w levels were 0.85 for growth of *A. carbonarius* isolates and 0.89 for OTA production. The main OTA producers in Turkish dried figs belong to *Aspergillus* division Nigri. The frequency of *A. ochraceus* and *P. verrucosum* is low (Karbancıoğlu-Güler and Heperkan, 2008). Black *Aspergillus* species are grouped into three according to their phenotypic characters as *A. niger* aggregate, *A. carbonarius*, and uniseriate *Aspergillus* (Abarca *et al.*, 2004). In general, the climatic conditions present during fig-fruit development, harvest, and the drying phases favor the development of black-spored *Aspergillus* species, which cause smut and may result in OTA formation (Aksoy, 2003; Bircan, 2009). However, despite quite similar climatic conditions, the prevalence of the *Aspergillus* species identified in Turkish and Californian fig orchards varied, possibly because of differences in terrain and the cultural practices applied (Doster *et al.*, 1996; Heperkan *et al.*, 2012b).

OTA contamination in dried figs is reported for almost every country of origin, either through the alert systems of the EU and other importing countries or from research results (Table 19.2). The frequency of OTA-contaminated samples varies between 0% and 54.3% due to either the origin or the quality of samples. The Aegean Exporters' Association in İzmir, Turkey, has a long-term monitoring program on OTA frequency and contamination levels in dried figs, which showed a high impact of the yearly conditions (unpublished data). However, OTA occurrence in dried whole-fig samples derived from different quality classes, fig paste, and moldy culled figs between the 1999 and 2001 seasons showed that a total of 80% of samples had no OTA contamination. However, due to the yearly conditions, the percentage of samples below the detection limit was 40% in 2000, and 38% of the samples were contaminated at levels ranging from 2 to 10 μg kg^{-1}. The ratio of OTA contamination was high (97%) in fig-paste samples, and the levels ranged between 0.5 and 50 μg kg^{-1}. In all 13 moldy culled figs, OTA was present but the contamination level was not regularly distributed and ranged between 0.9 and 2380 μg kg^{-1} (Aksoy *et al.*, 2003). Italian products made from dried fig such as molasses and salami had also higher OTA contamination compared with dried figs for table consumption due to the use of lower-quality figs in processing (di Sanzo *et al.*, 2018).

During processing and packaging of dried figs, moldy and culled figs are removed as they are defects and mostly not suitable for human consumption. In Turkey, there is a special system put

Table 19.2. Results of studies on the frequency and concentration of dried-fig samples contaminated with OTA.

Number of dried-fig samples analyzed	OTA-positive samples				Additional remarks	Reference
	Frequency (%)	Min OTA μg kg^{-1}	Mean OTA μg kg^{-1}	Max OTA μg kg^{-1}		
54	20.3	19	185	–		Sanlı *et al.* (1990)
282	5.7	0.5	–	62.9		Demir *et al.* (1990)
103	3.0	5.2	–	8.8		Özay and Alperden (1991)
4	100	60	–	120		Zohri and Abdel-Gawad (1993)
45	0	–	–	–		Pala *et al.* (1994)
15	40.0	–	–	9600		Doster *et al.* (1996)
20		0.2		0.8		Anon. (1999)
21	4.8	NA	NA	151		Anon. (2002)
20	–	–	–	–	Whole figs sampled between 1999 and 2001	Aksoy *et al.* (2003)
13	–	0.9	–	2380	Moldy culled figs	Aksoy *et al.* (2003)
6	0	0	0	0		Abdulkadar *et al.* (2004)
20	10	70.0	–	160.0		Alghalibi and Shater (2004)
	26	5.0	–	–		Iamanaka *et al.* (2005)
		0.03	0.33	1.42		Zinedine *et al.* (2007b)
15	46.7	1.01	1.35	4.33		Dündar (2008)
	47.8	–	–	–	96.5% of samples had <1 μg kg^{-1}	Karbancıoğlu-Güler and Heperkan (2008)
98	18.4	0.87	–	24.4		Bircan (2009)
115	48.0	0.1	–	15.3		Heperkan *et al.* (2012a)
35	54.3	3.15	–	277		Pavon *et al.* (2012b)
21	52.4	–	1.77	6.51	No samples above limits	Kaya and Tosun (2013)
	45.5	–	–	<10		Heshmati *et al.* (2017)
14	21.4	–	4.20	–		Iqbal *et al.* (2018)

NA, not available.

in place for the collection and destruction of mycotoxins, mainly for BGYF aflatoxin-contaminated figs, to prevent circulation in the market. Moreover, if culled figs are destined for the food industry, safety checks for the presence of mycotoxins are compulsory. Black-spored *Aspergillus* generally develops inside the fruit cavity of figs so is mostly invisible. During processing, dried figs are checked visually through the ostiole opening to see whether smut is present and if detected, they are separated as rejects.

19.3.3 Fumonisins

Fumonisins are an important group among mycotoxins and first identified in animal feed. Their chemical structure is based on a polyhydroxy alkylamine esterified with two carbon acids. The differences among fumonisins are due to the presence and position of the free hydroxyl groups. Currently, four series of fumonisins are identified: A, B, C, and P. The B group, FB_1, FB_2, and FB_3, are the most toxic fumonisins found in food and feed. The fumonisin B (FB) series of fumonisins have a free amine. The fumonisins were detected first in feed contaminated with *F. moniliforme* (*F. verticilloides*); however, the number of fumonisin-producing fungal species is increasing. The fumonisin-producing fungi are mainly *Fusarium* species (*F. verticillioides*, *F. proliferatum*, *F. fujikuroi*, *F. globosum*, *F. nygamai*, *F. subglutinans*, *F. napiforme*, *F. dlamini*, *F. anthophilum*, *F. beomiforme*, *F. polyphialidicum*, *F. thapsinum*, and *F. oxysporum*), but new fumonisin producers such as *Aspergillus* (*A. niger* and *A. welwitschiae*), and *Alternaria alternata* have been determined (Kösoğlu-Vural, 2008; Gil-Serna *et al.*, 2019). *Fusarium* species are among the most frequently found fungal populations, existing as both saprophytes hosted by decaying litter and plant pathogens causing various diseases such as wilt and rot. Fumonisin-producing fungal species are found widespread in fig orchards. *F. verticilloides* is the main cause of internal rot (pink rot) in figs, especially in varieties requiring caprification for fruit set. The male fig fruit is the main host, and the fungus is transmitted by the fig wasp during caprification. Together with *F. verticillioides*, *F. solani*, *Fusarium dimerum*, *F. proliferatum*, *F. subglutinans*, *Fusarium lacti*, and *Fusarium ramigenum* have also been isolated from figs (Moretti *et al.*, 2000). The *Fusarium* spores carried by the fig wasp first germinate on the wasp remains, and sporulation then moves toward the fruit tissue. The temperature range for growth of *F. verticillioides* is rather wide, but the optimal temperature for growth is 28°C. The temperature requirement for ascospore formation is 16°C. The optimum conditions for FB_1 production by *F. verticillioides* is 30°C and 0.97 a_w, whereas for *F. proliferatum*, 15°C and 0.97 a_w provide the optimum conditions. The value a_w is the critical factor, and below 0.92 a_w, both growth and toxin formation slow down, terminating at 0.88 a_w. *Fusarium* species attack the floral organs as an endophyte. Fig fruit infected by *Fusarium* species starts to rot with ripening, especially at the ostiole end, and a pink exudate starts dropping from the ostiole. Pest damage and physiological stress conditions promote growth of the fungus. After infection with *F. verticilloides*, there was a delay of 6–7 weeks for fungal growth due to the inhibitory effect of the fig latex; however, fig latex had no effect on *A. niger*, which started growing immediately (Kösoğlu-Vural, 2008).

The fumonisins are reported to create health problems in humans and animals. Some animals, as Equidae, are more sensitive. Fumonisins cause leucoencephalomalacia in horses and pulmonary edema in pigs. Studies have revealed that FB_1 does not pass to the animal products in the form of milk and meat. In humans, they are linked to esophageal cancer, especially in southern Africa and China where maize, a staple food of the poor, is contaminated with fumonisins. There are already limits set for fumonisins in maize and its products, and for baby food. There is no maximum limit for fumonisins in dried figs.

There have been a few surveys on fumonisin incidence in dried fig fruit in Turkey and Italy. Karbancıoğlu-Güler and Heperkan (2009) found FB_1 contamination in dried figs sampled during drying from different districts in western Turkey to be 71.8% in 2003 and 79.5% in 2004. The ranges were 0–3649 μg kg^{-1} and 0–2760 μg kg^{-1} in 2003 and 2004, respectively. However, in FB_1-positive samples, only 9.6% had concentrations exceeding 1 μg kg^{-1}. In studies carried out in the same region, the frequency of fumonisin-contaminated samples was found to differ in 2004 (52%) and 2005 (80%). In 2004, there were no dried figs contaminated with FB_1+FB_2 in the range of 0.21–0.40 μg g^{-1}, whereas 17 samples were within this range in 2005. The highest FB_1 (0.32 μg g^{-1}) and total FB_1+FB_2 (0.39 μg g^{-1}) levels were found in 2005. The yearly difference in fumonisin occurrence was attributed to higher rainfall and humidity in 2005, which favored sporulation and growth of *Fusarium* species. The critical periods were during the entrance of the fig wasp from the overwintering mamme to caprifig profichi, from caprifig profichi to the female main crop fruit (caprification), and during fruit ripening (Kösoğlu-Vural, 2008). Kaya and Tosun (2013) analyzed 19 organic certified dried fig samples and found the fumonisin incidence, without specifying the fumonisin type, to be 63.1% and the range to be 0–1645 μg kg^{-1}. Some *Aspergillus* section *Nigri* isolates known to produce OTA in dried figs may also produce fumonisin. When fumonisin-producing fungal isolates were evaluated for OTA production, *Aspergillus* section *Nigri* isolates considered to be potential OTA producers did not produce detectable levels of OTA (Daşkaya, 2011). In Italy, dried figs and fig-based product samples had FB_1 contamination; the frequency was five out of 28 and FB_1 contamination ranged between 153.81 and 5412.96 μg kg^{-1} in dried figs, while in fig-based products, the frequency was three out of 19 with a range of 249.97–1368.94 μg kg^{-1}. In eight baked dried-fig samples, no FB_1 was detected (di Sanzo, *et al.*, 2018).

19.4 Co-occurrence of Mycotoxins in Dried Figs

Mycotoxins may form as a single toxin or sometimes as a mixture of different mycotoxins. Co-occurrence has been the core objective in various studies in the field of mycotoxins, as the presence of multiple mycotoxins may create an additive effect and can be more detrimental for health. Analytical methods that can analyze several mycotoxins in one sample have contributed to enhancing these co-occurrence studies. In the processed products of figs, mycotoxin mixtures may also result from the ingredients used in processing, such as nuts, cocoa, and spices (di Sanzo *et al.*, 2018). Dried figs act as a suitable substrate during the senescence phase and may host various fungal species having the potential to form different toxins (Benlioğlu *et al.* 2008). This phenomenon stimulated various research studies to identify the presence of multiple mycotoxins in dried figs in Turkey, Iran, Italy, and China. Heperkan *et al.* (2012a) analyzed the mycoflora of dried-fig samples collected during drying. All fig samples were infected by fungi, with 94.7% having one or more toxigenic species. *Aspergillus* section *Nigri* members were by far the most prominent at 93.9%. The other fungal species identified were *Fusarium* species, *Aspergillus* section *Flavi*, and *Penicillium* species. The samples were analyzed for aflatoxins, OTA, FB_1, and cyclopiazonic acid. The percentage of samples contaminated with one of the four mycotoxins analyzed was 33.0%, while 5.2% contained three mycotoxins and 3.5% contained four mycotoxins. Kaya and Tosun (2013) investigated the co-occurrence of aflatoxins, OTA, and fumonisin in dried figs and molasses made from lower-quality dried figs. In Iran, the share of samples contaminated with aflatoxins and OTA was 18.2% (Heshmati *et al.*, 2017). Şenyuva *et al.* (2008) reported a lower probability of having both aflatoxin and OTA contamination in Turkish dried figs: 3.8% in 2003 and 4.9% in 2004. Di Sanzo *et al.* (2018) reported that 7.3% of dried figs or fig-product samples were contaminated with two, 5.5% by three, 10.9% by four, and 5.5% by five mycotoxins. The presence of fusaric acid and beauvericin were first reported in Italian dried figs. Analysis of 50 individual dried figs contaminated with black *Aspergillus* showed that OTA was present in 14% at a range of 0.66–250 μg kg^{-1}. When dried figs showing BGYF were analyzed for OTA, only one in nine figs was contaminated with OTA (Cankaya, 2018). Özer (1996) did not find OTA in non-fluorescent figs, and found OTA contamination at a level of 10 μg kg^{-1} in one of the BGY fluorescent fig fruits that was analyzed individually.

In dried-fig samples taken from production areas in the Menderes Basin in western Turkey in 2000 and 2001, OTA was found in 36% of the samples at detectable levels. In 2000, to determine the OTA incidence in BGYF samples, 74 samples were screened under long-wave UV light and 35 showed BGYF. Within this BGYF-positive group, OTA was detected in nearly half (49%). In 2001, 48 out of 74 samples had BGYF, but only 11 (23%) were found to be contaminated with OTA in the range 0.4–1710 μg kg^{-1}.

BGYF seen under long-wave UV light is due to kojic acid and does not confirm the presence of aflatoxins (Şahin, 2003). In another survey based on analysis of a large number of dried-fig samples, Şenyuva *et al.* (2008) reported that no significant relationship was found between AFB_1 and OTA levels. Taking into consideration the newly discovered mycotoxins and toxigenic fungi, co-occurrence in dried figs seems set to continue as a research topic.

19.5 Strategies for Prevention or Reduction of Mycotoxins

Control strategies applied in the preharvest period are important to reduce fungal infection and toxin formation (Aksoy, 2003; Amezqueta *et al.*, 2012). Most of the mycotoxigenic fungi are the most prevalent fungal species in fig orchards, and some such as *Fusarium* and *Alternaria* species also cause internal or external rots. Removing rotten, damaged, and moldy berries from healthy fruit before drying and processing, and avoiding excessive irrigation during cultivation, was shown to reduce the amount of OTA produced by black *Aspergillus* in grapes (Kozakiewicz *et al.*, 2004; Varga and Kozakiewicz, 2006; Meyvacı *et al.*, 2012). To develop and apply GAP as a preventive approach to reduce mycotoxins in fig orchards, the first step is long-term monitoring of mycotoxin frequency and concentrations, and of prevalent fungal populations. The next step could be to design specific GAP and HACCP protocols at critical points (Anon., 2001). The Code of Practice developed by the Codex Alimentarius (2008) suggests an approach. In varieties that require pollination for fruit set (Smyrna-type and the second crop of San Pedro-type varieties), transmission of fungal spores by the fig wasp can be restricted by removing unhealthy male fruits before bringing caprifigs to the orchard. The orchard floor management strategy should consider that tillage just prior to harvest may scatter toxigenic microbial populations, increasing the risks. Effective pest management against dried-fruit beetle (*Carpophilus* species) and fruit flies (*Drosophila* species) is important to prevent transmission of spores into the fruit cavity. Care at harvest is of the utmost importance, and moldy, rotten, and damaged fruits should be separated from healthy, sound ones starting from the first day of harvesting, and then throughout drying and while in storage. The defective fruit may create an inoculum for dispersal of toxigenic fungi so must be kept away from the sound, healthy figs.

Biocontrol is another strategy that has promising results. Although not widely applied in practice, biocontrol studies are especially targeted to suppress aflatoxigenic strains and reduce aflatoxin formation at the preharvest stages by using non-toxigenic strains as biocontrol agents. Commercial biocontrol products are also available for soil and foliar application. The biocontrol strategy is based on repeated introduction of non-toxigenic *A. flavus* strains to reduce the size of the toxigenic population, but long-term impacts on the soil microbiota need to be taken into account (Doster *et al.*, 2008; Ehrlich, 2014).

Artificial drying or structures such as well-ventilated high tunnels may also help to prevent toxin formation or lower the concentrations by shortening the period during which a_w levels favor fungal growth and toxin formation (Şahin, 2003; Benalia *et al.*, 2015; Mutlu-İngök and Karbancıoğlu-Güler, 2015; Petric *et al.*, 2018).

At the processing stage, removal of moldy, rotten, and damaged fruits, or BYGF fruit in the case of aflatoxins, manually or using sorting machines may reduce the final mycotoxin content, especially in combination with proper drying techniques. Automatic sorting machines are commercially available and are already used in processing plants for identification and removal of BGYF dried figs to lower aflatoxin contamination in traded lots. These fluorescent figs can only be identified under UV light as with the naked eye they look healthy and sound. Thus, a special program must be put in place to prevent circulation of these rejected contaminated figs in the market. Studies to identify and remove dried figs contaminated with mycotoxins are ongoing but are still at pilot level, except those for aflatoxins based on BGYF removal.

19.6 Concluding Remarks

Mycotoxigenic fungi are abundant in fig-producing regions worldwide, and fig fruit with its sugar-rich composition and soft and juicy peel provides favorable conditions as a complete substrate for fungal growth and toxin formation.

Many studies have been carried out and are still ongoing regarding the incidence of mycotoxins and the causal factors. As new mycotoxins are identified and/or health implications are revealed, fig has become a subject of interest. There are a number of mycotoxins found in dried figs, with aflatoxins, OTA, and fumonisins being found most frequently. Patulin, kojic acid, zearalenone, HT-2 toxin, tenuazonic acid, alternariol, cyclopiazonic acid, fusaric acid, beauvericin, fusaproliferin, and ergosterol have also been reported in studies. Fresh figs may have OTA and fumonisins, but no aflatoxins have been found in fresh figs (except for the study by Demir *et al.*, 1990) due to the a_w requirements of the toxigenic fungi. Although dried figs are among products that are very often contaminated with aflatoxins and OTA, exposure assessments that have been carried out have demonstrated that its contribution to total exposure is not significant due to the low consumption rates, even in Turkey (JECFA, 2008; Başeğmez, 2019; Schrenk *et al.*, 2020).

The co-occurrence of more than one mycotoxin is one of the major topics raising interest due to the additive effect of co-occurring mycotoxins and to more recent analytical methods that can concurrently analyze tens of mycotoxins within a short period of time in the same sample. Studies on the co-occurrence of mycotoxins, especially of aflatoxins, have not displayed high co-occurrence rates (Şenyuva *et al.*, 2005). However, these studies will remain a priority.

Based on diversity in the prevalence and toxigenicity of toxigenic fungi, it is recommended to carry out studies in existing fig regions or in areas where fig is being newly introduced. In this respect, climate change may also bring new risks. Aflatoxin-producing *Aspergillus* species are limited in number and are more adapted to milder climates in subtropical regions; however, new species are still being identified. In contrast, OTA and fumonisins can be produced by a higher number of fungal genera and species adapted to a wider climate range, making local assessment more important. Major dried-fig-producing regions such as the Aegean region (western Turkey) and California (USA) have black-spored *Aspergillus* as the dominant species, but there are differences regarding diversity and toxigenicity. Mycotoxins are highly stable compounds and detoxification is not allowed for food commodities, although there are some scientific studies on ozone or sulfur treatments. Therefore, preventive measures that limit the spread, growth, and toxin formation of toxigenic fungi at the orchard level (e.g. orchard hygiene, no-tillage, fast-drying techniques) and temperature and moisture control during processing and storage will help to reduce contamination levels. At the orchard level, biocontrol strategies focusing on non-toxigenic strains have some potential but require more exploration. Various studies on developing automated sorting machines based on hyperspectral images at various wavelengths to remove contaminated figs are ongoing. Among these, only those selecting and removing the BGYF fruit are in practice for reduction of aflatoxin contamination. Figs are known as products that are good for health and they have been used for medicinal purposes in many civilizations for many centuries. Consumers with more attention on health lead the market for organic figs. Thus, producing figs free of any mycotoxin contamination is one of the basic requirements for all fig-producing regions. Because fig is a suitable substrate and there are additional challenges such as climate change, more scientific work is expected regarding newly identified mycotoxins and in different fig-producing regions.

References

Abarca, M.L., Bragulat, M.R., Castella, G., and Cabanes, F.J. (1994) Ochratoxin A production by strains of *Aspergillus niger* var. niger. *Applied and Environmental Microbiology* 60, 2650–2652.

Abarca, M.L., Accensi, F., Cano, J., and Cabanes, F.J. (2004) Taxonomy and significance of black aspergilli. *Antonie van Leeuwenhoek* 86, 33–49.

Abdulkadar, A.H.W., Abdulla, A.A.A., Afrah, M.A., and Jassim, H.A. (2004) Mycotoxins in food products available in Qatar. *Food Control* 15, 543–548.

Abrunhosa, L., Paterson, R.R.M., Kozakiewicz, Z., Lima, N., and Venancio, A. (2001) Mycotoxin production from fungi isolated from grapes. *Letters in Applied Microbiology* 32, 240–242.

Adejumo, T.O. and Adejoro, D.O. (2014) Incidence of aflatoxins, fumonisins, trichothecenes and ochratoxins in Nigerian foods and possible intervention strategies. *Food Science and Quality Management* 31, 127–146.

Aish, J.L., Rippon, E.H., Barlow, T., and Hattersley, S.J. (2004) Ochratoxin A. In: Magan, N. and Olsen, M. (eds) *Mycotoxins in Food: Detection and Control*. Woodhead Publishing, Cambridge, UK pp. 307–338.

Aksoy, U. (1990) Aflatoksin. In: *Tarımsal Uygulama ve Araştırma Merkezi Yayım Bülteni*. Bornova, İzmir, Turkey.

Aksoy, U. (2003) Kuru incir üretiminde aflatoksinler ve okratoksin A'nın engellenmesine yönelik iyi tarım uygulamaları. In: *Mikotosinler Biyoteknolojisi; Sorunlar ve Çözümler Çalıştayı, June 17–20, 2003, İzmir, Turkey*.

Aksoy, U., Sabır, E., Eltem, R., Kıraç, S., Sarıgül, N., *et al.* (2003) Kuru incirlerde okratoksin a'nın potansiyel kontaminasyon riskinin araştırılması. In: *Ulusal Mikotoksin Sempozyumu*, September 18–19, 2003, İstanbul, Turkey.

Aksoy, U., Eltem, R., Meyvaci, K.B., Altindisli, A., and Karabat, S. (2007) Five-year survey of ochratoxin A in processed sultanas from Turkey. *Food Additives and Contaminants* 24, 292–296.

Aksoy, U., Can, H.Z., Meyvacı, K.B., Şen, F., İrget, E., *et al.* (2013) Kuru İncirde İyi Uygulamaların Aflatoksinlerin Önlenmesi/Azaltılması Üzerindeki Etkilerinin Belirlenmesi. In: *Ege İhracatçı Birlikleri Proje Sonuç Raporu*. İzmir, Turkey.

Alghalibi, S.M.S. and Shater, A.R.M. (2004) Microflora and mycotoxin contamination of some dried fruits in Yemen Republic. *Assiut University Bulletin Environmental Researches* 7, 19–27.

Altındişli, A., Aksoy, U., Eltem, R., Çakır, M., and Meyvacı, K.B. (2002) Çekirdeksiz kuru üzümde okratoksin A oluşumunun nedenleri ve azaltıcı önlemler üzerinde araştırmalar. In: *Türkiye V. Bağcılık ve Şarapçılık Sempozyumu, October 5–9, 2002, Cappadocia, Nevşehir, Turkey*, pp. 112–121.

Amezqueta, S., Schorr-Galindo, S., Murillo-Arbizu, M., González-Penas, E., Lopez de Cerain, A., and Guiraud, J.P. (2012) OTA-producing fungi in foodstuffs: a review. *Food Control* 26, 259–268.

Anlı, E. and Alkış, M.İ. (2010) Ochratoxin A and brewing technology: a review. *Journal of the Institute of Brewing* 116, 23–32.

Anon. (1996) *Aflatoxin Surveillance of Retail and Imported Nuts, Nut Products, Dried Figs and Fig Products*. Food Surveillance Information Sheet No. 81. Joint Food Safety and Standard Group, Ministry of Agriculture, Fisheries and Food, London.

Anon. (1999) *Survey of Retail Products for Ochratoxin-A*. Food Surveillance Information Sheet No. 185. Joint Food Safety and Standard Group, Ministry of Agriculture, Fisheries and Food, London.

Anon. (2001) *Manual on the Application of the HACCP System in Mycotoxin Prevention and Control*. Food and Nutrition Paper No. 73. Food and Agriculture Organization of the United Nations, Rome.

Anon. (2002) *Survey of nuts, nut products and dried tree fruits for mycotoxins*. Food Surveillance Information Sheet No. 21/02. Joint Food Safety and Standard Group, Ministry of Agriculture, Fisheries and Food, London.

Anon. (2020) Risk assessment of ochratoxin A in food. *EFSA Journal* 18, 6113–6262.

Asghar, M.A., Ahmed, A., Zahir, E., Asgher, M.A., Iqbal, J., and Walker, G. (2017) Incidence of aflatoxins contamination in dry fruits and edible nuts collected from Pakistan. *Food Control* 78, 169–175.

Ashiq, S. (2015) Natural occurrence of mycotoxins in food and feed: Pakistan perspective. *Comprehensive Reviews in Food Science and Food Safety*, 14, 159–175.

Atalla, M.M., Hassanein, N.M., El-Beih, A.A., and Youssef, Y.A.G. (2003) Mycotoxin production in wheat grains by different aspergilli in relation to different relative humidities and storage periods. *Nahrung* 47, 6–10.

Bakırcı, G.T. (2020) Investigation of aflatoxins levels in commercial dried figs from western Turkey. *International Food Research Journal* 27, 245–251.

Barkai-Golan, R. (2008) *Aspergillus* mycotoxins. In: Barkai-Golan, R. and Paster, N. (eds.) *Mycotoxins in Fruits and Vegetables*. Academic Press, San Diego, California, pp. 115–152.

Başeğmez, H.I.O. (2019) Dietary exposure assessment of aflatoxin from dried figs in Turkey. *Hittite Journal of Science and Engineering* 6, 173–177.

Bata, A., Rafai, P., and Kovacs, S. (2001) Investigation and a new evaluation method of the resistance of maize hybrids grown in Hungary to *Fusarium* moulds. *Phytopathology* 149, 107–111.

Battilani, P., Giorni, P., Bertuzzi, T., Formenti, S., and Pietri, A. (2006) Black aspergilli and ochratoxin A in grapes in Italy. *International of Food Microbiology* 111, S53–S60.

Bayman, P. and Baker, J.L. (2006) Ochratoxins: a global perspective. *Mycopathologia* 162, 215–223.

Bayman, P., Baker, J.L., Doster, M.A., Michailides, T.J., and Mahoney, N.E. (2002) Ochratoxin production by the *Aspergillus ochraceus* group and *Aspergillus alliaceus*. *Applied and Environmental Microbiology* 68, 2326–2329.

Belli, N., Bau, M., Marin, S., Abarca, M.L., Ramos, A.J., and Bragulat, M.R. (2006) Mycobiota and ochratoxin A producing fungi from Spanish wine grapes. *International Journal of Food Microbiology* 111, 40–45.

Benalia, S., Bernardi, B., Cubero, S., Leuzzi, A., Larizza, M., and Blasco, J. (2015) Preliminary trials on hyperspectral imaging implementation to detect mycotoxins in dried figs. *Chemical Engineering Transactions* 44, 157–162.

Benlioğlu, S., Yıldız, A., and Başpınar, N. (2008) Aydın İli'nden ihraç edilen kuru incirlerde fungal bulaşıklık. *ADÜ Ziraat Fakültesi Dergisi* 5, 3–8.

Bennett, J.W. and Klich, M. (2003) Mycotoxins. *Clinical Microbiology Reviews* 16, 497–516.

Betina, V. (1989) Biological effects of mycotoxins. In: Betina, V. (ed.) *Mycotoxins: Chemical, Biological and Environmental Aspects*. Elsevier Science Publishers, Amsterdam, pp. 42–58.

Bhatnagar, D., Payne, G.A., Cleveland, T.E., and Robens, J.F. (2004) Mycotoxins: current issues in USA. In: Barug, D., van Egmond, H.P., Lopez-Garcia, R., van Osenbruggen, W.A.,and Visconti, A. (eds) *Meeting the Mycotoxin Menace*. Wageningen Academic Publishers, Wageningen, The Netherlands, pp. 17–47.

Bircan, C. (2009) Incidence of ochratoxin A in dried fruits and co-occurrence with aflatoxins in dried figs. *Food Chemical Toxicology* 47, 1996–2000.

Bircan, C., Barringer, S.A., Ulken, U., and Pehlivan, R. (2008a) Aflatoxin levels in dried figs, nuts and paprika for export from Turkey. *International Journal of Food Science and Technology* 43, 1492–1498.

Bircan, C., Barringer, S.A., Ulken, U., and Pehlivan, R. (2008b) Increased aflatoxin contamination of dried figs in a drought year. *Food Additives and Contaminants* 25, 1400–1408.

Bottalico, A. (1998) Fusarium diseases of cereals: species complex and related mycotoxin profiles in Europe. *Journal of Plant Pathology* 80, 85–103.

Buchanan, J.R, Sommer, N.F., and Fortlage, R.J. (1975) *Aspergillus flavus* infection and aflatoxin in fig fruits. *Microbiology* 30, 1050–1051.

Cankaya, C.C. (2018) *Siyah küf ile bulaşık kuru incirlerde okratoksin A varlığının araştırılması (Research on presence of OTA in dried figs contaminated with black mold).* MSc thesis, Graduate School on Natural and Applied Sciences, Ege University, İzmir, Turkey.

Ciegler, A. (1978) Fungi that produce mycotoxins: conditions and occurrence. *Mycopathologia* 65, 5–11.

Codex Alimentarius (2008) *Code of Practice for the Prevention and Reduction of Aflatoxin Contamination in Dried Figs*. CAC/RCP 65-2008. Food and Agriculture Organization of the United Nations, Rome/World Health Organization, Geneva.

Codex Alimentarius (2009) General Standard for Contaminants and Toxins in Food and Feed CXS 193-1995. Adopted in 1995. Revised in 1997, 2006, 2008, 2009. Available at www.fao.org/fao-who-codexalimentarius (accessed December 16, 2021).

Daşkaya, C. (2011) Aspergillus Section Nigri üyelerinin fumonisin oluşturma potansiyelleri ve pozitif suşların okratoksin A üretimi özelliklerinin incelenmesi. MSc thesis, Istanbul Technical University Graduate School, Istanbul, Turkey.

Demir, T., Özar, A.İ., Gülsevim, O., Çoksöyler, N., Aksoy, U., and ve Düzbastılar, M. (1990) Ege Bölgesinde İncirlerde Görülen Aflatoksin, Okratoksin-A Oluşumu ile Önlenmesi Üzerinde Araştırma, Tarım ve Köyişleri Bakanlığı Projesi. Project No, KKGA/B/03/F/052. Project Final Report. İzmir, Turkey.

Desjardins, A., Maragos, C., Norred, W., Pestka, J., Philips, T., *et al.* (2003) *Mycotoxins: Risks in Plant, Animal and Human Systems*. Council for Agricultural Science and Technology, Ames, Iowa.

Desjardins, A.E. and Proctor, R.H. (2007) Molecular biology of *Fusarium* mycotoxins. *International Journal of Food Microbiology* 119, 47–50.

di Sanzo, R., Carabetta, S., Campone, L., Bonavita, S., Iaria, D., *et al.* (2018) Assessment of mycotoxins co-occurrence in Italian dried figs and in dried figs-based products. *Journal of Food Safety* 38: e12536.

di Stefano, V., Avellone, G., Pitonzo, R., Capocchiano, V.G., Mazza, A., *et al.* (2015) Natural co-occurrence of ochratoxin A, ochratoxin B and aflatoxins in Sicilian red wines. *Food Additives & Contaminants: Part A. Chemistry, Analysis, Control, Exposure & Risk Assessment* 32, 1343–1351.

Dickens, J.W. and Whitaker, T. (1981) Bright greenish-yellow fluorescence and aflatoxin in recently harvested yellow corn marketed in North Carolina, *Journal of the American Oil Chemists' Society* 58, 973–975.

Doko, M., Rapior, S., Visconti, A., and Schjoth, J.E. (1995) Incidence of levels of fumonisin contamination in maize by genotypes grown in Europe and Africa. *Journal of Agriculture and Food Chemistry* 43, 429–434.

Doster, M.A. and Michailides, T.J. (1998) Production of bright greenish yellow fluorescence in figs infected by *Aspergillus species* in California orchards. *Plant Disease* 82, 669–673.

Doster, M., Michailides, T., and Morgan, D. (1996) *Aspergillus* species and mycotoxins in figs from California orchards. *Plant Disease* 80, 484–489.
Doster, M.A., Michailides, T.J., Doster, L., and Cotty, P.J. (2008) Biocontrol of aflatoxins in figs. *Acta Horticulturae* 798, 223–226.
Duarte, S.C., Pena, A., and Lino, C.M. (2010) A review on ochratoxin A occurrence and effects of processing of cereal and cereal derived food products. *Food Microbiology* 27, 187–198.
Dündar, D. (2008) İstanbul İlinde Satışa Sunulan Kuru İncir ve Kuru Üzümlerde Aflatoksin ve Okratoksin A Varlığı. MSc thesis, Food Engineering Department, Namık Kemal University, Tekirdağ, Turkey.
EC (European Commission) (2002) Commission Regulation (EC) No 472/2002 of 12 March 2002 amending Regulation (EC) No 466/2001 setting maximum levels for certain contaminants in foodstuffs. *Official Journal of the European Union* L75, 18–20.
EC (European Commission) (2006a) Commission Regulation (EC) No 1881/2006 of 19 December 2006 setting maximum levels for certain contaminants in foodstuffs. *Official Journal of the European Union* L364, 5–24.
EC (European Commission) (2006b) Commission Regulation (EC) No 401/2006 of 23 February 2006 laying down the methods of sampling and analysis for the official control of the levels of mycotoxins in foodstuffs. *Official Journal of the European Union* L70, 12–34.
EC (European Commission) (2010) Commission Regulation (EU) No 105/2010 of 5 February 2010 amending Regulation (EC) No 1881/2006 setting maximum levels for certain contaminants in foodstuffs as regards ochratoxin A. *Official Journal of the European Union* L35, 7–8.
EC (European Commission) (2012) Commission Regulation (EU) No 1058/2012 of 12 November 2012 amending Regulation (EC) No 1881/2006 as regards maximum levels for aflatoxins in dried figs. *Official Journal of the European Union* L313, 14–15.
Ehrlich, K. (2014) Non-aflatoxigenic *Aspergillus flavus* to prevent aflatoxin contamination in crops: advantages and limitations. *Frontiers in Microbiology* 5: 50.
Esteban, A., Abarca, M.L., Bragulat, M.R., and Cabanes, F.J. (2006) Influence of water activity and temperature on growth of isolates of *Aspergillus* section Nigri obtained from grapes. *Food Microbiology* 23, 634–640.
Farjood, E. and Banihashemi, Z. (2013) Incidence of *Aspergillus* species and mycotoxins in dried figs in southern Iran. *Iranian Journal of Plant Pathology* 48, 183–189.
Frank, H.K. (1991) Food contamination by ochratoxin A in Germany. In: Castegnaro, M., Plestina, R., Dirheimer, G., Chernozemsky, I.N., and Bartsch, H. (eds) *Mycotoxins, Endemic Nephropathy and Urinary Tract Tumors*. International Agency for Research on Cancer, Lyon, France, pp. 77–81.
Frisvad, J.C., Larsen, T.O., Thrane, U., Meijer, M., Varga, J., *et al.* (2011) Fumonisin and ochratoxin production in industrial *Aspergillus niger* strains. *PLoS One* 6: e23496.
Gaddi, B.L. (1973) Mycotoxin-producing potential of fungi isolated from cheese. PhD thesis, University of Wisconsin-Madison, Madison, Wisconsin.
Galvan, A.I., Rodriguez, A., Martin, A., Serradilla, M.J., Martinez-Dorado, A., and Cordoba, M.G. (2021) Effect of temperature during drying and storage of dried figs on growth, gene expression and aflatoxin production. *Toxins* 13, 134–147.
Gilbert, J. and Şenyuva, H. (2008) Fungal and mycotoxin contamination of dried figs—a review. *Mycotoxins* 58, 73–83.
Gil-Serna, J., Vazquez, C., and Patino, B. (2019) Mycotoxins: toxicology. In: Batt, C.A. and Torterello, M.L. (eds) *Encyclopedia of Food Microbiology*, 2nd edn. Elsevier/Academic Press, Amsterdam, pp. 1539–1547.
Grenier, B. and Oswald, I. (2011) Mycotoxin co-contamination of food and feed: meta-analysis of publications describing toxicological interactions. *World Mycotoxin Journal* 4, 285–313.
Hassan, Z., Khan, M.Z., Khan, A., and Javed, J. (2010) Pathological responses of White Leghorn breeder hens kept on ochratoxin A contaminated feed. *Pakistan Veterinary Journal* 30, 118–123.
Hayat, A., Paniel, N., Rhouati, A., Marty, J., and Barthelmebs, L. (2012) Recent advances in ochratoxin A producing fungi detection based on PCR methods and OTA analysis in food matrices. *Food Control Journal* 26, 401–415.
Hedawoo, G.B., Bijwe, H.V., and Maggirwar, R.C. (2017) Occurrence of mycobiota associated with *Ficus carica* L. *International Journal of Researches in Biosciences, Agriculture and Technology* 2, 985–989.
Heperkan, D. (2006) The importance of mycotoxins and a brief history of mycotoxin studies in Turkey. Special issue "Mycotoxins: hidden hazards in food." *ARI Bulletin of Istanbul Technical University* 54, 18–27.
Heperkan, D., Karbancioglu-Guler, F., and Oktay, H.I. (2012a) Mycoflora and natural occurrence of aflatoxin, cyclopiazonic acid, fumonisin and ochratoxin A in dried figs. *Food Additives and Contaminants: Part A* 29, 277–286.

Heperkan, D., Moretti, A., Dikmen, C.D., and Logrieco, A. (2012b) Toxigenic fungi and mycotoxin associated with figs in the Mediterranean area. *Phytopathologia Mediterranea* 51, 119–130.

Heseltine, C.W. (1976) Conditions leading to mycotoxin contamination of food and feeds. In: Rodricks, J.V. (ed.) *Mycotoxins and Other Fungal Related Food Problems*. American Chemical Society, Washington, DC, pp. 1–22.

Heshmati, A., Zohrevand, T., Khaneghah, A.M., Amir Sasan, M.N., and Sant'Ana, A.S. (2017) Co-occurrence of aflatoxins and ochratoxin A in dried fruits in Iran: dietary exposure risk assessment. *Food and Chemical Toxicology* 106, 202–208.

Hocking, A., Su-Lin Leong, J., and Pitt, J.I. (2002) Ochratoxin A implications for the dried vine fruits industry. *Food Additives and Contaminants: Part B Surveillance* 7, 157–161.

Hussein, S.H. and Brasel, J.M. (2001) Toxicity metabolism and impact of mycotoxins on humans and animals. *Toxicology* 167, 101–134.

Iamanaka, B.T., Taniwaki, M.H., Menezes, H.C., Vicente, E., and Fungaro, M.H.P. (2005) Incidence of toxigenic fungi and ochratoxin A in dried fruits sold in Brazil. *Food Additives and Contaminants* 22, 1258–1263.

Iamanaka, B.T., Taniwaki, M.H., Vicente, E., and Menezes, H.C. (2006) Fungi producing ochratoxin in dried fruits. *Advances in Experimental Medicine and Biology* 571, 181–188.

Iamanaka, B.T., Menezes, H.C., Vicente, E., Leite, R.S.F., and Taniwaki, M.H. (2007) Aflatoxigenic fungi and aflatoxins occurrence in sultanas and dried figs commercialized in Brazil. *Food Control* 18, 454–457.

Ilic, Z., Bui, T., Tran-Dinh, N., Dang, M.N.V., Kennedy, I., and Carter, D. (2007) Survey of Vietnamese coffee beans for the presence of ochratoxigenic aspergilli. *Mycopathologia* 163, 177–182.

Iqbal, S.Z., Mehmood, Z., Asi, M.R., Shahid, M., Sehar, M., and Malik, N. (2018) Co-occurrence of aflatoxins and ochratoxin A in nuts, dry fruits, and nut products. *Journal of Food Safety* 38: e12462.

Isman, B. and Biyik, H. (2009) The aflatoxin contamination of fig fruits in Aydin City (Turkey). *Journal of Food Safety* 29, 318–330.

JECFA (Joint FAO/WHO Expert Committee on Food Additives) (2008) Evaluation of certain food additives and contaminants. World Health Organization Technical Report Series 995. World Health Organization, Geneva. Available at www.who.int/publications/i/item/9789241209953 (accessed December 16, 2021).

Jeswal, P. and Kumar, D. (2015) Mycobiota and natural incidence of aflatoxins, ochratoxin A, and citrinin in Indian spices confirmed by LC-MS/MS. *International Journal of Microbiology* 3: 242486.

Jonathan, S.G. and Esho, E.O. (2010) Fungi and aflatoxin detection in two stored oyster mushrooms (*Pleurotus ostreatus* and *Pleurotus pulmonarius*) from Nigeria. *Electronic Journal of Environmental, Agricultural and Food Chemistry* 9, 1722–1730.

Juan, C., Zinedine, A., Molto, J.C., Idrissi, L., and Manes, J. (2008) Aflatoxin levels in dried fruits and nuts from Rabat-Salé area, Morocco. *Food Control* 19, 849–853.

Kapetanakou, A.E., Panagou, E.Z., Gialitaki, M., Drosinos, E.H., and Skandamis, P.N. (2009) Evaluating the combined effect of water activity, pH and temperature on ochratoxin A production by *Aspergillus ochraceus* and *Aspergillus carbonarius* on culture medium and Corinth raisins. *Food Control* 20, 725–732.

Karaca, H. and Nas, S. (2008) Mycotoxins in dried figs. *Acta Horticulturae* 798, 307–312.

Karbancıoğlu-Güler, F. and Heperkan, D. (2008) Natural occurrence of ochratoxin A in dried figs. *Analytica Chimica Acta* 617, 32–36.

Karbancıoğlu-Güler, F. and Heperkan, D. (2009) Natural occurence of fumonisin B_1 in dried figs as an unexpected hazard. *Food and Chemical Toxicology* 47, 289–292.

Kaya, S.B. and Tosun, H. (2013) Occurrence of total aflatoxin, ochratoxin A and fumonisin in some organic food. *Journal of Pure and Applied Microbiology* 7, 2925–2932.

Khoury, A.E. and Atoui, A. (2010) Ochratoxin A: general overview and actual molecular status. *Toxins* 2, 461–493.

Kösoğlu-Vural, I. (2008) Research on occurrence of fumonisin in dried fig fruits of Sarilop fig (*Ficus carica* L.) variety. PhD thesis, Graduate School of Natural and Applied Sciences, Ege University, İzmir, Turkey.

Kowalska, A. and Manning, L. (2020) Using the rapid alert system for food and feed: potential benefits and problems on data interpretation. *Critical Reviews in Food Science and Nutrition* 61, 906–919.

Kozakiewicz, Z., Battilani, P., Javier Cabanes, J., Venancio, A., Mule, G., *et al.* (2004) Making wine safer: the case of ochratoxin. In: Barug, D., van Egmond, H., Lopez-Garcia, R., van Osenbruggen, T., and Visconti, A. (eds) *Meeting the Mycotoxin Menace*. Wageningen Academic Publishers, Wageningen, The Netherlands, pp. 133–142.

MacDonald, S., Wilson, P., Barnes, K., Damant, A., Massey, R., *et al.* (1999) Ochratoxin A in dried vine fruit: method development and survey. *Food Additives and Contaminants* 16, 253–260.

Magnoli, C., Astoreca, A., Ponsone, L., Fernández-Juri, M.G., Chiacchiera, S., and Dalcero, A. (2006) Ochratoxin A and the occurrence of ochratoxin A-producing black aspergilli in stored peanut seeds from Córdoba, Argentina. *Journal of Science of Food and Agriculture* 86, 2369–2373.
Marin, S., Ramos, A.J., Cano-Sancho, G., and Sanchis, V. (2013) Mycotoxins: occurrence, toxicology, and exposure assessment. *Food and Chemical Toxicology* 60, 218–237.
Meyvacı, K.B. (2003) Kuru İncir İşletmelerinde Etkin Aflatoksin Kontrolü. In: *Mikotosinler: Biyoteknolojisi; Sorunlar ve Çözümler Çalıştayı*. İzmir, Turkey.
Meyvacı, K.B., Aksoy, U., Eltem, R., Altindişli, A., Aşkun, T., and Taşkin, E. (2012) Effect of yearly conditions and management practices on ochratoxin A production in Sultana Seedless vineyards. *Food Additives & Contaminants: Part A: Chemistry, Analysis, Control, Exposure & Risk Assessment* 29, 1157–1167.
Mimoune, A.N., Arroyo-Manzanares, N., Gamiz-Gracia, L., García-Campana, A.M., Bouti, K., and Sabaou, N. (2018) *Aspergillus* section *Flavi* and aflatoxins in dried figs and nuts in Algeria. *Food Additives & Contaminants Part B* 11, 116–125.
Moller, T.E. and Nyberg, M. (2003) Ochratoxin A in raisins and currants: basic extraction procedure used in two small marketing surveys of the occurrence and control of the heterogeneity of the toxins in samples. *Food Additives and Contaminants* 20, 1072–1076.
Moreau, C. (1979) *Moulds, Mycotoxins and Foods*. Wiley, Chichester, New York.
Moretti, A., Ferracane, R., Ritieni, A., Frisullo, S., Lops, A., and Logrieco, A. (2000) *Fusarium* species from fig in Apulia: biological and toxicological characterization. *Mitteilungen aus der Biologischen Bundesanstalt für Land-Forstwirtsch* 377, 31–32.
Moss, M.O. (2008) Fungi, quality and safety issues in fresh fruits and vegetables. *Journal of Applied Microbiology* 104, 1239–1243.
Mutlu-İngök, A. and Karbancioglu-Güler, F. (2014) Effect of temperature on the growth and ochratoxin A production of the *Aspergillus* section *Nigri* members isolated from dried figs. *Journal of Food Safety* 34, 333–339.
Mutlu-İngök, A. and Karbancioglu-Güler, F. (2015) Sıcaklığın *Aspergillus* section *Nigri* üyelerinin okratoksin A oluşturması üzerine etkisinin incelenmesi. *Bilecik Şeyh Edebali Üniversitesi Fen Bilimleri Dergisi* 2, 1–7.
Mutlu-İngök, A., Elikoğlu, C., Temir, H.N., and Karbancıoğlu-Güler, F. (2020) Growth and ochratoxin A production by *Aspergillus carbonarius* isolated from dried figs in Aegean Region of Turkey affected by temperature and water activity. *Sakarya University Journal of Science* 24, 140–150.
Njumbe Ediage, E., van Poucke, C., and de Saeger, S. (2015) A multi-analyte LC-MS/MS method for the analysis of 23 mycotoxins in different sorghum varieties: the forgotten sample matrix. *Food Chemistry* 177, 397–404.
Noonim, P., Mahakarnchanakul, W., Nielsen, K.F., Frisvadi, J.C., and Samson, R.A. (2008) Isolation, identification and toxigenic potential of ochratoxin A-producing *Aspergillus* species from coffee beans grown in two regions of Thailand. *International Journal of Food Microbiology* 128, 197–202.
Northolt, M.D. and Soentoro, P.S.S. (1988) Fungal growth on foodstuffs related to mycotoxin contamination. In: Samson, R.A., van Reenen-Hoekstra, E.S. and van Dorsehot, C.A.N. (eds) *Introduction to Food Borne Fungi*. Centraalbureau voor Schimmelcultures, Baarn, The Netherlands, pp. 231–238.
Oksar, R.E., Şen, F., and Aksoy, U. (2017) Effect of drying under plastic tunnels on drying rate and quality of fig (*Ficus carica* L. 'Sarılop'). *Acta Horticulturae* 1173, 297–302.
Özar, A.G., Önder, P., Sarıbay, A., Özkut, S., Gündoğdu, M., *et al.* (1986) Ege Bölgesinde Görülen Hastalık ve Zararlılarla Savaşım Olanaklarının Saptanması ve Geliştirilmesi Üzerinde Araştırmalar. *Doğa, Türkiye Tarım ve Ormancılık Dergisi* 10, 263–277.
Özay, G. (1989) Kuru İncirde Aflatoksin Kontaminasyonu. In: *Gıdalarda Küfler ve Mikotoksinler Sempozyumu, Gebze, Kocaeli, Turkey*.
Özay, G. and Alperden, I. (1991) Kuru incirlerde (*Ficus carica* L.) aflatoksin ve okratoksin A olusumu. *Gıda Sanayi* 24, 57–62.
Özay, G. and Eke, D. (1988) Depolama Aşamasındaki Mısırlarda (*Zea mays* L.) Küf Gelişmesi ve Mikotoksin Oluşumu. *Gıda Sanayi* 8, 9–15.
Özay, G. and Özer, H. (2008) Mycotoxin problems in nuts and dried fruits from the Mediterranean basin. In: Leslie, J.F., Bandyopadhyay, R., and Visconti, A. (eds) *Mycotoxins: Detection Methods, Management, Public Health and Agricultural Trade*. CAB International, Wallingford, UK, pp. 133–138.
Özay, G., Aran, M., and Pala, M. (1995) Influence of harvesting and drying technique on mycoflora and mycotoxins of figs. *Nahrung* 39, 156–165.

Özer Meyvacı, K.B. (2009) Bahçe ürünlerinde mikotoksin problemi (Mycotoxin problem in horticultural products). In: Şen, F. (ed.) *Hasat Sonrası İyi Tarım Uygulamaları (Good Agricultural Practices After Harvest)*. Basım Yayım Matbaacılık San. Tic. Şti., İzmir, Turkey, pp. 207–221.

Özer, K.B. (1996) Kuru İncir Meyvelerinde Aflatoksin, Okratoksin A ve Psoralen Oluşumu Üzerinde Araştırmalar. MSc thesis, Graduate School of Natural and Applied Sciences, Ege University, İzmir, Turkey.

Özer, K.B. and Derici, B. (1998) A research on the relationship between aflatoxin and ochratoxin A formation and plant nutrients. *Acta Horticulturae* 480, 199–207.

Özkaya, Ş., Başaran, A., Elden, E., and Kaymak, T. (2002) Gıda ve Yemlerde Mikotoksin Düzeylerinin Tespiti. In: *Gıda Katkı-Kalıntı ve Bulaşanların İzlenmesi Proje Raporu*. Bursa, Turkey.

Pala, M., Özay, G., and Aran, N. (1994) Kuru İncir Üretiminde Küf ve Mikotoksin Oluşumunun İncelenmesi. In: *Alman Teknik İş birliği Teşkilatı Proje Raporu*. Bonn, Germany.

Palumbo, J.D., O'Keeffe, T.L., Vasquez, S.J., and Mahoney, N.E. (2011) Isolation and identification of ochratoxin A-producing *Aspergillus* section *Nigri* strains from California raisins. *Letters in Applied Microbiology* 52, 330–336.

Pavon, M.A., Gonzalez, I., Martin, R., and Garcia, T. (2012a) Competitive direct ELISA based on a monoclonal antibody for detection of ochratoxin A in dried fig samples. *Food and Agricultural Immunology* 23, 83–91.

Pavon, M.A., Gonzalez, I., de la Cruz, S., Martin, R., and Lacarra, T.G. (2012b) The use of high-performance liquid chromatography to detect ochratoxin A in dried figs from the Spanish market. *Journal of the Science of Food and Agriculture* 92, 74–77.

Petric, J., Sarkanj, B., Mujic, I., Mujic, A., Sulyok, M., *et al.* (2018) Effect of pretreatments on mycotoxin profiles and levels in dried figs. *Arhiv za Higijenu Rada i Toksikologiju* 69, 328–333.

Phillips, J.D., Purcell, S.L., and Stanley, G.I. (1980) Aflatoxins in almonds. *Agricultural Reviews and Manuals, Western Series* 20, 1–13.

Pitt, J.I. (2000) Toxigenic fungi and mycotoxins. *British Medical Bulletin* 56, 184–192.

Pohland, A.E., Nesheim, S., and Friedman, L. (1992) Ochratoxin A: a review. *International Union of Pure and Applied Chemistry* 64, 1029–1046.

Raiola, A., Meca, G., Mañes, J., and Ritieni, A. (2012) Bioaccessibility of deoxynivalenol and its natural co-occurrence with ochratoxin A and aflatoxin B_1 in Italian commercial pasta. *Food and Chemical Toxicology* 50, 280–287.

Reddy, S.V., Mayi, D.K., Reddy, M.U., Thirumala-Devi, K., and Reddy, D.V.R. (2001) Aflatoxins B_1 in different grades of chillies (*Capsicum annum* L.) in India as determined by indirect competitive-ELISA. *Food Additives and Contaminants* 18, 553–558.

Remiro, R., González-Peñas, E., Lizarraga, E., and López de Cerain, A. (2012) Quantification of ochratoxin A and five analogs in Navarra red wines. *Food Control* 27, 139–145.

Richard, J.L. (2007) Some major mycotoxins and their mycotoxicoses—an overview. *International Journal of Food Microbiology* 119, 3–10.

Rodricks, V. and Lovett, J. (1984) Toxigenic fungi. *Food Safety: Foodborne Illness*, 484–489.

Saadullah, A.A.M. and Abdullah, S.K. (2015) Contamination of dried figs with fungi and aflatoxigenic potential of some isolates of *Aspergillus* section *Flavi*. *Journal of Biology, Agriculture and Healthcare* 5, 76–80.

Şahin, E. (2003) Büyük ve Küçük Menderes Havzalarında Yetiştirilen Kurutmalık İncirlerde (*Ficus carica* L.) Aflatoksin ve Okratoksin A Varlığının, Dağılımının ve Kalite ile İlişkisinin Araştırılması. PhD thesis, Graduate School of Natural and Applied Sciences, Ege University, İzmir, Turkey.

Sanlı, Y., Yavuz, H., and Akar, F. (1990) Mycotoxin contaminations in dried figs. *Ankara Universitesi Veterinerlik Fakültesi Dergisi* 37, 293–308.

Schaafsma, A.W. and Hooker, D.C. (2007) Climatic models to predict occurrence of *Fusarium* toxins in wheat and maize. *International Journal of Food Microbiology* 119, 116–125.

Schrenk, D., Bignami, M., Bodin, L., Chipman, J.K., del Mazo, J., *et al.* (2020) Scientific opinion – risk assessment of aflatoxins in food. *EFSA Journal* 18, 6040–6151.

Şen, F., Aksoy, U., Ozer, K.B., Can, H.Z., Kosoglu, I., and Konak, R. (2017) Impact of yearly conditions on major physical and chemical properties of fresh, semi-dried and sun-dried fig (*Ficus carica* L., Sarılop) fruit. *Acta Horticulturae* 1173, 309–314.

Şen, L. and Nas, S. (2010) Kuru incir, üzüm ve kırmızıbiberlerde mikotoksin varlığı. *Akademik Gıda* 8, 24–32.

Şenyuva, H.Z., Gilbert, J., Ozcan, S., and Ulken, U. (2005) Survey for co-occurrence of ochratoxin A and aflatoxin B_1 in dried figs in Turkey by using a single laboratory-validated alkaline extraction method for ochratoxin A. *Journal of Food Protection* 68, 1512–1515.

Şenyuva, H.Z., Gilbert, J., Samson, R.A., Ozcan, S., Ozturkoglu, S., and Onal, D. (2008) Occurrences of fungi and their mycotoxins in individual Turkish dried figs. *World Mycotoxin Journal* 1, 79–86.

Serra, R., Abrunhosa, L., Kozakiewicz, Z., and Venancio, A. (2003) Black *Apergillus* species as ochratoxin A producers in Portuguese wine grapes. *International Journal of Food Microbiology* 88, 63–68.

Sharman, M., Patey, A.L., Bloomfield, D.A., and Gilbert, J. (1991) Surveillance and control of aflatoxin contamination of dried fig and fig paste imported into the United Kingdom. *Food Additives and Contaminants* 8, 299–304.

Steiner, W.E., Rieker, R.H., and Battaglia, R. (1988) Aflatoxin contamination in dried figs: distribution and association with fluorescence. *Journal of Agriculture and Food Chemistry* 36, 88–91.

Stoev, S.D. (2015) Foodborne mycotoxicoses, risk assessment and underestimated hazard of masked mycotoxins and joint mycotoxin effects or interaction. *Environmental Toxicology and Pharmacology* 39, 794–809.

Sweeney, M.J. and Dobson, A.D.W. (1998) Mycotoxin production by *Aspergillus*, *Fusarium* and *Penicillium* species. *International Journal of Food Microbiology* 43, 141–158.

Taylor, G., Petróczi, A., Nepusz, T., and Naughton, D.P. (2013) The procrustean bed of EU food safety notifications via the Rapid Alert System for Food and Feed: does one size fit all? *Food and Chemical Toxicology* 56, 411–418.

Techarat, S. and Dachoupakan, C. (2012) Growth and ochratoxin A production of black aspergilli isolated from Thai wine grapes. *Asian Journal of Food and Agro-Industry* 5, 172–182.

Topal, Ş. (1993) Gıdalarda Küf Kontaminasyon Riskleri ve Önlemleri. *Gıda Sanayinde Mikrobiyoloji ve Uygulamaları* 124, 162–174.

Trucksess, M.W. and Scott, P.M. (2008) Mycotoxins in botanicals and dried fruits: a review. *Food Additives and Contaminants* 25, 181–192.

UNECE (United Nations Economic Commission for Europe) (2016) *UNECE Standard DDP-14 Concerning the Marketing and Commercial Quality Control of Dried Figs*. United Nations, New York and Geneva. Available at https://unece.org/fileadmin/DAM/trade/agr/standard/dry/dry_e/14DriedFigs_E2016.pdf (accessed December 16, 2021).

van der Merwe, K.J., Steyn, P.S. and Fourie, L. (1965) Ochratoxin A, a toxic metabolite produced by *Aspergillus ochraceus* Wilh. *Nature* 265, 1112–1113.

van Egmond, H.P., Schothorst, R.C., and Jonker, M.A. (2007) Regulations relating to mycotoxins in food. *Annals of Bioanalytical Chemistry* 389, 147–157.

Varga, J. and Kozakiewicz, Z. (2006) Ochratoxin A in grapes and grape-derived products. *Trends in Food Science and Technology* 17, 72–81.

WHO/FAO (World Health Organization/Food and Agriculture Organization of the United Nations) (2018) *Safety Evaluation of Certain Contaminants in Food: Prepared by the Eighty-third Meeting of the Joint FAO/WHO Expert Committee on Food Additives. WHO Technical Report Series*. Available at https://apps.who.int/iris/handle/10665/254893 (accessed December 16, 2021).

Zinedine, A., González-Osnaya, L., Soriano, J.M., Moltó, J.C., Idrissi, L., and Mañes, J. (2007a) Presence of aflatoxin M_1 in pasteurized milk from Morocco. *International Journal of Food Microbiology* 114, 25–29.

Zinedine, A., Soriano, J.M., Juan, C., Mojemmı, M., Molto, J.C., *et al.* (2007b) Incidence of ochratoxin A in rice and dried fruits from Rabat and Salé area, Morocco. *Food Additives and Contaminants* 24, 285–291.

Zohri, A.A. and Abdel-Gawad, K.M. (1993) Survey of microflora and mycotoxins of some dried fruits in Egypt. *Journal of Basic Microbiology* 33, 279–288.

Zorlugenc, B., Zorlugenç, F.K., Oztekin, S., and Evliya, B. (2008) The influence of gaseous ozone and ozonated water on microbial flora and degradation of aflatoxin B_1 in dried figs. *Food and Chemical Toxicity* 46, 3593–3597.

Part V

Fig Omics Analysis

20 Fig Genetics, Genomics, and Transcriptomics

Hidetoshi Ikegami[1]* and Kenta Shirasawa[2]
[1]*Fukuoka Agriculture and Forestry Research Center Buzen Branch, Yukuhashi, Japan;* [2]*Kazusa DNA Research Institute, Kisarazu, Japan*

20.1 Introduction

Modern life sciences, including genetics, are currently undergoing a major change for animals, plants, and all other organisms. DNA is the blueprint of life, containing the necessary codes to construct the proteins and non-coding RNAs that affect all life systems. Understanding DNA sequences has made it possible to elucidate the structure and function of proteins and RNAs, and to understand the underlying causes of many life phenomena. However, genetics is not only an aspect of theoretical biology; it also applies to practical fields, such as medical diagnoses and breeding trials, and is now linked to many other fields, including general medicine, engineering, and agriculture.

Recent technological and analytical advances have shifted the focus of genetic approaches from hypothesis-linked to data-based. In particular, next-generation sequencing (NGS), the leader in data-based research, is a powerful technology that can sequence large numbers of DNA molecules simultaneously. For example, data on the whole genome and on sites of genomic mutations constructed from NGS reads have led to a plethora of gene discoveries in a very short time. Transcriptome analysis can focus on significant differences in gene expression that are calculated based on the numerous NGS reads. Both approaches are very powerful compared with pre-NGS approaches, which made use of only limited information resources. Fruit-tree breeding programs had been lagging behind those of cereals and vegetables because of the long generation times, long juvenile periods, and the large plant size. However, NGS technologies have led us to the new era of genomics-assisted fruit-tree breeding with genome-wide association studies (GWAS) and genomics selection, even in non-model plants including fruit trees (Iwata *et al.*, 2016).

In this chapter, we present an overview, starting with classical genetics based on mating experiments, through to entry into molecular genetics with the evaluation of fig genetic resources using DNA marker data, and then on to genomic science based on genome decoding with the advent of NGS. The chapter is arranged to provide a genome-centered overview of each omics layer along a time axis, and discusses the future of fig genetic research along with related issues and directions.

20.2 Fig Genetics: Early History

Genetics and breeding cannot be separated. Mendel's laws of inheritance were discovered in

* Email: ikegami@farc.pref.fukuoka.jp

 Advances in Fig Research and Sustainable Production (M.A. Flaishman and U. Aksoy)
DOI: 10.1079/9781789242492.0020

repeated crossing experiments. The appearance of a particular trait, such as fruit skin color or fruit weight, in current breeding crosses can be recognized as a genetic event. Therefore, breeding efforts over the years have often produced genetics results, while, in contrast, those efforts have been progressed on genetic backgrounds. Fig genetics was initially triggered by the onset of several breeding programs planned and implemented in the USA in the early 20th century (Ferguson *et al.*, 1990). Cultivated fig breeding studies led Storey (1955, 1975) to hypothesize that sex in *Ficus* species is genetically determined by a supergene consisting of two very tightly linked genes (Flaishman *et al.*, 2008). Saleeb (1965) then reported, in his doctoral dissertation, the inheritance of parthenocarpy. In each of these studies, experimental design was based on multiple planned cross combinations, with simple and clear genetic models for target traits. These findings provided the ideal basis for performing fig genetics. A few additional genetic studies were conducted based on breeding experiments until around 2000 (Awamura *et al.*, 1996a,b), when molecular biology techniques began to be applied to assess genetic resources. However, theorized genetic models have only been established for a few traits. Details of these models will be described in section 20.6.

20.3 Germplasm Evaluation and DNA Marker Development

Genetic diversity represents the heritable variation within and between populations of organisms and among plant species. This pool of genetic variation within an intermating population is the basis for selection, as well as for plant improvement. Thus, conservation of this plant genetic diversity is essential for humankind's present and future well-being (Ramanatha Rao and Hodgkin, 2002).

In this context, many attempts have been made to evaluate the genetic diversity and genetic relationships of *Ficus carica* using DNA markers. Specifically, fig genetic resources have been evaluated in the following countries or regions: Tunisia (Hedfi *et al.*, 2003; Salhi-Hannachi *et al.*, 2008; Trifi *et al.*, 2008; Chatti *et al.*, 2010; Baraket *et al.*, 2011; Saddoud *et al.*, 2011; Bendhifi *et al.*, 2013; Ahmed *et al.*, 2015; Essid *et al.*, 2015; Haffar *et al.*, 2017; Aljane and Essid, 2018), Italy (Chessa *et al.*, 1998; Laddomada *et al.*, 2008; Resta *et al.*, 2008; Oliveira *et al.*, 2012; Ciarmiello *et al.*, 2015; Ferrara *et al.*, 2017; de Masi *et al.*, 2017; Mafrica *et al.*, 2017; Vivaldi *et al.*, 2017; Rodolfi *et al.*, 2018), France and Morocco (Khadari *et al.*, 2003; Oukabli and Khadari, 2005; Achtak *et al.*, 2010; Haddou *et al.*, 2018), Spain (Giraldo *et al.*, 2008; Perez-Jiménez *et al.*, 2012), Turkey (Cabrita *et al.*, 2001; Ikten *et al.*, 2010; Caliskan *et al.*, 2012), the USA (Aradhya *et al.*, 2017; Knap *et al.*, 2018), Jordan (Sadder and Ateyyeh, 2006; Almajali *et al.*, 2012), Palestine (Basheer-Salimia *et al.*, 2012; Ali-Shtayeh *et al.*, 2014), Slovenia and Croatia (Knap *et al.*, 2018), Portugal (Elisiário *et al.*, 1998), Japan (Ikegami *et al.*, 2009), Brazil (Val *et al.*, 2013), and Algeria (Boudchicha *et al.*, 2018).

Genetic diversity can also be assessed using the phenotypic and chemical index values of traits, which are, in fact, significant criteria. However, phenotypic characteristics and chemical indices can be affected by environmental impacts. One advantage of genetic diversity assessment using molecular DNA markers is that they are not affected by any environmental factors. The following marker types (genetic regions/loci) were adopted in the above studies: random amplification of polymorphic DNA (RAPD) (Elisiário *et al.*, 1998; Cabrita *et al.*, 2001; de Masi *et al.*, 2003, 2005, 2017; Khadari *et al.*, 2003; Sadder and Ateyyeh, 2006; Salhi-Hannachi *et al.*, 2008; Ikegami *et al.*, 2009; Chatti *et al.*, 2010; Basheer-Salimia *et al.*, 2012; Caliskan *et al.*, 2012; Bendhifi *et al.*, 2013; Ali-Shtayeh *et al.*, 2014; Ciarmiello *et al.*, 2015; Aljane and Essid, 2018), inter-sequence simple repeat (ISSR) (Khadari *et al.*, 2003; Oukabli and Khadari, 2005; Salhi-Hannachi *et al.*, 2005, 2008; Ikegami *et al.*, 2009; Chatti *et al.*, 2010; Almajali *et al.*, 2012; Haddou *et al.*, 2018), random amplified microsatellite polymorphism (Chatti *et al.*, 2010), amplified fragment length polymorphism (AFLP) (Cabrita *et al.*, 2001; Laddomada *et al.*, 2008; Resta *et al.*, 2008), simple sequence repeat (SSR) (Khadari *et al.*, 2003; Oukabli and Khadari, 2005; Giraldo *et al.*, 2008; Trifi *et al.*, 2008; Achtak *et al.*, 2010; Aradhya *et al.*, 2010, 2017; Chatti *et al.*, 2010; Saddoud *et al.*, 2011; Caliskan *et al.*, 2012; Perez-Jiménez *et al.*, 2012; Val *et al.*, 2013; Ahmed *et al.*, 2015; Essid *et al.*, 2015; Ganopoulos *et al.*, 2015; Ferrara *et al.*, 2016, 2017; Knap *et al.*, 2018; Mafrica *et al.*, 2017; Vivaldi *et al.*, 2017; Aljane and Essid, 2018; Boudchicha *et al.*, 2018; Haddou *et al.*, 2018), sequence-related amplified polymorphism (SRAP) (Ikten *et al.*, 2010), internal transcribed spacer (Baraket *et al.*, 2011; Oliveira *et al.*, 2012; Ghada *et al.*, 2013), chloroplast DNA (Baraket *et al.*, 2011), conserved DNA-derived polymorphism (Haffar *et al.*, 2017), single-nucleotide polymorphism (SNP) (Mori *et al.*, 2017), and isozymes (Chessa *et al.*, 1998; Elisiário *et al.*, 1998; Hedfi *et al.*, 2003).

The isozyme marker is based on the amino acid sequence polymorphism of proteins, while the other markers are based on DNA sequence polymorphisms, and today the latter are mostly commonly used. Codominant markers (e.g. SSR and SNP) identify all the alleles at a particular locus, and with the codominant marker, heterozygotes can be distinguished from homozygotes. In contrast, dominant markers (e.g. RAPD, ISSR, AFLP, and SRAP) identify only a single dominant allele, and heterozygosity cannot be determined in these marker systems. In general, codominant markers are preferable to dominant markers due to the larger information content, but dominant markers continue to be very popular, mainly because of economic reasons (Piepho and Koch, 2000). When a single specific marker is applied, the genomic features corresponding to that marker type are informationally extracted (Powell *et al.*, 1996). Therefore, these markers can be used alone or in combination, where the application of multiple markers is expected to lead to evaluations that identify more various and comprehensive genomic polymorphisms. The decision about which marker(s) to use depends on the scientific purpose of the analysis, and on the cost and equipment requirements, but SSR markers seem to be the most commonly used to date.

SSR markers are characterized by codominance, high rates of polymorphism detection, stable reproducibility, high conservation in closely related species, and high translocational availability. There has been much development of new SSR markers in *Ficus* species. Khadari *et al.* (2001) succeeded in developing eight SSR markers by screening DNA fragments with repetitive sequences from a fig genome library (variety Violette de Sollies), and their applicability was confirmed in other varieties. Giraldo *et al.* (2005) developed 26 SSR markers from a genomic library of the fig variety Cuello Dama Blanco, and investigated SSR polymorphisms in 15 different varieties, confirming their reassignment to 13 fig species. Bandelj *et al.* (2007) also developed 15 SSR markers using common fig. Moe and Weiblen (2011) developed 10 SSR primer sets from a genomic library of three fig species and used populations of several origins for marker characterization. Heer *et al.* (2012) developed seven new SSR markers with expressed sequence tag libraries from two fig species and examined their applicability in 11 *Ficus* species. Knap *et al.* (2017) developed 39 new SSR markers from a genomic library and confirmed that 29 of them are polymorphic in fig. Ikten (2018) examined the transferability of 54 SSR markers developed previously for six *Ficus* species (*F. montana*, *F. septica*, *F. sycomorus*, *F. (Sycomorus) racemosa*, *F. (Urostigma) rubiginosa*, and *F. insipida*) (Zavodna *et al.*, 2005; Vignes *et al.*, 2006; Ahmed *et al.*, 2007; Crozier *et al.*, 2007) to *F. carica*. These reports indicated that there is no obstacle to the shared use of SSRs among different *Ficus* species.

The advent of NGS has made informative marker development possible. The first case was based on the pyrosequencing of five fig species (Garcia *et al.*, 2012). These authors designed 5723 potential microsatellite (SSR) primers designed from sequence reads of NGS, and investigated the performance of 20 of them. A similar microsatellite primer experiment was performed by Mori *et al.* (2017), with 3651 pairs of SSR primers designed and partially integrated into a high-density genetic linkage map. Meanwhile, the increase in NGS usage is facilitating accumulation of SNP information, and it is expected that the DNA markers used in fig (*Ficus*) species will gradually shift from SSRs to SNPs.

20.4 Basic Background on the Fig Genome

Most cultivated fig species (*F. carica*) are diploid, with $2n = 26$ and high heterogeneity. Condit (1969) and Storey (1977) described the basic number of chromosomes as 13. This was later confirmed indirectly based on the number of described DNA marker linkage groups (Mori *et al.*, 2017).

Ploidy is an important factor in understanding and handling the inheritance of a species. Through cytogenetic surveys of genetic resources in Italy (central Italy and Sardinia), figs were shown to be diploid ($2n = 26$), as well as triploid ($2n = 39$) and tetraploid ($2n = 52$) (Falistocco, 2009, 2016). In addition, large pollen, which is potentially a non-reducing gamete, was observed in triploids. The observations suggested that the triploids were derived from crosses between a diploid and a tetraploid (Falistocco, 2009, 2016). The author presented two possibilities for the origin of 3x plants: sexual polyploidization by means of the formation of $2n$ gametes, and hybridization between 2x and 4x genotypes (Rhoades, 1936; Müntzing, 1937), supporting the former theory due to the non-reducing gamete observation (Falistocco, 2016).

Because, in general, diploid species do not have a homogeneous genome set, there are fewer barriers to performing genetic analyses of diploid versus other polyploid or homoploid species. As an example, because the number of homologous genes is greater in a tetraploid than in a diploid, we can expect low discrimination among these homologs. Similarly, it is difficult to trace the inheritance of a parent allele in the progeny of the crossing if the experimental materials are isoploid. Thus, it may have been reasonable to target diploids as a first step in genetic studies of *F. carica*.

High-accuracy genome size studies have been conducted. The estimated *F. carica* genome size (1C) is equivalent to 356–357 Mb (Loureiro *et al.*, 2007; Bennett and Leitch, 2011). This value is supported by the genome-assembly results described in section 20.5 (Usai *et al.*, 2019; Shirasawa *et al.*, 2020). The fig genome size is close to that estimated for mulberry (345–366 Mb; Yamanouchi *et al.*, 2010). Remarkably, the diploid number of *F. carica* chromosomes is also close to that in mulberry (26 in *F. carica*, 28 in *Morus alba*), and high synteny between the two is expected, although this has not been verified. Although jujube (*Ziziphus jujuba*), also in the order Rosales but in an entirely different family, has a genome size of about 440 Mb, correspondence of SNP markers between fig and jujube also suggests high synteny (Mori *et al.*, 2017).

20.5 Genome Sequencing

To date, the genome sequences of three common figs and related species have been released (Table 20.1). Mori *et al.* (2017) reported the first-draft genome assembly for the common fig. The target variety was Houraishi. The assembly consisted of 27,995 contigs, with a length of 248 Mb and an N50 of 166 kb. Because only Illumina short-read sequence data (98×) was used for the assembly, the resulting sequences do not fully cover the fig genome (approx. 356 Mb), and contiguity is restricted to short regions. Nevertheless, 90% of the complete single-copy genes in benchmarking universal single-copy orthologs (BUSCO) are represented in the assembly, suggesting that the sequences cover large parts of gene-rich regions within the fig genome. A total 36,138 genes were predicted. Using this sequence as a reference, double-digest restriction-site-associated DNA sequencing (ddRAD-Seq) analysis was performed to detect SNPs in an F_2 mapping population derived from a cross between the Houraishi and Caprifig 6085 varieties. A high-density genetic map based on 8180 SNPs together with 64 SSRs was established; however, few sequences were assigned to chromosomes because of the aforementioned short sequence limitation.

A second common fig genome sequence was recently reported for the variety Dottato (Usai *et al.*, 2019). A total of 26.4 Gb of long-read sequences (74× genome coverage) from the PacBio Sequel system (PacBio, Menlo Park, CA) were employed for the assembly. As expected, genome coverage and sequence continuity were remarkably improved over the first report (Mori *et al.*, 2017). The total length of the assembly, 333 Mb, is close to the estimated genome size. The assembly consists of 905 contigs with an N50 length of 823 kb. Complete, single-copy BUSCO gene coverage is 93.3%. A total of 37,840 genes are predicted in the sequences. In accordance with genome and map information from Mori *et al.* (2017), 407 sequences (266 Mb) were anchored to fig chromosomes to establish 13 pseudomolecule sequences, corresponding to the appropriate number of fig chromosomes.

In parallel, the genome sequence of a wild fig relative, *Ficus erecta*, was also reported (Shirasawa *et al.*, 2020). This tree is known as a donor of resistance to *Ceratocystis* canker, which is a

Table 20.1. Genome sequences of common fig (*Ficus carica*).

Line	Sequence version	Total assembly length (Mb)	Number of sequences	N50 length (kb)	Gap (%)	Pseudomolecule sequences	Number of predicted genes	Reference
Houraishi	Houraishi_ver.1	248	27,995	166	14	Not available	36,138	Mori *et al.* (2017)
Dottato	FiCari_1.0	333	905	823	0	Available	37,840	Usai *et al.* (2020)
FE-Hiroshima-1	FER_r1.1	331	538	1856	0	Available	51,806	Shirasawa *et al.* (2020)

disruptive disease in common fig. Therefore, *F. erecta* is a useful genetic resource for resistance breeding programs in fig (Shimizu and Miyoshi, 1999; Morita *et al.*, 2011). A total of 21.3 Gb of long-read sequences (65× genome coverage) from the PacBio Sequel system were assembled into 538 contigs covering 331 Mb with an N50 length of 1856 kb. The complete, single-copy BUSCO gene coverage was 92.3%. A total of 51,806 genes were predicted in the sequences. In the process, 13 pseudomolecule sequences consisting of 204 sequences (273 Mb) were established in accordance with a genetic map of 13,320 SNPs for a BC_1F_1 population, in which *F. erecta* and common fig were employed as the donor and recurrent parent, respectively.

We compared sequence similarity of the two *Ficus* species *F. carica* and *F. erecta* using the two available chromosome-level genome sequences (Usai *et al.*, 2019; Shirasawa *et al.*, 2020). The resulting dot plot showed the comparisons and similarities of the sequences and structures of the two genomes (Fig. 20.1). Pseudomolecule sequences corresponded in a one-to-one manner, and potential structural variants disrupting the collinearity of the plot (e.g. inversions and translocations) were also observed, although this could be misassemblies or misalignments of sequences on the chromosomes.

20.6 Genetics and Gene Discovery

Genome analysis enables the identification of DNA sequences that control traits of interest. However, few insights into molecular mechanisms were revealed in fig because of a lack of genome information. Therefore, recent genome-sequencing analyses have shed light on many long-unsolved issues.

20.6.1 Sex determination

Fig gender is controlled by two tightly linked genetic loci, *A* and *G*, which have been proposed as genes associated with stamen formation and pistil length, respectively (Storey, 1975). Mori *et al.* (2017) reported on the candidate *A* gene through genetic and genomics approaches based on the Houraishi variety genome sequence discussed in section 20.5. Fig sex determinant-associated sequences, publicly available in GenBank under accession numbers AB331722–AB331728, showed similarity to the Houraishi genome contigs. These sequences were mapped on to a single genome position between 8.8 and 11.7 cM of chromosome Fc01a (Fig. 20.2). A GWAS with 119 divergent fig lines also detected the candidate region between 2.4 and 31.4 cM, which includes these sex determinant-associated sequences. To narrow down the candidate gene(s) even more, whole-genome resequencing analysis was performed on five representative lines to find 889 SNPs in the region. Based on possible SNP gene function effects, 39 possible deleterious mutations were selected and genotyped across 18 male lines to find two SNP genotyping scores that matched the phenotype. Subsequent genotyping analysis of the two SNPs across the 119 lines used in the GWAS confirmed that their genotypes correlated completely with the phenotypes. The two SNPs were located on a single gene—s00259g14131.t1 (Mori *et al.*, 2017), *RESPONSIVE-TO-ANTAGONIST1* (*RAN1*)—encoding a copper-transporting P-type ATPase (Hirayama *et al.*, 1999), and this *F. carica RAN1* gene, designated *FcRAN1*, was therefore identified as a candidate sex-determination gene. The molecular function of *RAN1* was proposed to involve copper delivery to create functional ethylene receptors (Hirayama *et al.*, 1999). In this scenario, male trees possessing functional *RAN1* are able to carry copper to the receptors, resulting in differentiation of the sex phenotype depending on the presence or absence of ethylene. However, in female figs, the constitutive response to the ethylene signal, due to lack of *RAN1* function, always leads to female flower formation, regardless of the presence or absence of ethylene. A functional cleaved amplified polymorphic sequence (CAPS) marker for identifying *FcRAN1* allele type and sex phenotype was developed (Fig. 20.2).

20.6.2 Parthenocarpy

The *F. carica* fig (female fig) is classified into three types: Smyrna, San Pedro, and common (Stover and Aradhya, 2008). The type depends on whether syconia require pollination or not, and how they persist on the tree. The common type

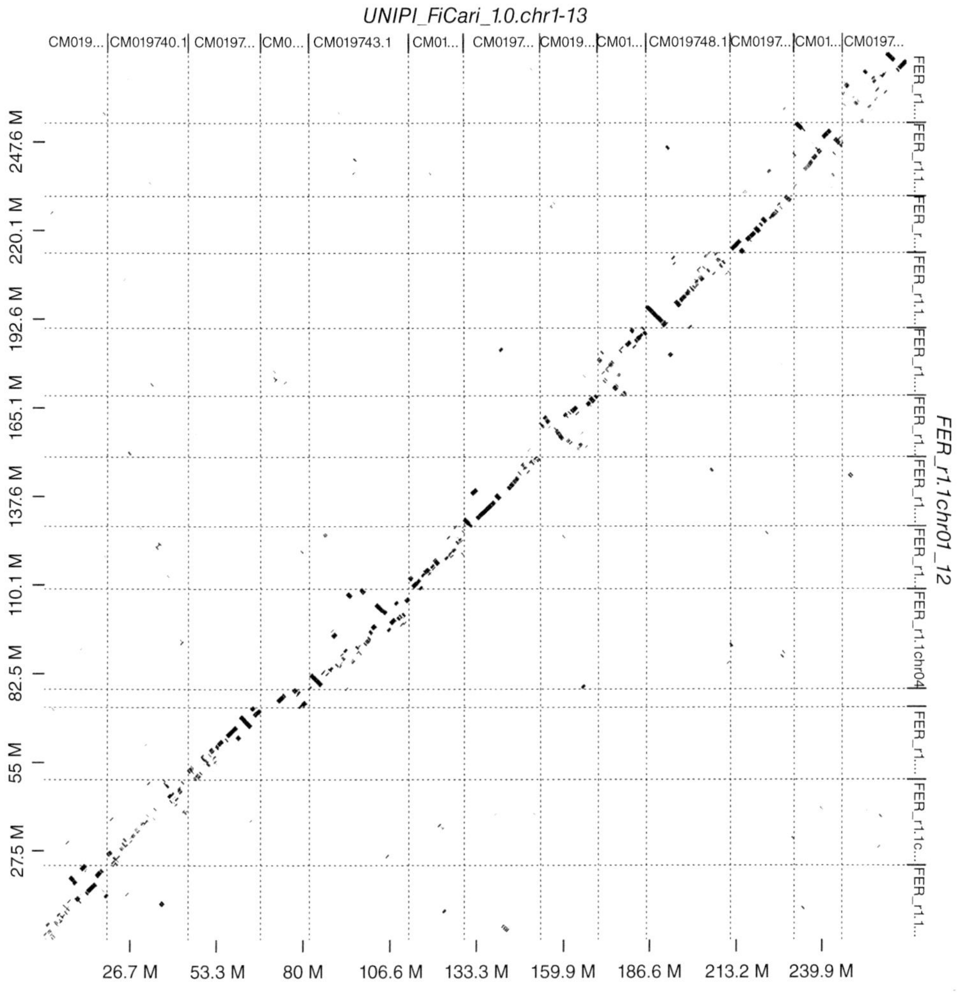

Fig. 20.1. Sequence similarity of the two *Ficus* species *F. carica* and *F. erecta* whose chromosome-level genome sequences are available. Dots indicate sequence similarity of the *F. carica* genome (UNIPI_FiCari_1.0) on the vertical axis versus those of *F. erecta* (FER_r1.1) on the horizontal axis.

has complete parthenocarpy, which means that all of the fruit crops persist on the tree without pollination. The San Pedro type has parthenocarpy only in its breba crop, and the Smyrna type has complete non-parthenocarpy. Parthenocarpy comes in two forms: stimulative, which requires external stimuli such as pollination or chemical treatment, and vegetative, which is genetically determined, independent of external stimuli. Although the syconium is not a true fruit, horticulturally speaking, "persistence" is scientifically accurate; however, in this chapter, we will mainly use the word "parthenocarpy."

Parthenocarpy is a fascinating scientific phenomenon. Saleeb (1965) first presented a genetic model for fig parthenocarpy in which he assumed a P factor—a causative factor of persistence that also induced female gamete lethality—based on the phenotype segregation of caducous-type and persistence-type in a diallel cross mating. The identity of the P factor locus remains unknown. However, the lethality of female gametes may be a clue worth considering as the determinant of parthenocarpy. Awamura *et al.* (1996a) investigated the effects of genetic differences between seed parents on

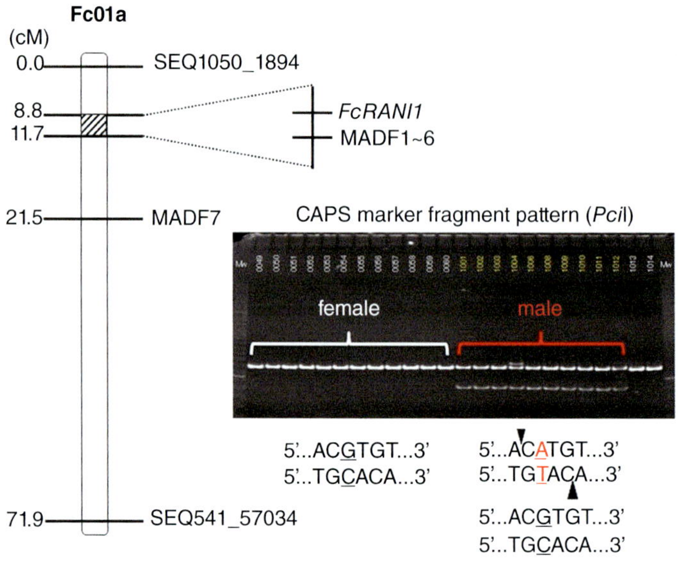

Fig. 20.2. Left: A schematic diagram of the Fc01a chromosome. "MADF" sequences show sexual phenotype-linked markers that are located near *FcRAN1*, which is strongly associated with sexual phenotype. The MADF and *FcRAN1* marker-mapped region is indicated by diagonal dotted lines. Right: Image and illustration of cleaved amplified polymorphic sequence (CAPS) marker genotyping. The genotype of the female plant is homozygous for G/G, but that of the male plant is heterozygous for G/A; the SNP position can be cleaved by the *Pci*I restriction enzyme.

the frequency of parthenocarpy in progeny for genders and crops separately. Heredity experiments using six cross combinations showed that the parthenocarpy attribute of seed parents actually affects the progeny parthenocarpy in both the first and second crop. More specifically, in a caprifig population (males), many individuals with a parthenocarpic first crop appeared in the progeny of San Pedro-type and common-type seed parents. In the female fig population, there were many individuals with a parthenocarpic second crop in the progeny of common-type seed parents. When Smyrna, the non-parthenocarpic type, was a seed parent, the frequency of appearance of parthenocarpic individuals in the progeny was very low. These results were consistent with the hypothesis of Saleeb's (1965) and Saleeb and Storey (1975) in that the genetic background of the seed parent affected the progeny's parthenocarpic characteristics. In addition, they showed that these parthenocarpic characteristics display quantitative annual variation, and suggested that environmental conditions (e.g. weather) might influence fig parthenocarpy independent of genetic factors (Awamura *et al.*, 1996a).

20.6.3 Fruit skin color

The skin of the fig fruit (syconium) contains multiple pigment types. A wide range of colors are seen among varieties, from green to black, depending on the compositional ratio of the different pigments (Solomon *et al.*, 2006; Dueñas *et al.*, 2008; Ercisli *et al.*, 2012; Yemiș *et al.*, 2012). Condit (1955) used the expressions Green, Yellow, Brown, Red, Violet, and Black for skin color classification, and Awamura *et al.* (1996b) applied the Munsell color chart method to classify the color types into nine groups.

Awamura *et al.* (1996b) reported the only study on the inheritance of syconium skin color. They observed the skin color of three seedling groups obtained by crossing three seed parents with one pollen parent (Caprifig 6085) using a color difference meter separately for each sex (first crop for the male plant, second crop for the female plant). The results—the frequency distribution of b* values (blue direction associated with purple-black or black color) in the progeny of the crosses—did not identify any clear inheritance of skin color, because this trait was broadly distributed in the progeny (Awamura *et al.*, 1996b). From this point of view, the existence of multiple genetic factors for skin color might be expected. Intriguingly, individuals with a low b* value were only obtained in the female populations, and the frequency of appearance of these individuals was higher than that of other individuals. This suggests that the expression of black skin color is suppressed by some factors in male plants, and that the male varieties used for crossing (e.g. Caprifig 6085) might have a major gene controlling black skin color.

20.6.4 *Ceratocystis* canker resistance

Ceratocystis canker resistance in *F. erecta* is controlled by a single dominant gene (Yakushiji *et al.*, 2012). To map the resistance locus on the genome sequence, Shirasawa *et al.* (2020) employed a BC_1F_1 population (Yakushiji *et al.*, 2019) and genome-wide genotypes that were analyzed with ddRAD-Seq technology. A total of 13,320 SNPs were obtained from the genotyping analysis and employed for GWAS. As expected, a single peak was observed in chromosome 2 of the *F. erecta* genome. Graphical genotyping of the BC_1F_1 population delimited the candidate locus to a 66.3 kb region. Nine genes are predicted to encode proteins in this region. Of these, only one, Fer_r1.1chr02_g012310.1—a predicted gene in the *F. erecta* genome—shows sequence similarity with the resistance genes. Transcriptome analysis revealed that this gene's expression pattern is stable in the stems, regardless of pathogen infection, whereas it is upregulated in the leaves. The counterpart sequence is conserved in susceptible fig genomes, but the corresponding gene is split into two parts, s01133g27051 and s01133g27052 (Mori *et al.* 2017), due to a single base deletion leading to a frameshift mutation in the first exon of Fer_r1.1chr02_g012310.1. This deletion could be expected to be a functional mutation; however, it is not conserved in other fig lines. Instead, many deleterious mutations are found in the gene sequences, suggesting that there are multiple susceptible alleles in fig lines.

20.7 Gene Expression and Transcriptome Analysis

Gene-expression and transcriptome information provide dynamic and quantitative gene-expression profiles and play an important role in exploring the mechanisms of various life phenomena. Here, we trace the progress of these studies' integration with physiology and ecology in figs.

20.7.1 Parthenocarpy

Two countries, Israel and China, have actively conducted parthenocarpy transcriptome analyses. The Israeli group aimed to understand the stimulative form of parthenocarpy. Their first study was a reverse transcriptase polymerase chain reaction (RT-PCR) analysis of the expression of 13 genes related to chlorophyll synthesis and degradation using four kinds of tissue (fruit skin, inflorescence, leaf, and root) from parthenocarpic varieties with different pollination status (Rosianskey *et al.*, 2016). They showed a strong negative correlation between the levels of pheophorbide a oxygenase and chlorophyll during the ripening process of pollinated fruit. They also found six genes (*FcGluTR1*, *FcGluTR2*, *FcCLS1*, *FcCLH1*, *FcCLH2*, and *FcCCR1*) that behave differently in the two forms of parthenocarpy (Rosianskey *et al.*, 2016). They then performed RNA sequencing (RNA-seq) analysis at five stages for each tissue from parthenocarpic-variety syconia of different pollination status. The 147,000 assembled transcripts revealed significantly different transcriptional patterns between the pulp and inflorescence in phase I and phase III, and a particularly strong difference in gene expression during fruit inflorescence. Gas chromatography analysis of hormone levels showed that both ethylene and auxin increase during fruit ripening with and without pollination, but no production of active gibberellins or cytokinins was observed. This indicated that the gene-expression levels of ethylene, auxin, and abscisic acid (ABA) are clearly different between pollinated and non-pollinated inflorescences. Finally, they suggested that fruit maturation is controlled by the reproductive part, and that differences in ABA production may control differences in the ripeness features of parthenocarpic varieties, depending on pollination (Rosianski *et al.*, 2016).

In contrast, the Chinese group analyzed the vegetative form of parthenocarpy using San Pedro-type fruit (Chai *et al.*, 2017, 2018). In the San Pedro type, the breba crop is parthenocarpic, and the main crop is non-parthenocarpic. In their experiments, they detected significant downregulation of indoleacetic acid–amino acid hydrolase and the gibberellic acid (GA) GA20ox, and significant upregulation of GA2ox, 9-*cis*-epoxycarotenoid dioxygenase, and 1-aminocyclopropane 1-carboxylic acid (ACC) synthase (ACS) in the main crop compared with the breba crop. The hormone levels of indoleacetic acid and gibberellin were lower in the younger main crops and remained lower until after flowering, but zeatin and ABA levels were higher in the main crop than in breba (Chai *et al.*, 2017). Based on a series of results, they suggested that auxin plays a major role in the early stages of syconium development and in the fate of parthenocarpy in San Pedro-type fruit (Chai *et al.*, 2017). Next, they studied parthenocarpy induced by GA (Chai *et al.*, 2018). GA treatment was applied to the main crop of the flowering stage of the San Pedro type, and a quantification of hormone levels and RNA-seq analysis were performed. Both indicated that GA treatment affects the control of auxin, ethylene, and ABA synthesis and signal transduction, and they showed that the properties of GA-induced parthenocarpy are controlled by cooperation of the female flowers and receptacle. They concluded that suppression of ABA and ethylene synthesis and catabolism of GA may be the driving forces for generating parthenocarpic properties (Chai *et al.*, 2018).

Previous studies have shown that plant hormones are involved in the expression of parthenocarpy (Crane, 1964; Lodhi *et al.*, 1969), and the above studies support this. Plant hormones may be involved in the control of trait expression by acting directly or indirectly as positive or negative regulators on a parthenocarpy master gene(s).

20.7.2 Flowering and reproductive development

Because *F. carica* has syconia and does not show any obvious flower organ, it is described as a "no-flower fruit," which is the Chinese phrase for fig; in reality, however, flower formation is induced. The *F. carica* syconium (inflorescence) can always differentiate during the elongation of a new shoot, and the timing of the differentiation depends on the nodal position of the shoot. This flowering style is unique to figs, and is of interest as an example of flowering diversity among plant species. Plant floral induction and flower organ development have been studied in figs, as well as in *Arabidopsis thaliana*.

The first flowering-related gene identified in *F. carica* was *FLOWERING LOCUS T* (*FcFT1*) (Ikegami *et al.*, 2013a). *FcFT1* belongs to the phosphatidylethanolamine-binding protein (PEBP) gene family, which also includes *FT*, *TERMINAL FLOWER 1* (*TFL1*), *TWIN SISTER OF FT* (*TSF*), *A. thaliana CENTRORADIALIS homolog* (*ATC*), *MOTHER OF FT AND TFL1* (*MFT*), and *BROTHER OF FT AND TFL1* (*BFT*) (Bradley *et al.*, 1997; Kardailsky *et al.*, 1999; Kobayashi *et al.*, 1999; Mimida *et al.*, 2001; Yoo *et al.*, 2004; Yamaguchi *et al.*, 2005). *FcFT1* is expressed in leaves, with high expression levels in the upper nodal leaves (mature leaves) and low levels in the lower nodal leaves (immature leaves). *FcFT1* also shows light-responsive, persistent expression throughout the growing season (Ikegami *et al.*, 2013a). This suggests that *FcFT1* expression is regulated by both light-dependent and autonomous pathways. The fact that *FcFT1* is under the control of a light-dependent pathway is consistent with the result that fruit yield decreases when the plant is growing in the shade (Matsuura and Araki, 1995). In addition, in a transgenic tobacco experiment, *FcFT1* overexpression resulted in earlier flowering in the transformants than in the wild type. There are no reports on PEBP genes other than *FcFT1* in *F. carica*, but a total of at least eight genes, including three *FT*, three *TFL*, and two *MFT* homologs in the *F. carica* genome has been confirmed (Mori *et al.*, 2017).

LEAFY (*LFY*) was also identified as a gene involved in fig flower bud differentiation (Ma *et al.*, 2020). The highest expression level of *FcLFY* was observed in axillary bud tissue, and flowering time in *Arabidopsis* overexpressing *FcLFY* was earlier than in the wild type. The phenotype was partially complementary to an *Arabidopsis lfy15* mutant, similar to experiments with *LFY* genes in other species. The expression of flowering-promoting genes in the transformants was upregulated, while, in contrast, the expression of flowering-suppressing genes was downregulated (Ma *et al.*, 2020). The authors suggested that *FcLFY* is involved in fig flower bud differentiation and apical meristem formation. No analyses of flowering-related genes other than *FcFT1* and *FcLFY* have been performed. Further analyses are expected to elucidate the physiological and ecological phenomena related to *F. carica* floral induction and flower development.

20.7.3 Fruit ripeness and softening

Ripened fresh figs are highly perishable. Softening is a characteristic property of fruit ripening and is a matter of fruit handling (Chessa, 1997; Owino *et al.*, 2004; Freiman *et al.*, 2014).

The first genetic report on ripening and softening analyzed the expression of genes encoding cell-wall-modifying enzymes (Owino *et al.*, 2004). Those authors studied the expression patterns of complementary DNA clones of 11 related genes encoding endopolygalacturonases (*FcPg1*, *FcPg2*), endoglucanases (*FcCel1*, *FcCel2*), β-galactosidases (*FcGal1*, *FcGal2*), xyloglucan endotransglycosylases (*FcXTH1*, *FcXTH2*, *FcXTH3*), expansins (*FcExp1*), and α-arabinofuranosidases (*FcArabf1*). Both receptacle and drupelet tissues were analyzed to help understand the rapid softening that occurs during fig ripening. They also studied the expression of four genes (*FcACS1*, *FcACS2*, *FcACS3*, and *FcACO1* (ACC oxidase 1)) in response to oil, propylene, auxin, or 1-methylcyclopropene (1-MCP) treatment, and confirmed that these genes were regulated by the treatments through a multi-responsive signal-transduction pathway during maturation and ripening (Owino *et al.*, 2004).

Freiman *et al.* (2014) reported the first NGS analysis of genes involved in fruit maturation. They used a Roche 454 system with common fruit types at four developmental stages to build 12,751 unigenes and 46,927 singletons. In their study, the expression of genes related to ethylene synthesis increased with maturation. Molecular resources from that study created a basis for research on the fruit-ripening process. Subsequently, Freiman *et al.* (2015) investigated the expression of maturation-related genes, including those related to ethylene synthesis, and those involved in signaling processes, in untreated and 1-MCP-treated fig fruit. They saw that drupelet ethylene synthesis and signaling pathway gene-expression patterns were similar to those of mature-stage tomato and apple. The conclusions were that *FcACS2*, *FcACS4*, and *FcACO3* are associated with the autoinhibition of ethylene synthesis in 1-MCP-treated fruit, and that *FcMADS8*, ethylene-insensitive 3-binding F-box 1 (*FcEBF1*), and ethylene-insensitive 3-like (*FcEIL3*) may be key genes for enhancing the storage of 1-MCP-treated fruit. They proposed that one of the ethylene-responsive factor genes (*FcERF*) may induce non-climacteric autoinhibition of ethylene production (Freiman *et al.*, 2015).

Cui *et al.* (2019) applied quantitative isobaric tags for relative and absolute quantitation (iTRAQ) and RNA-seq simultaneously with commercial-ripe (CR) and tree-ripe (TR) fruit to better understand expression profiles and metabolic changes during fruit ripening. They identified 1226 proteins with iTRAQ, 84 of which showed significantly varying levels between CR and TR fruit. RNA-seq revealed 2087 differentially expressed genes (at least twofold expression change) between CR and TR fruit. The results indicated that factors involved in sugar accumulation and texture are upregulated in both proteins and transcripts in TR fruit, and that 1-carboxylate oxidase and other ethylene-responsive factors are activated in the receptacle tissue, as reported previously (Freiman *et al.*, 2015). Much molecular information and knowledge of the fruit-ripening process has been accumulated over the last few decades. We expect that with more advanced analyses, using new temporal and spatial condition parameters, the complicated but integrated fig-maturation process will be elucidated.

20.7.4 Fruit skin color

Fig-fruit skin colors include green, yellow, brown, red, purple, black, and others. The coloring is controlled by pigments and is a major factor in the commercial value of the fruit. Therefore, elucidation of coloring mechanisms is important to agriculture. Fig-fruit skin variation is regulated by diverse gene functions, and by the upstream regulation and consequent expression of various pigment-synthesis pathways. Significant progress has been made in our understanding of these mechanisms in recent years.

Ikegami *et al.* (2013b) collected some sequence information on anthocyanin-synthesis pathway genes with NGS in fig, but their association with fruit skin color remained unclear. Cao *et al.* (2016) cloned the anthocyanin synthase gene *FcANS1* from the variety Brown Turkey, and found the pattern of gene expression and anthocyanin content to be consistent throughout the development of syconia. *FcANS1* expression and the amount of anthocyanin were dependent on light conditions in the study. Wang *et al.* (2017, 2019) used the purple skin mutant Zibao variety to conduct approaches based on transcriptome and metabolite analyses. In their 2017 report, they compared the transcriptomes of a green-skinned variety and its mutant (Zibao), and found differentially regulated expression of genes encoding enzymes in the phenylpropanoid- and flavonoid-synthesis pathways between young and mature stages. They inferred the possible activation of a transposable element, and its involvement in the mutation of various transcription factors, such as R2R3-MYB (Stracke *et al.*, 2001). Their 2019 report compared Zibao anthocyanin content with the transcriptome of female tissues under different light conditions. They found that the anthocyanin content of the fruit skin was affected by shade, but that the anthocyanin content of the female flower tissue was not, nor was expression of the phenylpropanoid- or flavonoid-synthesis pathways. This means that coloration of the skin and florets (female flowers) of the syconium by anthocyanins is independently controlled. In addition, they identified two candidate transcription factors, *FcHY5* and *FcMYB114*, which bind to the chalcone synthase (*FcCHS*) and dihydroflavonol-4-reductase (*FcDFR*) promoter regions. These latter two genes are the major transcripts of the flavonoid-synthesis pathway (Fig. 20.3) (Wang *et al.*, 2019).

Varietal differences

Simply obtaining and comparing genomic sequences or transcriptomes from different varieties and tissues, and then considering those differences, can significantly enhance fig genomics.

Ikegami *et al.* (2013b) compared the sequence reads from caprifig and common-type fig syconia, demonstrating the potential of a molecular approach for addressing some traits, including gender and parthenocarpy. An Italian group obtained transcriptome information for the variety Dottato and compared it with the transcriptome of the Japanese variety Houraishi (Zambrano *et al.*, 2017). They found 39,474 transcripts that were common to both cultivars, 2383 transcripts specific to Dotatto, and 4803 transcripts specific to Houraishi. Mori *et al.* (2017) obtained transcriptome information for three types of tissue (branch, leaf, and fruit (syconium)) separately from the variety Houraishi. They investigated the number of genes expressed in a tissue-specific manner. However, their results did not refer to the complete genome background, so updated genome information (Usai *et al.*, 2019; Shirasawa *et al.*, 2020) may yield more comprehensive analyses.

20.8 Issues Around and Beyond Genome

Fig genomics has made great strides due to improved accessibility of molecular genome data. However, there are cases for which such goals cannot be achieved by simply handling and understanding genomic information. Research in new directions may require incorporating and integrating other epigenetic factors. This section discusses some of those issues that may be addressed in connection with fig genomics research and peripheral technologies.

20.8.1 Gene functional validation and genome editing

Genetic recombination and genome editing are essential tools for proving the functions of new genes discovered in functional genomics research. When a target gene candidate has been found, the phenotype of a transformant into which the candidate gene has been introduced or modified provides strong evidence of target gene function.

Today's standard method for transforming figs was presented by Yancheva *et al.* (2005).

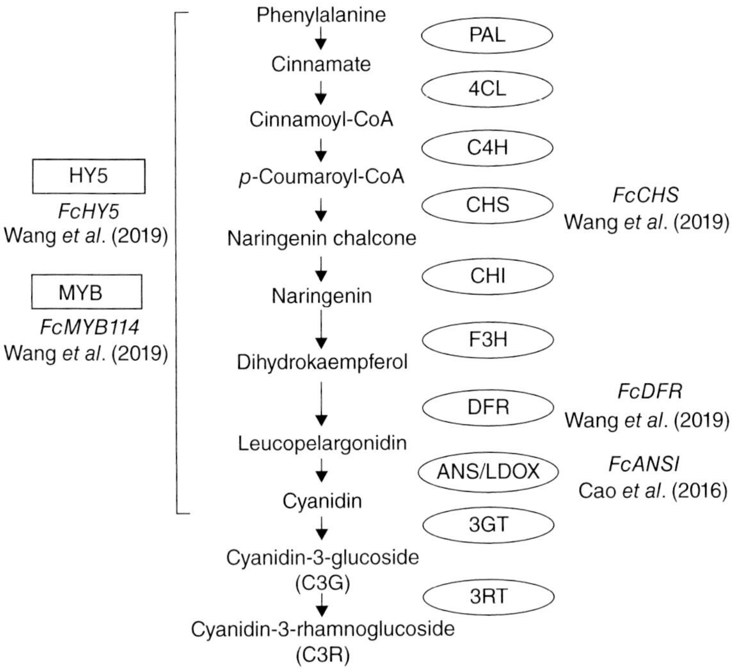

Fig. 20.3. A schematic diagram of the general anthocyanin biosynthetic pathway in fig. Enzymes for each step are indicated by ovals. Transcription factors are indicated by boxes. Bibliographic information is provided near each enzyme and transcription factor. PAL, phenylalanine ammonia lyase; 4CL, 4-coumarate-CoA ligase; C4H, cinnamate 4-hydroxylase; CHS, chalcone synthase; CHI, chalcone isomerase; F3H, flavanone-3β-hydroxylase; DFR, dihydroflavonol 4-reductase; ANS/LDOX, anthocyanidin synthase/ leucoanthocyanidin dioxygenase. Enzymes specific for various flavonoid compounds: 3GT, flavonoid 3-*O*-glucosyltransferase; 3RT, flavonoid 3-*O*-rhamnosyltransferase.

They showed the highest shoot-regeneration efficiency using a medium that included MS (Murashige and Skoog) basal salt mixture supplemented with 100 mg L^{-1} myoinositol, 1 mg L^{-1} thiamine HCl, 2.0 mg L^{-1} thidiazuron, and 2 mg L^{-1} indole-3-butyric acid, in 4% sucrose and 0.8% agar medium. Efficiency depended on the explant's orientation (adaxial side up was better). Soliman *et al.* (2010) obtained fig transformants derived from somatic embryos using an example of this method. They obtained the best plantlet differentiation with concentrations of 30 mg L^{-1} isopentenyl adenine, 7 mg L^{-1} thidiazuron and 0.25 mg L^{-1} naphthalene acetic acid, with a maximum regeneration efficiency of 83%.

In recent years, it has become relatively easy to adopt genome-editing technology (Kim *et al.*, 1996; Christian *et al.*, 2010; Carroll, 2011; Leeb and Wutz, 2011; Jinek *et al.*, 2012; Cong *et al.*, 2013; Mali *et al.*, 2013; Kim and Kim, 2014) for pinpoint gene knockout. Peer *et al.* (2015) first reported targeted mutagenesis using zinc-finger nucleases in *F. carica*. Mutations were subsequently introduced into the host endogenous phytoene desaturase gene in *F. carica* (*FcPDS*) through the CRISPR/Cas9 (clustered regularly interspaced short palindromic repeats/ CRISPR-associated protein 9) system (Flaishman *et al.*, 2020). The *FcPDS* gene encodes an important enzyme in the carotenoid-biosynthesis pathway, and loss of function of this gene in *Arabidopsis* results in an albino and dwarf phenotype (Qin *et al.*, 2007).

Woody plants, including *F. carica*, generally require a long time in culture after regeneration. In addition, fruit trees need a very long time for young individuals to mature and bear fruit; therefore, carrying out experiments in which fig transformants are created is quite burdensome. Consequently, one way to test these systems is to use more quickly maturing experimental model plants, such as *A. thaliana* or tomato, if it can be assumed that the target genes have common functions in the different species. However, it should be noted that this approach lacks validity if the target gene function or the gene itself is species specific.

20.8.2 Phenotyping technology

Obtaining large-scale phenotypic data can be challenging with some plant materials. The time and cost of phenotyping are bottlenecks in breeding trials (Furbank and Tester, 2011) and GWAS strategy. In genomic selection, for example, the accuracy of the prediction model depends on the size of the training population (Endelman *et al.*, 2014). Additionally, it takes more time and effort as the population size increases. In addition, the measurement of detailed morphological information is difficult in most cases, because complex traits are hard to convert to numerical form and need to be first modeled and quantified (Yabe *et al.*, 2016).

High-throughput phenotyping technology using imaging platforms and computer-based tools has attracted attention as a solution to such problems (Fahlgren *et al.*, 2015; Yabe *et al.*, 2016). In the case of *Arabidopsis*, the acquisition of quantitative data of multiple phenotypes involving various environmental factors has been automated (Fujita *et al.*, 2018). This system enables high-throughput and accurate phenotypic measurement by mechanical processing. Furthermore, genotype × environment interaction research is becoming increasingly important in agrigenetics, including GWAS and genomic-selection techniques (Lopez-Cruz *et al.*, 2015; Jarquín *et al.*, 2017; Sukumaran *et al.*, 2017a,b). Therefore, in the future, it will become necessary to perform phenotyping with a consideration of environmental data.

20.8.3 Trans-omics

All organisms, including *F. carica*, are controlled by an information network that spans multiple omics layers, including the genome, epigenome, transcriptome, proteome, and metabolome. The discipline of analyzing biological systems through the construction of a global molecular network for understanding living organisms as a whole, by hierarchically connecting multilayered omics data with quantitative information on science properties and statistical mathematical analyses, is called trans-omics or multi-omics (Yugi *et al.*, 2016). Trans-omics research usually involves the combination of a set of high-throughput technologies such as proteomics and metabolomics, with bioinformatics techniques that enable the processing, analysis, and integration of large amounts of data. Much information has been acquired at many omic levels in *F. carica*, and those cases have been described. In one study, two types of omic information were obtained simultaneously for the same material (Cui *et al.*, 2019). These data all provide a basis for driving trans-omics forward in *F. carica*. However, the existence of omics data where the cost is still relatively expensive (e.g. metabolome) still poses a hurdle for analyses, but this situation may eventually be resolved as the cost of the associated technology decreases. Another factor that narrows the gateway to trans-omics is information analysis. The learning of specialized knowledge and the performance of analyses with a high degree of freedom often remain challenging tasks for biologists. Collaboration with other people or groups specializing in the field is another option for progress.

20.9 Concluding Remarks

Although fig genetics was established in the 20th century (Storey, 1955, 1975; Saleeb, 1965), it has entered a new phase with the advent of molecular biology, genomics, and information science. The basic *F. carica* species is diploid ($2n$ = 26), and its genome is relatively compact (356–357 Mb); therefore, it has many advantages for use in various omics analyses compared with other plant species. Although *F. carica* is a non-model plant, its molecular resources have greatly accumulated in recent years. Today, we can access most of the necessary information for our research at a level comparable to model organisms. In addition, *F. carica* has many unique characteristics, and clarifying such characteristics may lead to the acquisition of knowledge that will impact not only this particular species, but also plant science in general. Macroscopically, we can say that fig genomics research and the further development of other fig omics research have reached the first stage, with the foundation finally set for takeoff. Although high-cost issues remain at some levels of analysis, it is likely that these will eventually decrease. An important key to advancing these studies in the future is the expected expansion of allied information science and associated technologies, not only for genomics research but for all fields.

References

Achtak, H., Ater, M., Oukabli, A., Santoni, S., Kjellberg, F., and Khadari, B. (2010) Traditional agroecosystems as conservatories and incubators of cultivated plant varietal diversity: the case of fig (*Ficus carica* L.) in Morocco. *BMC Plant Biology* 10: 28.

Ahmed, B.A., Ghada, B., Laila, E., Hafid, A., Bouchaib, K., and Amel, S.-H. (2015) Use of morphological traits and microsatellite markers to characterize the Tunisian cultivated and wild figs (*Ficus carica* L.). *Biochemical Systematics and Ecology* 59, 209–219.

Ahmed, S., Gilmartin, P.M., Dawson, D.A., and Compton, S.G. (2007) Characterization of microsatellite loci in the African fig *Ficus sycomorus* L. (Moraceae). *Molecular Ecology Notes* 7, 1175–1177.

Ali-Shtayeh, M.S., Jamous, R.M., Abu Zaitoun, S.Y., Mallah, O.B., and Mubaslat, A.K. (2014) Genetic diversity of the Palestinian fig (*Ficus carica* L.) collection by pomological traits and RAPD markers. *American Journal of Plant Sciences* 2014, 1139–1155.

Aljane, F. and Essid, A. (2018) Characterization and assessment of fig (*Ficus carica* L.) genetic resources in Tunisia: an overview. In: Kallel, A., Ksibi, M., Ben Dhia, H., and Khélifi, N. (eds) *Recent Advances in Environmental Science from the Euro-Mediterranean and Surrounding Regions*. EMCEI 2017. Advances in Science, Technology & Innovation (IEREK Interdisciplinary Series for Sustainable Development). Springer, Cham, Switzerland, pp. 1355–1356.

Almajali, Da., Abdel-Ghani, A.H., and Migdadi, H. (2012) Evaluation of genetic diversity among Jordanian fig germplasm accessions by morphological traits and ISSR markers. *Scientia Horticulturae* 147, 8–19.

Aradhya, M., Preece, J.E., and Velasco, D. (2017) Multivariate analysis of molecular and morphological diversity in fig (*Ficus carica* L.). *Acta Horticulturae* 1173, 23–28.

Aradhya, M.K., Stover, E., Velasco, D., and Koehmstedt, A. (2010) Genetic structure and differentiation in cultivated fig (*Ficus carica* L.). *Genetica* 138, 681–694.

Awamura, M., Shoda, K., and Yahata, D. (1996a) Effect of various seed parents on frequency distribution of parthenocarpy among seedling progenies of fig (*Ficus carica* L.). *Horticulture Journal* 65, 21–26.

Awamura, M., Yahata, D., and Shoda, K. (1996b) Classification of fig (*Ficus carica* L.) skin color by color measurement with a color diference meter. *Bulletin of the Fukuoka Agricultural Research Center* 15, 87–90.

Bandelj, D., Javornik, B., and Jakse, J. (2007) Development of microsatellite markers in the common fig, *Ficus carica* L. *Molecular Ecology Resources* 7, 1311–1314.

Baraket, G., Abdelkrim, A.B., Mars, M., and Salhi-Hannachi, A. (2011) Cyto-nuclear discordance in the genetic relationships among Tunisian fig cultivars (*Ficus carica* L.): evidence from non coding *trn*L–*trn*F and ITS regions of chloroplast and ribosomal DNAs. *Scientia Horticulturae* 130, 203–210.

Basheer-Salimia, R., Awad, M., and Ward, J. (2012) Assessments of biodiversity based on molecular markers and morphological traits among West-Bank, Palestine fig genotypes (*Ficus carica* L.). *American Journal of Plant Sciences* 3, 1241–1251.

Bendhifi, M., Baraket, G., Zourgui, L., Souid, S., and Salhi-Hannachi, A. (2013) Assessment of genetic diversity of Tunisian Barbary fig (*Opuntia ficus indica*) cultivars by RAPD markers and morphological traits. *Scientia Horticulturae.* 158, 1–7.

Bennett, M.D. and Leitch, I.J. (2011) Nuclear DNA amounts in angiosperms: targets, trends and tomorrow. *Annals of Botany* 107, 467–590.

Boudchicha, R.H., Hormaza, J.I., and Benbouza, H. (2018) Diversity analysis and genetic relationships among local Algerian fig cultivars (*Ficus carica* L.) using SSR markers. *South African Journal of Botany* 116, 207–215.

Bradley, D., Ratcliffe, O., Vincent, C., Carpenter, R., and Coen, E. (1997) Inflorescence commitment and architecture in *Arabidopsis*. *Science* 275, 80–83.

Cabrita, L.F., Aksoy, U., Hepaksoy, S., and Leitão, J.M. (2001) Suitability of isozyme, RAPD and AFLP markers to assess genetic differences and relatedness among fig (*Ficus carica* L.) clones. *Scientia Horticulturae* 87, 261–273.

Caliskan, O., Polat, A.A., Celikkol, P., and Bakir, M. (2012) Molecular characterization of autochthonous Turkish fig accessions. *Spanish Journal of Agricultural Research* 10, 130–140.

Cao, L., Xu, X., Chen, S., and Ma, H. (2016) Cloning and expression analysis of *Ficus carica* anthocyanidin synthase 1 gene. *Scientia Horticulturae* 211, 369–375.

Carroll, D. (2011) Genome engineering with zinc-finger nucleases. *Genetics* 188, 773–782.

Chai, L., Wang, Z., Chai, P., Chen, S., and Ma, H. (2017) Transcriptome analysis of San Pedro-type fig (*Ficus carica* L.) parthenocarpic breba and non-parthenocarpic main crop reveals divergent phytohormone-related gene expression. *Tree Genetics & Genomes* 13, 1–14.

Chai, L., Chai, P., Chen, S., Flaishman, M.A., and Ma, H. (2018) Transcriptome analysis unravels spatiotemporal modulation of phytohormone-pathway expression underlying gibberellin-induced parthenocarpic fruit set in San Pedro-type fig (*Ficus carica* L.). *BMC Plant Biology* 18: 100.

Chatti, K., Baraket, G., Abdelkrim, A.B., Saddoud, O., Mars, M., *et al.* (2010) Development of molecular tools for characterization and genetic diversity analysis in Tunisian fig (*Ficus carica*) cultivars. *Biochemical Genetics* 48, 789–806.

Chessa, I. (1997) Fig. In: Mitra, S. (ed.) *Postharvest Physiology and Storage of Tropical and Subtropical Fruits*. CAB International, Wallingford, UK, pp. 245–268.

Chessa, I., Nieddu, G., and Serra, P. (1998) Fig germplasm characterization using isozyme analysis. *Acta Horticulturae* 480, 143–148.

Christian, M., Cermak, T., Doyle, E.L., Schmidt, C., Zhang, F., *et al.* (2010) Targeting DNA double-strand breaks with TAL effector nucleases. *Genetics* 186, 757–761.

Ciarmiello, L.F., Piccirillo, P., Carillo, P., de Luca, A., and Woodrow, P. (2015) Determination of the genetic relatedness of fig (*Ficus carica* L.) accessions using RAPD fingerprint and their agro-morphological characterization. *South African Journal of Botany* 97, 40–47.

Condit, I.J. (1955) Fig varieties: a monograph. *Hilgardia* 11, 323–538.

Condit, I.J. (1969) *Ficus: The Exotic Species*. Division of Agricultural Sciences, University of California, Los Angeles, California.

Cong, L., Ran, F.A., Cox, D., Lin, S., Barretto, R., *et al.* (2013) Multiplex genome engineering using CRISPR/Cas systems. *Science* 339, 819–823.

Crane, J.C. (1964) Growth substances in fruit setting and development. *Annual Review of Plant Physiology* 15, 303–326.

Crozier, Y.C., Jia, X.C., Yao, J.Y., Field, A.R., Cook, J.M., and Crozier, R.H. (2007) Microsatellite primers for *Ficus racemosa* and *Ficus rubiginosa*. *Molecular Ecology Notes* 7, 57–59.

Cui, Y., Wang, Z., Chen, S., Vainstein, A., and Ma, H. (2019) Proteome and transcriptome analyses reveal key molecular differences between quality parameters of commercial-ripe and tree-ripe fig (*Ficus carica* L.). *BMC Plant Biology* 19: 146.

de Masi, L., Cipollaro, M., di Bernardo, G., and Simeone, A. (2003) Clonal selection and molecular characterisation by RAPD analysis of the fig (*Ficus carica* L.) “Dottato” and “Bianco del Cilento” cultivars in Italy. *Acta Horticulturae* 605, 65–68.

de Masi, L., Castaldo, D., Galano, G., and Laratta, B. (2005) Genotyping of fig (*Ficus carica* L) via RAPD markers. *Journal of the Science of Food and Agriculture* 85, 2235–2242.

de Masi, L., Vella, F.M., Laratta, B., Volpe, M.G., Tiseo, M., and la Cara, F. (2017) Biochemical and genetic characterization of a red fig cultivar (*Ficus carica*) from Southern Italy. *Acta Horticulturae* 1173, 81–86.

Dueñas, M., Pérez-Alonso, J.J., Santos-Buelga, C., and Escribano-Bailón, T. (2008) Anthocyanin composition in fig (*Ficus carica* L.). *Journal of Food Composition and Analysis* 21, 107–115.

Elisiário, P.J., Neto, M.C., Cabrita, L.F., and Leitão, J.M. (1998) Isozyme and RAPDs characterisation of a collection of fig (*Ficus carica* L.) traditional varieties. *Acta Horticulturae* 480, 149–154.

Endelman, J.B., Atlin, G.N., Beyene, Y., Semagn, K., Zhang, X., *et al.* (2014) Optimal design of preliminary yield trials with genome-wide markers. *Crop Science* 54, 48–59.

Ercisli, S., Tosun, M., Karlidag, H., Dzubur, A., Hadziabulic, S., and Aliman, Y. (2012) Color and antioxidant characteristics of some fresh fig (*Ficus carica* L.) genotypes from northeastern Turkey. *Plant Foods for Human Nutrition* 67, 271–276.

Essid, A., Aljane, F., Ferchichi, A., and Hormaza, J.I. (2015) Analysis of genetic diversity of Tunisian caprifig (*Ficus carica* L.) accessions using simple sequence repeat (SSR) markers. *Hereditas* 152: 1.

Fahlgren, N., Gehan, M.A., and Baxter, I. (2015) Lights, camera, action: high-throughput plant phenotyping is ready for a close-up. *Current Opinion in Plant Biology* 24, 93–99.

Falistocco, E. (2009) Presence of triploid cytotypes in the common fig (*Ficus carica* L.). *Genome* 52, 919–925.

Falistocco, E. (2016) Recurrent events of polyploidy in *Ficus carica* L. (Moraceae). *International Journal of Plant Sciences* 177, 319–325.

Ferguson, L., Michailides, T.J., and Shorey, H.H. (1990) The California fig industry. *Horticultural Reviews* 12, 409–490.

Ferrara, G., Mazzeo, A., Pacucci, C., Matarrese, A.M.S., Tarantino, A., *et al.* Characterization of edible fig germplasm from Puglia, southeastern Italy: Is the distinction of three fig types (Smyrna, San Pedro and Common) still valid? *Scientia Horticulturae* 205, 52–58.

Ferrara, G., Mazzeo, A., Gallotta, A., Pacucci, C., Matarrese, A.M.S., *et al.* (2017) Fruit-set and SSR markers of fig cultivars from Puglia region, Southeastern Italy. *Acta Horticulturae* 1173, 39–44.

Flaishman, M., Rodov, V., and Stover, E. (2008) Fig (*Ficus carica*): botany, horticulture and breeding. *Horticultural Reviews* 34, 113–196.

Flaishman, M.A., Peer, R., Raz, A., Cohen, O., Izhaki, K., *et al.* (2020) Advanced molecular tools for breeding in Mediterranean fruit trees: genome editing approach of *Ficus carica* L. *Acta Horticulturae* 1280, 1–10.

Freiman, Z.E., Doron-Faigenboim, A., Dasmohapatra, R., Yablovitz, Z., and Flaishman, M.A. (2014) High-throughput sequencing analysis of common fig (*Ficus carica* L.) transcriptome during fruit ripening. *Tree Genetics & Genomes* 10, 923–935.

Freiman, Z.E., Rosianskey, Y., Dasmohapatra, R., Kamara, I., and Flaishman, M.A. (2015) The ambiguous ripening nature of the fig (*Ficus carica* L.) fruit: a gene-expression study of potential ripening regulators and ethylene-related genes. *Journal of Experimental Botany* 66, 3309–3324.

Fujita, M., Tanabata, T., Urano, K., Kikuchi, S., and Shinozaki, K. (2018) RIPPS: a plant phenotyping system for quantitative evaluation of growth under controlled environmental stress conditions. *Plant and Cell Physiology* 59, 2030–2038.

Furbank, R.T. and Tester, M. (2011) Phenomics – technologies to relieve the phenotyping bottleneck. *Trends in Plant Science* 16, 635–644.

Ganopoulos, I., Xanthopoulou, A., Molassiotis, A., Karagiannis, E., Moysiadis, T., *et al.* (2015) Mediterranean basin *Ficus carica* L.: from genetic diversity and structure to authentication of a Protected Designation of Origin cultivar using microsatellite markers. *Trees* 29, 1959–1971.

Garcia, M., Bain, A., Tzeng, H.Y., Peng, Y.Q., Chou, L.S., and Kjellberg, F. (2012) Portable microsatellite primers for *Ficus* (Moraceae). *American Journal of Botany* 99, e187–e192.

Ghada, B., Ahmed, B.A., Messaoud, M., and Amel, S.-H. (2013) Genetic diversity and molecular evolution of the internal transcribed spacer (ITSs) of nuclear ribosomal DNA in the Tunisian fig cultivars (*Ficus carica* L.; Moracea). *Biochemical Systematics and Ecology* 48, 20–33.

Giraldo, E., López-Corrales, M., Roger, J.P., Khadari, B., Hochu, I., *et al.* (2008) Standardization of experimental protocols and SSR markers for the management of fig germplasm collections. *Acta Horticulturae* 798, 213–216.

Giraldo, E., Viruel, M.A., López-Corrales, M., and Hormaza, J.I. (2005) Characterisation and cross-species transferability of microsatellites in the common fig (*Ficus carica* L.). *Journal of Horticultural Science and Biotechnology* 80, 217–224.

Haddou, L.A., Charafi, J., Blenzar, A., Messaoudi, Z., and Outghouliast, H. (2018) Molecular characterization and study of genetic relationships among local cultivars of the Moroccan fig (*Ficus carica* L.) using microsatellite and ISSR markers. *International Journal of Environment Agriculture and Biotechnology* 3, 18–27.

Haffar, S., Ben Abdelkrim, A., Salhi-Hannachi, A., and Baraket, G. (2017) Genetic diversity assessment of Tunisian fig (*Ficus carica* L.) using conserved DNA-derived polymorphism. *Acta Horticulturae* 1173, 51–56.

Hedfi, J., Trifi, M., Salhi-Hannachi, A., Ould Mohamed Salem, A., and Marrakchi, M. (2003) Morphological and isoenzymatic polymorphism in Tunisian fig (*Ficus carica* L.) collection. *Acta Horticulturae* 605, 319–325.

Heer, K., Machado, C.A., Himler, A.G., Herre, E.A., Kalko, E.K., and Dick, C.W. (2012) Anonymous and EST-based microsatellite DNA markers that transfer broadly across the fig tree genus (*Ficus*, Moraceae). *American Journal of Botany* 99, e330–e333.

Hirayama, T., Kieber, J.J., Hirayama, N., Kogan, M., Guzman, P., *et al.* (1999) RESPONSIVE-TO-ANTAGONIST1, a Menkes/Wilson disease-related copper transporter, is required for ethylene signaling in *Arabidopsis*. *Cell* 97, 383–393.

Ikegami, H., Nogata, H., Hirashima, K., Awamura, M., and Nakahara, T. (2009) Analysis of genetic diversity among European and Asian fig varieties (*Ficus carica* L.) using ISSR, RAPD, and SSR markers. *Genetic Resources and Crop Evolution* 56, 201–209.

Ikegami, H., Nogata, H., Inoue, Y., Himeno, S., Yakushiji, H., Hirata, C., Hirashima, K., Mori, M., Awamura, M., Nakahara, T. (2013a) Expression of *FcFT1*, a *FLOWERING LOCUS T*-like gene, is regulated by light and associated with inflorescence differentiation in fig (*Ficus carica* L.). *BMC Plant Biology* 13, 216.

Ikegami, H., Habu, T., Mori, K., Nogata, H., Hirata, C., Hirashima, K., Tashiro, K., Kuhara, S. (2013b) *De novo* sequencing and comparative analysis of expressed sequence tags from gynodioecious fig (*Ficus carica* L.) fruits: caprifig and common fig. *Tree Genetics & Genomes* 9, 1075–1088.

Ikten, H., Mutlu, N., Gulsen, O., Kocatas, H., and Aksoy, U. (2010) Elucidating genetic relationships, diversity and population structure among the Turkish female figs. *Genetica* 138, 169–177.

Ikten, H., Solak, S.S., and Yilmaz, Y. (2018) Transferability of SSR markers from related *Ficus* species to *Ficus carica* L. and assessment of effectiveness of the markers. *Applied Ecology and Environmental Research* 16, 1909–1918.

Iwata, H., Minamikawa, M., Kajiya-Kanegae, H., Ishimori, M., and Hayashi, T. (2016) Genomics-assisted breeding in fruit trees. *Breeding Science* 66, 100–115.

Jarquín, D., Lemes da Silva, C., Gaynor, R.C., Poland, J., Fritz, A., *et al.* (2017) Increasing genomic-enabled prediction accuracy by modeling genotype × environment interactions in Kansas wheat. *Plant Genome* 10: 0130.

Jinek, M., Chylinski, K., Fonfara, I., Hauer, M., Doudna, J.A., and Charpentier, E. (2012) A programmable dual-RNA-guided DNA endonuclease in adaptive bacterial immunity. *Science* 337, 816–821.

Kardailsky, I., Shukla, V.K., Ahn, J.H., Dagenais, N., Christensen, S.K., *et al.* (1999) Activation tagging of the floral inducer FT. *Science* 286, 1962–1965.

Khadari, B., Hochu, I., Santoni, S., and Kjellberg, F. (2001) Identification and characterization of microsatellite loci in the common fig (*Ficus carica* L.) and representative species of the genus *Ficus*. *Molecular Ecology Resources* 1, 191–193.

Khadari, B., Hochu, I., Santoni, S., Oukabli, A., Ater, M., *et al.* (2003) Which molecular markers are best suited to identify fig cultivars: a comparison of RAPD, ISSR and microsatellite markers. *Acta Horticulturae* 605, 69–75.

Kim, H. and Kim, J.S. (2014) A guide to genome engineering with programmable nucleases. *Nature Reviews Genetics* 15, 321–334.

Kim, Y.G., Cha, J., and Chandrasegaran, S. (1996) Hybrid restriction enzymes: zinc finger fusions to Fok I cleavage domain. *Proceedings of the National Academy of Sciences USA* 93, 1156–1160.

Knap, T., Jakše, J., and Bandelj, D. (2017) Novel microsatellite markers for common fig, *Ficus carica* L. *Acta Horticulturae* 1173, 87–92.

Knap, T., Aradhya, M., Baruca Arbeiter, A., Hladnik, M., and Bandelj, D. (2018) DNA profiling of figs (*Ficus carica* L.) from Slovenia and Californian USDA collection revealed the uniqueness of some North Adriatic varieties. *Genetic Resources and Crop Evolution* 65, 1503–1516.

Kobayashi, Y., Kaya, H., Goto, K., Iwabuchi, M., and Araki, T. (1999) A pair of related genes with antagonistic roles in mediating flowering signals. *Science* 286, 1960–1962.

Laddomada, B., Gerardi, C., Mita, G., Lumare, D., Minonne, F., *et al.* (2008) Molecular characterization of Apulian fig (*Ficus carica* L.) germplasm collection using fluorescence-based AFLP markers. *Acta Horticulturae* 798, 205–211.

Leeb, M. and Wutz, A. (2011) Derivation of haploid embryonic stem cells from mouse embryos. *Nature* 479, 131–134.

Lodhi, F., Bradley, M.V., and Crane, J.C. (1969) Auxins and gibberellin-like substances in parthenocarpic and non-parthenocarpic syconia of *Ficus carica* L., cv. King. *Plant Physiology* 44, 555–561.

Lopez-Cruz, M., Crossa, J., Bonnett, D., Dreisigacker, S., Poland, J., *et al.* (2015) Increased prediction accuracy in wheat breeding trials using a marker × environment interaction genomic selection model. *G3: Genes, Genomes, Genetics* 5, 569–582.

Loureiro, J., Rodriguez, E., Doležel, J., and Santos, C. (2007) Two new nuclear isolation buffers for plant DNA flow cytometry: a test with 37 species. *Annals of Botany* 100, 875–888.

Ma, N., An, Y., Li, J., and Wang, L. (2020) Cloning and characterization of a homologue of the *FLORICAULA/LEAFY* gene in *Ficus carica* L., *FcLFY*, and its role in flower bud differentiation. *Scientia Horticulturae* 261: 109014.

Mafrica, R., Marchese, A., Bruno, M., Costa, F., Fretto, S., *et al.* (2017) Morphological and molecular variability within the fig cultivar 'Dottato' in the Italian protected designation origin area "Fichi di Cosenza". *Acta Horticulturae* 1173, 29–34.

Mali, P., Yang, L., Esvelt, K.M., Aach, J., Guell, M., *et al.* (2013) RNA-guided human genome engineering via Cas9. *Science* 339, 823–826.

Matsuura, K. and Araki, H. (1995) Effect of shading on the growth, fruit set, yield and fruit quality in fig trees of 'Masui Dauphine'. *Bulletin of the Hyogo Prefectural Agricultural Institute* 43, 21–26.

Mimida, N., Goto, K., Kobayashi, Y., Araki, T., Ahn, J.H., *et al.* (2001) Functional divergence of the *TFL1*-like gene family in *Arabidopsis* revealed by characterization of a novel homologue. *Genes to Cells* 6, 327–336.

Moe, A.M. and Weiblen, G.D. (2011) Development and characterization of microsatellite loci in dioecious figs (*Ficus*, Moraceae). *American Journal of Botany* 98, e25–e27.

Mori, K., Shirasawa, K., Nogata, H., Hirata, C., Tashiro, K., *et al.* (2017) Identification of RAN1 orthologue associated with sex determination through whole genome sequencing analysis in fig (*Ficus carica* L.). *Scientific Reports* 7: 41124.

Morita, T., Mise, D., and Jikumaru, S. (2011) Relative resistance of *Ficus erecta* and cultivars of *F. carica* to *Ceratocystis fimbriata* using their seedlings. *Annual Report of the Kansai Plant Protection Society* 53, 51–52.

Müntzing, A. (1937) The effects of chromosomal variation in *Dactylis*. *Hereditas* 23, 113–235.

Oliveira, A.P., Baptista, P., Andrade, P.B., Martins, F., Pereira, J.A., *et al.* (2012) Characterization of *Ficus carica* L. cultivars by DNA and secondary metabolite analysis: is genetic diversity reflected in the chemical composition? *Food Research International* 49, 710–719.

Oukabli, A. and Khadari, B. (2005) Caractérisation des variétés polyclonales marocaines de figuiers, *Ficus carica* L. *Fruits* 60, 47–54.

Owino, W.O., Nakano, R., Kubo, Y., and Inaba, A. (2004) Coordinated expression patterns of genes encoding cell wall modifying enzymes during ripening in distinct anatomical tissue regions of the fig (*Ficus carica* L.) fruit. *Postharvest Biology and Technology* 32, 253–261.

Peer, R., Rivlin, G., Golobovitch, S., Lapidot, M., Gal-On, A., *et al.* (2015) Targeted mutagenesis using zinc-finger nucleases in perennial fruit trees. *Planta* 241, 941–951.

Perez-Jiménez, M., López, B., Dorado, G., Pujadas-Salvá, A., Guzmán, G., and Hernandez, P. (2012) Analysis of genetic diversity of southern Spain fig tree (*Ficus carica* L.) and reference materials as a tool for breeding and conservation. *Hereditas* 149, 108–113.

Piepho, H.-P. and Koch, G. (2000) Codominant analysis of banding data from a dominant marker system by normal mixtures. *Genetics* 155, 1459–1468.

Powell, W., Morgante, M., Andre, C., Hanafey, M., Vogel, J., *et al.* (1996) The comparison of RFLP, RAPD, AFLP and SSR (microsatellite) markers for germplasm analysis. *Molecular Breeding* 2, 225–238.

Qin, G., Gu, H., Ma, L., Peng, Y., Deng, X.W., *et al.* (2007) Disruption of phytoene desaturase gene results in albino and dwarf phenotypes in *Arabidopsis* by impairing chlorophyll, carotenoid, and gibberellin biosynthesis. *Cell Research* 17, 471–482.

Ramanatha Rao, V. and Hodgkin, T. (2002) Genetic diversity and conservation and utilization of plant genetic resources. *Plant Cell, Tissue and Organ Culture* 68, 1–19.

Resta, P., Roselli, M., Palasciano, M.A., Lamaj, F., Fanizza, G., and Ferrara, E. (2008) Detection of multiple denominations of the same fig genotypes grown in southern Italy. *Acta Horticulturae* 798, 169–176.

Rhoades, M.M. (1936) Note on the origin of triploidy in maize. *Journal of Genetics* 33, 355–357.

Rodolfi, M., Ganino, T., Chiancone, B., and Petruccelli, R. (2018) Identification and characterization of Italian common figs (*Ficus carica*) using nuclear microsatellite markers. *Genetic Resources and Crop Evolution* 65, 1337–1348.

Rosianskey, Y., Dahan, Y., Yadav, S., Freiman, Z.E., Milo-Cochavi, S., *et al.* (2016) Chlorophyll metabolism in pollinated vs. parthenocarpic fig fruits throughout development and ripening. *Planta* 244, 491–504.

Rosianski, Y., Doron-Faigenboim, A., Freiman, Z.E., Lama, K., Milo-Cochavi, S., *et al.* (2016) Tissue-specific transcriptome and hormonal regulation of pollinated and parthenocarpic fig (*Ficus carica* L.) fruit suggest that fruit ripening is coordinated by the reproductive part of the syconium. *Frontiers in Plant Science* 7: 1696.

Sadder, M.T. and Ateyyeh, A.F. (2006) Molecular assessment of polymorphism among local Jordanian genotypes of the common fig (*Ficus carica* L.). *Scientia Horticulturae* 107, 347–351.

Saddoud, O., Baraket, G., Chatti, K., Trifi, M., Marrakchi, M., *et al.* (2011) Using morphological characters and simple sequence repeat (SSR) markers to characterize Tunisian fig (*Ficus carica* L.) cultivars. *Acta Biologica Cracoviensia Series Botanica* 53, 7–14.

Saleeb, W.F. (1965) *Genetics and cytology of syconium persistence in Ficus carica L.* PhD thesis, University of California, Riverside, California.

Saleeb, W.F. and Storey, W. (1975) The genetics of persistent vs. caducous synconia in fig. *HortScience* 10, 328.

Salhi-Hannachi, A., Khaled, C., Messaoud, M., Mohamed, M., and Mokhtar, T. (2005) Comparative analysis of genetic diversity in two Tunisian collections of fig cultivars based on random amplified polymorphic DNA and inter simple sequence repeats fingerprints. *Genetic Resources and Crop Evolution* 52, 563–573.

Salhi-Hannachi, A., Chatti, K., Saddoud, O., Marrakchi, M., Trifi, M., *et al.* (2008) Genetic diversity in Tunisian fig germplasm assessed with molecular markers: implication for its conservation. *Acta Horticulturae* 798, 191–197.

Shimizu, S. and Miyoshi, T. (1999) Occurrence of control of fig Ceratocystis canker caused by *Ceratocystis fimbriata*. *Plant Protection* 53, 25–27.

Shirasawa, K., Yakushiji, H., Nishimura, R., Morita, T., Jikumaru, S., *et al.* (2020) The *Ficus erecta* genome aids *Ceratocystis* canker resistance breeding in common fig (*F. carica*). *Plant Journal* 102, 1313–1322.

Soliman, H.I., Gabr, M.F., and Abdallah, N.A. (2010) Efficient transformation and regeneration of fig (*Ficus carica* L.) via somatic embryogenesis. *GM Crops* 11, 40–51.

Solomon, A., Golubowicz, S., Yablowicz, Z., Grossman, S., Bergman, M., *et al.* (2006) Antioxidant activities and anthocyanin content of fresh fruits of common fig (*Ficus carica* L.). *Journal of Agricultural and Food Chemistry* 54, 7717–7723.

Storey, W. (1975) Figs. In: Janick, J. and Moore, J. (eds) *Advances in Fruit Breeding*. Purdue University Press, West Lafayette, Indiana.

Storey, W.B. (1955) Sex inheritance in figs. California Fig Institute *Proceedings of the American Research Conference* 9, 15–17.

Storey, W.B. (1977) *The Fig (Ficus carica Linnaeus): Its Biology, History, Culture, and Utilization*. Jurupa Mountains Cultural Center, Riverside, California.

Stover, E. and Aradhya, M. (2008) Fig genetic resources and research at the US National Clonal Germplasm Repository in Davis, California. *Acta Horticulturae* 798, 57–68.

Stracke, R., Werber, M., and Weisshaar, B. (2001) The *R2R3-MYB* gene family in *Arabidopsis thaliana*. *Current Opinion in Plant Biology* 4, 447–456.

Sukumaran, S., Crossa, J., Jarquin, D., Lopes, M., and Reynolds, M.P. (2017a) Genomic prediction with pedigree and genotype × environment interaction in spring wheat grown in South and West Asia, North Africa, and Mexico. *G3: Genes, Genomes, Genetics* 7, 481–495.

Sukumaran, S., Crossa, J., Jarquin, D., and Reynolds, M. (2017b) Pedigree-based prediction models with genotype × environment interaction in multienvironment trials of CIMMYT wheat. *Crop Science* 57, 1865–1880.

Trifi, M., Chatti, K., Saddoud, O., Salhi-Hannachi, A., Marrakchi, M., and Mars, M. (2008) Survey of molecular markers in Tunisian figs: DNA fingerprintings. *Acta Horticulturae* 798, 159–164.

Usai, G., Mascagni, F., Giordani, T., Vangelisti, A., Bosi, E., *et al.* (2019) Epigenetic patterns within the haplotype phased fig (*Ficus carica* L.) genome. *Plant Journal* 102, 600–614.

Val, A.D., Souza, C.S., Ferreira, E.A., Salgado, S.M., Pasqual, M.K., and Cançado, G.M. (2013) Evaluation of genetic diversity in fig accessions by using microsatellite markers. *Genetics and Molecular Research* 12, 1383–1391.

Vignes, H., Hossaert-Mckey, M., Beaune, D., Fevre, D., Anstett, M.-C., *et al.* (2006) Development and characterization of microsatellite markers for a monoecious *Ficus* species, *Ficus insipida*, and cross-species amplification among different sections of *Ficus*. *Molecular Ecology Notes* 6, 792–795.

Vivaldi, G.A., Quartararo, A., Camposeo, S., Belloni, P., Volo, P. and Caruso, T. (2017) Genetic diversity of early ripening breba accessions (*Ficus carica* L.) found in the “Pomona” Apulian collection. *Acta Horticulturae* 1173, 121–126.

Wang, Z., Cui, Y., Vainstein, A., Chen, S., and Ma, H. (2017) Regulation of fig (*Ficus carica* L.) fruit color: metabolomic and transcriptomic analyses of the flavonoid biosynthetic pathway. *Frontiers in Plant Science* 8: 1990.

Wang, Z., Song, M., Li, Y., Chen, S., and Ma, H. (2019) Differential color development and response to light deprivation of fig (*Ficus carica* L.) syconia peel and female flower tissues: transcriptome elucidation. *BMC Plant Biology* 19: 217.

Yabe, S., Uehara, K., Yoshitsu, Y., Watanabe, K., and Noshita, K. (2016) How far have phenotyping systems been evolving? *Breeding Research* 18, 67–71.

Yakushiji, H., Morita, T., Jikumaru, S., Ikegami, H., Azuma, A., and Koshita, Y. (2012) Interspecific hybridization of fig (*Ficus carica* L.) and *Ficus erecta* Thunb., a source of Ceratocystis canker resistance. *Euphytica* 183, 39–47.

Yakushiji, H., Morita, T., and Jikumaru, S. (2019) Ceratocystis canker resistance in BC_1 populations of interspecific hybridization of fig (*Ficus carica*) and *F. erecta*. *Scientia Horticulturae* 252, 71–76.
Yamaguchi, A., Kobayashi, Y., Goto, K., Abe, M., and Araki, T. (2005) *TWIN SISTER OF FT* (*TSF*) acts as a floral pathway integrator redundantly with FT. *Plant Cell Physiology* 46, 1175–1189.
Yamanouchi, H., Koyama, A., Takyu, T., and Muramatsu, N. (2010) Nuclear DNA amounts in diploid mulberry species (*Morus* spp.). *Journal of Sericultural Science of Japan* 79, 1–8.
Yancheva, S.D., Golubowicz, S., Yablowicz, Z., Perl, A., and Flaishman, M.A. (2005) Efficient *Agrobacterium*-mediated transformation and recovery of transgenic fig (*Ficus carica* L.) plants. *Plant Science* 168, 1433–1441.
Yemiş, O., Bakkalbaşı, E., and Artık, N. (2012) Changes in pigment profile and surface colour of fig (*Ficus carica* L.) during drying. *International Journal of Food Science & Technology* 47, 1710–1719.
Yoo, S.Y., Kardailsky, I., Lee, J.S., Weigel, D., and Ahn, J.H. (2004) Acceleration of flowering by overexpression of *MFT* (*MOTHER OF FT AND TFL1*). *Molecules and Cells* 17, 95–101.
Yugi, K., Kubota, H., Hatano, A., and Kuroda, S. (2016) Trans-omics: how to reconstruct biochemical networks across multiple 'omic' layers. *Trends in Biotechnology* 34, 276–290.
Zambrano, L.S., Usai, G., Vangelisti, A., Mascagni, F., Giordani, T., *et al.* (2017) Cultivar-specific transcriptome prediction and annotation in *Ficus carica* L. *Data in Brief* 13, 64–66.
Zavodna, M., Arens, P., Dijk, P., and Vosman, B. (2005) Development and characterization of microsatellite markers for two dioecious *Ficus* species. *Molecular Ecology Notes* 5, 355–357.

21 Fig Flavor

Nesibe Ebru Kafkas[1]*, Muhammet Ali Gündeşli[2], Burhanettin Imrak[1], and Salih Kafkas[1]

[1]Department of Horticulture, Faculty of Agriculture, University of Çukurova, Balcali, Adana, Turkey; [2]Department of Plant and Animal Production, Nurdağı Vocational School, Gaziantep University, Gaziantep, Turkey

21.1 Introduction

Ficus carica L. (common fig) is a member of the family Moraceae, consisting of more than 850 species classified into about 40 genera, and is one of the oldest cultivated fruit trees among the horticultural crops (Mawa *et al.*, 2013). Although its origin is not yet precise, it is thought to have originated in western Asia and from there slowly spread through the Mediterranean region (Aksoy *et al.*, 2003; Stover *et al.*, 2007; Simsek, 2009; Gundesli *et al.*, 2020). Fruits can be consumed either fresh or processed dried, canned or in any other preserved form. It is mostly grown in the Mediterranean climate zone as it is adapted to subtropical and mild temperate climates (Eriten, 2005; Stover *et al.*, 2007). Located in the Mediterranean climate, Turkey is a major producer and exporter of figs with a 23% share of world production (Uzundumlu *et al.*, 2018). Fig production covers 315,000 ha in the world, which produced 1.15 million t in 2017. In total, 19% of the total fig area is in Morocco with 60 thousand ha of fig-production areas. Fig cultivation is carried out in a wide range of areas in the world. According to FAO data in 2019, fig production was carried out on an area of 290,000 ha throughout the world, with Morocco having the largest share (21.7%) of the fig-production area, followed by Turkey with a share of 18.0%, and Algeria in third place with a share of 13.6%. World fig production increased by 7.4% in 2019 compared with the previous year and amounted to approximately 1.3 million t. Turkey is ranked first in world fig production with 310,000 t, Egypt second with 225,000 t and Morocco third with 153,000 t of production. (FAO, 2020; Anon., 2021). These countries typically have mild winters and hot dry summers (Solomon *et al.*, 2006; Oliveira *et al.*, 2010a; Simsek, 2011; Mawa *et al.*, 2013; Gundesli *et al.*, 2020).

Nowadays, because of the research due to the beneficial compounds for human health, interest in figs among consumers and producers is increasing. Consumption of antioxidants from natural sources such as figs was suggested to be the "big secret" of the long life of Mediterranean populations (Solomon *et al.*, 2006; Calışkan and Polat, 2011). The presence of many secondary metabolites such as phenols, organic acids, sugars, vitamin E, volatile compounds, carotenoids, and dietary fiber, as well as the high amounts of amino acids with relatively low amounts of cholesterol, was reported for figs (Vinson, 1999; Grison-Pigé *et al.*, 2002; Solomon *et al.*, 2006;

* Email: ebruyasakafkas@gmail.com

 Advances in Fig Research and Sustainable Production (M.A. Flaishman and U. Aksoy)
DOI: 10.1079/9781789242492.0021

Trichopoulou *et al.*, 2006; Cuevas-Glory *et al.*, 2007; Veberic *et al.*, 2008; Oliveira *et al.*, 2009; Calışkan and Polat, 2011; Slatnar *et al.*, 2011; Mujić *et al.*, 2014). Epidemiological studies have revealed a positive association. Figs are well known in the traditional medical system for their healing and therapeutic potentials, and have traditionally been exploited for their respiratory, cardiovascular, laxative, antispasmodic, and anti-inflammatory therapeutic benefits. Fig fruits, roots, and leaves are known to be used in folk medicine in the treatment of different diseases such as gastrointestinal problems (colic, ulcer, indigestion, anorexia, and diarrhea), respiratory problems (sore throat and bronchial problems), inflammation, cancer, and cardiovascular diseases (Guarrera, 2005; Patil and Patil, 2012; Mawa *et al.*, 2013; Badhujar *et al.*, 2014; Soltana *et al.*, 2017). These compounds found in figs are also known for their preventive effects against many diseases such as cataracts, cancer, and cardiovascular diseases (Duke, 2002; Solomon *et al.*, 2006; Ribechini *et al.*, 2011; Mawa *et al.*, 2013). Therefore, figs are known to be associated with longevity and for their contribution to the quality of human life. The nutritional values and functional characteristics of figs are closely related to fruit quality and are usually influenced by genotype and ripening stage, as well as by environmental conditions and orchard management practices (Crisosto *et al.*, 2010). The fruits are perishable and the compounds regulating biological activity are sensitive to degradation. Therefore, protection of the quality and nutritive values throughout the distribution chains, from harvesting to marketing, via transportation and storage, is very important (Çakmak *et al.*, 2010).

Fruit quality is measured by four distinctive features, which consumers follow: (i) color; (ii) appearance; (iii) taste/aroma (i.e. flavor); and (iv) nutritional values (Barrett *et al.*, 1998; Ong *et al.*, 2008; Valero and Serrano, 2010; Pereira and Câmara, 2011; King *et al.*, 2012; Pereira *et al.*, 2017). Kramer (1965) stated that the color and appearance of a product are the most critical quality features in determining consumers' preferences. The nutritional value of the product is another important criterion and is increasingly more valued by consumers, scientists, and the medical sphere (Barrett *et al.*, 2010). The flavor of the fruit, consisting of its aroma and taste (Grison-Pigé *et al.*, 2002; Barrett *et al.*, 2010; Oliveira *et al.*, 2010b; Simsek *et al.*, 2010; Mohamed El Hadi *et al.*, 2013), along with its texture, are responsible for the repeated purchase (Beaulieu and Baldwin, 2002; Song and Forney, 2008; Gundesli *et al.*, 2020). However, many factors affect the production of the compounds responsible for flavor, such as biotic and abiotic stress factors, ecological factors, the production system, and pre- and postharvest processing conditions. Improving fruit flavor is not an easy task for researchers and breeders. As with other fruits, sugars, acids, and phenolic compounds contribute to the taste and color of figs and to its flavor characteristics, which depend mainly on the accurate balance between the volatile components (Oliveira *et al.*, 2010a,b; Pereira and Câmara, 2011; Mawa *et al.*, 2013; Ersoy *et al.*, 2015; Gundesli *et al.*, 2020). At the same time, in many fruits the aromatic compound profile is considered to be unique and has a great impact on organoleptic properties and thus consumer acceptance (Oliveira *et al.*, 2010a,b; Valero and Serrano, 2010; Mujić *et al.*, 2014). Most of the studies determining the quality characteristics of fig varieties state the link between quality properties and the physicochemical composition (Polat and Calışkan, 2008; Gözlekçi *et al.*, 2011), firmness, or aroma compounds (Grison-Pigé *et al.*, 2002; Oliveira *et al.*, 2010a; Gözlekçi *et al.*, 2011). Thus, cultivar breeding requires information regarding flavor constitution to improve fruit quality and modify its aroma-based flavor (Minh *et al.*, 2002; Kafkas *et al.*, 2009; Mujić *et al.*, 2014; Gundesli *et al.*, 2020). As only limited reviews have been published to date on fig flavor, this chapter aims to give a broad view by summarizing the most relevant and recent studies considering the importance of flavor.

21.2 Fig Taste

Among the fruit quality characteristics of fresh figs, fruit size and skin and flesh colors are part of the "external" quality parameters. There are also internal parameters, such as taste, sugar content, and acidity, which are considered to be parameters important in defining fig quality (Aksoy, 1998; Flaishman *et al.*, 2008). In recent years, the internal quality parameters have become more important for consumers who no longer buy fruit only for their appearance. Sugars and acids, along with various phenolic molecules, play a crucial role in the sensory and nutritional quality of the product and may vary according to the variety, maturity, storage conditions, pedo-climatic conditions, and cultural practices. Among many phytochemical compounds in figs, the sugar composition and level, especially of fructose, affect the level of sweetness (Setser, 1993; Calışkan and Polat, 2011). However, sugar and carboxylic acid contents according to varieties have a great influence on consumer acceptance due to their organoleptic properties (Veberic *et al.*, 2008; Calışkan and Polat, 2011; Slatnar *et al.*, 2011). The relative amounts of glucose, fructose, and sucrose measured in different fig varieties reveal significant differences in sugar content (Matthews *et al.*, 1987; Trad *et al.*, 2014; Pereira *et al.*, 2017). These differences could be attributed to variety-specific characteristics.

It is obvious that soluble sugars and their metabolites contribute significantly to fig-fruit quality. Naturally high sugar content is particularly desirable as a staple food for the development and growth of children. Fructose (56%) and glucose (43%) are the predominant saccharides in figs, whereas sucrose levels are relatively very low or do not exist in analyzed varieties (Melgarejo *et al.*, 2003; Genna *et al.*, 2005; Aljane *et al.*, 2007; Veberic *et al.*, 2008; Slatnar *et al.*, 2011; Simsek, 2016). Slatnar *et al.* (2011) also reported that fructose (52%) and glucose (46%) were the predominant sugars, according to the results of their study on different varieties. Matthews *et al.* (1987) reported that fresh figs contained 3.7–2.8 g of glucose and fructose 100 g^{-1} fresh weight (FW), and dried figs contained 28.6–26.0 g 100 g^{-1} dry weight (DW), in 100 g edible portions. Aljane *et al.* (2007) determined the ranges of glucose (1.21–6.13 g 100 g^{-1} FW) and fructose (1.91–4.65 g 100 g^{-1} FW) in 14 Tunisian fig varieties. Slatnar *et al.* (2011) reported the glucose content as 2.50–3.81 g 100 g^{-1} FW, fructose as 2.34–3.40 g 100 g^{-1} FW, and sucrose as 0.58–0.98 g 100 g^{-1} FW. Calışkan and Polat (2011) found glucose to be 7.2–10.7 g 100 g^{-1} FW, fructose 6.2–9.0 g 100 g^{-1} FW, and sucrose 0.05–0.36 g 100 g^{-1} FW in different fig varieties and genotypes. Pereira *et al.* (2017) analyzed nine different fig varieties and showed that the fructose content was 4.94–7.47 g 100 g^{-1} FW, glucose was 4.70–7.54 g 100 g^{-1} FW and sucrose 0.3–2.1 g 100 g^{-1} FW (Table 20.1). In contrast, for five varieties of Tunisian figs (i.e. Bouhouli, Zidi, Thgagli, Bidhi, and Khedri), Trad *et al.* (2014) found that fructose content (4.4–6.1 g 100 g^{-1} FW) was lower than glucose (5.4–7.7 g 100 g^{-1} FW) and that their sucrose content varied between 1.9 and 2.6 g 100 g^{-1} FW. Hssaini *et al.* (2020) analyzed 11 different fig varieties and reported that glucose content was 5.55–29.94 g kg^{-1} DW, fructose was 6.23–28.15 g kg^{-1} DW, and sucrose was detected between 1–5 g kg^{-1} DW. Previous studies showed that generally the relative sweetness of fructose was higher than that of glucose (Setser, 1993). Hence, the perception that sweetness can be determined by the total amount of sugars, the ratio of the titratable acidity, and relative amounts of fructose in fig varieties. Sugar content in figs is predominantly the hexoses glucose and fructose, almost at a 1:1 ratio; however, different studies reveal slightly higher shares of fructose or glucose.

Organic acid content varies according to the species, variety, and maturity stage of the fruits. The ratio of sugar:acid content is the most important determinant of taste. Organic acids in fruits mostly form compounds such as salts, esters, and glycosides (Shiraishi *et al.*, 1996; Veberic *et al.*, 2008; Gundogdu and Yılmaz, 2012; Trad *et al.*, 2014). Carboxylic acids such as malic acid, citric acid, oxalic acid, tartaric acid, fumaric acid, and succinic acid are immediately oxidized, so they do not have harmful effects on the body. It is known that carboxylic acids affect taste formation, as well as many different physiological processes. Different carboxylic acids were examined in figs and studies found significant variation in carboxylic acid composition in different fig varieties. These differences vary according to the different genetic characteristics, types, climatic conditions, sedimentology, and other environmental factors of the examined materials. Citric acid was defined as a major acid in different fig varieties. There was no quinic acid in the pulps and skin. A study by Melgarejo *et al.* (2003) determined that fig fruit contains malic, oxalic, and citric acids in concentrations ranging from

Table 21.1. Summary of mono- and disaccharide profiles of *Ficus carica* fruit and/or tissue.

Fruit type	Number/names of varieties	Place of origin	Relative amount: Glucose	Fructose	Sucrose	Reference
Dried	The sugar content values were developed from an extensive review of recent scientific literature		28.60 g 100 g^{-1} DW	25.98 g 100 g^{-1} DW	6.52 g 100 g^{-1} DW	Matthews *et al.* (1987)
Fresh			3.6 g 100 g^{-1} FW	2.8 g 100 g^{-1} FW	0.4 g 100 g^{-1} FW	
Fresh	14	Tunisia	1.21–6.13 g 100 g^{-1} FW	1.91–4.65 g 100 g^{-1} FW	–	Aljane *et al.* (2007)
Fresh	Bela Petrovka	–	2.50–3.81 g 100 g^{-1} FW	2.34–3.40 g 100 g^{-1} FW	0.58–0.98 g 100 g^{-1} FW	Slatnar *et al.* (2011)
Fresh	75	Turkey	7.2–10.7 g 100 g^{-1} FW	6.2–9.0 g 100 g^{-1} FW	0.05–0.36 g 100 g^{-1} FW	Calışkan and Polat (2011)
Fresh	Bouhouli, Zidi, Thgagli, Bidhi, Khedri	Tunisia	5.4–7.7 g 100 g^{-1} FW	4.4–6.1 g 100 g^{-1} FW	1.9–2.6 g 100 g^{-1} FW	Trad *et al.* (2014)
Fresh	9 breba varieties	Spain	4.7–7.54 g 100 g^{-1} FW	4.94–7.47 g 100 g^{-1} FW	0.3–2.1 g 100 g^{-1} FW	Pereira *et al.* (2017)
Fresh	Croatian Yellow Giant	Bulgaria	5.7 g 100 g^{-1} FW	5.8 g 100 g^{-1} FW	1.4 g 100 g^{-1} FW	Petkova *et al.* (2019)
Fresh	27 ecotypes	Tunisia	1.12–5.68 g 100 g^{-1} FW	0.86–5.43 g 100 g^{-1} fw	–	Aljane *et al.* (2020)
Dried	El Quoti Lbied, Nabout, Fassi, Noukali Ghoudan, Snowden, White Adriatic, Kadota, Palmeras, Cuello Dama Blanca, Breval Blanca	Morocco, USA, Italy, Spain	5.55–29.94 g kg^{-1} DW	6.23–28.15 g kg^{-1} DW	1.00–5.00 g kg^{-1} DW	Hssaini *et al.* (2020)

DW, dry weight; FW, fresh weight.

0.081% to 0.094% as the main carboxylic acids analyzed in fruit juice. Veberic *et al.* (2008) reported that citric acid (2.75–4.61 g kg^{-1}) was one of the main acids in juice made from figs from the first crop, while malic acid (1.06–2.17 g kg^{-1}) was one of the main acids in juice of the second crop. Oliveira *et al.* (2009) found that the carboxylic acid profile of fig leaves consists of six organic acids: oxalic, citric, malic, quinic, shikimic, and fumaric acids. Trad *et al.* (2014) analyzed different fig varieties and reported that the citric acid content ranged between 0.23 and 0.43 g 100 g^{-1} FW and malic acid content between 0.11 and 0.17 g 100 g^{-1} FW. Hssaini *et al.* (2020) reported that malic acid was the predominant organic acid in all fig samples and recorded high levels compared with citric acid (2.55–4.99 g kg^{-1} DW). Citric acid ranged between 0.31 and 1 g kg^{-1} DW in juice of the first-crop figs. Pereira *et al.* (2017) found that malic acid was detected in higher concentrations than succinic or citric acid in breba fruit of nine varieties. The acid contents for fruit ranged from 0.1 to 0.9 g kg^{-1} FW for citric acid, 0.8 to 3.6 g kg^{-1} FW for malic acid, and 1 to 4.3 g kg^{-1} FW for succinic acid. In general, citric acid is the major carboxylic acid in fig, namely of the second crop, whereas malic acid is dominant in breba, the first-crop fruit (Table 21.2). Fig varieties may display important differences in terms of sugar and carboxylic acid content, differences that vary depending on genetic characteristics of the study materials, climatic conditions, and other environmental factors.

21.3 Phenolic Compounds

Phenolic compounds are one of the largest and most widespread secondary metabolite groups found naturally in the plant. There is an increasing worldwide interest in antioxidant compounds that can delay or prevent oxidative stress by inactivating free radicals. Natural antioxidant compounds such as phenolic compounds,

Table 21.2. Carboxylic acid profiles of common fig (*Ficus carica* L.) whole fruit, skin, and leaves.

Part of the fruit analyzed	Names of varieties	Country of origin	Relative amount		Reference
			Citric acid	Malic acid	
Crop juice	Colar	Spain	0.0212%	0.0564%	Melgarejo *et al.* (2003)
	Florancha		0.0074%	0.0299%	
	Tío Antonio		0.0295%	0.0601%	
	Gobernad		0.0399%	0.0318%	
Fresh fruit	Bela Petrovka	Slovenia	1.36–1.83 g kg^{-1} FW	0.52–0.76 g kg^{-1} FW	Slatnar *et al.* (2011)
Fresh fruit	Bouhouli, Zidi, Thgagli, Bidhi, Khedri	Tunisia	0.23–0.43 g 100 g^{-1} FW	0.11–0.17 g 100 g^{-1} FW	Trad *et al.* (2014)
Fruit skin/fruit flesh	Colar Elche, Cuello Dama Blanco, Tres Voltas, San Antonio, Blanca Bétera, Brown Tiberio, San Antonio, Banane	Spain, Turkey	0.9 g kg^{-1} FW 0.9–2.6 g kg^{-1} FW	0.8–3.4 g kg^{-1} FW 1.2–4.0 g kg^{-1} FW	Pereira *et al.* (2017)
Fresh fruit	Croatian Yellow Giant	Bulgaria	0.19 g kg^{-1} FW	–	Petkova *et al.* (2019)
Dried fruit	El Quoti Lbied, Nabout, Fassi, Noukali, Ghoudan, Snowden, White Adriatic, Kadota, Palmeras, Cuello Dama Blanca, Breval Blanca	Morocco, USA, Italy, Spain	1–0.31 g kg^{-1} DW	2.55–4.99 g kg^{-1} DW	Hssaini *et al.* (2020)

DW, dry weight; FW, fresh weight.

organic acids, vitamin E, and carotenoids therefore have important benefits for human health. These compounds also play many physiological roles in plants, and are found in fruits and vegetables, including fig. Additionally, within each species, the nature of these compounds may vary from organ to organ, and may contribute to variability in phenolic composition such as crop plant and genetics, geographic origin, maturity, climate, tree location, agricultural practices, industrial processing, and storage conditions. Moreover, phenolic compounds inhibit free-radical-induced reactions, inhibiting the formation of many diseases such as Alzheimer's disease and cardiovascular diseases. The compounds are popular because of their well-known antioxidant capacity and their effect on color, aroma, and taste. The two main categories of phenolic compounds are phenolic acids and flavonoids. These compounds have also shown a wide range of important biological effects on human health, including antioxidant, anti-carcinogenic, anti-atherogenic, anti-inflammatory, and antimicrobial activities. Therefore, phenolic compounds have been subject to much research due to their health-promoting properties (Daniel *et al.*, 1999; Eberhardt *et al.*, 2000; Kim *et al.*, 2003; Vinson *et al.*, 2005; Del Caro and Piga, 2008; Crisosto *et al.*, 2010; Calışkan and Polat, 2011; Slatnar *et al.*, 2011; Ercişli *et al.*, 2012; Faleh *et al.*, 2012; Ergun and Bozkurt, 2020). Knowing the phenolic composition of figs can lead to improvement of quality and health properties in breeding studies, and especially through the improvement and standardization of the production process. Although there are many studies about the antioxidant activity of phenols in many fruit species in the literature (Ferreira *et al.*, 2002; Kim *et al.*, 2003; Erturk *et al.*, 2009; Devic *et al.*, 2010; Kamiloğlu *et al.*, 2015; Shahidi and Ambigaipalan, 2015; Gundesli *et al.*, 2019; Kafkas *et al.*, 2020), scientific interest has focused on the qualitative and quantitative analysis of phenolic acids and flavonoids in different varieties of fresh and dried figs and the distribution between pulp and skin. To study the potential health-promoting constituents of fig fruits, Solomon *et al.* (2006) examined six commercial fig varieties differing in color (Fig 21.1). Analysis of total polyphenols, flavonoids, antioxidant capacity, and amount and profile of anthocyanins showed various concentrations of the anthocyanins, and cyanidin-3-*O*-rhamnoglucoside as the main anthocyanin. The color of the fig extract correlated well with total polyphenols, flavonoids, anthocyanins, and antioxidant capacity. Extracts of darker varieties showed higher contents of phytochemicals compared with light-colored varieties. Fruit skins contributed most of the above phytochemicals and antioxidant activity compared with the fruit pulp. Antioxidant capacity correlated well with the amount of polyphenols and anthocyanins. Fruits of the Mission variety contained the highest levels of polyphenols, flavonoids, and anthocyanins, and exhibited the highest antioxidant capacity (Fig. 21.1).

Many studies have been conducted on the effects of color, fruit variety, harvest season, and drying processes on the concentration of phenolic compounds (Veberic *et al.*, 2008; Slatnar *et al.*, 2011; Bachir Bey and Louaileche, 2015; Ersoy *et al.*, 2015; Kamiloğlu *et al.*, 2015; Ajmal *et al.*, 2016; Harzallah *et al.*, 2016; Konak *et al.*, 2017; Maghsoudlou *et al.*, 2017; Arvaniti *et al.*, 2019).

Colorimetric tests are the best method for the determination of phenolic acids and flavonoids in figs. The total phenolic content of different tissue extracts in figs is usually determined by the Folin–Ciocalteu method. The method is reliable and has been applied by many researchers to determine total phenols from a variety of samples. Various studies have used whole fruits of fresh and dried-fig varieties, as well as their skin and pulp fractions, in quantifying total phenolics (TP) and total flavonoids (TF) through spectrophotometric tests. In recent years, more specific and selective analytical methods have also been used in many studies to identify individual phenol compounds in fig extracts. The most widely used instrumentation for the qualitative and quantitative determination of phenolic compounds in fig samples is high-performance liquid chromatography (HPLC) with ultraviolet visible or diode array detection (DAD) and mass spectrometry (MS) or tandem MS (MS/MS) detection in negative or positive ionization mode. In most cases, these analytical methods are useful for identifying these compounds from different parts of fresh and dried whole fruit, or from skin and pulp fractions and leaves, at very low detection limits. In addition to analyses of fresh and dried fruits using an HPLC-DAD system, it is the most important device for measuring trace substances by making

Fig. 21.1. A. Color appearance of six fig varieties. B. Relationship between (1) total polyphenols content, (2) total flavonoids content, and (3) total anthocyanins content, and Trolox equivalent antioxidant capacity (TEAC) of the six fig varieties expressed as TEAC 100 g^{-1} FW of fruits. Modified from Solomon *et al.* (2006).

precise measurements in fresh and dried fruits in addition to fresh and dried fruit analysis using the HPLC-DAD system. (Solomon *et al.*, 2006; Veberic *et al.*, 2008; Mujić *et al.*, 2012; Ouchemoukh *et al.*, 2012; Nakilcioğlu and Hısıl, 2013; Bachir Bey *et al.*, 2014; Debib *et al.*, 2014; Bachir Bey and Louaileche, 2015; Ersoy *et al.*, 2015; Hoxha *et al.*, 2015; Kamiloğlu *et al.*, 2015; Harzallah *et al.*, 2016; Amessis-Ouchemoukh *et al.*, 2017; Konak *et al.*, 2017; Pereira *et al.*, 2017; Arvaniti *et al.*, 2019).

Fig is rich in minerals, vitamins, and fiber. Moreover, it has antibacterial and antioxidant activity effects due to the presence of phenolic compounds (Solomon *et al.*, 2006; Calışkan and Polat, 2008; Veberic *et al.*, 2008). Different studies have estimated the phytochemical composition of whole fruits, fresh and dried-fig varieties, and tissues including pulp and skin of these compounds. Some studies report amounts of TP (Vinson, 1999; Vinson *et al.*, 2005; Solomon *et al.*, 2006; Calışkan and Polat, 2011; Ercişli *et al.*, 2012; Ersoy *et al.*, 2015; Maghsoudlou *et al.*, 2017), while others have examined individual phenolic components (Vinson, 1999; Slavin, 2006; Del Caro and Piga, 2008; Veberic *et al.*, 2008; Oliveira *et al.*, 2009; Slatnar *et al.*, 2011; Vallejo *et al.*, 2012; Harzallah *et al.*, 2016; Konak *et al.*, 2017; Pereira *et al.*, 2017).

TP, TF, and individual phenolic compounds have been determined using spectrophotometric tests and other techniques, and the concentrations of the detected phytochemical compounds are presented in Tables 21.3–21.6. There are differences in phenolic composition of different fig varieties, and in the regions where figs are grown. Over the past years, various groups have investigated the polyphenol composition of fresh and/or dried fruit or parts of different fig varieties. Most of these studies have been performed in major fig-producing countries; however, studies have covered almost all of the following: Turkey (Calışkan and

Table 21.3. Total phenolic content in fresh and sun-dried or artificially dried figs.

Country of origin	Number of varieties	Total phenolics: Fresh (mg GAE 100 g^{-1} FW)	Dried (mg GAE 100 g^{-1} DW): Sun dried	Oven dried	Reference
Albania	6	–	45.24–160.42	–	Hoxa *et al.* (2015)
Algeria	2	–	399.79–466.76	–	Bachir Bey *et al.* (2013)
	1	–	540	–	Ouchemoukh *et al.* (2012)
	2	–	500–756.6	–	Amessis-Ouchemoukh *et al.* (2017)
Croatia	5	930–1117	–	–	Mujić *et al.* (2012)
Georgia	1	72.7	–	–	Pande and Akoh (2010)
India	1	–	331	–	Reddy *et al.* (2010)
Israel	6	48.6–281.10		–	Solomon *et al.* (2006)
Turkey	10	198.81–307.64	81.77–158.77	–	Nakilcioğlu and Hısıl (2013)
	79	28.6–211.9	–	–	Calışkan and Polat (2011)
	4	25.47–37.29	–	–	Ersoy *et al.* (2015)
	4	55.54–144.32	245.60–459.98	272.54–553.32	Konak *et al.* (2017)
	2	211–493	193–417	–	Kamiloğlu *et al.* (2015)
Tunisia	3	50.57–74.16	–	–	Harzallah *et al.* (2016)

GAE, gallic acid equivalents; FW, fresh weight; DW, dry weight.

Table 21.4. Total flavonoid contents in different fresh and sun-dried or artificially dried figs.

Country of origin	Number of varieties	Total flavonoids: Fresh (mg GAE 100 g^{-1} FW)	Dried (mg C3G 100 g^{-1} DW): Sun dried	Oven dried	Reference
Albania	6	–	18.11–36.95	–	Hoxa *et al.* (2015)
Algeria	2	–	60–1374.4	–	Debib *et al.* (2014)
	1	–		–	Ouchemoukh *et al.* (2012)
	2	–	105.6	–	Amessis-Ouchemoukh *et al.* (2017)
Croatia	5	512–568	858.43	–	Mujić *et al.* (2012)
Israel	6	2.1–21.5	–	–	Solomon *et al.* (2006)
Turkey	10	67.33–147.51	–	–	Nakilcioğlu and Hısıl (2013)
	2	8–66	32.78–52.52	–	Kamiloğlu *et al.* (2015)
Tunisia	3	11.30–12.75	14–52	–	Harzallah *et al.* (2016)

GAE, gallic acid equivalents; FW, fresh weight; C3G, cyanidin-3-*O*-glucoside; DW, dry weight.

Polat, 2011; Yemis *et al.*, 2012; Nakilcioğlu and Hısıl, 2013; Ersoy *et al.*, 2015; Kamiloğlu *et al.*, 2015; Konak *et al.*, 2017), Algeria (Ouchemoukh *et al.*, 2012; Bachir Bey *et al.*, 2013; Bachir Bey *et al.*, 2014; Debib *et al.*, 2014; Bachir Bey and Louaileche, 2015; Amessis-Ouchemoukh *et al.*, 2017), Croatia (Mujić *et al.*, 2012), Italy (Del Caro and Piga, 2008), Portugal (Oliveira *et al.*, 2009), Albania (Hoxha *et al.*, 2015; Hoxha and Kongoli, 2016), Israel (Solomon *et al.*, 2006), Tunisia (Harzallah *et al.*, 2016), Georgia (Pande and Akoh, 2010), Pakistan (Ajmal *et al.*, 2016), Iran (Maghsoudlou *et al.*, 2017), Spain (Dueñas *et al.*, 2008), and Slovenia (Veberic *et al.*, 2008; Slatnar *et al.*, 2011).

According to the data released in scientific studies, TP content in whole fresh fig fruit ranges from 28.6 to 211.9 mg gallic acid equivalents (GAE) 100 g^{-1} FW (Calışkan and Polat, 2011) and 50.57 to 74.16 mg GAE 100 g^{-1} FW (Harzallah *et al.*, 2016), and TF from 2.1 (Solomon *et al.*, 2006) to 1130 (Harzallah *et al.*, 2016) mg

Table 21.5. Total phenolic and total flavonoid contents in different tissues (skin and pulp) of fresh figs.

		Total phenolics		Total flavonoids		
Country of origin	Number of varieties	Skin (mg GAE 100 g^{-1} FW)	Pulp (mg GAE 100 g^{-1} FW)	Skin (mg (+) C3G 100 g^{-1} FW)	Pulp (mg (+) C3G 100 g^{-1} FW)	References
Israel	6	65.5–463.0	36.5–100.6	2.2–45.6	1.6–5.7	Solomon *et al.* (2006)
Iran	2	4669–8079	4554–6535	589–1010	547–751	Maghsoudlou *et al.* (2017)
Pakistan	1	190	0.04	38.90	4.49	Ajmal *et al.* (2016)
Turkey	2	351–950	167–351	63–234	7–44	Kamiloğlu *et al.* (2015)
Tunisia	3	61.47	<6311	12.75	8	Harzallah *et al.* (2016)

GAE, gallic acid equivalents; FW, fresh weight; C3G, cyanidin-3-*O*-glucoside.

cyanidin-3-*O*-glucoside (C3G) 100 g^{-1} FW (Tables 21.3 and 21.4). The TP and TF of dried figs were determined to be between 81.77 (Nakilcioğlu and Hısıl, 2013) and 756.6 (Amessis-Ouchemoukh *et al.*, 2017) mg GAE 100 g^{-1} FW, and between 14 (Kamiloğlu *et al.*, 2015) and 1374.4 (Debib *et al.*, 2014) mg C3G 100 g^{-1} DW, respectively. As shown by the values in Tables 21.3 and 21.4, the TP and TF contents of fresh figs grown in Turkey, Algeria, and Tunisia were higher compared with those from other countries (Debib *et al.*, 2014; Ersoy *et al.*, 2015; Harzallah *et al.*, 2016). The TP and TF concentrations found in the skin and pulp of the fig fruit are shown in Table 21.5. TP and TF were found to be in the range of 65.5 (Solomon *et al.*, 2006) to 7416 (Harzallah *et al.*, 2016) mg GAE 100 g^{-1} FW, and 2.2 (Solomon *et al.*, 2006) to 1275 (Harzallah *et al.*, 2016) mg C3G 100 g^{-1} FW, respectively, in the fig skin (Table 21.5). TP and TF of the fig pulp were in the range of 0.04 (Ajmal *et al.*, 2016) to 6535 (Maghsoudlou *et al.*, 2017) mg GAE 100 g^{-1} FW, and 1.6 (Solomon *et al.*, 2006) to 8 (Harzallah *et al.*, 2016) mg C3G 100 g^{-1} FW, respectively (Table 21.5). All these results suggest that fig skin has higher amounts of phenolic compounds than fig pulp, although the color of the skin and pulp may create significant differences.

The major phenolic compounds in whole fresh and dried figs and their tissues are shown in Table 21.6. In terms of phenolic acids, gallic acid and syringic acid belonging to benzoic acid and hydroxylated derivatives of chlorogenic acid, ferulic acid, caffeic acid, and cinnamic acid were the most common compounds in all fresh and dried figs. The most predominant phenolic acids found in dried figs were gallic acid and chlorogenic acid in the range of 2.76 to 5.04 and 0.27 to 67.15 mg GAE 100 g^{-1} DW (Bachir Bey *et al.*, 2016; Kelebek *et al.*, 2018), respectively. Rutin content in fresh fig fruit ranged from 14.69 (Kelebek *et al.*, 2018) to 68.21 (Nakilcioğlu and Hısıl, 2013) mg GAE 100 g^{-1} FW. Gallic acid, chlorogenic acid, rutin, quercetin-3-*O*-rutinoside and epicatechin were the predominant phenolic acids and flavonoids in dried- and fresh-fig varieties. Additionally, gallic acid, ellagic acid, chlorogenic acid, psolarene, rutin, catechin, epicatechin, cyanidin-3-*O*-rutinoside, and C3G were predominant in all extracts examined in relation to the phytochemical composition in the skin and pulp as individual phenolic compounds (Table 21.6). The results showed that figs appear to be a rich source of phenolic compounds and flavonoids that have beneficial contributions to human health.

21.4 Fig Aroma

Aroma is known as a complex mixture of many volatile compounds, the composition of which is unique to the species and generally variable according to the various fruits. Although different fruits usually have many aromatic properties, each fruit's mixture of volatile substances has a unique flavor that depends on the concentration

Table 21.6. Summary of major phenolic compounds found in dried and fresh figs.

Country of origin	Variety	Major compounds	Dried fig Unit	Dried fig Content	Fresh fig Unit	Fresh fig Content	Skin Unit	Skin Content	Pulp Unit	Pulp Content	Whole fruit Unit	Whole fruit Content	Reference
Turkey	Sarılop (= Calimyrna)	Gallic acid	mg 100 g^{-1} DM	0.47–2.32	mg 100 g^{-1} DM	1.37–5.10		–		–		–	Nakilcioğlu and Hısıl (2013)
		Chlorogenic acid		0.40–1.81		2.97–6.54		–		–		–	
		(–)-Epicatechin		14.76–23.42		42.29–76.90		–		–		–	
		Syringic acid		0.41–2.83		3.03–4.45		–		–		–	
		Rutin		5.75–12.75		15.16–68.21		–		–		–	
Turkey	Sarılop	Gallic acid	mg 100 g^{-1} DM	0.1		–	mg 100 g^{-1} DW	0.2	mg 100 g^{-1} DW	0.3	mg 100 g^{-1} DW	0.3	Kamiloğlu *et al.* (2015)
		Chlorogenic acid		0.8		–		12		0.5		1.9	
		Epicatechin		0.6		–		2.7		0.9		1.1	
		Ellagic acid		0.2		–		0.3		0.2		0.2	
		Rutin		9		–		86		2		12	
		p-Coumaric acid		1.1		–		5.1		2.2		2.7	
	Bursa Black	Gallic acid	mg 100 g^{-1} DM	0.5		–	mg 100 g^{-1} DW	0.3	mg 100 g^{-1} DW	0.5	mg 100 g^{-1} DW	0.3	
		Chlorogenic acid		1.1		–		21		6		8	
		Epicatechin		2.2		–		29		4.8		6.9	
		Ellagic acid		0.3		–		0.7		0.1		0.2	
		Rutin		15		–		180		0.9		13	
		p-Coumaric acid		3.1		–		18		32		4.7	
Algeria	Aberkane	Gallic acid	mg 100 g^{-1} DM	4.32	mg 100 g^{-1} DM	1.76	mg 100 g^{-1} DM	–	mg 100 g^{-1} DM	–	mg 100 g^{-1} DM	–	Bachir Bey *et al.* (2016)
		Chlorogenic acid		0.27		0.44		–		–		–	
		Caffeic acid		ND		0.51		–		–		–	
		Ferulic acid		1.40		1.98		–		–		–	
		p-Coumaric acid		0.33		0.63		–		–		–	
		Syringic acid		0.43		1.03		–		–		–	
		Catechin		1.80		8.63		–		–		–	
		Quercetin		0.64		3.22		–		–		–	
		Salicylic acid		1.28		1.50		–		–		–	
		Vanillic acid		2.42		1.88		–		–		–	
		Benzoic acid		2.56		5.81		–		–		–	

Continued

Table 21.6. Continued.

Country of origin	Variety	Major compounds	Dried fig Unit	Dried fig Content	Fresh fig Unit	Fresh fig Content	Skin Unit	Skin Content	Pulp Unit	Pulp Content	Whole fruit Unit	Whole fruit Content	Reference
Algeria	Azandjar	Gallic acid	mg 100	2.76	mg 100	1.40		–		–		–	Bachir Bey
		Chlorogenic acid	g^{-1}	0.17	g^{-1}	0.93		–		–		–	*et al.*
		Caffeic acid	DM	ND	DM	0.14		–		–		–	(2016)
		Ferulic acid		1.01		3.51		–		–		–	
		p-Coumaric acid		0.45		0.31		–		–		–	
		Syringic acid		0.29		0.72		–		–		–	
		Catechin		1.80		1.72		–		–		–	
		Quercetin		1.62		16.53		–		–		–	
		Salicylic acid		2.04		3.68		–		–		–	
		Vanillic acid		3.32		3.90		–		–		–	
		Benzoic acid		1.10				–		–		–	
	Bouankik	Gallic acid	mg 100	5.04	mg 100	1.58		–		–		–	
		Chlorogenic acid	g^{-1}	0.34	g^{-1}	2.09		–		–		–	
		Caffeic acid	DM	ND	DM	0.37		–		–		–	
		Ferulic acid		2.56		3.37		–		–		–	
		p-Coumaric acid		0.44		0.51		–		–		–	
		Syringic acid		ND		1.20		–		–		–	
		Catechin		3.19		4.80		–		–		–	
		Quercetin		3.43		6.76		–		–		–	
		Salicylic acid		2.57		4.91		–		–		–	
		Vanillic acid		2.14		2.17		–		–		–	
		Benzoic acid		2.22		3.59		–		–		–	
Georgia	Brown Turkey	Gallic acid		–		–	mg 100	2.8	mg 100	2.6	mg 100	2.8	Pande and
		Ellagic acid		–		–	g^{-1}	0.4	g^{-1}	0.5	g^{-1}	0.2	Akoh
		Ferulic acid		–		–	FW	ND	FW	ND	FW	ND	(2010)
		Catechin		–		–		17.1		21.2		16.7	
		Epicatechin		–		–		31.8		28.6		32.1	

Slovenia	Skofjotka	Gallic acid		–		–	–	–	mg 100 g^{-1} FW	0.14–0.23	Veberic *et al.* (2008)
		Chlorogenic acid		–		–	–	–		0.46–1.44	
		(–)-Epicatechin		–		–	–	–		0.34–0.66	
		Syringic acid		–		–	–	–		0.022–0.033	
		Rutin		–		–	–	–		4.89–8.52	
		Catechin		–		–	–	–		1.07–1.49	
	Crna Petrovka	Gallic acid		–		–	–	–	mg 100 g^{-1} FW	0.30–0.38	
		Chlorogenic acid		–		–	–	–		0.71–1.24	
		(–)-Epicatechin		–		–	–	–		0.43–0.58	
		Syringic acid		–		–	–	–		0.079–0.104	
		Rutin		–		–	–	–		10.4–15.6	
		Catechin		–		–	–	–		2.29–4.03	
	Miljska Figa	Gallic acid		–		–	–	–	mg 100 g^{-1} FW	0.29–0.30	
		Chlorogenic acid		–		–	–	–		1.57–1.71	
		(–)-Epicatechin		–		–	–	–		0.94–0.97	
		Syringic acid		–		–	–	–		0.027–0.034	
		Rutin		–		–	–	–		27.3–28.7	
		Catechin		–		–	–	–		1.71–1.80	
Slovenia	Bela Petrovka	Chlorogenic acid	mg 100 g^{-1} DM	3.42–15.88	mg 100 g^{-1} FW	1.33–4.91	–	–		–	Slatnar *et al.* (2011)
		Catechin		5.88–11.46		1.36–2.88	–	–		–	
		Epicatechin		10.44–23.30		7.11–8.67	–	–		–	
		Rutin		1.38–12.06		ND	–	–		–	
		Kaempferol-3-*O*-glucoside		0.46–0.59		0.04–0.13	–	–		–	
		Luteolin-8-*C*-glucoside		0.13–0.16		ND	–	–		–	
		Quercetin-3-*O*-glucoside		0.56–2.40		0.18–0.60	–	–		–	

Continued

Table 21.6. Continued.

Country of origin	Variety	Major compounds	Dried fig		Fresh fig		Skin		Pulp		Whole fruit		Reference
			Unit	Content	Unit	Content	Unit	Content	Unit	Content	Unit	Content	
Portugal	Pingo De Mel	Ferulic acid		–		–	mg kg^{-1} FW	20.5	mg kg^{-1} FW	ND		–	Olivera *et al.* (2009)
		3-*O*-caffeoylquinic acid		–		–		3.2		ND		–	
		5-*O*-caffeoylquinic acid		–		–		43.8		32.9		–	
		Quercetin 3-*O*-glucoside		–		–		31.4		ND		–	
		Rutin		–		–		499.1		ND		–	
		Psoralen		–		–		2.6		3.7		–	
	Branca Tradicional	Ferulic acid		–		–	mg kg^{-1} FW	9.2	mg kg^{-1} FW	ND		–	
		3-*O*-caffeoylquinic acid		–		–		ND		ND		–	
		5-*O*-caffeoylquinic acid		–		–		8.3		2.8		–	
		Quercetin 3-*O*-glucoside		–		–		30.8		ND		–	
		Rutin		–		–		629.6		64.6		–	
		Psoralen		–		–		18.4		17.0		–	
Italy	Mattalona	Chlorogenic acid		–		–	mg kg^{-1} FW	69.33	mg kg^{-1} FW	ND		–	Del Caro and Piga (2008)
		Rutin		–		–		1071.41		ND		–	
		Cyan 3-*O*-rutinoside		–		–		901.03		4.92		–	
		Cyan 3-*O*-glucoside		–		–		28.24		ND		–	
	San Pietro	Chlorogenic acid		–		–	mg kg^{-1} FW	39.63	mg kg^{-1} FW	ND		–	
		Rutin		–		–		527.96		ND		–	
		Cyan 3-*O*-rutinoside		–		–		ND		17.60		–	
		Cyan 3-*O*-glucoside		–		–		ND		ND		–	

Turkey	Sarılop	Chlorogenic acid	mg 100 g^{-1} DM	28.59	mg 100 g^{-1} FW	115.57	–	–	–	Kelebek *et al.* (2018)
		Catechin		24.83		33.58	–	–	–	
		Epicatechin		3.90		6.06	–	–	–	
		Rutin		8.77		14.69	–	–	–	
		Vanillic acid		7.45		10.99	–	–	–	
		Quercetin		0.39		0.56	–	–	–	
		3-*O*-caffeoylquinic acid		21.33		34.67	–	–	–	
		5-*O*-caffeoylquinic acid		29.41		60.36	–	–	–	
	Bursa Black	Chlorogenic acid	mg 100 g^{-1} DM	67.15	mg 100 g^{-1} DM	115.05	–	–	–	
		Catechin		13.24		24.83	–	–	–	
		Epicatechin		3.22		4.48	–	–	–	
		Rutin		28.10		52.90	–	–	–	
		Vanillic acid		11.88		17.15	–	–	–	
		Quercetin		6.44		8.91	–	–	–	
		3-*O*-caffeoylquinic acid		6.77		10.87	–	–	–	
		5-*O*-caffeoylquinic acid		12.88		19.57	–	–	–	
		Cyan 3-*O*-glucoside		–		–	–	–	–	

ND, not detected.

of the individual volatile compounds and the detection threshold. Flavor is one of the most important features of fruit, and volatile flavor compounds in particular play an important role in determining the perception and acceptability of products by consumers. Determination of the essential volatile compounds that form the most important characteristics of fruit is of utmost importance, as it provides the basic sensory identity and characteristic flavor of the fruit. Due to the complex structure of volatile aroma profiles, volatile compounds in fresh fruits are constantly changing. To date, limited information is available on how these factors interact to determine the actual volatile composition of the fruit and the resulting flavor. Evaluation of volatiles at the germplasm level is useful for future breeding efforts aimed at improving fruit quality through effects on fruit flavor (Baldwin *et al.*, 2000; Gonçalves *et al.*, 2018).

Volatile flavor compounds in fruit are affected by many factors such as variety, cultural practices, ripeness, harvest maturity, and postharvest processing. Of these factors, the most important criterion that changes the amounts of volatile compounds is maturity (Lester, 2006). The fruit should be harvested in the most appropriate period to determine the optimum levels of volatile aroma compounds that affect the taste components. Many preharvest factors such as sunlight, water availability, fertilization, and chemical treatments can affect the growth and taste of the product, as well as the quality characteristics.

Several techniques are used to extend the shelf life of fruits after harvest. These techniques include optimizing storage and processing such as cold, heat, irradiation, modified storage environments, and chemical treatments. Postharvest treatments can also affect the amount of flavor compounds. Storage temperature is the most important factor affecting the color and taste of fruits. In particular, by lowering oxygen and increasing carbon dioxide levels, the quality of many fresh fruits can be maintained for a long time. This may also affect the taste in some chemical processes applied to fresh products (Fellman *et al.*, 2003; Dudareva *et al.*, 2004; Menager *et al.*, 2004; Peyrot des Gachons *et al.*, 2005; Harb *et al.*, 2008; de la Peña Moreno *et al.*, 2010; Ortiz *et al.*, 2010; Tietel *et al.*, 2012).

Aroma is one of the most important quality features of fig fruit and the most important quality parameter in both edible and processed products. Knowledge about the aroma composition composites of figs is necessary to improve fruit quality and change the aroma formation, and is important for cultivation and subsequent processing. Volatile compounds in figs occur mainly through the oxidation of fatty acids and are of great concern due to their large influence on sensory properties. The increasing demand for quality figs in recent years may be linked not only to potential health disadvantages but also to their distinctive organoleptic character. Many researchers have focused on the determination of aromatic compounds by trying different analysis methods from leaves, spirits, extracts, and others for the volatile aroma compounds in different organs of figs (Gibernau *et al.*, 1997; Oliveira *et al.*, 2010a,b; Gözlekçi *et al.*, 2011).

The aromatic profiles of figs are affected by various factors including genetic characteristics, biotic and abiotic stresses, cultural practices, fertilization, irrigation, planting systems, ecological conditions, and soil (Kafkas *et al.*, 2009; Gundesli *et al.*, 2020). Fig volatile flavor compounds have been studied extensively in the last decade. When determining fruit quality, the aromatic profile and individual volatile compounds are two of the important features that should be monitored and evaluated. Each product has a characteristic and unique composition of volatile ingredients. The aroma of most food products is usually made up of complex mixtures of several hundred compounds (Mujić *et al.*, 2014). Recently, much attention has been paid to methods that lead to aroma development in dried fruits and enhancement of aromatic properties in the production of superior dried and fresh fruit. However, there are very few studies on the volatile fraction of figs in different fig-producing countries in the world. Furthermore, the results show a large variability in the nature and concentration of the aroma compounds isolated from different varieties (Gibernau *et al.*, 1997; Grison *et al.*, 1999; Grison-Pigé *et al.*, 2002; Oliveira *et al.*, 2010a; Gözlekçi *et al.*, 2011; Li *et al.*, 2012; Ficsor *et al.*, 2013; Mujić *et al.*, 2014; Palassarou *et al.*, 2017; Russo *et al.*, 2017; Rodríguez-Solana *et al.*, 2018; Gundesli *et al.*, 2020). Key determinants of fig-fruit quality perceived by consumers include many compounds with varying degrees of volatility. Aroma compounds are detected using different techniques, even at very low concentrations. To date, more than 100 different volatile compounds have been identified in different parts of various fig varieties. The different tissues of fig (*F. carica*) are identified and distributed according to different chemical classes such as terpenes (monoterpenes and sesquiterpenes), alcohols, aldehydes, ketones, esters, and other compounds (Ware *et al.*, 1993; Mawa *et al.*, 2013; Trad *et al.*, 2014;). The contribution of each compound to the specific flavor profile of each variety depends on the activity of the relevant enzymes in the biosynthetic pathway and the substrate specificity, the presence of the substrate, the odor threshold on which the compound can be detected and the presence of other compounds. It is stated that aldehydes and esters are the main aroma compounds found in figs and contribute more to the general aroma of fig than other groups.

Recently, the importance of volatile compounds in the aromatic profile is increasing, and development of rapid analysis techniques makes it possible to determine the volatile compounds for different fig varieties. Solid-phase microextraction (SPME) is relatively easy for the determination of fruit volatiles (Kafkas *et al.*, 2005a,b; Dong *et al.*, 2006; Cuevas-Glory *et al.*, 2007; Gül and Tekeli, 2019), as it is fast, sensitive, reliable, simple, and solvent free. This method consists of a fused silica fiber coated with a polymeric constant excess that is fed to a liquid or gas sample (Hawthorne *et al.*, 1992; Grison-Pigé *et al.*, 2002; Kafkas *et al.*, 2005b; Mujić *et al.*, 2014). Because it is a solvent-free extraction alternative, no change in ingredients occurs due to solvent or temperature effect. The SPME technique is widely used for the analysis of volatile components from environmental, biological, and food samples by direct injecting into packed columns (Fig. 22.2). However this technique is outdated and has gradually been replaced by headspace techniques in capillary columns (Fig. 22.2) (Oliveira *et al.*, 2010b). Another downfall of the SPME method is that the choice of fiber coating, which depends mainly on the nature of the analytes, may bias the results. Polydimethylsiloxane (PDMS), Carboxen (CAR; a carbon molecular sieve), and divinylbenzene (DVB) fibers are mostly used for this purpose. Among these, PDMS is the most applied fiber for headspace and immersion techniques. In addition, combined fibers such as CAR/PDMS and DVB/CAR/PDMS provide the best extraction efficiencies for a wide range of analytes with different polarities and molecular weights. They are available in various coating combinations, blends or copolymers,

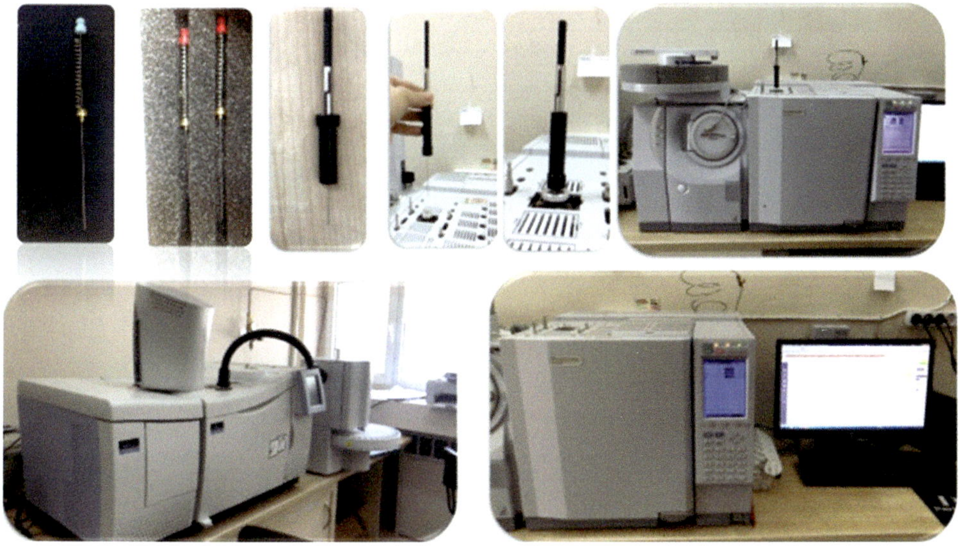

Figure 21.2. A view of headspace and SPME adsorption fibers, inlet apparatus, and desorption parts together with various gas chromatography/mass spectrometry instruments. Photographs from the laboratory of N.E. Kafkas.

film thicknesses, and fiber assemblies. PDMS fibers are better suited for the analysis of non-polar analytes, whereas polyacrylate fibers are more useful to extract polar analytes. DVB is also a polar porous solid coating and therefore efficient in extracting polar compounds such as disulfides and trisulfides. Polar compounds like alcohols, aldehydes, ketones, ethers, and carboxylic acids are often most efficiently extracted by fibers consisting of a combination of e.g. non-polar material (PDMS) and a polar material (DVB). Thus, the perception of aroma volatiles is directly related to their concentrations in the headspace (Kafkas *et al.*, 2005a,b, 2009; Taveira *et al.*, 2009). Headspace (HS)-SPME) with gas chromatography/mass spectrometry detector (GC-MS) is used for individual identification and measurement of volatile flavor compounds such as alcohols, aldehydes, ketones, esters, and terpenoids. Additionally, the presence and concentrations of these compounds depend on various factors, such as soil, climate, genotype, maturity, and technological aspects (Forney *et al.*, 2000).

Aldehydes are generally important for sensory perception in fruit and have a positive effect. Oliveira *et al.* (2010a) reported aldehydes such as 2-methylbutanal, 3-methylbutanal, (E)-2-pentanal, hexanal, heptanal, octanal, and nonanal, and alcohols such as 1-penten-3-ol, 3-methylbutanol, benzyl alcohol, (E)-2-nonenol, and phenylethyl alcohol in common fig.

Esters make the most important contribution to fruit flavor and are very important in determining the ripening stage of figs and the appropriate harvest time. They are produced by esterification of alcohols and acyl-CoAs derived from both fatty acid and amino acid metabolism in a reaction catalyzed by alcohol-O-acyltransferase. The esters present exclusively in figs are methyl butanoate, ethyl benzoate, methyl hexanoate, hexyl acetate, methyl salicylate, ethyl and methyl esters of 6-, 7-, 8-, 9-, 10-, 11-, 12-, 13-, 14-, 16-, and 20-carbon fatty acids, and ethyl butyrate, methyl acetate, and diethyl succinate (Oliveira *et al.*, 2010a,b; Barolo *et al.*, 2014; Russo *et al.*, 2014; Rodríguez-Solana *et al.*, 2018; Gundesli *et al.*, 2020).

Terpenes with highly complex structures are the largest class of plant secondary metabolites such as monoterpenes and sesquiterpenes. Limonene accounts for the major component in fig aroma. As well as limonene, terpinene, myrcene, and α-pinene are abundant in fig fruits. While α-pinene is one of the main monoterpenes mentioned in various studies found in different tissues of figs, the sesquiterpenes listed are β-elemene, β-cubebene, α-copaene, α-ylangene, β-bourbonene, caryophyllene, (+)-ledene (viridifloren), α-bulnesene, acoradien, α-muurolene, E-nerolidol, β-damascenone, cadalene, and α-gurjunene. Sesquiterpenes are specifically identified in fig leaves (Mawa *et al.*, 2013; Barolo *et al.*, 2014; Trad *et al.*, 2014; Palassarou *et al.*, 2017; Rodríguez-Solana *et al.*, 2018).

Fatty alcohols have been identified as the most important additives to fig flavor, especially terpene alcohols that have the characteristic aroma. Identified alcohols are described by a wide variety of sensory descriptors; some are reported with "positive" properties, particularly 1-hexanol, 1-heptanol, and 1-nonanol, while others may have "negative" sensory properties, such as methyl alcohol, ethyl alcohol, and isobutyl alcohol. Obviously, this strongly depends on their concentration and interaction with other volatiles, including the food matrix in which the compound was detected. The predominant alcohol is linalool. Alcohols also play an important role in ester biosynthesis. (Z)-3-hexanol, hexanol, (E)-2-hexanol and 2-ethylhexanol were detected in *tert*-butyl methyl ether extracts. Fruits and leaves contain (Z)-3-hexen-1-ol, 2-methyl-1-butanol, and 1-heptanol, while alcoholic beverages contain methanol, ethanol, 1-propanol, 1-butanol, 2-methyl-propanol (isobutyl alcohol), 1-hexanol, octanol and decanol, as well as 1-penten-3-ol, benzyl alcohol and (E)-2-nonen-1-ol, β-(methylamino)ethanol, epoxylinalol, and nonylol (Oliveira *et al.*, 2010a; Mawa *et al.*, 2013; Russo *et al.*, 2017; Rodríguez-Solana *et al.*, 2018; Gundesli *et al.*, 2020).

Ketones are found in relatively lower amounts in different organs of figs. The first ketone-bearing compound found in figs was 3-hydroxy-2-butanone (acetoin). Other ketones, such as 6-methyl-5-hepten-2-one and 3-pentanone have been identified in the fruits and leaves of fig varieties, respectively (Gözlekçi *et al.*, 2011; Mawa *et al.*, 2013; Trad *et al.*, 2014; Russo *et al.*, 2017).

21.5 Concluding Remarks

In horticultural crops, the quality of the products can be evaluated in many ways, including their flavor. Epidemiological studies have shown a clear positive relationship between the intake of many phytochemicals in fruits and reduced disease rate. Phenolic substances, which are flavor components, have shown beneficial effects for human health such as preventing many diseases. In the past, plant breeders made significant progress improving yield, pest and disease resistance, and nutritional values, with the pay-off being loss of flavors. Many primary and secondary metabolites play an important role in the formation of flavor. Mixtures of sweetness, sourness, bitterness, saltiness, and acidity are responsible for the fruit taste. Volatile compounds, fatty acids, sugars, and phenolic compounds have been evaluated in numerous studies using different methods under various conditions.

This chapter has reviewed the vast amount of information available on these properties in figs. Like other crops, cultivation, and breeding practices such as irrigation, fertigation, pruning, ecological conditions, variety choice, ripening stage, and technological factors, as well as post-harvest conditions, play an important role in the formation of fig flavor. Fig fruits are consumed both dried and fresh. Fig color, aroma, texture, and nutritional values are the most important factors of consumers' choice. However, all this information is still not fully sufficient and further studies are needed. It is essential to understand the processes behind the formation of compounds and the new advanced techniques used, but more importantly, how they will affect the perception of aroma and taste of consumers, and their propensity to buy such products. The results have indicated that figs can be used as a good source of phenolic compounds. Given the important antioxidant property of phenolic compounds and the high sugar and organic acid levels, fresh and dried figs are a healthy alternative for human consumption. The results of such studies can be used to characterize the volatile composition of figs, and to carry out quality control, genetic selection, and plant breeding programs, as well as to improve and standardize the production process. In addition, many aroma compounds provide convincing support for future clinical uses in contemporary therapy for human health and nutrition.

References

Ajmal, M., Arshad, M.U., Saeed, F., Ahmed, T., Khan, A.U., *et al.* (2016) Exploring the nutritional characteristics of different parts of fig in relation to hypoglycemic potential. *Pakistan Journal of Life and Social Sciences* 14, 115–122.

Aksoy, U. (1998) Why figs? An old taste and a new perspective. *Acta Horticulturae* 480, 25–27.

Aksoy, U., Balci, B., Can, H.Z., and Hepaksoy, S. (2003) Some significant results of the research work in Turkey on fig. *Acta Horticulturae* 605, 173–181.

Aljane, F., Toumi, I., and Ferchichi, A. (2007) HPLC determination of sugars and atomic absorption analysis of mineral salts in fresh figs of Tunisian cultivars. *African Journal of Biotechnology* 6, 599–602.

Aljane, F., Neily, M.H., and Msaddak, A. (2020) Phytochemical characteristics and antioxidant activity of several Fig (*Ficus carica* L.) ecotypes. *Italian Journal of Food Science* 32, 755–768.

Amessis-Ouchemoukh, N., Ouchemoukh, S., Meziant, N., Idiri, Y., Hernanz, D., *et al.* (2017) Bioactive metabolites involved in the antioxidant, anticancer and anticalpain activities of *Ficus carica* L., *Ceratonia siliqua* L. and *Quercus ilex* L. extracts. *Industrial Crops and Products* 95, 6–17.

Anon. (2021) Institute of Agricultural Economy and Policy Development (TEPGE), Agricultural Products Markets Fig Report in 2021. Available at: https://arastirma.tarimorman.gov.tr/tepge (accessed January 18, 2022).

Arvaniti, O.S., Samaras, Y., Gatidou, G., Thomaidis, N.S., and Stasinakis, A.S. (2019) Review on fresh and dried figs: chemical analysis and occurrence of phytochemical compounds, antioxidant capacity and health effects. *Food Research International* 119, 244–267.

Bachir Bey, M. and Louaileche, H. (2015) A comparative study of phytochemical profile and *in vitro* antioxidant activities of dark and light dried fig (*Ficus carica* L.) varieties. *Journal of Phytopharmacology* 4, 41–48.

Bachir Bey, M., Louaileche, H., and Zemouri, S. (2013) Optimization of phenolic compound recovery and antioxidant of light and dark fig (*Ficus carica* L.) varieties. *Food Science and Biotechnology* 22, 1613–1619.

Bachir Bey, M., Meziant, L., Benchikh, Y., and Louaileche, H. (2014) Deployment of response surface methodology to optimize recovery of dried dark fig (*Ficus carica* L. var. Azenjar) total phenolic compounds and antioxidant activity. *International Food Research Journal* 21, 1477–1482.

Bachir Bey, M., Richard, G., Meziant, L., Fauconnier, M., and Louaileche, H. (2016) Effects of sun-drying on physiochemical characteristics, phenolic composition, and *in vitro* antioxidant activity of dark fig varieties. *Journal of Food Processing and Preservation* 41: e13164.

Badhujar, S.B., Patel, V.V., Bandivdekar, A.H., and Mahajan, R.T. (2014) Traditional uses, phytochemistry and pharmacology of *Ficus carica*: a review. *Pharmaceutical Biology* 52, 1487–1503.

Baldwin, E.A., Scott, J.W., Shewmaker, C.K., and Shuch, W. (2000) Flavor trivia and tomato aroma: biochemistry and possible mechanisms for control of important aroma components. *HortScience* 35, 1013–1022.

Barolo, M.I., Mostacero, N.R., and López, S.N. (2014) *Ficus carica* L. (Moraceae): an ancient source of food and health. *Food Chemistry* 164, 119–127.

Barrett, D.M., Garcia, E., and Wayne, J.E. (1998) Textural modification of processing tomatoes. *Critical Reviews in Food Science and Nutrition* 38, 173–258.

Barrett, D.M., Beaulieu, J.C., and Shewfelt, R. (2010) Color, flavor, texture, and nutritional quality of fresh-cut fruits and vegetables: desirable levels, instrumental and sensory measurement, and the effects of processing. *Critical Reviews in Food Science and Nutrition* 50, 369–389.

Beaulieu, J. and Baldwin, E. (2002) Flavor and aroma of fresh-cut fruits and vegetables. In: Olusola Lamikanra (ed.) *Fresh-cut Fruit and Vegetables Science, Technology and Market*. CRC Press, Boca Raton, Florida, pp. 391–425.

Çakmak, B., Alayunt, F.N., Akdeniz, R.C., Aksoy, A., and Can, H.Z. (2010) Assessment of the quality losses of fresh fig fruits during transportation. *Journal of Agricultural Sciences* 16, 180–193.

Calışkan, O. and Polat, A.A. (2008) Fruit characteristics of fig cultivars and genotypes grown in Turkey. *Scientia Horticulturae* 115, 360–367.

Calışkan, O. and Polat, A.A. (2011) Phytochemical and antioxidant properties of selected fig (*Ficus carica* L.) accessions from the eastern Mediterranean region of Turkey. *Scientia Horticulturae* 128, 473–478.

Crisosto, C.H., Bremer, V., Ferguson, L., and Crisosto, G.M. (2010) Evaluating quality attributes of four fresh figs (*Ficus carica* L.) cultivars harvested at two maturity stages. *HortScience* 45, 707–710.

Cuevas-Glory, L.F., Pino, J.A., Santiago, L.S., and Sauri-Duch E. (2007) A review of volatile analytical methods for determining the botanical origin of honey. *Food Chemistry* 103, 1032–1043.

Daniel, O., Meier, M.S., Schlatter, J., and Frischknecht, P. (1999) Selected phenolic compounds in cultivated plants: ecologic functions, health implications, and modulation by pesticides. *Environmental Health Perspectives* 107, 107–114.

de la Peña Moreno, F., Monagas, M., Blanch, G.P., Bartolomé, B., and Ruiz del Castillo, M.L. (2010) Enhancement of phenolic and aroma compounds in strawberry fruit through methyl-jasmonate vapour treatment. *European Food Research and Technology* 230, 989–999.

Debib, A., Tir-Touili, A., Mothana, R.A., Meddah, B., and Sonnet, P. (2014) Phenolic content antioxidant and antimicrobial activities of two fruit varieties of Algerian *Ficus carica*. *Journal of Food Biochemistry* 38, 207–215.

Del Caro, A.H. and Piga, A. (2008) Polyphenol composition of peel and pulp of two İtalian fresh fig fruits cultivars (*Ficus carica* L.). *European Food Research and Technology* 226, 715–719.

Devic, E., Guyot, S., Daudin, J.D., and Bonazzi, C. (2010) Kinetics of polyphenol losses during soaking and drying of cider apples. *Food and Bioprocess Technology* 3, 867–877.

Dong, C., Mei, Y., and Chen, L. (2006) Simultaneous determination of sorbic and benzoic acids in food dressing by headspace solid-phase microextraction and gas chromatography. *Journal of Chromatography A* 1117, 109–114.

Dudareva, N., Pichersky, E., and Gershenzon, J. (2004) Biochemistry of plant volatiles. *Plant Physiology* 135, 1893–1902.

Dueñas, M., Pérez-Alonso, J.J., Santos-Buelga, C., and Escribano-Bailón, T. (2008) Anthocyanin composition in fig. *Journal of Food Composition and Analysis* 21, 107–115.

Duke, P.K. (2002) *Handbook of Medicinal Herbs*. CRC Press, Boca Raton, Florida.

Eberhardt, M.V., Lee, C.Y., and Liu, R.H. (2000) Antioxidant activity of fresh apples. *Nature* 405, 903–904.

Ercişli, S., Tosun, M., Karlidag, H., Dzubur, A., Hadziabulic, S., and Aliman, Y. (2012) Color and antioxidant characteristics of some fresh fig (*Ficus carica* L.) genotypes from northeastern Turkey. *Plant Foods for Human Nutrition* 67, 271–276.

Ergun, Z. and Bozkurt, T. (2020) Determination of fatty acid composition and antioxidant activity of fig seed oil. *International Journal of Agricultural and Natural Sciences* 13, 101–107.

Eriten, A. (2005) Estimates of demand relationships for figs and figs products in Turkey. MSc thesis, Middle East Technical University, Graduate School of Social Sciences, Ankara, Turkey.

Ersoy, N., Gozlekci, S., Gok, V., and Yilmaz, S. (2015) Fig (*Ficus carica* L.) fruit, some physical and chemical properties. *Acta Horticulturae* 1173, 329–334.

Erturk, Y., Ercisli, S., and Tosun, M. (2009) Physico-chemical characteristics of wild plum fruits (*Prunus spinosa* L.) *International Journal of Plant Production* 3, 89–92.

Faleh, E., Oliveira, P.A., Valentao, P., Ferchichi, A., Silva, B.M., and Andrade, P.B. (2012) Influence of Tunisian *Ficus carica* fruit variability in phenolic profiles and *in vitro* radical scavenging potential. *Brazilian Journal of Pharmacognosy* 22, 1282–1289.

FAO (Food and Agriculture Organization of the United Nations) (2020) FAOSTAT. Available at: www.fao.org/faostat/en/#home (accessed January 18, 2022).

Fellman, J.K., Rudell, D.R., Mattinson, D.S., and Mattheis, J.P. (2003) Relationship of harvest maturity to flavor regeneration after CA storage of “Delicious” apples. *Postharvest Biology and Technology* 27, 39–51.

Ferreira, D., Guyot, S., Marnet, N., Delgadillo, I., Renard, C., and Coimbra, M.A. (2002) Composition of phenolic compounds in a Portuguese pear (*Pyrus communis* L. var. S. Bartolomeu) and changes after sundrying. *Journal of Agricultural and Food Chemistry* 50, 4537–4544.

Ficsor, E., Szentmihályi, K., Lemberkovics, E., Blázovics, A., and Balázs, A. (2013) Analysis of *Ficus carica* L. – volatile components and mineral content. *European Chemical Bulletin* 2, 126–129.

Flaishman, M.A., Rodov, V., and Stover, E. (2008) The fig: botany, horticulture, and breeding. *Horticultural Reviews* 34, 113–197.

Forney, C.F., Kalt, W., and Jordan, M.A. (2000) The composition of strawberry aroma is influenced by cultivar, maturity, and storage. *HortScience* 35, 1022–1026.

Genna, A., Vecchi, P., de Bruno, M., and Maestrelli, A. (2005) Quality of Cosenza dried fig: a sensory and physicochemical approach. *Acta Horticulturae* 798, 319–323.

Gibernau, M., Buser, H.R., Frey, J.E., and Hossaert-McKey, M. (1997) Volatile compounds from extracts of figs of *Ficus carica*. *Phytochemistry* 46, 241–244.

Gonçalves, B., Oliveira, I., Bacelar, E., Morais, M.C., Aires, A., *et al.* (2018) Aromas and flavours of fruits. In: Vilela, A. (ed.) *Generation of Aromas and Flavors*. IntechOpen, London, pp. 9–31.

Gözlekçi, S., Kafkas, E., and Ercisli, S. (2011) Volatile compounds determined by HS/GC-MS technique in peel and pulp of fig (*Ficus carica* L.) cultivars grown in Mediterranean region of Turkey. *Notulae Botanicae Horti Agrobotanici Cluj-Napoca* 39, 105–108.

Grison, L., Edwards, A.A., and Hossaert-McKey, M. (1999) Interspecies variation in floral fragrances emitted by tropical *Ficus* species. *Phytochemistry* 52, 1293–1299.

Grison-Pigé, L., Hossaert-McKey, M., Greeff, J.M., and Bessière, J.M. (2002) Fig volatile compounds – a first comparative study. *Phytochemistry* 61, 61–71.

Guarrera, P.M. (2005) Traditional phytotherapy in Central Italy (Marche, Abruzzo, and Latium). *Fitoterapia* 76, 1–25.

Gül, H. and Tekeli, S.G. (2019) Inclusion of *Rosa damascena* Mill. powder into cookies: nutritional, antioxidant and quality characteristics. *International Journal of Agriculture Forestry and Life Science* 3, 301–306.

Gundesli, M.A., Korkmaz, N., and Okatan, V. (2019) Polyphenol content and antioxidant capacity of berries: a review. *International Journal of Agriculture, Forestry and Life Sciences* 3, 350–361.

Gundesli, M.A., Kafkas, N.E., Okatan, V., and Usanmaz, S. (2020) Identification and characterisation of volatile compounds determined by HS/GCMS technique in pulp of ‘Abbas’ fig (*Ficus carica* L.) variety. *Pakistan Journal of Agricultural Science* 57, 623–629.

Gundogdu, M. and Yılmaz, H. (2012) Organic acid, phenolic profile and antioxidant capacities of pomegranate (*Punica granatum* L.) cultivars and selected genotypes. *Scientia Horticulturae* 143, 38–42.

Harb, J., Bisharat, R., and Streif, J. (2008) Changes in volatile constituents of blackcurrants (*Ribes nigrum* L. cv. “Titania”) following controlled atmosphere storage. *Postharvest Biology and Technology* 47, 271–279.

Harzallah, A., Bhouri, A.M., Amri, Z., Soltana, H., and Hammami, M. (2016) Phytochemical content and

antioxidant activity of different fruit parts juices of three figs varieties grown in Tunisia. *Industrial Crops and Products* 83, 255–267.
Hawthorne, S., Miller, D., Pawliszyn, J., and Arthur, C. (1992) Solventless determination of caffeine in beverages using solid-phase micro extraction with fused silica fibers. *Journal of Chromatography A* 603, 185–191.
Hoxha, L. and Kongoli, R. (2016) Evaluation of antioxidant and potential of Albanian fig varieties “Kraps Zi” and “Kraps Bardhe” cultivated in the region of Tirana. *Journal of Hygienic Engineering and Design* 16, 70–74.
Hoxha, L., Kongoli, R., and Hoxha, M. (2015). Antioxidant activity of some dried autochthonous Albanian fig (*Ficus carica* L.) cultivars. *International Journal of Crop Science and Technology* 1, 20–26.
Hssaini, L., Charafi, J., Razouk, R., Hernández, F., Fauconnier, M.L., *et al.* (2020) Assessment of morphological traits and fruit metabolites in eleven fig varieties (*Ficus carica* L.). *International Journal of Fruit Science* 20, 8–28.
Kafkas, E., Kafkas, S., Koch-Dean, M., Schwab, W., Larkov, O., *et al.* (2005a) Comparison of methodologies for the identification of aroma compounds in strawberry. *Turkish Journal of Agriculture and Forestry* 29, 383–390.
Kafkas, E., Kafkas, S., Koch-Dean, M., and Schwab, W. (2005b) Comparison of methodologies for the identification of aroma compounds in strawberry. *Turkish Journal of Agriculture and Forestry* 29, 383–390.
Kafkas, E., Ercisli, S., Koc, N.K., Baydar, K., and Yilmaz, H. (2009) Chemical composition of blood orange varieties from Turkey: a comparative study. *Pharmacognosy Magazine* 5, 329–335.
Kafkas, E., Attar, S.H., Gundesli, M.A., Ozcan, A., and Ergun, M. (2020) Phenolic and fatty acid profile, and protein content of different walnut cultivars and genotypes (*Juglans regia* L.) grown in the USA. *International Journal of Fruit Science* 20, 1711–1720.
Kamiloğlu, S., Toydemir, G., Boyacioglu, D., Beekwilder, J., Robert, D., *et al.* (2015) A review on the effect of drying on antioxidant potential of fruits and vegetables. *Critical Reviews in Food Science and Nutrition* 56, 110–129.
Kelebek, H., Diblan, S., Kadiroglu, P., Kola, O., and Selli, S. (2018) Kurutma işlemlerinin incirlerin (*Ficus carica* L.) fenolik bileşikler, antioksidan kapasite ve diğer önemli bazi kalite kriterleri üzerine etkileri. *Çukurova Tarim Gida Bilimleri Dergisi* 33, 127–136.
Kim, D.O., Jeong, S.W., and Lee, C.Y. (2003) Antioxidant capacity of phenolic phytochemicals from various cultivars of plums. *Food Chemistry* 81, 321–326.
King, E.S., Hopfer, H., Haug, M.T., Orsi, J.D., Heymann, H., *et al.* (2012) Describing the appearance and flavor profiles of fresh fig (*Ficus carica* L.) cultivars. *Journal of Food Science* 77, 419–429.
Konak, R., Kosoglu, I., and Yemenicioglu, A. (2017) Effects of different drying methods on phenolic content antioxidant capacity and general characteristics of selected Turkish fig cultivars. *Acta Horticulturae* 1173, 335–340.
Kramer, A. (1965) Evaluation of quality of fruits and vegetables. In: Irving, G.W., Jr. and Hoover, S.R. (eds) *Food Quality*. American Association for the Advancement of Science, Washington, DC, pp. 9–18.
Lester, G. (2006) Consumer preference quality attributes of melon fruit. *Acta Horticulturae* 712, 175–181.
Li, J., Tian, Y., Sun, B., Yang, D., Chen, J.P., and Men, Q.M. (2012) Analysis on volatile constituents in leaves and fruits of *Ficus carica* by GC-MS. *Chinese Herbal Medicines* 4, 63–69.
Maghsoudlou, E., Kenari, R.E., and Amiri, Z.R. (2017) Evaluation of antioxidant activity of fig (*Ficus carica*) pulp and skin extract and its application enhancing stability of canola oil. *Journal of Food Processing and Preservation* 41, 1–11.
Matthews, R.H., Pehrsson, P.R., and Farhat-Sabet, M. (1987) Sugar content of selected foods. In: *Home Economics Research Report No. 48*. US Department of Agriculture, Agriculture Research Services, Beltsville, Maryland.
Mawa, S., Husain, K., and Jantan, İ. (2013) *Ficus carica* L. (Moraceae): Phytochemistry, traditional uses and biological activities. *Evidence-based Complementary and Alternative Medicine* 13: 974256.
Melgarejo, P., Hernandez, F.C.A., Martinez, J.J., Sanchez, M.J., and Salazar, D.M. (2003) Organic acids and sugars from first and second crop fig juices. *Acta Horticulturae* 605, 237–239.
Menager, İ., Jost, M., and Aubert, C. (2004) Changes in physicochemical characteristics and volatile constituents of strawberry (cv. Cigaline) during maturation. *Journal of Agricultural and Food Chemistry* 52, 1248–1254.
Minh Tu, N., Thanh, L.X., Une, A., Ukeda, H., and Sawamura, M. (2002) Volatile constituents of Vietnamese pummelo, orange, tangerine and lime peel oils. *Flavor and Fragrance Journal* 17, 169–174.

Mohamed El Hadi, M.A., Zhang, F.J., Wu, F.F., Zhou, C.H., and Tao, J. (2013) Advances in fruit aroma volatile research. *Molecules* 18, 8200–8229.
Mujić, İ., Kralj, M.B., Jokić, S., Jarni, K., *et al.* (2012) Changes in aromatic profile of fresh and dried fig – the role of pre-treatments in drying process. *International Journal of Food Science and Technology* 47, 2282–2288.
Mujić, İ., Kralj, M.B., Jokić, S., Jug, T., Šubarić, D., *et al.* (2014) Characterisation of volatiles in dried white varieties figs (*Ficus carica* L.). *International Journal of Food Science and Technology* 51, 1837–1846.
Nakilcioğlu, E. and Hısıl, Y. (2013) Research on the phenolic compounds in Sarilop (*Ficus carica* L.) fig variety. *Journal of Food* 38, 267–274.
Oliveira, A.P., Valentao, P., Pereira, J.A., Silva, B.M., Tavares, F., and Andrade, P.B. (2009) *Ficus carica* L.: metabolic and biological screening. *Food and Chemical Toxicology* 47, 2841–2846.
Oliveira, A.P., Silva, L.R., Andrade, P.B., Valentao, P., Silva, B.M., *et al.* (2010a) Determination of low molecular weight volatiles in *Ficus carica* using HS-SPME and GC/FID. *Food Chemistry* 121, 1289–1295.
Oliveira, A.P., Silva, L.R., Guedes de Pinho, P., Gil-İzquierdo, A., Valentao, P., *et al.* (2010b) Volatile profiling of *Ficus carica* varieties by HS-SPME and GC–IT-MS. *Food Chemistry* 123, 548–557.
Ong, B.T., Nazimah, S.A.H., Tan, C.P., Mirhosseini, H., Osman, A., *et al.* (2008) Analysis of volatile compounds in five jackfruit (*Artocarpus heterophyllus* L.) cultivars using solid-phase microextraction (SPME) and gas chromatography-time-of-flight mass spectrometry (GC-TOFMS). *Journal of Food Composition and Analysis* 21, 416–422.
Ortiz, A., Echeverría, G., Graell, J., and Lara, I. (2010) The emission of flavor-contributing volatile esters by “Golden Reinders” apples is improved after mid-term storage by postharvest calcium treatment. *Postharvest Biology and Technology* 57, 114–123.
Ouchemoukh, S., Hachoud, S., Boudraham, H., Mokrani, A., and Louaileche, H. (2012) Antioxidant activities of some dried fruits consumed in Algeria. *Food Science and Technology* 49, 329–332.
Palassarou, M., Melliou, E., Liouni, M., Michaelakis, A., Balayiannis, G., and Magiatis, P. (2017) Volatile profile of Greek dried white figs (*Ficus carica* L.) and investigation of the role of β-damascenone in aroma formation in fig liquors. *Journal of the Science of Food and Agriculture* 97, 5254–5270.
Pande, G. and Akoh, C.C. (2010) Organic acids, antioxidant capacity, phenolic content and lipid characterisation of Georgia-grown underutilized fruit crops. *Food Chemistry* 120, 1067–1075.
Patil, V.V. and Patil, V.R. (2012) Evaluation of anti-inflammatory activity of *Ficus carica* Linn. *Indian Journal of Natural Products and Resources* 2, 151–155.
Pereira, C., López Corrales, M., Martín, A., Villalobos, M. del C., Córdoba, M. de G., and Serradilla, M.J. (2017) Physicochemical and nutritional characterization of brebas for fresh consumption from nine fig varieties (*Ficus carica* L.) grown in Extremadura (Spain). *Journal of Food Quality* 2017: 6302109.
Pereira, J. and Câmara, J.S. (2011) Effectiveness of different solid-phase microextraction fibres for differentiation of selected Madeira Island fruits based on their volatile metabolite profile—identification of novel compounds profile. *Talanta* 83, 899–906.
Petkova, N., Ivanov, I., and Denev, P. (2019) Changes in phytochemical compounds and antioxidant potential of fresh, frozen, and processed figs (*Ficus carica* L.). *International Food Research Journal* 26, 1881–1888.
Peyrot des Gachons, C., Leeuwen, C.V., Tominaga, T., Soyer, J.P., Gaudillere, J.P., and Denis, D. (2005) Influence of water and nitrogen deficit on fruit ripening and aroma potential of *Vitis vinifera* L cv Sauvignon blanc in field conditions. *Journal of the Science of Food and Agriculture* 85, 73–85.
Polat, A.A. and Calışkan, O. (2008) Fruit characteristics of table fig (*Ficus carica* L) cultivars in subtropical climate conditions of the Mediterranean region. *New Zealand Journal of Crop and Horticultural Science* 36, 107–115.
Ribechini, E., Pérez-Arantegui, J., and Colombini, M.P. (2011) Gas chromatography/mass spectrometry and pyrolysis-gas chromatography/mass spectrometry for the chemical characterization of modern and archaeological figs (*Ficus carica*). *Journal of Chromatography A* 1218, 3915–3922.
Rodríguez-Solana, R., Ludovina, R.G., Pérez-Santín, E., and Romano, A. (2018) Production method and varietal source influence the volatile profiles of spirits prepared from fig fruits (*Ficus carica* L.). *European Food Research and Technology* 244, 2213–2229.
Russo, F., Caporaso, N., Antonello Paduano, A., and Sacchi, R. (2017) Characterisation of volatile compounds in Cilento (Italy) figs (*Ficus carica* L.) cv. Dottato as affected by the drying process, *International Journal of Food Properties* 20, 1366–1376.
Russo, F., Caporaso, N., Paduano, A., and Sacchi, R. (2014) Phenolic compounds in fresh and dried figs from Cilento (Italy), by considering breba crop and full crop, in comparison to Turkish and Greek dried figs. *Journal of Food Science* 79, C1278–C1284.

Setser, C.S. (1993) Sensory properties. In: Macrae, R., Robinson, R.K., and Sadler, M.J. (eds) *Encyclopedia of Food Science, Food Technology and Nutrition*. Academic Press, London.

Shahidi, F. and Ambigaipalan, P. (2015) Phenolics and polyphenolics in foods, beverages and spices: antioxidant activity and health effects – a review. *Journal of Functional Foods* 18, 820–897.

Simsek, M. (2009) Fruit performances of the selected fig types in Turkey. *African Journal of Agricultural Research* 4, 1260–1267.

Simsek, M. (2011) A study on selection and identification of table fig types in east edge of Firat River. *Asian Journal of Animal and Veterinary Advances* 6, 265–273.

Simsek, M. (2016) Chemical, mineral, and fatty acid compositions of various types of walnut (*Juglans regia* L.) in Turkey. *Bulgarian Chemical Communications* 1, 66–70.

Simsek, M., Yılmaz, K.U., and Demirkıran, A.R. (2010) Selection and determination of some significant properties of superior walnut genotypes. *Scientific Research and Essays* 5, 2987–2996.

Slatnar, A., Klancar, U., Stampar, F., and Veberic, R. (2011) Effect of drying of figs (*Ficus carica* L.) on the contents of sugars, organic acids, and phenolic compounds. *Journal of Agricultural and Food Chemistry* 59, 11696–11702.

Slavin, J.L. (2006) Figs: past, present and future. *Nutrition Today* 41, 180–184.

Solomon, A., Golubowicz, S., Yablowicz, Z., Kerem, Z., and Flaishman, M.A. (2006) Antioxidant activities and anthocyanin content of fresh fruits of common fig (*Ficus carica* L.). *Journal of Agricultural and Food Chemistry* 54, 7717–7723.

Soltana, H., Flamini, G., and Hammami, M. (2017) Volatile compounds from six varieties of *Ficus carica* from Tunisia. *Records of Natural Products* 11, 157–165.

Song, J. and Forney, C.F. (2008) Flavor volatile production and regulation in fruit. *Canadian Journal of Plant Science* 88, 537–550.

Stover, E., Aradhya, M., Ferguson, L., and Crisosto, C.H. (2007) The fig: overview of an ancient fruit. *HortScience* 42, 1083–1087.

Taveira, M., Guedes de Pinho, P., Gonçalves, R.F., Andrade, P.B., and Valentão, P. (2009) Determination of eighty-one volatile organic compounds in dietary *Rumex induratus* leaves by GC/IT-MS, using different extractive techniques. *Microchemical Journal* 93, 67–72.

Tietel, Z., Lewinsohn, E., Fallik, E., and Porat, R. (2012) Importance of storage temperatures in maintaining flavor and quality of mandarins. *Postharvest Biology and Technology* 64, 175–182.

Trad, M., Le Bourvellec, C., Gaaliche, B., Renard, C.M.G.C. and Mars, M. (2014) Nutritional compounds in figs from the southern Mediterranean region. *International Journal of Food Properties* 17, 491–499.

Trichopoulou, A., Vasilopoulou, E., Georga, K., Soukara, S., and Dilis, V. (2006) Traditional foods: why and how to sustain them. *Trends in Food Science and Technology* 17, 498–504.

Uzundumlu, A.S., Oksuz, M.E., and Kurtoglu, S. (2018) Future of fig production in Turkey. *Journal of the Tekirdag Agricultural Faculty* 15, 138–146.

Valero, D. and Serrano, M. (2010) *Postharvest Biology and Technology for Preserving Fruit Quality*. CRC Press, Boca Raton, Florida.

Vallejo, F., Marín, J.G., and Tomás-Barberán, F.A. (2012) Phenolic compound content of fresh and dried figs (*Ficus carica* L.). *Food Chemistry* 130, 485–492.

Veberic, R., Jakopic, J., and Stampar, F. (2008) Internal fruit quality of figs (*Ficus carica* L.) in the Northern Mediterranean Region. *Italian Journal of Food Science* 20, 255–262.

Vinson, J.A. (1999) The functional food properties of figs. *Cereal Foods World* 4, 82–87.

Vinson, J.A., Zubik, L., Bose, P., Sammam, N., and Proch, J. (2005) Dried fruits: excellent *in vitro* and *in vivo* antioxidants. *Journal of the American College of Nutrition* 24, 44–50.

Ware, A.B., Kaye, P.T., Compton, S.G., and van Noort, S. (1993) Fig volatiles: their role in attracting pollinators and maintaining pollinator specificity. *Plant Systematics and Evolution* 186, 147–156.

Yemis, O., Bakkalbas, E., and Artık, N. (2012) Changes in pigment profile and surface colour of fig (*Ficus carica* L.) during drying. *International Journal of Food Science and Technology* 47, 1710–1719.

22 The Role of Fig in Human Nutrition

Manuel Joaquín Serradilla[1]*, Cristina García-Viguera[2], Alberto Martín[3], Margarita López-Corrales[4], and Daniel Valero[5]

[1]*Área de Postcosecha, Instituto Tecnológico Agroalimentario de Extremadura (INTAEX), Centro de Investigaciones Científicas y Tecnológicas de Extremadura (CICYTEX), Junta de Extremadura, Badajoz, Spain;* [2]*Phytochemistry and Healthy Food Lab (LabFAS), Department of Food Science and Technology, CEBAS-CSIC, Campus de Espinardo, Murcia, Spain;* [3]*Nutrición y Bromatología, Escuela de Ingenierías Agrarias, Universidad de Extremadura, Badajoz, Spain;* [4]*Fruticultura Mediterránea, Centro de Investigación Finca La Orden-Valdesequera, Junta de Extremadura, Badajoz, Spain;* [5]*Department of Food Technology, EPSO, University Miguel Hernández, Alicante, Spain*

22.1 Introduction

Brebas and figs have been part of the Mediterranean diet for a long time, especially in their dried format, probably because this species has grown in this area since antiquity. In recent years, brebas and figs have been in great demand, first because of their attraction as subtropical fruit and secondly because of their high nutritional and bioactive values. Figs and brebas are nutritious fruits rich in fiber, potassium, calcium, and iron, as well as being an important source of vitamins, amino acids, and antioxidants. For this last reason, certain types of compounds present in both brebas and figs have acquired great attention due to their functional or therapeutic value.

The aim of this chapter, as well as discussing the nutritional aspects of brebas and figs, is to focus on the bioactive compounds present in brebas and figs and how these compositions are affected by the ripening process, as well as their subsequent postharvest life. Finally, the impact of these bioactive compounds on human health will be discussed.

22.2 Fruit Chemistry

The main quality traits of fresh fruits (brebas and figs) include fruit size, firmness, skin color, flesh color, total soluble solids (TSS, °Brix), titratable acidity (TA), and flavor (Flaishman *et al.*, 2008). All of these quality parameters are very important, as consumer acceptance depends on them (Flaishman *et al.*, 2008; Crisosto *et al.*, 2010; Çalişkan and Polat, 2011). The maturation index, which is the TSS:TA ratio, is positively correlated with consumer acceptability (Crisosto *et al.*, 2010; Valero and Serrano, 2010). TSS and TA are both related to fruit ripening and influence shelf life, and are therefore important parameters for fruit marketing. Fruits with a higher TSS and lower TA content (i.e. high maturation index) are the most sensitive to physical damage, and thus their shelf life is shorter (Pereira *et al.*, 2017a). The relevant quality parameters are affected by genotype, ripening stage at harvest, storage conditions, crop year, and orchard management practices (Crisosto *et al.*, 2010; Çalişkan and Polat, 2011). Furthermore, differences in quality parameters

* Email: manuel.serradilla@juntaex.es

 Advances in Fig Research and Sustainable Production (M.A. Flaishman and U. Aksoy)
DOI: 10.1079/9781789242492.0022

have been observed between brebas and figs (Pereira *et al.*, 2015, 2017b).

Several authors have reported that fresh figs from different varieties, grown in different countries, showed a different range of TSS with values ranging between 12.0°Brix and 30.1°Brix. TSS of varieties grown in California had higher values (Crisosto *et al.*, 2010) than those grown in Turkey (Çalişkan and Polat, 2011), Spain (Pereira *et al.*, 2015, 2017b) or Tunisia (Gaaliche *et al.*, 2012). Regarding TA, varieties grown in California had higher values (0.14–1.00%) than the varieties from Turkey (0.09–0.26%) (Çalişkan and Polat, 2011) or Spain (0.11–0.20%) (Pereira *et al.*, 2017b).

Breba crop quality parameters have been less studied. Pereira *et al.* (2017a) studied nine breba varieties whose TSS values ranged between 15.2°Brix (Brown Turkey) and 19.1°Brix (Cuello Dama Blanco). They showed significant differences among the varieties, allowing classification of the fruits into three groups according to their TSS content: high content (19°Brix or more), moderate content (around 18°Brix) and low content (15–17°Brix). Usually, figs show higher TSS content than brebas, and brebas are slightly more acidic than figs, with TA values ranging from 0.18% to 0.20% citric acid.

Fig-fruit growth can be described by a double-sigmoid curve comprising three phases, or periods. Phase I is characterized by rapid size growth due to cell division. During phase II, the fruit remains at nearly the same size. Rapid fruit growth in phase III is the outcome of cell expansion. During phase III, the fruit ripens, indicated by color change and softening of the fruit, along with alteration of the flesh texture to an edible state (Fig. 22.1). The final growth of breba figs coincides with phase I growth of the main commercial crop, increasing the sink demand, potentially resulting in reduced fruit size of the commercial crop (Ferguson *et al.*, 1990). Figure 22.1 shows the visual aspect of breba fruits at 12 different maturity stages.

The ripening stage at harvest has been shown to have a clear effect on TSS and TA. Crisosto *et al.* (2010) studied the quality attributes of four fresh-fig varieties harvested at two maturity stages: commercial and tree ripe. These authors showed that tree-ripe figs had higher levels of TSS, lower TA and a higher TSS:TA ratio than commercial figs. In another study of different varieties harvested at these two maturity stages, tree-ripe figs showed 22% higher levels of TSS, 30% lower TA and higher TSS:TA ratios than commercial figs (Crisosto *et al.*, 2017). For breba crop, differences between varieties at three ripening stages were observed in these quality parameters. For the same variety, the TSS value increased from stage 1 (15.6°Brix) to stage 3 (19.1°Brix), and the TSS:TA was higher as the ripening stage of the fruit increased, with values from ranging from 110 in stage 1 to 220 in stage 3 (Pereira *et al.*, 2017a). The preferred ratio will vary according to the use of the fruits, but will provide guidance in selecting varieties for specific uses (Can, 1993). Çalişkan and Polat (2011) indicated that genotypes with high soluble solids:acid ratios produce high quality dried fruits.

TSS and TA also are also affected by environmental conditions. One study evaluated the role of different agroecological zones on fruit quality of the Zidi variety grown in Tunisia. The results showed significant differences in quality parameters according to the growth area. The best fruit taste was achieved in north-east Tunisia with a high TSS rate of 19°Brix and a TA value of 0.18%. These differences reflect large agroecological adaptation in the Zidi variety, attributed to intra-varietal genetic variations and different climatic conditions between growing areas (Gaaliche *et al.*, 2012). These results are in accordance with

Fig. 22.1. Ripening stages of breba fruits. Visual aspect of breba fruits at 12 different maturity stages.

other studies (Küden *et al.*, 2008; Messaoudi and Haddadi, 2008; Crisosto *et al.*, 2010). TSS values for the varieties Cuello Dama Negro and Colar Elche grown in Spain (Pereira *et al.*, 2015) were higher than those achieved for the same varieties grown in California (Crisosto *et al.*, 2010). However, TA values were lower than those obtained for the same varieties under different climates (Crisosto *et al.*, 2010).

Aroma is an important criterion for fruit quality assessment. However, only a few studies are available regarding aroma constituents of ripe fruit and fig aroma profiles at ripeness are not yet defined. In three varieties studied, 51 aroma-related compounds were identified, dominated by esters, ketones, and alcohols. The main compounds among these were butyl acetate, isoamyl acetate, 3-hydroxy-2-butanone (acetoin), and hexyl acetate (Trad *et al.*, 2012).

Finally, the caprification process is another important factor affecting the quality of fresh fig fruits in San Pedro- and Smyrna-type varieties (Mars *et al.*, 2009). Some studies have reported that pollinated fruits were characterized by superior quality (higher soluble solids content, firmness, and more intense external and internal color) and could be better stored than non-pollinated ones (Aksoy *et al.*, 2003). Results obtained by Gaaliche *et al.* (2011) indicated that caprification carried out four times with an interval of 4 days between each occurrence could efficiently improve fruit quality TSS and other parameters on two fig varieties in Tunisia. These results accord with another report from Iran, where pollen source significantly affected fruit length, TSS, ostiole size, and percentage of seed germination (Rahemi and Jafari, 2008).

22.3 Nutritional Composition

22.3.1 Sugars

The main sugars in fresh figs are the hexose monosaccharides glucose and fructose with a constant glucose:fructose ratio of 1.1–1.2, independent of the variety (Melgarejo *et al.*, 2003; Veberic *et al.*, 2008; Slatnar *et al.*, 2011; Pisimisi *et al.*, 2012). The fructose and glucose content ranged from 4.7% to 8.29% and from 5.05% to 9.62% of fresh matter, respectively, in ten Tunisian varieties (Aljane and Ferchichi, 2009). Variation in sugar content in fresh fig was discovered to be genotype dependent, as suggested by the analysis of 12 fig genotypes (Çalişkan and Polat, 2011). In this study, fructose was described as the dominant sugar in the fresh matter (7.2–10.7%) followed by glucose (6.0–9.0%) and sucrose (0.14–0.53%). Melgarejo *et al.* (2003) reported sucrose and maltose as minor sugars in the first (breba) and second (main) crops of four varieties in Spain with concentrations lower than 1% in all cases. Commonly, the concentration of sugars is lower in breba crops than in the main crops, and there are also differences between the skin and flesh, and between the different ripening stages (Pereira *et al.*, 2017a). The accumulation of both glucose and fructose occurs during fruit ripening, whereas the amount of sucrose decreases due to the high invertase activity during this period, although it increases sharply at the end of ripening (Vemmos *et al.*, 2013). Other sugars found at minor levels in fig that are associated with the metabolism of the fruit are ribose, isomaltose, glucose 6-phosphate, fucose, and erythritol, among others (Allegra *et al.*, 2018). Sugar content is higher in dried figs than in fresh, highlighting the content of glucose over fructose (Slatnar *et al.*, 2011). This high sugar content makes dried figs among the most energy-rich dried fruits (Veberic and Mikulic-Petkovsek, 2016).

22.3.2 Organic acids

Citric acid is the major organic acid in fig fruits, with a mean value of 0.35 g 100 g^{-1} fresh weight (FW), followed by malic acid with 0.13 g 100 g^{-1} FW, as shown in five Tunisian fig varieties (Trad *et al.*, 2012). In three Slovenian varieties, citric and malic acid ranged from 2.75 to 4.61 g kg^{-1} and from 1.06 to 2.17 g kg^{-1}, respectively, while fumaric and shikimic acid were minor constituents (Veberic *et al.*, 2008). No differences associated with the skin color of the varieties were found for either of these organic acids (Veberic *et al.*, 2008). Other organic acids such as gluconic, succinic, pipecolic, ascorbic, lactic, and acetic acid have been described in fruits and juices from both first and second crops (Melgarejo *et al.*, 2003; Allegra *et al.*, 2018). A reduced shelf life of fig crops is the result of low TA, ranging from 0.18% to 0.24% for Turkish varieties with pH values between 4.6 and 5.2 (Çalişkan and Polat, 2011), and from 0.08% to 0.20% with pH values between 5.3 and 6 in both brebas and main-crop figs for Spanish varieties (Pereira *et al.*, 2015, 2017a). For dried figs, the citric acid concentration is 2.4–5.6-fold higher than in fresh fruits, while the malic acid content is 2.8–17.4-fold higher (Slatnar *et al.*, 2011).

22.3.3 Nitrogen compounds

Crude protein concentration in fresh fig also depends on both variety and ripening stage. The early ripening stage of fig is associated with higher protein content, with mean values of 4.4–5.9%. It is variety dependent, as measured for nine Spanish varieties (Pereira *et al.*, 2017a). Regarding amino acids, aspartic acid and glutamine are the most abundant amino acids found in figs, with levels above 5.5 mg g^{-1} dry weight (DW), whereas methionine is found in values lower than 0.5 mg g^{-1} DW. These authors also demonstrated the impact of growing conditions on amino acid concentrations, noting that plants grown in saline soil were expected to accumulate higher amino acid levels, in both fruit and leaves. In addition to conventional amino acids, other amino acid derivatives in fig fruit have been described such as α-ketoglutarate, oxoproline, *O*-acetylserine, and cyanoalanine (Allegra *et al.*, 2018).

22.3.4 Fiber and minerals

Figs and breba fruits are rich in fiber (5.8%, w/w, with more than 28% soluble fiber) and minerals (Vinson, 1999; Stover *et al.*, 2007). Potassium is the dominant mineral, followed by calcium and sodium. The mineral composition is mostly genotype dependent but is also affected by environmental conditions, fertilization, and the nutritional status of the plant. Aljane and Ferchichi (2009) described significant differences in the amounts of these minerals detected in ten Tunisian fig varieties, in the ranges 6.9–10.5, 1.7–4.1, and 0.4–1.8 mg g^{-1} DW for potassium, calcium, and sodium, respectively. Pereira *et al.* (2017a) found similar values for these major minerals in nine Spanish varieties, but also reported values above 1 mg g^{-1} DW for phosphorus and magnesium. In contrast, trace minerals such as iron, zinc, and copper were found in fresh fig in the ranges 0.03–0.1 mg g^{-1} for iron, 0.003–0.01 mg g^{-1} for zinc, and 0.002–0.004 mg g^{-1} for copper (Pereira *et al.*, 2017a; Khan *et al.*, 2011).

22.3.5 Vitamins

Brebas and figs contain vitamins A, C, E, K, thiamine (B_1), riboflavin (B_2), niacin (B_3), pantothenic acid (B_5), B_6, and total folate (B_9) (USDA Nutritional Database, 2019). The main vitamin is C, with average values of about 2 mg 100 g^{-1} FW. In addition, Pereira *et al.* (2017c) found that the concentration of vitamin C in skin was higher than in flesh, and that green-skinned varieties such as Banane were characterized by the highest concentrations of vitamin C in both skin and flesh. In dried figs, vitamin C is also characterized as the principal vitamin with an average value of 1.2 mg 100 g^{-1} FW (Khatib and Vaya, 2010). Also remarkable is the high content of vitamin K, as shown for dried figs with 15.6 μg 100 g^{-1} FW, which is much higher than that gained from fresh figs at only 4.5 μg 100 g^{-1} FW.

22.4 Phytochemical Composition and Antioxidant Activity

22.4.1 Carotenoids

The compounds determining pigments in fig fruits are carotenoids. Unlike anthocyanins, carotenoids are lipophilic and are responsible for the color spectrum of yellow, orange, and red. They can be divided into two groups: carotenes, which are non-oxygenated carotenoids, and xanthophylls, which are oxygenated derivatives of carotenes (Valero and Serrano, 2010). Regarding carotenes, lycopene has been identified as the most dominant, with a concentration of 0.32 mg 100 g^{-1} FW, followed by β-carotene with 0.04 mg 100 g^{-1} FW and α-carotene with 0.02 mg 100 g^{-1} FW (Veberic and Mikulic-Petkovsek, 2016). Regarding xanthophylls, fresh figs have high amounts of lutein, with a range of 0.614–0.715 mg 100 g^{-1} FW, followed by zeaxanthin 0.22–0.332 mg 100 g^{-1} DW, and cryptoxanthin at 0.185–0.204 mg 100 g^{-1} DW (Yemis *et al.*, 2012). Carotenoid concentrations are higher in skin (around 1.9 mg 100 g^{-1} FW) than in the flesh (1.1 mg 100 g^{-1} FW) (Veberic and Mikulic-Petkovsek, 2016). Although the initial amounts of these bioactive compounds in fresh figs are not very high compared with other fresh fruits, dried figs contain higher amounts of carotenoids compared with other dried fruits (Veberic and Mikulic-Petkovsek, 2016).

22.4.2 Phenolic compounds

Recently, a steady increase in consumers' demand for fruits and vegetables has been recognized, thanks to their health-promoting ingredients. In addition to their nutritional value, fruits and vegetables also produce a number of non-nutritional compounds or phytochemicals. These secondary metabolites synthesized by plant tissues through the process of photosynthesis ensure plant sustainability against biotic and abiotic stress (factors) such as drought, high soil salinity, environmental conditions, viruses, bacteria, fungi, and animal or insect attack. As well as these functions, it has been reported that these phenolic compounds play a crucial role in sensory properties such as the appearance, taste, aroma, and flavor of the fruit, either breba or main-crop figs (Lee and Kader, 2000; Tomás-Barberán and Espín, 2001; Piga *et al.*, 2008). Among the biological properties of these phenolic compounds is prevention or delay of degenerative diseases caused by oxidative stress, such as cancer, atherosclerosis, and cardiovascular disease (Watson and Preedy, 2010).

Breba and fig fruits are among the most popular fruits throughout the world, and in recent years, their consumption has gained attention due to their high content of phytochemicals; for this reason, they are considered to be a "superfruit." Both brebas and figs can be consumed fresh, either peeled or not, or dried and processed into jam, vinegar, or energy bars, increasing the economic benefits of this species compared with other fruits. The presence of phytochemicals in both figs and brebas is related to their antioxidant potential (Watson and Preedy, 2010). For this reason, several studies have been carried out to demonstrate the health-promoting abilities of figs and brebas, although the majority have focused on analysis solely at harvest time (Veberic *et al.*, 2008), with the exception of a few studies carried out by Crisosto *et al.* (2010) and Pereira *et al.* (2017c) who studied the influence of ripening stage on the content of bioactive compounds. Additionally, many other factors influence the phytochemicals in brebas and figs such as genotype, weather conditions, packaging, and storage (Vallejo *et al.*, 2012).

In general, phytochemicals or phenolic compounds can be divided into two groups: non-flavonoids and flavonoids. The non-flavonoids include, among others, phenolic acids, lignans, and stilbenes, while flavonoids are divided into anthocyanins, flavanols, procyanidins, flavans, flavones, and isoflavones (Fig. 22.2) (Khatib and Vaya, 2010).

Flavonoids have a generic structure called the flavan nucleus that is hydroxylated or methoxylated and/or glycosylated. Several studies on the distribution of flavonoids in the fruit have shown that fruit skins are the major source of phenolic compounds compared with flesh, which is lower in phenolic compounds. There are up to 70–80% more flavonoids in the skin than in the flesh (Vallejo *et al.*, 2012; Veberic and Mikulic-Petkovsek, 2016; Pereira *et al.*, 2017c). In addition, black-skinned brebas and figs contain higher levels of bioactive compounds, with 80–463 mg 100 g^{-1} FW compared with green-skinned ones, which have 42–78 mg 100 g^{-1} FW (Solomon *et al.*, 2006; Pereira *et al.*, 2017c).

Phenolic acids

Among the phenolic acids present in both brebas and figs are hydroxybenzoic acids, hydroxycinnamic acids, and ellagic acid. In the hydroxybenzoic acids, gallic acid is the dominant compound at 0.38–2.8 mg 100 g^{-1} FW (Veberic *et al.*, 2008; Khatib and Vaya, 2010). Within the hydroxycinnamic acids, chlorogenic acid is present at the highest concentration, especially in the skin, but is also considered the main phenolic acid. Its concentrations vary widely from 0.6 to 6.9 mg 100 g^{-1} FW, depending on the variety and ripening stage. According to Pereira *et al.* (2017c), chlorogenic acid seems to be produced at a higher concentration in green-skinned varieties, and its concentration increases slightly with the ripening process. Also noteworthy is the presence

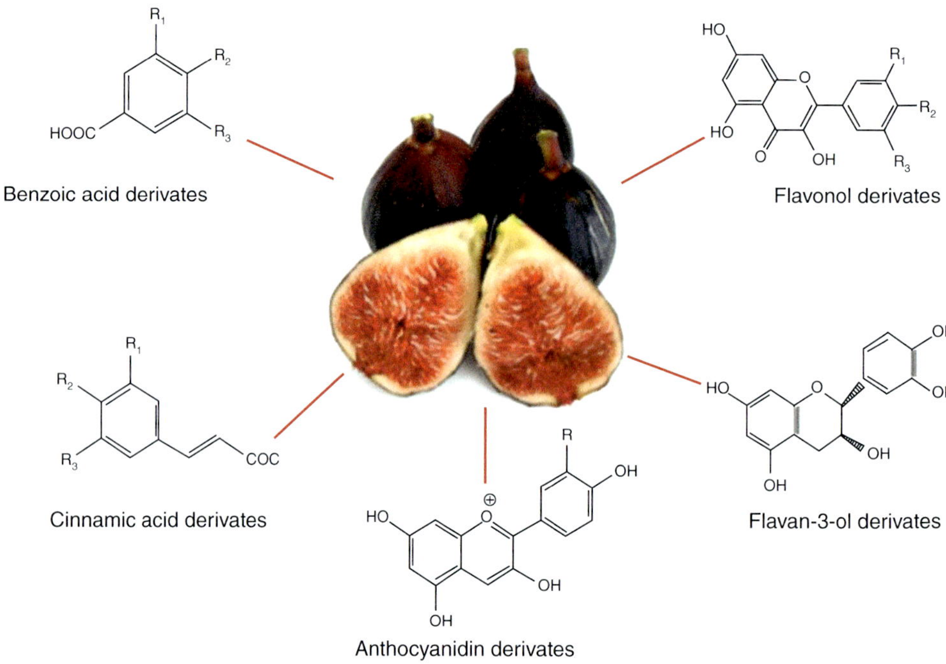

Fig. 22.2. The main phytochemical compounds present in breba and main-crop figs. Drawn using ACD/ChemSketch, ACDLabs).

of ellagic acid, at 0.4–2.7 mg 100 g^{-1} FW, depending on the variety and ripening stage, and whether skin or flesh (Veberic *et al.*, 2008; Pereira *et al.*, 2017c).

Flavonoids

ANTHOCYANINS Large groups of bioactive compounds or flavonoids are anthocyanins. Anthocyanins are hydrophilic pigments and are present in the vacuoles of the plant cell (Valero and Serrano, 2010). They determine the color of black-skinned fig varieties. Nine anthocyanins have been identified in the skin of dark varieties, with only two of them also present in green varieties. The main anthocyanins are cyanidin 3-*O*-rutinoside and cyanidin 3,5-diglucoside, which represent between 48% and 97% of the total anthocyanins identified in figs, depending on the variety and fruit part (Khatib and Vaya, 2010; Vallejo *et al.*, 2012). Cyanidin 3-*O*-glucoside and pelargonidin 3-*O*-rutinoside have also been identified (Pereira *et al.*, 2017c). The concentration of anthocyanins in general is higher in the skin than in the fig flesh, ranging from 12.19 to 97.4 mg 100 g^{-1} FW for cyanidin 3-*O*-rutinoside, while cyanidin 3-*O*-glucoside ranges from 0.12 to 6.39 mg 100 g^{-1} FW and pelargonidin 3-*O*-rutinoside from 0.09 to 5.64 mg 100 g^{-1} FW. The importance of these pigments, apart from determining the fruit color, is their strong antioxidant activity in metabolic reactions due to their great ability to neutralize free oxygen radicals as well as other reactive species (Solomon *et al.*, 2006). For dried figs, the cyanidin-3-*O*-rutinoside content is reduced during the sun-drying process (Slatnar *et al.*, 2011).

FLAVONOLS Another important group of flavonoids is the flavonols, which includes kaempferol 3-*O*-glucoside, kaempferol 3-*O*-rutinoside, quercetin 3-*O*-glucoside, quercetin 3-*O*-rutinoside (rutin), and quercetin 3-acetylglucoside. High concentrations of rutin were described for dark-skinned figs and it accumulated during ripening with values of 0.29–11.9 mg 100 g^{-1} FW in skin and values from 0.5–1 mg 100 g^{-1} FW in flesh (Veberic *et al.*, 2008; Pereira *et al.*, 2017c). Furthermore, Slatnar *et al.* (2011) found that sun drying leads to further accumulation of these compounds. Thus, dried figs have greater concentrations of rutin than fresh figs.

FLAVAN-3-OLS Brebas and figs also stand out for their flavan-3-ol content, especially their catechin and epicatechin content, as these compounds are well known for their beneficial biological activity (Johnston *et al.*, 2003; Yilmaz and Toledo, 2004; Singh *et al.*, 2011).

Like other phenolic compounds, concentrations depend on genotype and tissue. In the case of catechin, its concentration ranges from 0.08 to 17.1 mg 100 g^{-1} FW in the skin and from 2.0 to 21.2 mg 100 g^{-1} FW in the flesh. Epicatechin concentrations range from 4.8 to 16.9 mg 100 g^{-1} FW in skin, while the flesh values range from 2.2 to 6.8 mg 100 g^{-1} FW (Veberic *et al.*, 2008; Khatib and Vaya, 2010; Pereira *et al.*, 2017c). Again, the highest concentrations were observed in the skin of dark figs or brebas. Additionally, epicatechin has been shown to be the most abundant flavonoid in dried figs, with levels up to threefold higher than in fresh figs (Slatnar *et al.*, 2011).

22.4.3 Antioxidant activity

The total antioxidant capacity of bioactive compounds present in brebas and figs depends on their ability to scavenge reactive compounds generated during metabolic reactions. In general, dark-skinned figs and brebas have a higher total antioxidant capacity than green-skinned ones (Solomon *et al.*, 2006; Pereira *et al.*, 2017c). This is due to a higher concentration of bioactive compounds in these varieties. Thus, fig skin shows greater antioxidant activity than fig flesh, between two and ten times more, depending on the genotype and ripening stage. In this respect, it is advised to consume whole brebas and figs instead of peeled fruits (Pereira *et al.*, 2017c). Additionally, the hydrophilic fraction is characterized by a higher antioxidant capacity than the lipophilic fraction. This hydrophilic antioxidant activity is related to the presence of anthocyanins, while the lipophilic fraction is related to the presence of carotenoids (Valero and Serrano, 2010). According to Pereira *et al.* (2017c), the hydrophilic antioxidant activity in dark-skinned varieties such as Colar Elche and Cuello Dama Negro (Black Mission types) ranges between 177.4 and 109.4 mmol Trolox equivalents 100 g^{-1} FW, respectively, while brown- or green-skinned varieties, in general, show higher lipophilic antioxidant activity at the skin level, between 8.9 and 13.3 mmol Trolox equivalents 100 g^{-1} FW, than dark-skinned varieties.

22.5 Postharvest Factors Affecting Quality and Nutritional Compounds

Fresh figs present a challenge for postharvest technology. The fruit is highly perishable, even when harvested at the proper stage of ripening. Optimal harvest maturity is specific to each variety but is normally assigned to the beginning of phase III, as determined by changes in color and firmness (see discussion above, section 22.2). Fruit picked before optimal maturity will not reach its desired flavor. However, fruit harvested too late, after its prime state, has an extremely short postharvest life due to over-ripening, expressed as excessive softening.

The fig is categorized as a climacteric fruit with changing ethylene elevations throughout its growth phases. The ethylene production rate is relatively high at the beginning of phase I. It declines later although still during this stage, and rises again, before or in parallel with the onset of ripening at phase III. Before this spontaneous climacteric rise, in phase II, the fig fruit demonstrates an autocatalytic increase in ethylene production in response to on-tree ethylene application.

An inhibitor of ethylene action, 1-methylcyclopropene (1-MCP), binds to ethylene receptors so that ethylene perception and its effects on the treated tissue are blocked (Sisler and Serek, 2003). The use of 1-MCP has successfully delayed postharvest deterioration of various fruits, such as kiwifruit, apple, plum, and tomato (Martínez-Romero *et al.*, 2007). However, with figs picked at the commercial-maturity stage, postharvest 1-MCP application was ineffective or had only a limited effect, expressed as some retardation of softening (Gözlekçi *et al.*, 2005).

The application of preharvest 1-MCP (at 5 ppm) inside bags allowed concentration of the yield on the tree at the commercial-maturity stage, retarding the fruit advancement to full ripeness and thus reducing the need for labor-consuming frequent harvesting (Freiman *et al.*, 2012). The postharvest storage demonstrated that 1-MCP-treated figs improved keeping quality with up to a 7-day extension of storage life. During storage, senescence and spoilage of the 1-MCP-treated figs were slower and lower than in controls, as manifested by fruit color, firmness, internal texture, weight, size, shriveling, and decay. Thus, 1-MCP applied on fig trees at the fig's pre-climacteric stage can delay the senescence of Brown Turkey figs and improve their storage and transportation potential.

In the case of breba, some postharvest treatments such as 1-MCP and heat (45° for 5 min) with previous ethrel treatment or no treatment showed benefits in terms of quality after 14 days of storage at 2°C such as firmness retention and total acidity (Fig. 22.3A). In this experiment, during postharvest, levels of ethylene

A.

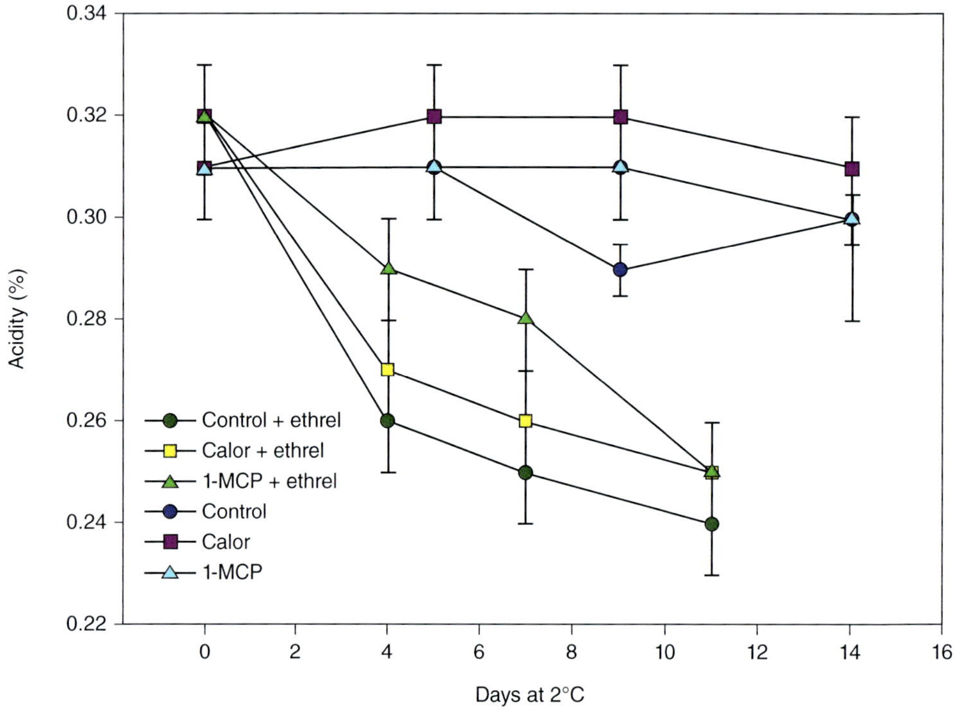

B.

Fig. 22.3. Heat and 1-methylcyclopropene (1-MCP) treatments in breba figs. A. Total acidity after heat or 1-MCP treatments in breba figs from control or ethrel-treated fruits. B. Visual aspect after 12 days of storage under control conditions (1), after heat treatment (calor) (2), and after 1-MCP treatment (3).

remained unchanged from those obtained at harvest (2 nl g^{-1} h^{-1}) (Fig. 22.3B).

An edible coating is an effective method to solve the problem of postharvest losses and provides a protective covering for fruit and vegetables. Edible coatings are of different types including hydrocolloids, lipids, and plasticizers, and they have good barrier properties to oxygen (O_2), carbon dioxide (CO_2), moisture, and water vapor. Chitosan, which is derived from chitin, is an edible polymer found in crustacean animal shells. Application of an alginate–chitosan edible film on figs led to higher levels of bioactive compounds and antioxidant capacity (Reyes-Avalos *et al.*, 2019). The senescence process was delayed (O_2 decreased and CO_2 increased) and total (poly)phenols were preserved in coated figs, as well as the content of anthocyanins, in particular cyanidin 3-*O*-rutinoside.

A controlled atmosphere was studied as a potential tool to extend the shelf life of Ottoman-it figs with several regimes of O_2/CO_2. Controlled-atmosphere conditions of 5 kPa O_2 and 5 kPa CO_2 maintained fruit quality better than an ambient atmosphere due to improved firmness, integrity, and control of decay (Bahar and Lichter, 2018). In addition, it is known that postharvest abiotic stresses, such as low O_2 concentration and high CO_2 concentration, induce a greater accumulation of bioactive compounds (Cisneros-Zevallos, 2003). Thus, Villalobos *et al.* (2015) were able to verify in figs of the Albacor variety that those stored under passive modified atmosphere conditions through microperforated films showed an increase in phenolic compound content and consequently an increase in antioxidant activity in both skin and flesh (see Chapter 13, this volume).

22.6 Bioactivity in Human Health: Current State and Future Prospects

Despite the fact that figs have been exploited in traditional medicine and studies have proven their potential therapeutic values, few scientific articles have focused on the importance of their ingestion in relation to nutrition or potential bioactivity for human health. In this aspect, an interesting review has been published, in which chemical analysis of fresh and dried figs in relation to their antioxidant capacity and health effects was studied in figs cultivated in Algeria and Turkey, as well as in European countries such as Greece, Italy, Portugal and Spain (Arvaniti *et al.*, 2019).

The nutritional importance of fig fruits has classically been attributed to its vitamins, minerals, and fiber (Russo *et al.*, 2012), indicating that they are also a good source of B vitamins. The nutritional composition is summarized in the USDA Nutritional Database (2019), showing that figs are rich in proteins, amino acids, fiber, and sugars, with low total fat content, and that they are also rich in calcium, iron, magnesium, phosphorus, potassium, zinc, copper, manganese, and selenium, and low in sodium, with vitamins A, C, E, B_6 and K, thiamine, riboflavin, niacin, and pantothenic acid also present.

The fatty acid composition has been described in dried Tunisian fig (Khadhraoui *et al.*, 2019) and their by-products (Conforti *et al.*, 2012), describing the presence of mainly linoleic, palmitic, and stearic methyl esters, which favorably affect atherosclerosis, coronary heart disease, inflammatory disease, and perhaps even behavioral disorders (Connor, 2000).

Other bioactive compounds, such as phenolics or carotenoids, found in fresh and dried figs of different varieties have been characterized, but the real effect on human health was not studied extensively. Most studies have focused on the antioxidant capacity of figs, as measured *in vitro*. Only a few studies have covered the potential effects on human health, including acetyl cholinesterase inhibition, and antifungal, antimicrobial, anti-helminthic, and anti-carcinogenic activities, as discussed below.

Regarding carotenoids, as mentioned previously, figs produce β-carotene, lutein, and zeaxanthin (Amessis-Ouchemoukh *et al.*, 2017; USDA Nutritional Database, 2019), which are antioxidants. Yemis *et al.* (2012) indicated that lutein, zeaxanthin, β-cryptoxanthin, and β-carotene are carotenoids present in yellow fig varieties. However, approximately 80% of carotenoid compounds in these fig varieties degrade when dried, significantly reducing their health properties.

Nevertheless, phenolic compounds such as phenolic acids and flavonoids are the predominant bioactive phytochemical compounds found in figs, with the most abundant being quercetin 3-*O*-rutinoside (rutin), followed by catechin, epicatechin, gallic acid, chlorogenic acid, and syringic acid, as well as tannins (Veberic *et al.*, 2008, Soltana *et al.*, 2018; Benmaghnia *et al.*, 2019). At the same time, anthocyanins have been described in black species, where cyanidin 3-*O*-rutinoside is the color flavonoid present in higher concentration (Dueñas *et al.*, 2008; Vallejo *et al.*, 2012). This rich (poly)phenolic composition suggests that figs can be good candidates to prevent certain non-communicable diseases (e.g. cardiovascular diseases, diabetes, cancer, and stroke), as there is much epidemiological evidence regarding the protective effects of fruit and vegetables attributed, in part, to phenolic secondary metabolites (Crozier *et al.*, 2009).

The (poly)phenolic compound levels are influenced by several (related) factors, such as the part of the fruit, color, and ripening stage, as well as the sun-drying process, which will affect the intake of these bioactive compounds.

In this aspect, it is remarkable that these compounds are distributed in different proportions in skin and flesh, as rutin is the major compound present in the skin extract, accounting for over 30% of its phenolic content, followed by chlorogenic acid, vanillic acid, and, in minor amounts, taxifolin and apigenin. In contrast, caffeic acid derivatives are the major components of the flesh, where rutin and apigenin derivatives are present only in trace amounts (Palmeira *et al.*, 2019). It has also been shown that the leaves and skin present richer qualitative profiles compared with other parts of the plants, with rutin the main compound in both although at different relative amounts (Ammar, *et al.*, 2015). In addition, 15 anthocyanins have been characterized in the skin of dark varieties (Dueñas *et al.*, 2008). This indicates that ingestion of the whole fruit (as in dried figs), or flesh/skin by-products incorporated into a functional food would have different nutritional activity, which needs to be studied carefully.

It has been pointed out that dark-colored varieties contain higher concentrations of phenolic compounds in comparison with light-colored ones (Veberic *et al.*, 2008), which could be a good indicator of the health-promoting compound composition or concentration. This is due to the different distribution of these bioactive compounds within the fruit, where the flavonoids, including anthocyanins (characteristics of dark varieties), are located mainly in the skin (Vallejo *et al.* 2012).

Another important point is the fact that some authors have indicated that dried fruits are a better source of these (poly)phenols when ingested (Vinson *et al.*, 2005), although others have reported a loss of around 30% during fig-drying processes (Bey *et al.*, 2017). Likewise, contradictory results have been found in the literature regarding the effect of air drying and sun drying on the total content of phytochemical compounds, as well as on the concentrations of individual phenolic compounds (Villalobos *et al.*, 2016). These aspects should be considered, in order to recommend the consumption of fresh or dry figs, as, although dried fruits might have more available bioactive compounds, a general loss of these health-promoting compounds can occur during the drying process.

The antioxidant capacity, based on the correlation of phenolic compounds and antioxidant activity, has been well established (Selby-Pham, *et al.*, 2017; Gu *et al.*, 2019). These studies have been mainly carried out *in vitro* (e.g. using 2,2-diphenyl-2-picrylhydrazyl (DPPH), 2,2′-azino-*bis*-3-ethylbenzothiazoline-6-sulfonic acid (ABTS), ferric reducing antioxidant power (FRAP) assays), in different species (Harzallah *et al.*, 2016; Benmaghnia *et al.*, 2019), which indicates only an approximation of what can really happen in the human body after ingestion of figs. However, this activity was demonstrated by Vinson *et al.* (2005), who administered figs together with high-fructose corn syrup in carbonated soft drinks to volunteers, and found that figs produced a significant increase in plasma antioxidant capacity for 4 h after consumption, and overcame the oxidative stress of the ingestion of these highly sugary drinks.

The antimicrobial activity of these fruits has also been demonstrated against some Gram-positive and -negative bacteria, where the use of figs or their extracts is recommended against some microbial infections (Benmaghnia *et al.* 2019; Palmeira *et al.*, 2019).

Regarding the mechanism of action of different parts of figs, only a few articles have been published and were summarized by Arvaniti *et al.* (2019), mainly in cancer cell lines or in rats with different induced disorders, indicating positive promising effects. Further bioaccessibility or phytochemical release approaches for simulations of *in vitro* gastrointestinal digestion have been performed (Kamiloglu and Capanoglu, 2013, 2015), indicating that phenolic compounds are available after fig intake, but that this possible bioavailability depended on the type of fruit and whether it was dried or fresh, with better results in fresh fruits than in dried ones.

Concerning the interesting potential bioactive composition, much more work needs to be done in order to prove the real health effects in humans, including studies *in vivo* or with volunteers, against some specific disorders such as cardiovascular diseases, cancer, and Alzheimer's disease.

The real activity of by-products/side-products should also be investigated, together with other parts of the tree (e.g. leaves, bark, or roots), which could validate the waste material as a good source of nutritional/bioactive ingredients, which could then be added to certain foods as functional food ingredients.

22.7 Concluding Remarks

The fig (*Ficus carica* L.) is famous for its nutritive values and is consumed both fresh and dried all over the world. Fig fruit is considered a functional food due to its unique chemical properties and rich nutritional content. Fig varieties show notable differences on the basis of the physicochemical and sensory properties. This confirms the relevance of characterizing fig varieties for the assessment of potential consumer acceptability. The differences are mainly related to the variety, environmental conditions, horticultural practices, and storage conditions. Further research is needed to isolate and characterize the functional components of fig. It will be valuable to make the connection between fig harvest time, storage, beneficial compounds, and medicinal modern science.

References

Aksoy, U., Balci, B., Can, H.Z., and Hepaksoy, B. (2003) Some significant results of the research-work in Turkey on fig. *Acta Horticulturae* 605, 173–180.

Aljane, F. and Ferchichi, A. (2009) Postharvest chemical properties and mineral contents of some fig (*Ficus carica* L.) cultivars in Tunisia. *Journal of Food, Agriculture and Environment* 7: e212.

Allegra, A., Gallotta, A., Carimi, F., Mercati, F., Inglese, P., and Marinelli, F. (2018) Metabolic profiling and post-harvest behavior of "Dottato" Fig (*Ficus carica* L.) fruit covered with an edible coating from *O. ficus-indica*. *Frontiers in Plant Science* 9: 1321.

Amessis-Ouchemoukh, N., Ouchemoukh, S., Meziant, N., Idiri, Y., Hernanz, D., *et al.* (2017) Bioactive metabolites involved in the antioxidant, anticancer and anticalpain activities of *Ficus carica* L., *Ceratonia siliqua* L. and *Quercus ilex* L. extracts. *Industrial Crops and Products* 95, 6–17.

Ammar, S., Del Mar Contreras, M., Belguith-Hadrich, O., Segura-Carretero, A., and Bouaziz, M. (2015) Assessment of the distribution of phenolic compounds and contribution to the antioxidant activity in Tunisian fig leaves, fruits, skins and pulps using mass spectrometry-based analysis. *Food and Function* 6, 3663–3677.

Arvaniti, O.S., Samaras, Y., Gatidou, G., Thomaidis, N.S., and Stasinakis, A.S. (2019) Review on fresh and dried figs: chemical analysis and occurrence of phytochemical compounds, antioxidant capacity and health effects. *Food Research International* 119, 244–267.

Bahar, A. and Lichter, A. (2018) Effect of controlled atmosphere on the storage potential of Ottomanit fig fruit. *Scientia Horticulturae* 227, 196–201.

Benmaghnia, S., Meddah, B., Tir-touil, A., and Gabaldón Hernández, J.A. (2019) Phytochemical analysis, antioxidant and antimicrobial activities of three samples of dried figs (*Ficus carica* L.) from the region of Mascara (western Algeria). *Journal of Microbiology, Biotechnology and Food Sciences* 9, 208–215.

Bey, M.B., Richard, G., Meziant, L., Fauconnier, M.L., and Louaileche, H. (2017) Effects of sun-drying on physicochemical characteristics, phenolic composition and *in vitro* antioxidant activity of dark fig varieties. *Journal of Food Processing and Preservation* 41: e13164.

Çalişkan, O. and Polat, A.A. (2011) Phytochemical and antioxidant properties of selected fig (*Ficus carica* L.) accessions from the eastern Mediterranean region of Turkey. *Scientia Horticulturae* 128, 473–478.

Can, H.Z. (1993) The investigation of same horticultural characteristics of same selected fig genotypes in Aegean Region. MSc Thesis, Ege University, Izmir, Tukey.

Cisneros-Zevallos, L. (2003) The use of controlled postharvest abiotic stresses as a tool for enhancing the nutraceutical content and adding-value of fresh fruits and vegetables. *Journal of Food Science* 68, 1560–1565.

Conforti, F., Menichini, G., Zanfini, L., Tundis, R., Statti, G.A., *et al.* (2012) Evaluation of phototoxic potential of aerial components of the fig tree against human melanoma. *Cell Proliferation* 45, 279–285.
Connor, W.E. (2000) Importance of n-3 fatty acids in health and disease. *American Journal of Clinical Nutrition* 71, 171S–175S.
Crisosto, C.H., Bremer, V., Ferguson, L., and Crisosto, G.M. (2010) Evaluating quality attributes of four fresh fig (*Ficus carica* L.) cultivars harvested at two maturity stages. *HortScience* 45, 707–710.
Crisosto, C.H., Ferguson, L., Preece, J.C., Michailides, T.J., Haug, M.T., *et al.* (2017) Developing the California fresh fig industry. *Acta Horticulturae* 1173, 285–292.
Crozier, A., Jaganath, I.B., and Clifford, M.N. (2009) Dietary phenolics: chemistry, bioavailability and effects on health. *Natural Product Reports* 26, 1001–1043.
Dueñas, M., Pérez-Alonso, J.J., Santos-Buelga, C., and Escribano-Bailón, T. (2008) Anthocyanin composition in fig (*Ficus carica* L.). *Journal of Food Composition and Analysis* 21, 107–115.
Ferguson, L., Michailides, T.J., and Shorey, H.H. (1990) The California fig industry. *Journal of the American Society for Horticultural Science* 12, 409–490.
Flaishman, M.A., Rover, V., and Stover, E. (2008) The fig: botany, horticulture, and breeding. *Horticultural Reviews* 34, 113–197.
Freiman, Z.E., Rodov, V., Yablovitz, Z., Horev, B., and Flaishman, M.A. (2012) Preharvest application of 1-methylcyclopropene inhibits ripening and improves keeping quality of ‘Brown Turkey’ figs (*Ficus carica* L.). *Scientia Horticulturae* 138, 266–272.
Gaaliche, B., Trad, M., and Mars, M. (2011) Effect of pollination intensity, frequency and pollen source on fig (*Ficus carica* L.) productivity and fruit quality. *Scientia Horticulturae* 130, 737–742.
Gaaliche, B., Trad, M., Hfaiedh, L., Lakhal, W., and Mars, M. (2012) Pomological and biochemical characteristics of fig (*Ficus carica* L.) cv. Zidi in different agro-ecological zones of Tunisia. *Pakistan Journal of Agricultural Sciences* 49, 425–428.
Gözlekçi, S., Erkan, M., Karaşahin, I., and Şahin, G. (2005) Effect of 1-methylcyclopropene (1-MCP) on fig (*Ficus carica* cv. Bardakci) storage. *Acta Horticulturae* 798, 325–330.
Gu, C., Howell, K., Dunshea, F.R., and Suleria, H.A. (2019) LC-ESI-QTOF/MS characterisation of phenolic acids and flavonoids in polyphenol-rich fruits and vegetables and their potential antioxidant activities. *Antioxidants Open Access* 8, 405–419.
Harzallah, A., Bhouri, A.M., Amri, Z., Soltana, H., and Hammami, M. (2016) Phytochemical content and antioxidant activity of different fruit parts juices of three figs (*Ficus carica* L.) varieties grown in Tunisia. *Industrial Crops and Products* 83, 255–267.
Johnston, K.L., Clifford, M.N., and Morgan, L.M. (2003) Coffee acutely modifies gastrointestinal hormone secretion and glucose tolerance in humans: glycemic effects of chlorogenic acid and caffeine. *American Journal of Clinical Nutrition* 78, 728–733.
Kamiloglu, S. and Capanoglu, E. (2013) Investigating the *in vitro* bioaccessibility of polyphenols in fresh and sun-dried figs (*Ficus carica* L.). *International Journal of Food Science and Technology* 48, 2621–2629.
Kamiloglu, S. and Capanoglu, E. (2015) Polyphenol content in figs (*Ficus carica* L.): effect of sun-drying. *International Journal of Food Properties* 18, 521–535.
Khadhraoui, M., Bagues, M., Artés, F., and Ferchichi, A. (2019) Phytochemical content, antioxidant potential, and fatty acid composition of dried Tunisian fig (*Ficus carica* L.) cultivars. *Journal of Applied Botany and Food Quality* 92, 143–150.
Khan, M.N., Sarwar, A., Adeel, M., and Wahab, M.F. (2011) Nutritional evaluation of *Ficus carica* indigenous to Pakistan. *African Journal of Food, Agriculture, Nutrition and Development* 11, 5187–5202.
Khatib, S. and Vaya, J. (2010) Fig, carob, pistachio, and health. In: Watson, R.R. and Preedy, V.R. (eds) *Bioactive Foods in Promoting Health: Fruits and Vegetables*. Academic Press, London, pp. 245–271.
Küden, A.B., Bayazit, S., and Çömlekcioglu, S. (2008) Morphological and pomological characteristics of fig genotypes selected from Mediterranean and South east Anatolia regions. *Acta Horticulturae* 798, 95–102.
Lee, S.K. and Kader, A.A. (2000) Preharvest and postharvest factors influencing vitamin C content of horticultural crops. *Postharvest Biology and Technology* 20, 207–220.
Mars, M., Gaaliche, B., Ouerfelli, I., and Chouat, S. (2009) Systèmes de production et ressources génétiques du figuier (*Ficus carica* L.) à Djebba et Kesra, deux villages de montagne au nord ouest de la Tunisie. *Revue des Régions Arides* 22, 33–45.
Martínez-Romero, D., Bailen, G., Serrano, M., Guillén, F., Valverde, J.M., *et al.* (2007) Tools to maintain postharvest fruit and vegetable quality through the inhibition of ethylene action: a review. *Critical Reviews in Food Science and Nutrition* 47, 543–560.

Melgarejo, P., Hernandez, F., Martínez Nicolás, J.J., Sánhez, J., and Salazar, D.M. (2003) Organic acids and sugars from first and second crop fig juices. *Acta Horticulturae* 605, 237–239.

Messaoudi, Z. and Haddadi, L. (2008) Morphological and chemical characterization of fourteen fig trees cultivated in Oulmès Area, Morocco. *Acta Horticulturae* 798, 83–86.

Palmeira, L., Pereira, C., Dias, M.I., Abreu, R.M.V., Correa, R.C.G., *et al.* (2019) Nutritional, chemical and bioactive profiles of different parts of a Portuguese common fig (*Ficus carica* L) variety. *Food Research International* 126, 108572–108581.

Pereira, C., Serradilla, M.J., Martín, A., Villalobos, M.C., Pérez-Gragera, F., and López-Corrales, M. (2015) Agronomic behaviour and quality of six fig cultivars for fresh consumption. *Scientia Horticulturae* 185, 121–128.

Pereira, C., López-Corrales, M., Martín, A., Villalobos, M.C., Córdoba, M.G., and Serradilla, M.J. (2017a) Physicochemical and nutritional characterization of brebas for fresh consumption from nine fig varieties (*Ficus carica* L.) grown in Extremadura (Spain). *Journal of Food Quality* 2017: 6302109.

Pereira, C., Serradilla, M.J., Pérez-Gragera, F., Martín, A., Villalobos, M.C., and López-Corrales, M. (2017b) Evaluation of agronomic and fruit quality traits of fig tree varieties (*Ficus carica* L.) grown in Mediterranean conditions. *Spanish Journal of Agricultural Research* 15: e0903.

Pereira, C., López-Corrales, M., Serradilla, M.J., Villalobos, M.C., Ruiz-Moyano, S., and Martín, A (2017c) Influence of ripening stage on bioactive compounds and antioxidant activity in nine fig (*Ficus carica* L.) varieties grown in Extremadura, Spain. *Journal of Food Composition and Analysis* 64, 203–212.

Piga, A., Del Caro, A., Milella, G., Pinna, I., Vacca, V., and Schirru, S. (2008) HPLC analysis of polyphenols in peel and pulp of fresh figs. *Acta Horticulturae* 798, 301–306.

Pisimisi, E., Vemmos, S.N., and Mili, E. (2012) The photosynthetic activity and evaluation of fruit quality in seven fig cultivars (*Ficus carica* L.). *Acta Horticulturae* 940, 341–348.

Rahemi, M. and Jafari, M. (2008) Effect of caprifig type on quantity and quality of Estahban dried fig *Ficus carica* cv.*Sabz.* *Acta Horticulturae* 798, 249–252.

Reyes-Avalos, M.C., Minjares-Fuentes, R., Femenia, A., Contreras-Esquivel, J.C., Quintero-Ramos, A., *et al.* (2019) Application of an alginate–chitosan edible film on figs (*Ficus carica*): effect on bioactive compounds and antioxidant capacity. *Food and Bioprocess Technology* 12, 499–511.

Russo, M., di Sanzo, R., Cefaly, V., Fuda, S., Postorino, S., and Suraci, F. (2012) Quality and nutritional composition of dried figs: a traditional Calabria Region (Italy) product. *Acta Horticulturae* 939, 189–196.

Selby-Pham, S., Cottrell, J., Dunshea, F., Ng, K., Bennett, L., and Howell, K. (2017) Dietary phytochemicals promote health by enhancing antioxidant defence in a pig model. *Nutrients* 9: 758.

Singh, B.N., Shankar, S., and Srivastava, R.K. (2011) Green tea catechin, epigallocatechin-3-gallate (EGCG): mechanisms, perspectives and clinical applications. *Biochemical Pharmacology* 82, 1807–1821.

Sisler, E.C. and Serek, M. (2003) Compounds interacting with the ethylene receptor in plants. *Journal of Plant Biology* 5, 473–480.

Slatnar, A., Klancar, U., Stampar, F., and Veberic, R. (2011) Effect of drying of figs (*Ficus carica* L.) on the contents of sugars, organic acids, and phenolic compounds. *Journal of Agricultural and Food Chemistry* 59, 11696–11702.

Solomon, A., Golubowicz, S., Yablowicz, Z., Grossman, S., Bergman, M., *et al.* (2006) Antioxidant activities and anthocyanin content of fresh fruits of common fig (*Ficus carica* L.). *Journal of Agricultural and Food Chemistry* 54, 7717–7723.

Soltana, H., de Rosso, M., Lazreg, H., Vedona, A.D., Hammami, M., and Flamini, R. (2018) LC-QTOF characterization of non-anthocyanic flavonoids in four Tunisian fig varieties. *Journal of Mass Spectrometry* 53, 817–823.

Stover, E., Aradhya, M., Ferguson, L., and Crisosto, C.H. (2007) The fig: overview of an ancient fruit. *Hortscience* 42, 1083–1087.

Tomás-Barberán, F.A. and Espín, J.C. (2001) Phenolic compounds and related enzymes as determinants of quality in fruits and vegetables. *Journal of Science and Food Agricultural* 81, 853–876.

Trad, M., Ginies, C., Gaaliche, B., Renard, C.M.G.C., and Mars, M. (2012) Does pollination affect aroma development in ripened fig (*Ficus carica* L.) fruit. *Scientia Horticulturae* 134, 93–99.

USDA Nutritional Database (2019) Figs, raw. Available at: https://fdc.nal.usda.gov/fdc-app.html#/food-details/173021/nutrients (accessed December 22, 2021).

Valero, D. and Serrano, M. (2010) Fruit ripening. In: Valero, D. and Serrano, M. (eds) *Postharvest Biology and Technology for Preserving Fruit Quality*. CRC Press, Boca Raton, Florida, pp. 7–47.

Vallejo, F., Marin, J.G., and Tomas-Barberan, F.A. (2012) Phenolic compound content of fresh and dried figs (*Ficus carica* L.). *Food Chemistry* 130, 485–492.

Veberic, R. and Mikulic-Petkovsek, M. (2016) Phytochemical composition of common fig (*Ficus carica* L.) cultivars. In: Simmonds, M.S.J. and Preedy, V.R. (eds) *Nutritional Composition of Fruit Cultivars*. Academic Press, London, pp. 235–255.

Veberic, R., Colaric, R., and Stampar, F. (2008) Phenolic acids and flavonoids of fig fruit (*Ficus carica* L.) in the northern Mediterranean region. *Food Chemistry* 106, 153–157.

Vemmos, S.N., Petri, E., and Stournaras, V. (2013) Seasonal changes in photosynthetic activity and carbohydrate content in leaves and fruit of three fig cultivars (*Ficus carica* L.). *Scientia Horticulturae* 160, 198–207.

Villalobos, M.C., Martín, A., Ruiz-Moyano, S., Martín, E., Córdoba, M.G., and Serradilla, M.J. (2015) Effect of modified atmosphere packaging on the antioxidant activity and total phenolic content in 'Albacor' figs. *Acta Horticulturae* 1079, 573–580.

Villalobos, M.C., Serradilla, M.J., Martín, A., Pereira, C., López-Corrales, M., and Córdoba, M.G. (2016) Evaluation of different drying systems as an alternative to sun drying for figs (*Ficus carica* L). *Innovative Food Science and Emerging Technologies* 36, 156–165.

Vinson, J. (1999) The functional food properties of figs. *Cereal Foods World* 44, 82–87.

Vinson, J.A., Zubik, L., Bose, P., Samman, N., and Proch, J. (2005) Dried fruits: excellent *in vitro* and *in vivo* antioxidants. *Journal of the American College of Nutrition* 24, 44–50.

Watson, R.R. and Preedy, V.R. (2010) *Bioactive Foods in Promoting Health: Fruits and Vegetables*. Academic Press, London.

Yemis, O., Bakkalbasi, E., and Artik, N. (2012) Changes in pigment profile and surface colour of fig (*Ficus carica* L.) during drying. *International Journal of Food Science and Technology* 47, 1710–1719.

Yilmaz, Y. and Toledo, R.T. (2004) Major flavonoids in grape seeds and skins: antioxidant capacity of catechin, epicatechin, and gallic acid. *Journal of Agricultural and Food Chemistry* 52, 255–260.

Part VI

Fig Varieties and Breeding

23 Fig Varieties

Uygun Aksoy[1]* and Moshe A. Flaishman[2]
[1]Department of Horticulture, Faculty of Agriculture, Ege University, Bornova, İzmir, Turkey; [2]Institute of Plant Sciences, Agricultural Research Organization, Rishon LeZion, Israel

23.1 Introduction

The common fig, *Ficus carica* L., is widely dispersed in wild areas around the Mediterranean as female or male forms that have progressed from natural seedlings. Both the male and the edible female figs display a wide diversity regarding fruit size, skin color, taste, and maturity period, and are named after these peculiarities or from where they have originated. Figs have received names from time immemorial, the earliest record being by Theophrastus (372–287 BCE). Pliny (23–79 CE) mentioned 29 fig varieties described with their skin color although without a full description (Condit, 1955). Furthermore, he explained caprification and mentioned fig trees that set fruit without pollination (Ülkümen *et al.*, 1948).

In all fig-producing regions of the world, you can find fig varieties named black fig, yellow fig, winter fig, Mission, Fichi di Cosenza, Fico Bianco del Cilento, Brown Turkey, or Bursa Black. As fig multiplies easily by cuttings, today's commercial varieties are mostly those that were selected by farmers and then propagated by others who appreciated their properties. As the years passed, these types were renamed either after the region of origin or after regions where they were introduced, leading to an abundance of synonyms. In addition, there are improved versions of these selections and varieties released from breeding programs. The commercial fig varieties today grown in or around these gene centers are fig varieties that require pollination, mostly Smyrna types, whereas toward the west and east, more common-type figs were introduced.

Starting from the 19th century, efforts were made to identify and classify fig varieties, and these were compiled in a monograph by Condit (1955), who described two important initiatives, one by Suffre de Salerno, who described 360 Italian fig varieties, and one by Geny, who compiled colored plates of French fig varieties, both of which stayed as drafts and were not published. Among the descriptive work since then was that by Columella (4–70 CE) mentioning eight Italian fig varieties with their local names, the French horticulturist La Quintinie (1626–1688), describing nine French fig varieties, and Noisette (1829) describing 37 French fig varieties. In England, where figs were grown in protected backyards, Hogg (1866) identified 66 figs with their synonyms and Barron (1891) described these 66 figs present in the Royal Horticultural Society hothouses in Chiswick, UK (Condit, 1941).

Eisen (1897) classified edible fig varieties under four groups: caprifigs, Smyrna figs, common

* Email: uygun.aksoy@gmail.com

©CAB International 2022. *Advances in Fig Research and Sustainable Production* (M.A. Flaishman and U. Aksoy)
DOI: 10.1079/9781789242492.0023

edible figs, and San Pedro figs. He also stated that the confusion about figs was chiefly due to incomplete description of the varieties, and he briefly described 27 fig varieties then present in the USA. Earle (1897) described confusion about names of fig varieties and referenced Celeste as the favorite variety in all parts of the US Gulf Coast states. He also listed 18 fig varieties that the American Pomological Society recommended for commercialization with their pomological characteristics such as ripening season, color, size, and quality of fruit. Eisen (1901) a few years later published a comprehensive book on figs with an introductory part on how to describe figs, and compiled a long list of fig varieties comprising 348 common figs, ten Smyrna figs, and 19 caprifigs. Some of the descriptions were detailed, some were cited only as a name, and some were accompanied by illustrations of the leaf or fruit. He tried to differentiate between distinct varieties and synonyms. The listed fig varieties were from France, the USA, Italy, Algeria, Portugal, Spain, Japan, Greece, and Dalmatia (Croatia and Montenegro). Where possible, he mentioned the region or locality where the variety originated. Vallese (1909) described nine caprifigs and 38 female fig varieties from Italy, and listed an additional 94 other Italian fig varieties (Condit, 1941).

Condit (1941), following in the footsteps of this previous work, published a book on descriptive terms and characterization of the varieties in 1941, and then published his monograph (Condit, 1955). The monograph included not only the fig varieties then present in the USA but also figs from Algeria, China, France, Greece, Japan, Portugal, Spain, and Turkey. In the monograph, the fig varieties were described under four categories: caprifig, Smyrna type, San Pedro type, and common type (Condit, 1955).

Ülkümen *et al.* (1948) identified 37 major Turkish fig varieties for fresh consumption or for sun drying, and provided information about the place of origin, fruit size, color, taste, and method of utilization. For caprifig varieties, seven Turkish and three Californian male varieties such as Roeding and Stanford (from the Aegean Region of Turkey), and Milco (from Italy), were described.

Arendt (1972) provided a pomological description of the caprifig and female fig varieties present at the Nikita Botanical Garden (Yalta, Crimea) and also those from Georgia and Central Asia. She also provided basic information regarding methods on how to carry out studies on varieties, disease resistance, floral biology, and their suitability for different ways of utilization including industrial use. The Appendix contains black and white photographs of leaves and fruits of 120 fig varieties.

Aksoy and Şahin (2001) published a catalog of Turkish fig varieties and included detailed descriptions and colored plates of ten caprifigs and 27 edible female fig varieties accepted as standard varieties. The female varieties were those found to be fit for recommendation out of a collection of 272 varieties.

In most fig-producing countries, fig is significant at regional level due to the strong link between the species or variety and the microclimatic conditions, traditions, culture, and gastronomy of the region. There are regional collections and descriptions in Puglia, Italy, and Majorca, Spain. The synonyms, botanical and fruit properties, and local distribution of 25 fig varieties in Puglia have been described and coupled with colored photographs of the leaf and fruit (Biscotti *et al.*, 2022). Boscana (2009) gave a comprehensive description of the tree growth, fruit characteristics, maturation, distribution, and history of 89 fig varieties present at the Son Mut Nou experiment station in Llucmajor, Majorca, Spain.

This chapter gives a brief overview of the efforts toward collection and description of fig varieties, and presents the major characteristics of fig varieties from different origins that are traded worldwide.

23.2 *Ex Situ* Collections

The diversity among wild and cultivated female and male fig populations enables variety improvement by screening the population and selecting better-performing plants for direct multiplication or for using in breeding. This selection generally has a specific objective such as high productivity, resistance to extreme conditions such as drought, high or low temperatures, pest and diseases, and/or fruit properties such as persistence, ripening period, nutritive and health properties, shelf life, and transportation. In most countries where fig has a commercial value, wild or planted common fig (*F. carica*) trees are being screened and evaluated in their natural habitats. The promising ones are also collected and preserved in ex situ collections and several papers and reports are released. These collection orchards are mostly created as a national repository; likewise, very few collections have occurred due to personal or communal interest. Sometimes, a core collection is enlarged with new introductions over time. The major problems with the early collections were proper identification, and too many homonyms, synonyms and/or misspelling of the variety names. However, with the development of new molecular tools, they can be screened, re-evaluated, renamed, and/or regenerated, leading to better management. Germplasm is an integral part of the genetic improvement program in any crop species. Fig germplasm collections with 25 or more accessions are listed in Table 23.1.

The Nikita Botanical Garden in Yalta, Crimea, established in 1812, is one of the oldest collections of fig trees. The old collection was renewed in 1925. It hosts 334 accessions collected by Nina K. Arendt from the local population and those brought from Azerbaijan, Bulgaria, Georgia, Turkey, Tunisia, France, Israel, Italy, the USA, Germany, and Yugoslavia. The Nikita Botanical Gardens has accessions from four species of the genus *Ficus*:—*F. carica* L., *F. palmata* Forssk., *F. pseudo-carica* Miq., and *F. virgata* Reinw. ex Blume—and hybrid forms obtained by breeding programs using *F. carica* either at the Botanical Garden or abroad (Yezhov *et al.*, 2005; Mitrofanova *et al.*, 2016).

In 1894, the US Department of Agriculture introduced scions of 66 fig varieties from the Royal Horticultural Society's collection in the UK and grafted these on to fig trees in the California Nursery Company (Condit, 1955). The National Clonal Germplasm Repository (NCGR) for Tree Fruit, Nut Crops, and Grapes in Davis (California) currently hosts 190 different accessions comprising "78 named fruiting cultivars, 44 regional selections from diverse locations, 40 advanced selections from plant breeders, 28 caprifigs, and a small number of species and hybrids" (www.ars-grin.gov/dav/; accessed December 23, 2021). The Repository carries out DNA microsatellite fingerprinting of fig accessions in order to provide proper identification (Stover and Aradhya, 2008).

In Spain, the fig collection in Badajoz (Extramadura, Spain) originally housed 220 fig accessions, 94% of which came from Spain. Later, accessions from the Canary Islands and Aragon were introduced, and upon screening 25 accessions having specificities were added. The fig germplasm is described according to the fig descriptors developed by the International Plant Genetic Resources Institute (IPGRI) and the International Centre for Advanced Mediterranean Agronomic Studies (CIHEAM) (IPGRI/CIHEAM 2003). Fingerprinting work continues as new accessions arrive.

In Portugal, there are currently two collections of fig trees, one located in Tavira, in the Algarve, belonging to the Regional Directorate for Agriculture and Fisheries in the Algarve, with 85 accessions, planted at 6 × 4 m, in 1994 and expanded in 2014 with the aim of focusing on material prospected in that region. The other collection is located in Alcobaça, belonging to the National Institute for Agricultural Research/Vieira Natividade National Fruit Research Station, with 35 accessions planted in 1992 and expanding as new accessions appear. The collection comprises some accessions from the Algarve region for fresh consumption (brebas and main crop), and also from the Torres Novas and Mirandela regions. Fig genetic material is characterized to solve the problems of homonymy and synonymy. For characterization, International Union for the Protection of New Varieties of plants (UPOV) descriptors are being used in the case of the Algarve collection and IPGRI descriptors in the case of the Alcobaça collection (www.ine.pt; accessed December 23, 2021).

The Turkish national fig germplasm collection is hosted at the Fig Research Institute in Aydın Province. The collection hosts 271 female fig accessions selected from different regions of Turkey, 86 clones of Sarılop (= Calimyrna), and 57 caprifigs. Selection studies were carried out in the 1970s and the collection was screened, and some were identified in detail through various research projects (Nalbant *et al.*, 1998). The genetic structure and diversity among cultigens and Sarılop clones present in the collection were subject to further studies (Cabrita *et al.*, 2001; İkten *et al.*, 2010). There are fig collections established by a few fig accessions generally selected from that region at Ege University (İzmir),

Table 23.1. Compilation of fig germplasm collections published with 25 or more accessions.

Location	Name	No. of accessions	Focus of collection	Reference
Castel St. Claire, France	Conservatoire Botanique National Méditerranéen de Porquerolles	277	France	Khadari *et al.* (2005); Aljane *et al.* (2018)
Winters, CA, USA	National Clonal Germplasm Repository	275	USA	
Rome, Italy	Istituto Sperimentale per la Frutticoltura	250	Italy	IPGRI (2006)
Guadajira, Spain	Junta de Extremadura. Servicio de Inv. y Desarrollo Tecnológico Finca la Orden	209	Spain	Giraldo *et al.* (2005)
St. Petersburg, Russian Federation	N.I. Vavilov All-Russian Scientific Research Institute of Plant Industry	187	Many from Ukraine	IPGRI (2006)
Moncada-Valencia, Spain	Instituto Valenciano de Investigaciones Agrarias	171	Spain	Amiri (2008)
Bet Dagan, Israel	Agricultural Research Organization, The Volcani Center	105	Israel	Amiri (2008)
Fig repository at the "P. Martucci" experimental station in Valenzano, University of Bari "Aldo Moro," Italy	University of Bari, Italy	100	Italy	Ferrara *et al.* (2016)
Tarragona, Spain	Centre de Mas Bove, Inst. Recerca i Tecnologia Agroalimen. (IRTA)	92	Spain	IPGRI (2006)
Meknes, Morocco	Aïn Taoujdate Experimental Station	75	Morocco and Algeria	Achtak *et al.* (2009)
Kalamata, Greece	Olive, Fruit and Vegetables Institute of Kalamata	66	Greece and Cyprus	Papadopoulou *et al.* (2002)
Tavira, Portugal	Direccao Regional de Agricultura do Algarve	60	Portugal	IPGRI (2006)
Boufarik, Algeria	Arboriculture Fruitière et de la Vigne	52	North Africa, S. Europe	IPGRI (2006)
Palma de Mallorca, Spain	Conselleria d'Agricultura i Pesca	52	–	IPGRI (2006)
Alcobaça, Portugal	Estacao Nacional de Fruticultura Vieira Natividade	50	–	IPGRI (2006)
Caserta, Italy	Istituto Sperimentale per la Futticoltura	50	Italy	Amiri (2008)
El Gordhab, Tataouine,Tunisia	The fig germplasm collection of the Arid Land Institute of Médenine	50	Tunisia	Essid *et al.* (2017)
Slovene Istria and Goriška Brda and Vipava Valley, Slovenia	University of Primorska, Slovenia	38	Slovenia and Croatia	Podgornik *et al.* (2010)
Nicosia, Cyprus	National (CYPARI) Genebank, Agricultural Research Institute	39	–	IPGRI (2006)
Sassari, Italy	Istituto di Coltivazioni Arboree (now Dipartimento di Universitá di Sassari)	31	Italy	Amiri (2008)

Contiuned

Table 23.1. Continued.

Location	Name	No. of accessions	Focus of collection	Reference
Chott Mariem, Médenine and Degache, Tunisia	–	35	Tunisia	Salhi-Hannachi *et al.* (2006)
Al-Mshaqer agricultural station, Jordan	National Center for Agricultural Research and Extension (NCARE)	31	Jordan	Almajali *et al.* (2012)
El Gordhab, Tunisia	Tunisian *ex situ* fig germplasm bank	30	Tunisia	Aljane (2016)
Neve-Yaar, Israel	Agricultural Research Organization, The Volcani Center	28	Israel	–
Potenza, Italy	Dipartimento di Produzione Vegetale, Universitá degli Studi della Basilicata	25	Italy	–
Tirana, Albania	Fig germplasm conservation in Albania	25	Albania	Koka (2008)

Atatürk Central Research Institute (Yalova) and Alata Horticultural Research Institute (Mersin).

In France, Porquerolles island (south-east of Toulon), hosts a fig collection comprising 277 accessions of caprifig and Smyrna, common and San Pedro types to display genetic diversity. The collection had six sources: (i) it started in 1981 with the transfer of 106 accessions of French and Mediterranean fig varieties that had been collected by Professor Rival over 30 years; (ii) 74 local varieties were collected from southern France and nurseries and planted in 1999; (iii) nine Portuguese accessions came from the Agronomical Research Institute of Alcobaça (Portugal); (iv) eight accessions came from the Nikita Botanical Garden (Yalta, Crimea); (v) 33 accessions came from the National Germplasm Depository (University of California, California); and (vi) 47 caprifigs came from French National Research Institute for Agriculture (INRA; Montpellier, France) in 1998. More than half of the accessions have been described pomologically (Roger and Khadari, 2003).

In Tunisia, the first attempts to collect fig genetic resources started in 1950, but these orchards were not maintained. Since the 1980s, studies have been undertaken to preserve the genetic resources, starting from central and southern Tunisia and five collection orchards were established at Ecole Supérieure d'Horticulture at Chott Mariem, the Institut des Régions Arides at Médenine, the Centre de Recherches Phoenicicoles at Degache, the Commissariat Régional au Développement Agricole at Gafsa, and the Groupement Interprofessionnels des Fruits at Sbikha (Mars *et al.*, 2008).

23.3 Descriptors

Proper identification is particularly important in figs as they come from local selections that are renamed, and hence traceability has been lost in most of the accessions. Sometimes, mutations may result in different phenotypic individuals that are propagated and multiplied and presented as a new variety, especially in vegetatively propagated species such as figs. In southern Italy, two fig varieties, Dottato grown around Cosenza and known as Fichi di Cosenza in Calabria and Fico Bianco del Cilento in the Salerno (Campania) area; both have protected designation of origin (PDO) status. Analysis of clones of these varieties revealed high similarity (de Masi *et al.*, 2003). In contrast, Sarızeybek, which has very similar fruit characteristics to Sarılop, was proven to be genetically different (Cabrita *et al.*, 2001).

All research on selection or varietal evaluation or on breeding of new varieties or cultivars requires criteria to assess performance either as a reference list of characteristics or as a descriptor. In the early years, this was done solely by evaluating morphological traits. Starting with Eisen (1901) and followed by Condit (1955), complete lists were prepared prior to any description.

Later, researchers such as Grassi and Pugliano (1984) developed and widened these descriptor lists according to their targets. Bioversity International (then IPGRI; www.bioversityinternational.org/; accessed December 23, 2021) released a "Descriptors for Fig (*Ficus carica*)" to benefit a wider scale. This Descriptor has an annex list of critical criteria for selection of female and male figs. These discriminating criteria are stated as "crop setting fruit, full maturity, harvest period, pollination requirement for fruit set, apical dominancy, rooting ability of the cuttings, leaf shape, fruit shape, fruit shape according to the location of the maximum width, fruit apex shape, fruit weight, ostiole width, ease of peeling, fruit skin cracks, resistance to ostiole-end cracks, fruit flesh thickness, fruit skin ground colour, pulp internal colour, fruit cavity, total soluble solids, number of dried fruits per kilogram, crops setting fruit, profichi fruit yield, profichi: amount of gall flowers, profichi: amount of male flowers, profichi: pollen maturation, mamme: date of Blastophaga's exit, profichi: date of Blastophaga's exit, cropping efficiency, breba: regularity of production, and estimated yield per tree" (IPGRI and CIHEAM, 2003). The UPOV (2010) released the guidelines for "harmonized examination of distinctness, uniformity and stability" to identify appropriate characteristics that will form the basis of this evaluation and "production of harmonized variety descriptions for figs (*Ficus carica* L.)." The UPOV guidelines recommend the following characteristics as useful in grouping figs: leaf, productive type, fruit shape, fruit skin ground and pulp color, and time of fruit maturity. As the basis of evaluation, rooted cuttings must be used. In UPOV descriptors, the characteristics are listed in English, French, German, and Spanish, and example(s) are given from some known varieties.

More recent studies are based on enhanced molecular methods for fingerprinting the accessions; however, agronomic and pomological characterization may guide the studies and give direct clues for variety selection. Morphological characteristics, especially those related to fruit and leaf shapes and sizes, and tree growth habit, may be used in the evaluation of fig genetic diversity (Fatahi *et al.*, 2017). DNA fingerprinting is used as a supplement to traditional morphology-based taxonomy. Currently, this is a widely accepted method for the characterization of genetic differences. DNA fingerprinting utilizes DNA markers for breeding and for variety identification. It is least affected by environmental interactions and spatial–temporal changes in gene expression. The most commonly used DNA fingerprinting techniques are polymerase chain reaction (PCR) based, such as random amplification of polymorphic DNA, amplified fragment length polymorphisms, simple sequence repeats, inter-simple sequence repeats, single-nucleotide polymorphisms, diversity arrays technology, and genotyping by sequencing (Nadeem *et al.*, 2018).

23.4 Major Fig Varieties

Edible female fig varieties are classified under three horticultural types according to their fruit setting: (i) Smyrna types (*F. carica* var. *smyrnica* Shinn.), which generally have the second crop as the main crop with caducous fruit, which requires caprification; (ii) common type (*F. carica* var. *hortensis* Shinn.), bearing two crops as breba (spring) and main crop (summer), which have persistent fruits without any caprification requirement; and (iii) San Pedro type (*F. carica* var. *intermedia* Shinn.), bearing two crops, the brebas being persistent and the second crop caducous, requiring caprification. The male figs or caprifigs (*F. carica* var. *sylvestris* Shinn.) can have two or three crops per year and may have persistent or caducous fruits (Crisosto *et al.*, 2011). The major female figs classified as Smyrna, common and San Pedro types, and male figs are described below (Fig. 23.1).

23.4.1 Smyrna-type varieties

Sarılop (Calimyrna)

This is the prominent Turkish variety known for its outstanding dried-fruit quality. According to 2019 figures, there are more than 8.5 million Sarılop fig trees grown as rainfed in the two western provinces of Turkey, Aydın and İzmir (https://biruni.tuik.gov.tr/medas/?kn=92&locale=tr; accessed January 18, 2022). Orchards are spread from the lowlands of the Big and Small Meander valleys toward the Aydın mountain

Fig. 23.1. Major commercial fig varieties. A. Sarilop. B. Sabz. C. Zidi. D. Bursa Black. E. Kalamata. F. Brown Turkey. G. Panache. H. Kadota. I. Black Mission. J. Toro Sentando. K. Evita. L. Dauphine.

range up to 850 m above sea level. The region has a typical Mediterranean climate with mild, mostly frost-free and rainy winters, and hot and dry summers. The long-term annual rainfall average is around 620–650 mm (https://izmir.mgm.gov.tr/; accessed December 23, 2021). Sarılop is a synonym of Calimyrna grown in California, USA, where it was renamed after Smyrna and California at the end of 19th century. The trees are all vegetatively propagated by rooting cuttings or replanting of rooted suckers. The tree growth is compact, but if not well trained, its canopy becomes spreading with drooping branches. The leaf is typical with five lobes and a deep sinus. It requires pollination by hanging caprifigs two to three times at 5–7-day intervals around mid-June for fruit set. Fruit ripening starts toward the end of July and continues till the end of September. The skin color of the fruit is greenish to light yellow and the internal color is pink to amber. The average fruit weight is 64–67 g, and has a wide, open ostiole sometimes filled with the sugary exudates. It has comparatively low titratable acidity (approx. 0.135% as citric acid) contributing to its sweet taste together with high soluble solids content (Aksoy and Şahin, 2001). Its fresh fruit are consumed locally, whereas its sun-dried fruit are exported widely. The dried fruits are large in size (average 40–50 dried fruit kg^{-1}), soft in texture, and light in color. Sarılop comprises nearly all Turkish dried-fig fruit exports.

Sabz

This is the main Iranian fig variety grown. It is grown under rainfed conditions in Estahban, the south-eastern part of Fars Province in Iran. It supplies 95% of the sun-dried figs in the region. The fig orchards are located on slopes of 45–50°. Arid climatic conditions prevail, with annual temperature averages of about 14.9°C, but extreme temperature differences may occur between −7°C in winter and 41°C in summer. Temperature increases during late August and September above 30–35°C may damage fig production. The average rainfall is 350 mm $year^{-1}$ and occurs mostly during late autumn and winter. Caprification is around March 20 and continues for 15–20 days. Sabz produces only the yellowish-green-colored summer crop, which dries partially on the tree, widely split at the eye, and is then collected and further dried on trays. The average dried-fruit weight is 6–7 g (160–170 dried figs kg^{-1}) (Rahemi and Jafari, 2008; Anon., 2011).

Zidi

This Tunisian fig variety is widely dispersed in south-east, center-east, north-west, and south-western parts (Mars *et al.*, 2008). The trees are vigorous, the leaves are large, and the fruits are very large, violet, black, very sweet, and juicy, with a brown sugar and berry flavor (www.fig-database.com/; accessed December 23, 2021).

Kalamata (Messenia)

This Greek variety is traditionally sun dried. The dried fruits are quite similar to those of Sarılop but with a thicker skin and larger seeds. The fruit is of medium size and spherical to obovate, with a very short and thick neck; the ribs are not prominent and the eye is large and open. The surface is somewhat glossy, with a delicate bloom, and with large, conspicuous white flecks. The skin is green, unattractive, often discolored, and has crisscross checks at maturity. The pulp is dark strawberry in color, flavor rich, and sweet (Condit, 1955).

Bursa Black

This Turkish fig variety originated from Bursa Province in the Marmara Region, in north-west Anatolia. Due to its high fruit quality as a fresh fig and its resistance to transportation, it was rapidly multiplied after the 1970s in the Marmara Region and also partially moved to the west, the Aegean, and south (Mediterranean) regions. The tree is vigorous forming a globe-shaped canopy. The leaves have three to five lobes, a deep sinus, and short, thick hairs. The main crop requires caprification. The harvest season is comparatively long due to the intervals between ripening of consecutive fruits. The harvest period begins in August and continues until the end of October. The fruits are purplish but sometimes have green sectors around the neck. The internal color is dark red when mature. The titratable acidity is approximately 0.21% as citric acid, and the soluble solids content is 18–21%. The fruit average weight is around 70 g with an oblong shape. The fruit is aromatic and resistant to cracking under high humidity conditions. The fruit performs well during transportation, especially if harvested before the full maturation stage. Bursa Black is a well-known variety in the fresh-fig trade for the late summer season (Aksoy and Şahin, 2001).

23.4.2 Common-type varieties

Brown Turkey (Aubique Noire, Negro Largo, San Piero, Masui Dauphine, Roxo de Valinhos)

This has medium-sized fruit. The skin is purplish-brown and the flesh pinkish-amber. It has a good flavor and is best when fresh. It produces a light breba crop. It grows as a small, hardy, vigorous tree, and should be pruned severely for the heaviest main crop. The number of fruit that set and ripen depends on pruning and the prevailing temperatures. It does best in many countries around the world: the USA, southern California, Brazil, China, Japan, and Israel. It ripens from July to December.

Panache (Striped Tiger, Tiger)

This has small to medium-sized fruit, with greenish-yellow skin with dark green strips. The flesh has a strawberry taste and is dry but sweet. It is best eaten fresh. It has no breba crop. It requires a long, warm growing season, and ripens from late August to October. It is grown in southern California, USA.

Kadota (Dottato, Ottato, Buttada, Florentine, White Kadota, Fichi di Cosenza)

This variety produces a medium-sized fruit; the skin is yellowish-green and the flesh is amber, tinged pink at the center. It has a rich flavor, and resists souring. There is little or no breba crop. The tree is upright and requires annual pruning to slow growth. It requires a hot, dry climate for the best quality, and is produced in Italy and the USA. In Italy, Dottato fig has local modified names of Ottato in the Puglia Region and Buttada in the Sardinia Region. In the European Community, Fichi di Cosenza has a PDO and is also utilized as sun-dried fruit (de Masi *et al.*, 2003).

Black Mission (Beers Black, Franciscan, Mission)

This variety originated in the Balearic Islands. The fruits are black purple all over and elongated, and the flesh is "watermelon" color to pink, with a fairly good taste. It is easily dried at home and is the single best all-round variety for the south, north, coast, and interior regions. The brebas are prolific and fairly rich. The tree is very large and should be planted at maximum spacing. After the tree reaches maturity, it should not be pruned. It commences growth mid-season.

Toro Sentando

This Spanish variety is grown in the south-eastern province of Alicante. It is now grown commercially in Peru for the export and domestic fresh market. It was selected by the Experimental Agricultural Station of Elche as promising for commercial production. The tree growth is not very vigorous so it is recommended for intensive orchards. The breba fruits are very large in size (63–73 g), while the main-crop fruits are smaller (average 47 g). The brebas are larger than the Black Mission variety. In both crops, the fruit is black or very dark purple color and may have longitudinal stripes, the skin is elastic, and the fruit has medium firmness. The fruit is sweet with a pale red pulp color and high organoleptic quality. It has high resistance to handling and transportation, and is commercially grown and exported from Peru. The smaller fruits are also dried locally (www.figdatabase.com/variety-details/2046/toro-sentado; accessed December 23, 2021).

Evita (Parisian, Barnistotte, Violette de Sollies)

The fruits of this variety are small to medium-sized, sometimes sold as "baby" figs, with a dark purple color, and a slightly flattened shape. When at peak maturity, they are very sweet, soft and juicy, and have a good balance of acids and sugars, enhancing the flavor. When not mature, the fruits are lighter in color and the taste is not good. They are grown in South Africa and France. It has a PDO in the European Union as Figue de Sollies (PDO-FR-0544). In South Africa, the harvest is between late January and May, supplying figs to the northern markets in winter (https://goodfruitguide.co.uk/product/evita/; accessed December 23, 2021).

Autumn Honey

This is a new cultivar that resulted from the Israeli breeding program. It is a medium- to large-sized fig (average 70–90 g), green–purple color, and with a slightly flattened shape. When at peak maturity, the fruit is sweet, soft, and juicy. This is a late-producing cultivar, harvested between September to December. It is grown in Israel and Portugal. This cultivar is used for fresh consumption and the tree has a heavy yield (30 t ha^{-1}).

24.4.3 San Pedro-type varieties

Dauphine (Grise de Tarascon, Adam, Rouge d'Argenteuil, Grosse de Juillet, Boule d'Or, Bouton d'Or)

This variety originated from Baud (Brittany, France), but was grown near Paris in the Argenteuil region and was thus also known as Rouge d'Argenteuil in the 19th century. It was also reported to be grown in the Palace of Versailles. It is cultivated extensively in southern France for the fresh market. It was introduced to California, USA, from England in 1883. The trees are vigorous and upright, with drooping outer branches. The leaves are medium to large, three- to five-lobed, and the upper surface is somewhat glossy; the upper sinuses are of medium depth and width, while the lower sinuses are shallow and broad, and the base is subcordate and the margins coarsely crenate. It produces a high-yielding breba and second crop. The breba fruit is large (average 92 g), sweet, and juicy, and has a red to dark brown glossy skin with white flecks and is shaded green toward the stalk. It has a short neck, broad ribs, a large eye, protruding violet scales around the eye, a firm and resistant texture, a light strawberry- to amber-colored pulp, large seeds, a rich flavor, and a fair to good quality, especially for shipping fresh. The second crop requires caprification for fruit set and ripens in autumn. It has a darker brown, purple- to black-colored fig with pink flesh and is a smaller oblate–spherical fruit (average 70 g) with a shorter neck compared with the breba. The pulp is juicy and good for fresh consumption (Condit, 1955; https://frenchfigfarm.com/fig-trees/dauphine; accessed December 23, 2021).

24.4.4 Caprifig varieties

Caprifigs or male varieties are an essential part of commercial production of Smyrna- or San Pedro-type female figs. There are two or three crops formed on male trees. The spring crop, called profichi, is used for pollination of the main-crop female fruit, whereas the other crops host the fig wasp and help its continuity. Especially in countries where Smyrna-type varieties are grown, caprifig varieties that have abundant pollen, withstand lower temperatures in winter and have profichi that ripen and shed pollen during the receptive period of the female variety are important. *Ex situ* collections host major caprifig varieties. Various studies have been performed in identification and selection of male fig varieties (Khadivi-Khub and Anjam, 2014; Anjam *et al.*, 2017; ÇalıŞkan *et al.*, 2017, 2018).

Roeding No. 3

This is the most valued caprifig variety in California. It is among a group of caprifigs introduced from the Meander valley of Turkey in 1887 by Fred Roeding, in Fresno. The numbers 1–3 indicate their place in the nursery row when the cuttings were first planted. Roeding No. 3 is the earliest in maturity, while No. 2 is the latest. The trees are vigorous. The dormant buds and the mamme-crop fig fruits are violet colored. The mamme crop is susceptible to frost injury. The profichi crop is very good and abundant in pollen; the mammoni crop is small but abundant. This variety is present in the National Germplasm Repository, Davis (Condit, 1955; www.figdatabase.com/variety-details/2046/toro-sentado; accessed December 23, 2021).

Roeding No. 4.

This variety is present in California. Its origin has not been clearly identified. The tree is vigorous and upright. The mamme crop is generally good. The fruit is spherical in shape without a neck, or pyriform in shape with a neck, the stalk is short, and the fruit is green with white flecks and a purple interior. The profichi crop is good, with medium or larger turbinate fruit, green in color with white flecks and an internal deep purple color. Its only importance is the late ripening of the profichi crop.

Stanford (Kara İlek)

Cuttings of this variety were imported from Turkey and stated as a synonym of Kara İlek in Turkey (Condit, 1955). It is a vigorous tree. The leaves are medium to large, five-lobed, and with medium to deep sinuses. It has a good mamme crop with green medium-sized fruit, and an excellent profichi crop, with pyriform-shaped light green fruit with white flecks, produced mid-season.

Hamza

This is a Turkish caprifig variety. Bud burst occurs from March 9 to 14 (in Aydın Province, western Turkey), with a few overwintering mamme, maturing between April 9 and 21. Maturation of the profichi is from June 12 to 19. The profichi fruit weigh 19–20 g, and are healthy fruits (not infected). The number of non-pollinating wasps per fruit is one, and they produce a medium amount of pollen. The pollen germination rate is 72%, and the average number of fig wasps per fruit is 115 (Aksoy and Şahin, 2001).

Ak İlek

This is a Turkish caprifig variety. Bud burst occurs from March 12 to 29 (in Aydın Province, western Turkey), with a few overwintering mamme, maturing between April 12 and 22. Maturation of the profichi is from June 10 to 17. The profichi fruit weigh 29–30 g and are healthy fruits (not infected). The number of non-pollinating wasps per fruit is one and they produce a medium amount of pollen. The pollen germination rate is 84%, and the average number of fig wasps per fruit is 462 (Aksoy and Şahin, 2001).

Çiçekli

This is a Turkish caprifig variety. Bud burst occurs from March 19 to 31 (in Aydın Province, western Turkey), with abundant overwintering mamme, maturing between April 10 and 21; Maturation of the profichi is from June 15 to 19. The profichi fruit weigh 17–18 g and the infected fruit ratio is 15%. The number of non-pollinating wasps per fruit is 33, and they produce a medium amount of pollen. The pollen germination rate is 93%, and the average number of fig wasps per fruit is 336 (Aksoy and Şahin, 2001).

23.5 Concluding Remarks

Fig production and trade is expanding to all continents from the traditional Mediterranean Basin, and new producers as well as importers are emerging for both fresh and dried produce. This is partially due to the health properties of figs uncovered by recent research; however, the increase in availability as a result of new varieties and almost year-round fresh production has exerted a marked impact. Cultivated and wild fig populations of the common fig (*F. carica* L.) remain rich sources of diversity on which future studies may focus. The focus is moving toward common-type varieties that avoid caprification, which requires manual labor, has increased production costs and creates higher fungal contamination risk. Furthermore, the thousands of *Ficus* species dispersed throughout the world can be utilized in breeding programs. In the past, varieties were distinguished based on morphological characters. However, over time, DNA-based markers were developed. DNA fingerprinting is now moving to next-generation technologies and genotyping is done by sequencing. Currently, varieties and cultivars can be identified and described by sequencing using new bioinformatics tools. DNA fingerprinting can be used for variety protection under Plant Breeders Rights rules. This will help broaden the genetic base of approved varieties of fig.

References

Achtak, H., Oukabli, A., Ater, M., Santoni, S., Kjellberg, F., and Khadari, B. (2009) Microsatellite markers as reliable tools for fig cultivar identification. *Journal of the American Society for Horticultural Science* 134, 624–631.

Aksoy, U. and Şahin, N. (2001) İncir ÇeŞit Kataloğu (Catalogue of Fig Varieties). Vocational Series (in Turkish and English). Ministry of Agriculture and Rural Affairs, DG Agricultural Production and Development, Merkez İkmal Müdürlüğü Matbaası.

Aljane, F. (2016) Analysis of genetic diversity in Tunisian fig (*Ficus carica* L.) germplasm bank revealed by RAPD markers and morphological characters. *European Journal of Scientific Research* 142, 172–192.

Aljane, F., Essid, A., and Nahdi, S. (2018) Improvement of fig (*Ficus carica* L.) by conventional breeding and biotechnology. In: Al-Khayri, J.M., Jain, M., and Johnson, D.V. (eds) *Advances in Plant Breeding Strategies: Fruits*. Springer, Cham, pp. 343–375.

Almajali, D., Abdel-Ghani, A.H., and Migdadi, H. (2012) Evaluation of genetic diversity among Jordanian fig germplasm accessions by morphological traits and ISSR markers. *Scientia Horticulturae* 147, 8–19.

Amiri, M.E. (2008) The status of genetic resources of deciduous, tropical, and subtropical fruit species in Iran. *Acta Horticulturae* 769, 159–167.

Anjam, K., Khadivi-Khub, A., and Sarkhosh, A. (2017) The potential of caprifig genotypes for sheltering *Blastophaga psenes* L. for caprification of edible figs. *Erwerbs Obstbau* 59, 45–49.

Anon. (2011) Case Study: Estahbanat Rainfed Fig System Iranian Agricultural Heritage System. FAO and Ministry of Jihad-e-Agriculture, Agricultural Planning, Economic and Rural Development Research Institute, Iran. Available at: www.fao.org/fileadmin/templates/giahs/PDF/Fig_traditional_system_Final_.pdf (accessed December 23, 2021).

Arendt, N.K. (1972) *Fig Cultivars*. The State Nikita Botanical Garden, Yalta, Proceedings LVI: 201 (in Russian).

Barron, A.F. (1891) Figs and their culture at Chiswick. *Royal Horticultural Society Journal* 13, 122–123.

Biscotti, N., Bonsanto, D., and Laghetti, G. (2021) Ethnobotanical study on traditional use of local fruit varieties in Gargano National Park (Apulia, Italy). *Genetic Resources and Crop Evolution (in press)* [https://doi.org/10.1007/s10722-021-01286-2].

Boscana, M.P.I. (2009) *Les Figueres a Les Illes Balears*. Consell de Mallorca, Ajuntament de Llucmajor, Obra Social la Caixa.

Cabrita, L.F., Aksoy, U., Hepaksoy, S., and Leitao, J.M. (2001) Suitability of isozyme, RAPD and AFLP markers to assess genetic differences and relatedness among (*Ficus carica* L.) clones. *Scientia Horticulturae* 87, 261–273.

Çalışkan, O., Bayazit, S., Ilgin, M., and Karatas, N. (2017) Morphological diversity of caprifig (*Ficus carica* var. *caprificus*) accessions in the eastern Mediterranean region of Turkey: potential utility for caprification. *Scientia Horticulturae* 222, 46–56.

Çalışkan, O., Bayazit, S., Ilgin, M., and Karatas, N. (2018) Characterization of caprifig (*Ficus carica* var. *caprificus*) accessions selected from various locations in the eastern Mediterranean region of Turkey. *Journal of Agricultural, Food and Environmental Sciences* 72, 203–207.

Condit I.J. (1941) Fig characteristics useful in the identification of varieties. *Hilgardia* 14, 1–69.
Condit, I.J. (1955) Fig varieties: a monograph. *Hilgardia* 23, 323–538.
Crisosto, H., Ferguson, L., Bremer, V., Stover, E., and Colelli, G. (2011) Fig (*Ficus carica* L.). In: Yahia, E.M. (ed.) *Postharvest Biology and Technology of Tropical and Subtropical Fruits: Cocona to Mango*. Woodhead Publishing Series in Food Science, Technology and Nutrition, Cambridge, UK, pp. 134–160.
de Masi, L., Cipollaro, M., di Bernardo, G., Galderisi, U., Galano, G., *et al.* (2003) Clonal selection and molecular characterisation by RAPD analysis of the fig (*Ficus carica* L.) "Dottato" and "Bianco del Cilento" cultivars in Italy. *Acta Horticulturae* 605, 65–68.
Earle, F.S. (1897) Fig culture in the Gulf States. *USDA Division of Pomology Bulletin* 5, 70–74.
Eisen, G. (1897) Fig culture, edible figs: their culture and curing. *USDA Division of Pomology Bulletin* 5, 1–53.
Eisen, G. (1901) The fig: its history, culture and curing with a descriptive catalogue of the known varieties of figs. *US Department of Agriculture Bulletin* 8, 201–282.
Essid, A., Aljane, F., and Ferchichi, A. (2017) Morphological characterization and pollen evaluation of some Tunisian *ex situ* planted caprifig (*Ficus carica* L.) ecotypes. *South African Journal of Botany* 111, 134–143.
Fatahi, S., Cheghamirza, K., Arji, I., and Zarei, L. (2017) Evaluation of genetic variation of common fig (*Ficus carica* L.) in west of Iran. *Journal of Medicinal Plants and By-products* 2, 229–240.
Ferrara, G., Mazzeo, A., Pacucci, C., Matarrese, A.M.S., Tarantino, A., *et al.* (2016) Characterization of edible fig germplasm from Puglia, southeastern Italy: is the distinction of three fig types (Smyrna, San Pedro and Common) still valid? *Scientia Horticulturae* 205, 52–58.
Giraldo, E., Viruel, M.A., López-Corrales, M., and Hormaza, J.I. (2005) Characterisation and cross-species transferability of microsatellites in the common fig (*Ficus carica* L.). *Journal of Horticultural Science and Biotechnology* 80, 217–224.
Grassi, G. and Pugliano, G. (1984) Schede per lo studio del fico, in uso presso l'istituto sperimentale per la frutticoltura di Roma (Forms for the fig, used at the Fruit Research Institute of Roma). *Annali Dell'istuto Sperimentale la Fruttticoltura* 15, 1–48.
Hogg, R. (1866) *The Fruit Manual*, 3rd edn. Journal of Horticulture, London.
İkten, H., Mutlu, N., Gulsen, O., Kocatas, H., and Aksoy, U. (2010) Elucidating genetic relationships, diversity and population structure among the Turkish female figs. *Genetica* 138, 169–177.
IPGRI (International Plant Genetic Resources Institute) (2006) Descriptors for Fig *(Ficus carica)*. Available at: www.bioversityinternational.org/fileadmin/_migrated/uploads/tx_news/Descriptors_for_fig__Ficus_carica__907.pdf (accessed January 18, 2022).
IPGRI/CIHEAM (2003) *Descriptors for Fig*. International Plant Genetic Resources Institute, Rome, and International Centre for Advanced Mediterranean Agronomic Studies, Paris.
Khadari, B., Oukabli, A., Ater, M., Mamouni, A., Roger, J.P., and Kjellberg, F. (2005) Molecular characterization of Moroccan fig germplasm using inter-simple sequence repeat and simple sequence repeat markers to establish a reference collection. *HortScience* 40, 29–32.
Khadivi-Khub, A. and Anjam, K. (2014) Characterization and evaluation of male fig (caprifig) accessions in Iran. *Plant Systematics and Evolution* 300, 2177–2189.
Koka, T. (2008) Fig germplasm conservation in Albania. *Acta Horticulturae* 798, 77–80.
Mars, M., Chatti, K., Saddoud, O., Salhi-Hannachi, A., Trifi, M., and Marakkchi, M. (2008) Fig cultivation and genetic resources in Tunisia, an overview. *Acta Horticulturae* 798, 27–32.
Mitrofanova, I.V., Mitrofanova, O.V., Lesnikova-Sedoshenko, N.P., Chelombit, S.V., Shishkina, E.L., and Chirkov, S.N. (2016) Phytosanitary status of *Ficus carica* collection orchards in Nikita Botanical Gardens and biotechnology of fig plants regeneration. In: *III Balkan Symposium on Fruit Growing, August 25, 2016, Belgrade, Serbia*, pp. 303–309.
Nadeem, M.A., Nawaz, M.A., Shahid, M.Q., Doğan, Y., Comertpay, G., *et al.* (2018) DNA molecular markers in plant breeding: current status and recent advancements in genomic selection and genome editing. *Biotechnology and Biotechnological Equipment* 32, 261–285.
Nalbant, M., Şahin, N., and Aydin, S. (1998) Fig genetic resources at the Fig Research Institute (Aydın-Turkey). In: *I International Symposium on Fig, December 1, 1998, Izmir, Turkey*, pp. 43–48.
Noisette, L. (1829) *Manuel Complet du Jardinier*. Wahlen & Tarlier, Brussels.
Papadopoulou, S.K., Papadopoulou, S.D., and Gallos, G.K. (2002) Macro- and micro-nutrient intake of adolescent Greek female volleyball players. *International Journal of Sport Nutrition and Exercise Metabolism* 12, 73–80.

Podgornik, M., Vuk, I., Vrhovnik, I., and Bandelj Mavsar, D. (2010) A survey and morphological evaluation of fig (*Ficus carica* L.) genetic resources from Slovenia. *Scientia Horticulturae* 125, 380–389.
Rahemi, M. and Jafari, M. (2008) Effect of caprifig type on quantity and quality of estahban dried fig *Ficus carica* cv. Sabz. In: *III International Symposium on Fig, September 30, 2008, Vilamoura, Portugal*, pp. 249–252.
Roger, J.P. and Khadari, B. (2003) Fig germplasm of CBNM Porquerolles displays an important genetic variability: evidence from pomological descriptors. In: *II International Symposium on Fig, April 28, 2003, Caceres, Spain*, pp. 51–59.
Salhi-Hannachi, A.M.E.L., Chatti, K., Saddoud, O., Mars, M., Rhouma, A., *et al.* (2006) Genetic diversity of different Tunisian fig (*Ficus carica* L.) collections revealed by RAPD fingerprints. *Hereditas* 143, 15–22.
Stover, E. and Aradhya, M. (2008) Fig genetic resources and research at the US national clonal germplasm repository in Davis, California. *Acta Horticulturae* 798, 57–68.
Ülkümen, L., Özbek, S., and İleri, M. (1948) *İncir ve Hastalıkları (Kitabi)*. Ankara Yüksek Ziraat Enstitüsü Basımevi, Ankara, Turkey.
UPOV (International Union for the Protection of New Varieties of Plants) (2010) *Fig. Guidelines for the Conduct of Tests for Distinctness, Uniformity and Stability, TG 265/1*. Geneva, Switzerland. Available at www.upov.int/edocs/tgdocs/en/tg265.doc (accessed December 23, 2021).
Vallese, F. (1909) *Il Fico*. F. Battiato, Catania, Italy.
Yezhov, V.N., Smykov, A.V., Smykov, V.K., Khokhlov, S.Y., Zaurov, D.E., *et al.* (2005) Genetic resources of temperate and subtropical fruit and nut species at the Nikita Botanical Gardens. *HortScience* 40, 5–9.

24 Conventional and Molecular Breeding in Fig

Reut Peer*, Dvory Namdar, Arpan Modi, and Moshe A. Flaishman
[1]*Institute of Plant Sciences, Agricultural Research Organization, Volcani Center, Rishon LeZion, Israel*

24.1 Introduction

Fig (*Ficus carica* L., Moraceae) trees are heterozygous diploid ($2n = 26$) plants. The DNA 2C values are rather low, from 1.37 pg, reflecting 270 million base pairs with an estimated 50,000 genes. Fig trees first spread from Persia, Asia Minor, and Syria, where they are endemic plants, and have been considered an important component of the Mediterranean diet for thousands of years (Solomon *et al.*, 2006). With human migration, figs became distributed throughout the Mediterranean area and further, to Europe, North and South Africa, the Americas, and many other Asian countries. A great number of fig varieties, mostly unnamed, have been developed to date.

The common fig has a monoecious ancestor (Machado *et al.*, 2001), which evolved later into a gynodioecious species with bisexual trees (functional male fig or caprifig) and unisexual female trees. Both female and male flowers are organized inside a fig-fruit structure termed a syconium. A female fruit contains hundreds of female-feature flowers. A male fig plant (caprifig) has hundreds of both male (staminate) and female (pistillate) flowers.

The differentiation between hermaphroditic “male” and female strains forms the basis for the close symbiotic relationship between fig plants and the *Blastophaga* species wasps (Galil and Neeman, 1977). Coevolution of *F. carica* and its fig wasp, *Blastophaga psenes*, has resulted in a complex breeding system involving two tree forms (caprifig and edible fig), three floral forms (female with long style, female with short style, and male), and the insect pollinator (Beck and Lord, 1988). This complex breeding system is a result of the gathering of diverse fig varieties growing in all Mediterranean countries into local fig collections, and their serving as sources for different breeding programs. Fruit breeding as an organized activity is a 19th-century innovation, where advances were made through intensification of the same forces that occurred during domestication, with increasing emphasis on adaptability, resistance to biotic stress, plant architectural modifications, and selection of color and taste mutations (Janick, 2012). Breeding strategies in fig are classified as either classical breeding or modern strategies. Modern strategies hold promise for increasing selection efficiency. They include marker-assisted selection (MAS), mutation breeding, transgenic plant development, and, recently, genome-editing approaches using the clustered regularly interspaced short palindromic repeats (CRISPR)/CRISPR-associated protein 9 (Cas9) system.

* Email: reutp@volcani.agri.gov.il

DOI: 10.1079/9781789242492.0024

In the present chapter, both strategies are detailed and compared. Prospects for the improvement of quality traits in different fig genotypes are also described.

24.2 Ethnological Fig Breeding

Most of the commercial fig varieties grown today are the results of unintended selections made by unknown growers. Cultivated figs vary in leaf morphology, plant vigor, fruit skin and flesh color, fruit flavor, soluble solids ratio and acidity level, seed and fruit morphology, skin thickness, ostiole diameter, and duration of the fruit-production season. Most of today's commercial fig varieties are selections from wild seedlings and open-pollinated seedlings of cultivated trees. Many countries have local fig collections that are used for the selection of superior clones (Mars, 2003). Numerous varieties expressing similar characteristics and similar genotype bear different names.

Reviewing the history of fig conservation demonstrates strong genetic corrosion in the last few decades. Therefore, it is extremely important to identify, conserve, and maintain the "genes" for future use and diversification (Aljane *et al.*, 2018). Two main approaches are applied for the conservation of genotypes: "*in situ*" or on-farm conservation, and "*in vitro*" conservation. On-farm conservation involves the maintenance of the qualities of varieties through vegetative propagation using classical methods, such as burrs, plunging, and grafting. Burrs are buds found on the roots with potential regenerative capacity. Propagation through burrs requires special care, as they are exposed to the soil surface and thus highly prone to nematode attack. In the plunging method, the vegetative part of the plant to be detached from the mother plant is removed only after it receives enough nutrients and water. Before its removal from the mother plant, roots are allowed to form, and the newly generated rooted plant is then transferred to the soil. Plunging and air layering are viable management practices for fig propagation. Grafting, however, requires two trees: the rootstock and the scion. The former contributes the root system and thus is generally screened for high water and nutrient uptake with disease resistance. The latter is the grafted tree on which fruiting will occur, and therefore plants with high fruit yield are selected for grafting (El-Borai and Duncan, 2005).

For *in vitro* conservation, many slow-growth protocols are available where plant tissues are nurtured under *in vitro* conditions to obtain the differentiation process in any plant part. In general, *in vitro* micropropagation techniques are divided into direct and indirect organogenesis. Protocols comprised of direct and indirect organogenesis have been optimized for commercial fig genotypes, as well as for endemic and endangered ones (Fráguas *et al.*, 2004; Yancheva *et al.*, 2005; Pasqual and Ferreira, 2007; Soliman *et al.*, 2010; Dhage *et al.*, 2012; Taha *et al.*, 2013; Shahcheraghi and Shekafandeh, 2016; Ling *et al.*, 2018).

Taphonomy is essential in figs, with the huge collection of germplasm; it would be beneficial to establish criteria for the identification and characterization of different accessions (Aradhya *et al.*, 2010). Traditionally, variety identification was based on morphological characteristics, such as leaf morphology, plant vigor, fruit color, fruit shape, skin diameter, and ostiole diameter. Highly diverse features, such as growth habits, leaves, life forms, position of staminate flowers, hairiness of tepals, and attachment of the style have been observed among *Ficus* species (Berg, 1989). Readily available biochemical and molecular markers have also been used to arrange local collections (Saddoud *et al.*, 2007; Perez-Jiménez *et al.*, 2012; Knap *et al.*, 2018).

24.3 Conventional Fig Breeding

Plant breeding is a dynamic area of applied horticultural science. It relies on genetic variation, and uses selection to improve plants with respect to the traits and characteristics that are of interest to both growers and consumers. The classical breeding system is highly efficient in producing new cultivars for almost all plant species. This breeding system consists of selecting diversified parents (either wild or cultivated) to produce new hybrids. This is followed by further hybrid selection and the fixation of new traits in subsequent generations. Therefore, beyond their conservation and identification, it is important to study the characteristics of the cultivars. Different genotypes of fig have been recognized since the 4th century BCE. Pliny listed 29 varieties of fig in the 1st century CE. Condit (1955) gathered them into a single list of 607 different fruit-producing varieties. Current commercial fig production in every region is based on only a few of these varieties. The fig industry in California, for example, is based on five main varieties.

To set up a breeding program, sex typing and flower phenology must be studied (Valdeyron and Lloyd, 1979). Figs demonstrate two tree forms (caprifig and edible) and three floral forms (long-styled females, short-styled females, and males). It has been noted that for *F. carica*, the two female floral forms differ only in style length and stigma shape. Breeding studies with this cultivated species led Storey to hypothesize that sex in *F. carica* is genetically determined by a supergene consisting of two very tightly linked genes (Storey, 1955, 1975). The first gene controls the style length of pistillate flowers (the *G* allele for short style is dominant over the *g* allele for long style), and the second controls the formation of staminate flowers (the *A* allele for male flowers is dominant over the *a* allele for absence of staminate flowers). Storey's experiments indicated that males are heterogametic (*GA*/*ga*) while females are homogametic (*ga*/*ga*) (Storey, 1955). As already noted, the males of dioecious *F. carica* also retain female flowers that have some ability to produce seeds in the male syconium (Storey, 1975; Neeman and Galil, 1978; Valdeyron and Lloyd, 1979). An artificial cross in which pollen was injected into the male syconium led to the production of fertile males and females (M.A. Flaishman, unpublished data). Thus, studies on fig floral biology, and an understanding of their coevolution with the fig wasp, have helped establish fig-breeding techniques (Storey, 1975; Weiblen, 2002).

24.3.1 Genetic resources

F. carica is known to have been domesticated from a diverse group of figs growing spontaneously in the south and east of the Mediterranean region some time in the Early Neolithic period (Zohary *et al.*, 2012). Large-fruited fig trees found in the deciduous forests of the Colchic district of northern Turkey and the Hyrcanian district of Iran and adjacent areas, which often intergrade into the Mediterranean figs, are considered by some botanists as a distinct ecotype of *F. carica* (Zhukovsky, 1962). According to Vavilov (1951), Transcaucasia is one of the centers of fig origin and diversity, as all phases of the domestication of fig can be seen in the southern Caucasus, where different types of fruit growing (i.e. wild, transitional, and modern) still exist. Domestication history and early migration along ancient trade routes strongly influenced the modern distribution, genetic diversity, and structure of fig (Storey, 1975). As fig cultivation spread across the world, introgression with local wild figs and landraces, along with human selection, resulted in numerous varieties and forms (Storey, 1975), but also in ambiguous nomenclature and descriptions. Condit (1955) listed more than 700 variety synonyms; however confusion persists in their identification and relationships. Morphological evaluation is often mislead by plasticity and due to environmental interactions, whereas molecular markers offer a stable and reliable alternative to the identification and characterization of germplasm to assess genetic diversity, structure, and differentiation in fig.

For several decades, the land area devoted to traditional village fig production has decreased significantly in many Mediterranean countries, and severe genetic erosion is threatening the local fig germplasms. Therefore, it is imperative to establish programs to preserve and characterize the genetic diversity of Mediterranean figs, a challenge that is being met by several countries. Proper variety identification is a key concern in many fig collections, mainly because individual varieties have been widely distributed with many synonyms, and the same name is often used for different varieties. The Institute of Kalamata, Greece, has 64 different fig varieties collected from Cyprus, Italy, Greece, Turkey, France, the USA, and Spain. The collection was characterized using random amplified polymorphic DNA markers, and the results were evaluated in conjunction with morphological and agronomical characteristics to determine the genetic relatedness of genotypes originating from diverse geographical regions. The results indicated that fig varieties have a rather narrow genetic base. Cluster analysis allowed the identification of groups based on geographical origin, phenotypic data, and pedigree. The US National Clonal Germplasm Repository (NCGR) in Davis, California, houses most of the Mediterranean-adapted fruit and nut crop collections in the USA, including fig. The NCGR fig collection currently contains 190 different accessions: 78 named fruiting varieties, 44 regional selections from diverse locations, 40 advanced selections from plant breeders (mainly from the UC Riverside breeding program), 28 caprifigs, and a small number of species and hybrids. The NCGR has completed DNA microsatellite fingerprinting of its fig accessions (Aradhya *et al.*, 2010). In Italy, a study of over 100 different fig genotypes, located in the fig repository at the P. Martucci experimental station, revealed a clear distinction between common and San Pedro-type figs (Ferrara *et al.*, 2016). A better understanding of fig biological caprification is important to distinguish fig types that require caprification from those that do not. To finalize the identification of fig varieties and cultivars from different collections around the world, it will be necessary to compare fingerprints with "typed" material from several collections. Microsatellite information and amplified fragment length polymorphism (AFLP) data will enable assessment of affiliations among fig genotypes and elucidation of evolutionary relationships within the genus *Ficus*, and conservation of fig plant material.

24.3.2 Conventional breeding methods

Traditional breeding techniques have been thoroughly described by Condit (1947) and Storey (1975). There are six main conventional breeding techniques, which involve the manipulation of natural growth, from flower to fruit, through pollen and seeds. The basic technical steps are: (i) selection of pollen and pollination intervention; (ii) building a seed collection; (iii) seedling evaluation and selection, i.e. parent selection, hybrid survey and phenotype analysis; (iv) shortening of the juvenile period; (v) interspecies breeding, and (vi) induction of mutations.

Pollen collection and pollination

Fig trees may produce one of two types of fruit—edible or non-edible—on the caprifig tree. The caprifig tree has three crop cycles a year; spring (called mamme), summer (called profici) and autumn (called mammoni). Only the caprifig profici crop develops stamens and generates pollen (Storey, 1975). To build a pollen collection, the caprifig profici crop is harvested and placed in a warm, dry place to dry for a few days. The anthers are then dissected and the clean pollen is shaken off the flower, collected, and stored until pollination. Pollen stored at 2°C remains viable for a year but before use, an artificial pollen germination test should be conducted. Pollination is the action of introducing pollen into the syconium. The pollen can be introduced into the summer fruit when they are light green and about 0.5–0.8 cm in size. The pollen is mixed in water and then injected into the fruit with a syringe. The pollinated syconium is then tagged with the date of pollination and the pollinating parent. The progeny from successful pollination are labeled, and their seeds are collected and germinated.

The selection of parent trees from which to collect pollen for breeding is based on their fruit's characteristics. Ideally, one wants all of the good qualities of two different trees, combining them into one supreme new cultivar with persistent syconia. Among the desired traits are inner and outer fruit color, fruit taste, and fruit size. The selection for caprifig fruit characteristics is similar to that for figs. Parent selection can also be made for fig-tree characteristics, rather than fruit quality alone. Preferred fig-tree traits might be persistence of syconia to ripeness, small compact trees, fruit productivity, and early or late crop maturation.

Seed collection

After maturation, the fruit contain fertile seeds, hundreds per syconium. The fruit are concealed and protected from birds that would otherwise feed on the seeds. For seed collection, the fruit are harvested and submerged in water for a few days to achieve fermentation, which cleans the seeds. After decanting the liquid with the floating fermented flesh, the clean seeds, which sink to the bottom of the container, can be collected. The seeds are then dried, and can be stored for several years at 2°C, pending germination and planting.

Shortening of the juvenile period

The process from germination to planted seedlings to fruitful tree can take 2–3 years. This juvenile period can be shortened to as little as 1 year by either grafting buds from seedlings on to adult trees (Storey, 1975), or suckering and staking seedlings in a vertical position during early development

(Ferguson, 1997), followed by intensive growth of the raised seedlings. Using the latter under the warm climatic conditions of Israel, Flaishman *et al.* (2008) obtained reproductive development and performed fruit evaluations as early as 1 year after seed extraction.

Upon seedling evaluation and selection, the undesired branches can be cut off and a new set of seedlings can use the vacant branch space in the following year. After 3–4 years, the fruit are evaluated for several parameters, such as size, shape, taste, and absence of splitting, as in other fruit, but also for the period of ripening during the crop season, seedling vigor, and anticipated productivity. The chosen seedlings are used for distribution. If they can be grown successfully under different environmental conditions, they can be distributed commercially, as well as used for further developments in the breeding chain.

24.3.3 Heredity of main fruit characteristics

We and others have shown that black skin color, red flesh color, smooth skin, and round fruit shape are dominant characteristics over green skin, pale flesh, cracked skin, and pear shape. Several hybrids have been produced using early-ripening varieties crossed with late-ripening varieties, resulting in prolonged fruit-harvesting time, lasting from April to mid-January. Moreover, new hybrids that can be grown in different climatic zones have been developed.

24.3.4 Classical breeding targets and achievements

Hybridization is a classical breeding strategy that is applied to achieve, for example, maximum productivity, resistance to pests and diseases, improved fruit-eating quality, improved shelf life, elimination of caprifigs, and persistence of syconia to ripeness. In addition to these fruit morphological characteristics, some health-promoting compounds have also been targeted. The fig fruit is a rich source of phytochemicals and has antioxidant activity (Vinson, 1999; Duenas *et al.*, 2008; Aljane and Ferchichi, 2009; Ercisli *et al.*, 2012). Moreover, polyphenol, anthocyanin, and flavonoid contents are dependent on the variety and harvest season (Solomon *et al.*, 2006; Vallejo *et al.*, 2012). The outcome of the research on phytochemical contents has shown that purple-blackish and red-greenish fruits have higher total polyphenol and anthocyanin contents, indicating a role for anthocyanins in the development of fruit skin color.

Fig breeding in California

Breeding programs were initiated in California at the beginning of the 20th century. In 1944, 141 combinations were made involving 49 female and 66 male parents. Several thousand seedlings were obtained. About 35 edible figs were selected and tested in commercial plots (Condit, 1947). The most significant achievement was the acceptance of four hybrid cultivars by commercial fig growers: Conadria, DiRedo, a sibling of Conadria, and Excel, mainly for dried-fruit consumption (Storey, 1975). The breeding program was revived in 1997 by James Doyle and Louise Ferguson using the germplasm established at University of California, Riverside, by Ira Condit and William Storey (Doyle and Ferguson, 1997). Breeding continued with the principal objective of developing a high-quality, large-size, light-colored dried fig, as similar as possible to the Calimyrna fig variety. In contrast to Calimyrna, this new fig would not need to be caprified (pollinated) and would have a small closed ostiole (eye), restricting insect access to the interior of the fig fruit (Doyle *et al.*, 2003). About 2400 hybrid fig seedlings were planted, and commercial plantings of three advanced fig selections were established. The California fig industry grows ten varieties, most of them of the "common" type. Seven are mainly for dried figs and fig paste: Calimyrna (the only Smyrna-type fig that needs to be pollinated); four groups identified as "Adriatics" but including Conadria, Adriatic, DiRedo, and Tena; and Kadota and Mission. Three are used in California primarily for the fresh-fig market: California Brown Turkey, the 2005 University of California release Sierra and the University of California 2013 release Sequoia with yellow-green skin color and reddish-amber pulp (Doyle *et al.*, 2003). Calimyrna, Mission, and Sierra are dual-purpose figs; all three dry well, but some growers direct part of the crop to the fresh market. Kadota is a multiple-use variety that can be successfully dried, canned, and picked for the fresh market.

The Israeli breeding program

A breeding program for fig cultivars for fresh and dry consumption was initiated in Israel in 2004 using diverse genetic resources, including more than 300 genotypes and hybrids of fig and other *Ficus* species. For improvement purposes, we assigned objective determinations for variety standards and developed selection procedures, targeting fruit quality and shelf life extension as the major objectives. Yearly assessment of about 8000 hybrid seedlings revealed the best parents and cross combinations for the selection of about 60 promising hybrids. The main fig-breeding objectives were: (i) high edible fruit quality; (ii) fig cultivars with distinct shape, size, and good flavor; (iii) parthenocarpic cultivars with a small ostiole that are resistant to splitting; (iv) maximum productivity; (v) extended shelf life; (vi) resistance to pests and diseases; and (vii) new dried-fig cultivars.

At the end of 2020, about 60,000 fig seedlings had been planted and screened at the Volcani Center in Rishon LeZion, Israel, and 180 cross combinations of 70 female and 26 male parents had been made. We preferred the use of commercial female varieties with high yield and good fruit quality, such as the local variety Hamadi and the introduced variety Brown Turkey for our crosses. In addition, we used superior cultivars from our breeding, such as the late-producing cultivar Autumn Honey, to further improvements. The phenotypic diversity of female figs screened by the broad Israeli cross-breeding program is shown in Fig. 24.1.

We found that selecting the right parental combination is a highly important stage. In addition, using the sex marker, we were able to separate male and female seedlings at an early stage (Sherman *et al.*, 2014). While breeding for distinct female cultivars, we also selected the superior caprifig. About 200 best-performing male cultivars were characterized and preserved for continuous breeding. Over the years, we identified superior male cultivars that, for example, were able to pass on a closed ostiole to the next generation. Superior males with diverse phenotypic combinations are presented in Fig. 24.2. The male collection contains a variety of colors and sizes. The purple cultivar 3-25, and the green cultivar 201-1, stand out in terms of size relative to the rest of the male fruit. In contrast, 4-28 and 13-01 are very small yet full of pollen. Cultivars 3-25, 147-7 and 145-2 are distinguished for their natural shine. Surprisingly, cultivars 200-3 and 1-36 accumulate sugar and soften toward ripening, unlike the characteristic male fruit. This collection of outstanding males has been used in many hybridizations over the years of research.

Several agronomic traits were evaluated within our segregating populations. We discovered that black skin color is dominant over green, while red flesh is dominant over pale color. Smooth skin is dominant over cracked fruit, and round fruit shape is dominant over pear shape. Fruit size is a quantitative trait, as two large-sized fruit parents will generate large-sized fruits.

We succeeded in developing diverse female cultivars with fruit of different sizes, colors, and flavors that produce high yield at different times of the year. We are currently growing five cultivars commercially in Israel from our breeding program: the early-producing dark cultivar Early FigARO, the late, highly productive strain Late FigARO, and three summer cultivars: the tasty green cultivar WOW, cultivar Shiny FigARO, and the small fruit cultivar Brown Candy. Using these fig cultivars in Israel, fig production starts in late May and ends in mid-January. In addition, we are evaluating 14 new fig hybrids from our breeding system in different climatic zones in Israel and worldwide. Among these are several cultivars with no ostiole opening that show extended pest resistance, mainly against the black fig fly (Fig. 24.3). The characteristics of the exceptional female progeny achieved by the Israeli breeding program are summarized in Table 24.1.

The superior female trees carry fig fruit in a variety of colors, shapes, and sizes. The dark cultivar Early FigARO has the largest fruit, 180–200 g. The smallest fruit is Brown Candy at only 14 g. In addition, there is wide variability in fruit sugar content. The taste of Brown Candy is unique and sweet, and it can be eaten directly from the tree, like candy. In taste and appearance, cultivars 44-26 and 43-53 had the highest indices. The green cultivar 44-26 is spectacular in the intense difference between its outer and

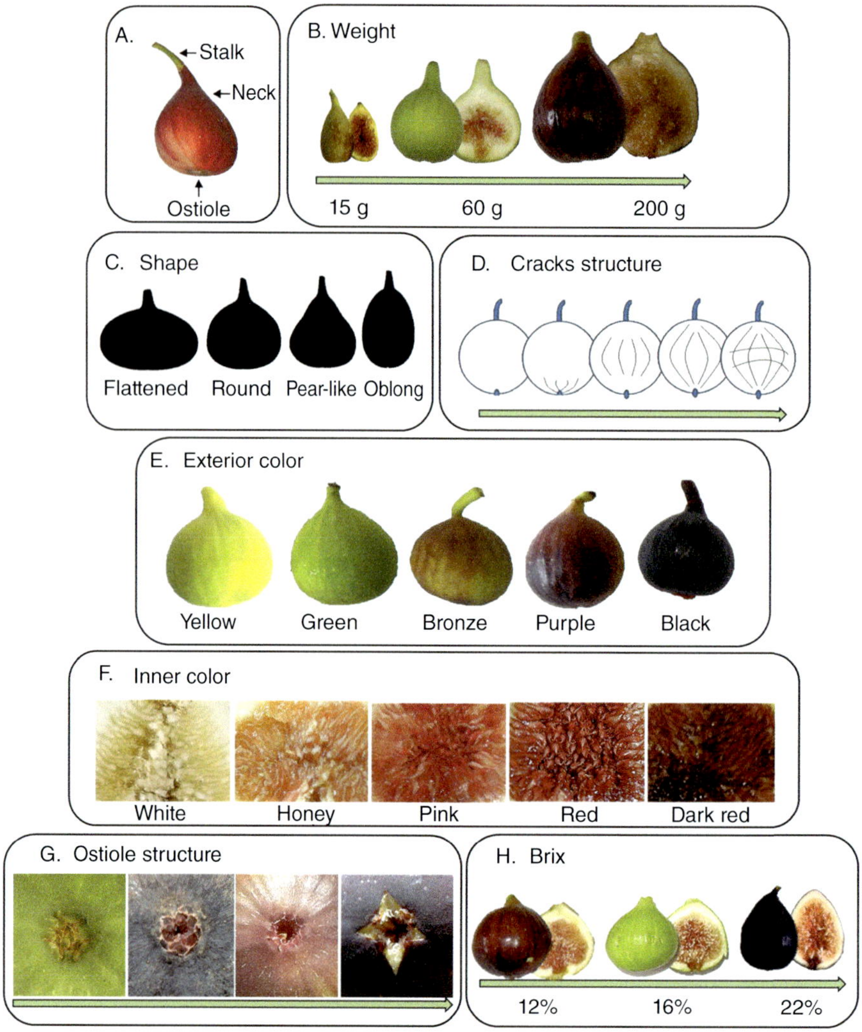

Fig. 24.1. Phenotypic diversity of female figs screened by the Israeli cross-breeding program. A. General structure. B. Weight variations. C. Fruit shape. D. Crack structure. E. Ripe fruit skin colors. F. Ripe fruit flesh colors. G. Ostiole structures. H. Sugar level indices in ripe fruit.

inner colors. Its taste is balanced with a delicate sweet acidity, giving the fruit its nickname WOW. The other popular standout is 43-53. Its skin color is dark and its flesh color is light, and it is the sweetest fruit in this outstanding collection with 25% sugar and a deep and unique taste.

24.3.5 Interspecific breeding

The genus *Ficus*, with over 750 species, contributes the highest number of species within the family Moraceae (Datwyler and Weiblen, 2004). The interaction between figs and fig-pollinating wasps is one of the most specific mutualisms known. Each fig wasp is specialized to pollinate only one *Ficus* species, developing a wall that prevents interspecific breeding of *Ficus* in nature. *Ficus* species show high diversity in their growth form; most are evergreen, and have typical syconia that vary in size, color, and taste (Fig. 24.4). Therefore, interspecific breeding could serve to increase diversity in *F. carica*. The first report of interspecific hybrids in *Ficus* was

Fig. 24.2. Outstanding males from the Israeli Agriculture Research Organization (ARO) fig-breeding program. Superior males (caprifig) with diverse phenotypic combinations achieved by the Israeli breeding program.

published by Condit (1950), who successfully produced an interspecific hybrid of *Ficus pumila* (as female parent) and *F. carica* (as male). The hybrid was more similar to *F. pumila*; it did not grow in a tree-like pattern, and its fruit morphology and leaf structures were also significantly altered. The intermediate *Ficus* hybrid can be used to study dioecious species genetics and sex determination in *Ficus* trees. In addition, natural hybrids between different *Ficus* species have been confirmed, suggesting that *Ficus* species can develop fertile interspecies (Parrish *et al.*, 2003). Using our *Ficus* collection in Israel, we made several interspecific crosses between the common fig and *F. pumila*, *Ficus palmata*, and *Ficus auriculata*. The leaves, shoots, and fruit of the hybrids were characterized. While some of the interspecific hybrid fruit showed fig characteristics, the others exhibited a new kind of fruit species. Progeny of common fig and *F. pumila* had a distinct taste and late ripening, and the hybrid plants were climbers. The fruit showed superb shelf life. A backcross of the F_1 cross of *F. carica* × *F. pumila* with *F. carica* produced segregation between the

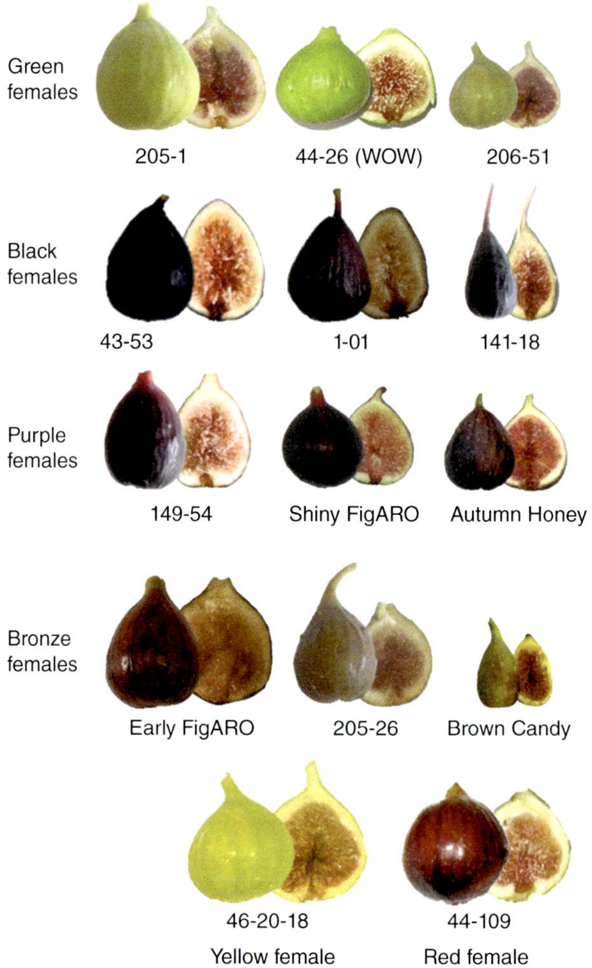

Fig. 24.3. Outstanding females from the Israeli Agriculture Research Organization (ARO) fig-breeding program. Superior females with diverse phenotypic combinations achieved by the Israeli breeding program.

climber and tree characteristics (Fig. 24.5). Crosses between the common fig and *F. auriculata* developed good-tasting red-colored fruit species. Finally, crosses between the common fig and *F. palmata* developed trees with a big trunk and small, tasty fruit (Fig. 24.6).

Interspecific crosses of various *Ficus* species were attempted to increase resistance to the fig fungal disease caused by *Ceratocystis fimbriata*. Yakushiji *et al.* (2012) developed an edible fig hybrid that is resistant to *Ceratocystis* canker, using interspecific crosses between *F. carica* and the *Ceratocystis* canker-resistant *Ficus erecta*. Seedlings of the fruit-bearing hybrid and the parents were infected with fungus directly on the shoots. There was no drying of the hybrid plants, even after 100 days of infection, whereas the seedlings of the susceptible species *F. carica* dried up within 70 days (Yakushiji *et al.*, 2012). Others have tried to achieve pest resistance in fig by using rootstocks of resistant varieties (Awamura and Nogata, 2005; Yakushiji *et al.*, 2012; Hosomi, 2015). We are currently testing the interspecific common fig × *F. palmata* hybrid as a rootstock that will confer resistance to *Batocera rufomaculata* (M.A. Flaishman *et al.*, unpublished data).

Table 24.1. Characteristics of the exceptional female progeny obtained in the Israeli breeding program.

Cultivar name	Weight (g)	Shape	Sugar content	Skin color	Flesh color	Production	Storage ability (days)	Fig type
205-1	90	Round	15%	Green	Dark red	July–August	14	Common
44-26	80	Round	21%	Green	Red	July–August	14	Common
206-51	32	Round	18%	Green	Red	July–August	14	Common
43-53	60	Pear-like	25%	Black	Red	July–August	14	Common
1-01	60	Pear-like	16%	Black	Pink	July–August	14	Common
141-18	30	Oblong	17%	Black	Dark red	July–August	14	Common
149-54	81	Pear-like	16%	Purple	Red	July–August	14	Common
Shiny FigARO 37-40	73	Round	12%	Purple	Pink	July–August	14	Common
Autumn Honey	55	Round	16%	Purple	Red	October–January	21	Common
Early FigARO 1008	180	Round with neck	18%	Bronze	Honey	May–June	21	San Pedro
205-26	45	Round with neck	19%	Bronze	Pink	July–August	14	Common
Brown Candy	14	Oblong	21%	Bronze	Dark red	July–August	14	Common
46-20-18	57	Round with neck	16%	Yellow	Honey	July–August	14	Common
44-109	60	Round	21%	Red	Pink	July–August	14	Common

Fig. 24.4. *Ficus* species with diverse fruit traits and growth forms.

B. rufomaculata is a species of beetle in the family Cerambycidae that feeds on the *F. carica* trunk and destroys the trees.

24.4 Mutational Breeding

Classical breeding strategies using conventional cross-breeding and induced mutations have played an important role in the development of new cultivars in fruit trees. Mutations can be induced by either physical or chemical mutagenic agents. These techniques are applied either *in vitro* or *ex vitro*. Mutation induction to improve traits and develop new cultivars has been studied in tropical fruit, including mango, citrus, papaya, avocado, and guava (Chan, 2009; Litz, 2009; Somsri *et al.*, 2009; Zamir *et al.*, 2009).

In fig, spontaneous mutants are also observed in commercial orchards. One such mutant from variety Kadota was isolated; its characteristics were similar to those of the mother plant but with red-colored fruit skin (Fig. 24.7A). It was registered under the name Red Kadota. Another type of spontaneous mutation was found in the variety Panache, sometimes called the "tiger" fig, producing a striking, yellow-green striped, variegated variety. This variegation mutation has developed several times with different fig varieties. Crosses with Panache confirmed that it is a vegetative, and thus non-inherited, mutation. We recently discovered another new variegated mutation in our breeding plot producing dark-colored fruit that are visually divided by pale red sectors (Fig. 24.7B) (Flaishman *et al.*, 2015).

In addition, mutational variations can be induced by specific treatments with physical and chemical mutagens, or by tissue culture. Among the mutagens, γ-rays are commonly used for effective mutation induction, and recently, a heavy-ion beam has been applied. γ-Irradiation of fig cuttings, seeds, and pollen has been reported to result in dwarfism and fruiting acceleration (Mars, 2003). Using such mutants, a new variety, Bol, was obtained by seed treatment with γ-irradiation. PI-189 is a mutant selection of Roxo de Valinhos, which originated from γ-irradiated buds at the Centro de Energia Nuclear na Agricultura da Universidade de São Paulo (CENA/USP) in São Paulo, Brazil (Rodrigues *et al.*, 2009). In comparison with Roxo de Valinhos, PI-189 stands out by showing good yields and a small, closed ostiole. Apart from these, in Turkey, the dried-fig variety Sarilop and the fresh-fig variety Bursa Black were subjected to cobalt γ-irradiation to produce a new genotype with different characteristics. The resultant varieties showed early maturation. Some of the trees also showed fruit with enhanced weight in both Sarilop (83 g versus 51 g in the control) and Bursa Black (111 g versus 81 g in the control) (Özen *et al.*, 2015).

Fig. 24.5. Progeny of interspecies crosses between common fig and *F. pumila* (F_1), and backcross (BC) of F_1 cross with *F. carica* (F_2).

24.5 New Strategies for Conventional Fig Breeding

It takes many breeding cycles to produce fruit-tree varieties that meet quality and environmental demands. The generation time in fig hinders the development of crosses, because the trees need to pass the juvenile period, which lasts 2–4 years. In addition, the large tree size makes it difficult to obtain large numbers of individuals for segregation families. Recent developments in high-throughput DNA-sequencing technologies have been exploited to identify the associations between genes and/or genomic intervals controlling important traits and phenotypes. Use of these technologies has enhanced the development of molecular markers for assisted selection and breeding for fruit quality, tree shape, and rootstock characteristics. Genomic methods provide a valuable approach for the selection of traits of interest in seedlings, and for the improvement of breeding efficiency through MAS and genome-wide association studies. Genomic analyses reveal the massive metabolome and transcriptome network changes underlying fig mutations.

24.5.1 Marker-assisted selection

MAS can be used in fig breeding because important traits, such as yield, quality, and disease resistance, are inherited quantitatively. Loci have

Fig. 24.6. Progeny of interspecies crosses between common fig and *F. palmata* (F_1).

Fig. 24.7. Spontaneous mutational variation of fig-fruit skin color. A. Kadota and mutated Red Kadota. B. Panache with darker color mutants.

been identified to target these multiple segregating genes, and these genotype markers can be combined with a phenotypic evaluation of individuals within the population, facilitating the identification of these quantitative trait loci (QTLs). Selection of appropriate varieties and cross-screening for a desired trait enable MAS. The joint approach of genotype marker segregation and phenotype evaluation of lines enables tracking of QTLs. DNA markers and powerful biometric methods have proven effective for plant QTL mapping. For fig breeding, QTLs can be used through the application of molecular markers, providing the basis for MAS and thus enhancing fig breeding and the development of new varieties.

Markers for early sexing in *F. carica* are economically important, although chromosome morphology and flow cytometry cannot be used for this purpose. Genetic control over sex postulates that if a single gene or a group of tightly linked genes controls sex determination, it should be possible to identify sex-specific DNA markers. So far, sex-related markers have been identified in a considerable number of dioecious plant species (Hormaza *et al.*, 1994; Zang *et al.*, 1998; Mandolino *et al.*, 1999; Deputy *et al.*, 2002). Parrish *et al.* (2004) reported a male-specific AFLP marker identified in the functionally dioecious fig species *Ficus fulva*. This study suggested that sex-specific AFLP markers are likely to be found in *F. carica*. To establish a molecular genetic sex marker in fig, we sequenced the transcriptomes of male and female figs. The sequence reads were assembled *de novo* and a set of single-nucleotide polymorphisms (SNPs) was discovered and compared with male and female DNA pools from our germplasm collection. Statistical and SNP analyses were applied to find two genetic markers that were linked to the sex-determination locus. This result supported the model for sex determination in fig, suggesting that the trait is controlled by one locus and that there are no sex chromosomes (Sherman *et al.*, 2014). To use sex-linked markers in fig breeding, fig seedlings were germinated and sampled. Robotic DNA extraction followed by real-time PCR was used to distinguish between the markers. Seedlings were classified and male and female progeny were planted separately in different plots (Fig. 24.8). As well as MAS in fig-breeding programs, identification of gender at the seedling stage plays a significant role in the production of fig fruit, as parthenocarpic varieties can be selected without pollination by the wasp originating on male varieties.

Recently, whole-genome sequencing of *F. carica* varieties Houraishi, Caprifig 6085, Masui Dauphine, Toyomitsuhime, and King was performed to identify sex-linked traits. *RESPONSIVE-TO-ANTAGONIST1* (*RAN1*), encoding a copper-transporting ATPase, was found to be associated with sex phenotypes. The SNP regions were targeted to determine early gender in figs (Mori *et al.*, 2017). Using this information, Ikten and Yilmaz (2019) developed cleaved amplified polymorphic sequence markers to check the sex type of 115 genotypes in the F_1 generation of Bursa Black × Ak Ilek. The amplified product (315 bp) was digested with *Pci*I and *Hpy*CH4IV, the former showing 100% efficiency at identifying gender, whereas *Hpy*CH4IV showed eight females as false negative (Ikten and Yilmaz, 2019).

24.5.2 Genome-wide sequencing for gene mining and transcriptome evaluation

Mori *et al.* (2017) generated the first draft of the fig genome sequence, using linkage analysis with a high-density genetic map established by a restriction-site-associated sequencing technique, and genome-wide association studies followed by whole-genome sequencing. Among other applications, genomic data can be used for *in silico* identification and characterization of genes and gene families related to important traits, thus enabling new tools for MAS and plant breeding. In the family Rosaceae, many gene families have been identified using this approach, among them numerous transcription factor (TF) families. RNA sequencing is a powerful tool for the analysis of transcriptomes due to the precise measurement of the expression level of each gene in a sample by quantifying short complementary DNA reads obtained by next-generation sequencing technologies, allowing comparison of entire transcriptomes between genotypes and conditions (Wang *et al.*, 2009). RNA reads can be mapped on a reference genome. In fruit trees, the study of transcriptomes has been widely used to characterize physiological issues, as well as tree and fruit development, fruit quality, tree pathogen infection, and resistance. Studying the pathways that lead to fig ripening may provide additional strategies for extending their storage life. We initiated a study dedicated to longer and higher-quality storage and shelf life, which has contributed to

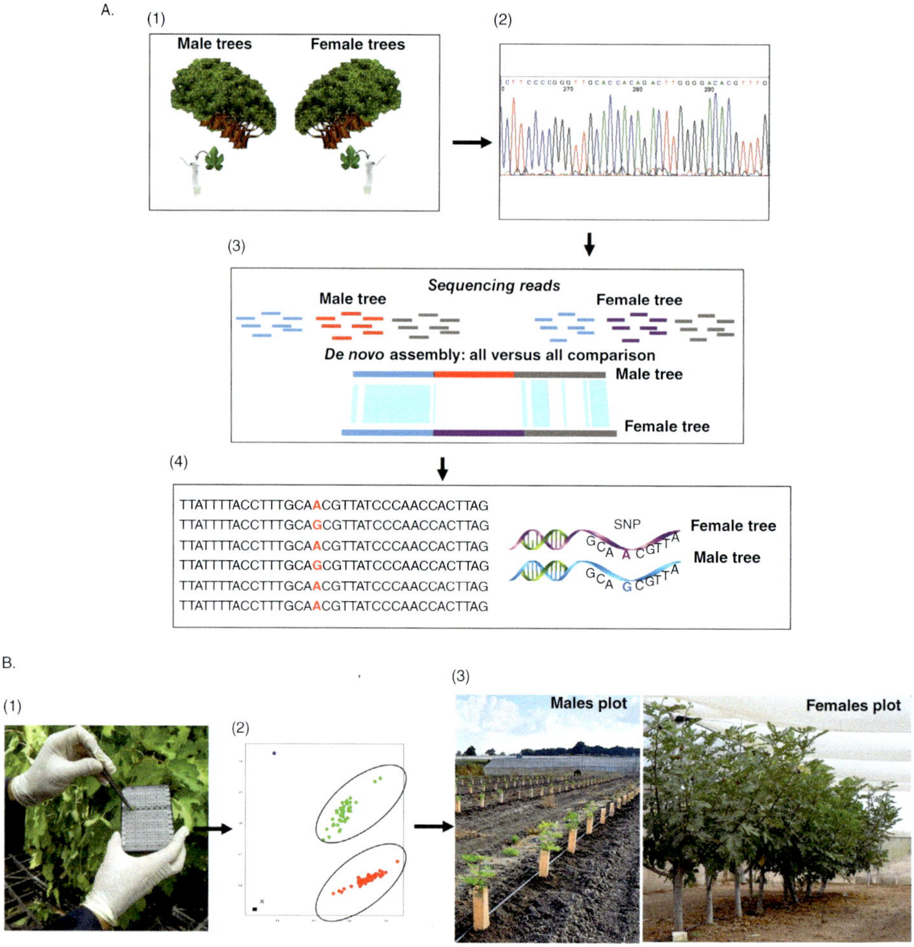

Fig. 24.8. A. Establishing a pathway in fig breeding. (1) Sample collection. (2) DNA extraction and sequencing. (3) *De novo* assembly. (4) Single-nucleotide polymorphism (SNP) analysis of the genetic marker linked to the sex-determination locus. B. Application of the molecular genetic sex marker in fig breeding. (1) Seedling notation and sampling. (2) Robotic DNA extraction followed by real-time PCR to distinguish between markers. (3) Classification and planting of males and females in separate plots.

our knowledge of ripening regulation by climacteric ethylene and TFs in fig (Freiman *et al.*, 2014). For transcriptome sequencing using 454 technology, we selected four different developmental stages in phase II and III fruit of the common fig Brown Turkey to identify the metabolic pathways and active TFs that are enriched during fruit ripening. This study provided the first expressed-gene data set for multiple developmental stages of fig fruit and an analysis of ripening metabolism, control, and regulation. A forward bioinformatics approach was adopted to locate the main changes during fig-fruit ripening (Freiman *et al.*, 2015). Several additional studies then described the metabolism in reproductive and non-reproductive parts of the fig fruit (Rosianskey *et al.*, 2016a; Lama *et al.*, 2019, 2020). The results identified candidate genes and provided new insights into the molecular process of regulating fruit ripening and fig-fruit manipulation using biotechnological approaches.

24.6 Molecular Fig Breeding

Emerging biotechnological approaches should be continuously evaluated for their potential to expedite fruit-tree breeding. With the growing availability of genome-wide sequence information on fruit trees, there is a greater need for methods that can use this vast amount of information to generate targeted and precise modifications of desired plant genes. Recent advances in the functional genomics of fruit trees have provided a basis for the development of precision crop genome engineering to improve fruit trees. Furthermore, targeted mutagenesis could facilitate the development of genetically modified crops and specific varieties without harboring transgenic DNA. Through the application of these new technologies, new fig varieties with novel traits can be generated.

24.6.1 Fig transformation

Progress in fig breeding has been achieved with the integration of biotechnology and the development of fig transformation systems. In general, the development of transgenic lines of any crop requires the optimization of two important steps. The first is a protocol for *in vitro* regeneration of the plant from non-meristematic, differentiated tissue, such as leaves or internodes. Micropropagation protocols have been established for various fig varieties (Yancheva *et al.*, 2005; Pasqual and Ferreira, 2007; Soliman *et al.*, 2010; Taha *et al.*, 2013; Moniruzzaman *et al.*, 2015; Al-Zahrani *et al.*, 2018; Ling *et al.*, 2018). The second step is the development of an efficient transformation protocol, with or without *Agrobacterium*. Differentiated tissue generated from *in vitro*- or *ex vitro*-grown plants is subjected to transformation, followed by its placement on regeneration medium. For *F. carica*, adventitious shoot regeneration in tissue culture has been reported by Yakushiji *et al.* (2003). An efficient and reproducible system for regeneration of the common fig variety Brown Turkey and Smyrna-type figs was reported by Yancheva *et al.* (2005). In addition, these authors reported an efficient and reproducible transformation system for both varieties (Yancheva *et al.*, 2005). Transgenic fig overexpressing β-glucuronidase (GUS) was obtained by selection on kanamycin (Fig. 24.9A). Soliman *et al.* (2010) reported efficient transformation and regeneration of fig (*F. carica*) via somatic embryogenesis of the fresh-consumed variety Sultani. Explants were transformed using *Agrobacterium* and microprojectile bombardment of plasmids harboring the *gus*-intron and *bar* genes. Recently, *Agrobacterium rhizogenes*-mediated transformation of *F. carica* was done for the efficient production of secondary metabolites (Amani *et al.*, 2020). The root-inducing plasmid-containing strain *A. rhizogenes* was used in combination with methyl jasmonate-induced elicitation to develop hairy roots, and thereby produce secondary metabolites such as gallic acid, caffeic acid, chlorogenic acid, coumaric acid, rosmarinic acid, cinnamic acid, rutin, quercetin, apigenin, and thymol. The primary objective of the study was to elicit the production of secondary metabolites instead of producing cell lines with altered characteristics. This experiment suggested additional scope for the fig plant in the pharmaceutical industry (Amani *et al.*, 2020).

24.6.2 Use of genome editing to produce desired mutations in fig

Genetic modification of crops to satisfy commercial needs is subject to strict regulations in many countries due to concern about transgenic introduction of foreign genes from bacteria or viruses into the organism of interest. Repairing mutated genes that already exist in the targeted organism can be an alternative to genetically modified crops. Genome-edited plants are not considered to be genetically modified organisms (GMOs) in countries such as the USA, Israel, and Japan, but are still under strict GMO regulation in Europe. The largest difference between genome-edited plants and GMOs is that the genomes of the edited plants can be freed of exogenous DNA sequences, which can be removed through genetic segregation (J. Xu *et al.*, 2019).

Targeted mutagenesis consists of introducing double-strand breaks, which are then repaired by the cell's own error-prone DNA-repair mechanism. Specific genomic regions are manipulated using sequence-specific nucleases that create double-strand breaks in targeted sites of the genomic DNA; these double-strand breaks are repaired by non-homologous end-joining

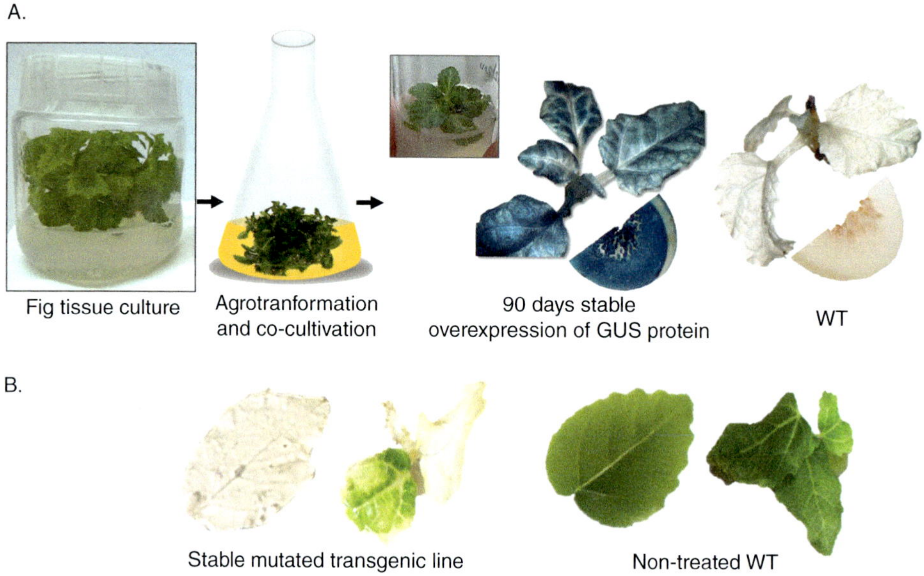

Fig. 24.9. Application of fig transformation and genome editing. A. Transgenic fig overexpressing β-glucuronidase (GUS) protein in fruit and leaves. B. Clustered regularly interspaced short palindromic repeats (CRISPR) and CRISPR-associated protein 9 (Cas9) system for genome editing in the endogenous fig gene phytoene desaturase.

(NHEJ) or homology-directed repair. In plants, the NHEJ process is far superior to the homology-directed repair process. During repair by NHEJ, various types of error, such as insertion or deletion of nucleotides by imperfect ligation at the tip of the double-strand break points, may be generated. Targeted mutagenesis can facilitate the development of genetically modified crops and specific varieties without harboring transgenic DNA.

Three systems have been developed for genome editing. A major advance was made with the development of zinc-finger nucleases in humans (Urnov *et al.*, 2005) and later also in plants (Townsend *et al.*, 2009). Zinc-finger nucleases were followed by a new generation of engineered double-strand break-inducing enzymes, such as *Xanthomonas*-derived transcription activator-like effector nucleases (TALENs) (Čermák *et al.*, 2015). Two types of TALEN are available: the standard TALEN, which has two DNA-recognition regions and is on the dimeric restriction enzyme *Fok*I, and compact TALEN, which is a monomeric DNA-cleaving protein. Successful generation of a much shorter endonuclease is based mainly on substitution of *Fok*I with I-*Tev*I homing endonuclease (Beurdeley *et al.*, 2013). A more recent genome-editing development, which has become the most practical editing tool, is the CRISPR/Cas9 system (Kim and Kim, 2014).

Studies that focus on adapting the CRISPR/Cas9 system to horticultural crops frequently use marker/reporter genes as targets, such as phytoene desaturase (*PDS*), whose mutation results in an albino phenotype. Among horticultural crops, tomato and potato have received a great deal of attention with respect to genome editing. In cereals, genome editing has been developed for rice and wheat (H. Xu *et al.*, 2019). Some floral and medicinal plants have also been successfully manipulated by genome editing. Recently, genome-editing technology has been developed for fruit trees, such as apple and grape.

A transgenic fig line carrying a ZFN under the control of a heat-shock promoter was developed. Editing of a mutated *uidA* gene, following expression of the ZFN gene by heat shock, was confirmed by GUS staining and PCR sequencing (Peer *et al.*, 2015). We further adopted the CRISPR/Cas9 system to create mutations in fig. We used the mutated GUS system and the marker/reporter gene *PDS* (Fig. 24.9B). The repair method for the mutated GUS was optimized and used to validate the CRISPR/Cas9 approach (Tovkach *et al.*, 2009). We used the CRISPR/Cas9 system to create mutations in fig genome. To confirm the transformation events in *FcPDS*-targeted samples, two main approaches were used: (i) to determine whether the albino phenotype due to the introduced double-strand break was followed by the plant's own (and error-prone) repair mechanism; and (ii) to isolate the DNA and amplify the PDS region with PDS-specific primers. The amplified product was sequenced and compared with the wild type to find the induced mutation. A complete albino phenotype was observed in the transgenic fig lines, whereas in control plants, clear green leaf color was observed. When the amplified product was sequenced and compared with the wild type, substitution mutations were observed at various places in the endogenous *FcPDS* gene (Fig. 24.9B).

24.6.3 Genome editing to improve fig storability

Optimally ripened fruit have high market value, warranting critical postharvest handling. For fig, ripening is an important process, as the fruit gains its size during this phase; premature picking will not allow it to gain the desired size, color, flavor, or texture. In contrast, late harvest results in overripened fruit that are susceptible to many pathogens (Flaishman *et al.*, 2008). Therefore, genetic modification of fig fruit is aimed at increasing their shelf life for postharvest handling (Armstrong and Lane, 2013; Gonsalves, 2014; Sidorova *et al.*, 2019). The rapid loss of firm texture determines the short postharvest life of this fruit. Disintegration of the cell wall and reduced cell adhesion due to dissolution of the middle lamella are the main factors controlling postharvest fruit softening (Quesada *et al.*, 2009). Genes involved in fruit ripening have been identified by sequencing the transcriptome of fig fruit at the ripening stage (Freiman *et al.*, 2015; Rosianskey *et al.*, 2016b; Lama *et al.*, 2019). These genes are related to the abscisic acid or ethylene processes and exhibit altered expression during ripening. Specifically, the abscisic acid-biosynthesis gene, *FcNCED2*, showed a ripening-related expression profile, exhibiting high expression levels at 10% ripening. Autoinhibition of ethylene production may be related to the direct function of the ethylene-biosynthesis gene *FcACS2* and could be used to alter fruit ripening. *FcNCED2* and *FcACS4* are thus candidate genes of choice for genome editing to alter the ripening process in fig fruit.

To study the gene-targeting efficacy of the CRISPR/Cas9 cassette, *FcNCED2* was used. The role of *FcNCED2* had been described in detail (Lama *et al.*, 2019), and knocking out this gene by short guide-RNA-directed mutagenesis was expected to result in delayed fruit softening. The transformation and genome-editing events were confirmed in a manner similar to that described for *FcPDS*. PCR sequencing demonstrated nucleotide substitution events that changed the amino acid sequence of the resultant protein. Phenotypic confirmation is not yet possible in the *in vitro* stage, and thus confirmation by sequencing was carried out (unpublished data). The phenotype of the genome-edited plants will be confirmed when they develop to produce fruit in the greenhouse.

24.7 Concluding Remarks

The common fig originated in the Middle East, and today wild forms are found in Mediterranean countries, Iran, Asia Minor, and Central Asia. For many years, fig fruit has been adopted and cultivated by many civilizations. Commercial fig varieties are grown around the world in different climatic regions. Fig is considered morphologically gynodioecious but functionally dioecious, and is pollinated by a specific pollinating wasp. Many varieties are the result of old selections that have been maintained by vegetative propagation, and numerous collections have been established in different countries. Phenotyping and molecular analysis have revealed high genetic diversity in cultivated fig germplasms. However, the threat of genetic erosion is evident, particularly for caprifig varieties. Efficient methods for genetic resource management and reference collections using universal descriptors will help maintain future variability, which will be useful for fig breeding. Perennial crops are affected by a large number of pathogens. The monogenic nature of many resistance genes has allowed the identification of homologs in wild relatives and varieties in germplasm collections, and has enabled the development of resistant varieties and cultivars and molecular markers linked to resistance loci.

There is increasing interest in fresh-fig marketing. Based on the consumer response to premium fresh figs, the potential market for new fresh figs is large and constantly growing. As the market for fresh figs expands, developments that are more extensive are required, such as development of postharvest technology, improved selection, and new varieties with higher quality. In addition, there is a need to develop new fig varieties with high eating quality, fruit with no cracking, and varieties adapted to environmental impacts associated with fig-fruit development. Many limitations are imposed by the tree and fruit characteristics, and new approaches must be developed for fig breeding. Numerous varieties and hybrids are available for breeding programs. Direct crossing of existing varieties or landraces may be very effective. Current breeding systems have succeeded to develop new fig varieties that are suitable for fresh consumption with advanced storage capacity, and for early breba and late summer production. Breeding programs have also isolated spontaneous mutations.

Combining genomic approaches with traditional breeding methods offers a promising prospect for improving perennial crops, including fig. As well as MAS, biotechnological interventions have been employed for the generation of transgenic lines in fig that exhibit diversified traits. This is because many genes associated with fruit and tree morphology and ripening have been characterized and their transcript levels have been measured. Considerable progress in recent years has enriched our knowledge of the molecular mechanisms underlying maturation and ripening of the fig fruit. This lays the groundwork for future development of new varieties and cultivars, using site-directed mutagenesis of key genes controlling ripening processes.

Today, there are around 700 identified species in the genus *Ficus*, which play an important role in the human diet. In general, interspecific crosses show intermediate characteristics, along with some new ones (Flaishman *et al.*, 2015). Interspecific hybridization may play an important role in the development of superior lines with improved fruit quality and higher disease resistance. The intermediate *Ficus* hybrid can also be used to study dioecious species genetics and sex determination in *Ficus* trees. Such programs have been carried out successfully in other fruit crops, such as strawberry, banana, mango, papaya, and citrus (Noguchi *et al.*, 2002; Vasugi *et al.*, 2009; Jayavalli *et al.*, 2011; Jenny *et al.*, 2011; Azad *et al.*, 2012; Smith *et al.*, 2013).

References

Aljane, F. and Ferchichi, A. (2009) Post-harvest chemical properties and mineral contents of some fig (*Ficus carica* L.) cultivars in Tunisia. *Journal of Food Agriculture and Environment* 7, 209–212.

Aljane, F., Essid, A., and Nahdi, S. (2018) Improvement of fig (*Ficus carica* L.) by conventional breeding and biotechnology. In: Al-Khayri, J.M., Jain, S.M., and Johnson, D.V. (eds) *Advances in Plant Breeding Strategies: Fruits*, Vol. 3. Springer, Cham, pp. 343–375.

Al-Zahrani, H.S., Almaghrabi, O.A., Fuller, M.P., Soliman, H.I., Farooq, M., and Metwali, E.M. (2018) Micropropagation of virus-free plants of Saudi fig (*Ficus carica* L.) and their identification through enzyme-linked immunosorbent assay methods. *In Vitro Cellular & Developmental Biology – Plant* 54, 626–636.

Amani, S., Mohebodini, M., Khademvatan, S., and Jafari, M. (2020) *Agrobacterium rhizogenes* mediated transformation of *Ficus carica* L. for the efficient production of secondary metabolites. *Journal of the Science of Food and Agriculture* 100, 2185–2197.

Aradhya, M.K., Stover, E., Velasco, D., and Koehmstedt, A. (2010) Genetic structure and differentiation in cultivated fig (*Ficus carica* L.). *Genetica* 138, 681–694.

Armstrong, J. and Lane, W.D. (2013) Genetically modified reduced-browning fruit-producing plant and produced fruit thereof, and method of obtaining such. US Patent No. 8563805.

Awamura, M. and Nogata, H. (2005) Effects of 'Celeste' rootstock on tree growth and fruit qualities. I. Young tree of fig cultivars '*Masui Dauphine*' and '*Houraishi*'. *Journal of the Japanese Society for Horticultural Science* 74 (Suppl. 2), 145.

Azad, M., Kalam, A., Rabbani, M., and Amin, L. (2012) Plant regeneration and somatic embryogenesis from immature embryos derived through interspecific hybridization among different *Carica* species. *International Journal of Molecular Sciences* 13, 17065–17076.

Beck, N.G. and Lord, E.M. (1988) Breeding system in *Ficus carica*, the common fig. II. Pollination events. *American Journal of Botany* 75, 1913–1922.

Berg, C.C. (1989) Classification and distribution of *Ficus*. *Experientia* 45, 605–611.

Beurdeley, M., Bietz, F., Li, J., Thomas, S., Stoddard, T., *et al.* (2013) Compact designer TALENs for efficient genome engineering. *Nature Communications* 4, 1–8.

Čermák, T., Baltes, N.J., Čegan, R., Zhang, Y., and Voytas, D.F. (2015) High-frequency, precise modification of the tomato genome. *Genome Biology* 16: 232.

Chan, Y.K. (2009) Radiation-induced mutation breeding of papaya. In: International Atomic Energy Agency (IAEA) (ed.) *Induced Mutation in Tropical Fruit Trees*. IAEA, Vienna, pp. 93–100.

Condit, I.J. (1947) *The Fig*. Chronica Botanica Company, Waltham, MA.

Condit, I.J. (1950) An interspecific hybrid in *Ficus*. *Journal of Heredity* 41,165–168.

Condit, I.J. (1955) Fig varieties: a monograph. *Hilgardia* 23, 323–538.

Datwyler, S.L. and Weiblen, G.D. (2004) On the origin of the fig: phylogenetic relationships of Moraceae from *ndh*F sequences. *American Journal of Botany* 91, 767–777.

Deputy, J.C., Ming, R., Ma, H., Liu, Z., Fitch, M.M.M., *et al.* (2002) Molecular markers for sex determination in papaya (*Carica papaya* L.). *Theoretical and Applied Genetics* 106, 107–111.

Dhage, S.S., Pawar, B.D., Chimote, V.P., Jadhav, A.S., and Kale, A.A. (2012) *In vitro* callus induction and plantlet regeneration in fig (*Ficus carica L.*). *Journal of Cell and Tissue Research* 12, 3395–3400.

Doyle, J. and Ferguson, L. (1997) Breeding persistent figs with Calimyrna quality. *Acta Horticulturae* 480, 259–264.

Doyle, J., Ferguson, L., Herman, K., López Corrales, M., and Bernalte García, M.J. (2003) Fig cultivar development and evaluation. *Acta Horticulturae* 605, 29–32.

Duenas, M., Perez-Alonso, J.J., Santos-Buelga, C. and Escribano-Bailon, T. (2008) Anthocyanin composition in fig (*Ficus carica* L.). *Journal of Food Composition and Analysis* 21,107–115.

El-Borai, F.E. and Duncan, L.W. (2005) Nematode parasites of subtropical and tropical fruit tree crops. In: Sikora, R., Coyne, D., Hallmann, J., and Timper, P. (eds) *Plant Parasitic Nematodes in Subtropical and Tropical Agriculture*. CAB International, Boston, MA, pp. 467–492.

Ercisli, S., Tosun, M., Karlidag, H., Dzubur, A., Hadziabulic, S., and Aliman, Y. (2012) Color and antioxidant characteristics of some fresh fig (*Ficus carica* L.) genotypes from Northeastern Turkey. *Plant Foods for Human Nutrition* 67, 271–276.

Ferguson, L. (1997) *Fig Breeding. Advanced Course on Fig Production*. EUFA, Turkey.

Ferrara, G., Mazzeo, A., Pacucci, C., Matarrese, A.M.S., Tarantino, A., *et al.* (2016) Characterization of edible fig germplasm from Puglia, southeastern Italy: is the distinction of three fig types (Smyrna, San Pedro and Common) still valid? *Scientia Horticulturae* 205, 52–58.

Flaishman, M.A., Rodov, V., and Stover, E. (2008) The fig: botany, horticulture, and breeding. *Horticultural Reviews* 34, 113–196.

Flaishman, M.A., Peer, R., Freiman, Z.E., Izhaki, Y., and Yablovitz, Z. (2015) Conventional and molecular breeding systems in fig (*Ficus carica* L.). *Acta Horticulturae* 1173, 1–10.

Fráguas, C.B., Pasqual, M., Dutra, L.F., and Cazetta, J.O. (2004) Micropropagation of fig (*Ficus carica* L.) 'Roxo de Valinhos' plants. *In Vitro Cellular & Developmental Biology – Plant* 40, 471–474.

Freiman, Z.E., Doron-Faigenboim, A., Dasmohapatra, R., Yablovitz, Z., and Flaishman, M.A. (2014) High-throughput sequencing analysis of common fig (*Ficus carica* L.) transcriptome during fruit ripening. *Tree Genetics and Genomes* 10, 923–935.

Freiman, Z.E., Rosianskey, Y., Dasmohapatra, R., Kamara, I., and Flaishman, M.A. (2015) The ambiguous ripening nature of the fig (*Ficus carica* L.) fruit: a gene-expression study of potential ripening regulators and ethylene-related genes. *Journal of Experimental Botany* 66, 3309–3324.

Galil, J. and Neeman, G. (1977) Pollen transfer and pollination in the common fig (*Ficus carica* L.). *New Phytologist* 79, 163–171.

Gonsalves, D. (2014) Hawaii's transgenic papaya story 1978–2012: a personal account. In: Ming, R. and Moore, P.H. (eds) *Genetics and Genomics of Papaya*. Springer, New York, pp. 115–142.

Hormaza, J.I., Dollo, L., and Polito, V.S. (1994) Identification of a RAPD marker linked to sex determination in *Pistacia vera* using bulked segregant analysis. *Theoretical and Applied Genetics* 89, 9–13.

Hosomi, A. (2015) Variation in graft compatibility of wild *Ficus* species as rootstock for common fig trees (*Ficus carica*). *Acta Horticulturae* 1173, 199–206.

Ikten, H. and Yilmaz, Y. (2019) Validation of gender specific CAPS marker in Turkish fig (*Ficus carica* L.) collection and F1 progenies. *Notulae Botanicae Horti Agrobotanici Cluj-Napoca* 47, 867–871.

Janick, J. (2012) Fruit breeding: past, present, and future. In: *XXII Congresso Brasilerio de Fruiticultura, Bento Goncalves-RS, 22a*, Vol. 26, pp. 1–22.

Jayavalli, R., Balamohan, T.N., Manivannan, N., and Govindaraj, M. (2011) Breaking the intergeneric hybridization barrier in *Carica papaya* and *Vasconcellea cauliflora*. *Scientia Horticulturae* 130, 787–794.

Jenny, C., Holtz, Y., Horry, J.P., and Bakry, F. (2011). Synthesis of new interspecific triploid hybrids from natural AB germplasm in banana (*Musa* sp.). In: *VII International Symposium on Banana: ISHS-ProMusa Symposium on Bananas and Plantains: Towards Sustainable Global Production and Improved Use, October 10–14, 2011, Salvador, Brazil*, pp. 209–217.

Kim, H. and Kim, J.S. (2014) A guide to genome engineering with programmable nucleases. *Nature Reviews Genetics* 15, 321–334.

Knap, T., Aradhya, M., Arbeiter, A.B., Hladnik, M., and Bandelj, D. (2018) DNA profiling of figs (*Ficus carica* L.) from Slovenia and Californian USDA collection revealed the uniqueness of some North Adriatic varieties. *Genetic Resources and Crop Evolution* 65, 1503–1516.

Lama, K., Yadav, S., Rosianskey, Y., Shaya, F., Lichter, A., *et al.* (2019) The distinct ripening processes in the reproductive and non-reproductive parts of the fig syconium are driven by ABA. *Journal of Experimental Botany* 70, 115–131.

Lama, K., Peer, R., Shlizerman, L., Meir, S., Doron-Faigenboim, A., *et al.* (2020) Tissue-specific organic acid metabolism in reproductive and non-reproductive parts of the fig fruit is partially induced by pollination. *Physiologia Plantarum* 168, 133–147.

Ling, W.T., Liew, F.C., Lim, W.Y., Subramaniam, S., and Chew, B.L. (2018) Shoot induction from axillary shoot tip explants of fig (*Ficus carica*) cv. Japanese BTM 6. *Tropical Life Sciences Research* 29, 165–174.

Litz, R.E. (2009) Recovery of mango plants with antrachnose resistance following mutation induction and selection *in vitro* with the culture filtrate of *Colletotrichum gloesporoides* Penz. In: International Atomic Energy Agency (IAEA) (ed.) *Induced Mutation in Tropical Fruit Trees*. IAEA, Vienna, pp. 7–16.

Machado, C.A., Jousselin, E., Kjellberg, F., Compton, S., and Herre, E.A. (2001) Phylogenetic relationships, historical biogeography and character evolution of fig-pollinating wasps. *Proceedings of the Royal Society B: Biological Sciences* 268, 685–694.

Mandolino, G., Carboni, A., Forapani, S., Faeti, V., and Ranalli, P. (1999) Identification of markers linked to the male sex in dioecious hemp (*Cannabis sativa* L.). *Theoretical and Applied Genetics* 98, 86–92.

Mars, M. (2003) Fig (*Ficus carica* L.) genetic resources and breeding. *Acta Horticulturae* 605, 19–27.

Moniruzzaman, M., Yaakob, Z., and Taha, R.A. (2015) *In vitro* production of fig (*Ficus carica* L.) plantlets. *Acta Horticulturae* 1173, 231–236.

Mori, K., Shirasawa, K., Nogata, H., Hirata, C., Tashiro, K., *et al.* (2017) Identification of RAN1 orthologue associated with sex determination through whole genome sequencing analysis in fig (*Ficus carica* L.). *Scientific Reports* 7: 41124.

Neeman, G. and Galil, J. (1978) Seed set in the 'male syconia' of the common fig *Ficus carica* L. (Caprificus). *New Phytologist* 81, 375–380.

Noguchi, Y., Mochizuki, T., and Sone, K. (2002) Breeding of a new aromatic strawberry by interspecific hybridization Fragaria × ananassa × F. nilgerrensis. *Journal of the Japanese Society for Horticultural Science* 71, 208–213.

Özen, M., Kocataş, H., Çobanoğlu, F., Ertan, B., Kösoğlu, İ., *et al.* (2015) Mutation breeding studies on fig. *Acta Horticulturae* 1173, 93–98.

Parrish, T.L., Koelewijn, H.P., van Dijk, P.J., and Kruijt, M. (2003) Genetic evidence for natural hybridization between species of dioecious *Ficus* on island populations. *Biotropica* 35, 333–343.

Parrish, R.L., Koelewijn, H.P., and van Dijk, P.J. (2004) Identification of a male-specific AFLP marker in a functionally dioecious fig, *Ficus fulva* Reinw. ex Bl. (Moraceae). *Sexual Plant Reproduction* 17, 17–22.

Pasqual, M. and Ferreira, E.A. (2007) Micropropagation of fig tree (*Ficus carica* sp). In: Jain, S.M. and Häggman, H. (eds) *Protocols for Micropropagation of Woody Trees and Fruits*. Springer, Dordrecht, pp. 409–416.

Peer, R., Rivlin, G., Golobovitch, S., Lapidot, M., Gal-On, A., *et al.* (2015) Targeted mutagenesis using zinc-finger nucleases in perennial fruit trees. *Planta* 241, 941–951.

Perez-Jiménez, M., López, B., Dorado, G., Pujadas-Salvá, A., Guzmán, G., and Hernandez, P. (2012) Analysis of genetic diversity of southern Spain fig tree (*Ficus carica* L.) and reference materials as a tool for breeding and conservation. *Hereditas* 149, 108–113.

Quesada, M.A., Blanco-Portales, R., Posé, S., García-Gago, J.A., Jiménez-Bermúdez, S., *et al.* (2009) Antisense down-regulation of the *FaPG1* gene reveals an unexpected central role for polygalacturonase in strawberry fruit softening. *Plant Physiology* 150, 1022–1032.
Rodrigues, M.G.F., Correa, L.D.S., and Boliani, A.C. (2009) Avaliação de seleções mutantes de figueira cv. *Roxo-de-Valinhos*. *Revista Brasileira de Fruticultura* 31, 771–777.
Rosianskey, Y., Dahan, Y., Yadav, S., Freiman, Z.E., Milo-Cochavi, S., *et al.* (2016a) Chlorophyll metabolism in pollinated vs. parthenocarpic fig fruits throughout development and ripening. *Planta* 244, 491–504.
Rosianskey, Y., Doron-Faigenboim, A., Freiman, Z.E., Lama, K., Milo-Cochavi, S., *et al.* (2016b) Tissue-specific transcriptome and hormonal regulation of pollinated and parthenocarpic fig (*Ficus carica* L.) fruit suggest that fruit ripening is coordinated by the reproductive part of the syconium. *Frontiers in Plant Science* 7: 1696.
Saddoud, O., Chatti, K., Salhi-Hannachi, A., Mars, M., Rhouma, A., *et al.* (2007) Genetic diversity of Tunisian figs (*Ficus carica* L.) as revealed by nuclear microsatellites. *Hereditas* 144, 149–157.
Shahcheraghi, S.T. and Shekafandeh, A. (2016) Micropropagation of three endemic and endangered fig (*Ficus carica* L.) genotypes. *Advances in Horticultural Science* 30, 129–134.
Sherman, A., Eshed, R., Peer, R., Rubinstein, M., Yablovitz, Z., *et al.* (2014) Discovery of sex linkage markers—a molecular proof for the genetic model for of sex determination in fig (*Ficus carica* L.). In: *Plant and Animal Genome XXII Conference, January 11–15, 2014, San Diego*.
Sidorova, T., Mikhailov, R., Pushin, A., Miroshnichenko, D., and Dolgov, S. (2019) *Agrobacterium*-mediated transformation of Russian commercial cv. 'Startovaya' (*Prunus domestica* L.) with virus-derived hairpin RNA construct confers durable resistance to PPV infection in mature plants. *Frontiers in Plant Science* 10: 286.
Smith, M.W., Gultzow, D.L., and Newman, T.K. (2013) First fruiting intergeneric hybrids between *Citrus* and *Citropsis*. *Journal of the American Society for Horticultural Science* 138, 57–63.
Soliman, H., Gabr, M., and Abdallah, N. (2010) Efficient transformation and regeneration of fig (*Ficus carica* L.) via somatic embryogenesis. *GM Crops and Food* 1, 47–58.
Solomon, A., Golubowicz, S., Yablowicz, Z., Grossman, S., Bergman, M., *et al.* (2006) Antioxidant activities and anthocyanin content of fresh fruits of common fig (*Ficus carica* L.). *Journal of Agriculture and Food Chemistry* 54, 7717–7723.
Somsri, S., Jompook, P., Kanhom, P., Thayamanont, P., Meecharoen, S., and Kumcha, U. (2009) Development of seedless fruits mutants in citrus including tangerine (*C. reticulata*) and pummelo (*C. grandis*) through induced mutations and biotechnology. In: International Atomic Energy Agency (IAEA) (ed.) *Induced Mutation in Tropical Fruit Trees*. IAEA, Vienna, pp. 33–40.
Storey, W.B. (1955) Sex inheritance in figs. *California Fig Institute, Proceedings of the Annual Research Conference* 9, 15–17.
Storey, W.B. (1975) Figs. In: Janick, J. and Moore, J.N. (eds) *Advances in Fruit Breeding*. Purdue University Press, West Lafayette, IN, pp. 568–589.
Taha, R.A., Mustafa, N.S., and Hassan, S.A. (2013) Protocol for micropropagation of two *Ficus carica* cultivars. *World Journal of Agricultural Sciences* 9, 383–388.
Tovkach, A., Zeevi, V., and Tzfira, T. (2009) A toolbox and procedural notes for characterizing novel zinc finger nucleases for genome editing in plant cells. *Plant Journal* 57, 747–757.
Townsend, J.A., Wright, D.A., Winfrey, R.J., Fu, F., Maeder, M.L., *et al.* (2009) High-frequency modification of plant genes using engineered zinc-finger nucleases. *Nature* 459, 442–445.
Urnov, F.D., Miller, J.C., Lee, Y.L., Beausejour, C.M., Rock, J.M., *et al.* (2005) Highly efficient endogenous human gene correction using designed zinc-finger nucleases. *Nature* 435, 646–651.
Valdeyron, G., and Lloyd, D.G. (1979) Sex differences and flowering phenology in the common fig, *Ficus carica* L. *Evolution* 33, 673–685.
Vallejo, F., Marin, J.G., and Tomas-Barberan, F.A. (2012) Phenolic compound content of fresh and dried figs (*Ficus carica* L.). *Food Chemistry* 130, 485–492.
Vasugi, C., Dinesh, M.R., and Sekar, K. (2009) Studies on inter and intraspecific crossability of *Mangifera* using cryopreserved pollens. *Journal of Genetics and Evolution* 2, 33–39.
Vavilov, N.I. (1951) *The Origin, Variation, Immunity and Breeding of Cultivated Plants*, Vol. 13. Chronica Botanica Company, Waltham, MA.
Vinson, J.A. (1999) The functional food properties of figs. *Cereal Foods World* 44, 82–87.
Wang, Z., Gerstein, M., and Snyder, M. (2009) RNA-Seq: a revolutionary tool for transcriptomics. *Nature Reviews Genetics* 10, 57–63.
Weiblen, G.D. (2002) How to be a fig wasp. *Annual Review of Entomology* 47, 299–330.

Xu, H., Xu, L., Yang, P., Cao, Y., Tang, Y., *et al.* (2019) Virus-induced phytoene desaturase (PDS) gene silencing using Tobacco rattle virus in *Lilium × formolongi*. *Horticultural Plant Journal* 5, 31–38.

Xu, J., Hua, K., and Lang, Z. (2019) Genome editing for horticultural crop improvement. *Horticulture Research* 6: 113.

Yakushiji, H., Mase, N., and Sato, Y. (2003) Adventitious bud formation and plantlet regeneration from leaves of fig (*Ficus carica* L.). *Journal of Horticultural Science and Biotechnology* 78, 874–878.

Yakushiji, H., Morita, T., Jikumaru, S., Ikegami, H., Azuma, A., and Koshita, Y. (2012) Interspecific hybridization of fig (*Ficus carica* L.) and *Ficus erecta* Thunb., a source of *Ceratocystis* canker resistance. *Euphytica* 183, 39–47.

Yancheva, S.D., Golubowicz, S., Yablowicz, Z., Perl, A., and Flaishman, M.A. (2005) Efficient *Agrobacterium*-mediated transformation and recovery of transgenic plants from fig (*Ficus carica* L.). *Plant Science* 168, 1433–1441.

Zamir, R., Ali, N., Shah, S.T., Mohammad, T., and Ahmad, J. (2009) Guava (*Psidium guajava* L.) improvement using *in vivo* and *in vitro* induced mutagenesis. In: International Atomic Energy Agency (IAEA) (ed.) *Induced Mutation in Tropical Fruit Trees*. IAEA, Vienna, pp. 101–112.

Zang, Y.H., Distilio, V.S., Rehman, F., Avery, A., Mulcahy, D., and Kesseli, R. (1998) Chromosome specific markers and the evolution of dioecy in the genus *Silene*. *Genome* 41, 141–147.

Zhukovsky, P.M. (1962) *Cultivated Plants and Their Wild Relatives*. Abridged translation by Hudson, P.S. CAB International, Wallingford, UK.

Zohary, D., Hopf, M., and Weiss, E. (2012) *Domestication of Plants in the Old World: The Origin and Spread of Domesticated Plants in Southwest Asia, Europe, and the Mediterranean Basin*. Oxford University Press (on demand), Oxford, UK.

Part VII

Fig Products and Trade

25 Fig Products from Tradition to Future

Antonio Piga[1]*, Christian Vera Ponce[2], and Uygun Aksoy[3]
[1]*Dipartimento di Agraria, Università degli Studi di Sassari, Sassari, Italy;* [2]*Ficatum-Higos con Historia, Miraflores, Lima, Peru;* [3]*Department of Horticulture, Faculty of Agriculture, Ege University, Bornova, İzmir, Turkey*

25.1 Introduction

Figs categorized as a functional food have gained popularity all over the world. The fitness branch in the West has adopted dried figs as an important source of nutrients as potassium, calcium, fiber, and carbohydrates to gain quick energy and repair tissues after a demanding physical activity (Stromberg, 2014). However, its gastronomical versatility (such as fresh and dried, but also in various forms of delicatessen such as Mediterranean fig bread) allows different kinds of culinary adaptations (Munar, 1999).

Ayurveda medicine considers figs a tonic with important qualities to increase vitality (Rosenberg, 2011). Functional foods with a long storage capacity and a balanced mixture of carbohydrates, minerals, and vitamins are one of the best options for outdoor and mountain sports. Other trends such as the paleo diet and even plant-based nutrition show different ways to adapt a very nutritious product like fig in this market development.

The focus on balanced food for sports, intellectual activity, and integral health opens up different opportunities for those who deal with dried figs and want to innovate, creating new products and eventually promoting consumption of local recipes using different regional fig varieties. One important fact is that figs have a natural storytelling as a sacred food (Chevalier and Gheerbrant, 2007) with remarkable references in mythology, literature, and art. Developing brands according to this trait represents a very powerful communication tool.

This market development has created new chances for the revival of old recipes that were very important in the past, like the use of dried-fig paste in the Mediterranean area or the production of fig coffee, which was a popular beverage in Austria in the 1950s. Fig paste is definitely a good alternative for those who want to develop functional food using small and irregular dried figs that are normally avoided due to the standards required in conventional markets.

In this chapter, a detailed description is given of traditional fig products from major producing regions such as the Mediterranean and Latin America, as well as commercial production of fig products through industrial processes.

25.2 Traditional Fig Products from the Mediterranean and Latin America

25.2.1 The fig paste tradition in the Mediterranean area

Smokvenjak (or Hib) in Croatia, *pan de higo* in Spain, *salami di fichi* in Italy, *sikomaida* in Greece,

* Email: pigaa@uniss.it

 Advances in Fig Research and Sustainable Production (M.A. Flaishman and U. Aksoy)
DOI: 10.1079/9781789242492.0025

and *figov hlevcek* in Slovenia are some expressions of an ancient use of fig paste in order to prepare a very consistent dough and mix it with herbs and/or nuts. Now these are available almost everywhere in Europe and are recognized as a real delicatessen, normally together with raw-milk cheeses or as a souvenir for tourists. Some are also supported by Slow Food (e.g. fig salami from the Marche region of Italy) and are linked to local gastronomic traditions. Fig bread from Elche and from Mallorca, both in Spain, are good examples of this.

Fig bread was very popular in ancient times because of its simplicity and benefits for human health. There are references in Mallorca, where fig bread was also an important source of energy for work in the field and as the main source of natural sugars. The combination bread with fig and local varieties of almonds offered a balanced mix of minerals, carbohydrates, and proteins, and was also a symbol of joy and sharing. References to dried figs as power foods are easy to find. The Spanish writer and cook Angeles Díaz Simón (2011), referred to the philosopher Plato in her book *Recetas con historia* (*Recipes with history*), who mentioned the importance of figs in Ancient Greece as a must for athletes in combination with oat and fermented milk.

Preparation of the fig paste requires a meat grinder (traditionally, the process was all done manually, but modern small factories use an electric grinder) and dried figs of different quality (normally the small or defect ones, e.g. torn or split, are used but there are also producers who prefer the larger ones). Some producers take off the peduncle before the paste production, but some just add the whole dried fig to the grinder. The result is a soft dough, which can be mixed with herbs (mainly anise and cinnamon in Spain) and nuts (almonds and walnuts). The Croatian tradition (e.g. on Vis Island) also includes *raki* and fennel, whereas in Corfu (Greece), *ouzo* will be used for the *sykomaida* and also a fig pie.

Fig bread was a very important source of nutrients in the Mediterranean area because of its long storage capacity. It is also a Christmas tradition in Spain and Croatia. This old product can be also commercialized as a power food of a very high quality and as a regional delicacy. The Fig Bread Festival in Barban, north Croatia, is a very good example. In Peru, fig bread is used in fitness sports, as activities such as running, cycling, or trekking require good sources of carbohydrates, potassium, and fiber. The long storage life of fig bread and its powerful consistency are also a good marketing tool for travelers.

25.2.2 Fig jams and preserves

Fig fruit rich in sugar and flavor is suitable for processing as jams. Traditionally, different jams and preserves were prepared using different kinds of figs, such as unripe male or female figs, and fresh mature or dried female fig fruits.

Unripe male figs

Male fig jam or green figs in syrup are typical in Greece and Turkey. The male fig (caprifig) tree bears three crops per year: the summer crop (profichi), autumnal crop (mammoni), and overwintering crop (mamme), which matures the following spring. The profichi crop is the one producing pollen, and is used to pollinate the summer crops of Smyrna- and San Pedro-type fig varieties (Armstrong, 2006). The male fig fruit suitable for jam is the unripe profichi crop collected before the fig wasps (*Blastophaga psenes* L.) enter the profichi after leaving the mamme in which they overwinter. The harvest period is early April in regions where unripe male fig fruits are traditionally harvested and marketed for jam making. There are different formulas ranging from very simple to more detailed. Figure 25.1 gives the two recipes used in making the jam.

In Turkey, calcium water is used while preparing male fig fruit jams to harden the fruit texture and make it glossy. It is prepared by dissolving calcium carbonate in water and leaving the solution for 12 h until the calcium settles at the bottom and a transparent layer of liquid is formed on top. This liquid is used to soak the fruit. In the old recipe for male fig jam, a minute amount of copper sulfate was also added into the boiling water to give a dark bluish-green color to the fruit. However, in recent years, copper sulfate has not been added. Jam prepared using unripe male fig fruits is also served as a preserve or candied, depending on the consistency of the syrup. Similar to male fig jam, the spring or summer crop fig fruits harvested as unripe (approximately 2 cm in diameter) may also be used for

Fig. 25.1. Traditional recipes for male fig jam (http://mamastaverna.com/baby-fig-spoon-sweet-sikalaki-gliko/; accessed January 3, 2022).

jam making. However, female fig fruits lack the flavor that the male profichi crop possesses.

Fresh ripe figs

The whole fruit, or quartered, sliced, or crushed fresh fruit, of all fig varieties can be utilized to make jam. The amount of sugar to be used or whether to preserve it prior to boiling or not depends on the sugar content and texture of the fruit. In most cases, 500–1000 g of sugar is used for 1 kg of fresh fig fruit. As fresh figs have a high water content, there is no need to add water. After removing the stem, the fruit is washed and then boiled in syrup for around 1 h. A tablespoon of lemon juice is added to thicken the syrup. Most of the traditional recipes recommend adding a teaspoon of butter to provide a shiny appearance. There are recipes where figs and sugar are combined with ginger, lemon, cinnamon, and vanilla, and the mixture is left for 30 min before boiling so that the flavors diffuse and the sugar dissolves (https://realfood.tesco.com/recipes/fig-and-ginger-jam.html; accessed January 3, 2022).

Dried fig

Dried fig fruit of all fig varieties can be utilized to make jam. First the dried figs are washed thoroughly and then left to drain. The hard part of the stem is cut and then the fruit is divided into cubes or quarters. Water and sugar are added and the mixture is boiled at medium heat. When the syrup starts boiling, the dried figs are added and after 10 min, the heat is lowered. One teaspoon of butter is added to give a shiny appearance as well as a tablespoon of lemon juice to increase the viscosity of the syrup.

Nuts or seeds can be added while preparing dried-fig jam. Walnut kernels can be added while boiling in syrup or walnut kernels can be stuffed into whole dried figs and then boiled in syrup. Similarly, sesame seeds can be roasted in oil and then added to the dried figs boiling in syrup (https://sedaninmutfagindan.wordpress.com/2013/01/04/cevizli-kuru-incir-receli/; www.eviminyemekleri.com/susamli-kuru-incir-receli/; accessed January 3, 2022).

25.2.3 Fig delight

Turkish delight, or “lokum” or “locum,” is a traditional confectionery known in Turkey and in the Middle Eastern countries for centuries and based on starch and sugar. It may contain nuts or may be prepared by mixing flavors such as mastic gum or rose water to add special aromas. Dried-fig paste is used as the basic raw material to prepare “fig delight” or “fig lokum.” Fig paste is obtained by grinding dried figs and is commercially available in two different forms: with seeds, with crushed seeds, or seedless. Fig delight, a modified version of Turkish delight, is prepared by mixing fig paste, sugar, modified starch, water, and citric acid. The thick mixture is baked and then cut into cubes and covered with shredded dried coconut (Şen *et al.*, 2015). Fig delight mostly contains walnuts; however, other nuts may also be used. Fig delight is consumed as a dessert or tea/coffee-break treat.

In fig-producing countries, there are many products processed from fig paste. The traditional types are mostly shaped as a sausage and filled or covered with nut kernels, chocolate, or candied citrus peel. At a worldwide level, fig paste is used to make fig bars and fig newtons.

25.2.4 Fig-seed oil

Fig varieties have different requirements regarding fruit set. Fruit of fig varieties that require caprification for fruit set have fertile seeds and thus the kernel contains oil, as in other seeds. However, some varieties may have hard endocarps—these are felt as seeds but no endosperm develops if they are not fertilized. Recently, oil has been extracted from dried fruit of the Sarılop (= Calimyrna) variety and sold commercially. Due to its high antioxidant capacity, fig-seed oil is accepted as an anti-aging compound and has additional medicinal properties such as being an edible oil or being used for skin treatments (Duman and Yazıcı, 2018). The commercial extraction method is cold extraction. Fig seeds obtained from the Sarılop variety contain 29.17% oil (Duman *et al.*, 2018). The other properties of fig-seed oil have been determined, including: acidity (4.6 mg potassium hydroxide g^{-1}), iodine value (176 g I_2 100 g^{-1}), peroxide value (4.5 mEq O_2 kg^{-1}), refractive index (1.480), density (0.925 mg ml^{-1}), viscosity (52.1 mPa), and oxidative stability (2.97 at 110°C h^{-1}). The major fatty acids were linolenic acid (41.8%), linoleic acid (29.0%), palmitic acid (6.7 %), and stearic acid (2.8%) in decreasing order. The oil extracted had a strong antimicrobial effect against Gram-positive bacteria. It also showed antifungal activity against tested molds and yeast. Joseph and Raj (2011) stated that dried seeds contain 30% edible oil that can be used as a lubricant. The fatty acid composition was reported as 33.72% linoleic acid, 32.95% linolenic acid, 18.99% oleic acid, 5.23% palmitic acid, 2.18% stearic acid, and 1.05% arachidic acid.

25.2.5 Fig tea

Fig leaves and fig fruits are consumed as herbal tea infusions, especially in far-eastern Asian countries. Products obtained from fig have been widely used not only as food but also for medicinal uses from antiquity. Many studies have been carried out on the bergapten (5-methoxypsoralen) and psoralen content of fig leaves, which are referred to as having anticancer effects. McGovern (2002) reported the results of two studies on fig leaves that displayed the presence of psoralens and bergapten throughout the growing season. Psoralen levels were three to eight times higher than bergapten levels. Fig leaves are also rich in pentosans, carotene, stigmasterol, tyrosine, ficusin, taraxasterol, β-sitosterol, rutin, sapogenin, calotropenyl acetate, lupeol acetate, and oleanolic acid (Joseph and Raj, 2011). Nakilcioğlu and Hışıl (2013) analyzed the phenolic composition (i.e. gallic acid, chlorogenic acid, epicatechin, syringic acid, rutin, and psoralen content) in dried fig fruit of the Sarılop variety collected from different orchards in the major fig-production region of Turkey. Epicatechin was the major phenolic compound ranging from 14.7 to 23.4 mg 100 g^{-1} dry matter. Rutin content showed higher variation ranging between 5.75 and 12.35 mg 100 g^{-1} dry matter. The content of analyzed phenolic compounds decreased in the order of psoralen, gallic acid, chlorogenic acid, and syringic acid.

The water extract and crude hot-water soluble polysaccharide from fig fruits were investigated for scavenging abilities against DPPH (2,2-diphenyl-1-picrylhydrazyl), superoxide, and hydroxyl radicals, and their reducing power, and the results demonstrated scavenging activity and immune system responses. The polysaccharide had a higher scavenging activity than the water extract on superoxide radicals and hydroxyl anion radicals (Yang *et al.*, 2009). As the medicinal properties unfold, herbal tea based on fig fruit and leaves are becoming more widespread.

25.2.6 Fig vinegar

Vinegar is one of the oldest methods of preservation, as well as a flavoring agent, and is still used all over the world. In addition to its use in gastronomy, vinegar has had medicinal usage for millennia. Vinegar can be produced from all crops or products rich in sugars and starch such as grape, apple, rice, malt, persimmon, and wine. It is produced through a fermentation process whereby yeast converts simple sugars to alcohol and the alcohol is then oxidized to acetic acid (Budak *et al.*, 2014).

There are various methods utilized in vinegar making. The methods used may require longer fermentation periods, as in the traditional methods, or may be produced within a day. Vinegar making is based on acetic acid bacteria belonging to the family *Acetobacteriaceae*, with *Acetobacter* and *Gluconobacter* being the most predominant genera. Their optimum growth temperature is 25–30°C and the optimum pH growth range is 5.0–6.5, although many are resistant to acidic pH levels below 5. These bacteria may produce different organic acids in vinegar, but acetic acid is the dominating one. The genus *Acetobacter* oxidizes alcohol firstly, while *Gluconobacter* preferably oxidizes glucose over ethanol, and therefore the choice of bacteria determines the method to be used. The two major methods of vinegar making are the "surface method" and the "submerged culture" method. The first method is used for traditional vinegar making and is based on wood shavings, which provide oxygen and allow bacterial growth on the interface between the air and liquid layers. In the "submerged culture" method, which is used more in commercial production, continuous oxygen is supplied to accelerate the fermentation process (Budak *et al.*, 2014).

Traditionally made fig vinegar is popular in countries where the fig tree is widespread, but fig vinegar has started to gain recognition in all markets due to its health benefits (Şengün and Kılıç, 2019). Fig vinegar can be made from fresh and/or dried figs. The research work on fig vinegar is rather limited. The western region of Turkey, mainly Aydın and İzmir Provinces, comprises the main area for commercial sun drying of figs destined for the export market. These two provinces alone supply 5% of the world dried-fig trade (INC, 2021).

Şengün (2013) collected home-made fig vinegar recipes (e.g. Fig. 25.2) from eight areas in the Aegean regions of Turkey and performed chemical and microbiological analyses on fig vinegar samples. Analyses of these eight traditionally made fig vinegars showed that the method of production affects the chemical composition. The yeast counts ranged between 0.00 and 6.49 log colony-forming units (CFU) ml^{-1}, the pH between 3.05 to 3.73, the volatile acidity between 1.97 and 6.46 g 100 ml^{-1}, and the total acidity between 2.1 and 6.97 g 100 ml^{-1}. The sample that was derived solely from fresh figs and that had the longest fermentation period had the highest numbers of lactic acid and acetic acid bacteria and thus the highest acidity value.

Fig. 25.2. Fig coffee production.

25.2.7 Photovoltaic-dried fig products

In sunny regions, it is possible to innovate with the use of solar power as an option for drying. The University of Hohenheim in Germany developed a solar dryer that is very practical for subtropical countries and was adopted in South Africa and Afghanistan to dry peach and spices (Lorenz-Ladener *et al.*, 2010). The potential of solar cells for the development of new products also brings the possibility of reaching new market groups, as use of solar energy is linked to more sustainable ways to produce food.

A small solar cell (10 Watts) activates a ventilator that maintains regular air circulation inside the dryer's structure, which looks like a tent and can be also adapted for self-consumption. This system is particularly affordable for small producers and can be a good option during rainy seasons. In Arequipa, Peru, one of the regions with high levels of sun radiation in the world, figs dry in 2 days.

As many market groups take an interest in the quality of food (e.g. whether it is healthy and sustainable, and the social aspect of its production), photovoltaic drying can be a good labeling option. Development of new products such as fig leather is recommended, or dried figs with a harder texture, which can be a good option for fig salami close to the hardness of a real dried meat, as experienced in Peru, offering a very hard version of the fig salami with Brazil nuts.

25.2.8 Solar fig coffee

Roasting is also another possibility for dried-fig processing. In the 19th century, there were fig coffee factories in Austria and south Germany (Boscana, 2009; Silla, 2019). Now there are also some companies in the USA and Argentina that commercialize this product as a caffeine-free beverage that replaces coffee without losing its energetic qualities. To develop alternatives for roasting and cooking with the sun, the solar oven ULOG, designed in Switzerland, has been used successfully in subtropical countries and Central Europe. This oven, made of wood, a recycled printing plate and lambswool, requires direct sunlight to work (Behringer and Götz, 2008) and can be used for fig coffee production. After 2 days of direct exposure to the sun, the dried figs will be hard enough but not burned and will be right for conversion to powder, with sweet and bitter notes. This product offers a good option for producers in sunny regions who want to develop new products with irregular, hard, and small dried figs. For this reason, it is also a "zero food waste" option that can easily be commercialized as a delicatessen product.

25.2.9 Chimbango and kombucha: fermented alternative dried-fig beverages

In South Peru, mainly in Arequipa and Moquegua, dried figs are traditionally used to prepare a fermented beverage called chimbango. For its production, dried figs, water, and (depending on the producer and their style) yeast or chicha de jora (a popular Peruvian fermented beverage made of corn and originating from Inca times) will be needed. In Santa Isabel de Siguas, a town located in Arequipa, this tradition is very old. A high concentration of old fig trees (between 100 and 150 years old) exists in this valley, with more than 1000 trees in the whole valley. Traditionally, the dried figs of best quality were reserved for home consumption and barter in exchange for goods such as Andean grains, Andean potatoes (with several varieties coming from a higher altitude), and dried meat. Those that were too hard or too small were used for the preparation of chimbango.

To make chimbango, to 1 kg of dried figs is added 5 l of cold water (with yeast or half a glass of chicha de jora added to accelerate the fermentation). It can be drunk the next day, but if the producer wants more intensity (i.e. more alcohol), it is necessary to replace the dried figs with others and continue with the fermentation process for a further 2 days. This beverage is normally consumed in celebrations and is a symbol of joy.

Kombucha is another fermented product made with tea, sugar, and a symbiotic culture of bacteria and yeast (also known as scoby). Its botanical name is *Medusomyces gisevi* and, depending on the proportions of sugars, tea, and yeast stock (scoby), the fermentation process will last between 15 and 30 days.

The fermentation process has three stages: 15 or 10 days with the scoby (also known as the "mushroom"), 3 or 5 days with fruits or herbs (without the mushroom or stock yeast), and 3 or 5 days in the bottle (R. Rossini, 2019, personal communication). In this final stage, carbonation occurs, which provides the "soda-style" liquid. Because of its probiotic qualities, this beverage is now becoming popular as a healthy drink.

The Brazilian chef Riders Rossini, based in Lima, Peru, has successfully experimented with dried figs for kombucha production. According to his experience, dried figs have enormous potential for the production of this beverage as they have important quantities of natural sugars that feed the bacteria and yeast and provide an aromatic flavor and carbonation stability once the kombucha is served in a glass.

25.2.10 Fig latex used in cheese making

Fig latex is rich in enzymes, namely rennin, protease, diastase, esterase, lipase, catalase, and peroxidase. Some of these are antioxidant enzymes, such as polyphenol oxidase, catalase, peroxidase, superoxide dismutase, glutathione *S*-tranferase, glutathione peroxidase, and glutathione reductase. Fig latex has been shown to demonstrate rennet-like milk-clotting protease activity that has been used historically in cheese making (Krishnaswamy and Johar, 1960; Badgujar *et al.*, 2014). This property makes fig latex an alternative plant coagulant to be used in case of any limitations occurring regarding calf rennet. Fig latex is mainly found in the shoots, leaf petiole, and unripe female and male fruits. In many countries, fig latex is among the plant rennets that were tested or used as alternatives to calf rennet many decades ago. Krishnaswamy and Johar (1960) reported that the enzyme from the latex of *Ficus carica* L. would be suitable for cheese making, and performed studies in India to investigate its purification and the physicochemical properties (including milk clotting) of the enzyme. Faccia *et al.* (2012) carried out a study on cacioricotta, a traditional Italian goat's cheese obtained from overheated milk (90°C) without the use of starter, and compared male fig latex (traditional) and calf rennet (industrial) as milk-clotting agents. The results showed that use of fig latex was linked to higher amounts of the nitrogen fractions and rich peptides compared with nearly no peptides formed with calf rennet. Caprifig clotting enzymes had strong proteolytic activity and gave a particular composition to the artisanal cacioricotta cheese. Öner and Akar (1993) examined milk-clotting enzymes from fig latex to produce traditional cheese from Gaziantep, Turkey, and compared it with the enzymes from calf rennet. The results showed no significant differences regarding acidity, bitterness, creaminess, off-flavor, and graininess, and supported using fig-tree latex as a chymosin substitute in cheese production.

25.3 Commercial Fig Products Other Than Dried Figs

Fig fruit (*F. carica* L.) is characterized by its high perishability, and this results in a very short fruit shelf life for fresh consumption, and thus it has mainly been destined for processing (Hardenburg *et al.*, 1986; Morton, 1987; Piga *et al.*, 1995). Nearly 50% of the global fig-fruit production is dried, while the remaining part is consumed as fresh or processed as canned fruits, jams and jellies, juices and syrup, liquors, and distillates (Polat and Siddiq, 2012). Drying is mostly carried out by exposure to the sun and requires very low capital, simple equipment, and low energy input. Reducing water activity below the threshold for microbial growth produces figs with excellent sensory properties resulting in food stabilization and reduced chemical and enzymatic changes. Dried fruit may be consumed directly as a whole or chopped to be mixed with other ingredients in muesli or fruit bars, or further processed as jams and pastes or used to obtain alcohol (Aksoy, 2015).

Figs may be processed by using methods other than drying, such as canning or juice extraction. In all cases, processors should consider that the fruits are characterized by high water-activity values and are in the low acidic range when applying different unit operations (Woodborn, 1982).

25.3.1 Fig canning

Canning foods in cans, jars, or plastic containers (trays and pouches) is based on the use of a heat sterilization process. Heat-processed foods should be both free of risk for public health and have an acceptable quality and extended shelf life with respect to fresh products. The main objective of the heat operation is to inactivate viable microorganisms at a desired level. First, pathogen growth should be avoided in thermal sterilized foods, and in this context *Clostridium botulinum* is the target microorganism being the most heat resistant (van Ermengem, 1897). Thus, the heat treatment should reach commercial sterility and render the food free from viable microorganisms, as undestroyed spores of resistant food-poisoning species can produce toxins fatal to humans. Some of these microorganisms including those of known public health significance are capable of growing at temperatures at which the food is likely to be held during distribution and storage (Department of Health, 1994). To avoid post-process contamination, cans and jars should maintain their hermetic closure, which is achieved by vacuum sealing containers. This operation involves removing air to reduce the in-container pressure, following heat treatment and closure, to a level lower than atmospheric pressure, thus helping closure and caps to perfectly adhere to the can and jar ends. Moreover, the headspace oxygen level is minimized, resulting in an extension of the food shelf life. Canned foods are, in general, stored and marketed at room temperature. When deciding to obtain canned figs, one has to consider that the can or jar headspace is under vacuum, that canned foods are generally stored at room temperature, and that figs are non-acidic foods (pH higher than 4.6), and that they have a high water-activity value. At the same time, these four parameters promote *C. botulinum* growth and rarely other *Clostridium* species (Smith *et al.*, 2015), and consequently, a heat sterilization process (i.e. using temperatures higher than 100°C) is required. Non-strict sterilization process, from home-canned figs, resulted in botulinum outbreaks (Woodborn, 1982; Fleck-Derderian *et al.*, 2018). The fig-fruit sterilization step will result in excessive quality loss, and therefore an acidification step is suggested to make heat pasteurization treatment (temperatures lower than 100°C) possible to obtain both a commercial sterility and a good-product quality. Canned figs are very popular in the USA, and a very good description of fruit canning has been given by Featherstone (2015a).

Fig harvest is the first critical point, as the ripening degree should be accurately selected. In fact, they should not be neither too green, as they lack flavor, nor too ripe, as they would be subjected to breaking during processing. Soon after harvesting and selection, figs should be gently water blanched at a temperature range of 66–77°C for 2–8 min to remove wax from the epidermis, and the blanched fruits should then be carefully selected to separate damaged fruits and unwanted matter, and finally the cans should be hand filled. If automatic fillers are used in canning, blanching is carried out inside the can by two different methods: dry or in water. In the dry exhaust mode, the cans move along a steam exhaust box at 98–99°C for a time ranging from 15 to 110 min, depending on the pack size, before adding the cut-out syrup without draining. In the liquid mode, the cans are filled with water and then passed inside a steam box at 98–99°C for 30–40 min, depending on the can size. The cans are then drained and syrup is added at temperatures ranging from 49°C to 66°C. Cut-out sucrose-syrup grades are required by the Standard of Identity and for certification by the US Department of Agriculture (USDA, 1977). A syrup made with a sweetener blend to produce diet canned figs resulted in a product that was shelf stable for at least 180 days without any significant difference compared with the standard sucrose-syrup canned figs (Caetano *et al.*, 2017a). The syrup should be acidified to reduce the fig pH to 4.6 or below. Woodborn (1982) suggested the addition of at least 4% (v/w of figs) of lemon juice, and the containers are then closed and processed with a combination of time/temperature that may vary based on the processor. In general, still retorts are used to pasteurize figs in glass jars, while continuous rotary cookers are preferred for tin-plate cans. The containers should be heated at least to 82°C at the cold spot, and after that are immediately water cooled to 25°C. These unit operations allow a commercial sterile product to be obtained that can be stored at ambient temperature and has a shelf life stability of at least 1 year.

25.3.2 Fig jams and jellies

One of the preferred methods to preserve fresh fruits is to process them as jams or jellies. The contemporary use of pectin and acidification, commonly with citric acid and cooking of fruits (jams and preserves) or fruit juices (jellies) with sugar, results in evaporation of water and a change in structure from a sol to a gel (Featherstone, 2015c). The total soluble solids should be greater than 65%. A proper jellification is obtained with a pH range of 2.8–3.2. The pectin acts as the gelling agent, while sugar and acid favor the change in structure, and water is the plasticizer. When the jam batch is cooked, water evaporates, the sugar content (from fruits and added sugar) increases, and the pectin precipitates. Moreover, the hydrogen ions derived from the added acids reduce the negative charge of the pectin molecules that tend to bond and form a network entrapping water and solutes. The formation of a good gel depends on correct values of pH and sugar and pectin concentration. The final product is characterized by a low pH and a water activity in the range of 0.80–0.85 (Bussiere and Serpelloni, 1985); thus, no pathogenic microbial proliferation is possible. The shelf life limiting factors are the possibility of moisture absorption from the surrounding atmosphere, mold growth on the surface, and oxidation, all of which can be prevented by producing a vacuum inside the glass jar as a packaging procedure. The quality of jams varies greatly, depending on the method used. For home-made jams, the open-kettle process is used by boiling the jam at atmospheric conditions, and thus the batch may exceed 107°C, causing an alteration in color and flavor of the fresh fruit. In the industrial vacuum cooking process, there are different beneficial effects with respect to the previously cited method, as the vacuum applied during cooking allows a maximum batch temperature of 80°C (except before filling). The main advantages are that the color, flavor, and wholeness of fruit pieces are maintained, there is less inverted sucrose, and sugar diffusion in the fruit is enhanced. Fig jams are very popular worldwide and are one of the preferred fruit products in the European market, which is the largest world market for this fruit specialty (CBI, 2018). This has stimulated some scientific research in recent years focused on improving the quality of the product, to assess the effect of processing on the main bioactive compounds of figs.

Levaj *et al.* (2010) studied the evolution of pectin fractions at different ripeness stages of Croatian fig on gel strength and sensory attributes of jams and low-sugar jams (preserves). The results showed that oxalate-soluble pectin fractions were the most abundant in figs. Ripeness of the fruit, although influencing the water and alkaline soluble pectin fractions, did not affect the gel strength of jams. As the fruit ripened, the strength of the preserves decreased. The fruit ripeness stage, however, did not have a significant impact on the sensory properties of either jams or preserves, whereas types of jams differed in preserving taste, firmness, and color (Levaj *et al.*, 2010).

Rababah *et al.* (2011) studied the influence of jam processing on the total phenolics, antioxidant activity, and anthocyanins of different fruit species, including fig, and also evaluated changes in these compounds at monthly intervals for up to 5 months of storage at 25°C. The results showed that the parameters studied decreased mainly because of processing and to a lesser extent during storage. Processing reduced the content of polyphenols, anthocyanins, and antioxidant activity by 76%, 60% and 39%, respectively. In contrast, storage had a minor effect during the first 3 months only on polyphenols and after 5 months on anthocyanins and antioxidant activity. These parameters maintained acceptable values during storage, despite their drastic reduction after processing.

Bof *et al.* (2012) evaluated the effect of processing and storage for 90 days on the antioxidant activity of jellies made with figs of the Roxo de Valinhos variety. Neither processing nor storage resulted in a decrease in antioxidant activity.

Caetano *et al.* (2017b) compared the physical and chemical attributes of regular and low-calorie jam, the latter made by substitution of sucrose with different commercial sweeteners. The results showed that it is possible to produce a light fig jam with all the sweeteners utilized, alone or in different combinations, as the measured parameters fell within the category of dietetic products. These results agreed in part with those found earlier by Hou *et al.* (1999), who optimized low-sugar fig jam by using cherry tomato pulp to turn high-methoxyl fig pectin into low-methoxyl pectin, a natural and synthetic colorant for improving the color and starch syrup, as a substitute for sucrose and found that the new formulation received better scores for color and the same judgment for the other sensory attributes, compared with conventional jams.

25.3.3 Fig juices and syrups

The term fruit juice can be used only to describe a product that is made of 100% fresh fruit or is eventually obtained by reconstituting it from a fruit concentrate (Featherstone, 2015b). The Codex Alimentarius (2005) General Standard for Fruit Juices states the following: "Fruit juice is the unfermented but fermentable liquid obtained from the edible part of sound, appropriately mature and fresh fruit or of fruit maintained in sound condition by suitable means including postharvest surface treatments applied in accordance with the applicable provisions of the Codex Alimentarius Commission."

The unit operations involved in juice processing should operate to keep the physical, chemical, nutritional, and sensory characteristics of the fresh fruit. It is possible, during juicing, to recover volatile substances that will be returned to the same fruit juice species. It is also possible to produce clear or cloudy juices.

Juice may be single if obtained from one fruit species or mixed if obtained by a blend of two or more fruit species. Juices may be obtained directly from fruit extraction or by reconstitution with water from a juice concentrate. Juice commercial sterility is industrially obtained by heat treatment followed by aseptic filling into a flexible material or plastic or glass bottles. Low-acid fruits are adequately stabilized by a simple pasteurization treatment, generally done for a time ranging from 30 s to several minutes in a plate or scraped surface heat exchanger with temperatures varying from 96°C to 104°C; the juice is then aseptically filled in pre-sterilized packages. For low-acid fruit juice, as for fig juice, the fruit is acidified to below pH 4.6 and then goes through pasteurization or a heat sterilization. A 5-log pathogen microbial reduction is required for all juices, as described by Mak *et al.* (2001) and Mazzotta (2001) and reported by the FDA (2001).

Fruit syrups are made from fruits and juices by concentrating to obtain a sweetened product.

Khezri *et al.* (2010) verified the possibility of using fig juice to produce a probiotic juice by fermenting it with three species of lactic acid bacteria (*Lactobacillus casei*, *Lactobacillus plantarum*, and *Lactobacillus delbrueckii*). *L. delbrueckii* was the lactic acid bacteria that grew better on fig juice, reaching nearly 9 log CFU ml^{-1} after 48 h of fermentation and with 6 log CFU ml^{-1} after 4 weeks of storage at 4°C. The use of probiotics negatively affected the sensory scores of taste, odor, consistency, and overall acceptability. *L. casei* was the most acceptable among the three lactobacilli used. The authors, however, affirmed that the hedonic quality of this probiotic juice may be improved by providing an adequate health information label and by using flavor-masking techniques.

Khezri *et al.* (2018) studied the possibility of developing both a probiotic and a symbiotic fig juice by fermenting it with a strain of *L. delbrueckii* or after inulin addition, respectively, and its weekly evolution over a 4-week storage period. The results showed that the fermented fig juices had a significantly higher total phenolic content and antioxidant capacity with respect to control samples and that the fig juice performed very well in terms of survival and proliferation of *L. delbrueckii* to have adequate health promotion. Again, the fermented fig juices had lower sensory scores than the control. The results showed that fig juice is a good candidate for producing a new functional food.

Gabriele *et al.* (2010) attempted, through an engineering approach using rheological analysis, to optimize the fig syrup production. The main process parameters considered were the grinding of fruit pulp and the addition of alkaline salts, to change the pH, improve the pectin extraction, and decrease the temperature of the process. Grinding gave a better structure to the material, and sodium bicarbonate resulted in a more efficient jellification, and when mild process temperatures and concentrations were applied, the result was a more structured product. The authors concluded that by changing process parameters and related variables, a wide range of functional products with different rheological properties may be obtained, ranging from syrup (liquid-like behavior) to sauce (liquid-like behavior) and up to a cream (solid-like behavior). This approach, based on analysis of rheological properties, may support the design of new products having controlled properties and may reduce the effect of raw material variability on the product quality.

Puoci *et al.* (2011) demonstrated that fig syrup may be used as a nutraceutical supplement. *In vitro* and high-performance liquid chromatography analyses gave a specific picture of the main antioxidant compounds in fig syrup. Moreover, results obtained for antioxidant activity, scavenging properties, and anticholinesterase activity showed that this fig specialty may surely be considered as a nutraceutical supplement due to its efficacy in preventing the negative effects caused by free radicals.

25.3.4 Fig distillates and liquors

Fruit spirits are derived from distillation of fermented mash or juice that contains mainly ethanol and water and many other volatile compounds known as congeners, which are very important in imparting the flavor and authenticity of spirits. The best-known fruit spirits are brandies derived from grape wines, but other fruits are traditionally used in different countries, such as apples in France and Italy (Versini *et al.*, 2009), apricots in Hungary, mulberries in Greece, and plums in the Balkan and Eastern

European countries (Tesević *et al.*, 2005). In the distillation process, the liquid is heat vaporized, collected, and re-collected in a container. The separation of distillate components is driven by the difference in their boiling points. Spirits become richer in ethanol with respect to the raw material because the boiling point of ethanol is lower than that of water, and thus heating the mixture to a temperature that is higher than the boiling point of ethanol and lower than that of water results in vapor richer in ethanol than water. Production of fruit spirits is normally carried out in batches with two types of equipment, a copper Charentais alembic (French style) and distillation columns (German style). The main difference between the two styles is the number of distillations needed to obtain an adequate ethanol concentration, which is multiple using the French style and single using the German style (Claus and Berglund, 2005).

Galego *et al.* (2011) used five different deodorized and concentrated fig distillates as an alcoholic base to produce a myrtle berry liquor. The results showed that improving processes of deodorization and concentrations of the fig distillates resulted in myrtle liquors with a significantly higher content of polyphenols and anthocyanins with respect to samples obtained by using the raw fig spirit, which obtained profiles very similar to those found for Sardinian myrtle berry liquors (Montoro *et al.*, 2006; Tuberoso *et al.*, 2007).

Miličević *et al.* (2017) found that the use of immobilized cell fermentation may influence the volatile profile and the sensory properties of spirits obtained from two Croatian commonly grown fig varieties. Aroma compounds were analyzed by gas chromatography (GC)/flame ionization detector and sensory analysis was performed using the German DLG model. About 40 different volatile compounds were found, and the use of immobilized yeast cells increased the ethanol content and decreased the methanol and esters amounts, with respect to traditional fermentation. The immobilized technology application, moreover, improved the sensory quality of the fig spirit, as odor was enhanced compared with control samples, and thus this technique is suggested as a good candidate for the production of high-quality fig-fruit distillates.

Rodríguez-Solana *et al.* (2018) explored the influence of varietal attributes and processing methods on the quality of fig distillates in terms of volatiles and sensory qualities of the distillates produced from eight Portuguese varieties of figs, the blends of laboratory sun-dried figs, and the variety of fig spirits found in the market. The authors detected a total of 130 volatile compounds by headspace solid-phase microextraction/GC-mass spectrometry (MS) and found that distillates obtained from fresh figs were better than those obtained from dried samples, due to a lower concentration of compounds associated with negative sensory attributes, such as ethyl acetate, ethyl lactate, and diethyl succinate, and thus proper storage of figs before distillation is recommended. Moreover, it was possible to better differentiate the spirits by using a unique variety, instead of more than one. The two white fig varieties, Burjassote Branco and Castelhana Branca, are recommended for the preparation of high-sensory quality fig distillates due to their better volatile profile.

Liquors are obtained by hydro-alcoholic infusion or decoction of fruits, with added sugar and sometimes added flavors that are usually derived from fruits, herbs, or nuts. Palassarou *et al.* (2017) prepared several liquors from dried Smyrna-type figs harvested in five producing zones in Greece and analyzed the volatile compounds of the obtained liquors by GC/MS, GC/MS/MS, and GC/time-of-flight-MS. The analyses revealed a total of 225 volatile and semi-volatile compounds and significant differences among the fig samples and their respective liquors. The results showed that there was significant aroma diversity among different fig samples and liquors, and that the figs from the Peloponnese geographical area produced liquors that were sensorially preferred for their "rum/brandy" odor profile. In particular, the Peloponnese Kalamon figs had a unique odor profile that distinguished them among the other fig samples. The aroma unicity of Kalamona liquor was attributed to the development of β-damascenone, which could thus be used as a marker. The authors concluded that Kalamon figs are the best for producing a commercially promising fig liquor.

25.4 Concluding Remarks

Pyke (1971) asserted that "The production of a new food commodity might seem to be a trivial matter unworthy of serious consideration; this is not necessarily so. The technological expertise upon which any one item depends may require the full depth of scientific understanding." It is undoubtedly true that the development of new food products by finding a new use for old items is of paramount importance for a company to be in the market. Every company needs to develop new products to increase their turnover. This continuous challenge is expensive, as it requires extensive research, adequate personnel, facilities, and financial availability. Unfortunately, there is no certainty that any of the newly developed products will be launched in the market or that the extra turnover given by its launch will cover the expenses afforded. In fact, of the thousands of new products launched worldwide each year, only a very minor percentage can be found on the shelves a year after their launch. Generally only one in 58 new product ideas arrives successfully on the market. However, the only way for a company to survive is to develop new products.

The development of a new product may be divided into several phases including idea generation, product design and process development, product marketing, and launch. Thus, what could be the main idea for proposing new fig products for the future? We undoubtedly have to consider the main important characteristics of fig fruit, which are its functionality and its flavor. Fresh fig fruit's functionality is due to its nutritional properties, as figs are rich in fiber, polyphenols, and phytosterols (Caliskan, 2015). Moreover, the fig flavor is unique and makes it one of the most delicious fruits in the world. Both of these merits are partially or totally lost during fig processing, due to thermal or physical alteration, as demonstrated earlier, and thus we can propose the development of new processed and stable fig products that are at the same time shelf stable and very similar to fresh ones and can thus meet the new consumer demand. The use of non-thermal or alternatives to traditional thermal technologies will surely help in fulfilling this aim. Among non-thermal technologies, the use of high hydrostatic pressure technology, by inactivating enzymes and microorganisms (mainly vegetative forms), could result in fig products that are fresher, more natural, and healthier. With this technology, it would be possible to develop jams or juices with nutritional and sensory properties very similar to fresh fruits, and in fact an important number of such fruit products are on the world market (Indrawati *et al.*, 2003). Ohmic heating is another promising advanced thermal food-processing technique that involves internal heat generation due to electrical resistance when an electric current is passed through the food; it provides fast and homogeneous heating, and thus thermal damage is minimized. The beneficial effect on different fruits, such as apple, apricot, grape, orange, peach, and others, has recently been reviewed (Kaur *et al.*, 2016), and so it would be interesting to study the effect of this technology on figs. Other emerging technologies, such as pulsed-electric fields, ultrasound, and microwave heating, could be of interest to the fig industry.

References

Aksoy, U. (2015) Dried figs. In: *UNECE Workshop, July 2015, Izmir, Turkey*. Available at: www.unece.org/fileadmin/DAM/trade/agr/meetings/ge.02/2015/Presentations-DDP-Workshop/03July-Aksoy-2-Figs.pdf (accessed December 28, 2021).

Armstrong, W. (2006) Sex Determination and Life Cycle of *Ficus carica*, www2.palomar.edu/users/warmstrong/pljun99b.htm (accessed December 28, 2021).

Badgujar, S.B., Patel, V.V., Bandivdekar, A.H., and Mahajan, R.T. (2014) Traditional uses, phytochemistry and pharmacology of *Ficus carica*: a review. *Pharmaceutical Biology* 52, 1487–1503.

Behringer, R. and Götz, M. (2008) *Kochen mit der Sonne: Solar kochen und backen in Mitteleuropa*. Ökobuch Verlag, Germany.

Bof, C.M.J., Fontana, R.C., Piemolini-Barreto, L.T., and Sandri, I.G. (2012) Effect of freezing and processing technologies on the antioxidant capacity of fruit pulp and jelly. *Brazilian Archives of Biology and Technology* 55, 107–114.

Boscana, M.P.I. (2009) *Ajuntament de Llucmajor*. Obra Social la Caixa, Barcelona, Spain.

Budak, N.H., Aykin, E., Seydim, A.C., Greene, A.K., and Güzel-Seydim, Z.B. (2014) Functional properties of vinegar. *Journal of Food Science* 79, R757–R764.

Bussiere, G. and Serpelloni, M. (1985) Confectionery and water activity determination of a_w by calculation. In: Simato, D. and Multon, J.L. (eds) *Properties of Water in Foods in Relation to Quality and Stability*. Martinus Nijhoff Publishers, Dordrecht, The Netherlands, pp. 627–645.

Caetano, P.K., Farinazzi-Machado, F.M.V., Nasser, F.M.C.M., Furlaneto, K.A., and Vieites, R.L. (2017a) Storage stability of standard and diet figs canned in syrup. *Food Science and Technology* 37, 154–159.

Caetano, P.K., Vieites, R.L, Daiuto, E.R., and Rolim de Moura, S.C.S (2017b) Processing and quality control of diet and conventional fig jams. *Brazilian Journal of Food Technology* 20: e2016026.

Caliskan, O. (2015) Mediterranean figs (*Ficus carica* L.) functional food properties. In: Preedy, V.R. and Watson, R.R. (eds) *The Mediterranean Diet: An Evidence-based Approach*. Academic Press, Cambridge, MA, pp. 629–637.

CBI (Center for the Promotion of Imports) (2018) Exporting Jams and Jellies to Europe. Ministry of Foreign Affairs, The Hague, The Netherlands. Available at: www.cbi.eu/market-information/processed-fruit-vegetables-edible-nuts/jams-jellies/europe/ (accessed December 28, 2021).

Chevalier, J. and Gheerbrant, A. (2007) *Diccionario de los Símbolos*. Herder, Barcelona, Spain.

Claus, M.J. and Berglund, K.A. (2005) Fruit brandy production by batch column distillation with reflux. *Journal of Food Process Engineering* 28, 53–67.

Codex Alimentarius (2005) General Standard for Fruit Juices and Nectars. CODEX STAN 247-2005. Food and Agriculture Organization of the United Nations, Rome/World Health Organization, Geneva.

Department of Health (1994) *Guidelines for the Safe Production of Heat Preserved Food*. Department of Health, Her Majesty's Stationery Office, London.

Díaz Simón, A. (2011) *Recetas con Historia*. Editorial Planeta, Barcelona, Spain.

Duman, E. and Yazıcı, A.S. (2018) Yaş incir (Mor Güz-Sarı Lop) çekirdek ve çekirdek yağlarının fiziko-kimyasal özellikleri (Physico-chemical properties of seeds and seed oil of fresh (Mor Güz-Sarılop) fig). *Anadolu Ege Tarımsal Araştırma Enstitüsü Dergisi* 2, 69–76.

Duman, E., Şimşek, M., and Özcan, M.M. (2018) Monitoring of composition and antimicrobial activity of fig (*Ficus carica* L.) fruit and seed oil. *Journal of Agroalimentary Processes and Technologies* 24, 75–80.

Faccia, M., Picariello, G., Trani, A., Loizzo, P., Gambacorta, G., *et al.* (2012) Proteolysis of Cacioricotta cheese made from goat milk coagulated with caprifig (*Ficus carica sylvestris*) or calf rennet. *European Food Research* and *Technology* 234, 527–533.

FDA (US Food and Drug Administration) (2001) Code of Federal Regulations—Title 21—Food and Drugs, Part 113: Thermally Processed Low-Acid Foods Packaged in Hermetically Sealed Containers, and Part 114: Acidified Foods. Available at: www.fda.gov/medical-devices/medical-device-databases/code-federal-regulations-title-21-food-and-drugs (accessed December 28, 2021).

Featherstone, S. (2015a) Canning of fruit. In: Featherstone, S. (ed.) *A Complete Course in Canning and Related Processes*, Vol. 3. *Processing Procedures for Canned Food Products*, 14th edn. Woodhead Publishing, Cambridge, UK, pp. 104–105.

Featherstone, S. (2015b) Canning of juices, fruit drinks, and water. In: Featherstone, S. (ed.) *A Complete Course in Canning and Related Processes*, Vol. 3. *Processing Procedures for Canned Food Products*, 14th edn. Woodhead Publishing, Cambridge, UK, pp. 135–168.

Featherstone, S. (2015c) Jams, jellies, and related products. In: Featherstone, S. (ed.) *A Complete Course in Canning and Related Processes*, Vol. 3. *Processing Procedures for Canned Food Products*, 14th edn. Woodhead Publishing, Cambridge, UK, pp. 313–349.

Fleck-Derderian, S., Shankar, M., Rao, A.K., Chatham-Stephens, K., Adjei, S., *et al.* (2018) The epidemiology of foodborne botulism outbreaks: a systematic review. *Clinical Infectious Diseases* 66, S73–S81.

Gabriele, D., Migliori, M., and de Cindio, B. (2010) Innovation in fig syrup production process: a rheological approach. *International Journal of Food Science and Technology* 45, 1947–1955.

Galego, L.R., Da Silva, J.P., Almeida, V.R., Bronze, M.R., and Boas, L.V. (2011) Preparation of novel distinct highly aromatic liquors using fruit distillates. *International Journal of Food Science & Technology* 46, 67–73.

Hardenburg, R.E., Watada, A.E., and Wang, C.Y. (1986) Fig. In: Hardenburg, R.E., Watada, A.E., and Wang, C.Y. (eds) *The Commercial Storage of Fruits, Vegetables and Nursery Stocks*. Agriculture Handbook No. 66. US Government Printing Office, Washington, DC, p. 40.

Hou, W.N., Kim, H.M., and Go, E.K. (1999) Processing of low sugar fig jam for marketable production. *Korean Journal of Food Science and Technology* 31, 651–657.

INC (International Nut and Dried Fruit Council) (2021) Dried figs. In: *Nuts and Dried Fruits Statistical Yearbook 2020/21*. INC, Reus, Spain, pp. 65–68.

Indrawati, van Loey, A., Smout, C., and Hendrickx, M. (2003) High hydrostatic pressure technology in food preservation. In: Zeuthen, P. and Bùgh-Sùrensen, L. (eds) *Food Preservation Techniques*. Woodhead Publishing, Cambridge, UK, pp. 428–448.
Joseph, B. and Raj, S.J. (2011) Pharmacognostic and phytochemical properties of *Ficus carica* Linn – an overview. *International Journal of PharmTech Research* 3, 8–12.
Kaur, R., Gul, K., and Singh, A.K. (2016) Nutritional impact of ohmic heating on fruits and vegetables – a review. *Cogent Food and Agriculture* 2, 1159000.
Khezri, S., Dehghan, P., Mahmoudi, R., and Jafarlou, M. (2010) Fig juice fermented with lactic acid bacteria as a nutraceutical product. *Pharmaceutical Sciences* 22, 260–266.
Khezri, S., Mahmoudi, R., and Dehghan, P. (2018) Fig juice fortified with inulin and *Lactobacillus delbrueckii*: a promising functional food. *Applied Food Biotechnology* 5, 97–106.
Krishnaswamy, M.A. and Johar, D.S. (1960) Preparation of Cheddar and Processed Cheese Using the Milk Clotting Enzyme from *Ficus carica* L. In: *Proceedings of the Symposium on Proteins, August 1960, Mysore, India*. Available at: http://ir.cftri.res.in/4953/ (accessed December 28, 2021).
Levaj, B., Bunić, N., Dragović-Uzelac, V., and Kovacević, D.B. (2010) Gel strength and sensory attributes of fig (*Ficus carica*) jams and preserves as influenced by ripeness. *Journal of Food Science* 75, S120–S124.
Lorenz-Ladener, C., Götz, M., Hoedt, H., and Zachl, E. (2010) *Trocknen und Dörren mit der Sonne*. Okobuch Verlag GmbH, Rastede, Germany.
Mak, P.P., Ingham, B.H., and Ingham, S.C. (2001) Validation of apple cider pasteurization treatments against *Escherichia coli* O157:H7, *Salmonella*, and *Listeria monocytogenes*. *Journal of Food Protection* 4, 1679–1689.
Mazzotta, A.S. (2001) Thermal inactivation of stationary-phase and acid-adapted *Escherichia coli* O157:H7, *Salmonella*, and *Listeria monocytogenes* in fruit juices. *Journal of Food Protection* 64, 315–320.
McGovern, T.W. (2002) Botanical briefs: the fig—*Ficus carica* L. *Cutis* 69, 339–340.
Miličević, B., Aćkar, Đ., Babić, J., Jozinović, A., Miličević, R., *et al.* (2017) Impact of the fermentation process with immobilized yeast cells on the aroma profile and sensory quality of distillates produced from two fig (*Ficus carica* L.) cultivars. *Poljoprivedra* 23, 49–55.
Montoro, P., Tuberoso, G.I.G., Perrone, A., Piacente, S., Cabras, P., and Pizza, C. (2006) Characterization by liquid chromatography electrospray tandem mass spectrometry of anthocyanins in extracts *Myrtus communis* L. berries used for the preparation of myrtle liquor. *Journal of Chromatography A* 1112, 232–240.
Morton, J.F. (1987) Fig. In: *Fruits of Warm Climates*. Creative Resource Systems, Winterville, North Carolina. Available at: www.hort.purdue.edu/newcrop/morton/fig.html (accessed December 28, 2021).
Munar, F. (1999) *De la Figuera a la Taula: Receptes de Cuina*. Edicions Documental Balear, Palma de Mallorca, Balearic Islands, Spain.
Nakilcioğlu, E. and Hışıl, Y. (2013) Research on the phenolic compounds in Sarilop (*Ficus carica* L.) fig variety. *Gıda* 38, 267–274.
Öner, M.D. and Akar, B.D. (1993) Separation of the proteolytic enzymes from fig tree latex and its utilization in Gaziantep cheese production. *LWT – Food Science and Technology* 26, 318–321.
Palassarou, M., Melliou, E., Liouni, M., Michaelakis, A., Balayiannisc, G., and Magiatis, P. (2017) Volatile profile of Greek dried white figs (*Ficus carica* L.) and investigation of the role of β-damascenone in aroma formation in fig liquors. *Journal of the Science of Food and Agriculture* 97, 5254–5270.
Piga, A., d'Aquino, S., Agabbio, M., and Papoff, C. (1995) Influenza del confezionamento con film plastici sulla conservazione del fico. *Italus Hortus* 2, 3–7.
Polat, A.A. and Siddiq, M. (2012) Figs. In: Siddiq, M., Ahmed, J., Lobo, M.G., and Ozdali, F. (eds) *Tropical and Subtropical Fruits: Postharvest Physiology, Processing and Packaging*. Wiley-Blackwell, Chichester, UK, pp. 455–478.
Puoci, F., Iemma, F., Spizzirri, U.G., Restuccia, D., Pezzi, V., *et al.* (2011) Antioxidant activity of a Mediterranean food product: "fig syrup". *Nutrients* 3, 317–329.
Pyke, M. (1971) A food scientist's reflections. *Food Science* 9, 10–12.
Rababah, T.M., Al-Mahasneh, M.A., Kilani, I., Yang, W., Alhamad, M.N., *et al.* (2011) Effect of jam processing and storage on total phenolics, antioxidant activity, and anthocyanins of different fruits. *Journal of the Science of Food and Agriculture* 91, 1096–1110.
Rodríguez-Solana, R., Galego, L.R., Pérez-Santín, E., and Romano, A. (2018) Production method and varietal source influence the volatile profiles of spirits prepared from fig fruits (*Ficus carica* L.). *European Food Research and Technology* 244, 2213–229.

Rosenberg, K. (2011) *Die Ayurveda-Ernährung*. Südwest Verlag, Munich, Germany.

Şen, F., Özer, K.B., Aksoy, U., and Koç Güler, S. (2015) Changes in quality of dried fig (*Ficus carica* l.) delight in different packages under cold and ambient storage, Ege Üniversitesi Ziraat Fakültesi Dergisi. *Journal of Ege University Faculty of Agriculture* 52, 235–242.

Şengün, I.Y. (2013) Microbiological and chemical properties of fig vinegar produced in Turkey. *African Journal of Microbiology Research* 7, 2332–2338.

Şengün, İ.Y. and Kılıç, G. (2019) Microflora, bioactive components and health effects of various kinds of vinegars. *Akademik Gıda* 17, 89–101.

Silla, T. (2019) *Feige*. Mandelbaum Wien, Berlin, Germany.

Smith, T.J., Hill, K.K., and Raphael, B.H. (2015) Historical and current perspectives on *Clostridium botulinum* diversity. *Research in Microbiology* 166, 290–302.

Stromberg, H. (2014) *Das Kochbuch der Nationalmannschaft*. Edel Books, Hamburg, Germany.

Tesević, V., Nikićević, N., Jovanović, A., Djoković, D., Vujsić, L.J., *et al.* (2005) Volatile components from old plum brandies. *Food Technology and Biotechnology* 43, 367–372.

Tuberoso, C.I.G., Melis, M., Angioni, A., Pala, M., and Cabras, P. (2007) Myrtle hydroalcoholic extracts obtained from different selections of *Myrtus communis* L. *Food Chemistry* 101, 806–811.

USDA (US Department of Agriculture) (1977) United States Standards for Grades of Canned Kadota Figs. Available at: www.ams.usda.gov/sites/default/files/media/Canned_Kadota_Figs_Standard%5B1%5D.pdf (accessed December 28, 2021).

van Ermengem, E. (1897) Ueber einen neuen anaëroben Bacillus und seine Beziehungen zum Botulismus. *Zeitschrift für Hygiene und Infektionskrankheiten* 26, 1–56.

Versini, G., Franco, M.A., Moser, S., Barchetti, P., and Manca, G. (2009) Characterisation of apple distillates from native varieties of Sardinia island and comparison with other Italian products. *Food Chemistry* 113, 1176–1183.

Woodborn, M. (1982) pH of Oregon-grown figs and their acidification for home-canning. *Journal of Food Protection* 45, 1245–1247.

Yang, X.M., Yu, W., Ou, Z.P., Ma, H.L., Liu, W.M., and Ji, X.L. (2009) Antioxidant and immunity activity of water extract and crude polysaccharide from *Ficus carica* L. fruit. *Plant Foods for Human Nutrition* 64, 167–173.

26 Traditional Ethnomedicinal Usage of *Ficus carica*

Uygun Aksoy*
Department of Horticulture, Faculty of Agriculture, Ege University, İzmir, Turkey

26.1 Introduction

The use of common fig (*Ficus carica* L.) and other *Ficus* species in traditional medicine dates back to time immemorial in many regions, whether found widespread in natural flora as a wild species or introduced later. Fig is not only a delicacy and a functional food, but all parts of the plant including unripe, fresh and dried fruit, leaves, roots, branches, bark, and latex are also traditionally utilized for medicinal purposes. The traditional treatments and medicines have played a more significant role in some regions of the world such as Africa and Far-eastern Asia. Factors such as rapid urbanization, access to modern medical care, and lack of written formulas have caused an erosion of knowledge on the use of fig in many parts of the world. However, during the last decades, traditional medicines and treatments have gained popularity in the western world, especially after the 1980s, which has resulted in many more recent publications. Based on this interest, the World Health Organization (WHO) started compiling work at a global level. The WHO defines traditional medicine as "the sum total of the knowledge, skill, and practices based on the theories, beliefs, and experiences indigenous to different cultures, whether explicable or not, used in the maintenance of health as well as in the prevention, diagnosis, improvement or treatment of physical and mental illness" (WHO, 2013). The WHO Strategy, developed first in 2002, focuses on the importance of widening the knowledge base in traditional medicine and strengthening quality assurance, safety, proper use, and effectiveness. The European Union has also given a mandate on herbal medicinal products to the European Medicines Agency and set up the application rules and procedure for traditional use: "no clinical tests and trials on safety and efficacy are required as long as sufficient safety data and plausible efficacy are demonstrated; involves assessment of mostly bibliographic safety and efficacy data; must have been used for at least 30 years, including at least 15 years within the European Union, and are intended to be used without the supervision of a medical practitioner and are not administered by injection" (EMA, 2004a,b). In traditional herbal medicine, some cases are tested and proven through scientifically accepted research, but orally transferred knowledge and practices still play a significant role.

Fig trees are accepted as holy species in monotheist religions signifying fig's beneficial role, possibly based on its therapeutic properties known and practiced by society. As well as the fig fruit, its roots, leaves, latex, seeds, and seed oil

* Email: uygun.aksoy@gmail.com

 Advances in Fig Research and Sustainable Production (M.A. Flaishman and U. Aksoy)
DOI: 10.1079/9781789242492.0026

also possess medicinal properties. There are many reviews that have presented a thorough analysis of the medicinal properties of various parts of *F. carica* (Oliveira *et al.*, 2009; Joseph and Raj, 2011; Patil and Patil, 2011; Chawla *et al.*, 2012; Barolo *et al.*, 2014; Al-Snafi, 2017; Li *et al.*, 2021). Many studies, including those carried out on animals, have proved that fig has biological activities such as antioxidant, anticancer, hepatoprotective (protecting the liver), immunomodulator (regulating the immune system), hypocholestrinemic (lowering cholesterol), antipyretic (lowering fever), anti-inflammatory (preventing inflammation), antibacterial, antihelmintic, and dissolving warts (Akçiçek, 2019). In Indian traditional medicine practice, *Ficus* species are used in the traditional systems of Ayurveda, Unani, Siddha, and homeopathy (Varanasi and Narayana, 2007; Subramanian *et al.*, 2013). Common fig (*F. carica*) fruits (unripe green, ripe, and dried), fig syrup, latex, leaves, and roots are rich in secondary metabolites that possess antioxidant, anticancer, cytotoxic, anti-inflammatory, and hypolipidemic effects. Due to this richness, fig is reported as a cure for a wide range of health problems related to digestive and dermatologic problems, constipation, loss of weight, and endocrine, reproductive, respiratory, and infectious diseases (Lansky and Paavilainen, 2010; Badjugar *et al.*, 2014; Al-Snafi, 2017). The edible fruits of *F. carica* were used traditionally in treating hemorrhoids, insect stings, gout, ulcers, and skin infections such as warts and viruses. Fig fruits were recommended as a laxative for those facing constipation problems, as nutrients for those active in sports, pregnant women or those requiring a high energy intake, and for mental and physical exhaustion. Its medicinal effects are considered to be "antipyretic, tonic, purgative, alexiteric, aphrodisiac, lithontriptic, anti-inflammatory, expectorant, diuretic, and used for treatment of pharyngitis, gastritis, bronchitis, irritative cough, weakness, paralysis, thirst, diseases of the liver and spleen, pain in the chest, to cure piles, to stimulate growth of hair, and for leprosy and nose bleeding" (Al-Snafi, 2017). The fig as a part of the Iranian agricultural heritage Estahbanat rainfed fig system is reported to be rich in calcium and protein and to have medicinal uses as a laxative and sedative, to reduce fever and thirst, and to relieve coronary thrombosis, inflammation of the spleen, and hemorrhoids (Anon., 2011). Despite being an allergen causing skin and eye irritation, fig latex is also used as a traditional medicine. Scientific results elucidating the phytochemical composition of different plant parts of the fig tree and the pharmacological properties of these compounds are increasing every year. In 2012, 85% of the 121 volatile constituents of fig leaves and 97% of those found in the fruits were first reported (Li *et al.*, 2012). More studies are needed to collect information on the ethnomedicinal use of fig and where possible to confirm its medicinal usages. Many new constituents of fig fruits and other organs will be revealed as analytical procedures are further enhanced.

This chapter provides insights into the traditional medicinal uses of *F. carica* and summarizes the studies reported in the literature that have investigated different organs used for specific purposes. Furthermore, articles that derive and collate the intangible heritage of fig used as a folk medicine are also discussed.

26.2 Traditional Medicinal Use

All over the world, there are thousands of *Ficus* species distributed mainly in the tropical, subtropical, and mild temperate climate zones that are used for medicinal use. It should be noted that *Ficus* species that are naturally present in the flora appear to be the main source for ethnomedical use. The species, variety, pedo-climatic factors, different organs (stem, bark, branch, latex, leaf, fruit), or the part of the fruit (pulp or skin), and the level of maturity (unripe, fresh, or dried) largely affect the composition of phytochemicals, their use in ethnomedicine, and the expected therapeutic effects. Although traditional uses date back centuries or even millennia, with the recent advances in analytical systems, many studies are being initiated to identify the bioactive compounds responsible for the therapeutic effect or to find out more about the methods of extraction, safe concentrations, or the mode of action. In the case of volatile compounds in fig fruits and leaves, which display a very wide diversity, each study finds new constituents (Oliveira *et al.*, 2010; Li *et al.*, 2012). In analysis of the volatiles of the *F. carica* leaves and fruit, only 18 compounds were in common out of the total of 210 volatiles quantified. The major volatile constituents in leaves were psoralen (10.12%), β-damascenone (10.17%), benzyl alcohol (4.56%), behenic acid (4.79%), and bergapten (1.99%), whereas furfural (10.55%), 5-methyl-2-furaldehyde (10.1%), and benzeneacetaldehyde (6.59%) were found in volatile oils of the fruit (Li *et al.*, 2012). The fruit volatiles in the light- and dark-colored Portuguese fig varieties Pingo de Mel, Branca Tradicional, Borrasota Tradicional, Verbera Preta, and Preta Traditional revealed an abundance of monoterpenes and aldehydes that differed from the leaves. The fruit peel and pulp volatile profiles were quite similar. Minor differences existed between the light- and dark-colored varieties. In total, 59 volatile compounds were identified and classified as aldehydes, alcohols, ketones, esters, monoterpenes, sesquiterpenes, and norisoprenoids. Among the 59 identified, 39 were first reported for *F. carica* (Oliveira *et al.*, 2010). Identification of new constituents in common fig, as well as other *Ficus* species, has unveiled new functions and medicinal usage. Classification of the *Ficus* taxon has undergone changes and been reclassified according to the symbiotic relationship with wasps of the family Agonidae (Ramírez, 1977; Berg, 1989), resulting in possible deviations in *Ficus* species names, creating controversy in the literature. However, *F. carica* including both the female and male figs remains the same.

Fig has been included in the widely used references such as pharmacopeias and the Physician's Desk Reference for Herbal Medicines. Indian, Chinese, and Iranian traditional medicine refer to use of figs as fresh or dried fruits, leaves, branches, trunks, roots, or latex. Many of the biological effects such as the antibacterial, antiviral, and antioxidant activity of *F. carica* have been studied and subjected to scientific studies that explore which chemical compound is responsible for a certain effect (Barolo *et al.*, 2014). Its uses comprise curing loss of appetite, colic, indigestion, constipation, dysentery, intestinal wounds, and intestinal parasite problems. It also has antispasmodic, antibacterial, anti-inflammatory, and natural laxative effects. Other uses include aiding in visual impairment, heart diseases, colon and lung problems, and sore throat and cough. Fig latex is applied in traditional treatments to heal boils and warts. In some European countries, fig syrup is widely used as a mild laxative (Başer, 2017). The compendium *Figs, the Genus Ficus* (Lansky and Paavilainen, 2010), based on a wide range of sources, refers to the traditional herbal medicinal usage of many *Ficus* species other than *F. carica* such as *F. asperifolia* Miq (Africa), *F. awkeotsang* L. (China, Singapore, Taiwan, and Thailand), *F. benghalensis* L. (India), *F. benjamina* L. (India and worldwide), *F. capensis* Thunb. (southern Africa, California), *F. chlamydocarpa* Mildbr. & Burret (Africa), *F. citrifolia* Mill. (Caribbean, Florida, and South America), *F. exasperata* Vahl (leaf) (tropical Africa, Yemen, and India), *F. hispida* L. (Southeast Asia, Australia, and the USA), *F. palmata* Forsk. (Middle East and south-west Asia), *F. pumila* L. (fruits and leaves) (China, Japan, and South-east Asia), *F. septica* Burm. f. (stems and roots) (Indonesia and Australia), and *F. sycomorus* L. (stem and bark) (Tanzania, Nigeria, Middle East, and tropical and South Africa), which have a wide range of therapeutic effects including antiviral, inhibition of smooth muscle contraction, smooth muscle relaxant, hepatoprotective, anticancer, anti-inflammatory, antiosteoporotic, antihelmintic, anticonvulsant, antimalarial, diuretic, antidiabetic, and antiparasitic. Lansky and Paavilainen (2010) reported that *F. carica* has digestive, immunological adjuvant, trace element deficienciy, antispasmodic, and antiplatelet uses. The ethnomedical uses of *F. carica* have been derived from various reviews and are summarized in Table 26.1 indicating the part of the fig plant utilized for therapy. Some sources specify the effect, such as inflammation of the breast or lymphatic system, whereas others report it as being effective to cure inflammation. Similarly, the use is attributed to different organs of the fig such as the fruit and leaves or leaves and latex, while others do not specify.

26.2.1 Medicinal use of the fig fruit

Fig fruit is a nutritious food consumed as fresh, dried, or processed. The unripe, ripe, and dried fig fruit can be further processed. In some producer countries, low-quality fig fruits may be used as animal feed. Varietal differences, stage of maturity, and preharvest factors affect the composition and thus the therapeutic effects. Furthermore, the storage conditions and processing methods utilized may affect the availability or bioaccessibility of phytochemicals. Sun drying

Table 26.1. Ethnomedical uses of various parts of *Ficus carica* L.

Therapeutic use	Part(s) utilized	Reference(s)
Various plant parts		
Anticancer	Fruit, leaf, latex	Lansky and Paavilainen (2010); Joseph and Raj (2011); Badjugar *et al.* (2014); Al-Snafi (2017)
Scavenging activity and immune response	Fruit (hot-water extract)	Joseph and Raj (2011); Badjugar *et al.* (2014)
Antiplatelet	Dried fruit (water extract)	Gilani *et al.* (2008)
Anemia	Fruit	Joseph and Raj (2011); Badjugar *et al.* (2014)
Ascites	Fruit	Badjugar *et al.* (2014)
Antipyretic	Ethanol extract of fruit and branches	Joseph and Raj (2011); Patil and Patil (2011); Badjugar *et al.* (2014); Al-Snafi (2017)
Antioxidant	Fruit, leaf	Badjugar *et al.* (2014); Barolo *et al.* (2014); Rahmani and Aldebasi (2017)
Nephro-protective	Fruit	Rahmani and Aldebasi (2017)
Eye vision problems (associated with *Amaranthus viridis*)	Fruit	Barolo *et al.* (2014)
Antispasmodic	Dried fruit	Menale *et al.* (2006); Gilani *et al.* (2008); Rahmani and Aldebasi (2017)
Breast inflammation and mastitis, externally	Fruit	Lansky and Paavilainen (2010)
Breast cancer	Fruit	Jasmine *et al.* (2015)
Emollient	Fruit	Mars (2003)
Diarrhea, orally	Fruit	Lansky and Paavilainen (2010); Barolo *et al.* (2014)
Warts, externally	Fresh fruit, sap	Lansky and Paavilainen (2010); Al-Snafi (2017)
Lung tuberculosis, orally	Fresh fruit	Lansky and Paavilainen (2010)
Liver diseases and hepatitis, orally, decoction	Fresh fruit	Lansky and Paavilainen (2010); Badjugar *et al.* (2014); Al-Snafi (2017)
Gangrene, orally	Fresh fruit	Lansky and Paavilainen (2010)
Adenitis, inflammation of the lymphatic system	Dried fruit	Lansky and Paavilainen (2010)
Boils, externally as warm poultice	Dried fruit	Lansky and Paavilainen (2010); Badjugar *et al.* (2014)
Bronchitis, roasted, powder/decoction, orally	Dried fruit	Lansky and Paavilainen (2010)
Expectorant	Fruit	Patil and Patil (2011); Mawa *et al.* (2013); Badjugar *et al.* (2014)
Demulcent	Fruit	Mars (2003)
Excess mucus secretion, roasted powder, orally	Dried fruit	Lansky and Paavilainen (2010)
Cold, decoction, orally	Dried fruit	Lansky and Paavilainen (2010)
Cough, suppressant, decoction of boiled fruit, orally	Fruit, dried fruit	Menale *et al.* (2006); Badjugar *et al.* (2014)
Cystitis and nephritis and urinary disorders, decoction, orally	Dried fruit	Joseph and Raj (2011)
Lenitive and antispasmodic, orally	Dried fruit	Menale *et al.* (2006); Lansky and Paavilainen (2010)
Dermatitis, externally	Dried fruit	Lansky and Paavilainen (2010)
Burns, externally as a paste	Dried fruit, fruit, latex	Lansky and Paavilainen (2010); Badjugar *et al.* (2014)
Maturation of skin pustules	Heated dried fruit	Barolo *et al.* (2014)

Continued

Table 26.1. Continued.

Therapeutic use	Part(s) utilized	Reference(s)
Gingivitis, decoction in milk, as a gargle	Dried fruit	Lansky and Paavilainen (2010)
Hemorrhoids, orally	Dried fruit	Lansky and Paavilainen (2010)
Eczema, externally as paste	Dried fruit	Lansky and Paavilainen (2010)
Anti-inflammation, orally or externally	Fruit, dried fruit	Lansky and Paavilainen (2010); Joseph and Raj (2011); Patil and Patil (2011)
Oral disinfectant for sore throat and aphthous complaints, as decoction, orally	Dried fruit	Joseph and Raj (2011)
Pleurisy (inflammation of the lining around the lungs), roasted powder	Dried fruit	Lansky and Paavilainen (2010)
Mouth cavity diseases, tooth abscess, decoction	Dried fruit	Lansky and Paavilainen (2010); Badjugar *et al.* (2014)
Measles, decoction	Dried fruit	Lansky and Paavilainen (2010)
Smallpox, decoction, orally	Dried fruit	Lansky and Paavilainen (2010)
Weakness (as energy source)	Dried fruit soaked in water	Anaç (2012); Badjugar *et al.* (2014)
Hemorrhage	Fruit juice with honey	Patil and Patil (2011)
Fodder for animals	Dried fruit (culled figs), leaf	Mars (2003); Lansky and Paavilainen (2010); Lazreg-Aref *et al.* (2010)
Digestant, orally	Dried fruit	Lansky and Paavilainen (2010); Anaç (2012)
Laxative, direct, hot-water extract or decoction, orally against constipation and abdominal cramping	Dried fruit, fruit paste	Mars (2003); Lazreg-Aref *et al.* (2010); Joseph and Raj (2011); Patil and Patil (2011); Mawa *et al.* (2013); Al-Snafi (2017)
Laxative, orally	Baked dried leaf, latex, or dried entire plant	Lansky and Paavilainen (2010)
Reproductive effect	Aqueous ethanol extract of fruit	Al-Snafi (2017)
Diuretic, hot-water extract or decoction, orally	Dried fruit, latex, fresh leaf	Mars (2003); Lansky and Paavilainen (2010); Joseph and Raj (2011); Patil and Patil (2011); Mawa *et al.* (2013)
Colic treatment	Fruit, root, and leaf	Mawa *et al.* (2013)
Indigestion	Fruit, root, and leaf	Mawa *et al.* (2013)
Antacid	Fruit ash	Joseph and Raj (2011)
Seeds and seed oil		
Antimicrobial, suppresses growth of selected bacteria, fungus, and yeast	Seeds and seed oil (of Sarılop, a caprified fig variety)	Duman *et al.* (2018); Küçükerdönmez *et al.* (2021)
Latex		
Antifungal	Fresh latex	Badjugar *et al.* (2014); Al-Snafi (2017)
Anticancer	Latex	Rahmani and Aldebasi (2017)
Antiviral (some human viruses)	Latex	Al-Snafi (2017); Rahmani and Aldebasi (2017)
Antiparasitic: earthworms, nematodes, and other larvae	Water and methanol extracts of fruit and latex	Badjugar *et al.* (2014); Al-Snafi (2017)
Warts, externally	Dried or fresh latex	Lansky and Paavilainen (2010); Joseph and Raj (2011); Badjugar *et al.* (2014); Barolo *et al.* (2014); Al-Snafi (2017)

Table 26.1. Continued.

Therapeutic use	Part(s) utilized	Reference(s)
Eye sore	Latex	Lagrez-Aref *et al.* (2010)
Skin diseases	Fruit and latex	Badjugar *et al.* (2014)
Abscess, promotes maturation, externally	Latex	Lansky and Paavilainen (2010)
Inhibition of intestinal absorption of cholesterol	Latex	Badjugar *et al.* (2014)
Digestive for domestic animals, decoction, orally mixed with feed	Latex	Menale *et al.* (2006); Lansky and Paavilainen (2010)
Boils, externally	Fresh latex	Lansky and Paavilainen (2010)
Bee sting, externally	Fresh latex	Yazıcıoğlu and Tuzlacı (1995)
Eruptions, externally	Fresh latex	Lansky and Paavilainen (2010)
Ulcers and wounds, externally, mixed with egg yolk or vegetable oil	Fresh latex, fruit, fruit juice	Lansky and Paavilainen (2010); Bandujar *et al.* (2014)
Pains, externally as paste	Latex from stem, dried fruit	Lansky and Paavilainen (2010)
Hemorrhoids, externally	Latex from stem	Yazıcıoğlu and Tuzlacı (1995); Lansky and Paavilainen (2010)
Sunburn (together with olive oil)	Latex (raw)	Menale *et al.* (2006)
Pain, scorpion bite	Latex (female and male trees)	Yazıcıoğlu and Tuzlacı (1995); Honda *et al.* (1996)
Antiangiogenic and antiproliferation	Latex, leaf	Al-Snafi (2017)
Anthelmintic, hot-water extract	Latex, fresh leaf	Lansky and Paavilainen (2010); Joseph and Raj (2011); Mawa *et al.* (2013)
Leaf, bark, and branches		
Demulcent, hot-water extract, orally	Fresh leaf	Lansky and Paavilainen (2010)
Skin problems, including hair, emollient, hot-water extract, orally	Fresh leaf	Yazıcıoğlu and Tuzlacı (1995); Lansky and Paavilainen (2010)
Diabetes, decoction as a supplement	Leaf	Lansky and Paavilainen (2010); Joseph and Raj (2011); Al-Snafi (2017)
Calcification in liver and kidneys, decoction	Leaf	Joseph and Raj (2011)
Anti-inflammatory including treatment of gout, pancreatitis, and prostatitis	Leaf	Patil and Patil (2011); Badjugar *et al.* (2014); Rahmani and Aldebasi (2017)
Antiviral	Leaf	Rahmani and Aldebasi (2017)
Antipyretic	Leaf	Rahmani and Aldebasi (2017)
Hypocholesterolemic, decoction	Leaf	Lazreg-Aref *et al.* (2010); Badjugar *et al.* (2014)
Hypolipidemic	Leaf	Badjugar *et al.* (2014)
Hepatoprotective, against jaundice	Leaf	Gond and Khadabadi (2008); Patil and Patil (2011)
Liver and spleen disorders	Leaf	Mohan *et al.* (2007)
Hypoglycemic for type I diabetes, decoction	Leaf	Lazreg-Aref *et al.* (2010)
Tuberculosis	Leaf	Mawa *et al.* (2013)
Nutritional anemia	Leaf	Mawa *et al.* (2013)
Cough	Leaf	Mawa *et al.* (2013)
Anthelmintic	Leaf	Mawa *et al.* (2013)

Continued

Table 26.1. Continued.

Therapeutic use	Part(s) utilized	Reference(s)
Antinematicidal	Leaf	Rahmani and Aldebasi (2017)
Antimutagenic	Branches	Al-Snafi (2017)
Edema, decoction, orally	Dried branchlets	Lansky and Paavilainen (2010)
Pharyngitis and pharyngeal problems, infusion, or direct orally	Dried shoots	Lansky and Paavilainen (2010)
Stomatitis, orally	Dried shoots	Lansky and Paavilainen (2010)
Buccal problems, infusion orally	Dried shoots	Lansky and Paavilainen (2010)
Wounds of head, externally	Ash from stem and twigs	Yazıcıoğlu and Tuzlacı (1995); Lansky and Paavilainen (2010)
Leukoderma, skin disorder	Roots	Patil and Patil (2011)
Ringworms	Roots	Patil and Patil (2011)
Part not specified		
Incontinence, hot-water extract, orally	Unspecified	Lansky and Paavilainen (2010)
Hematemesis, hot-water extract, orally	Unspecified	Lansky and Paavilainen (2010)
Birth, to ease delivery, decoction, orally	Unspecified	Lansky and Paavilainen (2010)
Emmenagogue, hot-water extract, orally/as a pessary, vaginal	Unspecified	Lansky and Paavilainen (2010)
Frostbite, externally	Unspecified	Lansky and Paavilainen (2010)

has been a very common practice in producer countries for thousands of years, supplying year-round dried figs not only at the site of origin but also in many other parts of the world, triggering its traditional use for medicinal purposes. Few studies have been carried out to investigate the effect of processing (e.g. producing jam or vinegar) on bioactive compounds. Petkova *et al.* (2019) examined the biochemical fraction in fresh, frozen, and processed (jam) fruit of Croatian Yellow Giant fig, a light yellow variety. Jam was prepared by adding 800 g of sugar, 250 g of water, 2.5 g of citric acid, and 4 g of methoxyl pectin to 2 kg of fresh figs. The analysis focused on moisture content, ash, titratable acidity, total chlorophylls, total carotenoids, total phenols, total flavonoids, and total anthocyanins, as well as the sugars fructose, glucose, and sucrose. The antioxidant capacity was determined by 2,2-diphenyl-1-picrylhydrazyl (DPPH) and ferric reducing antioxidant power (FRAP) assays. Fresh fruit contained the highest levels of bioactive compounds and antioxidant activity, but deep freezing was not effective in preserving these levels and after 4 months of storage at −18°C, significant decreases were registered in sucrose, total phenol, total flavonoid, and total anthocyanin contents. Processing figs as jam particularly preserved the carotenoids and phenolic compounds potential.

As some of the bioactive compounds are responsible for the color of the peel and the pulp, significant differences may be present among light- and dark-colored fig varieties (Kamiloğlu and Capanoğlu, 2015), or between the peel and the pulp. However, this difference is not limited to color. The edible fig fruit is botanically a composite fruit comprising vegetative and generative parts, and thus the composition of the skin may vary from that of the pulp, even in light-colored varieties (Elmagülü *et al.*, 2020). Similarly, the composition of persistent fruit that is set without fertilization as brebas and second-crop fruit of common-type varieties differs from the second crops of Smyrna and San Pedro types with caducous fruit that requires pollination for fruit set. In caprified second-crop fruit, pollination and fertilization result in "seeds" or achenes, with fully developed oil-rich endocarps that promote the flavor and at the same time contribute to some additional health benefits.

Fig fruit is a rich source of calcium, potassium, and iron, but is low in sodium. The calcium content of fresh figs is around 35 mg 100 g^{-1} but is 162 mg 100 g^{-1} in dried figs. As a rich calcium source, fig fruit is traditionally recommended for pregnant women and toddlers who are in the process of rapid tooth and bone formation. The potassium content is around 232 mg 100 g^{-1} in fresh fruit and 680 mg 100 g^{-1} in dried fruit. It is recommended as a rich potassium source in DASH (dietary approaches to stop hypertension) diets (Akçiçek, 2019).

Polysaccharides are among the bioactive compounds that have importance in traditional medicine, including Chinese medicine. They are long-chain polymers of monosaccharide units bound together by glycosidic linkages. Polysaccharides are known to have antioxidant, antitumor, anti-inflammatory, antidiabetic, and hypolipidemic effects, and regulate the immune system functions. Research focusing on the role of modulation played by fig polysaccharides on dendritic cells revealed that they are able to activate these cells, which are responsible for initiation of adaptive immune responses. This activation was "partially via the dectin-1/Syk pathway, and promotes their maturation, which leads to an increased production of inflammatory factors, augmenting T cell responses." Activation of dectin-1 induced the production of multiple cytokines. The outcome of the research work showed that fig "polysaccharides may be useful as an immune stimulant and a promising candidate as a potential medicine with low toxicity" (Tian *et al.*, 2014).

Kamiloğlu and Capanoğlu (2015) analyzed Sarılop (= Calimyrna) and Bursa Black fig varieties and evaluated the effects of sun drying on major phytochemicals with antioxidant capacity. Sarılop has a yellow skin and amber pulp color, whereas Bursa Black has dark purple skin and a red internal color. Bursa Black had a higher phenolic acid content, especially due to its purple color, rich in chlorogenic acid, ellagic acid, and *p*-coumaric acid compared with the pulp. In contrast, the pulp had a comparatively higher gallic acid content. In the Sarılop variety, the major phenolic compound was rutin. In Bursa Black, cyanidin 3-rutinoside was predominant. Sun drying at 31–34°C for 8 days reduced the chlorogenic acid and *p*-coumaric acid contents by more than 50% and increased the gallic acid content by 67% in both varieties. Both the total flavonoids and total proanthocyanidin contents were increased in Sarılop fruits after drying. Drying is thought to adversely influence the medicinal properties due to the possibility of enzymatic decomposition or degradation of polyphenols under high temperatures. However, the total phenolic content was increased in Sarılop fruit after sun drying. The two possible reasons for this were given as higher extractability of bioactive compounds due to breakdown of cell walls, and as a result of Maillard reaction products causing browning of the fruit color. In the case of Bursa Black, the total phenolic content decreased despite increases in major flavonoids such as rutin, apigenin, and kaempferol rutinoside (Kamiloğlu and Capanoğlu, 2015). Sun drying of fig fruit results in a reduction of volume and easy storage and distribution due to the low water content. In this regard, most of the dried figs traded worldwide come from yellow-colored varieties such as Sarılop, Sabz, and Kalamata, which seem to preserve most of their health properties after drying.

In general, fig fruit is rich in phytochemicals and its wider use is to help digestion. However, the effect of digestion on the targeted bioactive compounds is not known. Kamiloğlu and Capanoğlu (2015) carried out an *in vitro* simulation study on Sarılop and Bursa Black varieties. In a simulation of gastric digestion, the total anthocyanin content was increased in purple-colored Bursa Black. The reason was attributed to the pH, which dropped from pH 5 in the fresh fig to pH 1.7 after pepsin digestion and may be related to metabolization to "some non-colored forms, oxidized, or degraded into other chemicals."

Scientific studies aim to test and find evidence to support traditional medicinal uses. Aqueous ethanol (80%) extract of *F. carica* fruit was tested for two purposes: (i) to relieve spasms in the middle part of the small intestine, the jejunum, which helps digestion of food received from the stomach; and (ii) to prevent platelet formation and clotting *ex vivo* (Gilani *et al.*, 2008). When applied at a dose of 0.1–3.0 mg ml^{-1}, the *F. carica* extract produced relief of contractions in the rabbit jejunum, possibly through the activation of ATP-sensitive potassium channels. This result supports the role that fig fruit plays in the digestion and its laxative effects, facilitating movement of food through the gastrointestinal system. For its effect on prevention of platelets in humans, an extract of 0.6 and 0.12 mg ml^{-1}

inhibited platelet aggregation induced by adenosine 5′-diphosphate and adrenaline, signifying its important use in inflammatory disorders. Extracts of *F. carica* were found to have phytochemicals that have health properties such as alkaloids, flavonoids, coumarins, saponins, sterols, and terpenes.

A water extract of dried fig fruit was effective in relieving the symptoms of mild to moderate atopic dermatitis among children aged 4 months to 14 years and is recommended as a safe, efficient, and tolerable alternative to routine application of 1% hydrocortisone (Abbasi *et al.*, 2017). Unripe fig fruit is used as a homeopathic medicine and its uses include "aperient, emollient, demulcent, nutritive, for removal of stones in kidney or bladder, cures piles, gout, disorders of liver and removal of warts" (Subramanian *et al.*, 2013).

26.2.2 Medicinal use of the fig latex

Latex is the thick, milky, vascular exudate that flows when a branch, leaf, fruit, or trunk is cut in male or female trees of *F. carica*. Fig latex is known for its proteolytic activity, especially due to the cysteine protease ficin; however, it also contains collagenolytic protease, lipase, amylase, and chitinolytic enzymes. Latex also comprises amino acids, vitamins, carbohydrates, lipids, cardiac glycosides, and alkaloids. Latex is traditionally used for medicinal purposes, either fresh or dried. Latex can be collected from fig trees when they are abundant in early spring, dried under shade, and then stored for future applications. Application of latex can be directly externally or internally or mixed with honey, fenugreek flour, barley flour, starch, wheat, ground bitter almonds, mustard, egg yolk, grease, or wax. It is used locally as a coagulant agent, especially in making cheese or as a meat tenderizer (Lansky and Paavilainen, 2010). Fig latex is traditionally used for warts, which are small tumor-like benign protrusions caused by human papilloma virus. When compared with cryotherapy, applying fig latex had no adverse effects and cured within a shorter period of time (Bohlooli *et al.*, 2007).

Latex has powerful bactericidal and fungicidal effects and plays a protective role against pathogens that may attack the fig tree. The latex yield and composition vary according to the variety, organ, and time of sampling. Analysis of latex extracted from 20 Tunisian female figs (Smyrna, San Pedro, and common types) and male figs revealed high protease activity compared with lipase and amylase in all varieties. In Smyrna-type varieties and in male figs, the proteolytic activity increased during fruit maturation, whereas it was highest at the unripe fruit stage in the common- and San Pedro-type varieties. Lipase activity stopped in all fig varieties during the receptive period, except in male figs, which displayed an increase in lipase and protease activity. This similar trend in lipase and protease enzyme activities may indicate that it is related to the latex volatiles, which may be accepted as a sign of coevolution of the latex and enzyme activities (Lazreg-Aref *et al.*, 2018). The latex volatiles are the chemical signals for the "nursery pollination mutualism" whereby the fig tree hosts the fig wasp in its inflorescence for its reproduction, and the wasp in return pollinates the female flowers. Fig latex releases mainly mono- and sesquiterpenes as volatiles, which play a vital role in the attraction of fig wasps (Hossaert-McKey *et al.*, 2010; Lazreg-Aref *et al.*, 2018).

Latex from different organs shows divergence. Latex from unripe fruit, the petiole, and the trunk were compared to ascertain the latex-based defense mechanisms in fresh (unripe fruit and petiole) and lignified (trunk) tissues of *F. carica*. Petiole latex was rich in trypsin-inhibitor proteins, whereas the unripe fruit and trunk latex had high levels of cysteine proteases. The latex of all three organs had high levels of galloylglycerol, which is possibly a defense-related metabolite. The pathogenesis-related proteins were expressed at higher levels in the trunk latex compared with the petiole or unripe fruit latex, implying a protective role against microorganisms. In latex obtained from unripe fig fruits, the sesquiterpenoid and phenylpropanoid pathways, which are linked to defense volatiles, were expressed less (Kitajima *et al.* 2018). The protein concentration in fig latex shows a steady increase during fruit development, although the concentration of the specific constituents may vary. Ficin levels are higher at the beginning of fruit development, while the chitinolytic activity is higher at the beginning of flowering. Higher chitinolytic activity levels provide protection against insects and fungi, but also facilitate dissolving the fig wasps that have entered the fig fruit. The collagenolytic protease content does not show a marked divergence over time (Lazreg-Aref *et al.*, 2018).

26.2.3 Medicinal uses of the fig leaf

Fig leaf has been used traditionally as a tea or medicine, and as an infusion, it is still a highly valued herbal tea in many countries. Fig leaves are used to feed animals; however, for humans it is less edible and is utilized more as a medicine. Several studies on the composition of *F. carica* leaves have suggested a composition rich in secondary metabolites comprising flavonoids (quercetin and luteolin), esters and alcohols, volatile oil (psoralen and bergapten), phenolic compounds (chlorogenic acid, rutin, psoralen), triterpenoids (calotropenyl acetate, methyl maslinate, lupeol acetate), proteins (proteoses, peptone, amino acids containing tyrosine), and proteolytic enzymes (Lansky and Paavilainen, 2010).

One of the ancient uses of fig leaves was for the treatment of various cancers, which were tested on some animals or *in vitro* studies based on human cells. Anticancer activity tests were carried out on liver, cervical, breast, prostate, and colon cancers. The results obtained in *in vitro* studies showed that fig leaves could inhibit the growth of liver and cervical cancer cells. A water extract of fig leaves slowed down the movement as well as the viability of breast cancer cells by decreasing expression of the genes involved in the cell cycle. Various bioactive compounds found in fig leaves are reported to inhibit the growth of certain cancers, such as rutin inhibiting prostate cancer cells, δ-elemene inhibiting colon cancer cells, and β-caryophyllene showing a selective antiproliferative effect on colorectal cancer cells (Li *et al.*, 2021). Major treatments using fig leaves also include "treatment of diseases associated with inflammation, in the prevention of nutritional anemia and as anthelmintic" (Mahmoudi *et al.*, 2016).

The phenolic composition may display differences according to the sex, variety, pedoclimatic conditions, and cultural practices; however, in general, the levels of phenolic compounds in fig leaves are reported to be higher than in the fig fruit. The phenolic contents in fig leaves exceeded the levels found in red wine or tea (Li *et al.*, 2021). Analysis of the phenolic contents of the leaf, pulp, and peel of two Portuguese fig varieties, Pingo de Mel and Branca Tradicional, revealed a similar profile comprising 3-*O*- and 5-*O*-caffeoylquinic acids, ferulic acid, quercetin 3-*O*-glucoside, quercetin 3-*O*-rutinoside, psoralen, and bergapten. The leaf organic acid profile comprised oxalic, citric, malic, quinic, shikimic, and fumaric acids. Comparison of the antioxidant activity of the pulp, peel, and leaves showed that the leaves were the most effective organ and the only part that had the capacity to scavenge superoxide radicals due to their phenolic content and composition (Oliveira *et al.*, 2009). Fig leaves contain various phenolic compounds and for many years were reported to have diverse pharmacological effects. Every year the composition is widened, proposing new areas of research. The volatile compounds that define the aroma are also used in food processing and cosmetic industries.

The phenolic compounds have different scaffolds, chemical structures, and pharmacological activities. Li *et al.* (2021) referred to a total of 126 chemical constituents in the leaves of the common fig, *F. carica*, that had antioxidant, anticancer, antidiabetic, hepatoprotective, anticholinesterase, anti-herpes simplex virus type 1, antibacterial, anti-inflammatory, and renoprotective activities. The phenolic compounds in fig leaves can be evaluated under different categories such as hydroxybenzoic acids, hydroxycinnamic acids, flavonoids, coumarins, furanocoumarins, volatile constituents, triterpenoids, and others (Li *et al.*, 2021). Both hydroxybenzoic acids and hydrocinnamic acids are widely found in other foods as well as figs. Hydroxybenzoic acids are derivatives of benzoic acid containing the C6–C1 scaffold, whereas hydroxycinnamic acid has a scaffold of C6–C3. They have been shown to have health benefits. Hydroxybenzoic acids have antioxidant, anti-inflammatory, antibacterial, and hepatoprotective activities, and directly affect the antioxidant activity of the fig leaf. In fig leaves, about ten hydroxycinnamic acids have been reported showing antioxidant, anti-inflammatory, antitumor, hypoglycemic, and neurodegenerative properties. "Flavonoids contain two benzene rings (A and B) linked via a heterocyclic pyran ring (C)" (Li *et al.*, 2021). In fig leaves, 34 flavonoids have been reported, which are known to possess anticancer, antidiabetic, antioxidant, hepatoprotective, anti-inflammatory, renoprotective, and antibacterial activities.

Flavonoid content exerts a marked effect on the antioxidant activity. Rutin, as the major flavonoid in fig leaves, is reported to possess diverse biological activities including anticancer, anti-inflammatory, antidiabetic, antioxidant, and hepatoprotective. It is also reported to have a positive impact on some neurodegenerative processes (Li *et al.*, 2021). To determine the compounds responsible for the biological activity, fig leaves collected from Mount Meron in Israel were oven-dried at 60°C and extracted with 70% ethanol. A total of 22 flavonoids were identified in the extract determined with high-performance liquid chromatography coupled with mass spectrometry. The results revealed that luteolin was the main free flavonoid, not bound in the form of a glycoside (187 mg kg^{-1} extract). In fig leaves, the leading flavonoid was luteolin (680 mg kg^{-1} extract), followed by quercetin (630 mg kg^{-1} extract). Free flavonoids comprise less than 10% of the total flavonoids present in *Ficus* species, as in the case of quercetin, where the free level was 53 mg kg^{-1} compared with the total of 630 mg kg^{-1} extract. Biochanin A was present mainly as a free aglycone, although at low concentrations (Vaya and Mahmood, 2006; Jeong *et al.*, 2009). The variable concentrations or constituents in these studies are mainly due to the sample analyzed and the extraction methods. Many other studies are still required to determine the metabolic functions of the defined constituents and the conditions under which they are formed or become active.

In figs, the highest total phenolic content is in the leaves, followed by the fruit peel and pulp. The major reason for high levels in the leaves is attributed to their screening function and the protection against ultraviolet (UV) radiation by phenolic compounds, namely flavonoids (Mahmoudi *et al.*, 2016). However, this topic requires further studies in *F. carica* to determine the mechanisms involved. Stezner *et al.* (2019) in their search to identify the role of flavonoids and hydroxycinnamic acid derivatives in filtering UV radiation and protecting sunflower leaf tissue found that both are located in the upper epidermis layers and that under low light conditions, these two compounds act in a complementary way. In contrast, high light conditions triggered accumulation of hydroxycinnamic acids in the vacuoles, and hydroxycinnamic acids contributed solely to UV-A protection of the leaf tissue.

An analysis of fig leaves taken from Algerian male (one tree) and female (uniferous = bearing one crop and biferous, two crops) trees, dried at room temperature (20°C) and extracted with methanol, determined the levels of flavonoids, alkaloids, coumarins, and saponins. The total phenolic content varied slightly among the analyzed group of varieties and the average concentrations decreased in the order of biferous > uniferous > male within the range of 52.3 down to 46.1 mg gallic acid equivalents g^{-1} dry plant extract. The varieties that had the highest phenolic content displayed the highest capacity to scavenge free-radical DPPH. The leaf extracts proved to be more effective against Gram-positive compared with Gram-negative bacteria. Although the complete explanation is unknown, this difference in sensitivity was attributed to the outer membrane of Gram-negative bacteria, which contains peptidoglycan and lipopolysaccharide, creating a strong hydrophilic surface acting as a barrier (Mahmoudi *et al.*, 2016).

Fig leaves are rich in coumarins and furanocoumarins. "Coumarin contains a benzene ring fused with an α-pyrene ring, whereas furanocoumarins are tricyclic aromatic compounds containing an α-benzopyrone fused furan ring system" (Li *et al.*, 2021). In fig leaves, six coumarins have antioxidant, antibacterial, and antidiabetic activities, and the main coumarins are umbelliferone, 4′,5′-dihydropsoralen, and marmesin, with umbelliferone being the most abundant. Furanocoumarins have anticancer, antioxidant, and anticholinesterase activities, and the main furanocoumarins in fig leaves are reported to be psoralen and bergapten, with psoralen being the most abundant furanocoumarin possessing anticancer and anticholinesterase activities (Li *et al.*, 2021). The psoralen concentration is always higher than that of the other furanocoumarins including bergapten. The ratio of psoralen to bergapten also varied during the season and ranged from 2.8 to 7.7 (Innocenti *et al.*, 1982). There is a highly significant positive correlation between the psoralen content and antioxidant activity. Psoralen is also known as the cause of phototoxic reactions of fig leaves (Li *et al.*, 2021). Continuous contact with fig leaf or unripe fruit (latex) followed by exposure to sunlight or artificial UV-A light may result in phytotoxic reactions on the skin called "phytophotodermatitis". This occurs due to the presence of phototoxic compounds such as psoralen and bergapten, which are at higher levels in spring and summer (Özdamar *et al.*, 2003; Polat *et al.*, 2008; Al-Snafi, 2017). In studies carried out at Ege University (İzmir, Turkey) to determine the effect of drying conditions on the psoralen content of dried Sarılop fig leaves, mature seasonal fig leaves were collected during June and dried in an oven, in outdoor conditions under shade, or in outdoor conditions exposed to direct sun during the day. The results showed that the leaves dried under direct sun had five times more psoralen compared with oven- or shade-dried leaves (U. Aksoy and F. Şen, unpublished data). Thus, while preparing herbal tea or for use as traditional medicine, it is important to know the psoralen and bergapten contents in fig leaves.

In addition to the intrinsic factors linked to the sampled leaf, the extraction method applied also affects the phytochemical compounds derived from fig leaves, as well as the biological activity of the extract. The chemical composition of water and methanol extracts were found to be different. However, in the search for inhibition of enzymes related to diabetes mellitus and Alzheimer's disease, both water and methanol extracts were found to exert strong α-glucosidase and α-amylase inhibition accompanied by high antioxidant activity. The inhibition potential of the water extract was higher than that of the methanol extract against acetylcholinesterase and butyrylcholinesterase enzymes. However, the methanol extract exerted anticancer activity against the human breast cancer cell line MDA-MB-231 cells (Ergül *et al.*, 2019).

Jeong *et al.* (2009) repeatedly extracted dried and powdered fig leaf with methanol for 4 h at 80°C and the collected extract was tested against 11 oral bacteria. The results showed that the efficacy differed according to the tested bacteria, with strong antibacterial activity against *Streptococcus gordonii*, *Streptococcus anginosus*, *Prevotella intermedia*, *Prevotella gingivalis* and *Aggregatibacter actinomycetemcomitans* (minimum inhibitory concentration 0.156–0.625 mg ml^{-1}; minimum bactericidal concentration to kill 99% of bacteria 0.313–0.625 mg ml^{-1}). The activity was linked to the anticaries activity of phenolic compounds by inhibiting growth of *Streptococcus mutans* or of glucosyltransferases (Jeong *et al.*, 2009). The antibacterial effects against *Escherichia coli* and *Staphylococcus aureus* were moderate (Ergül *et al.*, 2019).

Volatile compounds in fig leaves are also of interest from a medicinal point of view, as they show anticancer, anti-inflammatory, and anticholinesterase properties. In total, 38 volatile compounds have been reported with medicinal properties in fig leaves (Li *et al.*, 2021). Fig leaves also have phenolic acids such as ferulic acid and phytosterols such as taraxasterol with medicinal properties (Jeong *et al.*, 2009).

26.2.4 Medicinal use of the woody parts of fig

In traditional medicine, the trunk, branches, roots, or bark of the fig tree are used due to their phytochemical composition, which is also rich in phenolic compounds. The fig roots are used as a tonic in treating leukoderma, which is the formation of white patches on the skin due to the partial or total loss of skin pigmentation or depigmentation. A tonic prepared from fig roots has also been used against ringworm (Al-Snafi, 2017; Lazreg-Aref *et al.*, 2018).

26.3 Fig as a Folk Medicine and Historical Recipes

Figs were known in the Near East prior to the Greek and Roman periods, but there are nearly no written sources. The Sumerian civilization that dominated southern Mesopotamia during the Chalcolithic and early Bronze Ages (4500–1900 BCE) identified figs together with apples, grapes, honey, cheese, onions, and garlic as their daily food. A prescription dating back to *c.*2750 BCE recognized the medicinal benefits of figs with pears and thyme (Sutton, 2014). Although the fruit, as the most easily accessible, was utilized more, different parts of the fig such as the branches, fruit, trunk, bark, or suckers were also used directly. For easier storage, the fruits, leaves, or other parts of the fig were dried and powdered or burned to obtain ash to be used as a medicine either through external application or for internal use. There are many ancient formulas based on fig ashes. Fig ash is mixed with wax and rose oil and applied externally to heal burns or wounds and prevent scarring. Fig wood ash mixed with olive oil was used externally to clean ulcers or scars on the skin, to decrease cramps or nerve pains, to whiten teeth, and to induce sweating. The ash and olive oil mixtures are also taken internally with water for bruises and cramps or to prevent the formation of platelets. Fig ash when mixed with honey is recommended for cold sores (Lansky and Paavilainen, 2010).

Turkey is currently the prime producer of figs; however, the importance of figs in Anatolia dates back a few thousand years (Sutton, 2014). The traditional use of figs among the rural communities inherited by word of mouth was subject to a few studies. Anaç (2012) quotes from Derviş Mehmed Zillî (1611–1682), known as Evliya Çelebi, an Ottoman explorer who traveled throughout the Empire, who mentions that figs were used in Köşk (a township in Aydın, Turkey) as a digestive and energy source and to strengthen eyesight. Figs and bread were the staple food and energy source during the economic crisis period in rural areas. During the same period, "in Western herbal, Nicholas Culpeper (1616–1654) wrote: 'A syrup made of the fig leaves is extremely good for the treatment of cough, lung diseases and it dissolves congealed blood caused by bruises or falls'" (Li *et al.*, 2021).

Özatalay (2014), in her interviews, reported that figs were used in the following ways: (i) to help digestion and prevent constipation, to increase blood circulation, and to heal wounds if dried fig fruits are cooked in hot water for half an hour, mashed, and then applied to the wound; the latex also helps healing of wounds; (ii) fig latex can cure the pain of scorpion bites; (iii) fig leaf infusions when brewed as a black tea can help patients with eczema or bone pains, and can help cure wounds; and (iv) dried fig fruits soaked in olive oil for 8–10 h help to cure stomach ache, or are fed to animals to clean up their stomachs to help them recover quickly after labor.

Başer (2017) gave two traditional recipes for a decoction using figs:

1. A 10% syrup or decoction is prepared from dried fig fruit and used as a laxative. For preparation, 120 g of dried figs is ground and soaked in 600 ml of water for 3 h, then drained through a muslin cloth, and 400 g of sugar is added and it is boiled for few minutes.

2. Macerated fresh-fig leaves or a 5% decoction can be used to ripen or puncture an abscess by making a compress with a clean cloth wetted with the decoction. Drinking two to three cups of the decoction helps to suppress hemorrhoids.

26.4 Concluding Remarks

As one of the first cultivated perennials with high nutritive properties known to humans and having a rich bioactive composition, fig has been used as a traditional medicine for thousands of years. There is still a need to collect the traditional knowledge on ethnomedical use of *F. carica* from different regions of the world. The ethnomedicinal usage of *Ficus* is mostly linked to the species that dominates the natural flora. Similarly, traditional use of *F. carica* is more concentrated in areas where it is cultivated or found in the natural flora. The reviews taken as references in this chapter also state that ethnomedicinal use of *F. carica*, the common fig is more concentrated in the Middle East or Asia, chiefly in Iran, Turkey, Pakistan, and India.

Because of the diversity in biologically active phytochemicals, different *Ficus* species have been subject to screening for thousands of years and the best-performing ones are reflected in practice. Lansky and Paavilainen (2010) compiled the medicinal properties of various *Ficus* species including *F. carica*. Additionally, during the last decade, various reviews have been released on the medicinal properties of *F. carica*, most of them overlapping regarding their sources. Honda *et al.* (1996) linked the use of herbal medicines with isolation of the region or difficulties in accessing modern medical care. When exposed to modern medical care, knowledge about herbal medicines was more preserved and transferred by the elderly. Thus, it is important to carry on this knowledge to future generations and to guide scientific studies (Barolo *et al.*, 2014). The scientific findings related to the phytochemicals and their pharmacological values found in different parts of *F. carica* are increasing. This trend will surely continue, enlarging the knowledge pool and disclosing scientific results on the use of common fig, thus supporting the shift toward traditional medicine and alternative therapeutic sources. This accumulated knowledge should go hand in hand with scientific research determining the active compounds possessing health benefits or analyzing whether there are others that could be detrimental to health. The research findings may also present contradictory results, which then require additional research work to provide an explanation (Nishimura *et al.*, 2021). In this respect, the plant material and extraction methods should be scrutinized first due to the vast diversity used, as these may be the major sources leading to controversial results. The plant material should be identified in detail in terms of the variety, location, time of sampling, moisture content, cultural practices and conditions (e.g. fresh/dried), or the drying conditions. Fig fruit is included in various pharmacopeia as a safe herbal medicine. However, some of the constituents that are biologically active at very low concentrations, especially those in fig leaves and latex, can be toxic if they exceed certain concentrations. Therefore, these active biocompounds need to be identified for wider use. Contact with fig leaves possessing furanocoumarins followed by exposure to sunlight or artificial UV-A light may cause phytodermatitis, which can be a professional problem or if it accumulates more to reach toxic levels, it may create problems when fig is used in a home-made recipe. The risks need to be assessed and safe use should always be considered when making recommendations. Cooperation among different disciplines and regions may help to widen the scope of the research work and bring valuable contributions to relevant fields.

References

Abbasi, S., Kamalinejad, M., Babaie, D., Shams, S.M., Sadr, Z., *et al.* (2017) A new topical treatment of atopic dermatitis in pediatric patients based on *Ficus carica* L. (fig): a randomized, placebo-controlled clinical trial. *Complementary Therapies in Medicine* 35, 85–91.

Akçiçek, E. (2019) İncirin tıbbi özellikleri (Medicinal properties of fig). In: *İncirin Akdenizdeki Yolculuğu (Travel of Figs in the Mediterranean). Proceedings of the Conference*. Dinç Ofset Printing, İzmir, Turkey, pp. 15–22.

Al-Snafi, A.E. (2017) Nutritional and pharmacological importance of *Ficus carica* – a review. *Journal of Pharmacy* 7, 33–48.

Anaç, H. (2012) *Geçmişten Günümüze İncir*. Aydın Ticaret Odasi, Efeler, Turkey.

Anon. (2011) *Estahbanat Rainfed Fig System Iranian Agricultural Heritage System*. Food and Agricultural Research Organization, Iran. Available at www.fao.org/fileadmin/templates/giahs/PDF/Fig_traditional_system_Final_.pdf (accessed December 31, 2021).

Badjugar, S.B., Patel, V.V., Bandivdekar, A.H., and Mahajan, R.T. (2014) Traditional uses, phytochemistry and pharmacology of *Ficus carica*: a review. *Pharmaceutical Biology* 52, 1487–1503.

Barolo, M.I., Mostacero, N.R., and Lopez, S.N. (2014) *Ficus carica* L. (Moraceae): an ancient source of food and health. *Food Chemistry* 164, 119–127.

Başer, H.C. (2017) İncir (*Ficus carica* L., Moraceae). *Bağ Bahçe Dergisi* 73, 30–32.

Berg, C.C. (1989) Classification and distribution of *Ficus*. *Experientia* 45, 605–611.

Bohlooli, A., Mohebipoor, A., Mohammadi, S., Kouhnavard, M., and Pashapoor, S. (2007) Comparative study of fig tree efficacy in the treatment of common warts (*Verruca vulgaris*) vs. cryotherapy. *International Journal of Dermatology* 46, 524–526.

Chawla, A., Kaur, R., and Sharma, A.K. (2012) *Ficus carica* Linn, a review on its pharamacognostic, phytochemical and pharmacologic aspects. *International Journal of Pharmaceutical and Phytopharmacological Research* 1, 215–232.

Duman, E., Şimşek, M., and Özcan, M.M. (2018) Monitoring of composition and antimicrobial activity of fig (*Ficus carica* L.) fruit and seed oil. *Journal of Agroalimentary Processes and Technologies* 24, 75.

Elmagülü, G., Aksoy, U., and Şen, F. (2020). Sugars in 'Sarılop' fig fruits. *Acta Horticulturae* 1280, 115–122.

Ergül, M., Ergül, M., Eruygur, N., Ataş, M., and Uçar, E. (2019) *In vitro* evaluation of the chemical composition and various biological activities of *Ficus carica* leaf extracts. *Turkish Journal of Pharmaceutical Science* 16, 401–409.

EMA (European Medicines Agency) (2004a) Herbal Medicinal Products Regulation (EC) No. 726/2004. Available at: www.ema.europa.eu/en/human-regulatory/herbal-medicinal-products (accessed December 31, 2021).

EMA (European Medicines Agency) (2004b) Directive 2004/24/EC of the European Parliament and of the Council of 31 March 2004, amending, as regards traditional herbal medicinal products, Directive 2001/83/EC on the Community code relating to medicinal products for human use. Available at: www.ema.europa.eu/en/human-regulatory/herbal-medicinal-products (accessed December 31, 2021).

Gilani, A.H., Mehmood, M.H., Janbaz, K.H., Khan, A., and Saeed, S.A. (2008) Ethnopharmacological studies on antispasmodic and antiplatelet activities of *Ficus carica*. *Journal of Ethnopharmacology* 119, 1–5.

Gond, N.Y. and Khadabadi, S.S. (2008), Hepatiprotective activity of *Ficus carica* leaf extract on rifamicin-induced hepatic damage in rats. *Indian Journal of Pharmaceutical Sciences* 70, 364–366.

Honda G., Yeşilada, E., Tabata, M., Sezik, E., Fujita, T., *et al.* (1996) Traditional medicine in Turkey VI. Folk medicine in West Anatolia: Afyon, Kütahya, Denizli, Muğla, Aydın provinces. *Journal of Ethnopharmocology* 53, 75–87.

Hossaert-McKey, M., Soler, C., Schatz, B., and Proffit, M. (2010) Floral scents: their roles in nursery pollination mutualisms. *Chemoecology* 20, 75–88.

Innocenti, G., Bettero, A., and Caporale, G. (1982) Determination of the coumarinic constituents of *Ficus carica* leaves by HPLC. *Il Farmaco; Edizione Scientifica* 37, 475–485.

Jasmine, R., Manikandan, K., and Karthikeyan, K. (2015) Evaluating the antioxidant and anticancer property of *Ficus carica* fruits. *African Journal of Biotechnology* 14, 634–641.

Jeong, M.R., Kim, H.Y., and Cha, J.D. (2009) Antimicrobial activity of methanol extract from *Ficus carica* leaves against oral bacteria. *Journal of Bacteriology and Virology* 39, 97–102.

Joseph, B. and Raj, S.J. (2011) Pharmacognostic and phytochemical properties of *Ficus carica* Linn – an overview. *International Journal of PharmaTech Research* 3, 8–12.

Kamiloğlu, S. and Capanoğlu, E. (2015) Polyphenol content in figs (*Ficus carica* L.): effect of sun-drying. *International Journal of Food Properties* 18, 521–535.

Kitajima, S., Aoki, W., Shibata, D., Nozomu, D., Nakajima, D., *et al.* (2018) Comparative multi-omics analysis reveals diverse latex-based defense strategies against pests among latex-producing organs of the fig tree (*Ficus carica*). *Planta* 247, 1423–1438.

Küçükerdönmez, O., Baygeldi, N., Akder, R.N., and Çağındı, Ö. (2021) Medicinal and nutritional analysis of fig (*Ficus carica*) seed oil; a new gamma tocopherol and omega-3 source. *Progress in Nutrition* 23: 2021052.

Lansky, E.P. and Paavilainen, H.M. (2010) *Figs: The Genus Ficus*. CRC Press, Boca Raton, Florida.

Lazreg-Aref, H., Hadj Salah, K.B., Chaumont, J.P., Fekih, A.W., Aouni, M., and Said, K. (2010) *In vitro* antimicrobial activity of four *Ficus carica* latex fractions against resistant human pathogens (antimicrobial activity of *Ficus carica* latex). *Pakistan Journal of Pharmaceutical Sciences* 23, 53–58.

Lazreg-Aref, H., Gaaliche, B., Ladhari, A., Hammami, M., and Hammami, S.O. (2018) Co-evolution of enzyme activities and latex in fig (*Ficus carica* L.) during fruit maturity process. *South African Journal of Botany* 115, 143–152.

Li, J., Tian, Y., Sun, B., Yang, B., Chen, J., and Men, Q. (2012) Analysis on volatile constituents in leaves and fruits of *Ficus carica* by GC-MS. *Chinese Herbal Medicines* 4, 63–69.

Li, Z., Yang, Y., Liu, M., Zhang, C., Shao, J., *et al.* (2021) A comprehensive review on phytochemistry, bioactivities, toxicity studies, and clinical studies on *Ficus carica* Linn. *Biomedicine & Pharmacotherapy* 137: 111393.

Mahmoudi, S., Khali, M., Benkhaled, A., Benamirouche, K., and Baiti, I. (2016) Phenolic and flavonoid contents, antioxidant and antimicrobial activities of leaf extracts from ten Algerian *Ficus carica* L. varieties. *Asian Pacific Journal of Tropical Biomedicine* 6, 239–245.

Mars, M. (2003) Fig (*Ficus carica* L.) genetic resources and breeding. *Acta Horticulturae* 605, 19–27.

Mawa, S., Husain, K., and Jantan, İ. (2013) *Ficus carica* L. (Moraceae): phytochemistry, traditional uses and biological activities. *Evidence-Based Complementary and Alternative Medicine* 2013: 974256.

Menale, B., Amato, G., Di Prisco, C., and Muoio, R. (2006) Traditional uses of plants in North-Western Molise (Central Italy). *Delpinoa* 48, 28–36.

Mohan, G.K., Pallavi, E., Ravi Kumar, B., Ramesh, M., and Venkatesh, S. (2007) Hepatoprotective activity of *Ficus carica* Linn. leaf extract against carbon tetrachloride-induced hepatotoxicity in rats. *DARU Journal of Pharmaceutical Sciences* 15, 162–166.

Nishimura, K., Higashiya, K., Ueshima, N., Kojima, K., Takita, T., *et al.* (2021) Insight into the collagen-degrading activity of a serine protease in the latex of *Ficus carica* cultivar Masui Dauphine. *Bioscience, Biotechnology and Biochemistry* 85, 1147–1156.

Oliveira, A.P., Valentão, P., Pereira, J.A., Tavares, F., Silva, B.M., and Andrade, P.B. (2009) *Ficus carica* L.: metabolic and biological screening. *Food and Chemical Toxicology* 47, 2841–2846.
Oliveira, A.P., Silva, L.R., de Pinho, P.G., Gil-Izquierdo, A., Valentão, P., *et al.* (2010) Volatile profiling of *Ficus carica* varieties by HS-SPME and GC–IT-MS. *Food Chemistry* 123, 548–557.
Özatalay, G.Z. (2014) Aydın yöresi halk hekimliğinde incirin kullanımı (The use of fig in folk medicine of Aydın region). *KMU Journal of Social and Economic Research* 16, 151–154.
Özdamar, E., Özbek, S., and Akın, S. (2003) An unusual cause of burn injury: fig leaf decoction used as a remedy for a dermatitis of unknown etiology. *Journal of Burn Care and Rehabilitation* 24, 229–233.
Patil, V.V. and Patil, V.R. (2011) *Ficus carica* Linn. an overview. *Research Journal of Medicinal Plant* 5, 246–253.
Petkova, N., Ivanov, I., and Denev, P. (2019) Changes in phytochemical compounds and antioxidant potential of fresh, frozen, and processed figs (*Ficus carica* L.). *International Food Research Journal* 26, 1881–1888.
Polat, M., Öztaş, P., Cik Dikilitaş, M., and Allı, N. (2008) Phytophotodermatitis due to *Ficus carica*. *Dermatology Online Journal* 14: 9.
Rahmani, A.H. and Aldebasi, Y.H. (2017) *Ficus carica* and its constituent's role in management of diseases. *Asian Journal of Pharmaceutical and Clinical Research* 10, 49–53.
Ramírez, B.W. (1977) A new classification of *Ficus*. *Annals of the Missouri Botanical Garden* 64, 296–310.
Stezner, J., Roemhild, R., Garibay-Hernandez, A., Harbaum-Piayda, B., Mock, H.P., and Bilger, W. (2019) Hydroxycinnamic acids in sunflower leaves serve as UV-A screening pigments. *Photochemical and Photobiological Sciences* 18, 970–988.
Subramanian, P., Padma Rao, P., Sheshashena Reddy, T., Sudhakar, P., Reddy, P.R., *et al.* (2013) Standardization of *Ficus carica* L.: in homoeopathic perspective. *Indian Journal of Research in Homoeopathy* 7, 153–157.
Sutton, D.C. (2014) *Figs, A Global History*. Reaktion Books, London, pp. 40–43.
Tian, J., Zhang, Y., Yang, X., Rui, K., Tang, X., *et al.*. (2014) *Ficus carica* polysaccharides promote the maturation and function of dendritic cells. *International Journal of Molecular Sciences* 15, 12469–12479.
Varanasi, S. and Narayana, A. (2007) Medico-historical review of Nysgrŏdha (*Ficus bengalensis* Linn.). *Bulletin of the Indian Institute of History Medicine (Hyderabad)* 37, 167–178.
Vaya, J. and Mahmood, S. (2006) Flavonoid content in leaf extracts of the fig (*Ficus carica* L.), carob (*Ceratonia siliqua* L.) and pistachio (*Pistacia lentiscus* L.). *Biofactors* 28, 169–175.
WHO (World Health Organization) (2013) *WHO Traditional Medicine Strategy: 2014–2023*. WHO, Geneva. Available at: www.who.int/publications/i/item/9789241506096 (accessed December 31, 2021).
Yazıcıoğlu, A. and Tuzlacı, E. (1995) Folks medicinal plants of Trabzon (Turkey). *Journal of Pharmacy Marmara University* 11, 333–342.

27 World Fig Production and Trade

Ece Tırkaz[1]*, Belit Balcı[2] and Uygun Aksoy[3]
[1]Aegean Exporters' Associations (EİB), İzmir, Turkey; [2]ALARA Tarım, Bursa, Turkey; [3]Department of Horticulture, Faculty of Agriculture, Ege University, Bornova, İzmir, Turkey

27.1 Introduction

Fig (*Ficus carica* L.) fruit, as one of the species widely dispersed in the natural flora of the Mediterranean and one of the oldest cultivated plants in the world, has been consumed by humans from time immemorial. Its consumption is not limited to fresh or dry fruit; it is also used for processing and medicinal purposes. Trading of fig fruit started first locally and became more globalized with the onset of the 20th century. Sun-dried fig fruits were the main subject of international trade for decades; however, new varieties and improved storage and transportation facilities paved the way for furthering fresh figs in global trade. In international trade, dried figs have competitors or substitutes such as dried grapes, apricots, and dates, but fresh fig is unique. When fig is introduced to new markets as a rare and "exotic" fruit with rich health properties, it triggers demand in these international markets, which in turn makes fig production appealing for growers in many regions of the world. Scientific studies on its nutritional and health properties, breeding and variety development, intensification of orchards, soil-less production systems, ripening, drying, and processing have also contributed to the emergence of new fig-production regions such as Central and South America, Saudi Arabia, and India.

Major global data sources include figs or dried figs in their statistics; however, there are still controversies or gaps in figures. In 2019, the world total fig production on a fresh basis totaled 1,332,813 t produced in 52 countries at quantities ranging from 310,000 t (Turkey) to 8 t (Qatar) (FAO, 2021). The top five producers are Turkey, Egypt, Algeria, Morocco, and Iran, supplying 70% of the world's global fig-fruit production. Although it is the prime fig producer and exporter, the United Nations Food and Agriculture Organization (FAO) trade statistics do not include Turkey as a fresh-fig exporter country. Trade Map, a tool developed by the International Trade Center for trade statistics (www.trademap.org; accessed January 4, 2022), shows the traded amounts of figs as fresh and dried together, and reports the total (dried and fresh) global trading capacity of figs as 150,000 t representing a value of $570,000,000 in 2019 (Trade Map, 2021). The top ten importing countries evaluated as the main fig-consuming markets or acting as transits are France, India, Germany, the USA, the UK, Canada, Austria, Italy, the Netherlands, and Switzerland. These countries represent 66% of world total trade value for

* Email: ece.tirkaz@eib.org.tr

 Advances in Fig Research and Sustainable Production (M.A. Flaishman and U. Aksoy)
DOI: 10.1079/9781789242492.0027

fresh and dried figs. Reliable data for fresh and dried figs separately is not fully available, and some figures require estimations based on calculation of the available data. Collection and processing of updated, reliable, and comparative data for the fresh- and dried-fig trade independently worldwide is necessary to have an accurate understanding of the market flows, especially with new fig-producing countries entering the global trade scene. There are some global institutions collating and providing global data on dried-fig trade. Data on the fresh-fig trade is rare and of low quality in most databases. Low-quality data results in conflicting figures for the same data set. There could be various reasons for this. Fig is grown as dispersed trees and traded locally or in low quantities as a minor crop. As stated in Eurostat documents, being a minor or underutilized species, fig is grouped with other fruit species that are classified under tropical, subtropical, or temperate fruit. The Eurostat guiding document asks the question "Are figs tropical fruit?" and makes a proposal to answer this. Fruits are typically distinguished into three groups:

1. Fruit from temperate climate zones: fruit that requires chilling (a certain number of hours of cold during the winter season) to flower/fructify.
2. Fruit from Mediterranean/subtropical climate zones: fruit that is are not hardy to extreme cold, and is mostly indifferent to chilling, although some may have chilling requirements.
3. Tropical fruit: fruit that can grow in all habitats but are intolerant to frost.

Figs, as well as persimmons, guava, avocados, and pomegranates, have in common this indifference to chilling (although it has been shown that for figs, production can be higher when there are more hours of chilling). Considering them under "subtropical" fruit is in line with FAO/Indicative Crop Classification (ICC) classification and with Nomenclature des Activités Économiques dans la Communauté Européenne (NACE). As a product, in the Combined Nomenclature, figs are in the same class as dates, avocados, guavas, and mangoes (code 0804). Therefore, even if they are not "tropical," they should be classified under F2000, which also includes the "subtropical" crops" (EU, 2020). The fig, 0804 20, is subdivided into groups as 0804 20 10—Fresh and 0804 20 90—Dried figs when subject to trade. For fig-paste nomenclature, there was confusion and it was recently classified under the dried-fig Combined Nomenclature code. Trade Map figures for the quantity and value of the fig trade do not differentiate between fresh and dried figs. However, dried figs can be produced from the same variety consumed as fresh (e.g. Black Mission) but requires a conversion factor for calculating the quantity of fresh utilized as dried. FAO (2021), Trade Map (2021), and the International Nut and Dried Fruit Council (INC, 2021) are major data sources for this chapter regarding global fig and dried-fig production and trade. Figures particularly from the Aegean Exporters' Association database (AEA, 2021) in Turkey for the dried-fig trade, and some information collected through personal communication, have helped to draw a more realistic global picture of the fig trade. This chapter aims to collate the available information on the current fresh- and dried-fig trade to inform for the future. Similarly, to serve this aim, the chapter also discusses the major historical shifts in fig trade routes linked to problems faced in production as lessons have been learned.

27.2 World Fig Production

27.2.1 Current dried- and fresh-fig production

Although global fig cultivation is more concentrated around the Mediterranean Basin, it is dispersed across all continents and is carried out in more than 50 countries, with data provided (Table 27.1). Fig production, which faced a decline toward the 1970s, has been increasing steadily in recent years and on a fresh-fruit basis exceeded 1.3 million t in 2019. However, this is not solely due to rapid enlargement of the areas under fig cultivation but more to intensification of the production systems (Fig. 27.1).

The world production of fig fruit was almost stable at around 1.1 million t year^{-1} for more than a decade (Fig. 27.1). The global fig area is calculated to be around 290,000–300,000 ha, and the global average yield as 4.5 t ha^{-1} according to 2019 data (FAO, 2021).

In global fig production, Asia supplies 43.8%, followed by Africa (39.4%), Europe (9.7%), and the Americas (7.0%) according to

Table 27.1. World fig production in 2019. From FAO (2021).

Rank	Country	Production (t)	Rank	Country	Production (t)	Rank	Country	Production (t)
1	Turkey	310,000	**16**	India	14,914	**31**	Colombia	2,413
2	Egypt	225,295	**17**	Japan	14,175	**32**	Israel	2,387
3	Morocco	153,472	**18**	Azerbaijan	12,074	**33**	South Africa	1,972
4	Iran	130,328	**19**	Italy	11,830	**34**	Cyprus	1,600
5	Algeria	114,092	**20**	Libya	9,673	**35**	Tajikistan	1,367
6	Spain	51,600	**21**	Mexico	9,466	**36**	Armenia	1,234
7	Syrian Arab Republic	43,015	**22**	Iraq	9,265	**37**	Bosnia and Herzegovina	980
8	USA	28,174	**23**	Yemen	5,593	**38**	Argentina	868
9	Tunisia	24,619	**24**	Uzbekistan	4,242	**39**	Croatia	810
10	Afghanistan	24,319	**25**	Lebanon	4,048	**40**	Bolivia	686
11	Brazil	22,320	**26**	Montenegro	3,921	**41**	Jordan	641
12	Albania	22,024	**27**	Palestine	3,459	**42**	North Macedonia	626
13	Greece	19,730	**28**	Portugal	3,380	**43**	United Arab Emirates	462
14	China	17,225	**29**	France	3,340	**44**	Pakistan	225
15	China, mainland	17,225	**30**	Peru	3,095	**45**	Chile	104

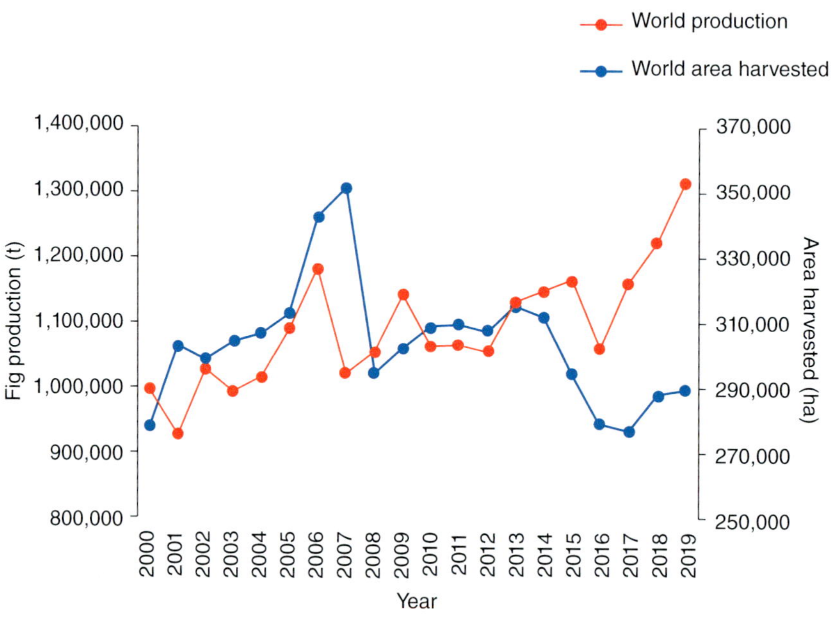

Fig. 27.1. Global fig production (t) and the area of fig cultivation (ha) between 2000 and 2019. Modified from FAO (2021).

the average for the last 18 years. Most fig production comes from regions located around the Mediterranean Sea and the Near East. However, other countries such as the USA, Brazil, India, Japan, and China where a Mediterranean climate prevails are becoming important producers in the market, with a stable or increasing fig production since 2000 (FAO, 2021).

Fresh figs are highly perishable, and thus most of the figs produced are used for drying to extend their shelf life by lowering water activity levels and to facilitate storage, transportation, and trade. Sun drying is a millennia-old practice, and sun-dried figs have been the subject of international trade for centuries. There are a few varieties known in the international dried-fig trade with commercial importance such as Sarılop (= Calimyrna), Sabz, Zidi, Kadota, and Black Mission. In contrast, most fresh-fig fruits are consumed locally and comprise a diverse palette of fig varieties as brebas and second crops, small or large, from globular to spherical in shape, and with white, yellow, black, purple, or striped skins. The quantities traded as fresh are quite limited at both the worldwide and country level. Among the top-ranking fig-producing countries (Table 27.1), only Brazil, located in the tropical zone, is a fresh-fig producer and trader, whereas other countries marketing fresh figs are prominent dried-fig producers, as well. Israel, Mexico, Peru, and Argentina are comparatively new in the world fresh-fig trade.

27.2.2 Long-term fig-production trends

Evaluation of global fig production within a time frame of around 60 years, between 1961 and 2019, has revealed a major shift from developed countries to the developing ones. In 1961, the leaders in fig production were the developed countries of the Mediterranean—Italy, Spain, Portugal, and Greece—which during the latter decades faced drastic declines. The production figures of these countries were reduced from a few hundred thousand tons to a few ten thousand tons (Table 27.2). A similar decrease but at a much lower rate was observed in California, USA. Developing countries such as Afghanistan, Morocco, and Iran have doubled or tripled their fig production during this timespan, and Egypt reported a 40-fold increase in annual fig production. Turkey showed an increase but at a much lower rate than the aforementioned developing countries. Turkey was the only major fig producer maintaining its production capacity and became the leader with a slow but steady increase.

In 2019, world fig production was close to the total quantity obtained 60 years ago. The decline in global fig production starting from the 1970s and until 2000 was around 40%. This was due to major reductions from developed countries. To balance this loss and recover the reduced quantities, developing countries required a few decades to establish new fig plantations and obtain accountable yields. Based on the FAO statistical data for 1961, 85% of global fig production was supplied by the Mediterranean countries. Italy was the major producer, followed by Portugal,

Table 27.2. Long-term fig production in the major fig-producing countries in 1961, 1971, 1981, 2000, 2009, and 2019. From FAO (2021).

	Fig production (t)					
Country	1961	1971	1981	2000	2009	2019
Afghanistan	8,000	11,400	11,700	15,000	11,544	24,319
Algeria	62,000	51,136	70,589	54,326	83,801	114,092
Egypt	5,358	15,106	10,737	187,698	286,682	225,295
Greece	169,259	139,459	115,397	29,117	20,838	19,730
Iran	10,000	15,000	21,000	57,201	58,836	130,328
Italy	319,100	175,800	73,700	25,000	12,106	11,830
Morocco	77,000	67,000	66,000	68,400	109,200	153,472
Portugal	270,000	155,300	49,500	3,496	3,003	3,380
Spain	204,000	101,400	48,800	56,014	29,120	51,600
Turkey	203,700	195,000	250,000	240,000	244,351	310,000
USA	57,000	41,367	34,650	50,712	39,689	28,174
Total production in leading countries (t)	1,385,417	967,968	752,073	786,964	899,170	1,072,220
Share of leading countries globally (%)	88.1	83.9	78.2	78.9	78.7	81.5
Total world fig production (t)	1,572,296	1,153,694	961,590	997,043	1,142,611	1,315,588

Spain, Turkey, and Greece. The bulk of fig production was commercialized as sun-dried figs and traded within the territory but some was also exported. The reduction trend among these countries started in the 1960s, and the production quantities steadily decreased with slight fluctuations linked to the climatic conditions of the season. As fig is a perennial species, the trees were not uprooted in 1 or 2 years (except 2007–2008), as could be seen in the areas under fig cultivation. In the mid-1990s, Portugal still had large areas covered with fig trees not harvested but left for animal grazing.

Increased demand and high prices in the market promoted fig production in developing countries, which started filling the gap, especially in the dried-fig trade. Sun drying was the best solution to preserve and market the perishable fig fruit (Aksoy, 2021). A short summary starting at the beginning of the 20th century may help to understand the link between dried-fig production and regional conditions, the reasons for change, and solutions brought through competitive advantages in fig production and trade (Eisen, 1901). In Italy, fig was produced throughout the country; however, northern production was destined for fresh consumption, while southern regions focused on sun drying. Different fig varieties were dried, but the quality was not always good, and therefore dried figs were consumed mainly by the poor. Thus, the whole packaging process was ignored to reduce production costs. Fig was grown all over Greece, including the islands. The highest-quality dried figs came from Kalamata. Greek figs were exported to Russia, Austria, France, and Central Europe, with a few to the USA. The lower-quality figs were used in making "coffee" and brandy. In Spain, most of the dried figs were sent to France and Austria for distilleries, and export to the USA started at the beginning of the 20th century. In Portugal, figs were grown around Lisbon for local consumption, either fresh or dried; however, those from the southern Algarve region that were of high quality were mainly exported to Europe. Fresh or semi-dried figs were collected by merchants for distilling as brandy due to their lower quality. Although distributed throughout the country, the Peloponnese Region was the center of fig production in Greece, and high-quality dried figs were exported to many countries. Egyptian and Kabylian (Algeria) figs were also noteworthy. Tizi Ouzou, a town in north-central Algeria in the Great Kabylie mountain region, was the major fig-producing region outside Smyrna in İzmir, Turkey. White or black fig varieties were dried and exported and helped the livelihoods of the local populations.

Among the current leading fig producers, only Turkey maintained substantial fig production from the early 1900s until the 1970s. Fig drying is an old tradition in Turkey, which continues today with important economic returns in the two western provinces, Aydın and İzmir. Eisen (1901) mentioned that "Smyrna" (İzmir, Turkey) figs were well known not only in his era but also during the first century, with the best-quality figs brought to the Roman Emperor from Smyrna. Figs were separated into different quality classes visually and manually according to the size, color, and thinness of the skin. The dried-fig value chain was then established between the grower and the packer. Specialized camel owners were among the important actors in the chain by carrying dried figs of different quality in different bags—high quality figs in bags made from goat or camel hair, and the culled figs in ordinary yellow sacks (Eisen, 1901). These reflections dating back a century show that dried figs were subject to global trade while fresh figs were consumed locally. The dried-fruit quality determined the price and care given to the product by all actors in the chain from growers to processors.

Portugal was one of the major suppliers of dried figs to the USA until the late 1920s with a share of 13%. The major producing regions were Algarve and Torres Novas, with a smaller contribution from the Douro region. The figs from Torras Novas were not edible but were used for distillation. The major variety, Cotio, was smaller in size compared with the Sarılop (Turkish) and Kalamata (Greek) fig varieties. The farm sizes were small, and the fig trees were interspersed with other fruit trees. However, as dried-fig production increased in California, US imports fell sharply. This was coupled to the decreases in Algarve production due to the introduction of irrigation and replacement of figs with mainly citrus and almonds. Due to lower quality, Portugal started processing fig paste as a solution to maintain markets after 1954–1955, which also coincided with lower production in

Greece and Turkey. The price of Portuguese paste in the USA was lower than that imported from Turkey or produced in California (Mehr and Ferguson, 1958). Fig trees were planted on the borders or mixed with grapes; however, in the 20th century, fig orchards were established in Greece. During the period between the 1920s and 1940s, fresh- and dried-fig production started to receive more care; for example, a pruning campaign was launched in 1930. Some of the male and female fresh-fig fruits were canned and marketed. The major producing regions in Greece were Peloponnese, the central region around Athens, and Crete. However, the highest-quality dried figs were obtained around Kalamata, Mesini, Pilia, and Trifilia in Messenia Province where every village had a farmers' cooperative. Moreover, the Greek Fig Protection Office was founded to promote fig production and support the sector. A major problem in international trade was fig moth infestation, and in cases where fig lots were rejected from the USA, the fig insurance company reimbursed the cost incurred. The property of the fig load was then transferred to the Fig Protection Office (Aday, 1938). Europe was the major market for Mediterranean dried figs; however, World War II affected the dried-fig market and thus the fig-producing countries.

In the USA, as today, California was the major fig region, and the fig area underwent a 177% increase between 1920 and 1930. The average dried-fig production (1923–1928) was 10,470 t and figs for fresh and canning comprised 4368 t. Fig production was for fresh, canning, and dried fruits, and the major varieties were Mission, White Adriatic, Brown Turkey, Calimyrna (Sarılop), and Kadota (canning) (Condit, 1946). In 1959, the total dried-fig production in California was 19,140 t (Valley Fig Growers, 2009).

In 1961, world fig production was 1.57 million t, and the leaders were the Mediterranean countries (Table 27.1). During the period between 1970 and the 1990s, the annual average world fig production was reduced to around 1 million t, and dried-fig production was around 100,000 t. Turkey supplied 51% of the global dried-fig production, followed by Greece (17.5%) and the USA (10.9%). In Egypt, during the periods 1985–1989 to 1995–1999, average fig production on a fresh basis increased from 25,800 t to 216,011 t according to FAO statistics (Eriten, 2005). The boost in Egypt could have been due to the starting of fruit bearing in new fig plantations but also because of revisions to the national statistics. Spanish fig production was reduced by 50% between 1990 and 2015, and the reasons stated were lack of information on commercial varieties and research supporting fig production to adapt to changing conditions (Pereira *et al.*, 2017).

When discussing the change in top positions of global trade among fig-producing countries, the first factor to be considered is that fig is a long-lived perennial species adapted to a certain climatic zone and that it is not a staple food but a delicacy. Competition exists not only among different producer countries but also within a country. In terms of market demand, other dried fruits such as raisins, dried apricots, or dates may substitute for dried figs in some of its uses. In this respect, global competition among producers was limited to countries with Mediterranean climate zones. Turkey, which carried fig production onward from the 1960s toward the 2000s, had a number of advantages with respect to production and trade: the variety Sarılop, which is well adapted to the conditions in the Meander Basin of the western Aegean Region and produces the best-quality sun-dried figs, and the availability of cheap and qualified labor required during harvest and processing. The agricultural policies after 1950 prioritized exports and targeted the post-war European market. Moreover, Turkey increased its global competitiveness through a radical change within its production zone. The lowland fig orchards moved uphill in the same area as irrigation networks were established in the lowland. This competition resulted in annual cash crops such as cotton and vegetables replacing lowland fig orchards. This induced the establishment of new orchards on slopes with poor soil quality under rainfed conditions. Higher number of trees per unit area helped the fig growers to maintain productivity and continue their livelihoods. Moreover, figs have a competitive advantage on such marginal land and slopes, especially when sun dried, due to their extended shelf life, enhancing elasticity in marketing. Extension of traditional production systems and the use of varieties well adapted to the local conditions provide a comparative advantage for commercial dried-fig production. Thus, the enhanced competitiveness of Iran, Afghanistan, Syria, Algeria, and Morocco helped them to become global actors in the dried-fig trade, despite existing challenges.

27.2.3 Dried-fig production

The annual dried-fig production is estimated to be around 130,000–160,000 t and may vary slightly according to the climatic conditions prevailing in the major fig-producing countries. If the fresh:dried fig ratio is accepted roughly as 4:1 (the conversion rate may vary according to the variety or site-specific conditions), annually 520,000–640,000 t or nearly 50% of the total fig production is destined for drying and then marketed as dried figs. As of 2019, artificially dried figs are at negligible quantities and therefore all dried figs are assumed to be sun dried in open-field conditions.

In 2019, world dried-fig production increased by 17% compared with 2018, reaching 158,500 t, the highest level of the last decade. In the 2019/20 season, Turkey kept its rank as the top dried-fig producer, accounting for 56% of the world share, followed by Iran with a 19% share. The next 21% of the production was distributed among the USA, Greece, Spain, Italy, and Afghanistan, with shares ranging between 3% and 6% (INC, 2021). According to the 5-year average between 2015 and 2019, the annual dried-fig production and their shares were, in decreasing order: Turkey, 76,741 t (54%); Iran, 26,800 t (19%); USA, 9040 t (6%), Greece, 7700 t (6%); Spain, 7080 t (5%); Afghanistan, 5900 t (4%); Italy, 4160 t (3%), and others, 4750 t (3%). Countries other than the major producers do not have significant shares, so global fig trade is sourced from a limited number of countries. Dried-fig production in Turkey, Iran, Greece, and Afghanistan mostly benefits from the dominating variety suitable for sun drying and that is well adapted to the low-input production systems. Although there are rich fig genetic resources in all countries, commercial dried-fig production depends on very few varieties.

In Turkey, dried-fig production is based solely on the Sarılop (= Calimyrna) variety. It is a Smyrna-type fig requiring pollination for fruit set and characterized by a light-colored soft and thin skin and a large, sweet, fleshy fruit. According to the Turkish Statistics Institute (TUIK; www.tuik.gov.tr/; accessed January 4, 2022), two neighboring provinces Aydın (36,902 ha under fig cultivation; 6.4 million trees; 62,000 t of dried-fig production) and İzmir (8196 ha; 1.6 million fig trees; 18,000 t) in western Turkey supply all the sun-dried figs.

In Iran, Estahban, a semi-arid region located in the south-east of Fars Province is the major region for dried figs, suppling nearly 90% of Iranian dried figs, and Sabz is the major variety used for sun drying (Jafari *et al.*, 2017).

In Afghanistan, figs are produced in Herat, Balkh, Farah, Zabul, and Jawzjan Provinces; however, southern Kandahar Province is the center of dried-fig production, and a yellow variety that bears two crops per year from the Shah Wali Kot District is the major commercialized variety (www.fondazioneslowfood.com/en/ark-of-taste-slow-food/shah-wali-kot-dried-yellow-fig/; accessed January 4, 2022).

In Morocco, the area under fig cultivation comprises 46,000 ha and production is estimated as 57,000 t. Fig orchards are located mainly in mountainous regions with poor soils, and the main production regions are Taounate (22,230 ha), Chefchaouen (7050 ha), Al Hoceima (5000 ha), Ouazzane (3150 ha) and Tetouan (2000 ha). In Moroccan dried-fig production, there is no dominating variety but there are more than ten varieties that are sun dried. Varietal diversity creates problems in terms of marketing standards, storage-pest problems, mycotoxins, and lack of care at postharvest stages and these are the major challenges faced by the industry (Salih, 2020).

Fig production in Egypt is mainly (50%) located along the north-western coast of Alexandria and in Sinai Governorates. These regions are characterized by a very arid climate with an average rainfall of 120 mm. Sultani is the major variety grown; however, Fayoumi, Ramadi, Barshoumy, Sidi Gaber, and Hegazi are also used as synonyms (Khamis *et al.*, 2006). Mosaic virus has created significant losses in productivity and quality. Egypt ranks among the top dried-fig producers; however, there are no data for its role in international trade and information on the quantity traded is not documented.

Dried-fig production in Greece centers around Kalamata with the Kalamata variety, the main variety used for drying. It is very rich in taste, sugars, and pectin, with an oblong shape, and a thick greenish peel that becomes yellowish to beige after drying. The fresh fruit reaches 5.5 cm in diameter. The dried figs are marketed mostly as whole figs. Kimis is the second commercial variety used for sun drying and has a thin peel and is cut in half (Eleni Tsantili, University of Athens, Athens, Greece, personal communication).

The main region for sun drying figs in Spain is Extremadura, the western region bordering Portugal. The variety grown is Calabacita, which has been well adapted to the microclimate in the Caceres area for decades. The fruit is also used as fresh and as processed. A developed processing industry has helped to add value and increase the income generated within the region.

In Italy, dried figs are produced from Dottato and white fig varieties, and the fruit size is comparatively smaller than Turkish or Greek dried figs. The major producing areas are, in decreasing order, Campania, Calabrita, Puglia, and Sicily. The best-quality product is sold in Italy and overseas markets and is marketed as plain (plaits, rings) or processed with nuts, citrus fruit peels, and various aromatic herbs, and sometimes coated with chocolate. The dried-fig sector in Italy helped in the creation of confectionery firms processing ancient traditional dried-fig products (Nesci *et al.*, 2004).

Dried-fig production and processing in the USA is located mainly in the San Joaquin Valley, California. According to the Agricultural Resource Marketing Center, total fig production in 2017 on a fresh basis was 31,200 t, and 87% were processed as Black Mission for dried fruit, paste, or juice concentrate, Calimyrna for dried fig and paste, and Kadota for canning, preserving, and fig paste (www.agmrc.org/commodities-products/fruits/figs; accessed January 4, 2021).

27.3 International Fig Trade

27.3.1 Overview of the global fig trade

In total, figs (fresh and dried combined) have a global trading capacity of 160,000 t annually, which represents a value of $570,000,000 (Trade Map, 2021). The figures show that the dried-fig trade dominates over fresh figs (Tables 27.3–27.7). The dried-fig trade is estimated to have a total volume ranging between 120,000 and 135,000 t. Taking these figures into consideration, estimates of the total fresh-fig trade

Table 27.3. Global fig exports as quantity (t) between 2016 and 2019 (listed in decreasing order as of 2019). From Trade Map (2021).

Rank	Country	2016	2017	2018	2019
–	**World**	**144,413**	**162,786**	**162,444**	**158,183**
1	Turkey	69,696	79,302	78,452	85,037
2	Afghanistan	7,585	11,952	9,602	12,091
3	Spain	6,412	7,525	7,871	7,568
4	Saudi Arabia	7,142	10,487	6,788	6,724
5	Syrian Arab Republic	4,702	4,463	8,020	4,153
6	Austria	3,373	3,212	3,125	3,907
7	Germany	3,403	3,178	3,507	3,872
8	Netherlands	3,814	3,884	3,708	3,774
9	Greece	3,168	4,484	3,135	3,610
10	United Arab Emirates	657	255	1,974	3,551
11	Belarus	2,177	2,597	2,708	3,124
12	USA	4,721	4 520	3,639	2,301
13	Iran, Islamic Republic of	8,886	11,093	13,307	2,299
14	Italy	3,569	3,384	3,423	1,923
15	France	1,910	1,649	1,473	1,626
16	Brazil	1,191	1,313	1,394	1,377

Table 27.4. Global fig exports as value (1000 US$) between 2016 and 2020 (listed in decreasing order as of 2019). From Trade Map (2021).

Rank	Country	2016	2017	2018	2019	2020
–	**World**	**490,753**	**587,788**	**583,629**	**568,467**	–
1	Turkey	249,570	298,673	289,143	284,492	293,691
2	Afghanistan	33,861	81,581	67,782	85,737	–
3	Spain	16,160	16,736	19,911	20,613	20,677
4	Germany	15,259	15,575	18,199	19,230	21,514
5	Netherlands	17,760	20,020	20,369	18,768	21,186
6	Greece	11,276	15,770	12,606	16,474	14,509
7	Austria	14,726	13,263	12,959	15,569	31,116
8	USA	19,225	16,859	17,477	13,472	11,036
9	Syrian Arab Republic	7,462	6,133	10,239	10,720	–
10	France	9,666	9,000	8,861	9,104	8,848
11	Brazil	6,304	6,627	6,940	6,671	4,839
12	Saudi Arabia	4,986	8,021	5,994	6,437	–
13	Mexico	4,458	5,410	6,047	6,395	–
14	Iran, Islamic Republic of	36,063	27,632	33,612	4,987	–
15	Peru	1,596	1,567	5,011	4,815	5,450

Table 27.5. Average unit export price of fig (US$ kg^{-1}) in the world and in some selected countries in 2016–2018. Data from Trade Map (2021).

Country	2016	2017	2018
World	**3.40**	**3.61**	**3.59**
Turkey (majority dried)	3.58	3.77	3.68
Iran (dried)	4.06	2.49	3.96
Syria (dried)	1.59	1.37	1.28
Afghanistan (dried)	4.46	6.80	7.06
Brazil (fresh figs)	5.29	5.04	4.98

global volume range between 25,000 and 40,000 t annually. In both the fresh- and dried-fig trade, there is nearly no overstock in producing countries, so annual production can be assumed to be sold within a year. In the Turkish dried-fig sector, a small quantity (<5000 t) is stored under cold conditions and may be marketed within 2 years without any quality deterioration. However, as discussed above, the figures quoted in major data sources often give contradictory figures. In some cases, data from Turkey, the major supplier of dried and fresh figs, are given as derived from the Aegean Exporters' Associations to better visualize international fig trade.

Turkey is the major supplier of figs for international trade with a share of 48.2–53.7% of the market between 2016 and 2019. According to the 2019 figures, Afghanistan, Spain, Syria, Greece, the USA, Iran, Italy, France, and Brazil are the other important fig-producing countries. Austria, Germany, the Netherlands, Saudi Arabia, United Arab Emirates, and Belarus which are also listed among top exporters are transit countries, which import and then re-export. These countries are also listed among major importers.

Among the fig-exporting countries, Turkey, Iran, Syria, and Afghanistan are the major dried-fig producers. When the unit price of fig as US$ kg^{-1} is calculated based on the quantity exported and value gained, the results show that there is a marked difference between prices (Table 27.5). The world average is around US$3.4–3.6 kg^{-1}. The average unit prices for dried figs range between US$1.28 and US$7.06 kg^{-1}, varying according to the quality, quantity, and origin. The highest prices are in Afghanistan and the lowest in Syria. In Iran, the unit fig prices showed fluctuations. The Turkish dried-fig prices averaged US$3.58–3.77 kg^{-1}. It should be noted that the yearly conditions, especially during fruit maturation and drying, affect the overall quality. In years when cull ratios are high, the overall exported quantity may remain the same but the average price drops because of the higher weight of low-unit priced culled figs compared with the lower share of high-quality figs. Supporting the price–quality relationship, as of 2019 and 2020, the average export prices per unit (kg) of Turkish dried figs according to

Table 27.6. Major fig importers as quantity (t) between 2016 and 2019 (listed in decreasing order as of 2019). From Trade Map (2021).

Rank	Country	2016	2017	2018	2019
–	**World**	**145,816**	**149,185**	**150,463**	–
1	Germany	15,244	17,011	16,949	17,988
2	France	18,911	16,909	18,249	17,476
3	USA	11,170	10,527	11,394	12,262
4	India	8,411	12,726	9,725	11,940
5	UK	9,845	8,138	5,940	7,496
6	Austria	4,703	5,110	5,196	7,234
7	Russian Federation	2,958	4,282	5,428	5,795
8	Italy	5,314	4,785	4,299	5,237
9	Netherlands	4,579	4,566	4,523	4,168
10	Canada	3,993	4,368	4,136	4,120
11	United Arab Emirates	2,997	3,100	3,539	3,876
12	Saudi Arabia	3,791	4,452	3,520	3,600
13	Switzerland	3,820	3,681	3,554	3,590
14	Belarus	2,283	2,730	2,882	3,272
15	Belgium	3,152	2,974	2,918	2,969
16	China	1,922	654	1,009	2,768
17	Australia	2,068	1,776	1,876	1,773
18	Hong Kong, China	1,661	1,960	1,617	1,736

Table 27.7. Major fig importers as value (1000 US$) between 2016 and 2020 (listed in decreasing order as of 2019). From Trade Map (2021).

Rank	Country	2016	2017	2018	2019	2020
–	**World**	**514,109**	**598058**	**601529**	**604197**	–
1	India	59,675	101,231	81,484	96,147	–
2	Germany	53,144	66,342	64,032	63,176	–
3	France	60,347	58,348	67,040	60,770	70,147
4	USA	40,926	41,752	47,066	44,480	46,430
5	UK	29,674	30,112	26,753	28,447	24,551
6	Austria	17,929	20,398	21,741	27,993	38,161
7	Canada	20,377	21,733	24,087	23,383	24,757
8	Italy	22,151	21,873	19,769	22,562	21,395
9	Netherlands	17,359	20,572	21,052	18,481	25,921
10	Switzerland	17,693	19,301	19,194	18,083	21,796
11	Japan	9,677	11,502	11,671	11,909	10,849
12	Hong Kong, China	13,217	14,858	11,759	11,354	–
13	Belgium	11,933	11,517	12,669	11,287	12,337
14	United Arab Emirates	7,074	8,590	9,951	10,170	–
15	Russian Federation	4,876	7,720	8,361	9,721	11,837

the different qualities or processing types were as follows: whole dried figs, US$4.35–4.13 kg^{-1}; diced/sliced dried figs US$2.75–2.59, fig paste US$1.44–1.43, and culled figs US$1.60–1.73 (Aegean Exporters' Associations, unpublished data). Although Egypt, Morocco, and Algeria are among the major fig producers, they are not listed in the top rankings in terms of international trade, mainly due to problems in quality and value-chain management.

Mexico, Peru, and Brazil are among the fresh-fig exporters from the southern hemisphere. Brazil, located in the tropical zone, is known for exporting fresh Roxo de Valinhos (Brown Turkey) figs. The average unit price of fresh figs from Brazil was around US$5.0–5.3 kg^{-1} (Table 27.5).

In terms of total import value of fig fruit, the top six countries—India, Germany, France, the USA, the UK, and Austria—spend over 53% of the total amount traded, globally. When it comes to the quantity of figs imported, these same countries are listed as the top six. However, India, although it imports less than Germany, France, and the USA in terms of value, is listed at the top because of the range of quality imported and thus relative prices are higher than the other countries. According to the statistical data, in 2016, 2017, and 2018, India imported 7,450, 12,710, and 9,730 t of dried figs, respectively, mainly from Afghanistan and Turkey (www.statista.com/statistics/872368/india-dried-fig-import-volume/; accessed January 5, 2022). These figures are in line with those in Table 27.7, and in addition confirm that all figs imported to India were dried, and that the unit prices were higher than the other importing countries (Table 27.8). The unit export price of Afghanistan, a major supplier of figs to India, was the highest unit export price for figs. The average import prices for the selected 3 years are rather stable with a slight increase. There is a significant difference among importing countries, as with exporting countries. The highest average import prices (US$ kg^{-1}) in India (7.09–8.38) were followed in decreasing order by the USA (3.66–4.13), Germany (3.48–3.90), and the Russian Federation (1.54–1.80). Average unit import and export prices of fig (US$ kg^{-1}) in some selected countries show coupling of certain producers with their major markets, as in the case of Afghanistan–India, Turkey–Germany, USA–Canada, and Syria–Russian Federation. Dried-fig prices increased after the 1980s and up until then had averaged around US$1 kg^{-1}, lower than the current values. Between 1980 and 2000, the dried-fig unit (kg) prices were highest in Hong Kong SAR (around US$3), followed by Switzerland (US$2.54), Israel (US$2.36), and Canada (US$2.31). The lowest export unit prices were for Austria at US$0.67 (Eriten, 2005). The differences in unit price are directly linked to the quality of the dried figs traded. Austria had a history of importing culled figs for industrial use and therefore the lowest values were reported for Austria. The major reasons for the increase in dried-fig prices exported from Turkey after the 1990s were the increased costs incurred because of additional human labor required for the removal of bright greenish-yellow fluorescent (BGYF) figs to lower aflatoxin contamination, the cost of the removed BGYF product, and destroying the contaminated figs from exported lots (see Chapter 18, this volume).

Table 27.8. Average unit import price of fig (US$ kg^{-1}) in the world and in some selected countries in 2016–2018. Data from Trade Map (2021).

Country	2016	2017	2018
World	**3.52**	**4.00**	**4.02**
India	7.09	7.95	8.38
Germany	3.48	3.90	3.78
USA	3.66	3.97	4.13
Russian Federation	1.65	1.80	1.54

Many countries apply an import tariff, which also affects fig prices and product flows in international trade. The average import tariff for fresh or dried figs was 18.1% in 2018. The import tariff on figs was highest in Iran at 90%; other countries with the highest tariffs were Cyprus (72%), Angola (50%), Syria (46%), and Turkey (45.4%) (OEC, 2021). Turkey applies different tariff rates for figs from different countries as follows: Belarus (2001), 5.00%; China (2020), 30.00%; the European Union (EU; 2020), 0.001%; the Russian Federation (2020), 5.00%; Saudi Arabia (2020), 0.00%; and Vietnam (2018), 30.00%. The EU applies tariffs for dried figs only to the USA as US$7.9 cents kg^{-1} (2020), and the US applies a 20% tariff for Mexico (2020) (www.nutfruit.org/files/tariffs/; accessed January 5, 2022).

27.3.2 The dried-fig trade

Sun drying of figs is a well-known drying process and has been practiced in fig-producing regions from time immemorial. Drying has many advantages for preserving product quality for longer and facilitating trade. It decreases water activity, reduces microbiological activity, and minimizes physical and chemical changes. Drying is the most popular and effective way of preserving figs. Dried figs can be consumed directly, processed partially as fig paste, or utilized as a raw material in food and the beverage processing industry. Important advantages of this method are that the product is durable and low cost, involves low-technology production, is energy efficient, has a low carbon footprint, and the final product does not depend on refrigeration and can be stored and transported under ambient conditions. One of the major problems faced in international trade starting from the early phases is storage pests. In this respect, official measures were implemented in major fig-producing countries, such as making sulfur treatment, methyl bromide fumigation, or any other method compulsory for controlling pests and pest damage (Aday, 1938; Mehr and Ferguson, 1958; Erakay and Özar, 1979). For decades, methyl bromide was the unique fumigant used until it was phased out globally due to its ozone-layer-depletion effect. Therefore, alternative methods that are effective in controlling storage pests and exert no negative effects on dried-fig quality have been investigated and implemented (Meyvacı *et al.*, 2010; Şen *et al.*, 2015) (see Chapter 14, this volume).

Sun drying of figs utilizes about 50% of the world fig production on a fresh basis. Dried figs are distinguished by their high nutritional value and functional food properties, which are triggering the demand worldwide. Every year, new studies are carried out suggesting additional nutritional and health benefits (Sridhar *et al.*, 2019). Global dried-fig exports exhibit an increasing trend. In 2018, global dried-fig shipments remained at the same record level reached in 2017, adding up to around 132,600 t; however in 2019, total exports reached 145,331 t (Fig. 27.2.) Turkey continued to be the top world supplier of dried figs in 2019, amounting to 44.5% of the global exports. Turkey's top destinations were the EU countries plus the UK (54% of the country share), among which Germany (13%), France (13%), and the UK (6%) ranked as the leading importers, with 9% destined for the USA. Similarly, top importers of Spanish figs were the EU and UK (79%) and the USA (19%). For Afghanistan, 95% of its shipments were bound for India. The INC data showed that in 2020, the dried-fig trade (0804 20) originating from Afghanistan totaled 12,697 t, of which, India with 11,588 t had the major share. Spain supplies 7566 t to the global market, with 6751 t going to the EU/UK/European Free Trade Association (EFTA) markets. Dried-fig exports from Iran, totaling 75,935 t, are destined for the EU/UK/EFTA (18,525 t), Turkey (14,093 t), the Russian Federation (11,099 t), and United Arab Emirates (10,200 t) markets. Turkey marketed 84,702 t of dried figs as 46,472 t to EU/UK/EFTA, 10,200 t to the USA, and 3495 t to China. The USA's 2019 export of dried figs totaled 2294 t, and 55.3% (1268 t) was sent to Canada followed by China (352 t), Mexico (265 t), and Japan (248 t), with other countries importing less than 50 t (www.nutfruit.org; accessed January 5, 2022).

Turkey is the leading country in dried-fig production and trade in the world and exports approximately 80% of its dried figs to more than 100 countries. Turkey's dried-fig export amounts between the 2004/05 and 2019/20

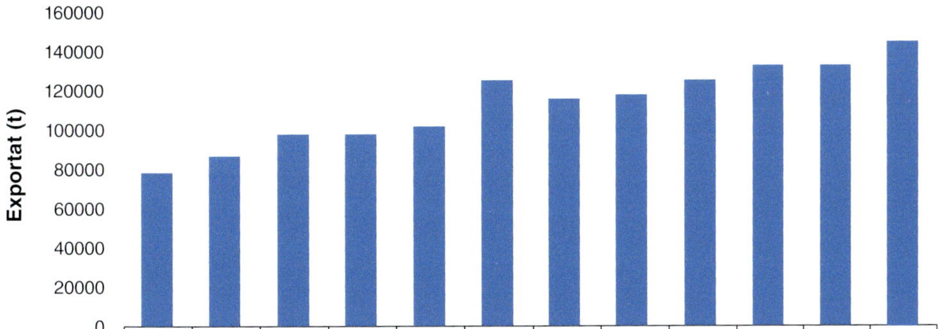

	2008	2009	2010	2011	2012	2013	2014	2015	2016	2017	2018	2019
■ Export (t)	78,709	86,921	98,155	97,792	102,135	125,775	116,252	118,223	125,408	132,819	132,585	145,331

Fig. 27.2. Annual global dried-fig exports (t) between 2008 and 2019. From INC (2021) and www.nutfruit.org (accessed January 5, 2022) with permission of INC.

seasons showed that export quantity is related to the total production of dried figs of that year but is more linked to the overall quality. The lowest exported quantity was in the 2007/08 season (38,359 t) due to the shortage of the 2007 crop and was highest was in the 2019/20 season with 64,708 t (Table 27.9). There is nearly no stock of dried figs in Turkey, as in many countries, and the dried-fig crop is sold out until the next year's crop arrives. When it comes to dried-fig substitutes such as dried apricots or grapes, there are always overstocks to compensate years with short crops. The climatic conditions, especially high humidity and strong winds during the ripening and drying period, trigger some of the defects leading to a higher share of culled figs, which are destined for industrial use. However, climatic conditions in general do not affect the quantity of internationally traded figs, as all quality classes including culled figs are subject to international trade, although the unit price and total value of the exports may vary (Çobanoğlu *et al.*, 2007).

EU countries are the major export markets for Turkish dried figs. Total fig fruit imported to the EU as fresh or dried was 80,851 t in 2019 (www.trademap.org; accessed January 5, 2022). Turkey provides 56% of this volume with 45,226 t. The breakdown of fig exportation from Turkey to the EU is 65.6% (29,650 t) of dried figs and 34.4% (15,576 t) of fresh figs. Some of the European importer countries (e.g. Germany, the Netherlands, France, Belgium) import quantities more than the national consumption because they are transit countries partially re-exporting imported quantities to third countries (Table 27.10).

The quality required for the dried-fig market is generally determined according to national or international standards. A large size, soft texture, light color (for white varieties), and being free from any defects, mechanical or pest damage, or visible mold are the major quality attributes. In this respect, higher-quality figs are sold as Extra or Class I grade (UN ECE Standard) destined for table consumption as ready-to-eat. Dried figs with slight defects are sliced, cut, or ground, and processed as fig paste. Dried figs for direct consumption are also shaped manually into different types. Whole figs are consumed as a snack and as an ingredient in home baking and cooking; softened figs are consumed as snacks and desserts; chopped figs are used in muesli, breakfast cereals, and fruit bars; clipped figs are used for jam and marmalade; and dried-fig paste is used in fig bars, biscuits, and cakes. Substandard low-grade culled figs are used for industrial use (alcohol, aroma) and for animal feed. The breakdown of these types in Turkish dried-fig exports in 2019 were as follows: whole figs, 72%; diced or sliced, 18%; fig paste, 8%; and culled figs, 2% (Aegean Exporters' Associations, unpublished data). The climatic conditions prevailing, especially during fruit maturation and the drying period, affect the share of classes of dried figs of different quality. In extreme years, the substandard quality may be up to 40% of the total production. Having different types allows

Table 27.9. Quantity (t) of annual dried-fig (total of all types) exports from Turkey to different markets between 2014 and 2019. From AEA (2021).

	2014	2015	2016	2017	2018	2019
Total	**63,889**	**61,659**	**64,779**	**62,979**	**61,327**	**64,708**
USA	7,818	7,421	8,820	7,948	7,075	10,100
France	8,724	8,013	8,719	8,262	8,486	7,877
Germany	7,526	7,157	7,881	6,933	7,023	7,660
Italy	2,849	3,470	2,985	2,140	2,637	2,831
Belarus	119	280	1,922	2,726	2,658	2,778
Mexico	1,194	2,476	1,998	1,525	1,333	2,711
UK	3,375	3,299	3,289	3,239	2,807	2,688
China	2,070	2,880	1,245	479	954	2,408
Canada	702	982	866	1,413	1,551	1,569
Switzerland	1,940	1,406	1 752	1,736	1,572	1,489
10 countries, total	36,317	37,384	39,477	36,403	36,094	42,111
Other countries	27,572	24,275	25,302	26,576	25,233	22,597

Table 27.10. Fig (dried in 2018 and dried and fresh in 2019) exports from Turkey destined for Europe and consumption (total and per capita) with a breakdown of the major importer countries (including the UK and Switzerland). Data from Trade Map (2021) and INC (2021).

Country	2018 (dried figs)					2019 (dried + fresh figs)				
	Total import (t)	Share of import from Turkey (%)	National consumption (t)	Consumption per capita (g $year^{-1}$)[a]	Estimated consumption per capita (g $year^{-1}$)[b]	Total import(t)	Share of imports from Turkey (%)	National consumption (t)	Consumption per capita (g $year^{-1}$)[a]	Estimated consumption per capita of special groups (g $year^{-1}$)[b]
Germany	19,297	36	15,809	190	761	17,077	86	13,990	168	673
France	14,862	57	12,871	198	792	17,476	52	15,135	233	931
UK	7,003	40	6,617	99	394	7,496	63	7,083	106	422
Netherlands	6,285	24	2,580	151	605	4,168	82	1,711	100	401
Italy	4,394	60	5,272	87	348	5,237	58	6,283	104	415
Switzerland	3,323	47	3,297	387	773	3,590	61	3,562	418	835
Belgium	2,333	24	1,619	141	564	2,969	25	2,060	179	718
Poland	1,784	26	1,775	47	94	1,642	52	1,634	43	87
Greece	869	30	4,656	420	885	612	24	3,279	296	623
EU										
Total import (dried+ fresh)			77,213					80,851		

[a]Total consumption expressed in g per person. Population data from United Nations, Department of Economic and Social Affairs, Population Division (2019). World Population Prospects 2019, online edn., revision 1. [b]Based on the estimated percentage of the population consuming dried figs.

marketing of all quality classes of dried figs for different purposes, in which unit prices may vary significantly. The availability of high-quality whole figs also affects the unit price of figs.

Turkey's whole-fig export market is mainly to France, Germany, and the USA. US import quantities increased with exportation of organic dried figs certified according to the National Organic Program of the US Department of Agriculture (USDA). Asia-Pacific countries, mainly China, are the fastest-growing market for Turkish dried figs. Whole dried-fig exports have varied between 45,000 and 50,000 t since 2014. Turkish whole figs are well known in export markets with their largest size, light color, and soft texture (Table 27.11).

Culled figs are the lowest-quality dried figs due to the high percentage of defective fruit. Culled figs can be utilized as a mixture for coffee after grinding, for preparing fig molasses, or as a raw material to extract fig aroma. Culled fig exports have been increasing since 2014 and reached 1339 t in 2019. The top three export markets for Turkish culled figs are Russia, China, and France (Table 27.12).

Diced or sliced figs can be used in cookies, cakes, muffins, breads, pastries, fruit cakes, frozen products, confectionery, muesli, granola, energy bars, rice mixes, and trail mixes. They can be either natural or certified organic, and diced fig exports have been increasing since 2014, reaching 4999 t in 2019. The top three export markets for Turkish diced figs are France, Germany, and the Netherlands (Table 27.13).

Fig paste is generally prepared from dried fruit of secondary quality and is destined for the food industry. Fig paste is an old ingredient used in the biscuit industry, such as fig rolls. The paste is prepared by grinding dried figs, either crushing the seeds or keeping seeds intact according to the specifications demanded. It is very popular in the biscuit industry as a filling thanks to its taste and high nutritional value. Table 27.14 shows Turkey's fig-paste export by country. The major dried-fig paste markets are the USA, Mexico, and the UK. In 2014, Turkey's fig-paste

Table 27.11. Turkey's whole-fig exports (t) between 2014 and 2019 by country. From AEA (2021).

	2014	2015	2016	2017	2018	2019
Total	**50,279**	**45,717**	**49,191**	**49,449**	**46,766**	**46,429**
France	6,845	6,300	6,954	6,697	6,698	6,106
Germany	5,589	5,001	5,606	4,699	4,746	5,772
USA	4,551	3,486	4,230	5,057	3,939	4,400
Belarus	119	164	1,749	2,706	2,540	2,555
Italy	2,652	3,272	2,690	1,940	2,385	2,508
China	2,070	2,643	966	319	719	2,053
Canada	674	916	831	1,361	1,483	1,494
Switzerland	1,774	1,248	1,624	1,633	1,438	1,330
Australia	1,104	1,341	1,421	1,318	1,364	1,242
Spain	1,312	1,024	954	881	914	931
Netherlands	1,264	1,286	1,272	1,140	1,027	886
Other countries	22,325	19,036	20,894	21,697	19,513	17,152

Table 27.12. Turkey's culled fig exports (t) by country. From AEA (2021).

	2014	2015	2016	2017	2018	2019
Importing countries	**620**	**756**	**367**	**650**	**1308**	**1339**
Russia	199	214	–	81	186	376
China	–	7	–	25	–	252
France	174	280	220	48	244	202
Belarus	–	–	20	–	98	164
Belgium	–	64	16	104	64	68
Italy	–	16	6	23	14	47

Table 27.13. Turkey's diced or sliced fig export (t) by countries. From AEA (2021).

Diced or sliced figs	2014	2015	2016	2017	2018	2019
Total	**4039**	**5186**	**5423**	**4822**	**5016**	**4999**
France	1091	1038	1306	1302	1274	1316
Germany	1194	1418	1578	1554	1501	1295
Netherlands	269	286	262	303	463	381
Japan	300	232	199	279	295	358
Italy	171	158	259	147	190	245
Australia	163	217	336	166	171	225
UK	36	102	122	143	133	146
Greece	0	10	65	57	109	125
Switzerland	159	144	112	91	101	119
Saudi Arabia	1	0	1	73	61	74
Other countries	657	1,578	1,184	707	718	714

Table 27.14. Fig-paste exports (t) from Turkey by country. From AEA (2021).

Country	2014	2015	2016	2017	2018	2019
Total	**8,177**	**9149**	**9,450**	**7,590**	**7,824**	**11,485**
USA	3,155	3,235	4,204	2,804	3,084	5,585
Mexico	1,191	2,431	1,986	1,496	1,323	2,699
UK	1,818	1,662	1,764	1,883	1,848	1,727
Germany	700	737	697	665	767	572
France	577	367	193	188	225	205

exports were 8177 t and, due to high demand, reached 11,485 t in 2019.

Commercial market quality standards

The commercial quality standards developed by the United Nations Economic Commission for Europe (UNECE) Working Party on Agricultural Quality Standards help facilitate international trade, encourage high-quality production, improve profitability, and protect consumer interests. UNECE standards are used voluntarily by governments, producers, traders, importers and exporters, and other international organizations. Many countries accept UNECE STANDARD DDP-14 concerning the marketing and commercial quality control of dried figs and participate in the activities of the Working Party and align their national standard on dried figs (UNECE, 2016). This standard applies only to whole dried figs grown from common fig (*F. carica* L.) and intended for direct consumption, and excludes those destined for industrial processing. The standard defines the product, defects, and the allowed maximum moisture content for figs that are untreated or treated with preservatives or any other methods such as pasteurization. The quality classes are Extra, Class I, and Class II, and the defects allowed in each quality class are defined. Sizing is optional and can be performed based on the number of fruit per kilogram or diameter. There are certain tolerances for size range in each quality class. When sizing according to diameter, the minimum is 18 mm for black varieties and 22 mm for white varieties (UNECE, 2016).

In Turkey, sizing according to the number of fruit per kilogram is similar to the UNECE (2016) standard; however, the TS 541 standard concerning dried figs is compulsory for export. Quality class and size information must be given on the package label. Size 1 starts with up to 40 dried fruit kg^{-1} and goes up to size 11 with 121 or more fruits kg^{-1} (TS 541, 2006). The Turkish national quality standard for dried figs, Turkish Standard 541, also covers processed and unprocessed dried figs as well as industrial (cull) dried figs and gives information on the different styles. It does not cover dried figs that are dipped in flour, sugar, honey, etc. either whole or fragmented, in or out of which products such as sesame seeds, almonds, walnuts, etc. have gone through intensive production. The commercial quality specifications of fig paste are also defined in another standard (TS 542, 2015).

The US Standards for Grades of Dried Fig (USDA, 2001) covers quality characteristics, sizes, and styles (Style I; whole as loose, pulled, or layered; Style II: pieces) of white and black dried figs. Dried figs may or may not be treated with sulfur or any other bleaching methods. Sizes are determined by the diameter and classified as Jumbo, Extra fancy, Fancy, Extra choice, Choice, and Standard. Quality grades are stated as US Grade A, US Grade B, and substandard. The maximum allowed moisture content for untreated dried figs is 24% compared with 26% in the UNECE Standard. The prerequisite quality factors on which the standard evaluates dried figs are stated as "flavor and odor, varietal characteristics, defects, maturity, uniformity of color and uniformity of size." Furthermore, sugaring is accepted as a defect in the US Standard (USDA, 2001).

Dried-fig types in trade

Dried whole figs are either packed as loose figs, termed "natural," or manually shaped into different forms or tied on strings. Some of these forms are known to have existed a century ago, while others developed during the last 50 years. Fig-producing countries may have one or more types dominating in their trade. In the USA, pulled and layered are the two types cited in the dried-fig standard. In Afghanistan and Central Asian countries, putting dried figs on a string like a necklace is the popular type, whereas in southern Italy, traditionally two figs are torn apart from their ostiole end up to the stem and the internal sides are then pasted together. The nine different types found dominantly in international trade are as follows:

- Chocolate: each dried fig is cut off at the stem and ostiole end and then wrapped individually with a packaging material such as aluminum foil or cellophane.
- Turkish delight (locum): the dried figs are shaped into squares and their corners torn and placed in special molds. Two to four layers must be placed on top of each other and pressed, thus giving them the appearance of Turkish delight (locum). This is one of the oldest forms known in the dried-fig trade.
- Layered: the dried figs are cut or torn longitudinally toward the stem and placed in special molds. Two to three layers are placed on top of each other and then pressed.
- Pulled: the dried figs must be shaped by concealing their stems, with domed tops and square bottoms.
- Protoben: the dried figs must be aligned, superposing each other by concealing their stems and ostioles.
- Lerida: the dried figs are pressed from the stem and ostiole side to give a flat and round shape and are then aligned side by side or one on top of the other.
- Macaroni: the dried figs must be prepared in the form of a roller manually pressed with their stems and ostioles at each end.
- Tied: the dried figs must be individually tied at their stems with a thread or raffia and placed in boxes in different forms.
- Garland: the dried figs are first shaped as lerida and then put in the form of a round bread in special round molds by connecting with thread, raffia, etc. through their mid-section or not.

When the 6-year average of export quantities is considered, the breakdown of different types of Turkish dried fig exported is: 32% as natural, 19% as lerida, 14% as fig paste, 8% as garlands, 8% as diced/sliced, 7% as layered, 5% as pulled, 4% as protoben, 1% as tied, and the rest as culled or other types (Aegean Exporters' Associations, unpublished data). Due to the high manual labor costs and risks of contamination, unshaped "natural" packing using automated machines is becoming widespread and its share in total exports is increasing steadily.

Dried-fig consumption

Dried-fig production is concentrated in a few countries, but the market is increasing to include new destinations. The International Nut and Dried Fruit Council consumption data state that dried-fruit consumption is, as of 2019, highest in the Middle East (29%), followed by Europe (24%), Asia (20%), North America

(15%), Africa (6%), and others (6%) (INC, 2021). The INC database calculates the global consumption of dried figs in two categories based on available data (Table 27.10). Dried-fig consumption per capita as calculated by dividing the population by the amount of figs present in the domestic market (production plus import, or production–export) is 26 g per person per year. The second data set is calculated based on estimation of the percentage of the population consuming dried figs in a specific country. The highest consumption levels were for Iran as 431 and 1307 g, respectively, for per capita and per capita of the specific consumer group in 2019. The second consumption data set assumes that not every person from all ages consumes dried figs. The average per capita consumption for the specific group shows that dried-fig consumption decreases after Iran in the following order: Germany (842 g), France (753 g), Switzerland (747 g), United Arab Emirates (553 g), Italy (549 g), Russian Federation (491 g), Netherlands (458 g), Canada (432 g), UK (407 g), Greece (353 g), Australia (254 g) and the USA (229 g).

Dried fig is a traditional product and is mostly consumed during religious festivities or as processed fruit. It is popular among the elderly either as an energy source or to help the digestive tract. More recently, dried figs have become popular among younger people, especially during sports activities, and the demand for dried figs has been increasing with changes in the type of consumer.

27.3.3 The fresh-fig trade

Fig trees are dispersed throughout fig-producing countries and were mostly consumed locally for centuries due to the delicate nature of the fruit. Fresh-fig fruits can be stored only for limited periods, and long-distance transport lowers the fig-fruit quality and palatability very rapidly. Compared with dried figs, the fresh-fig trade started to become more pronounced globally only after the mid-1980s. Initially, major trade routes for fresh fig were to neighboring countries in the vicinity. In North America, this was between the USA and Canada, and in Europe from Spain and Italy to Germany or the UK. The introduction of the Bursa Black variety to the global fresh market marked a milestone. The total quantity of fresh figs exported globally between 1980 and 1984 was still very little, calculated as 3,217 t, with 50% supplied by Turkey with Bursa Black variety, while the other suppliers were Italy and Greece. The total value was US$2,392,000. Between 2000 and 2003, the total exported volume increased to 18,440 t and the value to US$26,269,000. Turkey had a share of 29.7% followed by Spain with 17.4%, and Italy with 8.02 % (Eriten, 2005). As there are no specific data on the fresh-fig trade, export data presented for fresh and dried combined by the Observatory of Economic Complexity (OEC) for 2019 was used for the major fresh-fig producers. However, for Spain, Italy, and the USA, fresh and dried combined figures are presented. To process available data further, official fresh-fig export figures for Turkey were obtained from the Aegean Exporters' Associations, and all other countries are accepted as fresh-fig producers and exporters (Table 27.15).

Fresh-fig exports from Turkey increased from 2015 to 2020. Bursa Black is the fig variety for the fresh market, and the harvest season starts in August, September, and October. There are price differences in the importing markets ranging from US$1.17 kg^{-1} in Russia to US$4.26

Table 27.15. Total value of fresh-fig exports from major fig-producing countries in 2019. From OEC (2021).

Rank	Country	Export value (1000 US$)
1	Turkey	48,945[a]
2	Italy	74,772[b]
3	Spain	23,894[b]
4	USA	14,284[b]
5	France	8,210[b]
6	Brazil	8,206
7	Saudi Arabia	6,869
8	Mexico	6,196
9	Peru	5,271
10	United Arab Emirates	4,327
11	South Africa	3,772
12	Slovakia	3,450
13	Israel	2,917
14	China	1,041

[a]Aegean Exporters' Association's figure for Turkish fresh-fig exports in 2019. [b]Spain, Italy, and the USA also export dried figs, which is included in the total export value.

in Hong Kong. The overall average unit export price is calculated as US$2.76 kg^{-1}. Germany is the leading market for Turkish fresh figs comprising 33.4% of the exported amount in 2019 (Table 27.16). In 2018, Brazil exported 6,946 t of fresh figs at a value of US$1,383,015, with the Netherlands being the major importer receiving 58% of the tonnage (http://wits.worldbank.org; accessed January 10, 2022). Almost one-third of the Brazilian fresh-fig production is exported (www.tridge.com/intelligences/fig/BR; accessed January 5, 2022). The share of the other southern countries is rather low but shows an increasing trend. The fresh-fig export quantity and value of these countries, as of 2018, are: Saudi Arabia (6788 t, US$5,994,000), Peru (722 t, US$5,032,780), Israel (528 t, US$2,341,000), South Africa (US$149 t, US$705,000), India (146 t, US$752,000) and Argentina (65 t, US$357,000). Fresh figs from Peru are sent on cargo flights to the Netherlands, the UK, and the USA, whereas Saudi Arabian figs are marketed in neighboring countries. Israeli exports are destined for the UK, Belgium, the Netherlands, and Hong Kong (http://wits.worldbank.orgaccessed; January 10, 2022).

In international trade, fresh-fig quality is evaluated by voluntary national and international standards. The UNECE (2017) FFV-17 standard "applies to fresh figs of varieties (cultivars) grown from *Ficus carica* L. to be supplied fresh to the consumer, fresh figs for industrial processing being excluded." Fig fruit is classified according to their quality as Extra, Class I, and Class II, and certain tolerances for defects are accepted. However, in all classes, fresh figs should be able to withstand handling and transportation and reach the final market destination in good condition. Moreover, they should be "intact, sound, produce not affected by rotting or deterioration such as to make it unfit for consumption is excluded, clean, practically free of any visible foreign matter, fresh in appearance, practically free from pests, free from damage caused by pests affecting the flesh, free of abnormal external moisture, free of any foreign smell and/or taste" (UNECE, 2017). The OECD (2015) Standard for fresh figs comprises quality classification similar to that of UNECE (2017); however is compiled in English and French and has illustrations showing some varieties as well as defects.

The overarching aim of the fresh-fig trade is to supply fig fruit of high quality and tasty figs throughout the year. The major marketing channel for the global fresh-fig trade is the supermarkets, not only the high-end stores but also discounters selling fresh figs. As fresh-fig fruit cannot be stored for long periods, availability in the market is maintained by growing breba and main-crop-producing varieties in northern and southern hemispheres. However, the number of varieties subject to global trade is still limited in number and is dominated by the purple skin-colored varieties. White or light-colored

Table 27.16. Fresh-fig exports as quantity (1000 t) and value (FOB 1000 USD) from Turkey (AEA, 2021).

	2015		2017		2020	
	Quantity (1000 t)	Value (FOB 1000 US$)	Quantity (1000 t)	Value (1000 US$)	Quantity (1000 t)	Value (1000 US$)
Total	**14,523**	**38,149**	**16,131**	**46,245**	**21,476**	**60,004**
Germany	5,345	13,949	5,512	15,734	7,169	20,715
Austria	570	1,473	1,006	2,989	1,908	6,430
Netherlands	1,710	4,559	1,850	5,461	2,051	5,642
Russian Federation	671	1,093	1,417	2,567	2,277	4,960
UK	1,822	5,646	1,668	5,128	1,830	4,730
France	1,331	3,159	1,326	3,836	1,559	4,589
Hong Kong	236	777	683	3216	802	3,422
Switzerland	500	1,426	507	1,527	662	1,661
Poland	360	1,211	226	840	549	1,557
Italy	157	487	197	572	351	949

FOB, freight on board.

varieties are more susceptible to bruises and discoloration/browning that lowers marketability. Turhan *et al.* (2013) stated that color is the major determinant of price in the Bursa Black variety among the other fruit characteristics analyzed. The major fresh-fig varieties reaching global markets from different sources of origin and their presence in the market are summarized in Fig. 27.3.

In parallel with demand, the global fresh-fig trade is increasing with the entrance of new production regions outside the traditional fig-producing countries. The support of innovative production systems and postharvest technology help to extend the availability of fresh figs. Supermarket chains experienced in value-chain management are the major outlets forcing producers to supply higher quantities with well-defined quality and safety specifications throughout the year. Double-cropping and common-type varieties partially serve this aim; however, fig varieties with high yields, bearing parthenocarpic and tasty fruit with longer harvest season and shelf life may help further development of the global fresh-fig market (see Chapter 23, this volume).

27.4 Concluding Remarks

Data on the fresh- and dried-fig trade is rather scarce as fig is a minor fruit and the global fresh market is still limited. Planning production and trade based on accurate data is of the utmost importance for a perennial species such as fig, with a comparatively longer juvenile period. As trade volumes increase and sourcing origins diversify, detailed breakdown of trade data will become more decisive. The historical data show that dried-fig production has faded away in the major producers of the 1950s—Italy, Portugal, Greece, and Spain—within a timespan of 60 years. As these countries have decreased their production, fig-producing countries from the developing world have emerged and filled the gap in global fig production. Thus, it is important to understand the past and current trade flows and to focus on major factors that influence the sustainability of the value chains.

Except in the USA or newly emerging countries, fig production is mostly realized by smallholder farms where available family labor is used for harvesting, drying, and other farm operations. Despite some attempts to develop mechanical harvesting, manual harvesting is still widespread and remains at almost 100% in fresh-fig production. As manual shaping is required in the dried-fig industry and mostly packing is done manually, fig processing and packing generates employment and income, mostly for women, which can be considered while developing policies. Işın *et al.* (2003) stated that the social role of fig production should not be undermined as it engages around 35,000 families and employs around 30,000 seasonal workers for harvest in just two Turkish provinces, supplying the whole national dried-fig production. Furthermore, fig production concentrates in certain areas where production, processing, and trade

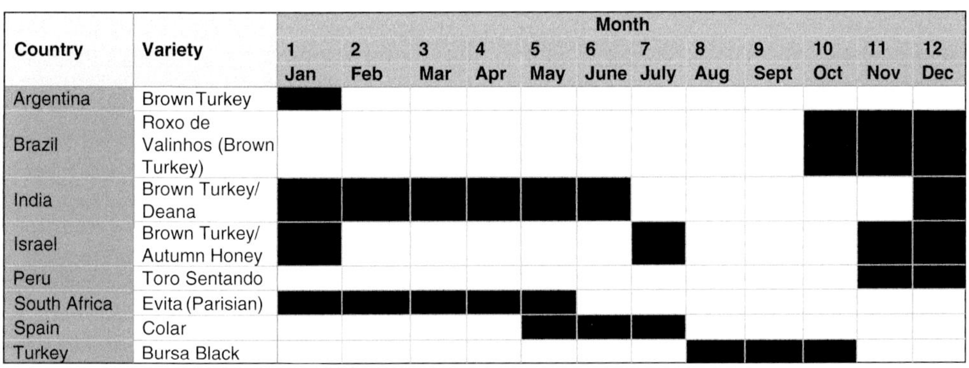

Fig. 27.3. Major fresh-fig exporting countries, varieties, and availability period (may vary according to the yearly conditions) in the global markets (Author's own figure).

may bring high economic viability to the region. Current data show that all classes of dried figs can be marketed as whole, diced/sliced, paste or cull; however, the quality determines the price. Thus, the aim is always to have the highest share of Extra class whole figs.

The leadership of Turkey in the dried-fig trade seems set to continue for a while, with forecasts showing a slight increase in dried-fig production (Uzundumlu *et al.*, 2018; Çakan, 2020). One of the factors that enabled Turkey to become the prime producer and leader in this trade was the comparative advantage of low-input sun-dried fig production on marginal land and the transfer of knowledge and experience accumulated over centuries to the coming generations. Moreover, with the high demand for organic produce, during the last decade low-input Turkish dried figs were certified as organic and directed toward new and organic marketing channels in Europe and the USA. Countries that produce significant quantities of dried figs but are not subjected to global trade yet such as Egypt, Algeria, Morocco, and Syria may get higher shares as the global demand increases. The high requirement for hand labor is an opportunity to create jobs in developing countries. However, to develop the fig sector and trade, quality and safety issues should be focused throughout the value chain. In the dried-fig trade, in addition to low market quality, two major problems exist globally: storage pests (the species may vary) and mycotoxins, both of which require integration of basic scientific and local knowledge. In this regard, fig-producing countries should focus on solving the existing problems, and increase the quantities exported by improving dried-fig value chains.

One of the major threats in traditional fig-producing regions is climate change. Thus, sustainability and carbon emissions are becoming of major concern among consumers in importing countries demanding implementation of regenerative practices to help carbon sequestration and minimize carbon emissions.

The breaking point for all fig-producing countries could be to further process and create high value, new and healthy snacks, or food based on figs according to consumer demand. The global dried-fig trade is dominated by a few varieties such as Sarılop (= Calimyrna), Black Mission, and Sabz, all of which have different quality attributes. New varieties suitable for drying (e.g. small, round, and black) could be introduced. Moreover, artificial drying used to obtain low-moisture fig products will help generate new products and tastes.

Among the fig varieties traded globally, very few varieties (Black Mission, Kadota (Dottato)) are utilized both as dried and fresh, as the quality parameters demanded for the fresh and dried industry may differ. However, the rich diversity within common fig and related *Ficus* species offers a high potential for variety development with the support of molecular techniques. The recent interest in many parts of the world suggests that the future fig trade will encompass new varieties, new products, and new uses.

There are many research results outlining the health and nutritional properties of the fig fruit, chiefly linked to the phenolic content. This aspect makes fig a product in demand by health-conscious consumers. In many fig-producing countries, national consumption figures are very low; however, local consumption can be the driving force to increase production. Certified production for organic, geographic indications, fair-trade only, or as multi-certification is gaining popularity among importers and consumers. Thus, fig products, whether dried, fresh, or processed, can be an element in rural development and tourism introducing "exotic," perishable pleasure, taste, and flavor. The emerging markets, among which Asia-Pacific is the fastest growing, will provide new chances for all fig-producing countries. Due to the perishable nature of fresh figs, trade is mostly led by neighboring countries. However, there are a few countries such as the Netherlands and Hong Kong that act as transits. Intensive systems of fig production, such as soil-less production under protection, aiming to supply the fresh-fruit trade, create opportunities for many countries even if the climatic conditions are not optimal. In conclusion, sustainable agri-food production and trade policies granting production and trade of safe and high-quality fig fruit or its products is a key issue to nourish the global fig trade.

References

Aday, N. (1938) Yunan Kuru İncirleri (Greek dried figs). In: *1st Rural and Agricultural Development Congress*. Ministry of Agriculture of the Turkish Republic Publications, Ankara, Turkey.

AEA (Aegean Exporters' Associations) (2021) Dried Fig Statistics, E-Birlik database. Available at: http://istanbul.ebirlik.net/ebnet/app (accessed December 31, 2021).

Aksoy, U. (2021) Dried fig industry: progress and challenges. *Acta Horticulturae* 1310, 157–164.

Çakan, V.A. (2020) Forecasts for Turkey fresh fig production and dried fig export: ARIMA model approach. *Journal of Tekirdag Agricultural Faculty* 17, 357–368.

Çobanoğlu, F., Kocatas, H., Ozen, M., Tutmus, E., and Konak, R. (2007) An evaluation of approach in determination of climate factors impact on dried fig exportation in Turkey. *Journal of the Faculty of Agriculture Ondokuz Mayıs University* 22, 11–19.

Condit, I. (1946) *Fig Production in California*. İzmir Ege Publishing Co., Ministry of Agriculture Publications Office, Izmir, Turkey (translated into Turkish).

Eisen, G. (1901) *The Fig: Its History, Culture and Curing*. Kessinger Publishing, Whitefish, Montana.

Erakay, S. and Özar, A.I. (1979) Ege bölgesinde kuru incirlerde zararli böceklerin bulaşma oranlari ve savaş yöntemleri üzerinde ön araştirmalar (Preliminary works on the infestation rates and the control measures of insects in dried figs in Aegean region). *Bitki Koruma Bülteni* 19, 159–170.

Eriten, A. (2005) Estimates of demand relationships for figs and fig products in Turkey. MSc thesis, Graduate School of Social Sciences, Middle East Technical University, Ankara, Turkey.

EU (European Union) (2020) Commission Implementing Regulation (EU) 2020/1577 of 21 September 2020 amending Annex I to Council Regulation (EEC) No 2658/87 on the tariff and statistical nomenclature and on the Common Customs Tariff. *Official Journal of the European Union, L* 361. Available at: https://eur-lex.europa.eu/legal-content/EN/TXT/?uri=CELEX:32020R1577 (accessed December 31, 2021).

FAO (Food and Agriculture Organization of the United Nations) (2021) FAOSTAT. Available at: www.fao.org/faostat/en/#data/QCL (accessed December 31, 2021).

INC (2021) Dried figs. In: *Nuts and Dried Fruits Statistical Yearbook 2020/21*. International Nut and Dried Fruit Council, Reus, Spain, pp. 65–68. Available at: www.nutfruit.org (accessed December 31, 2021).

Işın, F., Cukur, T., and Armagan, G. (2003) An evaluation of dried fig production and marketing in Turkey from dried fig exportation standpoint. *Acta Horticulturae* 605, 183–189.

Jafari, M., Rahemi, M., and Zare, H. (2017) Increasing the tolerance of fig tree to drought stress by trunk thinning. *Acta Horticulturae* 1173, 189–195.

Khamis, M.A., Bakry, K.H.A., El-Sayed, M.E., and El-Husseiny, A.M. (2006) Evaluation of some fig cultivar in Egypt 2—fruiting and fruit quality. *Journal of Biological Chemistry and Environmental Science* 1, 105–116.

Mehr, S. and Ferguson, H.K. (1958) *Dried Fig Industry of Portugal*. Foreign Agricultural Service, US Department of Agriculture, Washington, DC.

Meyvacı, K.B., Sen, F., and Aksoy, U. (2010) Optimization of magnesium phosphide treatment used to control major dried fig storage pests. *Horticulture, Environment, and Biotechnology* 51, 33–38.

Nesci, F.S., Nicolosi, A., and Privitera, D. (2004). Regional marketing of dried fig-based products: an exploratory consumer study. *New Medit* 3, 55–60.

OEC (Observatory of Economic Complexity) (2021) Figs, Fresh or Dried. Available at: https://oec.world/en/profile/hs92/figs-fresh-or-dried (accessed December 31, 2021).

OECD (Organisation for Economic Co-operation and Development) (2015) Fresh Figs, International Standards for Fruits and Vegetables. Available at www.oecd-ilibrary.org/docserver/9789264234086-en-fr.pdf?expires=1626754114&id=id&accname=guest&checksum=4BDE096739B7F2E2D576B43C734D4770 (accessed December 31, 2021).

Pereira, C., Serradilla, M.J., Perez-Gragera, F., Martin, A., Villalobos, M.C., and Lopez-Corrales, M. (2017) Evaluation of agronomic and fruit quality traits of fig varieties (*Ficus carica* L.) grown in Mediterranean conditions. *Spanish Journal of Agricultural Research* 15: e0903.

Salih, G. (2020) Assessment of production and marketing practices for dried figs: surveys of Moroccan fig producers and sellers. *Moroccan Journal of Agricultural Sciences* 1, 84–90.

Şen, F., Aksoy, U., Emekci, M., and Ferizli, A.G. (2015) Determining effect of phosphine ($ECO_2FUME^{®}$) fumigation under atmospheric and vacuum conditions on dried fig quality. *Fresenius Environmental Bulletin* 24, 2046–2051.

Sridhar, D., Venugopaga, S., and Swamysekhar, R. (2019) Different drying methods for fig. In: Dar, S.A. (ed.) *Research Trends in Horticulture Sciences*. AkiNik Publications, Delhi, India, pp. 63–69.

Trade Map (2021) Trade Statistics for International Business Development. Available at: www.trademap.org/Country_SelProductCountry_TS.aspx?nvpm= (accessed December 31, 2021).
TS 541 (2006) *Kuru İncir (Dried Fig)*. Turkish Standards, Türk Standartları Enstitüsü, Ankara.
TS 542 (2015) *İncir Ezmesi (Fig Paste)*. Turkish Standards, Türk Standartları Enstitüsü, Ankara.
Turhan, S., Nargeleçekenler, M., and Cetin, B. (2013) A hedonic analysis of Bursa's Black Fig bid prices and product quality characteristics in Turkey. *Journal of Food, Agriculture and Environment* 11, 264–267.
UNECE (United Nations Economic Commission for Europe) (2016) *UNECE Standard DDP-14 Concerning the Marketing and Commercial Quality Control of Dried Figs*. United Nations, New York/Geneva.
UNECE (United Nations Economic Commission for Europe) (2017) *UNECE Standard FFV-17 Concerning the Marketing and Commercial Quality Control of Fresh Figs*. United Nations, New York/Geneva.
USDA (US Department of Agriculture) (2001) United States Standards for Grades of Dried Figs. In: *USDA Agricultural Marketing Service, Fruit and Vegetable Programs*. Processed Products Branch, Washington, DC.
Uzundumlu, A.S., Öksüz, M.E., and Kurtoğlu, S. (2018) Future of fig production in Turkey. *Journal of Tekirdag Agricultural Faculty* 15,138–146.
Valley Fig Growers (2009) *Valley Fig Growers 1959–2009. 50th Anniversary Celebration: A Pictoral History of Valley Fig Growers*. Available at: www.valleyfig.com/our-story (accessed December 31, 2021).

28 Fig Marketing Strategies

Yanlei Zhai and Huiqin Ma*
College of Horticulture, China Agricultural University, Beijing, China

28.1 Introduction

The fig has been cultivated as an edible commodity for thousands of years throughout the Mediterranean regions from where it became widespread in warm and dry climates around the world. Fig is a dioecious gynodioecious tree, with two tree morphologies, the usually non-eaten pollinator caprifig and the edible fig. According to the Food and Agriculture Organization of the United Nations (FAO), the world production of fig fruit is stable, with a decade average of approximately 1.1 million t year^{-1}. Most of the global production of figs is harvested in Turkey, followed by Egypt, Algeria, Iran, and Morocco. Other major producers are Syria, the USA, Brazil, Spain, and Tunisia. Fig production in China has increased dramatically in the last decade, with estimated annual production of 200,000–300,000 t (i.e. around 15–28% of world production).

Fresh and dried figs have become a staple of Middle Eastern and Mediterranean cuisine. Both fresh and dried figs are especially rich in nutrients such as sugar, dietary fiber, and proteins, and have high levels of potassium, calcium, vitamins E and B, and minerals, but also organic acids and polyphenols. There are a number of different edible fig varieties, with different fruit flavors; some varieties are parthenocarpic, while others require pollination via the fig wasp. Fruit drying is the easiest way of fig-fruit preservation. Dried fruits have a shelf life of over 1 year and can be shipped and stored easily. In addition, the fruits can be eaten, canned, or kept in other preserved forms. The consumption of figs should be encouraged as they have high beneficial effects and are potential healthy alternatives for sweets.

Although fig is an ancient and traditional fruit, marketing and promotion are still a necessity for new markets and new consumers, and to compete with other fresh/dried fruits. The marketing of figs mainly includes fresh figs, dried figs, and other processed fig products such as cakes and jams. A few merits can be shared in all fig-product marketing, such as the high dietary fiber content, which has an effect in promoting gastrointestinal peristalsis. Lifestyle changes make people sit for longer, and have less exercise/sports time, in addition to changed food structure, so consumers need more dietary fiber to maintain their gastrointestinal health. The main sugars in fresh and dried figs are hexoses, which can be used directly by our body, and the low acidity makes fresh and dried figs fit for people of different ages and a wide taste preference.

Fig has a rich culture and is mentioned many times in the Bible and in poems and legends.

* Email: hqma@cau.edu.cn

DOI: 10.1079/9781789242492.0028

The fruits and leaves can be observed in paintings, sculptures, and various daily decorations such as plates, bricks, blankets, carpets, and clothes. Beyond the basic function of providing energy and fighting hunger, fig is also an integrated part of culture and living style. Promoting the culture of fig, its history, specific ways of fig consumption, and the use of figs in cuisine will increase consumer stickiness.

There are also significant differences between fresh- and dried-fig marketing. Fresh figs are perishable and need cold-chain transportation. Many consumers are unfamiliar with fresh figs but have a positive attitude, indicating that education promotion can play a crucial role in marketing (Crisosto *et al.*, 2010). Marketing of fresh figs is similar in many ways to the marketing of table grapes, whereas marketing of dried figs could be compared to that of raisins. Freshness, high quality, exotic flavor and taste, healthy benefits, and Mediterranean fashion could be the aims for fresh-fig marketing, similar to cherries, strawberries, and other fresh eating berries. Dried figs could be promoted as sweets or for home baking enjoyment.

A number of challenges are also faced by fig marketing. For fresh figs, the short shelf life is a key limiting factor. Controlling the storage and transportation conditions of fresh figs, such as temperature, humidity, and vehicle vibration, is crucial to reducing quality loss and extending shelf life. Research on fig quality formation also revealed that pollination led to prolonged shelf life for parthenocarpic varieties (Rosianski *et al.*, 2016). Beyond the cultivation and postharvest measures, fundamental and strategic chance could lie in high-tech-facilitated breeding, as was achieved in the last three decades with tomatoes. In history, fresh figs were consumed locally in the production area, while the major commodity form is dried figs. Consequently, fig variety selection was more focused on yield, size, and other extrinsic attributes such as color, rather than storability. With the increase in the fresh consumption market, shelf life is considered an important selection criterion in fig breeding.

For dried figs, improvement, innovation, and market positioning are top factors that need to be considered for sustainable development. Natural drying, by exposure to the sun, is the traditional processing method for dried figs. Insufficient drying and storage conditions can lead to problems with insects, *Aspergillus*, and dust pollution, which negatively affect the quality of dried figs. Sorting dried figs based on appearance (e.g. color and fruit size) before sale helps obtain high-quality products and increase sales prices. As well as sun drying, modern drying technology provides alternatives for areas where the fig harvest season is not naturally dry enough. In China, most dried figs are processed by freeze-dry technology, which provides better nutrition maintenance, especially of second metabolites, with the cost that the figs must be cut into pieces for the process.

In many traditional production regions, dried figs usually lack packaging and labels showing the origin, variety, or expiration date in the retail market, which is not conducive to market promotion and communication with consumers. For packaged dried figs, the package is often in a rather old style and does not fit well for the modern family size, fast-paced city life, and the preferences of young consumers. Wine and beer are both historical drinks, and their package size and label design have been constantly upgraded when new materials, new closures, and new lifestyles have emerged. Moreover, since the start of the new century, much business has moved online, especially to mobile terminals. E-commerce and social media are widely used in fresh and dried agricultural product promotions. This trend is expected to become more prevalent, and figs will not be an exception.

In this chapter, the targeted market, distribution channels, and major promotion practices for fig and fig products are addressed. This is the first marketing review in a fig book as this topic has long been neglected. As more data, especially that from e-commerce, become available together with new publications, this chapter can be revised and updated in a few years.

28.2 Target Market

28.2.1 Market segmentation

The target market refers to the specific market that production or distribution units plan to enter. Customers in a particular market have different preferences and purchasing behaviors regarding price, the quality of the produce, product origin, brand, and other factors (Chang *et al.*, 2016). Market segmentation is the process of dividing the overall market into submarkets based on geographic, demographic, behavioral, or psychographic characteristics (Deepak and Jeyakumar, 2019). This is of particular importance in the marketing of fresh fig and fig products. With the aid of market studies, fig producers and distributors can obtain or follow up on specific data-based information regarding the needs and preferences of a certain market, identify business opportunities, and provide the appropriate products and services. Market segmentation can also help the fig industry identify competitors and potential partners in the market.

Previous studies have suggested several criteria for market segments, including profitability, variability, accessibility, market size, market growth, and competitive forces (McQueen and Miller, 1985; Loker and Perdue, 1992; Simkin and Dibb, 1998). In general, the range of the market segment should be clear and extant, such that its capacity can be measured. Through media, distribution channels and other measures, the target market can be entered by producers and distributors, and product information and products can be delivered effectively to consumers. The market segment needs to be large enough to amass purchasing power, making the business profitable.

Fig producers and distributors can use a variety of factors to define their potential market. Commonly applied criteria include domestic or foreign markets, based on legislative and geographical factors. The market of a country or region can be further segmented by demographic factors, such as age, gender, family background, income, occupation, educational background, and others (Rajagopal and Bansal, 2013; Gindi *et al.*, 2018). Among these factors, household income is a very strong influencer for fruit consumption (Park *et al.*, 1996), while knowledge and awareness of the health benefits of fruit, as well as other factors, play minor but noticeable roles (James *et al.*, 2009; Nandi *et al.*, 2017).

Consumers' psychological profiles and lifestyles can also be used in market segmentation, such as personality, motivation, and lifestyle-related purchasing and consumption behaviors. Some consumers demonstrate a strong curiosity to try novel fruits and processed products, while others may be conservative in their food choices (Shim *et al.*, 2001). Antioxidants can be highly valued by some consumers (Romano *et al.*, 2015), providing a targeted market for fresh figs, which are rich in these, as well as in phenolic compounds (Wojdyło *et al.*, 2016). In addition, the consumers' cultural environment can be used as a reference for target-market choices. Globalization, intercontinental businesses, tourists, and immigration establish new neighborhood communities, and the purchasing behavior and preferences of such social groups can be target markets for certain fruit and distribution channels, a good example being the fruit businesses in Dubai in the United Arab Emirates (Rajagopal and Bansal, 2013) and the popular tourist destination Bali, in Indonesia (Korry *et al.*, 2017). Globalization creates new market possibilities for figs in regions where they are not traditionally consumed.

Fig fruit can be consumed fresh or after processing into dried figs, jam, alcoholic drinks, and other products. Traditionally, the fig market is in the Mediterranean region: among the world's main producers are Turkey, Egypt, Iran, and Morocco (Crisosto *et al.*, 2011). Fresh figs are highly perishable and not suited to long-distance transport without modern postharvest technology, so figs in these traditional markets are in the form of dried fruit.

There is a linear positive fit between local growth and local consumption of figs. Outside the traditional fig-consuming regions, commercial production is found in the USA, Japan, China, Australia, South Africa, and other New World countries (FAO, 2018). In China, Sichuan, Shandong, and Xinjiang are the main fig-producing provinces (Weibin *et al.*, 2001; Zhang *et al.*, 2020). In the New World markets (i.e. North and South America, Australia, New Zealand, South Africa and East Asia countries), fresh figs have been redefined as an exotic, high-quality, perishable pleasure for the high-end market. For the grower, fresh fruit production is often more important than, or at least as important as, the processed goods. In the New World markets, consumers also buy dried figs, fig jams, and other products. Processed fig products are sold to the food industry in bulk to produce biscuits, pastries, energy bars, and other products (Shamin-Shazwan *et al.*, 2019). In China, processed figs are estimated at around 10% of the total production.

28.2.2 Market surveys

A market survey is a process whereby enterprises or service agencies systematically collect, sort, and analyze market information. The results can be purchased to understand market demand, to draft and adjust market positioning, and to revise product structure, marketing channels, pricing, and promotion measures (Zhou *et al.*, 2015). Market surveys can also help enterprises in market prediction and new-product development. A market survey can be qualitative or quantitative in nature. A qualitative survey provides a holistic study of the market, explaining market phenomena or consumer behavior by induction. A quantitative survey investigates and analyzes quantifiable indicators, such as demographic attributes, consumer behaviors, and so on. Specific goals need to be clarified before launching a survey; then the required information, and how it will be collected, are defined. The investigators must devise a survey plan, organize an investigation team, and collect the information/data accordingly. Finally, the raw data are analyzed using statistical tools, the necessary validations are made, and the survey report is written up.

Market survey methods

Appropriate methods and measures are the primary components of a successful market survey. There are several widely used methods for market surveys of figs and their products.

OBSERVATIONAL STUDY This method is aimed at elucidating the cause-and-effect relationships when it is difficult to conduct controlled experiments (Cochran and Chambers, 1965), and is well used in market studies, for example to investigate the effect of treatments when subjects are not randomly assigned (Rosenbaum, 2005). This method is direct, mostly regarded as objective, and easy to operate. An observational study has no questions or communication with consumers, which could avoid discrepancies caused by investigator induction or unsuitable question settings. This method can also follow a dynamic recording model to understand changes in consumer behavior.

INTERVIEWS This refers to the use of structured, semi-structured, or open-ended questions used to conduct interviews with respondents about the survey. The interview method is usually characterized as direct and in depth. Before the interview, the investigators should make adequate preparations by learning as much as possible about the subject of the survey, creating an interview outline, and establishing preliminary expectations of the interview results. During the interview, the investigators should create an atmosphere of relaxed and free communication, and ask questions clearly and appropriately; they should fully consider the differences in the respondents' form of expression, their ability to understand, and other communications aspects, and adjust the interaction accordingly. A semi-structured interview was successfully performed by a telephone questionnaire survey on consumers' choice of cactus pear quality and credence attributes (Migliore *et al.*, 2015), supporting the use of this interviewing method in fig marketing studies. An interview survey in Mostar, in Bosnia and Herzegovina, studied consumers' satisfaction with the fresh- and dried-fig supply (Knežović *et al.*, 2017). It reported that 95% of consumers like fresh figs, which was higher than the 65% liking dried figs. Consumers attach importance to organoleptic properties (i.e. shape, size of fruit, variety, color, firmness), but with differences according to their socio-demographic characteristics. When purchasing dried fruits, consumers prefer local products with high intrinsic quality rather than a good appearance.

EXPERIMENTAL STUDIES In this method, investigators purposely change one or a few factors, such as the price and package of figs, dried figs, and processed fig products, and observe the market's or respondents' responses to determine the influence of the changed factors. By controlling the numerous factors and the experimental environment, investigators can obtain first-hand data directly and reveal the correlations between different market factors.

ONLINE MARKETING STUDIES This is a convenient way for investigators to communicate with respondents through the Internet. The psychological buffer effect of cyberspace can offer a relaxed environment for respondents. Computer- and smartphone-based data collection can increase the scale and efficiency of the survey. Online methods also have the advantages of low cost and less need for human and material resources. The methods can be further divided into online surveys, email surveys, search-engine surveys, etc. Using tools such as web crawlers (e.g. Python), investigators can capture and analyze public data pertaining to online sales (Kausar *et al.*, 2013; Thomas and Mathur, 2019).

The contents of the market survey should be closely related to the preset goals, and the questions should be arranged in an order that will avoid confusion. In comprehensive surveys, questions pertaining to different topics should be organized into modules. Questions should be reasonable, specific, and clear, to avoid ambiguity. For cross-cultural surveys, common-sense errors and culture conflicts should be carefully avoided to ensure the respondents' cooperation. The surveys should be objective, with no obvious hints or implied content. The survey should not be too lengthy, to avoid respondent resentment. A pre-investigation is always advisable to determine whether the plan is feasible and to enable correction of potential flaws in time.

Asking for cooperation from respondents is a prominent issue in market surveys. In general, investigators should inform the respondents of the purpose of the survey, while emphasizing the significance and importance of the respondents' participation. During the survey itself, the investigators should be warm, friendly, and polite to establish a good relationship with the respondent. The investigators can offer small rewards for completing the survey, such as souvenirs, coupons, or subway tickets (Olsho *et al.*, 2015), or they can promise to share the final study results with the respondents. The investigators must ensure secure privacy of the respondents' personal information.

The main contents of fig marketing surveys include but are not limited to the following:

- **Product supply information**. This includes the fig-growing acreage, yield, and yearly production-related information of a certain region. From the marketplace, information should be collected on varieties, region of origin, grade, quality, brand, packaging, and prices of the fig fruit (or processed products). It may be necessary to investigate the number, business scale, and operational performance of potential competitors, as well as the main competing products in the market. A study of the Italian market found that fresh figs are used for end consumption, a small part of which is used to produce other minor products. Local dried-fig products, usually made from the Dottato and white fig varieties, face competition from Turkish and Greek figs which are derived from varieties with large-seeded and bulky syconiums (Nesci *et al.*, 2004).
- **Consumer behavior information**. This contains information on purchasing capacity, purchasing motivation, consumption scenarios, consumption habits, and product preferences of consumers in the target market. The aim is to differentiate the target group, assess the market size, and reveal consumers' behavior patterns and consumption preferences. A study in southern Italy found that the fruit's sensory qualities, namely appearance, odor, texture, taste, and flavor, are the main drivers for purchasing fresh and cold-stored figs; this information could be useful in optimizing the period of cold storage, prolonging fresh-fig shelf life, and enhancing its market value (Ingrassia *et al.*, 2015). Consumers in Spain were tested using a set of physico-chemical and sensory indicators, namely maturation index, sensory and flavor-related flesh color parameters, and terpenes in their preference decision making. Cuello Dama Negro and San Antonio gained the highest consumer preference in the nine tested varieties (Pereira *et al.*, 2020). A study on fresh figs in California demonstrated that consumer acceptance of tree-ripe figs is higher than that of commercial-ripe figs, suggesting that varieties with high soluble solids content and firmness should be developed to support the fresh-fig industry (Crisosto *et al.*, 2010).
- **Market price information**. This contains the price points of figs and processed products, pricing models, and price differences among different distribution channels and regions, distributors' responses to wholesale prices, consumers' responses to retail prices, and the effects of price promotions, among others.
- **Distribution channels and promotion information**. This includes information on

upstream wholesale markets, and the number, location, sales scale, and service level of wholesalers, retailers, and agents in a certain region. It also includes surveys on transportation methods, distribution efficiency, storage facilities, and preservation and quality regulation of figs and processed products. Investigators can also evaluate different promotion methods and their effects, such as point-of-sale promotions, sales promotions, public relations, advertising, and so on.

28.2.3 Packaging

Packaging has four basic functions: protection, communication, convenience, and containment (Han, 2005). Packaging is a vital component of product image and quality in marketing, directly affecting the consumers' first impressions of a product. Packaging provides information on product quantity, quality, region of origin, and so on, and is an important means for companies to compete in the marketplace and to guide consumers to purchase their product (Krishna *et al.*, 2017). Packaging styles—for convenience, beauty, novelty, or to indicate a high-end product—can attract consumer attention and generate consumer engagement. Packaging also serves as a promotion device by reflecting the brand's culture and the value proposition of the producers and distributors. An effective packaging strategy must consider factors such as product characteristics, market demand and consumer psychology, and selection of the appropriate packaging materials and designs to promote sales. Four fresh figs (usually dark-skinned varieties harvested at one-third of the coloring stage) in a square plastic box with around 200 g in total for 10 renminbi (RMB) (around US$1.5) is the most common package in supermarkets in China, which is suited to the target market of young persons, in a package for convenient consumption by one to two people. Similarly, bulk-imported dried figs are repacked into 250 g or 500 g plastic bags and sold for about RMB 45 (around US$7) (Fig. 28.1).

Packaging strategies in marketing

GREEN PACKAGING Different packaging materials are perceived as having varied environmental impacts (van Dam and Apeldoorn, 1996; Lindh *et al.*, 2016). The green packaging concept advocates environmental protection and coping with resource shortages. It includes reduction and thinning of packaging materials to avoid excessive packaging, reuse and recyclability of packaging materials, and the use of degradable packaging materials. Plastic packages are regarded as harmful to the environment by organic fruit consumers (van Herpen *et al.*, 2016). Green packaging materials commonly include bamboo, cotton, linen, and other natural materials, as well as fiber, starch, and synthetic polymer materials that can be biodegraded, chemically degraded, or recycled. Figs have a long tradition in many regions where green packaging materials are primarily used for the transportation and sales of fresh and dried products. The sales of these products are traditionally based on local markets. Newly developed green packaging materials support the long-distance transportation of figs, making fig products available to more consumers.

Fig. 28.1. Plastic and carton fig packaging.

CONVENIENCE PACKAGING In sales, the convenience strategy is reflected in package design, such as comfortable handles, a suitable size and easy opening. Diversity in the demand for convenience packaging in different countries, regions, markets, and consumer groups must be considered. Traditional regions and areas with relatively larger households may need larger consumption units, whereas for emerging markets related to more experiential consumption, the packaging unit may be smaller. A smaller package might increase consumption due to the lower price per package, and prevent the consumer-handling losses that occur in self-service sales (Asioli *et al.*, 2016). China's fresh-fig market offers different packaging units for different consumers, containing four, six, 12, or 24 fresh figs per package.

GIFT PACKAGING Fruits are frequently given as gifts in East Asia, South-east Asia, and other regions (Shim *et al.*, 2001; Gindi *et al.*, 2018). Gift packaging has developed in response to this demand. A gift of fresh figs is common in emerging markets where fresh figs are rare and sold as high-end fruit. Fig wine, fig spirits, processed fig chocolates, and fig desserts are also frequently gifted. Gift packaging needs to comprehensively reflect the brand's image and product characteristics, and needs to be suited to consumer psychology and consumption purposes. In terms of material selection and design, these should attract the consumers' attention and mesh with their value comprehension, while enhancing their experience of novelty and quality.

DIVERSIFIED AND CONTEXTUALIZED PACKAGING Figs are sold through a variety of channels, from the orchard door and farmers' markets to supermarkets and high-end or organic food stores, and even online. Highly diversified consumption and packing requirements are observed for consumer groups using different distribution channels. Packaging must adapt correspondingly to the demand of consumption scenarios. Consumers who purchase from orchards or farmers' markets prefer convenience of consumption and tasting, and want to determine the freshness of the figs, and therefore packages should clearly display the fresh figs and processed products and enable on-the-spot tasting. Some consumers prefer affordable products, and for them, the packaging should be as simple as possible. Gift consumption scenarios often require higher-end and display packaging.

TRANSPARENT PACKAGING This refers to packaging that is partially or completely transparent, such as transparent plastic, baskets, nets, and open-window boxes. Transparent packaging provides visual cues to freshness, integrity, and other appearance attributes of figs and their processed products, enhancing the consumers' desire to buy them (Fernqvist *et al.*, 2015). Moreover, in comparison to opaque packaging materials, transparent packaging can efficiently decrease their opening by customers who wish to check the product, thereby improving the figs' shelf life.

COMBINATION PACKAGING Diverse kinds or varieties of fruit can be packaged in the same sales unit, as commonly seen with gift fruit baskets (Nasir *et al.*, 2017) and mixed fresh-cut fruit for lunch boxes (Sortino *et al.*, 2017). Fresh-fig varieties come in different peel and flesh colors. Fruits of two to four varieties can be packaged in the same unit to enhance the consumers' desire to purchase them. It should be noted that, in combination packaging, fig fruits of different varieties need to be of equivalent size and shape to facilitate packaging standardization. Orchards that grow several fig varieties can classify figs according to standards such as size and shape, and develop a few packaging combinations. Processing factories, stores, and online stores with different fig products can also combine cross-category products for sale. Brochures, small gifts, and other items can be packed with fig fruit or processed products as a form of combination packaging.

DIGITAL PACKAGING Also called intelligent packing, this is an extension of packaging functions, adding perception, communication, and feedback (Ghaani *et al.*, 2016). For example, the information systems established by the QR (quick response) code, NFC (near-field communication), and other technologies enable consumers to interact with the packaging using smartphones to obtain production, processing, transportation, and sales information. Digital packaging makes information transmission more efficient, and promotes the interaction between consumers and brands. Two-dimensional barcodes are often added to agri-product packages; by scanning with a smartphone, consumers can join the brand's club and receive regular e-promotional materials, enhancing product loyalty.

28.2.4 Pricing

Price is one of the main factors driving consumer purchase and consumption of fruit (Asioli *et al.*, 2016; Deliens *et al.*, 2016). Appropriate pricing strategies can improve the market competitiveness of producers and distributors. Pricing is affected by many factors, such as cost, supply and demand, market competition, and consumer perceptions. A US survey revealed that consumers are willing to pay more for locally grown agri-products than products originating from other regions (James *et al.*, 2009). A survey in India found that consumers are willing to pay a premium price for better-quality and organically grown fruit (Nandi *et al.*, 2017). Therefore, different strategies can be used in the pricing of figs and their processed products for different target markets.

Pricing strategies

PRICE SKIMMING Price skimming is a product-pricing strategy whereby a relatively high price is set at the preliminary stages of a product's release to recover costs as quickly as possible and obtain higher profits. It is used with newly released fruit cultivars and club cultivars (Asioli *et al.*, 2016). The premise for this strategy is that there are enough consumers with strong purchasing power in target markets who have low sensitivity to prices and are willing to pay higher prices for novel products. In the fruit business, growing and marketing of the table grape variety Sweet Sapphire, the apple cultivar Cosmic Crisp, and the kiwi cultivar SunGold are good examples of the skimming strategy. The success of this strategy requires the condition that there are few related products on the market, giving the novel products a strong competitive advantage. In the case of figs, innovative cultivars with uniqueness in shelf life, flavor, size, color, or other consumer high-valued attributes could try a skimming strategy at the cultivar release.

MARKET-PENETRATION PRICING In contrast to price skimming, market-penetration pricing involves setting a lower price compared with similar products at the preliminary stage of product release. This can generate price advantages and quickly increase the market share. Studies have revealed that reduced prices are consistently associated with higher selection (French *et al.*, 1997; Deliens *et al.*, 2016). Market-penetration pricing is suitable for markets with high price sensitivity, and uses the scale effect of sales to turn a profit.

DIFFERENTIAL PRICING This involves differentiating product prices according to differences in potential consumers, time, and place. It is based on differences in consumer purchasing power, demand, and preferences in segmented markets. A study revealed that foods, including fruits and vegetables, are 10–54% more expensive in small stores than in supermarkets (Caspi *et al.*, 2017). Differential pricing strategies can fine-tune the relationship between supply and demand to obtain optimal market return. For example, fresh figs can be divided into different grades according to their extrinsic and intrinsic qualities, and higher prices can be set for larger, higher-quality figs. Based on differences in purchasing locations, higher prices can be applied for high-end distribution channels. Fresh figs are produced seasonally, and setting different prices according to the time to market can also produce better returns. On a given day, prices for perishable fruit may be set differently for peak periods and low periods to optimize sales and reduce losses to decay (Chang *et al.*, 2016).

PROMOTIONAL PRICING Promotional pricing makes products seem affordable and promotes consumer purchases through price discounts, both direct and indirect. In direct discounts, the price is reduced at purchase. Common forms are quantity discounts, coupons, seasonal discounts, and functional discounts. The quantity discount involves giving a discount that corresponds to the total amount spent by the consumer, such as a discount on the next purchase. Coupons are widely used to encourage fruit and vegetable purchases and consumption, as the demand rate is exponentially dependent on price (Tripathi *et al.*, 2019). Seasonal discounts refer to setting discounts during the off-season for sales, encouraging consumers to buy and stimulating market demand. The functional discount involves setting different prices based on the market function of buyers, for example different discounts for wholesalers and retailers. In general, wholesalers enjoy higher price discounts.

28.3 Distribution Channels

28.3.1 Conventional channels

Distribution channels confer availability to consumers; they constitute the bridge between producers of fig-related products and services and consumers, through a series of intermediate links, to realize value. The production and sale of fresh and dried figs can be quite dispersed. Scattered origins and small scale are usually found at the production end, and inconsistent demand and low volume of single purchases characterize the consumer end. A study in Slovenia found fruit purchases to be most significantly influenced by their availability in retail outlets (Kuhar and Juvancic, 2010). An efficient distribution channel can overcome the limitations of dispersion, expand the market area, and broaden the scope of distribution.

Distribution channel models

In the international and regional trade of fresh fig and fig products, distribution channels involve the growers, agents, wholesalers, retailers, and consumers. Based on their different compositions and structure, distribution channels for figs and fig products can be summarized in the following models:

- **Channel 1: Growers + Consumers**. This is also referred to as direct marketing, where producers sell products to consumers with no intermediate links, such as farm stores, booth sales at farmers' markets, and online sales. Direct marketing connects the farmers with consumers. Shorter distribution chains usually mean lower intermediate costs. However, this model puts growers in the position of juggling both production and marketing demands, and the scale of sales is limited, which could result in high transportation costs for distribution (Clark and Inwood, 2016).
- **Channel 2: Growers + Retailers + Consumers**. Retailers buy products directly from producers and sell them to consumers. This model can scale up the business and improve the efficiency of moving the product, which is important for perishable fruit.
- **Channel 3: Growers + Wholesalers + Retailers + Consumers**. Wholesalers purchase products from producers and then wholesale them to retailers. Products are distributed across regions through wholesale markets.
- **Channel 4: Growers + Agents + Wholesalers + Retailers + Consumers**. This model is often used for large-scale sales and international fruit businesses. Production and sales companies set up regional agents in different areas, who are responsible for all product sales in the region. Other models are also possible, such as Producers + Agents + Retailers + Consumers.

A study in Maharashtra, India, found that channel 4 has the maximum marketing cost compared with the other channels. The local fig growers mainly sold their products through channel 2. The producer's share in consumer's rupees was lowest in channel 4 but highest in channel 1 (Ghulghule *et al.*, 2009).

In Morocco's dried-fig market, distribution starts with farmers entering the market and providing products to traders. Intermediaries act as bridges between producers and wholesalers in landlocked areas. The wholesalers sort fig batches based on extrinsic quality before selling. Retailers get dried figs from wholesalers and sell them in shops that sell spices, aromatic plants, and dried fruits. Dried figs are usually consumed directly. Some of them with reduced quality can be ground as raw materials in wine, ground coffee, fruit juices, and cake production (Salih, 2020). In the Turkish dried-fig market, TARIS (TARIS Fig Agricultural Sales Cooperatives Union) is an important participant, which owns 15 agricultural sales cooperatives, 5000 farmers, and 6000 m^2 of fig processing and storage facility (www.tarisincir.com.tr/; accessed January 5, 2022). As a government-authorized intervention organization, TARIS stabilizes market prices and provides technical support to producers. A study indicated that 8.33% of farmers surveyed sold all their production to TARIS (Isin *et al.*, 2003).

The main purchase places for figs

Fresh fig and fig products are sold in various places, including supermarkets, farmers' markets, convenience stores, and wholesale markets, making them available and convenient for consumers. Supermarkets are the leading purchase place for food products in the USA (Chrisinger *et al.*, 2018), and they are also major fruit-distribution targets in urban areas of many countries. Aside from the conventional supermarkets, comparable outlets are found with discount/limited assortment retailers (e.g. Aldi in Europe), and warehouse-type stores that charge a membership fee (e.g. Costco and Sam's Club in the USA). For fresh-fig and fig-product suppliers, supermarkets and their equivalents have certain thresholds, and unified procurement and balanced supply are usually required. The supply must be on a large scale for a certain period and meet the supermarket's quality standards. The advantage of supermarkets lies in their vast experience and management protocols in packaging and sales, mature logistics, and well-equipped storage facilities, which are conducive to the distribution of figs, as well as other perishable fresh products.

Farmers' markets are direct distribution channels, offering fresh, local fruits and vegetables. They play a key role in regional agricultural sustainability and enhance communities' intake of healthy food products (DeWitt *et al.*, 2017). Producers or retailers ship fig products to farmers' markets to sell them directly to the consumer. There are no intermediate links, leading to higher profits. However, this distribution channel is geographically restricted and can suffer from unstable sales and relative low transaction efficiency.

Wholesale markets are professional distribution markets established in the regions of production or center of distribution; they play a major role in large-scale, rapid, and widespread commodity circulation. The infrastructure of wholesale markets includes cold-chain storage, sorting and packing lines, service logistics services, and high-quality human resources, which are listed as the most crucial factors for centralized storage, extensive distribution, and cost control of wholesale markets (Li *et al.*, 2015; Kowalczyk *et al.*, 2016). Wholesale markets target sub-wholesalers, distributors, retailers, off-site markets, and processing companies. Fixed pricing and auctions are widely practiced in wholesale markets for fruit (Ahmad and Siddiqui 2015; Reid *et al.*, 2019). Wholesale markets are usually inevitable for fresh-fig and fig-product exporters, importers, and large-scale regional producers.

28.3.2 Agritourism

The improvement in living standards has led to the need for diversified leisure activities; rural tourism, as an alternative form of tourism in developed countries, gained popularity in the 1970s (Lane and Kastenholz, 2015). Nature and the countryside have long been recognized as a source of spiritual, physical, and emotional benefits (Jepson and Sharpley, 2015). Agritourism relies on the integration of agricultural resources, tourism, education, and services.

Fig agritourism is based on fig cultivation, supported by restaurants and entertainment facilities to achieve the overall goals of fruit production, sightseeing, entertainment, and activities such as picking, cooking, and tasting. Fig agritourism has a few advantages over other fruit. Many of the traditional fig-production regions in the Mediterranean area are rich in culture and national heritage, such as the Napoli area in southern Italy and the area surrounding the historical city of Bursa in Turkey. Recreational activities and displays of intangible and tangible agricultural heritage are two major factors in successful agritourism (Farsani *et al.*, 2019). Agritourism is also often established in areas surrounding cities with advantageous natural conditions; orchards and farms can become popular recreational sites supported by the short distance for touring families from the cities.

Agritourism can directly promote the sale and profit of agricultural products, enhance the popularity of the specific agricultural area or orchard, and help regional products expand their market. Agritourism is highly dependent on agricultural resources and is inseparable from the rural environment, landscape, and traditional customs. The more it is able to tie together local/regional historical, cultural, and landscape-based resources, the better the interplay of this network (Saxena and Ilbery, 2010).

Visitor participation is a key component of agritourism, and visitor satisfaction can be enhanced by providing unique entertainment activities, high-quality services, and marketing promotions.

Esthetics play a strong role in arousal, memory, and satisfaction from agritourism (Lane and Kastenholz, 2015). Orchards with agritourism functions should pay close attention to the ecosystem and environmental protection. Indigenous plants are encouraged in gardening and environmental decor. A complex ecosystem with trees, shrubs, and herbaceous plants can be designed and planted to make the environment suitable for sightseeing and hiking. Multiple fig varieties are usually required to provide visitors with far more choices than in supermarkets or grocery stores. By using protected cultivation in the form of greenhouses, high tunnels, and summer pruning, the production period can be extended, which will prolong the period of tourism attractiveness and experience. In addition to figs, other fruits suitable for picking can be grown to extend the season and enjoyment. In orchards set up for agritourism, folk entertainment and activities such as singing and dancing performances, fishing, fun activities for children and families, and dinner parties can be organized to enhance the visitor's experience.

Driving is usually the most common way for visitors to approach agritourism locations. Fig towns/farms should set up eye-catching signs along the roads and mark the location on electronic maps. In many cases, a lack of road signage in the last mile can be a cause of frustration for people seeking to reach a tourist farm. Even with the convenient electronic map apps, fig-featured interesting and attractive road signs are an integrated part in fig-promoting agritourism. Adequate and convenient parking is required, especially on weekends and holidays, when the number of visitors can rise significantly.

Due to the proteases in fig latex, visitors need to be cautioned in advance, and gloves need to be provided to protect self-pickers. Additional training on fig maturity may be needed to reduce fruit losses during picking. Note that the visitors' facilities in agritourism sites must be adequate, including sightseeing, shopping, and dining facilities, and a playground for children should be considered as well.

A farm store can be a good place for sales and a key promotion site for the orchards. In addition to fresh figs, farm stores can include processed figs, such as dried fruit, fig wine, baked products, desserts; books and cookbooks featuring figs; decorations such as paintings, ceramics, metalwork, and wooden articles; and tableware with fig images and kitchen utensils; and fig-patterned jewelry. In addition, in considering the visitors' basic needs, the stores should also provide drinking water and simple food.

Reputation and credits support premium added value and competitiveness of agritourism orchards. A series of concepts and practices can help fig tourist farms build their reputation and increase visitor intimacy, for example by exploring the meaning of local and seasonal food materials, reconnecting with nature, satisfying the search for freshness, taste, and authenticity, and addressing environmental issues (Sidali *et al.*, 2015). The historical, cultural, and lifestyle connotations of figs should be integrated into the orchard's promotion strategy. In regions with a favorable climate, year-round activities such as fig-themed painting, photography, baking, and fruit and salad tasting can be organized to enhance the visitors' experience and involvement, and to increase their awareness and loyalty.

28.3.3 E-commerce

E-commerce or e-business offers an important opportunity for cost reduction and enhanced demand for agricultural products (Zeng *et al.*, 2017). The last decade has seen internet connections change rapidly from a computer terminal-based model to a smartphone-based one, with a significant rise in the number of people connected to the internet. The Internet has provided a new platform for businesses with its wide coverage, extremely high-speed information dissemination, diverse forms of content, and direct feedback. E-commerce of fresh agricultural products refers to the direct sale of fresh products to meet the needs of consumers on internet platforms by means of electronic business, which could save greatly on the circulation costs of traditional business models, where four or even five circulation links are usually needed (Zhang *et al.*, 2016).

Taking China as an example, according to the 48th *Statistical Report on Internet Development in China* published by the China Internet Network Information Center (CNNIC), the number of internet users in China reached 1011 million in June 2021, with a penetration rate of 71.6%. The number of online shoppers reached 812 million. According to the *China Digital Rural Development Report* published in 2019 by the Ministry of Agriculture and Rural Affairs of China, the online retail sales of agricultural products in China reached RMB 230.5 billion in 2018, recording a 33.8% increase over 2017. Business models have been established and continue to improve for e-commerce of fresh agricultural products in terms of management operations, logistics, and other aspects (Lu, 2015). A very rapid increase in online fruit businesses has been recorded (Tan and Gao, 2017). A survey conducted in China found that consumers of different ages differed in online purchase intention of dried fruits: consumers under the age of 24 years regarded dried fruits as a new and healthy snack, and they intended to conduct high-quality communication with companies; consumers aged 25–29 took dried fruits as a healthy eating habit; consumers over 30 valued more convenient marketing strategies (Sun and Liang, 2020).

Fig e-commerce relies on the production base and distribution-system logistics. A set of tools have been developed and integrated to improve and supervise the supply chain (Ruan and Shi, 2016); brand promotion, customer service, and transactions are all Internet-based. E-commerce tools make it easier for consumers to communicate with sellers, and sellers can also provide their customers with personalized products and services. Producers and distributors can efficiently collect market feedback through online sales. E-commerce also has a variety of other functions, such as market surveys, branding, advertisement, and product promotions.

E-commerce models

In recent years, increased numbers of smallholders in developing countries such as China have begun to sell agri-products directly to consumers via online shops on large-scale, third-party trade platforms (Zeng *et al.*, 2017). A good example of an e-commerce service platform is Alibaba, where "Network (platform) + Company (network operators) + Farmers" is the most common business model (Lu, 2015). Producers and distributors can also build their own homepages, use professional platforms for specific products, and publish information about their products on websites of industrial associations or regional government departments (Huang *et al.*, 2017). Consumers shopping online learn about products and services via pictures, texts, videos, and reviews, choose what they like, order online, and pay electronically. The sellers receive the orders and send the products to the consumers through a third-party logistics system (Bao *et al.*, 2012). The consumers click "received" and electively give evaluations online, the sellers get the payment from the third-party bank system, and the transaction circle is complete.

There are five major e-commerce models: Business to Business (B2B); Business to Customer (B2C); Customer to Customer (C2C); Customer to Business (C2B); and Online to Offline (O2O) (Zeng *et al.*, 2017), and more recently the Customer to Manufacturer (C2M) model. The B2B model refers to product transactions between businesses, which is applied in trade among fig-processing companies, wholesalers, and retailers. The B2C model refers to businesses selling to customers. The Tmall platform of Alibaba is of a typical B2C nature, where B is required to be a registered legal corporation (i.e. a fig farm, a process company or an intermediate company for figs/fig products) and C refers to individual customers. B2C has a significantly higher sale scale than Alibaba's C2C platform Taobao; moreover, online flagship stores are only allowed on the Tmall platform of Alibaba. In other words, if someone wants to open a flagship store selling PDO (protected designation of origin) dried figs from Italy, the applicant needs to be a corporation.

The C2C model refers to transactions between individuals, such as those between fig growers and consumers via the Internet. An example is Alibaba's Taobao platform, which offers direct selling opportunities to natural persons in the sense of law. Individual fig growers, intermediate people, or companies involved in fig production, processing, and business can open an online shop on the Taobao platform and provide their products to individual consumers. Compared with B2C, the entrance criteria and running costs of C2C are lower. Small players selling fresh figs, dried figs, and fig products usually use the C2C model, while companies that have stores on the B2C Tmall usually have Taobao stores to gain more traffic for their business.

The C2M model has become popular for agricultural product retail in the last few years in China. In this model, intermediates are eliminated, and thus it can offer customers lower prices than other e-business models. Buy Together is representative of the C2M model, a platform that directly connects consumers and growers/producers, and has become a strong competitor of Alibaba in e-business in China. Fresh figs and fig products are as popular on Buy Together as on Alibaba's platforms.

The O2O model achieves seamless docking and a closed loop between online and offline by drawing potential customers from online channels and conducting physical store pick-up. This model has been adopted and analyzed in the fruit and vegetable business (Qin and Hou, 2016; Sturiale and Scuderi, 2017). Freshippo of Alibaba is of the O2O model. Young white collar workers usually order online and collect the fresh products from a convenient Freshippo on their way home, but in-shop purchasing is also offered. The model to some extent solved the third-party delivery problems and is called "new retail" in China. At present, Freshippo has become a strong competitor of the traditional fresh stores.

Tactics of the online shop

The structure of fruit e-commerce businesses usually consists of an online sales platform and a social media platform, which is true for fresh figs, dried figs, and other fig products. The online sales platform requires the establishment of online stores, which can provide product information and data through pictures, text, and videos on specifications, packaging, region of origin, taste, and logistics services. A trained customer service team should be established to provide online consulting and post-sale service, leading to positive evaluations from the consumers. Online stores often need to cooperate with the online store platforms for major marketing events, such as Black Friday and the November 11 Online Shopping Festival. A series of marketing interactions with the platform can improve the product's position on a list of similar products. Interactions with consumers, such as rebates (consumers usually get US$0.5 in cash after the first online purchase and join the online shop consumer group to receive promotion materials), best-feedback competitions (for those leaving a 5-star review and comments), discounts (after a certain purchase, free delivery or other discounts are provided), and other incentives can be used to promote customer loyalty.

Social media is an important channel to promote e-commerce. It has low thresholds and strong interactivity with its audience. Many food industries and brands have achieved prominent levels of engagement through social media, with direct links to purchasable items (Klassen *et al.*, 2018). The preferred social media can differ in different countries and in different segmented markets, and the relevant adjustments may be necessary. For example, WeChat public accounts, Weibo (microblog), Tik Tok, and others are important for promotion and marketing in China, whereas Facebook, Twitter, Instagram, and others are widely used in many countries (Niedermeier *et al.*, 2016). Social media marketing contents need to be released at regular intervals to maintain contact and interactions with potential customers. It should be noted that with the quick advance from third-generation (3G) telephony to 4G telephony, and the start of 5G telephony usage, the form of marketing content for mobile devices is gradually transitioning from text and photos to short video feeds that have visual and intuitive advantages.

28.4 Promotion

28.4.1 Brand management

A brand is defined as the "name, term, sign, symbol, design or a combination of these used to identify and differentiate products" (AHC, 1997). Traditionally, fruit is perceived as an undifferentiated commodity, a category of sales that relies more on volume than on quality differentiation, marketing, or promotion efforts. However, branding can indicate an agricultural product's region of origin, provide its identification with quality and trust, and improve and deepen communication with consumers.

In the last few decades, brands have emerged and taken on importance for fruit (Asioli *et al.*,

2016). ENZA for New Zealand apples and pears, ZESPR for New Zealand kiwi, Dole for fruits and vegetables, Cape for South African fruit, and Sunkist for Californian citrus are good examples. Successful fruit brands have become badges of honor or trust and, ultimately, an indication of value (Meads and Sharma, 2008). Investment is required for brand establishment and promotion (Pearson, 2003); in return, the brand owner benefits from product price and market share.

With high-pressure competition from diverse types of fruit supplied year-round in the marketplace, fresh figs and processed products need to maintain and expand their market through active brand management and promotion, to build product differentiation, transmit the concept of producer value (Kilian *et al.*, 2012), and establish price point support. A study revealed that labeling and brand image have a positive and significant influence on consumers' buying intentions for fruit purchases, indicating benefits in market share, if not supporting a better price (Korry *et al.*, 2017).

Brands can be regional or commercial. The establishment of regional brands can significantly improve the popularity of the main fig-producing areas, which is conducive to promotion and sales. Regional brands have public value and exclusivity; they are promoted and protected by industrial associations, clubs, and regional governments. The PDO is a good example of the strengthening, promotion, and protection of regional brands. Fig production in Italy is concentrated in the south, especially in Calabria (Cosenza area) and Campania (Cilento, in the Province of Salerno). The variety Dottato is used to produce dried figs and is labeled Fichi di Cosenza and Fico Bianco del Cilento, each with their respective PDO. Similarly, with the support of the FAO and the European Bank for Reconstruction and Development (EBRD), the Bursa region in Turkey has set up a geographic indication for the unique black figs in the region, to protect the "Bursa" brand and gain an advantage in the market.

Many food products have comparatively lower levels of branding (Pay *et al.*, 1996), due to their perishability and incapacity for year-round supply; consequently, it is difficult to conduct year-round promotions, as the gap of time during which the products are missing from the market makes it difficult for consumers to remember the brand (Cook, 2013). Compared with other fruits, under suitable growth conditions and management, fresh figs have a longer period of harvest and supply, which provides an advantage for brand building. Fig-selling companies can achieve annual sales by purchasing from different production areas to ensure that the brand appears in the market sustainably throughout the year, a practice that is followed in China. With regard to the building of commercial brands, the fig industry can learn from the operations of companies dealing with similar agricultural products, such as cherry, kiwi, and blueberry. Branding dried figs and other processed products is easier than fresh figs from the annual supply point of view, and the principles of branding agricultural products should be followed in this case. Besides branding the specific fig products, branding shops/stores, names of distributors, regions of production, and other partners in the production chain can be alternatives and supplements in building trust and promoting fig business.

28.4.2 Promotion methods

Promotion is the combination of selling incentives and methods used by producers and operators to disseminate product information to target audiences, thereby persuading them or influencing their purchases (Totten and Block, 1994). Promotional activities can improve the middleman's or consumer's understanding of the products, stimulate consumer demand, and maintain or expand sales. At the same time, market feedback from promotions helps marketers adjust or strengthen specific promotion measures.

Promotion consists of a combined series of tools, including personal promotions, sales promotions, public relations, and advertising. Different promotional activities focus on transmitting different perceived values to consumers and are intended to trigger different behavioral responses. Fig producers and marketers should define the purpose of the promotion when formulating promotional strategies and adopt appropriate promotion methods with full consideration of the advantages and characteristics of the products, within a planned budget.

Personal promotion

Personal promotion is a direct promotion strategy. The salesperson or sales promoter makes suitable purchase recommendations, through education, display materials, recipe cards, and so on, to increase consumers' awareness and knowledge of the product. Increased sales are usually achieved for fruit and vegetables by point-of-sale, onsite personal promotion (Miller and Pollard, 2005).

Briefing consumers is a key process of personal promotion. The salesperson should be familiar with figs and fig products, understand the different types of consumers in the market and their needs, motivations, and consumption habits, and should be aware of the main competitors' situations, to cope with comparisons by consumers. In emerging markets, considering consumers' overall lack of experience with figs, a salesperson can inform consumers of the healthy aspects of figs and provide them with flyers carrying consumption recommendations and menu ideas, or with tips on judging quality and selecting fruit.

Sales promotion

Sales promotion is a short-term market-stimulating method. Common sales promotion tactics include gift giving, price discounts, tasting, and drawings. Gift giving is available in the form of "free samples," "increased quantity," "buy one, get one free," coupons, and other forms that can provide consumers with a sense of additional value and stimulate their purchases. Distributors should carefully choose the timing and methods for the price discounts; the range of discounts should be within the framework of the overall price strategy for the products, to avoid confusion of consumers with respect to price, value, and the positioning of the product/brand in the market.

In the practice of fruit sales promotion, tasting is an important means for new types of fruit or new fruit varieties or cultivars. The salesperson can invite consumers to taste fresh figs or fig-related products to establish their first experience. Successful tasting can stimulate consumers' desire to purchase and promote transactions. A store coupon can be remarkably effective for some market segments; discount coupons worth 50 cents were found to be highly used, and to attract new shoppers to the farmers' markets for fruits and vegetables (Glanz and Yaroch, 2004).

Moreover, sales promotions also include incentives for channels such as wholesalers, agents, and retailers, as well as the salesperson. Practices for the sales promotion include granting a channel as the exclusive agent, and support of distribution and display facilities, such as store signage and shelf banners. Managers of supermarkets and grocery stores are the gatekeepers in the distribution channels, as they have the power to select product placement at the points of sale in large retail chains (Asioli *et al.*, 2016). Sales goals can be set for them, with a reward for those who meet those goals.

Public relations

Production regions and production and sales companies for figs should establish good public relations, and maintain a good social and public image, to achieve understanding, trust, and support from the public while creating a beneficial environment for marketing. Especially in this era of globalization and frequent trade wars, public relations management can be used to expand publicity, promote images of production regions and brands, and improve popularity and reputation. Trade barriers can be eliminated or reduced through government cooperation. The import of fresh fruit and fruit products is usually strictly regulated by countries, and public relation campaigns can help place the region of origin of agri-products on an importers' whitelist with a favorable tariff. Public relations activities should provide the contents of the region, create the producers' and brands' social image, and build a friendly and integrated environment by creating social credibility, which will support long-term development. Enterprises can also make use of public hotspots and major events to conduct marketing activities that will increase consumer attention and participation.

Advertising

Fig production and sales companies can adopt a variety of advertising strategies, among them product advertising, brand advertising, and public service advertising. Advertising can introduce the characteristics of figs and fig products to consumers, convey the joy and satisfaction of fig consumption, create a positive emotional experience, and establish an attractive image.

Industry reports and news in public media can be used to display products and corporate images from a soft and relatively objective perspective. When choosing which media type to use for product advertising, companies should consider the convenience and effectiveness of dissemination to the targeted audience/readers, be it through television, radio, press, or billboard advertising. In recent years, an Internet-based personalized advertising/recommendation system, using big data and algorithms, has entered wide use (Kumar and Balakrishnan, 2019). In terms of the content and expression of the advertising style, it is also necessary to consider the characteristics of the public environment and consumption habits, and the specific content and style needs that will suit the culture and consumption concept of the target market.

28.4.3 Festival and event marketing

Festival and event marketing are thematic promotion methods based on a series of industrial, economic, cultural, and touristic resources. Freshness and product origin are the most crucial factors in consumers' choice of fruit (Girgenti *et al.*, 2016), supporting the attractiveness of festivals and events that are usually organized during the harvest season in the central town of a production region. Festivals and events can attract massive attention and participants from both the media and the public through activities such as parades, parties, performances, competitions, exhibitions, and commerce and trade activities, achieving the goals of brand promotion and sales.

For fig festivals and events organized in some countries and major production regions, the activities include fig-themed performances, fig dishes and cooking festivals, fig picking, beauty contests for "fig princess" or "fig queen", fresh fig and processed product competitions, fig auctions, painting and photography competitions, etc. Fig exhibitions and tastings can be organized, and fireworks and performances offered in the evening. The resultant media reports will enhance exposure of the region and promote the local fig industry (Fig. 28.2).

Large-scale festivals and events can often become the focus of regional media, which can

Fig. 28.2. Fig festivals and exhibition promotion in China.

widely attract people's attention and exponentially improve the reputation of the place. For the host cities/towns, festivals and events will not only improve the popularity and influence of the fig industry but will also generate additional benefits for other sectors of the economy, such as hotels, restaurants, bars, and transportation services, as well as improving the comprehensive environment of the venue.

The level of regional economic development is the basis for festival and event marketing; complete infrastructures and adequate tourism resources are the necessary material conditions. Festivals and events are also part of agritourism, facilitating the sustainable development of rural areas, and therefore government agencies are usually involved in promoting the festival or event (Amir *et al.*, 2015). The government should create a good public environment and provide favorable policies and security guarantees that are conducive to successful festivals and events.

Participating companies should set clear marketing goals, design activity plans in response to market demand, and provide high-quality products. Enterprises and local government branches involved in urban transportation, telecommunications, entertainment, and finance should also be fully prepared to provide complete supportive services.

Geographical location is an important external factor for the organization and promotion of festivals and events. Inconvenient transportation and poor accessibility are often limiting factors for the development of these events in many regions. In addition to the main production areas, festivals and events can be held in neighboring cities or major cities in the area, where better site conditions will attract more visitors, attention, and effective promotion.

The success of festival and event marketing is closely related to the choice of theme. Strong cultural, heritage, or hedonistic themes can increase visitation. Theme selection should make full use of local resources; for example, it can be combined with local history, culture, and customs to reflect the area's diverse connotations. If possible, different themes and distinctive activities should be set up on an annual basis to increase the attraction and repeated participation of consumers.

Consumers who participate in festivals and events are driven by experiential need; they are willing to take part in the various activities set up by the organizers. One investigation showed that all respondents engaging in rural tourism activities had highly emotive experiences (Jepson and Sharpley, 2015). Fig-production regions in the northern hemisphere cannot easily organize festivals and events during the traditional holiday seasons, such as Christmas, New Year's Day and Easter due to the fig ripening and harvesting periods. However, the fig season in the summer coincides with school holidays and the most popular annual vacation times, suiting the launch of family-oriented summer activities to attract visitors. Fig harvest season is good for both learning-oriented and recreation-oriented outdoor activities (Huang *et al.*, 2017). Festivals and events create a strong entertainment atmosphere, increasing the interest of those who are taking part in the festivities. Consumers tend to extend their participation time, resulting in an increased level of consumption.

A wide range of publicity channels should be adopted for festival and event marketing; candidate collaborative partners include government departments, industrial associations, international organizations, and well-known corporations. The publicity can be carried out through newspapers and magazines, television, online media, travel agencies, and other channels to achieve satisfactory outcomes. The organizers can hold a press conference before the festival or event, invite journalists, internet celebrities, and key consumers to visit the venue, or send them promotional materials. Outdoor advertising can also be set up in important areas along the way, such as airports, train stations, and landmarks. Through these efforts, festivals and events can obtain stable public attention and consumer flow.

28.5 Concluding Remarks

Here we have summarized the applicability of different marketing aspects and methods on fresh and dried figs (Table 28.1). Distributors and sellers of fresh figs and fig products are well advised to explore the connotations of figs from the perspective of geographical conditions, and historical and humanistic backgrounds, integrate cultural characteristics into the brand, and identify product differences. In cross-regional trade, it

Table 28.1. Applicability of different marketing aspects and methods to fresh and dried figs.

Marketing scheme	Marketing method	Fresh figs	Dried figs	Comments
Market analysis	Market segmentation	++++	++++	Fresh and dried figs could share the same or be targeted at different consumers
	Market survey	+++	+++	
	Packing	+++++	++++	Extremely important for fresh-fig postharvest quality and market positioning
	Pricing	++++	++++	
Distribution channel	Conventional channel	+++++	+++++	
	Agritourism	++++	+++	Agritourism is more applicable for seasonal fresh products, but dried figs and other products can also be promoted by agritourism
	E-commerce	+++	++++	Dried figs and processed fig products are more applicable for e-commerce than perishable fresh figs, mainly in the B2C, C2C and C2M models
Promotion	Branding	+++	++++	
	Personal promotion	++++	++++	Offer tastings and recommend consumption scenarios
	Sales promotion	+++	+++	
	Public relations	++++	++++	Health benefits and culture should be addressed
	Advertising	+++++	+++++	
	Festival and event marketing	++++	++++	Regional food and harvest festivals, and traditional festivals where people eat fresh and dried figs

B2C, Business to Customer; C2C, Customer to Customer; C2M, Customer to Manufacturer.

is necessary to conduct market surveys and formulate cultural adaptation strategies for the target market to reduce the impact of cultural differences and maintain good public relations.

Fig production and sales companies can use various channels to promote brand popularity and influence. Regional governments, industrial associations, and companies can build platforms for publicity, participate in trade fairs, and organize promotion activities. Intermediaries and distribution channels with good social images are also supportive in joint promotion. Traditional media, such as newspapers, television, and radio, can be used to increase exposure. It is becoming increasingly important to use social media for brand recommendations and to stimulate consumer conversations, increase consumer loyalty, and acquire new customers (Liu and Lopez, 2016).

Social media influencers, such as popular lifestyle bloggers, video bloggers, internet celebrities, and key consumers, have a broad readership. Companies can invite them to visit production regions, plantations, farms, and factories, and experience relevant activities or tours. Social media influencers are recognized and trusted by their followers and readers, and have a strong impact on their purchasing decisions. Many food-focused social media influencers, such as gourmet writers, well-known chefs, and food anchors, may play important roles in promoting figs, fig products, and their brands. Key consumers are the core consumers who are continually active in social media. They are willing to share consumption experiences and repost relevant articles and videos online, and have an influence on the small social or community group to which they belong.

Figs have been cloned, selected, cultivated, and consumed for several thousand years, and the Mediterranean area and western Asian regions developed a rich fig culture and integrated figs into the local lifestyle during this prolonged period. In the past century, modern technology has provided this historical fruit crop with great opportunities to achieve high yield by improved processing and sorting techniques, high-speed logistics, and cold-chain services for fresh figs. In the past decade, we have experienced the profound impact of the Internet, with changed lifestyles, and the rapid evolution of e-business and marketing. Today, innovation and marketing are necessities for the survival and development of every industry, and fig is no exception. Understanding the true need of consumers, communicating with them, building trust, and promoting the value of fresh figs, dried figs, and various fig products will support the sustainable development of the industry and offer thousands of jobs. Innovation is especially required for historical products, as "the end of Stone Age was not because of running out of stones." Fig marketing can be viewed in the same light.

References

AHC (Australian Horticultural Corporation) (1997) *Produce Branding*. Australian Horticultural Corporation in conjunction with Good Fruit and Vegetable Magazine, Sydney, Australia.

Ahmad, M.S. and Siddiqui, M.W. (2015) Supply chain and marketing channels of fruits. In: *Postharvest Quality Assurance of Fruits*. Springer, Cham, Switzerland, pp. 207–215.

Amir, A.F., Ghapar, A.A., Jamal, S.A., and Ahmad, K.N. (2015) Sustainable tourism development: a study on community resilience for rural tourism in Malaysia. *Procedia Social and Behavioral Sciences* 168, 116–122.

Asioli, D., Canavari, M., Malaguti, L., and Mignani, C. (2016) Fruit branding: exploring factors affecting adoption of the new pear cultivar 'Angelys' in Italian large retail. *International Journal of Fruit Science* 16, 284–300.

Bao, L., Huang, Y., Ma, Z., Zhang, J., and Lv, Q. (2012) On the supply chain management supported by e-commerce service platform for agreement based circulation of fruits and vegetables. *Physics Procedia* 33, 1957–1963.

Caspi, C., Pelletier, J., Harnack, L., Erickson, D., Lenk, K., and Laska, M.N. (2017) Pricing of staple foods at supermarkets versus small food stores. *International Journal of Environmental Research and Public Health* 14: 915.

Chang, X., Li, J., Rodriguez, D., and Su, Q. (2016) Agent-based simulation of pricing strategy for agri-products considering customer preference. *International Journal of Production Research* 54, 3777–3795.

Chrisinger, B.W., Kallan, M.J., Whiteman, E.D., and Hillier, A. (2018) Where do US households purchase healthy foods? An analysis of food-at-home purchases across different types of retailers in a nationally representative dataset. *Preventive Medicine* 112, 15–22.

Clark, J.K. and Inwood, S.M. (2016) Scaling-up regional fruit and vegetable distribution: potential for adaptive change in the food system. *Agriculture and Human Values* 33, 503–519.

Cochran, W.G. and Chambers, S.P. (1965) The planning of observational studies of human populations. *Journal of the Royal Statistical Society: Series A (General)* 128, 234–266.

Cook, R.A. (2013) *Trends in the Marketing of Fresh Produce and Fresh-cut/Value-added Produce*. University of California, Davis, California.

Crisosto, C.H., Bremer, V., Ferguson, L., and Crisosto, G.M. (2010) Evaluating quality attributes of four fresh fig (*Ficus carica* L.) cultivars harvested at two maturity stages. *HortScience* 45, 707–710.

Crisosto, H., Ferguson, L., Bremer, V., Stover, E., and Colelli, G. (2011) Fig (*Ficus carica* L.). In: Yahia, E.M. (ed.) *Postharvest Biology and Technology of Tropical and Subtropical Fruits*. Woodhead Publishing, Cambridge, UK, pp. 134–160.

Deepak, R.K.A. and Jeyakumar, S. (2019) *Marketing Management*. Educreation Publishing, Delhi, India.

Deliens, T., Deforche, B., Annemans, L., de Bourdeaudhuij, I., and Clarys, P. (2016) Effectiveness of pricing strategies on French fries and fruit purchases among university students: results from an on-campus restaurant experiment. *PLoS One* 11: e0165298.

DeWitt, E., McGladrey, M., Liu, E., Peritore, N., Webber, K., *et al.* (2017) A community-based marketing campaign at farmers markets to encourage fruit and vegetable purchases in rural counties with high rates of obesity, Kentucky, 2015–2016. *Preventing Chronic Disease* 14: E72.

FAO (Food and Agriculture Organization of the United Nations) (2018) FAOSTAT. Available at: www.fao.org/faostat/en/#data/QCL (accessed January 5, 2022).

Farsani, N.T., Ghotbabadi, S.S., and Altafi, M. (2019) Agricultural heritage as a creative tourism attraction. *Asia Pacific Journal of Tourism Research* 24, 541–549.

Fernqvist, F., Olsson, A., and Spendrup, S. (2015) What's in it for me? Food packaging and consumer responses, a focus group study. *British Food Journal* 117, 1122–1135.

French, S.A., Story, M., Jeffery, R.W., Snyder, P.A.T., Eisenberg, M., *et al.* (1997) Pricing strategy to promote fruit and vegetable purchase in high school cafeterias. *Journal of the Academy of Nutrition and Dietetics* 97, 1008–1010.

Ghaani, M., Cozzolino, C.A., Castelli, G., and Farris, S. (2016) An overview of the intelligent packaging technologies in the food sector. *Trends in Food Science & Technology* 51, 1–11.

Ghulghule, J.N., Asmatoddin, M.O.H.D., Thombre, A.R., and Shinde, R.B. (2009) Price spread in marketing of fig (*Ficus carica* Linn.) in Maharashtra. *Asian Journal of Horticulture* 4, 25–28.

Gindi, A.A., Abdullah, A.M., Ismail, M.M., and Nawi, N.M. (2018) Segmentation of fresh fruits consumers by product and store attributes at Klang valley, Malaysia. *International Journal of Food and Agricultural Economics* 6, 73–84.

Girgenti, V., Massaglia, S., Mosso, A., Peano, C., and Brun, F. (2016) Exploring perceptions of raspberries and blueberries by Italian consumers. *Sustainability* 8: 1027.

Glanz, K. and Yaroch, A.L. (2004) Strategies for increasing fruit and vegetable intake in grocery stores and communities: policy, pricing, and environmental change. *Preventive Medicine* 39, 75–80.

Han, J.H. (2005) New technologies in food packaging: overview. In: Han, J.H. (ed.) *Innovations in Food Packaging*. Elsevier/Academic Press, Amsterdam, p. 3–11.

Huang, Y., Zhou, C., Cao, N., and Zhou, L. (2017) Research on application of local fruit E-commerce based on Internet of Things. In: *2017 IEEE International Conference on Computational Science and Engineering (CSE) and IEEE International Conference on Embedded and Ubiquitous Computing (EUC)*, Vol. 2. IEEE, Piscataway, New Jersey, pp. 191–194.

Ingrassia, M., Chironi, S., Allegra, A., and Sortino, G. (2015) Consumer preferences for fig fruit (*Ficus carica* L.) quality attributes and postharvest storage at low temperature by in-store survey and focus group. *Acta Horticulturae* 1173, 383–388.

Isin, F., Cukur, T., and Armagan, G. (2003) An evaluation of dried fig production and marketing in Turkey from dried fig exportation standpoint. *Acta Horticulturae* 605, 183–189.

James, J.S., Rickard, B.J., and Rossman, W.J. (2009) Product differentiation and market segmentation in applesauce: using a choice experiment to assess the value of organic, local, and nutrition attributes. *Agricultural and Resource Economics Review* 38, 357–370.

Jepson, D. and Sharpley, R. (2015) More than sense of place? Exploring the emotional dimension of rural tourism experiences. *Journal of Sustainable Tourism* 23, 1157–1178.

Kausar, M.A., Dhaka, V.S., and Singh, S.K. (2013) Web crawler: a review. *International Journal of Computer Applications* 63, 31–36.

Kilian, B., Hettinga, J., Jiménez, G.A., Molina, S., and White, A. (2012) Case study on Dole's carbon-neutral fruits. *Journal of Business Research* 65, 1800–1810.

Klassen, K.M., Borleis, E.S., Brennan, L., Reid, M., McCaffrey, T.A., and Lim, M.C. (2018) What people "like": analysis of social media strategies used by food industry brands, lifestyle brands, and health promotion organizations on Facebook and Instagram. *Journal of Medical Internet Research* 20: e10227.

Knezović, Z., Šaravanja, P., and Micic, N. (2017) Survey on customer satisfaction with fresh and dried figs supply in Mostar. *АГРОЗНАЊЕ* 18, 47–56.

Korry, P.D.P., Yulianti, N.M.D.R., and Yunita, P.I. (2017) Increase the attractiveness of local fruits to buying intention of hedonic consumers in Bali. *International Research Journal of Management, IT and Social Sciences* 4, 10–16.

Kowalczyk, Z., Szelag-Sikora, A., and Grotkiewicz, K. (2016) The infrastructure and organization of the trade of fruits and vegetables based on the agri-food market in Kraków-Rybitwy. *Infrastruktura i Ekologia Terenów Wiejskich* 4, 1821–1832.

Krishna, A., Cian, L., and Aydınoğlu, N.Z. (2017) Sensory aspects of package design. *Journal of Retailing* 93, 43–54.

Kuhar, A. and Juvancic, L. (2010) Determinants of purchasing behaviour for organic and integrated fruits and vegetables in Slovenia. *Agricultural Economics Review* 11, 70–83.

Kumar, M.S. and Balakrishnan, K. (2019) Development of a model recommender system for agriculture using Apriori algorithm. In: Mallick, P.K., Balas, V.E., Bhoi, A.K., and Zobba, A.F. (eds) *Cognitive Informatics and Soft Computing*. Springer, Singapore, pp. 153–163.

Lane, B. and Kastenholz, E. (2015) Rural tourism: the evolution of practice and research approaches—towards a new generation concept? *Journal of Sustainable Tourism* 23, 1133–1156.

Li, L., Lichter, A., Kenigsbuch, D., and Porat, R. (2015) Effects of cooling delays at the wholesale market on the quality of fruit and vegetables after retail marketing. *Journal of Food Processing and Preservation* 39, 2533–2547.

Lindh, H., Olsson, A., and Williams, H. (2016) Consumer perceptions of food packaging: contributing to or counteracting environmentally sustainable development? *Packaging Technology and Science* 29, 3–23.

Liu, Y. and Lopez, R.A. (2016) The impact of social media conversations on consumer brand choices. *Marketing Letters* 27, 1–13.

Loker, L.E. and Perdue, R.R. (1992) A benefit-based segmentation of a nonresident summer travel market. *Journal of Travel Research* 31, 30–35.

Lu, J. (2015) The entire e-commerce application in Guangxi fruit industry. In: Lui, L. and Shi, J. (eds) *Proceedings of the 1st International Symposium on Social Science, August 29–30, 2015, Wuhan, China*. Atlantis Press, Beijing. Available at: www.atlantis-press.com/proceedings/ISSS-15/publishing (accessed January 1, 2022).

McQueen, J. and Miller, K.E. (1985) Target market selection of tourists: a comparison of approaches. *Journal of Travel Research* 24, 2–6.

Meads, C.A.A. and Sharma, P. (2008) The concept of "brand" in business value creation. In: *International DMI Education Conference*. ESSEC Business School, Cergy, France. Available at: www.forbes.com/sites/larrylight/2019/03/06/the-economics-of-brand-value-creation/?sh=c650ce644913 (accessed January 1, 2022).

Migliore, G., Galati, A., Romeo, P., Crescimanno, M., and Schifani, G. (2015) Quality attributes of cactus pear fruit and their role in consumer choice: the case of Italian consumers. *British Food Journal* 117, 1637–1651.

Miller, M. and Pollard, C. (2005) Health working with industry to promote fruit and vegetables: a case study of the Western Australian Fruit and Vegetable Campaign with reflection on effectiveness of intersectoral action. *Australian and New Zealand Journal of Public Health* 29, 176–182.

Nandi, R., Bokelmann, W., Gowdru, N.V., and Dias, G. (2017) Factors influencing consumers' willingness to pay for organic fruits and vegetables: empirical evidence from a consumer survey in India. *Journal of Food Products Marketing* 23, 430–451.

Nasir, N.A., Suboh, N.S., Ghani, F.A., and Yusoff, N.M. (2017) Purchasing fruits at retail store: assessing customer satisfaction. *Journal of Tourism, Hospitality and Culinary Arts* 9, 505–516.

Nesci, F.S., Nicolosi, A., and Privitera, D. (2004) Regional marketing of dried fig based products: an exploratory consumer study. *New Medit* 3, 55–60.

Niedermeier, K.E., Wang, E., and Zhang, X. (2016) The use of social media among business-to-business sales professionals in China: how social media helps create and solidify Guanxi relationships between sales professionals and customers. *Journal of Research in Interactive Marketing* 10, 33–49.

Olsho, L.E., Payne, G.H., Walker, D.K., Baronberg, S., Jernigan, J., and Abrami, A. (2015) Impacts of a farmers' market incentive programme on fruit and vegetable access, purchase and consumption. *Public Health Nutrition* 18, 2712–2721.

Park, J.L., Holcomb, R.B., Raper, K.C., and Capps, O., Jr. (1996) A demand systems analysis of food commodities by US households segmented by income. *American Journal of Agricultural Economics* 78, 290–300.

Pay, C.J., White, M.R., and Zwart, A.C. (1996) *The Role and Importance of Branding in Agricultural Marketing*. Department of Economics and Marketing, Lincoln University, Canterbury, New Zealand.

Pearson, D. (2003). Australia fresh fruits and vegetables: why do so many of them remain unbranded? *Agribusiness Review* 11: 132540.

Pereira, C., Martín, A., López-Corrales, M., Córdoba, M.D.G., Galván, A.I., and Serradilla, M.J. (2020) Evaluation of the physicochemical and sensory characteristics of different fig cultivars for the fresh fruit market. *Foods* 9, 619–634.

Qin, B. and Hou, H. (2016) Research on O2O model of fresh products. In: *International Conference on Logistics, Informatics and Service Sciences (LISS), July 24–27, 2016, Sydney, Australia*. IEEE. Piscataway, New Jersey. Available at: https://ieeexplore.ieee.org/document/7854518 (accessed January 1, 2022).

Rajagopal, S. and Bansal, I. (2013) Cross cultural analysis of purchasing behaviour of fresh fruits and vegetables in the United Arab Emirates (UAE). *International Journal of Management and Computing Sciences* 3, 2231–3303.

Reid, J., Simmonds, D., and Newbold, E. (2019) Wholesale produce auctions and regional food systems: the case of Seneca produce auction. *Renewable Agriculture and Food Systems* 34, 259–267.

Romano, K.R., Rosenthal, A., and Deliza, R. (2015) How do Brazilian consumers perceive a non-traditional and innovative fruit juice? An approach looking at the packaging. *Food Research International* 74, 123–130.

Rosenbaum, P.R. (2005) Observational study. In: Everitt, B. and Howell, D. (eds) *Encyclopedia of Statistics in Behavioral Science*. Wiley, Hoboken, New Jersey.

Rosianski, Y., Doron-Faigenboim, A., Freiman, Z.E., Lama, K., Milo-Cochavi, S., *et al.* (2016) Tissue-specific transcriptome and hormonal regulation of pollinated and parthenocarpic fig (*Ficus carica* L.) fruit suggest that fruit ripening is coordinated by the reproductive part of the syconium. *Frontiers in Plant Science* 7, 1696–1709.

Ruan, J. and Shi, Y. (2016) Monitoring and assessing fruit freshness in IOT-based e-commerce delivery using scenario analysis and interval number approaches. *Information Sciences* 373, 557–570.

Salih, G. (2020) Assessment of production and marketing practices for dried figs: surveys of Moroccan fig producers and sellers. *Moroccan Journal of Agricultural Sciences* 1. Available at: https://techagro.org/index.php/MJAS/article/view/834/880?download=pdf#info (accessed January 1, 2022).

Saxena, G. and Ilbery, B. (2010) Developing integrated rural tourism: actor practices in the English/Welsh border. *Journal of Rural Studies* 26, 260–271.

Shamin-Shazwan, K., Shahari, R., Amri, C.N.A.C., and Tajuddin, N.S.M. (2019) Figs (*Ficus carica* L.): cultivation method and production based in Malaysia. *Engineering Heritage Journal* 3, 6–8.

Shim, S., Gehrt, K., and Lotz, S. (2001) Export implications for the Japanese fruit market: fruit-specific lifestyle segments. *International Journal of Retail & Distribution Management* 29, 298–314.

Sidali, K.L., Kastenholz, E., and Bianchi, R. (2015) Food tourism, niche markets and products in rural tourism: combining the intimacy model and the experience economy as a rural development strategy. *Journal of Sustainable Tourism* 23, 1179–1197.

Simkin, L. and Dibb, S. (1998). Prioritising target markets. *Marketing Intelligence and Planning* 16, 407–417.

Sortino, G., Caviglia, V., Liguori, G., de Pasquale, C., Gianguzzi, G., and Farina, V. (2017) Quality changes of tropical and subtropical fresh-cut fruits mix in modified atmosphere packaging. *Chemical Engineering Transactions* 58, 397–402.

Sturiale, L. and Scuderi, A. (2017) The marketplaces and the integration between physic and virtual in the business models of fruit and vegetables e-commerce. In: *8th International Conference on Information and Communication Technologies in Agriculture, Food and Environment, September 21–24, 2017, Chania, Crete, Greece*. Available at: https: https://2017.haicta.gr/ (accessed January 1, 2022).

Sun, Y. and Liang, C. (2020) Effects of determinants of dried fruit purchase intention and the related consumer segmentation on e-commerce in China. *British Food Journal* 123, 1133–1154.

Tan, L. and Gao, Y. (2017) The analysis of consumer behavior of fresh-products e-business-based on research in Chengdu. In: *Wuhan International Conference on E-Business 2017 Proceedings, Chengdu, China*. Available at: https://aisel.aisnet.org/whiceb2017/47/ (accessed January 1, 2022).

Thomas, D.M. and Mathur, S. (2019) Data analysis by web scraping using python. In: *3rd International conference on Electronics, Communication and Aerospace Technology (ICECA), Coimbtore, India*. Available at: https://ieeexplore.ieee.org/document/8822022 (accessed January 1, 2022).

Totten, J.C. and Block, M.P. (1994) *Analyzing Sales Promotion: Text & Cases: How to Profit from the New Power of Promotion Marketing*, 2nd edn. The Dartnell Corporation, Chicago, Illinois.

Tripathi, R.P., Singh, D., Singh, D., and Rao, P. (2019) Inventory model with quantity discount, pricing and partial backlogging for a deteriorating items. *International Journal of Operational Research* 35, 208–223.

van Dam, Y.K. and Apeldoorn, P.A. (1996) Sustainable marketing. *Journal of Macromarketing* 16, 45–56.

van Herpen, E., Immink, V., and van den Puttelaar, J. (2016) Organics unpacked: the influence of packaging on the choice for organic fruits and vegetables. *Food Quality and Preference* 53, 90–96.

Weibin, J., Kai, M., Zhifeng, L., Yelin, W., and Lianju, W. (2001) The production and research of fig (*Ficus carica* L.) in China. *Acta Horticulturae* 605, 191–196.

Wojdyło, A., Nowicka, P., Carbonell-Barrachina, Á.A., and Hernández, F. (2016) Phenolic compounds, antioxidant and antidiabetic activity of different cultivars of *Ficus carica* L. fruits. *Journal of Functional Foods* 25, 421–432.

Zeng, Y., Jia, F., Wan, L., and Guo, H. (2017) E-commerce in agri-food sector: a systematic literature review. *International Food and Agribusiness Management Review* 20, 439–460.

Zhang, H.W., Zeng, Y.T., Wan, Z.Q., Zhang, W.T., and Wang, Q. (2016) Analysis on the development of e-business of fresh agricultural products in China. In: *DEStech Transactions on Social Science, Education and Human Science, December 9–11, 2016, Shanghai, China*. Available at: www.destechpub.com/product/destech-transactions-social-science-education-human-science/ (accessed January 1, 2022).

Zhang, X., Wang, X., Liu, L., Wang, W., Liu, Y., *et al.* (2020) Evaluation of the comparative morphological and quality in fig (Sichuan, China) with different colors under different ripening stages. *Scientia Horticulturae* 265: 109256.

Zhou, J., Cao, L., Chen, S., Perl, A., and Ma, H. (2015) Consumer-assisted selection: the preference for new tablegrape cultivars in China. *Australian Journal of Grape and Wine Research* 21, 351–360.

29 Conclusions and Future Research

Moshe A. Flaishman[1]* and Uygun Aksoy[2]
[1]*Institute of Plant Sciences, Agricultural Research Organization, Rishon LeZion, Israel;* [2]*Department of Horticulture, Faculty of Agriculture, Ege University, Bornova, İzmir, Turkey*

29.1 Introduction

The fig tree and fruit are one of the oldest trees cultivated and consumed by human kind. The high nutritious value, the sweet flavor, and the ability to consume it in fresh, dry, and different industrial forms ensure that fig will be following human footsteps for many generations to come, as in the past. Compared with many other fruits, fig production is limited. The annual global fig production is rather stable and is estimated at 1.2 million t, of which 60% is used for dried consumption. In the last few years, many fig products have become available. Fig products such as syrup, jam, paste, coffee, wine, beverages, vinegar, bread, fig bars, and chocolate-coated figs have become popular in many countries worldwide.

The fig is a very nourishing food; it is rich in antioxidants, phenolic compounds, vitamins, and minerals. Fresh figs are one of the highest plant sources of calcium and fiber. Dried figs are one of the richest fruits for fiber, copper, magnesium, potassium, and calcium. Figs contain many antioxidants and are a good source of flavonoids, polyphenols, and some other bioactive compounds with healing properties.

Worldwide, the main fig producers are traditionally the Mediterranean countries, as fig is adapted to zones where Mediterranean climate conditions—hot and dry summers and mild and rainy winters—prevail. Outside this zone, fig has moved to countries with similar climatic conditions such as California in the USA, South Africa, Central Asia, and Iran. The more recent fig production in Latin America, China, Israel, and Japan was built on fig varieties with persistent fruit set and appropriate training systems. As is being observed, there are more windows of opportunity for other regions or countries to start or commercialize their fig production. The diversity within common fig, *Ficus carica* L., and other *Ficus* species enables the use of already available new scientific and technological tools in fig breeding and variety development to create a wider impact. Environmental, economic, and social sustainability issues, the climate crisis, stringent regulations shaping the whole value chain, globalization, and enhanced competitiveness will together shape future fig production and trade. One historic example is when major producers on northern Mediterranean shores—Italy, Spain, and Portugal—started losing their competitiveness in the global fig trade around the mid-20th century due to limitations in manual labor, either availability or high cost. This wave affected Greece and the USA at a later phase. Turkey replaced these countries and became the leader in the global dried-fig market

* Email: vhmoshea@agri.gov.il

DOI: 10.1079/9781789242492.0029

followed by Iran and southern Mediterranean countries such as Egypt and Morocco. These countries produce sun-dried figs under rainfed conditions, mostly as low-input produce. There are projections that fig production and sun drying may be affected by the rising temperatures and changing rainfall regimes in the coming decades. As with many crops, new cultivars that are well adapted to climate as well as other biotic and abiotic stress factors will mimic the current fig diversity. To overcome the negatives, alternative solutions are also undertaken by practitioners such as cultivating figs under nets or drying fruit undercover or using artificial drying methods. A balance between high production and sustainable production should be capped in order to maintain future fig production, together with conservation of natural variety resources.

Over the last five decades, globalization and targets to increase market completion have created new production areas in eastern Asia and the southern hemisphere. These new fig-production areas or opportunities have stimulated new research directions. New biochemical and molecular methods have established a platform to develop new cultivars, and protected cultivation in soil or soil-less systems has provided a high and efficient growth cycle with improved fruit quality.

This chapter will give an overview of the current research on different topics, and discuss the major outputs of the research work to identify the existing knowledge gaps. The focus will be on research gaps, aiming to stimulate new directions of research and innovation, possibly for the short and medium term, as advances in science and technology may divert the targets. In addition, proposals are made to stimulate research cooperation between industry, academia, and producers in order to increase applied and participatory research and fast transfer of knowledge, and to create wider impact. Finally, directions to encourage fig education by development of research networks are discussed.

29.2 Development of New Fig Cultivars

29.2.1 Functional genomics

The recent revolution in omics technology such as genomic, transcriptomic, metabolomics and bioinformatics can accelerate our knowledge of unrecognized genes, proteins, and metabolites resulting in an understanding of the molecular determinants of key traits. With the recent release of the draft genome for *F. carica* (Mori *et al.*, 2017), increased efforts have been made to pursue functional genomics studies in fig. The estimated genome size of fig is 356 Mb, with a total of 36,138 gene loci and 25,011 annotated genes (Mori *et al.*, 2017). Many large-scale studies have identified candidate genes related to horticulturally important traits of fig such as reproductive transition, induction and initiation of flowers, fruit growth and ripening, fruit coloration, development of fruit flavor, and other traits. Functional genomics studies require enriched gene data resources and advanced technologies. The availability of next-generation sequencing, proteome and metabolome analysis, fig transformation, and, more recently, fig genome editing are all applicable tools for functional genomics studies in fig. In addition, the development of new technologies that compare whole transcriptomes among genotypes grown under different environmental conditions and with other species, small RNA sequencing, exome sequencing, and epigenetic technologies will allow further functional genomics studies in fig. This will contribute to a better understanding of the fig tree and fruit growth and development, and will contribute to the development of new cultivars.

29.2.2 Future breeding strategies

To breed better fig fruit, or provide the best quality for the targeted conditions, it is crucial to understand customer and market needs. To achieve this, the first step is to have a deep knowledge of the global fig industry, as well as an understanding of the networks between breeders, growers, industry, and marketing. This information is critical to build specific and multi-purpose breeding targets. Currently, many Mediterranean countries are growing local fig cultivars and developing cultivar conservation programs. In addition, local growers are continually selecting and growing superior fig cultivars derived from spontaneous crosses and mutations. Those programs are expanding the genetic pool, which is important for the breeding programs. When it comes to the long-term designed breeding programs, there are only a limited number. The two largest breeding programs in California and Israel have been ongoing for more than 20 years and have examined several thousand progeny. These fig-breeding programs have involved a few generations of breeders focusing on combining superior fig fruit quality, high productivity, long postharvest life, and pest resistance. In the USA, parthenocarpic fig varieties with light skins were developed to overcome some of the problems such as caprification, internal rot, and dried-fruit beetles faced in growing Calimyrna (= Sarılop), the main variety used for sun drying. Several new cultivars were developed through these long-term breeding programs, some of which are currently grown in commercial orchards, helping to expand production and widen the availability period in global markets.

Modern breeding program goals could be achieved using different methods such as conventional breeding combined with molecular breeding. The use of omics technologies for selection of superior lines during the early stage of tree development will greatly enhance the output of the breeding program. In addition, genome-editing technology using the clustered regularly interspaced short palindromic repeats (CRISPR)/CRISPR-associated protein 9 (Cas9) for gene silencing has been developed and will have a significant impact in future fig breeding. However, there are some limitations for genome-editing technology in certain regions of the world, mainly in the European countries, while in other regions such as North and South America and Asia, this technology has started to gain acceptance.

The availability of DNA markers and powerful biometric methods has led to considerable progress in quantitative trait locus (QTL) mapping in plants. Marker-assisted selection has become a valuable tool for fruit trees where the target species has a long generation cycle. To plan and screen the progeny, larger areas and longer periods are needed, which result in a high cost of maintenance. In fig-breeding programs, QTLs can be used by the application of molecular markers, providing the basis for marker-assisted selection. This can enhance efforts and accelerate the creation of cultivars. Markers for early sexing in *F. carica* are economically important. Statistical analysis and single-nucleotide polymorphism (SNP) analysis were applied to find two genetic markers that were linked to the sex-determination locus (Sherman *et al.*, 2014; Mori *et al.*, 2017). These sex-linkage markers were successfully used in the Israeli fig-breeding program for early classification of male and female seedlings in order to eliminate the male progeny and grow only the female seedlings in the breeding plot.

In figs, as in other plants, quantitative inheritance is a feature of many important traits such as yield, quality, and disease resistance. Means of analyzing quantitative variation, and especially of uncovering its potential genetic basis, are therefore of major importance for breeding purposes. Such quantitative variations result from the combined action of multiple segregating genes and environmental factors. The joint analysis of genotype marker segregation and phenotypic values of individuals or lines enables the detection and localization of loci affecting quantitative traits. Associations between genome-wide markers and phenotypes have enabled new genomic-based strategies such as genome-wide association studies (GWAS). The GWAS approach does not require establishment of a segregating population, which is time-consuming and costly. A GWAS approach has been used successfully in apples to identify the association between fruit color and the *MYB1* locus (Migicovsky *et al.*, 2016). Similarly, it can be used in figs to select fruit quality and to develop evergreen fig cultivars and disease-resistant progeny. An alternative strategy called genomic selection, which allows selection of superior genotypes, aims to predict the genetic potential and breeding value of progeny and breeding lines to be further used in the breeding program. Genomic selection based on genomic estimated breeding values derives information based on genome-wide markers and therefore is more effective for traits controlled by a large number of genes. Both GWAS and genomic selection would be useful in accelerating genetic improvement of fig cultivars. However, the accumulation of large data sets sufficient for conducting such analyses will be required.

29.2.3 Interspecies crosses

Recent viable fertile progeny derived from crosses between fig and several *Ficus* species have revealed the close relationship within the *Ficus* family and opened up a new field of interspecies crosses. Determination of whole-genome sequences from different *Ficus* species is needed. *Ficus* genome sequences will be used to identify functional traits and to increase diversity in *F. carica* in order to develop new species with better adaptability to environmental conditions, such as evergreen trees. The ability to develop viable progeny in a cross between deciduous and evergreen trees is currently limited. Most of the species in the *Ficus* family are evergreen trees, with fig being one of a limited number of deciduous trees. The development of an evergreen fig–*Ficus* interspecies tree has already started in the Israeli breeding program. It is expected that non-deciduous, evergreen fig–*Ficus* interspecies trees will have better adaptability to climatic conditions, which will allow growth of this new species in new locations around the world. In addition, as has already been seen in interspecies progeny between *F. carica* and *Ficus pumila* in the Israeli breeding program, new fruits with enriched health functional compounds and better storability can be obtained. Interspecies trees expand the horizon of fig cultivation and its fruit by development of combinations with increased fruit quality and productivity and new abilities to overcome biotic and abiotic stress.

29.2.4 The use of grafted trees in fig orchards

Grafting, a vegetative propagation technique that connects the shoot (scion) of one plant with the root system (rootstock) of the other plant, is a powerful old technique that can be used to create a dual plant system. Grafting has been widely used to improve horticultural traits, especially in fruit trees. The success of the graft depends on the compatibility between the rootstock and scion. Studies have indicated that grafts using different genera of the same family are rarely compatible, but grafts of different species within the same genus can survive by forming an effective graft union (Goldschmidt, 2014). Grafting can be used to assist plants in the adaptation to biotic stress, such as resistance to below- or above-ground pests and abiotic stress conditions including drought, salinity, and mineral deficiency. Grafting can be a tool to modify plant architecture, induce precocious flowering and juvenility in perennial seedlings, change fruit quality, improve yield, or overcome other physiological disorders.

Although propagation by cutting is still the most used system in fig due to its easy rooting, grafting can be used to increase fig production, modify tree architecture and fruiting habits, and, mainly, acquire pest and nematode resistance. Several wild *Ficus* species and different fig varieties have been evaluated as rootstock (Hosomi, 2017). The performance of *Ficus* species rootstock was weaker than their own rooted trees, while fig rootstock showed good performance. The fig variety Zidi is used as vigorous rootstock, with partial resistance to fig nematodes. A few studies have described the wild species *Ficus racemosa* as exhibiting nematode resistance (Condit, 1947). In order to advance *Ficus* species as rootstocks in the Israeli breeding program, the progeny of interspecies crosses were tested as a rootstock. In field experiments, several rootstocks showed similar performances to their own rooted trees. Currently, interspecies rootstock for *Batocera rufomaculata* resistance is being tested (M.A. Flaishman, unpublished data). *F. erecta* and the *F. carica* × *F. erecta* F_1 were found to be extremely resistant to the soil-borne fungus canker that is a serious disease in fig (Yakushiji *et al.*, 2019). Rootstock originated from *F. erecta* has extremely low grafting compatibility; however, the BC_1 progeny exhibited sufficient graft compatibility with fig, indicating that *Ceratocystis* canker-resistant BC_1 would be a practical fig rootstock source (Yakushiji *et al.*, 2019). Beneficial new rootstock–scion combinations such as fig–*Ficus* combinations will help the development of graft-induced phenotypic changes to combat abiotic stress and environmental challenges.

29.3 Sustainable Fig-production Systems and Climate Change

The great challenge of global environmental climate change calls for adaptation and precision of sustainable farming systems. A sustainable system involves conservation of energy and water and environmental protection by the use of ecologically and environmentally friendly methods and technologies. Today, it is still possible to adopt practices that can regenerate soil health, water sources and agri- and natural biodiversity.

29.3.1 Climate change and fig production

Climate change is one of the most prominent global environmental problems, predicting improved conditions for food production in mid to high latitude areas such as northern USA, Canada, northern Europe, and Russia, while low latitudes such as the main fig-production area in the Mediterranean could experience declining productivity. Agriculture contributes to climate change while simultaneously being affected by it (Thornton and Lipper, 2014). The main effect of climate change to the agricultural system includes the increase in water demand, and many fig orchards around the world are rainfed with no alternative water sources to compensate for loss of rain. In many cases, water harvesting could provide only partial solutions. Changes in precipitation regimes may also threaten sun drying, which has been a traditional practice for centuries. In addition, the stress due to water scarcity may increase the loss of fig biodiversity. Non-conventional water resources such as saltwater and greywater can be used for fig-production systems after proper treatment combined with selected fig varieties and advanced management. The effects of climate change are predicted to increase pest and disease pressure, and frequent exposure of fruits to high or low temperatures can also cause deterioration of fruit quality. Hence, the effects of climatic variations are predicted to worsen fig-production systems until and unless appropriate adaptation measures are implemented. New production areas will appear in time and replace the current major fig producers. Therefore, the effort toward sustainable fig production requires development of technological and social innovations that can improve farm management practices, enable long-term health of the soil and the environment, bring new partnership models, and maintain consumer demand. In this respect, better cooperation among food-sector industries, breeders, and farmers is required to develop diversified nutritious cultivars grown under agro-ecological management systems that reduce soil degradation and dependence on external inputs, such as fertilizers and pesticides, and enhance adaptation to climate change and resistance to emerging pests and diseases.

29.3.2 Agri-nanotechnology for sustainable agriculture

Traditional methods of farming have restricted the full potential utilization of farmlands, including fig orchards. Nanotechnology, the new field of the 21st century, has a significant impact on the world's economy and industries and on people's lives (Fayaz *et al.*, 2021). Agriculture nanotechnology, with its possible applications, can significantly enhance crop productivity. Nanotechnology can contribute novel tools to disease diagnosis and management, nanopesticide delivery including green pesticides and biopesticides, and slow and sustained release of micronutrients and fertilizers. Site-directed and controlled release of encapsulated fertilizers and pesticides is a revolutionary change for crop improvement and environment. New nano-agrochemicals have been developed with specific nutrients released in a controlled fashion, leading to maintaining soil fertility and meeting environmental challenges.

The most widely investigated and developed factor related to fruit trees is nanofertilizers, which play very important roles in increasing vegetative growth and improving reproductive growth and flowering, thereby increasing productivity and product quality and ultimately increasing shelf-life and decreasing fruit waste. These nanomaterials, which are generally sprayed at low concentrations on trees at different time intervals and in frequent sessions, are also considered to be growth stimulants. Macro- and micro-scale nanofertilizers such as zinc, boron, chitosan, and fertilizer nanocomposites such as ZnFeMnB (zinc, iron, manganese, and boron), NPKMg (nitrogen, phosphorus, potassium, and magnesium), and calcite have been shown to significantly improve the vegetative and reproductive traits of fruit trees such as pomegranate, strawberry, mango, date, coffee, and grape (Zahedi *et al.*, 2020). The use of agriculture nanotechnology in fig is currently limited. Nanoparticles in fig-fruit coatings have been used to enhance fig storability. The quality assessment of fig fruit coated with zinc oxide nanoparticles showed that the application of such a coating is a promising approach toward extending the shelf life of fig fruits (Lakshmi *et al.*, 2018). Begum *et al.* (2020) made an attempt to prepare a biocompatible silver nanoparticle in different molar concentrations by using supernatant prepared from fresh leaves of *F. carica*. Transmission electron microscopy showed that the silver nanoparticles interacted well and enhanced the antibiotic effect. The antibacterial studies showed that the silver-coated organic molecules contained proteins and flavonoids and effectively increased the inhibition zone up to ninefold. Knowledge of the effects of nanoparticles on fruit trees including fig and the biological effects on different traits is partial and there is a need for extensive research on these topics.

29.3.3 Using sensors for sustainable smart orchards

Reducing crop loss due to environmental and pathogen-related stresses, improving resource use efficiency, and selecting optimal plant traits are major challenges in plant agriculture industries worldwide. The current practices in fig orchards are based on the observation of the tree and fruit development. Quantitative approaches, which provide precise ways of determining crucial elements such as irrigation, fertilization, pest management, and other horticultural applications that enhance fruit growth and harvest management, are limited. Precision and sustainable agriculture has allowed farmers to be connected to their farm from anywhere and at any time. Sensors are used to monitor the farm conditions, while microcontrollers are used to automate related process. Smart phones can update growers on the ongoing conditions of their farm, resulting in increased farm productivity.

New technologies are required to accurately monitor, in real time and with high spatial and temporal resolution, the plant physiological and developmental responses to their microenvironment. Nanosensors have the potential to enable smart plant sensors that communicate with actuate electronic devices to improve plant productivity, optimize and automate water and agrochemical allocation, and enable high-throughput plant chemical phenotyping. Nanomaterials are allowing the translation of plant chemical signals into digital information that can be monitored by standoff electronic devices (Giraldo *et al.*, 2019).

A smart farming system for fig cultivation using the Internet of things approach was developed (Mohamood *et al.*, 2019). Smart farming includes the capability to deliver the functional requirements of monitoring and controlling tasks. The developed monitoring system is capable of tracking the surrounding temperature and humidity, soil humidity and pH, and fertilizer electrical conductivity value changes with the help of mobile devices. When looking at the outcomes from this project, all of the mentioned elements were successfully implemented, and executed their tasks in performing the surrounding humidity, irrigation, and fertilization control duties.

Although studies have recognized smart farming systems, research has yet to systematically investigate the effect of smart farming in fig plant application. We call for further development and utilization of smart plant sensors that can be used for (i) nanoformulations of agrochemicals for applying pesticides and fertilizers; (ii) plant protection for the identification of diseases and residues of agrochemicals; and (iii) nanodevices for postharvest management. Accurate electronic devices for monitoring and optimizing individual fig-tree productivity in fig orchards will counterbalance their adverse effects on crop productivity.

29.4 Exploring the Benefits of Fig Products to Control Human Diseases

The health benefits of fruit consumption as revealed through scientific studies pave the way toward educating consumers to consume fruit and vegetables daily in increasing portions. Respectively, the rising demand for fresh fig is attributed to consumers' preferences motivated by health issues. Recently, the number of fruits that have potential health benefits has grown tremendously and led to the recent term "superfruits." The medical usage of fig fruits, leaves, branches, and the milky latex was reported as early as 2900 BCE by Sumerians in Mesopotamia.

Figs and fig latex were the earliest known ancient drugs for cutaneous anthrax (Ben-Noun, 2003). The latex released upon picking the fruit is used by different cultures to treat skin tumors and warts and is also a meat tenderizer. A leaf decoction is taken as a remedy for diabetes and calcifications in the kidneys and liver. Fresh and dried figs have long been appreciated for their laxative property. Fig fruit is accepted among functional foods or ingredients of foods as providing an additional physiological benefit beyond its basic nutritional value. Fig fruits are a rich source of macronutrients and essential micronutrients (i.e. minerals and vitamins). They have high calcium and potassium content and phytochemicals such as polyphenols, anthocyanins, and carotenoids. The different phytochemicals in the fruits exhibit various bioactivities such as antioxidant, antihypertensive, antimicrobial, hepatoprotective, antispasmodic, antiplatelet, and other health impacts. Chapters 25 and 26 (this volume) have described various uses of fig fruit, leaves, branches, latex, and other organs in ethnomedicine traditionally and in the medicinal properties today disclosed by the current studies.

Different polarity crude extracts of *F. carica* have been screened for antioxidant activity. Solomon *et al.* (2006) analyzed the polyphenols, anthocyanins and flavonoids, and antioxidant properties of six different commercial fig varieties. Another study showed that the latex of *F. carica* is very rich in flavonoids and polyphenols, which have been widely shown to be potent antioxidants (Bouyahya *et al.*, 2016). Vinson *et al.* (2005) conducted research on fresh and dried fruits, and their results support the recommendation of including the dried fruits in the daily diet, as they are rich in antioxidants with a phenolic profile, nutrients and fiber. In addition to the fruit, *F.* carica leaves and stems have pharmacological effects and are used for the treatment of different diseases in traditional medicine and in contemporary treatment of various disorders (Salehi *et al.*, 2021). Methanol extracts of the leaves of *F. carica* exhibit strong *in vitro* antibacterial activity, and extracts from the leaves of *F. carica* also demonstrate an anti-inflammatory effect. Ethanol extracts from the fruits of *F. carica* have shown *in vitro* cytotoxic inhibitory effect on human breast cancer cells and on skin cancer cells. Limited clinical studies have also demonstrated the positive effect of *F. carica* extracts. In Korea, a double-blind, randomized study was carried out to investigate oral supplementation effects of *F. carica* paste to treat constipation, showing a substantial improvement compared with the placebo (Baek *et al.*, 2016). In another randomized study, a water extract of edible dried *F. carica* was successfully tested as a treatment for atopic dermatitis (Abbasi *et al.*, 2017).

There is considerable interest in developing food or well-being products from plants rich in protective vitamins or other compounds with potential health benefits. The introduction of a single stilbene synthase gene is sufficient to synthesize resveratrol in heterologous plants such as fig (Flaishman *et al.*, 2008). Currently, genetic engineering allows the development of fig cultivars that are highly enriched with respect to a specific beneficial compound.

The current growing world population requires increasing demand for nutritious, safe, and healthy food. The medicinal value of *F. carica* calls for better cooperation among fig breeders, fig growers, and medical scientists and pharmacologists in order to determine active plant ingredients from fig fruits, leaves, and stems to further test for human, animal, and plant therapy.

29.5 Productivity, Quality, and Safety Nexus

There are many market standards that decide the quality issues and international and national regulations on quality and safety that affect the trade of figs or products made from fig. In preparation of these rules, the risk posed to the consumer should be in balance with the producer. Everyone agrees that the marketed products should be safe and should respond to the expectations of the consumer. During the last decade, safety issues have gained more weight compared with quality standards. Within this context, slight irregularities in shape or mixing of small-sized fruit have become acceptable if the product is safe. Obligatory sizing of dried fig fruit has become optional. However, there are more stringent rules to guarantee not only the safety of the final product but also the production process. The safety issues could be linked to mycotoxins, as they are today, or pesticides, or perhaps some phytochemicals that have health benefits but also act as natural toxins. Many issues could arise in the future, but the productivity, quality, and safety nexus will always be of the utmost importance. In the world market, where the producer is far distant from the consumer, third-party certification in marketing, labeling for conformity, logos, and codes were introduced. As these systems became more diversified, the consumer's recognition or understanding of the differences has become harder. Therefore, consumers' preferences for locally produced fruits or for those with a known origin, with geographic indications, are increasing. Research work on marketing and consumers' preferences must be carried out and the information delivered to the producer. Producers' main concerns are high yields and good quality, bringing them higher revenues. Farmers are not quick to adapt to changes and innovations, so models that facilitate adoption by the farmer have to be included in the research work.

29.6 Applied Research, and Interaction with the Producers and the Industry

Fig production, which is considered a minor crop, is small compared with other fruit commodities such as apples, bananas, and dates. Most commercial efforts have been focused on dried-fig production, as it has a centuries-long history. However, the integration of technology is still very low in the dried-fig industry. However, great growth potential for fig appears to be in high-quality fresh figs. In many countries, consumers already enjoy fresh figs. Consumer prices for fresh figs are quite high, providing greater grower profits than dried figs. However, fresh-fig production is seasonal. There are significant opportunities to develop cultivars with enhanced fig production in both the early and late season. Brebas are the first figs of the season, typically maturing in June in the northern hemisphere. With the development of late-ripening cultivars in Israel and California, fresh figs could be harvested between June and December. This diversity currently also exists among the fig types present in many countries. With southern-hemisphere production, or through protected cultivation in suitable areas, it is almost possible to bridge the gap and market year-round fresh figs. The limitation to expansion of fresh-fig sales is the very short shelf life and the required manual labor for selective fig harvesting. Application of abscisic acid and ethephon enhances fruit ripening and reduces labor during harvesting (Lama *et al.*, 2019). However, short shelf life of 14 days from harvest to sale is typical, and advances in postharvest handling and/or development of varieties with a better shelf life are urgently needed. Recently, postharvest use of fruit coatings with nanoparticles was shown to enhance the shelf life of the Brown Turkey variety to up to 4 weeks. This is a good example for applied research with the industry in order to expand fig markets. In addition, the development of new commercial fig cultivars balancing acidity and flavor will advance the fresh-fig industry. Fig color is an additional trait to improve in both dried and fresh figs. Some producers or consumers favor green/yellow figs, while others prefer highly pigmented dark figs. Finally, different alcoholic beverages derived from the fruits and leaves of the fig plant, such as wine, liquor, and spirit, could provide additional high-quality fig products.

29.7 Research Networks and Education and Training

When evaluated globally, it is obvious that if the crop has a long history in a region, there will be more research leading to accumulation of knowledge and experience. Evaluation of global fig production and trade in the past decades has shown that production is shifting toward new countries/regions. The environmental or socio-economic conditions may be quite different; however, it is always easier when building upon an existing knowledge base. In the last few years, the number of scientists working in applied research in horticulture, especially in perennial fruit trees, and the number of institutional long-term research programs have decreased and financial support has been reduced. The opportunities have increased more in cross-cutting areas linked to strategic research priorities such as information and communication technology, sustainable farming systems, climate change mitigation, health benefits, innovative healthy and high-value products, and regional/rural planning and development. The advances in basic and applied research should result in horticulture practices for management and postharvest treatments. There are issues that need to be solved at a specific location where figs create an economic benefit, which may require national, regional, or local funding and cooperation. Transfer of new scientific advances in technologies for plant production, transportation, postharvest, and moving from the laboratory and greenhouses into commercially available and applicable products seems to be one of the most crucial needs for the relevant industry. Over the last years, university and government extension services have been reduced and the links between industry and research faded out, resulting in less efficient information flow, especially for resource-poor farming communities. In this regard, timely dissemination of applied research results toward appropriate targets is crucial, and currently is mostly a part of the project proposal. However, studies to assess the impact are also required.

International research networks and collaboration are important tools to support development of the fig industry. International and national symposia and conferences, many of which are held under the guidance of the International Society for Horticultural Sciences (ISHS), allow the meeting and exchange of ideas and scientific results where fig is the core subject. In addition, detailed reviews on figs focusing on advances achieved and giving future perspectives for research and development will stimulate cooperation to develop multidisciplinary research teams and to be included among the priority and policies of governments and/or industries. Scientific societies such as the ISHS that conduct a symposium on fig every 4 years should assemble farmers, corporate companies, and the researchers, and provide suitable platforms to develop networking for exchanging information among all stakeholders. These meetings can be a good start to develop mutual visits and to engage industry in the development and support of new research activities. Organization of similar meetings at the national or regional level could also be promoted.

References

Abbasi, S., Kamalinejad, M., Babaie, D., Shams, S., Sadr, Z., *et al.* (2017) A new topical treatment of atopic dermatitis in pediatric patients based on *Ficus carica* L. (fig): a randomized, placebo-controlled clinical trial. *Complementary Therapies in Medicine* 35, 85–91.

Baek, H.-I., Ha, K.-C., Kim, H.-M., Choi, E.-K., Park, E.-O., *et al.* (2016) Randomized, double-blind, placebo-controlled trial of *Ficus carica* paste for the management of functional constipation. *Asia Pacific Journal of Clinical Nutrition* 25, 487– 496.

Begum, M.Y., Alhamhoom, Y., and Roy, A. (2020) Study of antimicrobial and DNA cleavage property of biocompatible silver nanoparticles prepared by using *Ficus carica* L. *Materials Research Innovations* 25, 147–154.

Ben-Noun, L.L. (2003) Figs—the earliest known ancient drug for cutaneous anthrax. *Annals of Pharmacotherapy* 37, 297–300.

Bouyahya, A., Bensaid, M., Bakri, Y., and Dakka, N. (2016) Phytochemistry and ethnopharmacology of *Ficus carica*. *International Journal of Biochemistry Research & Review* 14, 1–12.

Condit, I.J. (1947) *The Fig*. Chronica Botanica Co., Waltham, Massachusetts.

Fayaz, M., Rabani, M.S., Wani, S.A., and Thoker, S.A. (2021) Nano-agriculture: a novel approach in agriculture. In: Hakeem, K.R., Dar, G.H., Mehmood, M.A., and Bhat, R.A. *Microbiota and Biofertilizers*. Springer, Cham, Switzerland, pp. 99–122.

Flaishman, M.A., Rodov, V., and Stover, E. (2008) The fig: botany, horticulture, and breeding. *Horticultural Reviews* 34, 113–197.

Giraldo, J.P., Wu, H., Newkirk, G.M., and Kruss, S. (2019) Nanobiotechnology approaches for engineering smart plant sensors. *Nature Nanotechnology* 14, 541–553.

Goldschmidt, E.E. (2014) Plant grafting: new mechanisms, evolutionary implications. *Frontiers in Plant Science* 5: 727.

Hosomi, A. (2017) Variation in graft compatibility of wild *Ficus* species as rootstock for common fig trees (*Ficus carica*). *Acta Horticulturae* 1173, 199–206.

Lakshmi, S.J., Roopa Bai, R.S., Sharanagouda, H., Ramachandra, C.T., Nadagouda, S., and Nidoni, U. (2018) Effect of biosynthesized zinc oxide nanoparticles coating on quality parameters of fig (*Ficus carica* L.) fruit. *Journal of Pharmacognosy and Phytochemistry* 7, 10–14.

Lama, K., Modi, A., Peer, R., Izhaki, Y., and Flaishman, M.A. (2019) On-tree ABA application synchronizes fruit ripening and maintains keeping quality of figs (*Ficus carica* L.). *Scientia Horticulturae* 253, 405–411.

Migicovsky, Z., Gardner, K.M., Money, D., Sawler, J., Bloom, J.S., *et al.* (2016) Genome to phenome mapping in apple using historical data. *Plant Genome* 9: 0113.

Mohamood, N., Zainal, N., Kadmin, A.F., Abd Gani, S.F., and Tengku Wook, T.M.F. (2019) Integrated smart farming based on Internet of Things (Iot) system for figs cultivation. *International Journal of Innovative Technology and Exploring Engineering* 9, 2278–3075.

Mori, K., Shirasawa, K., Nogata, H., Hirata, C., Tashiro, K., *et al.* (2017) Identification of *RAN1* orthologue associated with sex determination through whole genome sequencing analysis in fig (*Ficus carica* L.). *Scientific Reports* 7: 41124.

Salehi, B., Mishra, A.P., Nigam, M., Karazhan, N., Shukla, I., *et al.* (2021) *Ficus* plants: State of the art from a phytochemical, pharmacological, and toxicological perspective. *Phototherapy Research* 35, 1187–1217.

Sherman, A., Eshed, R., Peer, R., Rubinstein, M., Yablovitz, Z., *et al.* (2014) Discovery of sex linkage markers—a molecular proof for the genetic model for of sex determination in fig (*Ficus carica* L.). In: *Plant and Animal Genome XXII Conference, January 11–15, 2014, San Diego, California*.

Solomon, A., Golubowicz, S., Yablowicz, Z., Grossman, S., Bergman, M., *et al.* (2006) Antioxidant activities and anthocyanin content of fresh fruits of common fig (*Ficus carica* L.). *Journal of Agricultural and Food Chemistry* 54, 7717–7723.

Thornton, P.K. and Lipper, L. (2014) How does climate change alter agricultural strategies to support food security? IFPRI Discussion Paper 1340. International Food Policy Research Institute and Food and Agriculture Organization of the United Nations, Washington, DC.

Vinson, J.A., Zubik, L., Bose, P., Samman, N., and Proch, J. (2005). Dried fruits: excellent *in vitro* and *in vivo* antioxidants. *Journal of the American College of Nutrition* 24, 44–50.

Yakushiji, H., Morita, T., and Jikumaru, S. (2019) Ceratocystis canker resistance in BC1 populations of interspecific hybridization of fig (*Ficus carica*) and *F. erecta*. *Scientia Horticulturae* 252, 71–76.

Zahedi, S.M., Karimi, M., and Teixeira da Silva, J.A. (2020) The use of nanotechnology to increase quality and yield of fruit crops. *Journal of the Science of Food and Agriculture* 100, 25–31.

Index

Note: Page numbers in *italics* denote tables and figures.